Basic College Mathematics:
An Applied Approach

Instructor's Annotated Edition

Eighth Edition

Basic College Mathematics
An Applied Approach

Instructor's Annotated Edition

Richard N. Aufmann
Palomar College, California

Vernon C. Barker
Palomar College, California

Joanne S. Lockwood
Plymouth State University, New Hampshire

Houghton Mifflin Company
Boston   New York

Editor-in-Chief: Jack Shira
Senior Sponsoring Editor: Lynn Cox
Associate Editor: Melissa Parkin
Editorial Assistant: Noel Kamm
Senior Project Editor: Carol Merrigan
Editorial Assistant: Eric Moore
Manufacturing Manager: Karen Banks
Senior Marketing Manager: Ben Rivera
Marketing Assistant: Lisa Lawler

Cover photo: © David Muir/Masterfile

Photo Credits
Chapter 1: *p. 1:* Photodisc Blue/Getty Images; *p. 2:* Photodisc Green/Getty Images; *p. 12:* Alan Schein Photography/CORBIS; *p. 15:* Laura Dwight/PhotoEdit, Inc.; *p. 20:* Hulton-Deutsch Collection/CORBIS; *p. 24:* Rachel Epstein/PhotoEdit, Inc.; *p. 25:* CORBIS; *p. 32:* Blaine Harrington III/CORBIS; *p. 39:* Michelle D. Bridwell/PhotoEdit, Inc.; *p. 44:* CORBIS; *p. 54:* CORBIS; *p. 56:* Jonathan Nourak/PhotoEdit, Inc.; *p. 60:* Pete Saloutos/CORBIS; *p. 62:* Ed Young/CORBIS. **Chapter 2:** *p. 63:* AP/Wide World Photos; *p. 64:* Duomo/CORBIS; *p. 68:* Johnny Buzzerio/CORBIS; *p. 84:* Westlight Stock/OZ Productions/CORBIS; *p. 91:* Reuters/CORBIS; *p. 92:* Timothy A. Clary/Getty Images; *p. 100:* Joe Sohm/The Image Works; *p. 107:* David Young-Wolff/PhotoEdit, Inc.; *p. 108:* Tom McCarthy/PhotoEdit, Inc.; *p. 109:* Bill Aron/PhotoEdit, Inc.; *p. 120:* AP/Wide World Photos; *p. 124:* Kevin Lee/Getty Images. **Chapter 3:** *p. 125:* Michael Newman/PhotoEdit, Inc.; *p. 132:* AFP/Getty Images; *p. 134:* Mug Shots/CORBIS; *p. 148:* Richard Cummins/CORBIS; *p. 149:* CORBIS; *p. 157:* Eric Fowke/PhotoEdit, Inc.; *p. 164:* Tony Freeman/PhotoEdit, Inc.; *p. 168:* Ariel Skelley/CORBIS; *p. 172:* Dana White/PhotoEdit, Inc. **Chapter 4:** *p. 173:* Joel Simon/STONE/Getty Images; *p. 176:* Charles O'Rear/CORBIS; *p. 178:* Mike Powell/Allsport Concepts/Getty Images; *p. 180:* John Madere/CORBIS; *p. 181:* AP/Wide World Photos; *p. 182:* Warner Brothers Television/Getty Images; *p. 189:* Michael Newman/PhotoEdit, Inc.; *p. 190:* Digital Image © 1996 CORBIS, Original Image Courtesy of NASA/CORBIS; *p. 191:* Reproduced by Permission of The State Hermitage Museum, St. Petersburg, Russia/CORBIS; *p. 192:* Dallas & John Heaton/CORBIS; *p. 200:* Bob Daemmrich/PhotoEdit, Inc. **Chapter 5:** *p. 201:* David Sacks/The Image Bank/Getty Images; *p. 206:* David Chasey/Photodisc/Getty Images; *p. 209:* Galen Rowell/CORBIS; *p. 210:* Joe McBride/STONE /Getty Images; *p. 214:* Ulrike Welsch/PhotoEdit, Inc.; *p. 217:* Ariel Skelley/CORBIS; *p. 218:* Chuck Savage/CORBIS; *p. 224:* Ariel Skelley/CORBIS; *p. 228:* CORBIS. **Chapter 6:** *p. 233:* CORBIS; *p. 234:* Ainaco/CORBIS; *p. 236:* Myrleen Ferguson Cate/PhotoEdit, Inc.; *p. 240:* David Madison/STONE/Getty Images; *p. 247:* Todd A. Gipstein/CORBIS; *p. 248:* Dave Bartruff/CORBIS; *p. 255:* Richard Cummins/CORBIS; *p. 257:* Knut Platon/STONE/Getty Images; *p. 263:* Paul Conklin/PhotoEdit, Inc.; *p. 271:* Jeff Greenberg/PhotoEdit, Inc. **Chapter 7:** *p. 293:* Frederic J. Brown/AFP/Getty Images; *p. 294:* Roy Morsch/CORBIS; *p. 295:* Ethan Miller/CORBIS; *p. 314 (top):* CORBIS, *(bottom):* CORBIS; *p. 317:* Jason Reed/Reuters NewMedia, Inc./CORBIS; *p. 318:* Bettmann/CORBIS; *p. 325:* Tony Freeman/PhotoEdit, Inc. **Chapter 8:** *p. 339:* CORBIS; *p. 350:* Tony Freeman/PhotoEdit, Inc.; *p. 354:* Michael Newman/PhotoEdit, Inc.; *p. 359:* CORBIS; *p. 367:* Tony Freeman/PhotoEdit, Inc. **Chapter 9:** *p. 371:* Robert W. Ginn/PhotoEdit, Inc.; *p. 379:* Michael Newman/PhotoEdit, Inc.; *p. 380:* Robert Essel NYC/CORBIS; *p. 381:* Rachel Epstein/PhotoEdit, Inc.; *p. 385:* CORBIS; *p. 387:* Cleve Bryant/PhotoEdit, Inc.; *p. 388:* Dick Reed/CORBIS; *p. 392:* Digital Vision/Getty Images; *p. 398:* Randy M. Ury/CORBIS; *p. 400:* S. Carmona/CORBIS. **Chapter 10:** *p. 403:* Peter Johnson/CORBIS; *p. 410:* Tannen Maury/The Image Works; *p. 418:* DPA/The Image Works; *p. 426:* AP/Wide World Photos; *p. 444:* CORBIS. **Chapter 11:** *p. 455:* E. Dygas/TAXI/Getty Images; *p. 475:* Olson Scott/CORBIS SYGMA; *p. 476:* James Leynse/CORBIS SABA; *p. 478:* The Granger Collection; *p. 482:* Sandor Szabo/EPA/Landov; *p. 500:* Kaku Kurita/Time Life Pictures/Getty Images; *p. 501:* AP/Wide World Photos; *p. 508:* Ralf-Finn Hestoft/CORBIS; *p. 510:* Peter Beck/CORBIS. **Chapter 12:** *p. 513:* Kevin Flemming/CORBIS; *p. 527:* CORBIS; *p. 543:* CORBIS; *p. 553:* Kevin R. Morris/CORBIS; *p. 561:* Gianni Giansanti/CORBIS; *p. 582:* Dwayne Newton/PhotoEdit, Inc.

Printed in the U.S.A.

Library of Congress Control Number: 2004113918

ISBNs:
Student Edition: 0-618-50305-6
Instructor's Annotated Edition: 0-618-50680-2

123456789-WC-09 08 07 06 05

Contents

4   Ratio and Proportion   173

5   Percents   201

Preface

The eighth edition of *Basic College Mathematics: An Applied Approach* provides mathematically sound and comprehensive coverage of the topics considered essential in a basic college mathematics course. The text has been designed not only to meet the needs of the traditional college student but also to serve the needs of returning students whose mathematical proficiency may have declined during years away from formal education.

In this new edition of *Basic College Mathematics: An Applied Approach*, we have continued to integrate approaches suggested by AMATYC. Each chapter opens by illustrating and referencing a mathematical application within the chapter. At the end of each section there are Applying the Concepts exercises, that include writing, synthesis, critical thinking, and challenge problems. At the end of each chapter there is a "Focus on Problem Solving" that introduces students to various problem-solving strategies. This is followed by "Projects and Group Activities" that can be used for cooperative learning activities.

One of the main challenges for students is the ability to translate verbal phrases into mathematical expressions. One reason for this difficulty is that students are not exposed to verbal phrases until later in most texts. In *Basic College Mathematics: An Applied Approach*, we introduce verbal phrases for operations as we introduce the operation. For instance, after addition concepts have been presented, we provide exercises which say "Find the sum of . . ." or "What is 6 more than 7?" In this way, students are constantly confronted with verbal phrases and must make a mathematical connection between the phrase and a mathematical operation.

NEW! Changes to This Edition

In this textbook, as always, students are provided with ample practice in solving application problems requiring addition, subtraction, multiplication, or division. In this edition, however, we have increased the number of application problems that require a student first to determine which operation is required to solve the problem. See, for example, the Applying the Concept exercises on pages 44 and 45. Here is an opportunity for students to apply their understanding of the four basic operations and the words in an application problem that indicate each. This "mixed practice" will improve their ability to determine what operation to use to solve an application problem, a skill that is vital to their becoming good problem solvers. For further examples, see also the Applying the Concept exercises on pages 108 and 109 of Chapter 2 and pages 158 and 159 of Chapter 3.

We have updated the material in Chapter 1 on estimation. And to emphasize the importance of using estimation to check the result of a calculation performed on a calculator, exercises now ask students first to perform the operation on a calculator and then to use estimation to determine if that answer is reasonable. See, for example, the calculator exercises on pages 14, 23, 31, and 43. In this text, estimation is presented as a problem-solving tool; see, for example, the exercises on page 223 of Chapter 5.

In Chapter 2, we choose to present addition and subtraction of fractions prior to multiplication and division of fractions. For instructors who prefer to teach multiplication first, an alternative sequencing of the sections in the chapter is provided on page 77.

The exposition in Section 1 of Chapter 3 includes more instruction on the relationship between decimals and fractions, and the corresponding exercise set includes problems asking students to convert fractions to decimals and decimals to fractions.

Chapter 8 now has a section on converting units of time, and Chapter 11 now includes an objective in which students use a formula to convert degrees Fahrenheit to degrees Celsius.

The in-text examples are now highlighted by a prominent HOW TO bar. Students looking for a worked-out example can easily locate one of these problems.

Throughout the text, data problems have been updated to reflect current data and trends. Also, titles have been added to the application exercises in the exercise sets. These changes emphasize the relevance of mathematics and the variety of problems in real life that require mathematical analysis.

The Chapter Summaries have been remodeled and expanded. Students are provided with definitions, rules, and procedures, along with examples of each. An objective reference and a page reference accompany each entry. We are confident that these will be valuable aids as students review material and study for exams.

Margin notes entitled Integrating Technology provide suggestions for using a scientific calculator. An annotated illustration of a scientific calculator appears on the inside back cover of this text.

1 Whole Numbers

How much would it cost to take a vacation that looks as relaxing as this one? What would you have to pay for travel, accommodations, and food? Do you have enough money in the bank, or would you need to earn and save more money? If you are considering taking a vacation, you will need to know the answers to these and other questions. The **Focus on Problem Solving on page 54** illustrates how we use operations on whole numbers to determine costs associated with planning a vacation.

Need help? For online student resources, such as section quizzes, visit this textbook's website at **math.college.hmco.com/students.**

OBJECTIVES

Section 1.1

A To identify the order relation between two numbers

B To write whole numbers in words and in standard form

C To write whole numbers in expanded form

D To round a whole number to a given place value

Section 1.2

A To add whole numbers

B To solve application problems

Section 1.3

A To subtract whole numbers without borrowing

B To subtract whole numbers with borrowing

C To solve application problems

Section 1.4

A To multiply a number by a single digit

B To multiply larger whole numbers

C To solve application problems

Section 1.5

A To divide by a single digit with no remainder in the quotient

B To divide by a single digit with a remainder in the quotient

C To divide by larger whole numbers

D To solve application problems

Section 1.6

A To simplify expressions that contain exponents

B To use the Order of Operations Agreement to simplify expressions

Section 1.7

A To factor numbe[r]

B To find the prime[s] a number

Page 1

Chapter Opening Features

NEW! Chapter Opener

New, motivating chapter opener photos and captions have been added, illustrating and referencing a specific application from the chapter.

The at the bottom of the page lets students know of additional online resources at math.college.hmco.com/students.

Objective-Specific Approach

Each chapter begins with a list of learning objectives that form the framework for a complete learning system. The objectives are woven throughout the text (i.e., Exercises, Prep Tests, Chapter Review Exercises, Chapter Tests, Cumulative Review Exercises) as well as through the print and multimedia ancillaries. This results in a seamless learning system delivered in one consistent voice.

Page 2

Prep Test and Go Figure

Prep Tests occur at the beginning of each chapter and test students on previously covered concepts that are required in the coming chapter. Answers are provided in the Answer Section. Objective references are also provided if a student needs to review specific concepts.

The **Go Figure** problem that follows the *Prep Test* is a playful puzzle problem designed to engage students in problem solving.

 PREP TEST • • •

Do these exercises to prepare for Chapter 1.

1. Name the number of ♦s shown below.

♦ ♦ ♦ ♦ ♦ ♦ ♦ ♦

2. Write the numbers from 1 to 10.

1 __ __ __ __ __ __ __ __ 10

3. Match the number with its word form.

a. 4	**A.** five
b. 2	**B.** one
c. 5	**C.** zero
d. 1	**D.** four
e. 3	**E.** two
f. 0	**F.** three

 GO FIGURE • • •

Five adults and two children want to cross a river in a canoe. The canoe can hold one adult or two children or one child. Everyone is able to paddle the canoe. What is the minimum number of trips that will be necessary for everyone to get to the other side?

Aufmann Interactive Approach (AIM)

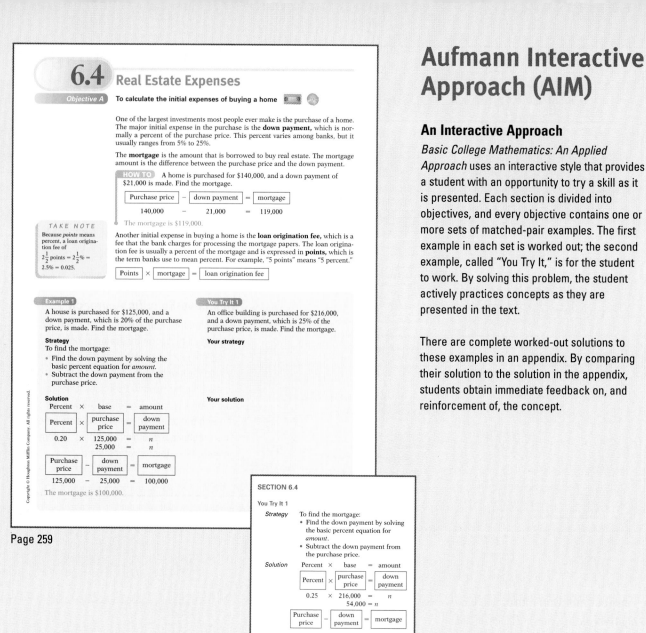

Page 259

Page S17

An Interactive Approach

Basic College Mathematics: An Applied Approach uses an interactive style that provides a student with an opportunity to try a skill as it is presented. Each section is divided into objectives, and every objective contains one or more sets of matched-pair examples. The first example in each set is worked out; the second example, called "You Try It," is for the student to work. By solving this problem, the student actively practices concepts as they are presented in the text.

There are complete worked-out solutions to these examples in an appendix. By comparing their solution to the solution in the appendix, students obtain immediate feedback on, and reinforcement of, the concept.

AIM for Success Student Preface

This student 'how to use this book' preface explains what is required of a student to be successful and how this text has been designed to foster student success, including the Aufmann Interactive Method (AIM). *AIM for Success* can be used as a lesson on the first day of class or as a project for students to complete to strengthen their study skills. There are suggestions for teaching this lesson in the *Instructor's Resource Manual.*

Page xxv

AIM for Success

Welcome to *Basic College Mathematics: An Applied Approach.* As you begin this course, we know two important facts: (1) We want you to succeed. (2) You want to succeed. To do that requires an effort from each of us. For the next few pages, we are going to show you what is required of you to achieve that success and how you can use the features of this text to be successful.

Motivation One of the most important keys to success is motivation. We can try to motivate you by offering interesting or important ways mathematics can benefit you. But, in the end, the motivation must come from you. On the first day of class, it is easy

Problem Solving

Focus on Problem Solving

At the end of each chapter is a Focus on Problem Solving feature which introduces the student to various successful problem-solving strategies. Strategies such as drawing a diagram, applying solutions to other problems, working backwards, inductive reasoning, and trial and error are some of the techniques that are demonstrated.

Focus on Problem Solving

Common Knowledge An application problem may not provide all the information that is needed to solve the problem. Sometimes, however, the necessary information is common knowledge.

HOW TO You are traveling by bus from Boston to New York. The trip is 4 hours long. If the bus leaves Boston at 10 A.M., what time should you arrive in New York?

What other information do you need to solve this problem?

You need to know that, using a 12-hour clock, the hours run

10 A.M.
11 A.M.
12 P.M.
1 P.M.
2 P.M.

Four hours after 10 A.M. is 2 P.M.

You should arrive in New York at 2 P.M.

HOW TO You purchase a 37¢ stamp at the Post Office and hand the clerk a one-dollar bill. How much change do you receive?

What information do you need to solve this problem?

You need to know that there are 100¢ in one dollar.

Your change is 100¢ − 37¢.

$$100 - 37 = 63$$

You receive 63¢ in change.

What information do you need to know to solve each of the following problems?

1. You sell a dozen tickets to a fundraiser. Each ticket costs $10. How much money do you collect?

2. The weekly lab period for your science course is 1 hour and 20 minutes long. Find the length of the science lab period in minutes.

3. An employee's monthly salary is $3750. Find the employee's annual salary.

4. A survey revealed that eighth graders spend an average of 3 hours each day watching television. Find the total time an eighth grader spends watching TV each week.

5. You want to buy a carpet for a room that is 15 feet wide the amount of carpet that you need.

Problem-Solving Strategies

The text features a carefully developed approach to problem solving that emphasizes the importance of *strategy* when solving problems. Students are encouraged to develop their own strategies—to draw diagrams, to write out the solution steps in words—as part of their solution to a problem. In each case, model strategies are presented as guides for students to follow as they attempt the "You Try It" problem. Having students provide strategies is a natural way to incorporate writing into the math curriculum.

Example 6

A $2\frac{2}{3}$-inch piece is cut from a $6\frac{5}{8}$-inch board. How much of the board is left?

Strategy
To find the length remaining, subtract the length of the piece cut from the total length of the board.

Solution
$$6\frac{5}{8} = 6\frac{15}{24} = 5\frac{39}{24}$$
$$-2\frac{2}{3} = 2\frac{16}{24} = 2\frac{16}{24}$$
$$3\frac{23}{24}$$

$3\frac{23}{24}$ inches of the board are left.

You Try It 6

A flight from New York to Los Angeles takes $5\frac{1}{2}$ hours. After the plane has been in the air for $2\frac{3}{4}$ hours, how much flight time remains?

Your strategy

Your solution

Example 7

Two painters are staining a house. In 1 day one painter stained $\frac{1}{3}$ of the house, and the other stained $\frac{1}{4}$ of the house. How much of the job remains to be done?

Strategy
To find how much of the job remains:
• Find the total amount of the house already stained $\left(\frac{1}{3} + \frac{1}{4}\right)$.
• Subtract the amount already stained from 1, which represents the complete job.

Solution
$$\frac{1}{3} = \frac{4}{12} \qquad 1 = \frac{12}{12}$$
$$+\frac{1}{4} = \frac{3}{12} \qquad -\frac{7}{12} = \frac{7}{12}$$
$$\frac{7}{12} \qquad\qquad \frac{5}{12}$$

$\frac{5}{12}$ of the house remains to be stained.

You Try It 7

A patient is put on a diet to lose 24 pounds in 3 months. The patient lost $7\frac{1}{2}$ pounds the first month and $5\frac{3}{4}$ pounds the second month. How much weight must be lost the third month to achieve the goal?

Your strategy

Your solution

Solutions on p. S6

Real Data and Applications

Applications

One way to motivate an interest in mathematics is through applications. Wherever appropriate, the last objective of a section presents applications that require the student to use problem-solving strategies, along with the skills covered in that section, to solve practical problems. This carefully integrated applied approach generates student awareness of the value of algebra as a real-life tool.

Applications are taken from many disciplines including agriculture, business, carpentry, chemistry, construction, education, finance, nutrition, real estate, sports, and weather.

Page 12 (inset)

Objective B   To solve application problems

To solve an application problem, first read the problem carefully. The **strategy** involves identifying the quantity to be found and planning the steps that are necessary to find that quantity. The **solution of an application problem** involves performing each operation stated in the strategy and writing the answer.

The table below displays the Wal-Mart store counts as reported in the Wal-Mart 2003 Annual Report.

	Discount Stores	Supercenters	SAM'S CLUBS	Neighborhood Markets
Within the United States	1568	1258	525	49
Outside the United States	942	238	71	37

HOW TO   Find the total number of Wal-Mart stores in the United States.

Strategy   To find the total number of Wal-Mart stores in the United States, read the table to find the number of each type of store in the United States. Then add these numbers.

Solution
```
   1568
   1258
    525
 +   49
   3400
```
Wal-Mart has a total of 3400 stores in the United States.

Example 4

According to the table above, how many discount stores does Wal-Mart have worldwide?

Strategy
To determine the number of discount stores worldwide, read the table to find the number of discount stores within the United States and the number of discount stores outside the United States. Then add the two numbers.

Solution
```
   1568
 +  942
   2510
```
Wal-Mart has 2510 discount stores worldwide.

You Try It 4

Use the table above to determine the total number of Wal-Mart stores that are outside the United States.

Your strategy

Your solution

Page 12

Real Data

Real data examples and exercises, identified by , ask students to analyze and solve problems taken from actual situations. Students are often required to work with tables, graphs, and charts drawn from a variety of disciplines.

Page 222

Page 222 (inset)

27.   **Lodging**   The graph at the right shows the breakdown of the locations of the 53,500 hotels throughout the United States. How many hotels in the United States are located along highways?

28.   **Poultry**   In a recent year, North Carolina produced 1,300,000,000 pounds of turkey. This was 18.6% of the U.S. total in that year. Calculate the U.S. total turkey production for that year. Round to the nearest billion.

29.   **Mining**   During 1 year, approximately 2,240,000 ounces of gold went into the manufacturing of electronic equipment in the United States. This is 16% of all the gold mined in the United States that year. How many ounces of gold were mined in the United States that year?

30.   **Demography**   The table at the right shows the predicted increase in population from 2000 to 2040 for each of four counties in the Central Valley of California.
a. What percent of the 2000 population of Sacramento County is the increase in population?
b. What percent of the 2000 population of Kern County is the increase in population? Round to the nearest tenth of a percent.

31.   **Police Officers**   The graph at the right shows the causes of death for all police officers killed in the line of duty during a recent year. What percent of the deaths were due to traffic accidents? Round to the nearest tenth of a percent.

32.   **Demography**   According to a 25-city survey of the status of hunger and homelessness by the U.S. Conference of Mayors, 41% of the homeless in the United States are single men, 41% are families with children, 13% are single women, and 5% are unaccompanied minors. How many homeless people in the United States are single men?

APPLYING THE CONCEPTS

33.   **The Federal Government**   In the 108th Senate, there were 51 Republicans, 48 Democrats, and 1 Independent. In the 108th House of Representatives, there were 229 Republicans, 205 Democrats, and 1 Independent. Which had the larger percent of Republicans, the 108th Senate or the 108th House of Representatives?

Most Hotels on Highways
Of the 53,500 hotels throughout the USA, most are found along highways. The breakdown:
Highways 42.2%
Suburban 33.6%
Resort 6.3%
Urban 10.2%
Airport 7.7%

Source: American Hotel and Lodging Association

County	2000 Population	Projected Increase
Sacramento	1,200,000	900,000
Kern	651,700	948,300
Fresno	794,200	705,800
San Joaquin	562,000	737,400

Source: California Department of Finance

Job-related illness 19
Other 6
Violent attacks 58
Traffic accidents 73

Causes of Death for Police Officers Killed in the Line of Duty
Source: International Union of Police Associations

Student Pedagogy

Icons

The 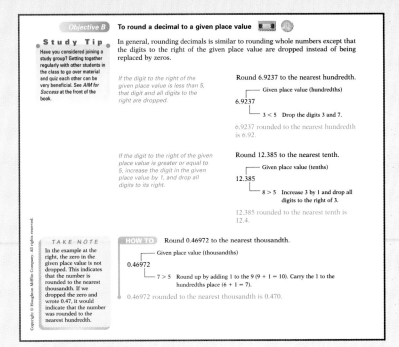 at each objective head remind students that both a video and a tutorial lesson are available for that objective.

Study Tips

These margin notes remind students of study skills presented in the *AIM for Success;* some notes provide page references to the original descriptions. They also provide students with reminders of how to practice good study habits.

Take Note

These margin notes alert students to a point requiring special attention or are used to amplify the concept under discussion.

Page 129

Page 260

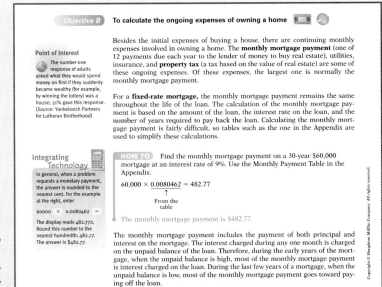

Page 111

The Order of Operations Agreement

Step 1. Do all the operations inside parentheses.
Step 2. Simplify any number expressions containing exponents.
Step 3. Do multiplications and divisions as they occur from left to right.
Step 4. Do additions and subtractions as they occur from left to right.

Point of Interest

These margin notes contain interesting sidelights about mathematics, its history, or its application.

Integrating Technology

These margin notes provide suggestions for using a scientific calculator.

HOW TO Examples

HOW TO examples use annotations to explain what is happening in key steps of the complete, worked-out solutions.

Key Terms and Concepts

Key terms, in bold, emphasize important terms. The key terms are also provided in a **Glossary** at the back of the text.

Key Concepts are presented in orange boxes in order to highlight these important concepts and to provide for easy reference.

For Exercises 61 to 64, use a calculator to add. Then round the numbers to the nearest ten-thousand, and use estimation to determine whether the sum is reasonable.

61.	67,421	**62.**	21,896	**63.**	281,421	**64.**	542,698
	82,984		4,235		9,874		97,327
	66,361		62,544		34,394		7,235
	10,792		21,892		526,398		73,667
	+ 34,037		+ 1,334		+ 94,631		+ 173,201

For Exercises 65 to 68, use a calculator to add. Then round the numbers to the nearest million, and use estimation to determine whether the sum is reasonable.

65.	28,627,052	**66.**	1,792,085	**67.**	12,377,491	**68.**	46,751,070
	983,073		29,919,301		3,409,723		6,095,832
	+ 3,081,496		+ 3,406,882		7,928,026		280,011
					+ 10,705,682		+ 1,563,897

Objective B **To solve application problems**

69. Demographics In a recent year, according to the U.S. Department of Health and Human Services, there were 110,670 twin births in this country, 6919 triplet births, 627 quadruplet deliveries, and 79 quintuplet and other higher-order multiple births. Find the total number of multiple births during the year.

70. Demographics The Census Bureau estimates that the U.S. population will grow by 296 million people from 2000 to 2100. Given that the U.S. population in 2000 was 281 million, find the Census Bureau's estimate of the U.S. population in 2100.

The Film Industry The graph at the right shows the domestic box-office income from the first four *Star Wars* movies. Use this information for Exercises 71 to 73.

71. Estimate the total income from the first four *Star Wars* movies.

72. Find the total income from the first four *Star Wars* movies.

73. a. Find the total income f...
the lowest box-office i...
b. Does the total income ...
the lowest box-office i...
from the 1977 *Star War*...

$500,000,000
$400,000,000
$300,000,000
$200,000,000
$100,000,000

BOX OFFICE

Page 15

APPLYING THE CONCEPTS

80. a. How many two-digit numbers are there? **b.** How many three-digit numbers are there?

81. If you roll two ordinary six-sided dice and add the two numbers that appear on top, how many different sums are possible?

82. If you add two *different* whole numbers, is the sum always greater than either one of the numbers? If not, give an example.

83. If you add two whole numbers, is the sum always greater than either one of the numbers? If not, give an example. (Compare this with the previous exercise.)

84. Make up a word problem for which the answer is the sum of 34 and 28.

85. Call a number "lucky" if it ends in a 7. How many lucky numbers are less than 100?

Page 16

Exercises and Projects

Exercises

The exercise sets of *Basic College Mathematics: An Applied Approach* emphasize skill building, skill maintenance, and applications. Concept-based writing or developmental exercises have been integrated with the exercise sets. Icons identify appropriate writing , data analysis , and calculator exercises.

Included in each exercise set are **Applying the Concepts** that present extensions of topics, require analysis, or offer challenge problems. The writing exercises ask students to explain answers, write about a topic in the section, or research and report on a related topic.

Projects and Group Activities

The Projects and Group Activities featured at the end of each chapter can be used as extra credit or for cooperative learning activities. The projects cover various aspects of mathematics, including the use of calculators, collecting data from the Internet, data analysis, and extended applications.

Projects and Group Activities

The Golden Ratio There are certain designs that have been repeated over and over in both art and architecture. One of these involves the **golden rectangle.**

A golden rectangle is drawn at the right. Begin with a square that measures, say, 2 inches on a side. Let *A* be the midpoint of a side (halfway between two corners). Now measure the distance from *A* to *B*. Place this length along the bottom of the square, starting at *A*. The resulting rectangle is a golden rectangle.

2 in.
1 in.
A
B

Golden Rectangle

The **golden ratio** is the ratio of the length of the golden rectangle to its width. If you have drawn the rectangle following the procedure above, you will find that the golden ratio is approximately 1.6 to 1.

The golden ratio appears in many different situations. Some historians claim that some of the great pyramids of Egypt are based on the golden ratio. The drawing at the right shows the Pyramid of Giza, which dates from approximately 2600 B.C. The ratio of the height to a side of the base is approximately 1.6 to 1.

Height
Side

1. There are instances of the golden rectangle in the Mona Lisa painted by Leonardo da Vinci. Do some research on this painting and write a few paragraphs summarizing your findings.
2. What do 3 × 5 and 5 × 8 index cards have to do with the golden rectangle?
3. What does the United Nations Building in New York City have to do with the golden rectangle?
4. When was the Parthenon in Athens, Greece, built? What does the front of that building have to do with the golden rectangle?

End of Chapter

Chapter Summary

At the end of each chapter there is a Chapter Summary that includes Key Words, Essential Rules and Procedures, and an example of each. Each entry includes an objective reference and a page reference indicating where the concept is introduced. These chapter summaries provide a single point of reference as the student prepares for a test.

Chapter 3 Summary

Key Words	Examples
A number written in *decimal notation* has three parts: a *whole-number part*, a *decimal point*, and a *decimal part*. The decimal part of a number represents a number less than 1. A number written in decimal notation is often simply called a *decimal*. [3.1A, p. 127]	For the decimal 31.25, 31 is the whole-number part and 25 is the decimal part.

Essential Rules and Procedures	Examples
To write a decimal in words, write the decimal part as if it were a whole number. Then name the place value of the last digit. The decimal point is read as "and." [3.1A, p. 127]	The decimal 12.875 is written in words as twelve and eight hundred seventy-five thousandths.
To write a decimal in standard form when it is written in words, write the whole-number part, replace the word *and* with a decimal point, and write the decimal part so that the last digit is in the given place-value position. [3.1A, p. 128]	The decimal forty-nine and sixty-three thousandths is written in standard form as 49.063.
To round a decimal to a given place value, use the same rules used with whole numbers, except drop the digits to the right of the given place value instead of replacing them with zeros. [3.1B, p. 129]	2.7134 rounded to the nearest tenth is 2.7. 0.4687 rounded to the nearest hundredth is 0.47.

Page 165

Chapter Review Exercises

Chapter Review Exercises are found at the end of each chapter. These exercises are selected to help the student integrate all of the topics presented in the chapter.

Page 167

Chapter 3 Review Exercises 167

Chapter 3 Review Exercises

1. Find the quotient of 3.6515 and 0.067.

2. Find the sum of 369.41, 88.3, 9.774, and 366.474.

3. Place the correct symbol, < or >, between the two numbers.
0.055   0.1

4. Write 22.0092 in words.

Chapter Test

Each Chapter Test is designed to simulate a possible test of the material in the chapter.

Page 169

Chapter 3 Test 169

Chapter 3 Test

1. Place the correct symbol, < or >, between the two numbers.
0.66   0.666

2. Subtract: $13.027 - 8.94$

3. Write 45.0302 in words.

4. Convert $\frac{9}{13}$ to a decimal. Round to the nearest thousandth.

Cumulative Review Exercises

Cumulative Review Exercises, which appear at the end of each chapter (beginning with Chapter 2), help students maintain skills learned in previous chapters.

The answers to all Chapter Review Exercises, all Chapter Test exercises, and all Cumulative Review Exercises are given in the Answer Section. Along with the answer, there is a reference to the objective that pertains to each exercise.

Page 171

Cumulative Review Exercises 171

Cumulative Review Exercises

1. Divide: $89\overline{)20{,}932}$

2. Simplify: $2^3 \cdot 4^2$

3. Simplify: $2^2 - (7 - 3) \div 2 + 1$

4. Find the LCM of 9, 12, and 24.

Page A8

CUMULATIVE REVIEW EXERCISES

1. 235 r17 [1.5C]   2. 128 [1.6A]   3. 3 [1.6B]   4. 72 [2.1A]   5. $4\frac{2}{5}$ [2.2B]   6. $\frac{37}{8}$ [2.2B]

Instructor's Annotated Edition—Format and Features

This edition offers a complete Instructor's Annotated Edition. Student pages are reduced, creating a margin for the following instructor-only features.

- **Instructor Notes**
Instructor notes include teaching ideas, warnings about common student errors, or historical notes.

- **New Vocabulary**
A list of new vocabulary introduced within a lesson is provided for appropriate objectives. There are similar lists for *New Symbols, Formulas, Rules, Properties,* and *Equations*.

- **Vocabulary to Review**
A list of vocabulary introduced in a previous objective that students will need to recall in order to understand the material in the present lesson is provided for appropriate objectives. There are similar lists for *Symbols, Formulas, Rules, Properties,* and *Equations*.

- **In-Class Examples**
For *every* objective, extra examples are offered that can be used during the presentation of the lesson. These extra examples do not duplicate the examples presented in the student textbook.

- **Discuss the Concepts**
These questions, or requests for an explanation, can be used for class discussion or for writing exercises. They require students to verbalize the basic concepts presented in the lesson.

- **Concept Check**
These questions or exercises can be used after the presentation of a lesson to test student understanding of the concepts developed.

- **Optional Student Activity**
These exercises can be assigned at the conclusion of the lesson. They can serve as a class activity, individual work, or cooperative learning projects. In general, activities were written to be accomplished within about a five-minute period.

- **Suggested Assignment**
At the beginning of *every* exercise set, there is a suggested homework assignment that covers the essential topics of the section.

- **Quick Quiz**
For *every* objective, there is a short quiz that can be given to students. These quizzes are designed to check basic concepts. These quizzes can also be downloaded from our website at math.college.hmco.com/instructors and are available on the Instructor *Class Prep* CD.

- **Answers to Writing Exercises, Focus on Problem Solving, and Projects and Group Activities**
Suggested answers to all the writing exercises are given in the Instructor's Annotated Edition. Also included, where appropriate, are answers to Focus on Problem Solving exercises and exercises in the Projects and Group Activities feature.

- **PowerPoints**
Next to many of the graphs or tables in the text, there is a P that indicates that a PowerPoint of that graph is available. These PowerPoints can be downloaded from our web site at math.college.hmco.com/instructors and are available on the Instructor *Class Prep* CD.

- **Answers to Exercises**
Answers to all the exercises are given in the Instructor's Annotated Edition. Only answers to selected exercises are given in the Answer Section in the student edition. Answers to writing exercises are not included in the student edition.

Instructor Resources

Basic College Mathematics: An Applied Approach has a complete set of support materials for the instructor.

Instructor's Annotated Edition This edition contains a replica of the student text and additional resources just for the instructor. These include: *Instructor Notes, New Vocabulary/Symbols, etc., Vocabulary/Symbols, etc. to Review, In-Class Examples, Discuss the Concepts, Concept Checks, Optional Student Activities, Suggested Assignments, Quick Quizzes, Answers to Writing Exercises/Focus on Problem Solving/Projects and Group Activities,* and *PowerPoints*. Answers to all exercises are also provided.

Instructor's Solutions Manual The *Instructor's Solutions Manual* contains worked-out solutions for all exercises in the text.

Instructor's Resource Manual with Testing This resource includes eight ready-to-use printed *Chapter Tests* per chapter, suggested *Course Sequences*, and a printout of the *AIM for Success* PowerPoint slide show. All resources are also available on the Instructor website and *Class Prep CD*.

HM ClassPrep with HM Testing CD-ROM *HM ClassPrep* contains a multitude of text-specific resources for instructors to use to enhance the classroom experience. These resources can be easily accessed by chapter or resource type and can also link you to the text's website. *HM Testing* is our computerized test generator and contains a database of algorithmic test items as well as providing **online testing** and **gradebook** functions.

Instructor Text-Specific Website The resources available on the *Class Prep CD* are also available on the instructor website at math.college.hmco.com/instructors. Appropriate items are password protected. Instructors also have access to the student part of the text's website.

WebCT ePacks *WebCT ePacks* provide instructors with a flexible, Internet-based education platform providing multiple ways to present learning materials. The *WebCT ePacks* come with a full array of features to enrich the online learning experience.

Blackboard Cartridges The *Houghton Mifflin Blackboard cartridge* allows flexible, efficient, and creative ways to present learning materials and opportunities. In addition to course management benefits, instructors may make use of an electronic grade book, receive papers from students enrolled in the course via the Internet, and track student use of the communication and collaboration functions.

NEW! **HM Eduspace®** is a powerful course management system powered by Blackboard that makes preparing, presenting, and managing courses easier. You can use this distance-learning platform to customize, create, and deliver course materials and tests online, and easily maintain student portfolios using the grade book, where grades for all assignments are automatically scored, averaged, and saved.

Student Resources

Student Solutions Manual The *Student Solutions Manual* contains complete solutions to all odd-numbered exercises in the text.

Math Study Skills Workbook by Paul D. Nolting This workbook is designed to reinforce skills and minimize frustration for students in any math class, lab, or study skills course. It offers a wealth of study tips and sound advice on note taking, time management, and reducing math anxiety. In addition, numerous opportunities for self-assessment enable students to track their own progress.

HM Eduspace® Online Learning Environment *Eduspace* is a text-specific, web-based learning environment which combines an algorithmic tutorial program with homework capabilities. Specific content is available 24 hours a day to help you further understand your textbook.

HM mathSpace® Tutorial CD-ROM This tutorial CD-ROM allows students to practice skills and review concepts as many times as necessary by providing algorithmically-generated exercises and step-by-step solutions for practice.

SMARTHINKING™ Live, Online Tutoring Houghton Mifflin has partnered with SMARTHINKING to provide an easy-to-use and effective online tutorial service. **Whiteboard Simulations** and **Practice Area** promote real-time visual interaction.

Three levels of service are offered.

- **Text-specific Tutoring** provides real-time, one-on-one instruction with a specially qualified "e-structor."
- **Questions Any Time** allows students to submit questions to the tutor outside the scheduled hours and receive a reply within 24 hours.
- **Independent Study Resources** connect students with around-the-clock access to additional educational services, including interactive websites, diagnostic tests, and Frequently Asked Questions posed to SMARTHINKING e-structors.

Houghton Mifflin Instructional Videos and DVDs Text-specific videos and DVDs, hosted by Dana Mosely, cover all sections of the text and provide a valuable resource for further instruction and review.

Student Text-Specific Website Online student resources can be found at this text's website at math.college.hmco.com/students.

Acknowledgments

The authors would like to thank the people who have reviewed this manuscript and provided many valuable suggestions.

Dorothy A. Brown, *Camden County College, NJ*
Kim Doyle, *Monroe Community College, NY*
Kimberly A. Gregor, *Delaware Technical and Community College, DE*
Allen Grommet, *East Arkansas Community College, AR*
Anne Haney
Rose M. Kaniper, *Burlington County College, NJ*
Mary Ann Klicka, *Bucks County Community College, PA*
Helen Medley
Dr. James R. Perry, *Owens Community College, OH*
Susan Wessner, *Tallahassee Community College, FL*

With special thanks to Dawn Nuttall for her contributions in the developmental editing of three editions of this textbook series.

AIM for Success

INSTRUCTOR NOTE
See the *Instructor's Resource Manual* or *Class Prep* CD for suggestions on how to teach this lesson.

Welcome to *Basic College Mathematics: An Applied Approach*. As you begin this course, we know two important facts: (1) We want you to succeed. (2) You want to succeed. To do that requires an effort from each of us. For the next few pages, we are going to show you what is required of you to achieve that success and how you can use the features of this text to be successful.

Motivation

One of the most important keys to success is motivation. We can try to motivate you by offering interesting or important ways mathematics can benefit you. But, in the end, the motivation must come from you. On the first day of class, it is easy to be motivated. Eight weeks into the term, it is harder to keep that motivation.

To stay motivated, there must be outcomes from this course that are worth your time, money, and energy.

List some reasons you are taking this course.

_____

_____

_____

TAKE NOTE

Motivation alone will not lead to success. For instance, suppose a person who cannot swim is placed in a boat, taken out to the middle of a lake, and then thrown overboard. That person has a lot of motivation but there is a high likelihood the person will drown without some help. Motivation gives us the desire to learn but is not the same as learning.

Although we hope that one of the reasons you listed was an interest in mathematics, we know that many of you are taking this course because it is required to graduate, it is a prerequisite for a course you must take, or because it is required for your major. Although you may not agree that this course is necessary, it is! If you are motivated to graduate or complete the requirements for your major, then use that motivation to succeed in this course. Do not become distracted from your goal to complete your education!

Commitment

To be successful, you must make a commitment to succeed. This means devoting time to math so that you achieve a better understanding of the subject.

List some activities (sports, hobbies, talents such as dance, art, or music) that you enjoy and at which you would like to become better.

ACTIVITY	TIME SPENT	TIME WISHED SPENT
_____	_____	_____
_____	_____	_____
_____	_____	_____

Thinking about these activities, put the number of hours that you spend each week practicing these activities next to the activity. Next to that number, indicate the number of hours per week you would like to spend on these activities.

Whether you listed surfing or sailing, aerobics or restoring cars, or any other activity you enjoy, note how many hours a week you spend doing it. To succeed in math, you must be willing to commit the same amount of time. Success requires some sacrifice.

The "I Can't Do Math" Syndrome

There may be things you cannot do, such as lift a two-ton boulder. You can, however, do math. It is much easier than lifting the two-ton boulder. When you first learned the activities you listed above, you probably could not do them well. With practice, you got better. With practice, you will be better at math. Stay focused, motivated, and committed to success.

It is difficult for us to emphasize how important it is to overcome the "I Can't Do Math" Syndrome. If you listen to interviews of very successful atheletes after a particularly bad performance, you will note that they focus on the positive aspect of what they did, not the negative. Sports psychologists encourage athletes to always be positive—to have a "Can Do" attitude. Develop this attitude toward math.

Strategies for Success

Textbook Review   Right now, do a 15-minute "textbook review" of this book. Here's how:

First, read the table of contents. Do it in three minutes or less. Next, look through the entire book, page by page. Move quickly. Scan titles, look at pictures, notice diagrams.

A textbook reconnaissance shows you where a course is going. It gives you the big picture. That's useful because brains work best when going from the general to the specific. Getting the big picture before you start makes details easier to recall and understand later on.

Your textbook reconnaissance will work even better if, as you scan, you look for ideas or topics that are interesting to you. List three facts, topics, or problems that you found interesting during your textbook reconnaissance.

_____

_____

_____

The idea behind this technique is simple: It's easier to work at learning material if you know it's going to be useful to you.

Not all the topics in this book will be "interesting" to you. But that is true of any subject. Surfers find that on some days the waves are better than others, musicians find some music more appealing than other music, computer gamers find some computer games more interesting than others, car enthusiasts find some cars more exciting than others. Some car enthusiasts would rather have a completely restored 1957 Chevrolet than a new Ferrari.

Know the Course Requirements   To do your best in this course, you must know exactly what your instructor requires. Course requirements may be stated in a *syllabus*, which is a printed outline of the main topics of the course, or they may be presented orally. When they are listed in a syllabus or on other printed pages, keep them in a safe place. When they are presented orally, make sure to take complete notes. In either case, it is important that you understand them completely and follow them exactly. Be sure you know the answer to each of the following questions.

1. What is your instructor's name?
2. Where is your instructor's office?
3. At what times does your instructor hold office hours?
4. Besides the textbook, what other materials does your instructor require?
5. What is your instructor's attendance policy?
6. If you must be absent from a class meeting, what should you do before returning to class? What should you do when you return to class?

7. What is the instructor's policy regarding collection or grading of homework assignments?

8. What options are available if you are having difficulty with an assignment? Is there a math tutoring center?

9. If there is a math lab at your school, where is it located? What hours is it open?

10. What is the instructor's policy if you miss a quiz?

11. What is the instructor's policy if you miss an exam?

12. Where can you get help when studying for an exam?

Remember: Your instructor wants to see you succeed. If you need help, ask! Do not fall behind. If you are running a race and fall behind by 100 yards, you may be able to catch up but it will require more effort than had you not fallen behind.

Time Management We know that there are demands on your time. Family, work, friends, and entertainment all compete for your time. We do not want to see you receive poor job evaluations because you are studying math. However, it is also true that we do not want to see you receive poor math test scores because you devoted too much time to work. When several competing and important tasks require your time and energy, the only way to manage the stress of being successful at both is to manage your time efficiently.

Instructors often advise students to spend twice the amount of time outside of class studying as they spend in the classroom. Time management is important if you are to accomplish this goal and succeed in school. The following activity is intended to help you structure your time more efficiently.

List the name of each course you are taking this term, the number of class hours each course meets, and the number of hours you should spend studying each subject outside of class. Then fill in a weekly schedule like the one printed below. Begin by writing in the hours spent in your classes, the hours spent at work (if you have a job), and any other commitments that are not flexible with respect to the time that you do them. Then begin to write down commitments that are more flexible, including hours spent studying. Remember to reserve time for activities such as meals and exercise. You should also schedule free time.

> **TAKE NOTE**
>
> Besides time management, there must be realistic ideas of how much time is available. There are very few people who can *successfully* work full-time and go to school full-time. If you work 40 hours a week, take 15 units, spend the recommended study time given at the right, and sleep 8 hours a day, you will use over 80% of the available hours in a week. That leaves less than 20% of the hours in a week for family, friends, eating, recreation, and other activities.

	Monday	Tuesday	Wednesday	Thursday	Friday	Saturday	Sunday
7–8 a.m.							
8–9 a.m.							
9–10 a.m.							
10–11 a.m.							
11–12 p.m.							
12–1 p.m.							
1–2 p.m.							
2–3 p.m.							
3–4 p.m.							
4–5 p.m.							
5–6 p.m.							
6–7 p.m.							
7–8 p.m.							
8–9 p.m.							
9–10 p.m.							
10–11 p.m.							
11–12 a.m.							

We know that many of you must work. If that is the case, realize that working 10 hours a week at a part-time job is equivalent to taking a three-unit class. If you must work, consider letting your education progress at a slower rate to allow you to be successful at both work and school. There is no rule that says you must finish school in a certain time frame.

Schedule Study Time   As we encouraged you to do by filling out the time management form above, schedule a certain time to study. You should think of this time the way you would the time for work or class—that is, reasons for missing study time should be as compelling as reasons for missing work or class. "I just didn't feel like it" is not a good reason to miss your scheduled study time.

Although this may seem like an obvious exercise, list a few reasons you might want to study.

_____

_____

_____

Of course we have no way of knowing the reasons you listed, but from our experience one reason given quite frequently is "To pass the course." There is nothing wrong with that reason. If that is the most important reason for you to study, then use it to stay focused.

One method of keeping to a study schedule is to form a ***study group***. Look for people who are committed to learning, who pay attention in class, and who are punctual. Ask them to join your group. Choose people with similar educational goals but different methods of learning. You can gain insight from seeing the material from a new perspective. Limit groups to four or five people; larger groups are unwieldy.

There are many ways to conduct a study group. Begin with the following suggestions and see what works best for your group.

1. Test each other by asking questions. Each group member might bring two or three sample test questions to each meeting.
2. Practice teaching each other. Many of us who are teachers learned a lot about our subject when we had to explain it to someone else.
3. Compare class notes. You might ask other students about material in your notes that is difficult for you to understand.
4. Brainstorm test questions.
5. Set an agenda for each meeting. Set approximate time limits for each agenda item and determine a quitting time.

And now, probably the most important aspect of studying is that it should be done in relatively small chunks. If you can only study three hours a week for this course (probably not enough for most people), do it in blocks of one hour on three separate days, preferably after class. Three hours of studying on a Sunday is not as productive as three hours of paced study.

Text Features That Promote Success   There are 12 chapters in this text. Each chapter is divided into sections, and each section is subdivided into learning objectives. Each learning objective is labeled with a letter from A to D.

Preparing for a Chapter   Before you begin a new chapter, you should take some time to review previously learned skills. There are two ways to do this. The first is to complete the ***Cumulative Review Exercises,*** which occurs after every chapter (except Chapter 1). For instance, turn to page 231. The questions in this review are taken from the previous chapters. The answers for all these exercises can be found on page A11. Turn to that page now and locate the answers for the Chapter 5 Cumulative Review Exercises. After the answer to the first exercise, which is 4, you will see the objective reference [1.6B]. This means that this question was taken from Chapter 1, Section 6, Objective B. If you missed this question, you should return to that objective and restudy the material.

A second way of preparing for a new chapter is to complete the ***Prep Test***. This test focuses on the particular skills that will be required for the new chapter. Turn to page 202 to see a Prep Test. The answers for the Prep Test are the first set of answers in the answer section for a chapter. Turn to page A10 to see the answers for the Prep Test for Chapter 5. Note that an objective reference is given for each question. If you answer a question incorrectly, restudy the objective from which the question was taken.

Before the class meeting in which your professor begins a new section, you should read each objective statement. Next, browse through the objective material, being sure to note each word in bold type. These words indicate important concepts that you must know in order to learn the material. Do not worry about trying to understand all the material. Your professor is there to assist you with that endeavor. The purpose of browsing through the material is that when it is presented to you, your brain will be prepared to accept and organize the new information.

Turn to page 3. Write down the title of the first objective in Section 1.1. Under the title of the objective, write down the words on that page that are in bold print. It is not necessary for you to understand the meaning of these words. You are in this class to learn their meaning.

_____   _____   _____   _____

_____   _____   _____   _____

_____   _____   _____   _____

Math Is Not a Spectator Sport   To learn mathematics you must be an active participant. Listening and watching your professor do mathematics is not enough. Mathematics requires that you interact with the lesson you are studying. If you filled in the blanks above, you were being interactive. There are other ways this textbook has been designed to help you be an active learner.

Annotated Examples   The HOW TO feature indicates an example with explanatory remarks to the right of the work. Using paper and pencil, you should work along as you go through the example.

When you complete the example, get a clean sheet of paper. Write down the problem and then try to complete the solution without referring to your notes or the book. When you can do that, move on to the next part of the objective.

Leaf through the book now and write down the page numbers of two other occurrences of a HOW TO feature.

$$\frac{5}{6} = \frac{10}{12}$$
$$-\frac{1}{4} = \frac{3}{12}$$
$$\overline{\quad\quad \frac{7}{12}}$$

Page 85

You Try Its   One of the key instructional features of this text is the paired examples. Notice that in each example box, the example on the left is completely worked out and the "You Try It" example on the right is not. Study the worked-out example carefully by working through each step. Then work the You Try It. If you get stuck, refer to the page number at the end of the example which directs you to the page on which the You Try It is solved—a complete worked-out solution is provided. Try to use the given solution to get a hint for the step you are stuck on. Then try to complete your solution.

Solution on p. S6

Page 86

When you have completed your solution, check your work against the solution we provided. (Turn to page S6 to see the solution of You Try It 2.) Be aware that frequently there is more than one way to solve a problem. Your answer, however, should be the same as the given answer. If you have any question as to whether your method will "always work," check with your instructor or with someone in the math center.

Browse through the textbook and write down the page numbers of two other occurrences of the paired example feature.

Remember: Be an active participant in your learning process. When you are sitting in class watching and listening to an explanation, you may think that you understand. However, until you actually try to do it, you will have no confirmation of the new knowledge or skill. Most of us have had the experience of sitting in class thinking we knew how to do something only to get home and realize that we didn't.

Word Problems   Word problems are difficult because we must read the problem, determine the quantity we must find, think of a method to do that, and then actually solve the problem. A short summary of this process is to formulate a *strategy* to solve the problem and then devise a *solution*.

Note in the paired example below that part of every word problem is a strategy and a solution. The strategy is a written description of how we will solve the problem. In the corresponding You Try It, you are asked to formulate a strategy. Do not skip this step and be sure to write it out.

Example 6

A $2\frac{2}{3}$-inch piece is cut from a $6\frac{5}{8}$-inch board. How much of the board is left?

Strategy
To find the length remaining, subtract the length of the piece cut from the total length of the board.

Solution

$$6\frac{5}{8} = 6\frac{15}{24} = 5\frac{39}{24}$$
$$-2\frac{2}{3} = 2\frac{16}{24} = 2\frac{16}{24}$$
$$\overline{\qquad\qquad\qquad 3\frac{23}{24}}$$

$3\frac{23}{24}$ inches of the board are left.

You Try It 6

A flight from New York to Los Angeles takes $5\frac{1}{2}$ hours. After the plane has been in the air for $2\frac{3}{4}$ hours, how much flight time remains?

Your strategy
To find how much flight time remains, subtract the amount of time the plane has been in the air $\left(2\frac{3}{4}\right)$ from the total time for the trip $\left(5\frac{1}{2}\right)$.

Your solution

$$5\frac{1}{2} = 5\frac{2}{4} = 4\frac{6}{4}$$
$$-2\frac{3}{4} = 2\frac{3}{4} = 2\frac{3}{4}$$
$$\overline{\qquad\qquad\qquad 2\frac{3}{4}}$$

$2\frac{3}{4}$ hours of flight time remains.

Page 88

Rule Boxes Pay special attention to rules placed in boxes. These rules give you the reasons certain types of problems are solved the way they are. When you see a rule, try to rewrite the rule in your own words.

Objective B   **To use the Order of Operations Agreement to simplify expressions**

More than one operation may occur in a numerical expression. The answer may be different, depending on the order in which the operations are performed. For example, consider $3 + 4 \times 5$.

Multiply first, then add.

$$3 + \underbrace{4 \times 5}$$
$$\underbrace{3 + 20}$$
$$23$$

Add first, then multiply.

$$\underbrace{3 + 4} \times 5$$
$$\underbrace{7 \times 5}$$
$$35$$

An Order of Operations Agreement is used so that only one answer is possible.

The Order of Operations Agreement

Step 1. Do all the operations inside parentheses.

Step 2. Simplify any number expressions containing exponents.

Step 3. Do multiplication and division as they occur from left to right.

Step 4. Do addition and subtraction as they occur from left to right.

Parentheses
Exponents
Multiplication and division
Addition and subtraction

TAKE NOTE

If you are working at home and need assistance, there is online help available at math.college.hmco.com/ students, at this text's website.

Chapter Exercises   When you have completed studying an objective, do the exercises in the exercise set that correspond with that objective. The exercises are labeled with the same letter as the objective. Math is a subject that needs to be learned in small sections and practiced continually in order to be mastered. Doing all of the exercises in each exercise set will help you master the problem-solving techniques necessary for success. As you work through the exercises for an objective, check your answers to the odd-numbered exercises with those in the back of the book.

Preparing for a Test   There are important features of this text that can be used to prepare for a test.

- Chapter Summary
- Chapter Review Exercises
- Chapter Test

After completing a chapter, read the Chapter Summary. (See pages 116–118 for the Chapter 2 Summary.) This summary highlights the important topics covered in the chapter. The page number following each topic refers you to the page in the text on which you can find more information about the concept.

Following the Chapter Summary are Chapter Review Exercises (see page 119) and a Chapter Test (see page 121). Doing the review exercises is an important way of testing your understanding of the chapter. The answer to each review exercise is given at the back of the book along with its objective reference. After checking your answers, restudy any objective from which a question you missed was taken. It may be helpful to retry some of the exercises for that objective to reinforce your problem-solving techniques.

The Chapter Test should be used to prepare for an exam. We suggest that you try the Chapter Test a few days before your actual exam. Take the test in a quiet place and try to complete the test in the same amount of time you will be allowed for your exam. When taking the Chapter Test, practice the strategies of successful test takers: 1) scan the entire test to get a feel for the questions; 2) read the directions carefully; 3) work the problems that are easiest for you first; and perhaps most importantly, 4) try to stay calm.

When you have completed the Chapter Test, check your answers. If you missed a question, review the material in that objective and rework some of the exercises from that objective. This will strengthen your ability to perform the skills in that objective.

Is it difficult to be successful? YES! Successful music groups, artists, professional athletes, chefs, and <u>Write your major here</u> have to work very hard to achieve their goals. They focus on their goals and ignored distractions. The things we ask you to do to achieve success take time and commitment. We are confident that if you follow our suggestions, you will succeed.

1 Whole Numbers

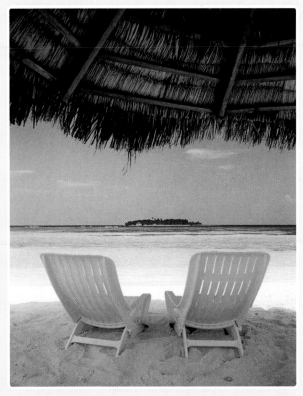

How much would it cost to take a vacation that looks as relaxing as this one? What would you have to pay for travel, accommodations, and food? Do you have enough money in the bank, or would you need to earn and save more money? If you are considering taking a vacation, you will need to know the answers to these and other questions. The **Focus on Problem Solving on page 54** illustrates how we use operations on whole numbers to determine costs associated with planning a vacation.

Need help? For online student resources, such as section quizzes, visit this textbook's website at **math.college.hmco.com/students.**

Features of the IAE

Among the features included in the margins of this Instructor's Annotated Edition are the following:

Instructor Note

These notes offer suggestions for presenting the material in the objective or information that may be helpful.

In-Class Examples

These exercises can be used during your presentation of the lesson. They do not duplicate examples presented in the textbook.

Discuss the Concepts

These questions or requests for an explanation can be used for class discussion or for writing exercises. They require students to verbalize the basic concepts presented in the lesson.

Concept Check

These questions or exercises can be used after the presentation of a lesson to test student understanding of the concepts developed.

Optional Student Activity

This exercise can be assigned at the conclusion of the lesson. It can serve as a class activity, as individual work, or as a cooperative learning project. In general, activities were written to be accomplished within about a five-minute period.

PREP TEST

Do these exercises to prepare for Chapter 1.

1. Name the number of ♦s shown below.

   ♦ ♦ ♦ ♦ ♦ ♦ ♦ ♦
   8

2. Write the numbers from 1 to 10.

   1 ___ ___ ___ ___ ___ ___ ___ ___ 10
   1 2 3 4 5 6 7 8 9 10

3. Match the number with its word form.
   a. 4   A. five
   b. 2   B. one
   c. 5   C. zero
   d. 1   D. four
   e. 3   E. two
   f. 0   F. three
   a and D; b and E; c and A; d and B; e and F; f and C

GO FIGURE

Five adults and two children want to cross a river in a canoe. The canoe can hold one adult or two children or one child. Everyone is able to paddle the canoe. What is the minimum number of trips that will be necessary for everyone to get to the other side?   21 trips

Instructor Note

Margin notes entitled *Point of Interest, Take Note, Study Tip,* and *Integrating Technology* are printed in the student text. The *Point of Interest* feature provides a historical note or mathematical fact of interest. The *Take Note* feature flags important information or provides assistance in understanding a concept. The *Study Tip* feature offers suggestions for using this text and approaches for creating good study habits. The *Integrating Technology* feature describes some of the functions of a calculator.

1.1 Introduction to Whole Numbers

Objective A   To identify the order relation between two numbers

The **whole numbers** are 0, 1, 2, 3, 4, 5, 6, 7, 8, 9, 10, 11, 12, 13, 14, ...

The three dots mean that the list continues on and on and that there is no largest whole number.

Just as distances are associated with the markings on the edge of a ruler, the whole numbers can be associated with points on a line. This line is called the **number line.** The arrow on the number line below indicates that there is no largest whole number.

The **graph of a whole number** is shown by placing a heavy dot directly above that number on the number line. Here is the graph of 7 on the number line:

The number line can be used to show the order of whole numbers. A number that appears to the left of a given number **is less than (<)** the given number. A number that appears to the right of a given number **is greater than (>)** the given number.

Four is less than seven.
4 < 7

Twelve is greater than seven.
12 > 7

Example 1   Graph 11 on the number line.

Solution

Example 2   Place the correct symbol, < or >, between the two numbers.
   a. 39   24
   b. 0   51

Solution   **a.** 39 > 24
   b. 0 < 51

You Try It 1   Graph 6 on the number line.

Your solution

You Try It 2   Place the correct symbol, < or >, between the two numbers.
   a. 45   29
   b. 27   0

Your solution   **a.** 45 > 29
   b. 27 > 0

Solutions on p. S1

Objective 1.1B

New Vocabulary

standard form
place value
place-value chart
period

Discuss the Concepts

Explain how we use the words written in red in the place-value chart when reading a number.

Concept Check

Write the following numbers in standard form and then arrange them in order from smallest to largest.

1. Eighteen thousand nine hundred thirty-two

2. Three thousand four hundred seventy-five

3. Five million three thousand seven hundred ninety

4. Four hundred twenty-six thousand one hundred eight

3475; 18,932; 426,108; 5,003,790

Point of Interest

The Babylonians had a place-value system based on 60. Its influence is still with us in angle measurement and time: 60 seconds in 1 minute, 60 minutes in 1 hour. It appears that the earliest record of a base-10 place-value system for natural numbers dates from the 8th century.

Objective B   **To write whole numbers in words and in standard form**

When a whole number is written using the digits 0, 1, 2, 3, 4, 5, 6, 7, 8, and 9, it is said to be in **standard form.** The position of each digit in the number determines the digit's **place value.** The diagram below shows a **place-value chart** naming the first 12 place values. The number 37,462 is in standard form and has been entered in the chart.

In the number 37,462, the position of the digit 3 determines that its place value is ten-thousands.

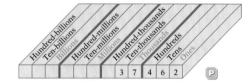

When a number is written in standard form, each group of digits separated from the digits by a comma (or commas) is called a **period.** The number 3,786,451,294 has four periods. The period names are shown in red in the place-value chart above.

To write a number in words, start from the left. Name the number in each period. Then write the period name in place of the comma.

3,786,451,294 is read "three billion seven hundred eighty-six million four hundred fifty-one thousand two hundred ninety-four."

To write a whole number in standard form, write the number named in each period, and replace each period name with a comma.

Four million sixty-two thousand five hundred eighty-four is written 4,062,584. The zero is used as a place holder for the hundred-thousands place.

Example 3   Write 25,478,083 in words.

Solution   Twenty-five million four hundred seventy-eight thousand eighty-three

Example 4   Write three hundred three thousand three in standard form.

Solution   303,003

You Try It 3   Write 36,462,075 in words.

Your solution   Thirty-six million four hundred sixty-two thousand seventy-five

You Try It 4   Write four hundred fifty-two thousand seven in standard form.

Your solution   452,007

Solutions on p. S1

Objective C   **To write whole numbers in expanded form**

The whole number 26,429 can be written in **expanded form** as

20,000 + 6000 + 400 + 20 + 9.

The place-value chart can be used to find the expanded form of a number.

In-Class Examples (Objective 1.1B)

Write the number in words.

1. 703   Seven hundred three

2. 4865   Four thousand eight hundred sixty-five

3. 560,491
Five hundred sixty thousand four hundred ninety-one

4. 4,205,312
Four million two hundred five thousand three hundred twelve

Write the number in standard form.

5. Six hundred forty-two   642

6. Forty-seven thousand eight hundred sixty   47,860

7. Seven hundred ten thousand two hundred two   710,202

8. Five million sixteen thousand four hundred thirty-one   5,016,431

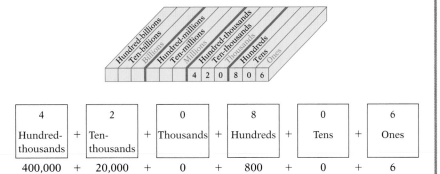

The number 420,806 is written in expanded form below. Note the effect of having zeros in the number.

or simply 400,000 + 20,000 + 800 + 6.

Example 5 Write 23,859 in expanded form.

Solution   20,000 + 3000 + 800 + 50 + 9

You Try It 5 Write 68,281 in expanded form.

Your solution 60,000 + 8000 + 200 + 80 + 1

Example 6 Write 709,542 in expanded form.

Solution   700,000 + 9000 + 500 + 40 + 2

You Try It 6 Write 109,207 in expanded form.

Your solution 100,000 + 9000 + 200 + 7

Solutions on p. S1

Objective D   **To round a whole number to a given place value**

When the distance to the moon is given as 240,000 miles, the number represents an approximation to the true distance. Taking an approximate value for an exact number is called **rounding**. A rounded number is always rounded to a given place value.

Objective 1.1D

Vocabulary to Review
place value   [1.1B]

New Vocabulary
rounding

Discuss the Concepts
Explain how to round a four-digit number to the nearest hundred.

Optional Student Activity
Determine whether the numbers are exact or approximations.

1. The population of Dallas, Texas, is 1,100,000.   Approximation

2. In 1975 there were 19 female delegates to the U.S. House of Representatives.   Exact

3. At Sotheby's auction of Jackie Kennedy Onassis's estate, a bidder paid $700,000 for the woods in JFK's set of golf clubs.   Exact

4. The projected population of people aged 100 and over in the United States in 2020 is 235,000.   Approximation

5. In a recent year, General Motors spent $2,121,040 on advertising.   Exact

---

37 is closer to 40 than it is to 30. 37 rounded to the nearest ten is 40.

673 rounded to the nearest ten is 670. 673 rounded to the nearest hundred is 700.

A whole number is rounded to a given place value without using the number line by looking at the first digit to the right of the given place value.

If the digit to the right of the given place value is less than 5, that digit and all digits to the right are replaced by zeros.

HOW TO   Round 13,834 to the nearest hundred.

Given place value
13,834
└── 3 < 5

13,834 rounded to the nearest hundred is 13,800.

If the digit to the right of the given place value is greater than or equal to 5, increase the digit in the given place value by 1, and replace all other digits to the right by zeros.

HOW TO   Round 386,217 to the nearest ten-thousand.

Given place value
386,217
└── 6 > 5

386,217 rounded to the nearest ten-thousand is 390,000.

Example 7   Round 525,453 to the nearest ten-thousand.

Solution

Given place value
525,453
└── 5 = 5

525,453 rounded to the nearest ten-thousand is 530,000.

You Try It 7   Round 368,492 to the nearest ten-thousand.

Your solution   370,000

Example 8   Round 1972 to the nearest hundred.

Solution

Given place value
1972
└── 7 > 5

1972 rounded to the nearest hundred is 2000.

You Try It 8   Round 3962 to the nearest hundred.

Your solution   4000

Solutions on p. S1

In-Class Examples (Objective 1.1D)
Round the number to the given place value.

1. 356 Tens   360
2. 150 Hundreds   200
3. 4060 Hundreds   4100
4. 2369 Thousands   2000
5. 35,099 Thousands   35,000
6. 228,560 Ten-thousands   230,000
7. 1,485,000 Millions   1,000,000

1.1 Exercises

Objective A   **To identify the order relation between two numbers**

For Exercises 1 to 4, graph the number on the number line.

1. 3

2. 5

3. 9

4. 0

For Exercises 5 to 13, place the correct symbol, < or >, between the two numbers.

5. 37 < 49

6. 58 > 21

7. 101 > 87

8. 16 > 5

9. 245 > 158

10. 2701 > 2071

11. 0 < 45

12. 107 > 0

13. 815 < 928

Objective B   **To write whole numbers in words and in standard form**

For Exercises 14 to 17, name the place value of the digit 3.

14. 83,479
Thousands

15. 3,491,507
Millions

16. 2,634,958
Ten-thousands

17. 76,319,204
Hundred-thousands

For Exercises 18 to 25, write the number in words.

18. 2675
Two thousand
six hundred
seventy-five

19. 3790
Three thousand
seven hundred
ninety

20. 42,928
Forty-two thousand
nine hundred
twenty-eight

21. 58,473
Fifty-eight thousand
four hundred
seventy-three

22. 356,943
Three hundred fifty-six
thousand nine hundred
forty-three

23. 498,512
Four hundred ninety-
eight thousand five
hundred twelve

24. 3,697,483
Three million
six hundred
ninety-seven thousand
four hundred
eighty-three

25. 6,842,715
Six million
eight hundred
forty-two
thousand seven
hundred fifteen

For Exercises 26 to 31, write the number in standard form.

26. Eighty-five
85

27. Three hundred fifty-seven
357

28. Three thousand four hundred fifty-six
3456

29. Sixty-three thousand seven hundred eighty
63,780

Quick Quiz (Objective 1.1A)
Place the correct symbol between the two numbers.
1. 6857   8675   6857 < 8675
2. 36,294   32,694   36,294 > 32,694

Quick Quiz (Objective 1.1B)
1. Write 27,902 in words.
Twenty-seven thousand nine hundred two
2. Write four million eight thousand fifty-one in
standard form.   4,008,051

30. Six hundred nine thousand nine hundred forty-eight
609,948

31. Seven million twenty-four thousand seven hundred nine
7,024,709

Objective C **To write whole numbers in expanded form**

For Exercises 32 to 39, write the number in expanded form.

32. 5287
5000 + 200 + 80 + 7

33. 6295
6000 + 200 + 90 + 5

34. 58,943
50,000 + 8000 + 900 + 40 + 3

35. 453,921
400,000 + 50,000 + 3000 + 900 + 20 + 1

36. 200,583
200,000 + 500 + 80 + 3

37. 301,809
300,000 + 1000 + 800 + 9

38. 403,705
400,000 + 3000 + 700 + 5

39. 3,000,642
3,000,000 + 600 + 40 + 2

Objective D **To round a whole number to a given place value**

For Exercises 40 to 51, round the number to the given place value.

40. 926 Tens
930

41. 845 Tens
850

42. 1439 Hundreds
1400

43. 3973 Hundreds
4000

44. 43,607 Thousands
44,000

45. 52,715 Thousands
53,000

46. 389,702 Thousands
390,000

47. 629,513 Thousands
630,000

48. 647,989 Ten-thousands
650,000

49. 253,678 Ten-thousands
250,000

50. 36,702,599 Millions
37,000,000

51. 71,834,250 Millions
72,000,000

APPLYING THE CONCEPTS

Answer true or false for Exercise 52a and 52b. If the answer is false, give an example to show that it is false.

52. a. If you are given two distinct whole numbers, then one of the numbers is always greater than the other number. True
b. A rounded-off number is always less than its exact value.
False; 8270 rounded to the nearest hundred is 8300.

53. What is the largest three-digit whole number? What is the smallest five-digit whole number?
999; 10,000

54. If 3846 is rounded to the nearest ten and then that number is rounded to the nearest hundred, is the result the same as what you get when you round 3846 to the nearest hundred? If not, which of the two methods is correct for rounding to the nearest hundred?
No. Round 3846 to the nearest hundred.

Quick Quiz (Objective 1.1C)
Write the number in expanded form.
1. 29,048   20,000 + 9000 + 40 + 8
2. 670,153   600,000 + 70,000 + 100 + 50 + 3

Quick Quiz (Objective 1.1D)
Round the number to the given place value.
1. 4298 Hundreds   4300
2. 29,074 Tens   29,070
3. 67,524 Thousands   68,000

1.2 Addition of Whole Numbers

Objective A To add whole numbers

Addition is the process of finding the total of two or more numbers.

> **TAKE NOTE**
> The numbers being added are called **addends.** The result is the **sum.**

By counting, we see that the total of $3 and $4 is $7.

$$\$3 \;+\; \$4 \;=\; \$7$$

Addend   Addend   Sum

Addition can be illustrated on the number line by using arrows to represent the addends. The size, or magnitude, of a number can be represented on the number line by an arrow.

The number 3 can be represented anywhere on the number line by an arrow that is 3 units in length.

> **Point of Interest**
> The first use of the plus sign appeared in 1489 in *Mercantile Arithmetic*. It was used to indicate a surplus, not as the symbol for addition. That use did not appear until about 1515.

To add on the number line, place the arrows representing the addends head to tail, with the first arrow starting at zero. The sum is represented by an arrow starting at zero and stopping at the tip of the last arrow.

$$3 + 4 = 7$$

More than two numbers can be added on the number line.

$$3 + 2 + 4 = 9$$

Some special properties of addition that are used frequently are given below.

Addition Property of Zero

Zero added to a number does not change the number.

$$4 + 0 = 4$$
$$0 + 7 = 7$$

Commutative Property of Addition

Two numbers can be added in either order; the sum will be the same.

$$4 + 8 = 8 + 4$$
$$12 = 12$$

Associative Property of Addition

> **TAKE NOTE**
> This is the same addition problem shown on the number line above.

Grouping the addition in any order gives the same result. The parentheses are grouping symbols and have the meaning "Do the operations inside the parentheses first."

$$(3 + 2) + 4 = 3 + (2 + 4)$$
$$5 + 4 = 3 + 6$$
$$9 = 9$$

New Vocabulary
addition
addend
sum
carrying
approximately equal to ($\approx$)

New Properties
Addition Property of Zero
Commutative Property of Addition
Associative Property of Addition

Discuss the Concepts
1. Provide at least three examples of situations in which we add numbers.
2. What is the difference between the Commutative Property of Addition and the Associative Property of Addition?

Concept Check
Which property justifies the statement?
1. $9 + 16 = 16 + 9$
   Commutative Property of Addition
2. $23 + 0 = 23$
   Addition Property of Zero
3. $6 + (8 + 5) = (6 + 8) + 5$
   Associative Property of Addition

Optional Student Activity
Use a number line to illustrate the sum of 5, 3, and 6. Label the addends and the sum.

In-Class Examples (Objective 1.2A)
Add.
1. $9831 + 2066$ 11,897
2. $1453 + 668 + 78,736$ 80,857
3. $29 + 6538 + 35,724 + 89$ 42,380

Optional Student Activity

How many numbers that are greater than 9 and less than 100 are increased by 9 when their digits are reversed?

8: 12, 23, 34, 45, 56, 67, 78, 89

Instructor Note

Carrying can be modeled with money. For instance, to add $87 + $45, think $7 + $5 is $12, which can be exchanged for 1 ten-dollar bill and 2 one-dollar bills. Add the 1 ten-dollar bill to the 8 tens and 4 tens. The result is 13 ten-dollar bills, which can be exchanged for 1 one-hundred-dollar bill and 3 ten-dollar bills.

Integrating Technology

Most scientific calculators use *algebraic logic:* the add (+), subtract (−), multiply (×), and divide (÷) keys perform the indicated operation on the number in the display and the next number keyed in. For instance, for the example at the right, enter 24 + 71 = . The display reads 95.

The number line is not useful for adding large numbers. The basic addition facts for adding one digit to one digit should be memorized. Addition of larger numbers requires the repeated use of the basic addition facts.

To add large numbers, begin by arranging the numbers vertically, keeping the digits of the same place value in the same column.

HOW TO   Add: $321 + 6472$

$$\begin{array}{r} 321 \\ + 6472 \\ \hline 6793 \end{array}$$

• Add the digits in each column.

There are several words or phrases in English that indicate the operation of addition. Here are some examples:

added to	3 added to 5	$5 + 3$
more than	7 more than 5	$5 + 7$
the sum of	the sum of 3 and 9	$3 + 9$
increased by	4 increased by 6	$4 + 6$
the total of	the total of 8 and 3	$8 + 3$
plus	5 plus 10	$5 + 10$

HOW TO   What is the sum of 24 and 71?

$$\begin{array}{r} 24 \\ + 71 \\ \hline 95 \end{array}$$

• The phrase *the sum of* means to add.

The sum of 24 and 71 is 95.

When the sum of the digits in a column exceeds 9, the addition will involve **carrying.**

HOW TO   Add: $487 + 369$

$$\begin{array}{r} \overset{1}{4}87 \\ + 369 \\ \hline 6 \end{array}$$

• Add the ones column.
  $7 + 9 = 16$ (1 ten + 6 ones).
  Write the 6 in the ones column and carry the 1 ten to the tens column.

$$\begin{array}{r} \overset{1}{4}\overset{1}{8}7 \\ + 369 \\ \hline 56 \end{array}$$

• Add the tens column.
  $1 + 8 + 6 = 15$ (1 hundred + 5 tens).
  Write the 5 in the tens column and carry the 1 hundred to the hundreds column.

$$\begin{array}{r} \overset{1}{4}\overset{1}{8}7 \\ + 369 \\ \hline 856 \end{array}$$

• Add the hundreds column.
  $1 + 4 + 3 = 8$ (8 hundreds).
  Write the 8 in the hundreds column.

Example 1 Find the total of 17, 103, and 8.

Solution
$$\begin{array}{r} 1 \\ 17 \\ 103 \\ +\ \ 8 \\ \hline 128 \end{array}$$

• 7 + 3 + 8 = 18
Write the 8 in the ones column. Carry the 1 to the tens column.

You Try It 1 What is 347 increased by 12,453?

Your solution 12,800

Example 2 Add: 89 + 36 + 98

Solution
$$\begin{array}{r} 2 \\ 89 \\ 36 \\ +\ 98 \\ \hline 223 \end{array}$$

• 9 + 6 + 8 = 23
Write the 3 in the ones column. Carry the 2 to the tens column.

You Try It 2 Add: 95 + 88 + 67

Your solution 250

Example 3 Add:
$$\begin{array}{r} 41,395 \\ 4,327 \\ 497,625 \\ +\ 32,991 \end{array}$$

Solution
$$\begin{array}{r} 112\ 21 \\ 41,395 \\ 4,327 \\ 497,625 \\ +\ 32,991 \\ \hline 576,338 \end{array}$$

You Try It 3 Add:
$$\begin{array}{r} 392 \\ 4,079 \\ 89,035 \\ +\ 4,992 \end{array}$$

Your solution 98,498

Solutions on p. S1

Solutions on p. S1

● E S T I M A T I O N ●

Estimation and Calculators

At some places in the text, you will be asked to use your calculator. Effective use of a calculator requires that you estimate the answer to the problem. This helps ensure that you have entered the numbers correctly and pressed the correct keys.

For example, if you use your calculator to find 22,347 + 5896 and the answer in the calculator's display is 131,757,912, you should realize that you have entered some part of the calculation incorrectly. In this case, you pressed ✕ instead of + . By estimating the answer to a problem, you can help ensure the accuracy of your calculations. We have a special symbol for **approximately equal to (≈).**

For example, to estimate the answer to 22,347 + 5896, round each number to the same place value. In this case, we will round to the nearest thousand. Then add.

$$\begin{array}{r} 22,347 \approx\ \ \ 22,000 \\ +\ \ 5,896 \approx\ +\ \ 6,000 \\ \hline 28,000 \end{array}$$

The sum 22,347 + 5896 is approximately 28,000. Knowing this, you would know that 131,757,912 is much too large and is therefore incorrect.

To estimate the sum of two numbers, first round each whole number to the same place value and then add. Compare this answer with the calculator's answer.

Integrating Technology
This example illustrates that estimation is important when one is using a calculator.

Optional Student Activity

In word sums, each letter stands for a single-digit number. Find the numbers that fit the sum shown below.

$$\begin{array}{r} FOUR \\ +\ FOUR \\ \hline EIGHT \end{array}$$

There is more than one answer. For example: 8532 + 8532 = 17,064.

Instructor Note

Estimation is an important skill. Students should estimate every time they use a calculator.

Discuss the Concepts

Explain how to estimate the sum of 31,287 and 4916.

Objective 1.2B

New Vocabulary

strategy
solution of an application problem

Instructor Note

There are many everyday instances when a table or a graph is used to present information. We have included numerous tables and graphs in this text and have asked students to answer questions based on the data.

Optional Student Activity

List five words or phrases that indicate the operation of addition. Use each one with numbers, and then show the translation of your phrase into mathematical symbols.
See page 10 for a list of six such phrases, their use with numbers, and their translations into mathematical symbols.

Instructor Note

Another major pedagogical feature of this text is written *strategies* that accompany every application problem. For the paired You Try It, we ask students to provide their own written strategy. A suggested strategy, along with a complete solution to the problem, is given in the Solutions section at the back of the text.

Objective B   **To solve application problems**

To solve an application problem, first read the problem carefully. The **strategy** involves identifying the quantity to be found and planning the steps that are necessary to find that quantity. The **solution of an application problem** involves performing each operation stated in the strategy and writing the answer.

The table below displays the Wal-Mart store counts as reported in the Wal-Mart 2003 Annual Report.

	Discount Stores	Supercenters	SAM'S CLUBS	Neighborhood Markets
Within the United States	1568	1258	525	49
Outside the United States	942	238	71	37

HOW TO   Find the total number of Wal-Mart stores in the United States.

Strategy   To find the total number of Wal-Mart stores in the United States, read the table to find the number of each type of store in the United States. Then add these numbers.

Solution
$$\begin{array}{r} 1568 \\ 1258 \\ 525 \\ +49 \\ \hline 3400 \end{array}$$

Wal-Mart has a total of 3400 stores in the United States.

Example 4

According to the table above, how many discount stores does Wal-Mart have worldwide?

Strategy
To determine the number of discount stores worldwide, read the table to find the number of discount stores within the United States and the number of discount stores outside the United States. Then add the two numbers.

Solution
$$\begin{array}{r} 1568 \\ +942 \\ \hline 2510 \end{array}$$

Wal-Mart has 2510 discount stores worldwide.

You Try It 4

Use the table above to determine the total number of Wal-Mart stores that are outside the United States.

Your strategy

Your solution   1288 stores

Solution on p. S1

In-Class Examples (Objective 1.2B)

1. A hospital emergency staff treated 64 people on Friday, 88 people on Saturday, and 73 people on Sunday. How many people did the emergency room staff treat on Friday, Saturday, and Sunday?   225 people

2. A software company had revenues of $1,560,752, $2,964,003, and $4,500,491 during its first three years. Find the software company's total revenue for these three years. $9,025,246

1.2 Exercises

Objective A **To add whole numbers**

For Exercises 1 to 32, add.

1.
$$\begin{array}{r} 17 \\ + 11 \\ \hline 28 \end{array}$$

2.
$$\begin{array}{r} 25 \\ + 63 \\ \hline 88 \end{array}$$

3.
$$\begin{array}{r} 83 \\ + 42 \\ \hline 125 \end{array}$$

4.
$$\begin{array}{r} 63 \\ + 94 \\ \hline 157 \end{array}$$

5.
$$\begin{array}{r} 77 \\ + 25 \\ \hline 102 \end{array}$$

6.
$$\begin{array}{r} 63 \\ + 49 \\ \hline 112 \end{array}$$

7.
$$\begin{array}{r} 56 \\ + 98 \\ \hline 154 \end{array}$$

8.
$$\begin{array}{r} 86 \\ + 68 \\ \hline 154 \end{array}$$

9.
$$\begin{array}{r} 658 \\ + 831 \\ \hline 1489 \end{array}$$

10.
$$\begin{array}{r} 842 \\ + 936 \\ \hline 1778 \end{array}$$

11.
$$\begin{array}{r} 735 \\ + 93 \\ \hline 828 \end{array}$$

12.
$$\begin{array}{r} 189 \\ + 50 \\ \hline 239 \end{array}$$

13.
$$\begin{array}{r} 859 \\ + 725 \\ \hline 1584 \end{array}$$

14.
$$\begin{array}{r} 637 \\ + 829 \\ \hline 1466 \end{array}$$

15.
$$\begin{array}{r} 470 \\ + 749 \\ \hline 1219 \end{array}$$

16.
$$\begin{array}{r} 427 \\ + 690 \\ \hline 1117 \end{array}$$

17.
$$\begin{array}{r} 36{,}925 \\ + 65{,}392 \\ \hline 102{,}317 \end{array}$$

18.
$$\begin{array}{r} 56{,}772 \\ + 51{,}239 \\ \hline 108{,}011 \end{array}$$

19.
$$\begin{array}{r} 50{,}873 \\ + 28{,}453 \\ \hline 79{,}326 \end{array}$$

20.
$$\begin{array}{r} 34{,}872 \\ + 46{,}079 \\ \hline 80{,}951 \end{array}$$

21.
$$\begin{array}{r} 878 \\ 737 \\ + 189 \\ \hline 1804 \end{array}$$

22.
$$\begin{array}{r} 768 \\ 461 \\ + 669 \\ \hline 1898 \end{array}$$

23.
$$\begin{array}{r} 319 \\ 348 \\ + 912 \\ \hline 1579 \end{array}$$

24.
$$\begin{array}{r} 292 \\ 579 \\ + 315 \\ \hline 1186 \end{array}$$

25.
$$\begin{array}{r} 9409 \\ 3253 \\ + 7078 \\ \hline 19{,}740 \end{array}$$

26.
$$\begin{array}{r} 8188 \\ 8020 \\ + 7104 \\ \hline 23{,}312 \end{array}$$

27.
$$\begin{array}{r} 2038 \\ 2243 \\ + 3139 \\ \hline 7420 \end{array}$$

28.
$$\begin{array}{r} 4252 \\ 6882 \\ + 5235 \\ \hline 16{,}369 \end{array}$$

29.
$$\begin{array}{r} 67{,}428 \\ 32{,}171 \\ + 20{,}971 \\ \hline 120{,}570 \end{array}$$

30.
$$\begin{array}{r} 52{,}801 \\ 11{,}664 \\ + 89{,}638 \\ \hline 154{,}103 \end{array}$$

31.
$$\begin{array}{r} 76{,}290 \\ 43{,}761 \\ + 87{,}402 \\ \hline 207{,}453 \end{array}$$

32.
$$\begin{array}{r} 43{,}901 \\ 98{,}301 \\ + 67{,}943 \\ \hline 210{,}145 \end{array}$$

Suggested Assignment
Exercises 1–65, every other odd
Exercises 69–79, odds
Exercises 84, 85
More challenging problems:
Exercises 80, 81

Quick Quiz (Objective 1.2A)
Add.
1. 905 + 1781 2686
2. 3976 + 491 + 27,885 32,352

For Exercises 33 to 40, add.

33. 20,958 + 3218 + 42
24,218

34. 80,973 + 5168 + 29
86,170

35. 392 + 37 + 10,924 + 621
11,974

36. 694 + 62 + 70,129 + 217
71,102

37. 294 + 1029 + 7935 + 65
9323

38. 692 + 2107 + 3196 + 92
6087

39. 97 + 7234 + 69,532 + 276
77,139

40. 87 + 1698 + 27,317 + 727
29,829

41. What is 9874 plus 4509?
14,383

42. What is 7988 plus 5678?
13,666

43. What is 3487 increased by 5986?
9473

44. What is 99,567 added to 126,863?
226,430

45. What is 23,569 more than 9678?
33,247

46. What is 7894 more than 45,872?
53,766

47. What is 479 added to 4579?
5058

48. What is 23,902 added to 23,885?
47,787

49. Find the total of 659, 55, and 1278.
1992

50. Find the total of 4561, 56, and 2309.
6926

51. Find the sum of 34, 329, 8, and 67,892.
68,263

52. Find the sum of 45, 1289, 7, and 32,876.
34,217

For Exercises 53 to 56, use a calculator to add. Then round the numbers to the nearest hundred, and use estimation to determine whether the sum is reasonable.

53. 1234 + 9780 + 6740
Cal.: 17,754
Est.: 17,700

54. 919 + 3642 + 8796
Cal.: 13,357
Est.: 13,300

55. 241 + 569 + 390 + 1672
Cal.: 2872
Est.: 2900

56. 107 + 984 + 1035 + 2904
Cal.: 5030
Est.: 5000

For Exercises 57 to 60, use a calculator to add. Then round the numbers to the nearest thousand, and use estimation to determine whether the sum is reasonable.

57.   32,461
              9,844
         + 59,407
Cal.: 101,712
Est.: 101,000

58.   29,036
             22,904
         +   7,903
Cal.: 59,843
Est.: 60,000

59.   25,432
             62,941
         + 70,390
Cal.: 158,763
Est.: 158,000

60.   66,541
             29,365
         + 98,742
Cal.: 194,648
Est.: 195,000

For Exercises 61 to 64, use a calculator to add. Then round the numbers to the nearest ten-thousand, and use estimation to determine whether the sum is reasonable.

61. 67,421
    82,984
    66,361
    10,792
 + 34,037
Cal.: 261,595
Est.: 260,000

62. 21,896
    4,235
  62,544
  21,892
 + 1,334
Cal.: 111,901
Est.: 100,000

63. 281,421
    9,874
  34,394
 526,398
 + 94,631
Cal.: 946,718
Est.: 940,000

64. 542,698
  97,327
   7,235
  73,667
+ 173,201
Cal.: 894,128
Est.: 890,000

 For Exercises 65 to 68, use a calculator to add. Then round the numbers to the nearest million, and use estimation to determine whether the sum is reasonable.

65. 28,627,052
   983,073
 + 3,081,496
Cal.: 32,691,621
Est.: 33,000,000

66. 1,792,085
 29,919,301
 + 3,406,882
Cal.: 35,118,268
Est.: 35,000,000

67. 12,377,491
  3,409,723
  7,928,026
+ 10,705,682
Cal.: 34,420,922
Est.: 34,000,000

68. 46,751,070
  6,095,832
   280,011
 + 1,563,897
Cal.: 54,690,810
Est.: 55,000,000

Objective B   **To solve application problems**

69. **Demographics** In a recent year, according to the U.S. Department of Health and Human Services, there were 110,670 twin births in this country, 6919 triplet births, 627 quadruplet deliveries, and 79 quintuplet and other higher-order multiple births. Find the total number of multiple births during the year.
118,295 multiple births

70. **Demographics** The Census Bureau estimates that the U.S. population will grow by 296 million people from 2000 to 2100. Given that the U.S. population in 2000 was 281 million, find the Census Bureau's estimate of the U.S. population in 2100.   577 million people

The Film Industry The graph at the right shows the domestic box-office income from the first four *Star Wars* movies. Use this information for Exercises 71 to 73.

71. Estimate the total income from the first four *Star Wars* movies. $1,500,000,000

72. Find the total income from the first four *Star Wars* movies. $1,491,400,000

73. **a.** Find the total income from the two movies with the lowest box-office incomes.
  b. Does the total income from the two movies with the lowest box-office incomes exceed the income from the 1977 *Star Wars* production?
  a. $599,300,000   **b.** Yes

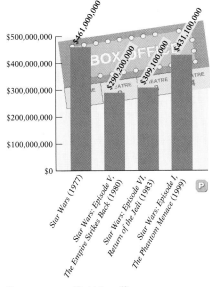

Source: **www.worldwideboxoffice.com**

Quick Quiz (Objective 1.2B)

1. You had a balance of $753 in your checking account before making deposits of $158, $269, and $374. What is your new checking account balance? $1554

74. **Geometry** The perimeter of a triangle is the sum of the lengths of the three sides of the triangle. Find the perimeter of a triangle that has sides that measure 12 inches, 14 inches, and 17 inches. 43 inches

75. **Travel** The odometer on a moving van reads 68,692. The driver plans to drive 515 miles the first day, 492 miles the second day, and 278 miles the third day.
a. How many miles will be driven during the three days? 1285 miles
b. What will the odometer reading be at the end of the trip? 69,977 miles

Investments The table below shows the average amounts of money that all Americans have invested in selected assets and the average amounts invested for Americans between the ages of 16 and 34. Use this table for Exercises 76 to 79.

76. What is the total average amount for all Americans in checking accounts, savings accounts, and U.S. Savings Bonds? $4527

77. What is the total average amount for Americans aged 16 to 34 in checking accounts, savings accounts, and U.S. Savings Bonds? $1796

78. What is the total average amount for Americans aged 16 to 34 in all categories except home equity and retirement? $7838

	All Americans	Ages 16 to 34
Checking accounts	$487	$375
Savings accounts	$3,494	$1,155
U.S. Savings Bonds	$546	$266
Money market	$10,911	$4,427
Stock/mutual funds	$4,510	$1,615
Home equity	$43,070	$17,184
Retirement	$9,016	$4,298

79. Is the sum of the average amounts invested in home equity and retirement for all Americans greater than or less than the sum of all categories for Americans between the ages of 16 and 34? Greater than

APPLYING THE CONCEPTS

80. **a.** How many two-digit numbers are there? **b.** How many three-digit numbers are there? 90; 900

81. If you roll two ordinary six-sided dice and add the two numbers that appear on top, how many different sums are possible? 11 different sums

82. If you add two *different* whole numbers, is the sum always greater than either one of the numbers? If not, give an example. No. 0 + 2 = 2

83. If you add two whole numbers, is the sum always greater than either one of the numbers? If not, give an example. (Compare this with the previous exercise.) No. 0 + 0 = 0

84. Make up a word problem for which the answer is the sum of 34 and 28.

85. Call a number "lucky" if it ends in a 7. How many lucky numbers are less than 100? 10 numbers

Answers to Writing Exercises

84. Answers will vary. For example:

A part-time instructor is teaching two classes this term, with 34 students in one class and 28 students in the other. How many students is the part-time instructor teaching this term?

62 students

1.3 Subtraction of Whole Numbers

Objective A   **To subtract whole numbers without borrowing**

Subtraction is the process of finding the difference between two numbers.

TAKE NOTE

The **minuend** is the number from which another number is subtracted. The **subtrahend** is the number that is subtracted from another number. The result is the **difference.**

By counting, we see that the difference between $8 and $5 is $3.

$$\underset{\text{Minuend}}{\$8} - \underset{\text{Subtrahend}}{\$5} = \underset{\text{Difference}}{\$3}$$

The difference 8 − 5 can be shown on the number line.

Note from the number line that addition and subtraction are related.

    Subtrahend   5
  + Difference   + 3
  = Minuend   8

Point of Interest

The use of the minus sign dates from the same period as the plus sign, around 1515.

The fact that the sum of the subtrahend and the difference equals the minuend can be used to check subtraction.

To subtract large numbers, begin by arranging the numbers vertically, keeping the digits that have the same place value in the same column. Then subtract the digits in each column.

HOW TO   Subtract 8955 − 2432 and check.

    8 9 5 5
  − 2 4 3 2
    6 5 2 3

Check: Subtrahend   2432
          + Difference   + 6523
          = Minuend   8955

Example 1   Subtract 6594 − 3271 and check.

Solution
        6594   *Check:*   3271
      − 3271   + 3323
        3323   6594

You Try It 1   Subtract 8925 − 6413 and check.

Your solution   2512

Example 2   Subtract 15,762 − 7541 and check.

Solution
       15,762   *Check:*   7,541
      − 7,541   + 8,221
        8,221   15,762

You Try It 2   Subtract 17,504 − 9302 and check.

Your solution   8202

Solutions on p. S1

Objective 1.3A

New Vocabulary
subtraction
minuend
subtrahend
difference

Discuss the Concepts
1. Provide at least three examples of situations in which we subtract numbers.
2. Explain why addition can be used to check the answer to a subtraction problem.

Concept Check
For each subtraction problem, find the difference and then write a related addition problem.
1. 68 − 25   43; 25 + 43 = 68
2. 1539 − 722
        817; 722 + 817 = 1539

Optional Student Activity
Use a number line to illustrate the difference between 12 and 7. Label the minuend, the subtrahend, and the difference.

In-Class Examples (Objective 1.3A)
Subtract.
1. 744 − 31   713
2. 5629 − 625   5004
3. 8427 − 6306   2121

Objective 1.3B

Copyright © Houghton Mifflin Company. All rights reserved.

New Vocabulary
borrowing

Instructor Note
Borrowing can be related to money. For instance, if Kelly has $27 as 2 ten-dollar bills and 7 one-dollar bills and Chris wants to borrow $9, then Kelly can exchange a ten-dollar bill for 10 one-dollar bills. Kelly then has 1 ten-dollar bill and 17 one-dollar bills. Kelly now can give Chris 9 one-dollar bills. This leaves Kelly with 1 ten-dollar bill and 8 one-dollar bills.

Instructor Note
The phrases that indicate subtraction are more difficult for students, especially the phrase "2 less than 7," which means 7 − 2.

Discuss the Concepts
Explain the process of borrowing when subtracting 43 from 81.

Objective B To subtract whole numbers with borrowing

In all the subtraction problems in the previous objective, for each place value the lower digit was not larger than the upper digit. When the lower digit *is* larger than the upper digit, subtraction will involve **borrowing.**

HOW TO Subtract: 692 − 378

Because 8 > 2, borrowing is necessary.
9 tens = 8 tens + 1 ten.

Borrow 1 ten from the tens column and write 10 in the ones column.

Add the borrowed 10 to 2.

Subtract the digits in each column.

The phrases below are used to indicate the operation of subtraction. An example is shown at the right of each phrase.

minus	8 minus 5	8 − 5
less	9 less 3	9 − 3
less than	2 less than 7	7 − 2
the difference between	the difference between 8 and 2	8 − 2
decreased by	5 decreased by 1	5 − 1

HOW TO Find the difference between 1234 and 485, and check.

"The difference between 1234 and 485" means 1234 − 485.

```
  2 14          1 12 14          0 11 12 14
1 2 3 4       1 2 3 4          X 2 3 4        Check:     11
−   4 8 5     −   4 8 5        −   4 8 5              485
        9             4 9            7 4 9          + 749
                                                     1234
```

Subtraction with a zero in the minuend involves repeated borrowing.

Study Tip
The HOW TO feature indicates an example with explanatory remarks. Using paper and pencil, you should work through the example. See *AIM for Success* at the front of the book.

HOW TO Subtract: 3904 − 1775

```
    8 10                    9                    9
3 9 0 4             8 10 14          8 10 14
− 1 7 7 5          3 9 0 4          3 9 0 4
                   − 1 7 7 5        − 1 7 7 5
                                      2 1 2 9
```

5 > 4
There is a 0 in the tens column. Borrow 1 hundred (= 10 tens) from the hundreds column and write 10 in the tens column.

Borrow 1 ten from the tens column and add 10 to the 4 in the ones column.

Subtract the digits in each column.

In-Class Examples (Objective 1.3B)
Subtract.
1. 351 − 69   282
2. 6402 − 517   5885
3. 40,824 − 6917   33,907

Example 3 Subtract 4392 − 678 and check.

Solution

$$
\begin{array}{r}
\overset{3}{\cancel{4}}\;\overset{13}{\cancel{3}}\;\overset{8}{\cancel{9}}\;\overset{12}{\cancel{2}} \\
-\quad 6\;7\;8 \\
\hline
3\;7\;1\;4
\end{array}
$$

Check:
$$
\begin{array}{r}
678 \\
+\,3714 \\
\hline
4392
\end{array}
$$

Example 4 Find 23,954 less than 63,221 and check.

Solution

$$
\begin{array}{r}
\overset{5}{\cancel{6}}\;\overset{12}{\cancel{3}},\overset{11}{\cancel{2}}\;\overset{11}{\cancel{2}}\;\overset{11}{\cancel{1}} \\
-\,2\;3,9\;5\;4 \\
\hline
3\;9,2\;6\;7
\end{array}
$$

Check:
$$
\begin{array}{r}
23{,}954 \\
+39{,}267 \\
\hline
63{,}221
\end{array}
$$

Example 5 Subtract 46,005 − 32,167 and check.

Solution

$$
\begin{array}{r}
4\;\;\overset{5}{\cancel{6}},\overset{10}{\cancel{0}}\;\;0\;\;5 \\
-3\;\;2,1\;\;6\;\;7
\end{array}
$$

• There are two zeros in the minuend. Borrow 1 thousand from the thousands column and write 10 in the hundreds column.

$$
\begin{array}{r}
4\;\;\overset{5}{\cancel{6}},\overset{\overset{9}{\cancel{10}}}{\cancel{0}}\;\overset{10}{\cancel{0}}\;\;5 \\
-3\;\;2,1\;\;6\;\;7
\end{array}
$$

• Borrow 1 hundred from the hundreds column and write 10 in the tens column.

$$
\begin{array}{r}
4\;\;\overset{5}{\cancel{6}},\overset{\overset{9}{\cancel{10}}}{\cancel{0}}\;\overset{\overset{9}{\cancel{10}}}{\cancel{0}}\;\overset{15}{\cancel{5}} \\
-3\;\;2,1\;\;6\;\;7 \\
\hline
1\;\;3,8\;\;3\;\;8
\end{array}
$$

• Borrow 1 ten from the tens column and add 10 to the 5 in the ones column.

Check:
$$
\begin{array}{r}
32{,}167 \\
+\,13{,}838 \\
\hline
46{,}005
\end{array}
$$

You Try It 3 Subtract 3481 − 865 and check.

Your solution 2616

You Try It 4 Find 54,562 decreased by 14,485 and check.

Your solution 40,077

You Try It 5 Subtract 64,003 − 54,936 and check.

Your solution 9067

Solutions on pp. S1–S2

Concept Check

Subtract

$$
\begin{array}{r}
932 \\
-\,475
\end{array}
$$

by first writing each number in expanded form.

$$
\begin{array}{r}
\overset{120}{} \\
800\quad \overset{}{\cancel{20}}\quad 12 \\
\cancel{900} + \cancel{30} + 2 \\
-400 - 70 - 5 \\
\hline
400 + 50 + 7 = 457
\end{array}
$$

Note: Students will find this exercise difficult. They need to realize they are borrowing 1 *ten* from 30 and 1 *hundred* from 900.

Optional Student Activity

Write down any seven-digit number. Use the same digits, but scramble them to make a different number. Subtract the smaller number from the larger number. Add the digits of the answer. If you end up with a two-digit number, add them to produce a one-digit number. What does the resulting number have to be?   9

Discuss the Concepts

Explain how to estimate the difference between 61,978 and 32,405.

ESTIMATION

Estimating the Difference Between Two Whole Numbers

Calculate 323,502 − 28,912. Then use estimation to determine whether the difference is reasonable.

Subtract to find the exact difference. To estimate the difference, round each number to the same place value. Here we have rounded to the nearest ten-thousand. Then subtract. The estimated answer is 290,000, which is very close to the exact difference 294,590.

$$
\begin{array}{rcr}
323{,}502 & \approx & 320{,}000 \\
-\;28{,}912 & \approx & -\;30{,}000 \\
\hline
294{,}590 & & 290{,}000
\end{array}
$$

Objective 1.3C

Instructor Note

In reference to the data displayed at the right, you might discuss with the students the fact that the large difference between the number of active military personnel in 1940 and the number in 1945 was a result of America's involvement in World War II. America entered the war in 1941.

Optional Student Activity

1. What is the least number of plus signs that need to be put between the digits of the number 123,456,789 so that the sum is 126?

   Five: 12 + 34 + 56 + 7 + 8 + 9

2. Now change three of the plus signs in the expression you wrote to minus signs so that the expression simplifies to 78.

   12 + 34 + 56 − 7 − 8 − 9

Objective C   **To solve application problems**

The table at the right shows the number of personnel on active duty in the branches of the U.S. military in 1940 and 1945. Use this table for Example 6 and You Try It 6.

Branch	1940	1945
U.S. Army	267,767	8,266,373
U.S. Navy	160,997	3,380,817
U.S. Air Force	51,165	2,282,259
U.S. Marine Corps	28,345	474,680

Source: Dept. of the Army, Dept. of the Navy, Air Force Dept., Dept. of the Marines, U.S. Dept. of Defense

Example 6

Find the difference between the number of U.S. Army personnel on active duty in 1945 and the number in 1940.

Strategy
To find the difference, subtract the number of U.S. Army personnel on active duty in 1940 (267,767) from the number on active duty in 1945 (8,266,373).

Solution

$$\begin{array}{r} 8,266,373 \\ -\ \ \ 267,767 \\ \hline 7,998,606 \end{array}$$

There were 7,998,606 more personnel on active duty in the U.S. Army in 1945 than in 1940.

You Try It 6

Find the difference between the number of personnel on active duty in the Navy and the number in the Air Force in 1945.

Your strategy

Your solution   1,098,558 personnel

Example 7

You had a balance of $415 on your student debit card. You then used the card, deducting $197 for books, $48 for art supplies, and $24 for theater tickets. What is your new student debit card balance?

Strategy
To find your new debit card balance:
- Add to find the total of the three deductions (197 + 48 + 24).
- Subtract the total of the three deductions from the old balance (415).

Solution

$$\begin{array}{r} 197 \\ 48 \\ +\ 24 \\ \hline 269 \end{array} \text{ total deductions}$$

$$\begin{array}{r} 415 \\ -269 \\ \hline 146 \end{array}$$

Your new debit card balance is $146.

You Try It 7

Your total weekly salary is $638. Deductions of $127 for taxes, $18 for insurance, and $35 for savings are taken from your pay. Find your weekly take-home pay.

Your strategy

Your solution $458

Solutions on p. S2

In-Class Examples (Objective 1.3C)

Note: Example 1 is a one-step problem. Example 2 is a two-step problem.

1. How much larger is Alaska than Texas? Alaska is 615,230 square miles in area, and Texas is 276,277 square miles in area.   338,953 square miles

2. You drove a car 25,950 miles in a three-year period. You drove 8070 miles the first year and 9759 miles the second year. How many miles did you drive the third year?   8121 miles

1.3 Exercises

Objective A   To subtract whole numbers without borrowing

For Exercises 1 to 35, subtract.

1. $\begin{array}{r} 9 \\ -5 \\ \hline 4 \end{array}$ **2.** $\begin{array}{r} 8 \\ -7 \\ \hline 1 \end{array}$ **3.** $\begin{array}{r} 8 \\ -4 \\ \hline 4 \end{array}$ **4.** $\begin{array}{r} 7 \\ -3 \\ \hline 4 \end{array}$ **5.** $\begin{array}{r} 10 \\ -0 \\ \hline 10 \end{array}$

6. $\begin{array}{r} 11 \\ -4 \\ \hline 7 \end{array}$ **7.** $\begin{array}{r} 12 \\ -8 \\ \hline 4 \end{array}$ **8.** $\begin{array}{r} 19 \\ -8 \\ \hline 11 \end{array}$ **9.** $\begin{array}{r} 15 \\ -6 \\ \hline 9 \end{array}$ **10.** $\begin{array}{r} 16 \\ -7 \\ \hline 9 \end{array}$

11. $\begin{array}{r} 25 \\ -3 \\ \hline 22 \end{array}$ **12.** $\begin{array}{r} 55 \\ -4 \\ \hline 51 \end{array}$ **13.** $\begin{array}{r} 68 \\ -8 \\ \hline 60 \end{array}$ **14.** $\begin{array}{r} 77 \\ -3 \\ \hline 74 \end{array}$ **15.** $\begin{array}{r} 89 \\ -23 \\ \hline 66 \end{array}$

16. $\begin{array}{r} 54 \\ -21 \\ \hline 33 \end{array}$ **17.** $\begin{array}{r} 88 \\ -57 \\ \hline 31 \end{array}$ **18.** $\begin{array}{r} 1202 \\ -701 \\ \hline 501 \end{array}$ **19.** $\begin{array}{r} 1305 \\ -404 \\ \hline 901 \end{array}$ **20.** $\begin{array}{r} 1763 \\ -801 \\ \hline 962 \end{array}$

21. $\begin{array}{r} 1497 \\ -706 \\ \hline 791 \end{array}$ **22.** $\begin{array}{r} 8974 \\ -3972 \\ \hline 5002 \end{array}$ **23.** $\begin{array}{r} 2836 \\ -1711 \\ \hline 1125 \end{array}$ **24.** $\begin{array}{r} 8976 \\ -7463 \\ \hline 1513 \end{array}$ **25.** $\begin{array}{r} 9273 \\ -6142 \\ \hline 3131 \end{array}$

26. $77 - 36$ 41   **27.** $129 - 82$ 47   **28.** $132 - 61$ 71   **29.** $969 - 44$ 925   **30.** $1347 - 103$ 1244

31. $4865 - 304$ 4561   **32.** $1525 - 702$ 823   **33.** $9999 - 6794$ 3205   **34.** $7806 - 3405$ 4401   **35.** $8843 - 7621$ 1222

36. What is 3795 minus 1092?   2703   **37.** What is 9071 minus 6050?   3021

38. Find the difference between 9763 and 541.   9222   **39.** Find the difference between 6094 and 3072.   3022

40. What is 3701 less than 6932?   3231   **41.** What is 2031 less than 5071?   3040

42. Find 6509 decreased by 3102.   3407   **43.** Find 7994 decreased by 7782.   212

44. Find 23,907 less 12,705.   11,202   **45.** Find 65,986 less 5741.   60,245

Quick Quiz (Objective 1.3A)

Subtract.

1. $936 - 25$ 911

2. $6993 - 1821$ 5172

Objective B **To subtract whole numbers with borrowing**

For Exercises 46 to 89, subtract.

46. 71 − 18 = 53   **47.** 93 − 28 = 65   **48.** 47 − 18 = 29   **49.** 44 − 27 = 17

50. 37 − 29 = 8   **51.** 50 − 27 = 23   **52.** 70 − 33 = 37   **53.** 993 − 537 = 456

54. 250 − 192 = 58   **55.** 840 − 783 = 57   **56.** 768 − 194 = 574   **57.** 770 − 395 = 375

58. 674 − 337 = 337   **59.** 3526 − 387 = 3139   **60.** 1712 − 289 = 1423   **61.** 4350 − 729 = 3621

62. 1702 − 948 = 754   **63.** 1607 − 869 = 738   **64.** 5933 − 3754 = 2179   **65.** 7293 − 3748 = 3545

66. 9407 − 2918 = 6489   **67.** 3706 − 2957 = 749   **68.** 8605 − 7716 = 889   **69.** 8052 − 2709 = 5343

70. 80,305 − 9176 = 71,129   **71.** 70,702 − 4239 = 66,463   **72.** 10,004 − 9306 = 698   **73.** 80,009 − 63,419 = 16,590

74. 70,618 − 41,213 = 29,405   **75.** 80,053 − 27,649 = 52,404   **76.** 70,700 − 21,076 = 49,624   **77.** 80,800 − 42,023 = 38,777

78. 2600 − 1972 = 628   **79.** 8400 − 3762 = 4638   **80.** 9003 − 2471 = 6532   **81.** 6004 − 2392 = 3612

82. 8202 − 3916 = 4286   **83.** 7050 − 4137 = 2913   **84.** 7015 − 2973 = 4042   **85.** 4207 − 1624 = 2583

86. 7005 − 1796 = 5209   **87.** 8003 − 2735 = 5268   **88.** 20,005 − 9,627 = 10,378   **89.** 80,004 − 8,237 = 71,767

Quick Quiz (Objective 1.3B)
Subtract.
1. 9344 − 793   8551
2. 75,068 − 9499   65,569

90. Find 10,051 less 9027.   1024

91. Find 17,031 less 5792.   11,239

92. Find the difference between 1003 and 447.
556

93. What is 29,874 minus 21,392?   8482

94. What is 29,797 less than 68,005?   38,208

95. What is 69,379 less than 70,004?   625

96. What is 25,432 decreased by 7994?   17,438

97. What is 86,701 decreased by 9976?   76,725

For Exercises 98 to 101, use the relationship between addition and subtraction
to complete the statement.

98. ___ + 39 = 104
65

99. 67 + ___ = 90
23

100. ___ + 497 = 862
365

101. 253 + ___ = 4901
4648

For Exercises 102 to 107, use a calculator to subtract. Then round the numbers to
the nearest ten-thousand and use estimation to determine whether the difference is
reasonable.

102.   80,032
          − 19,605
    Cal.: 60,427
    Est.: 60,000

103.   90,765
          − 60,928
    Cal.: 29,837
    Est.: 30,000

104.   32,574
          − 10,961
    Cal.: 21,613
    Est.: 20,000

105.   96,430
          − 59,762
    Cal.: 36,668
    Est.: 40,000

106.   567,423
          − 208,444
    Cal.: 358,979
    Est.: 360,000

107.   300,712
          − 198,714
    Cal.: 101,998
    Est.: 100,000

Objective C   **To solve application problems**

108. **Banking**   You have $304 in your checking account. If you write a check
for $139, how much is left in your checking account? $165

109. **Identity Theft**   Use the graph at the right to
determine the increase in the number of identity theft complaints from 2001 to 2002.
75,621 complaints

110. **Consumerism**   The tennis coach at a high school
purchased a video camera that costs $1079 and
made a down payment of $180. Find the amount
that remains to be paid. $899

111. **Consumerism**   Rod Guerra, an engineer, purchased a used car that cost $11,225 and made a
down payment of $950. Find the amount that
remains to be paid. $10,275

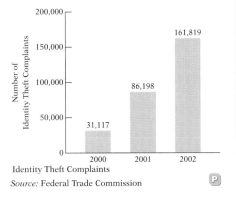

Identity Theft Complaints

Source: Federal Trade Commission

Quick Quiz (Objective 1.3C)

Note: Question 1 is a one-step problem. Question 2 is
a two-step problem.

1. After a trip of 728 miles, the odometer of your car
read 65,412 miles. What was the odometer
reading at the beginning of your trip?
64,684 miles

2. You had a bank balance of $843. You then wrote
checks for $192, $65, and $19. Find the new bank
balance. $567

112. Earth Science   Use the graph at the right to find the difference between the maximum height to which Great Fountain geyser erupts and the maximum height to which Valentine erupts.   15 feet

113. Earth Science   According to the graph at the right, how much higher is the eruption of the Giant than that of Old Faithful?   25 feet

114. Education   The National Center for Education Statistics estimates that 850,000 women and 587,000 men will earn a bachelor's degree in 2012. How many more women than men are expected to earn a bachelor's degree in 2012?   263,000 more women than men

Demographics   The graph at the right shows the expected U.S. population aged 100 and over, every 2 years from 2010 to 2020. Use this information for Exercises 115 and 116.

115. What is the expected growth in the population aged 100 and over during the 10-year period?   106,000

116. a. Which 2-year period has the smallest expected increase in the number of people aged 100 and over?   2010 to 2012

   b. Which 2-year period has the greatest expected increase?   2018 to 2020

117. Expense Accounts   A sales executive has a monthly expense account of $1500. This month the executive has already spent $479 for transportation, $268 for food, and $317 for lodging. Find the amount remaining in this month's expense account. $436

118. Finances   You had a credit card balance of $409 before you used the card to purchase books for $168, CDs for $36, and a pair of shoes for $97. You then made a payment to the credit card company of $350. Find your new credit card balance. $360

APPLYING THE CONCEPTS

119. Answer true or false.
   a. The phrases "the difference between 9 and 5" and "5 less than 9" mean the same thing.   True
   b. $9 - (5 - 3) = (9 - 5) - 3$.   False
   c. Subtraction is an associative operation. *Hint:* See part b of this exercise.   False

120. Make up a word problem for which the difference between 15 and 8 is the answer.

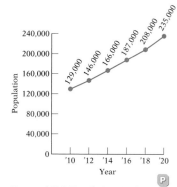

The Maximum Heights of the Eruptions of Six Geysers at Yellowstone National Park

Expected U.S. Population Aged 100 and Over
Source: Census Bureau

Answers to Writing Exercises

120. Answers will vary. For example:

Pat has earned 15 college credits, and Leslie has earned 8 college credits. How many more college credits has Pat earned?

7 college credits

1.4 Multiplication of Whole Numbers

Objective A To multiply a number by a single digit

Six boxes of toasters are ordered. Each box contains eight toasters. How many toasters are ordered?

This problem can be worked by adding 6 eights.

$$8 + 8 + 8 + 8 + 8 + 8 = 48$$

This problem involves repeated addition of the same number and can be worked by a shorter process called **multiplication.** Multiplication is the repeated addition of the same number.

8 + 8 + 8 + 8 + 8 + 8 = 48

The numbers that are multiplied are called **factors.** The result is called the **product.**

or

6 × 8 = 48

Factor Factor Product

The product of 6 × 8 can be represented on the number line. The arrow representing the whole number 8 is repeated 6 times. The result is the arrow representing 48.

The times sign "×" is only one symbol that is used to indicate multiplication. Each of the expressions that follow represent multiplication.

$$7 \times 8 \qquad 7 \cdot 8 \qquad 7(8) \qquad (7)(8) \qquad (7)8$$

As with addition, there are some useful properties of multiplication.

Multiplication Property of Zero

The product of a number and zero is zero.

$0 \times 4 = 0$
$7 \times 0 = 0$

Multiplication Property of One

The product of a number and one is the number.

$1 \times 6 = 6$
$8 \times 1 = 8$

Commutative Property of Multiplication

Two numbers can be multiplied in either order. The product will be the same.

$4 \times 3 = 3 \times 4$
$12 = 12$

Associative Property of Multiplication

Grouping the numbers to be multiplied in any order gives the same result. Do the multiplication inside the parentheses first.

$(4 \times 2) \times 3 = 4 \times (2 \times 3)$
$8 \quad \times 3 = 4 \times \quad 6$
$24 = 24$

Study Tip

Some students think that they can "coast" at the beginning of this course because the topic of Chapter 1 is whole numbers. However, this chapter lays the foundation for the entire course. Be sure you know and understand all the concepts presented. For example, study the properties of multiplication presented in this lesson.

Objective 1.4A

New Vocabulary
multiplication
factors
product

New Properties
Multiplication Property of Zero
Multiplication Property of One
Commutative Property of Multiplication
Associative Property of Multiplication

Discuss the Concepts
1. Provide at least three examples of situations in which we multiply a number by a single digit.
2. What is the difference between the Multiplication Property of Zero and the Multiplication Property of One?

Concept Check
Which property justifies the statement?
1. 8 × 12 = 12 × 8
   Commutative Property of Multiplication
2. 39 × 0 = 0
   Multiplication Property of Zero
3. 2 × (3 × 9) = (2 × 3) × 9
   Associative Property of Multiplication
4. 1 × 45 = 45
   Multiplication Property of One

Optional Student Activity
Solve the multiplication problem below. Each letter stands for a single-digit number.

```
    STRAW
  ×     4
    WARTS
```
21,978 × 4 = 87,912

In-Class Examples (Objective 1.4A)
Multiply.

1. 83
      × 9
     747

2. 245
      × 5
     1225

3. 7894
      × 6
     47,364

The basic facts for multiplying one-digit numbers should be memorized. Multiplication of larger numbers requires the repeated use of the basic multiplication facts.

HOW TO   Multiply: 37×4

$$\begin{array}{r} \overset{2}{} \\ 3\ 7 \\ \times\quad 4 \\ \hline 8 \end{array}$$

- $4 \times 7 = 28$ (2 tens + 8 ones). Write the 8 in the ones column and carry the 2 to the tens column.

$$\begin{array}{r} \overset{2}{} \\ 3\ 7 \\ \times\quad 4 \\ \hline 14\ 8 \end{array}$$

- The 3 in 37 is 3 tens.

  $\begin{array}{r} 4 \times 3 \text{ tens} = \ 12 \text{ tens} \\ \text{Add the carry digit.} \quad +\ 2 \text{ tens} \\ \hline 14 \text{ tens} \end{array}$

- Write the 14.

The phrases below are used to indicate the operation of multiplication. An example is shown at the right of each phrase.

times	7 times 3	$7 \cdot 3$
the product of	the product of 6 and 9	$6 \cdot 9$
multiplied by	8 multiplied by 2	$2 \cdot 8$

Example 1   Multiply: 735×9

Solution

$$\begin{array}{r} ^{3\ 4} \\ 735 \\ \times\quad 9 \\ \hline 6615 \end{array}$$

- $9 \times 5 = 45$
  Write the 5 in the ones column. Carry the 4 to the tens column.
  $9 \times 3 = 27, 27 + 4 = 31$
  $9 \times 7 = 63, 63 + 3 = 66$

You Try It 1   Multiply: 648×7

Your solution   4536

Solution on p. S2

Objective B   **To multiply larger whole numbers**

Note the pattern when the following numbers are multiplied.

Multiply the nonzero part of the factors.

Now attach the same number of zeros to the product as the total number of zeros in the factors.

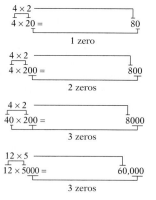

$$4 \times 2$$
$$4 \times 20 = \qquad\qquad 80$$
1 zero

$$4 \times 2$$
$$4 \times 200 = \qquad\qquad 800$$
2 zeros

$$4 \times 2$$
$$40 \times 200 = \qquad\qquad 8000$$
3 zeros

$$12 \times 5$$
$$12 \times 5000 = \qquad\qquad 60,000$$
3 zeros

Objective 1.4B

Discuss the Concepts

1. Provide at least three examples of situations in which we multiply by larger whole numbers.

2. Explain how to multiply a number by 100.

Concept Check

What will be the last digit in the product?

1. 438×617 6
2. 9152×306 2
3. $32,480 \times 579$ 0

In-Class Examples (Objective 1.4B)

Multiply.

1. $\begin{array}{r}15 \\ \times\ 29 \\ \hline 435\end{array}$	2. $\begin{array}{r}935 \\ \times\ 46 \\ \hline 43{,}010\end{array}$	3. $\begin{array}{r}4198 \\ \times\quad 24 \\ \hline 100{,}752\end{array}$	4. $\begin{array}{r}6572 \\ \times\quad 294 \\ \hline 1{,}932{,}168\end{array}$	5. $\begin{array}{r}728 \\ \times 306 \\ \hline 222{,}768\end{array}$

HOW TO   Find the product of 47 and 23.

Multiply by the ones digit.	Multiply by the tens digit.	Add.

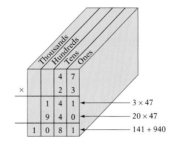

$$\begin{array}{r} 47 \\ \times\ 23 \\ \hline 141 \end{array} (= 47 \times 3)$$

$$\begin{array}{r} 47 \\ \times\ 23 \\ \hline 141 \\ 940 \end{array} (= 47 \times 20)$$

$$\begin{array}{r} 47 \\ \times\ 23 \\ \hline 141 \\ 940 \\ \hline 1081 \end{array}$$

Writing the 0 is optional.

3 × 47
20 × 47
141 + 940

The place-value chart illustrates the placement of the products.

Note the placement of the products when we are multiplying by a factor that contains a zero.

HOW TO   Multiply: 439×206

$$\begin{array}{r} 439 \\ \times\ 206 \\ \hline 2634 \\ 000 \\ 878 \\ \hline 90{,}434 \end{array}$$

0×439

When working the problem, we usually write only one zero. Writing this zero ensures the proper placement of the products.

$$\begin{array}{r} 439 \\ \times\ 206 \\ \hline 2634 \\ 8780 \\ \hline 90{,}434 \end{array}$$

Example 2   Find 829 multiplied by 603.

Solution

$$\begin{array}{r} 829 \\ \times\ 603 \\ \hline 2487 \\ 49740 \\ \hline 499{,}887 \end{array}$$

• $3 \times 829 = 2487$
• Write a zero in the tens column for 0×829.
• $6 \times 829 = 4974$

You Try It 2   Multiply: 756×305

Your solution   230,580

Solution on p. S2

● E S T I M A T I O N ●

Estimating the Product of Two Whole Numbers

Calculate 3267×389. Then use estimation to determine whether the product is reasonble.

Multiply to find the exact product.

$3267 \times 389 = 1{,}270{,}863$

To estimate the product, round each number so there is one nonzero digit. Then multiply. The estimated answer is 1,200,000, which is very close to the exact product 1,270,863.

$$\begin{array}{r} 3267 \approx 3000 \\ \times\ 389 \approx \times\ 400 \\ \hline 1{,}200{,}000 \end{array}$$

Optional Student Activity

Assume the pattern continues, and find the next three numbers in the pattern.

1. 4, 8, 12, 16, . . .   20, 24, 28

2. 17, 170, 1700, . . .
17,000; 170,000; 1,700,000

3. 10, 200, 4000, . . .
80,000; 1,600,000; 32,000,000

Objective 1.4C

Optional Student Activity

Shown below is the cost per person for bus trips organized by a travel agency. How much more would it cost for you and four members of your family to take the New York trip than to take the Montreal trip?

2 days	Foxwood	$139
3 days	New York	$479
4 days	Montreal	$389
4 days	Niagara Falls	$409

$450

Note: Have students describe their solutions. Some students will have found the difference between 5 × 479 and 5 × 389; others will have found the product of 5 and the difference between 479 and 389.

Objective C   To solve application problems

Example 3

An auto mechanic receives a salary of $1050 each week. How much does the auto mechanic earn in 4 weeks?

Strategy

To find the mechanic's earnings for 4 weeks, multiply the weekly salary (1050) by the number of weeks (4).

Solution

```
  1050
×    4
  4200
```

The mechanic earns $4200 in 4 weeks.

You Try It 3

A new-car dealer receives a shipment of 37 cars each month. Find the number of cars the dealer will receive in 12 months.

Your strategy

Your solution   444 cars

Example 4

A press operator earns $640 for working a 40-hour week. This week the press operator also worked 7 hours of overtime at $26 an hour. Find the press operator's total pay for the week.

Strategy

To find the press operator's total pay for the week:

- Find the overtime pay by multiplying the hours of overtime (7) by the overtime rate of pay (26).
- Add the weekly salary (640) to the overtime pay.

Solution

```
   26                 640
×   7                +182
  182 overtime pay    822
```

The press operator earned $822 this week.

You Try It 4

The buyer for Ross Department Store can buy 80 men's suits for $4800. Each sports jacket will cost the store $23. The manager orders 80 men's suits and 25 sports jackets. What is the total cost of the order?

Your strategy

Your solution $5375

Solutions on p. S2

In-Class Examples (Objective 1.4C)

Note: Example 1 is a one-step problem. Example 2 is a two-step problem.

1. The Environmental Protection Agency estimates that a motorcycle gets 43 miles per gallon of gasoline. How many miles can it get on 8 gallons of gasoline?   344 miles

2. A gasoline storage tank contains 66,000 gallons of gasoline. A valve is opened that lets out 30 gallons each minute. How many gallons remain in the tank after 40 minutes?   64,800 gallons

1.4 Exercises

Objective A   To multiply a number by a single digit

Suggested Assignment
Exercises 1–3, odds
Exercises 5–85, every other odd
Exercises 89–105, odds
Exercises 106, 108, 109

For Exercises 1 to 4, write the expression as a product.

1. $2 + 2 + 2 + 2 + 2 + 2$
6×2 or $6 \cdot 2$

2. $4 + 4 + 4 + 4 + 4$
5×4 or $5 \cdot 4$

3. $7 + 7 + 7 + 7$
4×7 or $4 \cdot 7$

4. $18 + 18 + 18$
3×18 or $3 \cdot 18$

For Exercises 5 to 39, multiply.

5. $\begin{array}{r} 3 \\ \times 4 \\ \hline 12 \end{array}$
6. $\begin{array}{r} 2 \\ \times 8 \\ \hline 16 \end{array}$
7. $\begin{array}{r} 5 \\ \times 7 \\ \hline 35 \end{array}$
8. $\begin{array}{r} 6 \\ \times 4 \\ \hline 24 \end{array}$
9. $\begin{array}{r} 5 \\ \times 5 \\ \hline 25 \end{array}$

10. $\begin{array}{r} 7 \\ \times 7 \\ \hline 49 \end{array}$
11. $\begin{array}{r} 0 \\ \times 7 \\ \hline 0 \end{array}$
12. $\begin{array}{r} 8 \\ \times 0 \\ \hline 0 \end{array}$
13. $\begin{array}{r} 8 \\ \times 9 \\ \hline 72 \end{array}$
14. $\begin{array}{r} 7 \\ \times 6 \\ \hline 42 \end{array}$

15. $\begin{array}{r} 66 \\ \times 3 \\ \hline 198 \end{array}$
16. $\begin{array}{r} 70 \\ \times 4 \\ \hline 280 \end{array}$
17. $\begin{array}{r} 67 \\ \times 5 \\ \hline 335 \end{array}$
18. $\begin{array}{r} 127 \\ \times 9 \\ \hline 1143 \end{array}$
19. $\begin{array}{r} 623 \\ \times 4 \\ \hline 2492 \end{array}$

20. $\begin{array}{r} 802 \\ \times 5 \\ \hline 4010 \end{array}$
21. $\begin{array}{r} 607 \\ \times 9 \\ \hline 5463 \end{array}$
22. $\begin{array}{r} 300 \\ \times 5 \\ \hline 1500 \end{array}$
23. $\begin{array}{r} 600 \\ \times 7 \\ \hline 4200 \end{array}$
24. $\begin{array}{r} 906 \\ \times 8 \\ \hline 7248 \end{array}$

25. $\begin{array}{r} 703 \\ \times 9 \\ \hline 6327 \end{array}$
26. $\begin{array}{r} 127 \\ \times 5 \\ \hline 635 \end{array}$
27. $\begin{array}{r} 632 \\ \times 3 \\ \hline 1896 \end{array}$
28. $\begin{array}{r} 559 \\ \times 4 \\ \hline 2236 \end{array}$
29. $\begin{array}{r} 632 \\ \times 8 \\ \hline 5056 \end{array}$

30. $\begin{array}{r} 524 \\ \times 4 \\ \hline 2096 \end{array}$
31. $\begin{array}{r} 337 \\ \times 5 \\ \hline 1685 \end{array}$
32. $\begin{array}{r} 841 \\ \times 6 \\ \hline 5046 \end{array}$
33. $\begin{array}{r} 6709 \\ \times 7 \\ \hline 46,963 \end{array}$
34. $\begin{array}{r} 3608 \\ \times 5 \\ \hline 18,040 \end{array}$

35. $\begin{array}{r} 8568 \\ \times 7 \\ \hline 59,976 \end{array}$
36. $\begin{array}{r} 5495 \\ \times 4 \\ \hline 21,980 \end{array}$
37. $\begin{array}{r} 4780 \\ \times 4 \\ \hline 19,120 \end{array}$
38. $\begin{array}{r} 3690 \\ \times 5 \\ \hline 18,450 \end{array}$
39. $\begin{array}{r} 9895 \\ \times 2 \\ \hline 19,790 \end{array}$

40. Find the product of 5, 7, and 4.   140

41. Find the product of 6, 2, and 9.   108

42. Find the product of 5304 and 9.   47,736

43. Find the product of 458 and 8.   3664

Quick Quiz (Objective 1.4A)
Multiply.

1. $\begin{array}{r} 78 \\ \times 4 \\ \hline 312 \end{array}$
2. $\begin{array}{r} 649 \\ \times 6 \\ \hline 3894 \end{array}$
3. $\begin{array}{r} 3724 \\ \times 5 \\ \hline 18,620 \end{array}$

44. What is 3208 multiplied by 7? 22,456   **45.** What is 5009 multiplied by 4? 20,036

46. What is 3105 times 6? 18,630   **47.** What is 8957 times 8? 71,656

> **Objective B**   **To multiply larger whole numbers**

For Exercises 48 to 79, multiply.

48.
$$16 \times 21$$
336

49.
$$18 \times 24$$
432

50.
$$35 \times 26$$
910

51.
$$27 \times 72$$
1944

52.
$$693 \times 91$$
63,063

53.
$$581 \times 72$$
41,832

54.
$$419 \times 80$$
33,520

55.
$$727 \times 60$$
43,620

56.
$$8279 \times 46$$
380,834

57.
$$9577 \times 35$$
335,195

58.
$$6938 \times 78$$
541,164

59.
$$8875 \times 67$$
594,625

60.
$$7035 \times 57$$
400,995

61.
$$6702 \times 48$$
321,696

62.
$$3009 \times 35$$
105,315

63.
$$6003 \times 57$$
342,171

64.
$$809 \times 530$$
428,770

65.
$$607 \times 460$$
279,220

66.
$$800 \times 325$$
260,000

67.
$$700 \times 274$$
191,800

68.
$$987 \times 349$$
344,463

69.
$$688 \times 674$$
463,712

70.
$$312 \times 134$$
41,808

71.
$$423 \times 427$$
180,621

72.
$$379 \times 500$$
189,500

73.
$$684 \times 700$$
478,800

74.
$$985 \times 408$$
401,880

75.
$$758 \times 209$$
158,422

76.
$$3407 \times 309$$
1,052,763

77.
$$5207 \times 902$$
4,696,714

78.
$$4258 \times 986$$
4,198,388

79.
$$6327 \times 876$$
5,542,452

80. What is 5763 times 45? 259,335   **81.** What is 7349 times 27? 198,423

82. Find the product of 2, 19, and 34. 1292   **83.** Find the product of 6, 73, and 43. 18,834

Quick Quiz (Objective 1.4B)

Multiply.

1.
$$75 \times 42$$
3150

2.
$$495 \times 37$$
18,315

3.
$$724 \times 831$$
601,644

4.
$$392 \times 407$$
159,544

84. What is 376 multiplied by 402? 151,152

85. What is 842 multiplied by 309? 260,178

86. Find the product of 3005 and 233,489.
701,634,445

87. Find the product of 9007 and 34,985.
315,109,895

For Exercises 88 to 95, use a calculator to multiply. Then use estimation to determine whether the product is reasonable.

88.   8745
   $\times$ 63
Cal.: 550,935
Est.: 540,000

89.   4732
   $\times$ 93
Cal.: 440,076
Est.: 450,000

90.   2937
   $\times$ 206
Cal.: 605,022
Est.: 600,000

91.   8941
   $\times$ 726
Cal.: 6,491,166
Est.: 6,300,000

92.   3097
   $\times$ 1025
Cal.: 3,174,425
Est.: 3,000,000

93.   6379
   $\times$ 2936
Cal.: 18,728,744
Est.: 18,000,000

94.   32,508
   $\times$ 591
Cal.: 19,212,228
Est.: 18,000,000

95.   62,504
   $\times$ 923
Cal.: 57,691,192
Est.: 54,000,000

Objective C   **To solve application problems**

96. **Fuel Efficiency**   Rob Hill owns a compact car that averages 43 miles on 1 gallon of gas. How many miles could the car travel on 12 gallons of gas?   516 miles

97. **Fuel Efficiency**   A plane flying from Los Angeles to Boston uses 865 gallons of jet fuel each hour. How many gallons of jet fuel were used on a 6-hour flight?   5190 gallons

98. **Geometry**   The perimeter of a square is equal to four times the length of a side of the square. Find the perimeter of a square whose side measures 16 miles.   64 miles

16 mi

99. **Geometry**   To find the area of a square, multiply the length of one side of the square times itself. What is the area of a square whose side measures 16 miles? The area of the square will be in square miles.
256 square miles

100. **Geometry**   The area of a rectangle is equal to the product of the length of the rectangle times its width. Find the area of a rectangle that has a length of 24 meters and a width of 15 meters. The area will be in square meters.   360 square meters

15 m
24 m

Interior Design   A lighting consultant to a bank suggests that the bank lobby contain 43 can lights, 15 high-intensity lights, 20 fire safety lights, and one chandelier. The table at the right gives the costs for each type of light from two companies. Use this table for Exercises 101 and 102.

	Company A	Company B
Can lights	$2 each	$3 each
High-intensity	$6 each	$4 each
Fire safety	$12 each	$11 each
Chandelier	$998 each	$1089 each

101. Which company offers the lights for the lower total price?   Company A

102. How much can the lighting consultant save by purchasing the lights from the company that offers the lower total price? $84

Quick Quiz (Objective 1.4C)

Note: Question 1 is a one-step problem. Question 2 is a two-step problem.

1. A mechanic has a car payment of $197 each month. What is the total of the car payments over a 12-month period? $2364

2. A baker can buy 1000 pounds of flour for $300 and one 100-pound bag of sugar for $64. The baker orders 1000 pounds of flour and fifteen 100-pound bags of sugar. What is the total cost of the order? $1260

Construction   The table at the right shows the hourly wages of four different job classifications at a small construction company. Use this table for Exercises 103 to 105.

Type of Work	Wage per Hour
Electrician	$34
Plumber	$30
Clerk	$16
Bookkeeper	$20

P

103. The owner of this company wants to provide the electrical installation for a new house. On the basis of the architectural plans for the house, it is estimated that it will require 3 electricians, each working 50 hours, to complete the job. What is the estimated cost for the electricians' labor? $5100

104. Carlos Vasquez, a plumbing contractor, hires 4 plumbers from this company at the hourly wage given in the table. If each plumber works 23 hours, what are the total wages paid by Carlos? $2760

105. The owner of this company estimates that remodeling a kitchen will require 1 electrician working 30 hours and 1 plumber working 33 hours. This project also requires 3 hours of clerical work and 4 hours of bookkeeping. What is the total cost for these four components of this remodeling? $2138

APPLYING THE CONCEPTS

106. Determine whether each of the following statements is always true, sometimes true, or never true.
   a. A whole number times zero is zero.   Always true
   b. A whole number times one is the whole number.   Always true
   c. The product of two whole numbers is greater than either one of the whole numbers.   Sometimes true

107. **Safety**   According to the National Safety Council, in a recent year a death resulting from an accident occurred at the rate of 1 every 5 minutes. At this rate, how many accidental deaths occurred each hour? Each day? Throughout the year? Explain how you arrived at your answers.   12 deaths each hour; 288 deaths each day; 105,120 deaths each year

108. **Demographics**   According to the Population Reference Bureau, in the world today, 261 people are born every minute and 101 people die every minute. Using this statistic, what is the increase in the world's population every hour? Every day? Every week? Every year? Use a 365-day year. Explain how you arrived at your answers.
   9600 people every hour; 230,400 people every day; 1,612,800 people every week; 84,096,000 people every year

109. Pick your favorite number between 1 and 9. Multiply the number by 3. Multiply that product by 37,037. How is the product related to your favorite number? Explain why this works. (*Suggestion:* Multiply 3 and 37,037 first.)

110. There are quite a few tricks based on whole numbers. Here's one about birthdays. Write down the month in which you were born. Multiply by 5. Add 7. Multiply by 20. Subtract 100. Add the day of the month on which you were born. Multiply by 4. Subtract 100. Multiply by 25. Add the year you were born. Subtract 3400. The answer is the month/day/year of your birthday.

Answers to Writing Exercises

109. $3 \times 37{,}037 = 111{,}111$. By the Multiplication Property of One, a number times 1 equals the number. Because the product of 3 and 37,037 is 111,111, the product of 111,111 and a number between 1 and 9 will be a six-digit number in which all digits equal the number between 1 and 9.

1.5 Division of Whole Numbers

Objective A — To divide by a single digit with no remainder in the quotient

Division is used to separate objects into equal groups.

A store manager wants to display 24 new objects equally on 4 shelves. From the diagram, we see that the manager would place 6 objects on each shelf.

The manager's division problem can be written as follows:

TAKE NOTE
The **divisor** is the number that is divided into another number. The **dividend** is the number into which the divisor is divided. The result is the **quotient.**

Number of shelves — **Divisor**
Number on each shelf — **Quotient**
$4\overline{)24}$
Number of objects — **Dividend**

Note that the quotient multiplied by the divisor equals the dividend.

$4\overline{)24}^{6}$ because $\boxed{6 \atop \text{Quotient}} \times \boxed{4 \atop \text{Divisor}} = \boxed{24 \atop \text{Dividend}}$

$9\overline{)54}^{6}$ because $6 \times 9 = 54$

$8\overline{)40}^{5}$ because $5 \times 8 = 40$

Here are some important quotients and the properties of zero in division:

Properties of One in Division

Any whole number, except zero, divided by itself is 1.
$8\overline{)8}^{1}$ $14\overline{)14}^{1}$ $10\overline{)10}^{1}$

Any whole number divided by 1 is the whole number.
$1\overline{)9}^{9}$ $1\overline{)27}^{27}$ $1\overline{)10}^{10}$

Properties of Zero in Division

Zero divided by any other whole number is zero.
$7\overline{)0}^{0}$ $13\overline{)0}^{0}$ $10\overline{)0}^{0}$

Division by zero is not allowed.
$0\overline{)8}^{?}$ There is no number whose product with 0 is 8.

Integrating Technology
Enter 8 ÷ 0 = on your calculator. An error message is displayed because division by zero is not allowed.

Objective 1.5A

New Vocabulary
division
divisor
quotient
dividend

New Properties
Properties of zero and one in division

Discuss the Concepts
1. Explain the difference between 0 ÷ 9 and 9 ÷ 0.
2. Explain why multiplication can be used to check the answer to a division problem.

Concept Check
For each division problem, find the quotient and then write a related multiplication problem.
1. 98 ÷ 7   14; 14 × 7 = 98
2. 1968 ÷ 6
   328; 328 × 6 = 1968

Instructor Note
One method to help students understand that division by zero is not allowed is to relate it to the problem of the store manager above. Ask how the manager can display 24 items on 4 shelves; on 3 shelves; on 2 shelves; on 1 shelf; on 0 shelves!

In-Class Examples (Objective 1.5A)
Divide.
1. $9\overline{)711}$ 79
2. $8\overline{)6728}$ 841
3. $4\overline{)78,384}$ 19,596

Optional Student Activity

Divide 111 by the sum of the digits (3). Divide 222 by the sum of the digits (6). Divide 333 by the sum of the digits (9). Continue this pattern through division of 999 by the sum of the digits (27). Describe the pattern. Why does this occur?

The quotient is always 37. Both the dividend and the divisor are multiplied by the same number.

When the dividend is a larger whole number, the digits in the quotient are found in steps.

HOW TO   Divide $4\overline{)3192}$ and check.

$$
\begin{array}{r}
7 \\
4\overline{)3192} \\
-28 \\
\hline
39
\end{array}
$$

- Think $4\overline{)31}$.
- Subtract 7×4.
- Bring down the 9.

$$
\begin{array}{r}
79 \\
4\overline{)3192} \\
-28 \\
\hline
39 \\
-36 \\
\hline
32
\end{array}
$$

- Think $4\overline{)39}$.
- Subtract 9×4.
- Bring down the 2.

$$
\begin{array}{r}
798 \\
4\overline{)3192} \\
-28 \\
\hline
39 \\
-36 \\
\hline
32 \\
-32 \\
\hline
0
\end{array}
$$

- Think $4\overline{)32}$.
- Subtract 8×4.

Check:
$$
\begin{array}{r}
798 \\
\times 4 \\
\hline
3192
\end{array}
$$

The place-value chart can be used to show why this method works.

$$
\begin{array}{r}
\text{HUNDREDS} \ \text{TENS} \ \text{ONES} \\
7 \ \ 9 \ \ 8 \\
4\overline{)3 \ 1 \ 9 \ 2} \\
-2 \ 8 \ 0 \ 0 \quad \text{7 hundreds} \times 4 \\
\hline
3 \ 9 \ 2 \\
-3 \ 6 \ 0 \quad \text{9 tens} \times 4 \\
\hline
3 \ 2 \\
-3 \ 2 \quad \text{8 ones} \times 4 \\
\hline
0
\end{array}
$$

There are other ways of expressing division.

54 divided by 9 equals 6.

54   ÷   9 equals 6.

$\dfrac{54}{9}$ equals 6.

Example 1   Divide $7\overline{)56}$ and check.

Solution
$$\frac{8}{7\overline{)56}}$$

Check: $8 \times 7 = 56$

You Try It 1   Divide $9\overline{)63}$ and check.

Your solution   7

Example 2   Divide $2808 \div 8$ and check.

Solution
$$\begin{array}{r} 351 \\ 8\overline{)\,2808} \\ -24 \\ \hline 40 \\ -40 \\ \hline 08 \\ -8 \\ \hline 0 \end{array}$$

Check: $351 \times 8 = 2808$

You Try It 2   Divide $4077 \div 9$ and check.

Your solution   453

Example 3   Divide $7\overline{)2856}$ and check.

Solution
$$\begin{array}{r} 408 \\ 7\overline{)\,2856} \\ -28 \\ \hline 05 \\ -0 \\ \hline 56 \\ -56 \\ \hline 0 \end{array}$$
   • Think $7\overline{)5}$. Place 0 in quotient.
   • Subtract 0×7.
   • Bring down the 6.

Check: $408 \times 7 = 2856$

You Try It 3   Divide $9\overline{)6345}$ and check.

Your solution   705

Solutions on pp. S2–S3

Objective B   **To divide by a single digit with a remainder in the quotient**

Sometimes it is not possible to separate objects into a whole number of equal groups.

A baker has 14 muffins to pack into 3 boxes. Each box holds 4 muffins. From the diagram, we see that after the baker places 4 muffins in each box, there are 2 left over. The 2 is called the **remainder.**

Objective 1.5B

New Vocabulary
remainder

36

Discuss the Concepts

In what situation does a division problem have a remainder?

Instructor Note

Some students have difficulty with the concept of remainder. Have these students try to give 15 pennies to 4 students so that each student has the same number of pennies.

Concept Check

Which of the following numbers divide evenly into 144? (That is, there is no remainder.)

1	10
2	11
3	12
4	13
5	14
6	15
7	16
8	17
9	18

1, 2, 3, 4, 6, 8, 9, 12, 16, 18

Optional Student Activity

The number 10,981 is not divisible by 4. By rearranging the digits, find the largest possible number that is divisible by 4.
91,180

The clerk's division problem could be written

$$
\begin{array}{r}
4 \quad \leftarrow \textbf{Quotient (Number in each box)} \\
\textbf{Divisor (Number of boxes)} \rightarrow 3\overline{)\ 14} \quad \leftarrow \textbf{Dividend (Total number of objects)} \\
-12 \\
\overline{2} \quad \leftarrow \textbf{Remainder (Number left over)}
\end{array}
$$

The answer to a division problem with a remainder is frequently written

$$3\overline{)14}^{\,4\ \text{r}2}$$

Note that $\boxed{\begin{array}{c}4\\\text{Quotient}\end{array}} \times \boxed{\begin{array}{c}3\\\text{Divisor}\end{array}} + \boxed{\begin{array}{c}2\\\text{Remainder}\end{array}} = \boxed{\begin{array}{c}14\\\text{Dividend}\end{array}}$.

Example 4   Divide $4\overline{)2522}$ and check.

Solution

$$
\begin{array}{r}
630\ \text{r}2 \\
4\overline{)\ 2522} \\
-24 \\
\overline{12} \\
-12 \\
\overline{02} \\
-\ 0 \\
\overline{2}
\end{array}
$$

• Think $4\overline{)2}$. Place 0 in quotient.
• Subtract 0×4.

Check: $(630 \times 4) + 2 =$
     $2520 \ + 2 = 2522$

You Try It 4   Divide $6\overline{)5225}$ and check.

Your solution   870 r5

Example 5   Divide $9\overline{)27,438}$ and check.

Solution

$$
\begin{array}{r}
3,048\ \text{r}6 \\
9\overline{)\ 27,438} \\
-27 \\
\overline{04} \\
-\ 0 \\
\overline{43} \\
-36 \\
\overline{78} \\
-72 \\
\overline{6}
\end{array}
$$

• Think $9\overline{)4}$.
• Subtract 0×9.

Check: $(3048 \times 9) + 6 =$
     $27,432 \ + 6 = 27,438$

You Try It 5   Divide $7\overline{)21,409}$ and check.

Your solution   3058 r3

Solutions on p. S3

In-Class Examples (Objective 1.5B)

Divide.

1. $8\overline{)547}$ 68 r3

2. $6\overline{)3743}$ 623 r5

3. $7\overline{)65,412}$ 9344 r4

Objective C   **To divide by larger whole numbers**

When the divisor has more than one digit, estimate at each step by using the first digit of the divisor. If that product is too large, lower the guess by 1 and try again.

Copyright © Houghton Mifflin Company. All rights reserved.

HOW TO   Divide $34\overline{)1598}$ and check.

$$\begin{array}{r} 5 \\ 34\overline{)\,1598} \\ -170 \end{array}$$ • Think $3\overline{)15}$.
• Subtract 5×34.

$$\begin{array}{r} 4 \\ 34\overline{)\,1598} \\ -136 \\ \hline 238 \end{array}$$ • Subtract 4×34.

170 is too large. Lower the guess by 1 and try again.

$$\begin{array}{r} 47 \\ 34\overline{)\,1598} \\ -136 \\ \hline 238 \\ -238 \\ \hline 0 \end{array}$$ • Think $3\overline{)23}$.
• Subtract 7×34.

Check: $\begin{array}{r} 47 \\ \times 34 \\ \hline 188 \\ 141 \\ \hline 1598 \end{array}$

The phrases below are used to indicate the operation of division. An example is shown at the right of each phrase.

| the quotient of | the quotient of 9 and 3 | $9 \div 3$ |
| divided by | 6 divided by 2 | $6 \div 2$ |

Study Tip

One of the key instructional features of this text is the Example/You Try It pairs. Each Example is completely worked. You are to solve the You Try It problems. When you are ready, check your solution against the one in the Solutions section. The solution for You Try It 6 below is on page S3 (see the reference at the bottom right of the You Try It.) See *AIM for Success* at the front of the book.

Example 6   Find 7077 divided by 34 and check.

Solution

$$\begin{array}{r} 208 \text{ r5} \\ 34\overline{)\,7077} \\ -68 \\ \hline 27 \\ -0 \\ \hline 277 \\ -272 \\ \hline 5 \end{array}$$

• Think $34\overline{)27}$.
• Place 0 in the quotient.
• Subtract 0×34.

Check: $(208 \times 34) + 5 =$
    $7072 \quad + 5 = 7077$

You Try It 6   Divide $4578 \div 42$ and check.

Your solution   109

Solution on p. S3

Objective 1.5C

Vocabulary to Review
divisor   [1.5A]
quotient   [1.5A]

Discuss the Concepts
Explain how to check the answer to a division problem that has a remainder.
Quotient × divisor + remainder = dividend

Optional Student Activity
Find a three-digit whole number such that, when it is divided by 11, the quotient is equal to the sum of its digits.   198

In-Class Examples (Objective 1.5C)
Divide.
1. $69\overline{)741}$ 10 r51
2. $96\overline{)6525}$ 67 r93
3. $73\overline{)29,645}$ 406 r7

Example 7   Find the quotient of 21,312 and 56 and check.

Solution

$$
\begin{array}{r}
380 \text{ r}32 \\
56\overline{)21{,}312} \\
-16\ 8 \\
\hline
4\ 51 \\
-4\ 48 \\
\hline
32 \\
-\ 0 \\
\hline
32
\end{array}
$$

Think 5)21.
4 × 56 is too large. Try 3.

Check: (380 × 56) + 32 =
21,280   + 32 = 21,312

You Try It 7   Divide 18,359 ÷ 39 and check.

Your solution   470 r29

Example 8   Divide 427)24,782 and check.

Solution

$$
\begin{array}{r}
58 \text{ r}16 \\
427\overline{)24{,}782} \\
-21\ 35 \\
\hline
3\ 432 \\
-3\ 416 \\
\hline
16
\end{array}
$$

Check: (58 × 427) + 16 =
24,766   + 16 = 24,782

You Try It 8   Divide 534)33,219 and check.

Your solution   62 r111

Example 9   Divide 386)206,149 and check.

Solution

$$
\begin{array}{r}
534 \text{ r}25 \\
386\overline{)206{,}149} \\
-193\ 0 \\
\hline
13\ 14 \\
-11\ 58 \\
\hline
1\ 569 \\
-1\ 544 \\
\hline
25
\end{array}
$$

Check: (534 × 386) + 25 =
206,124   + 25 = 206,149

You Try It 9   Divide 515)216,848 and check.

Your solution   421 r33

Solutions on p. S3

● E S T I M A T I O N ●

Estimating the Quotient of Two Whole Numbers

Calculate 36,936 ÷ 54. Then use estimation to determine whether the quotient is reasonable.

Divide to find the exact quotient.

36,936 ÷ 54 = 684

To estimate the quotient, round each number so there is one nonzero digit. Then divide. The estimated answer is 800, which is close to the exact quotient 684.

36,936 ÷ 54 ≈
40,000 ÷ 50 = 800

Objective D   To solve application problems

The **average** of several numbers is the sum of all the numbers divided by the number of those numbers.

$$\text{Average test score} = \frac{81 + 87 + 80 + 85 + 79 + 86}{6} = \frac{498}{6} = 83$$

HOW TO The table at the right shows what an upper-income family can expect to spend to raise a child to the age of 17 years. Find the average amount spent each year. Round to the nearest dollar.

Expenses to Raise a Child	
Housing	$89,580
Food	$35,670
Transportation	$32,760
Child care/education	$26,520
Clothing	$13,770
Health care	$13,380
Other	$30,090

Source: Department of Agriculture, *Expenditures on Children by Families* [P]

Strategy
To find the average amount spent each year:

- Add all the numbers in the table to find the total amount spent during the 17 years.
- Divide the sum by 17.

Solution

```
  89,580
  35,670
  32,760
  26,520
  13,770
  13,380
+ 30,090
---------
 241,770
```
Sum of all the costs

```
        14,221
   17) 241,770
      -17
      ----
       71
      -68
      ----
        3 7
       -3 4
       -----
         37
        -34
        ----
         30
        -17
        ----
         13
```

- When rounding to the nearest whole number, compare twice the remainder to the divisor. If twice the remainder is less than the divisor, drop the remainder. If twice the remainder is greater than or equal to the divisor, add 1 to the units digit of the quotient.

- Twice the remainder is 2 × 13 = 26. Because 26 > 17, add 1 to the units digit of the quotient.

The average amount spent each year to raise a child to the age of 17 is $14,222.

Objective 1.5D

New Vocabulary
average

Instructor Note
Ask students why the rounding rule given at the left works. You want them to discover that if twice the remainder is less than the divisor, the next digit is less than 5; if twice the remainder is greater than or equal to the divisor, the next digit is greater than or equal to 5.

In-Class Examples (Objective 1.5D)

1. A lottery prize of $857,000 is divided equally among 4 winners. What amount does each winner receive? $214,250

2. A shipment of 9810 diodes requires testing. The diodes are divided equally among 15 employees. How many diodes must each employee test?
654 diodes

40

Optional Student Activity

A community organization is planning a bus trip to the city. Members have signed up either to go to the Museum of Fine Arts or to see *Phantom of the Opera.* The number attending each is shown below. Buses will be ordered for the event, and each bus will carry 44 members. It is too expensive per person to have fewer than 20 people on a bus. Vans seating 11 people can be ordered through the bus company. How many buses are needed for the trip? Should any vans be ordered? If so, how many?

Museum of Fine Arts: 82 people
Phantom of the Opera: 109 people

4 buses; 2 vans (for 15 extra people)

Example 10

Ngan Hui, a freight supervisor, shipped 192,600 bushels of wheat in 9 railroad cars. Find the amount of wheat shipped in each car.

Strategy
To find the amount of wheat shipped in each car, divide the number of bushels (192,600) by the number of cars (9).

Solution

$$
\begin{array}{r}
21{,}400 \\
9)\overline{\,192{,}600} \\
-18 \\
\hline
12 \\
-9 \\
\hline
36 \\
-36 \\
\hline
0
\end{array}
$$

Each car carried 21,400 bushels of wheat.

You Try It 10

Suppose a Michelin retail outlet can store 270 tires on 15 shelves. How many tires can be stored on each shelf?

Your strategy

Your solution 18 tires

Example 11

The used car you are buying costs $11,216. A down payment of $2000 is required. The remaining balance is paid in 48 equal monthly payments. What is the monthly payment?

Strategy
To find the monthly payment:
- Find the remaining balance by subtracting the down payment (2000) from the total cost of the car (11,216).
- Divide the remaining balance by the number of equal monthly payments (48).

Solution

$$
\begin{array}{r}
11{,}216 \\
-\ 2{,}000 \\
\hline
9{,}216
\end{array}
$$
Remaining balance

$$
\begin{array}{r}
192 \\
48)\overline{\,9216} \\
-48 \\
\hline
441 \\
-432 \\
\hline
96 \\
-96 \\
\hline
0
\end{array}
$$

The monthly payment is $192.

You Try It 11

A soft-drink manufacturer produces 12,600 cans of soft drink each hour. Cans are packed 24 to a case. How many cases of soft drink are produced in 8 hours?

Your strategy

Your solution 4200 cases

Solutions on pp. S3–S4

1.5 Exercises

Objective A   To divide by a single digit with no remainder in the quotient

Section 1.5

Suggested Assignment
Exercises 1–101, every other odd
Exercises 105–111, odds
More challenging problem:
  Exercise 112

For Exercises 1 to 20, divide.

1. $4\overline{)8}$ 2
2. $3\overline{)9}$ 3
3. $6\overline{)36}$ 6
4. $9\overline{)81}$ 9

5. $7\overline{)49}$ 7
6. $5\overline{)80}$ 16
7. $6\overline{)96}$ 16
8. $6\overline{)480}$ 80

9. $4\overline{)840}$ 210
10. $3\overline{)690}$ 230
11. $7\overline{)308}$ 44
12. $7\overline{)203}$ 29

13. $9\overline{)6327}$ 703
14. $4\overline{)2120}$ 530
15. $8\overline{)7280}$ 910
16. $9\overline{)8118}$ 902

17. $3\overline{)64,680}$ 21,560
18. $4\overline{)50,760}$ 12,690
19. $6\overline{)21,480}$ 3580
20. $5\overline{)18,050}$ 3610

21. Find the quotient of 1446 and 3.   482
22. Find the quotient of 4123 and 7.   589

23. What is 7525 divided by 7?   1075
24. What is 32,364 divided by 4?   8091

For Exercises 25 to 28, use the relationship between multiplication and division to complete the multiplication problem.

25. ___ × 7 = 364   52
26. 8 × ___ = 376   47
27. 5 × ___ = 170   34
28. ___ × 4 = 92   23

Objective B   To divide by a single digit with a remainder in the quotient

For Exercises 29 to 51, divide.

29. $4\overline{)9}$ 2 r1
30. $2\overline{)7}$ 3 r1
31. $5\overline{)27}$ 5 r2
32. $9\overline{)88}$ 9 r7
33. $3\overline{)40}$ 13 r1

34. $6\overline{)97}$ 16 r1
35. $8\overline{)83}$ 10 r3
36. $5\overline{)54}$ 10 r4
37. $7\overline{)632}$ 90 r2
38. $4\overline{)363}$ 90 r3

Quick Quiz (Objective 1.5A)
Divide.
1. $6\overline{)270}$ 45
2. $7\overline{)2667}$ 381
3. $9\overline{)25,677}$ 2853

Quick Quiz (Objective 1.5B)
Divide.
1. $9\overline{)415}$ 46 r1
2. $8\overline{)7787}$ 973 r3
3. $6\overline{)85,300}$ 14,216 r4

230 r1
39. 4)921

120 r5
40. 7)845

204 r3
41. 8)1635

309 r3
42. 5)1548

1347 r3
43. 7)9432

1160 r4
44. 7)8124

1720 r2
45. 3)5162

708 r2
46. 5)3542

409 r2
47. 8)3274

3825 r1
48. 4)15,301

6214 r2
49. 7)43,500

9044 r2
50. 8)72,354

8708 r2
51. 5)43,542

52. Find the quotient of 3107 and 8. 388 r3

53. Find the quotient of 8642 and 8. 1080 r2

54. What is 45,738 divided by 4? Round to the nearest ten. 11,430

55. What is 37,896 divided by 9? Round to the nearest hundred. 4200

56. What is 3572 divided by 7? Round to the nearest ten. 510

57. What is 78,345 divided by 4? Round to the nearest hundred. 19,600

Objective C   **To divide by larger whole numbers**

For Exercises 58 to 85, divide.

3 r15
58. 27)96

1 r38
59. 44)82

2 r3
60. 42)87

1 r26
61. 67)93

21 r36
62. 41)897

21 r21
63. 32)693

34 r2
64. 23)784

30 r22
65. 25)772

8 r8
66. 74)600

5 r40
67. 92)500

4 r49
68. 70)329

9 r17
69. 50)467

200 r25
70. 36)7225

200 r21
71. 44)8821

203 r2
72. 19)3859

303 r1
73. 32)9697

35 r47
74. 88)3127

67 r13
75. 92)6177

271
76. 33)8943

176 r13
77. 27)4765

4484 r6
78. 22)98,654

1086 r7
79. 77)83,629

608
80. 64)38,912

403
81. 78)31,434

Quick Quiz (Objective 1.5C)

Divide.
1. 34)693   20 r13
2. 28)3518   125 r18
3. 94)79,683   847 r65

82. $\overset{15\ r7}{206\overline{)3097}}$ 83. $\overset{12\ r456}{504\overline{)6504}}$ 84. $\overset{1\ r563}{654\overline{)1217}}$ 85. $\overset{4\ r160}{546\overline{)2344}}$

86. Find the quotient of 5432 and 21.   258 r14

87. Find the quotient of 8507 and 53.   160 r27

88. What is 37,294 divided by 72?   517 r70

89. What is 76,788 divided by 46?   1669 r14

90. Find 23,457 divided by 43. Round to the nearest hundred.   500

91. Find 341,781 divided by 43. Round to the nearest ten.   7950

For Exercises 92 to 103, use a calculator to divide. Then use estimation to determine whether the quotient is reasonable.

92. Cal.: 5129 / Est.: 5000 $76\overline{)389,804}$

93. Cal.: 2225 / Est.: 2000 $53\overline{)117,925}$

94. Cal.: 21,968 / Est.: 20,000 $29\overline{)637,072}$

95. Cal.: 11,016 / Est.: 10,000 $67\overline{)738,072}$

96. Cal.: 24,596 / Est.: 22,500 $38\overline{)934,648}$

97. Cal.: 26,656 / Est.: 30,000 $34\overline{)906,304}$

98. Cal.: 2836 / Est.: 3000 $309\overline{)876,324}$

99. Cal.: 504 / Est.: 500 $642\overline{)323,568}$

100. Cal.: 3024 / Est.: 3000 $209\overline{)632,016}$

101. Cal.: 541 / Est.: 500 $614\overline{)332,174}$

102. Cal.: 32,036 / Est.: 30,000 $179\overline{)5,734,444}$

103. Cal.: 20,621 / Est.: 20,000 $374\overline{)7,712,254}$

Objective D   **To solve application problems**

Finance   The graph at the right shows the annual expenditures, in a recent year, of the average household in the United States. Use this information for Exercises 104 to 106. Round answers to the nearest whole number.

104. What is the total amount spent annually by the average household in the United States? $35,535

105. What is the average monthly expense for housing? $976

106. What is the difference between the average monthly expense for food and the average monthly expense for health care? $242

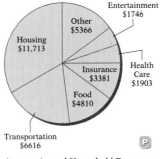

Other $5366
Entertainment $1746
Housing $11,713
Insurance $3381
Health Care $1903
Food $4810
Transportation $6616

Average Annual Household Expenses

Source: Bureau of Labor Statistics Consumer Expenditure Survey

Quick Quiz (Objective 1.5D)

Note: Question 1 is a one-step problem. Question 2 is a two-step problem.

1. A management consultant received a check for $1755 for 45 hours of work. What is the consultant's hourly fee? $39

2. A tannery produces and packages 320 briefcases each hour. Ten briefcases are put in each package for shipment. How many packages of briefcases can be produced in 8 hours?   256 packages

Insurance The table at the right shows the sources of insurance claims for losses of laptop computers in a recent year. Claims have been rounded to the nearest ten thousand dollars. Use this information for Exercises 107 and 108.

Source	Claims
Accidents	$560,000
Theft	$300,000
Power surge	$80,000
Lightning	$50,000
Transit	$20,000
Water/flood	$20,000
Other	$110,000

Source: Safeware, The Insurance Company

107. What was the average monthly claim for theft? $25,000

108. For all sources combined, find the average claims per month. $95,000

Work Hours The table at the right shows, for different countries, the average number of hours per year that employees work. Use this information for Exercises 109 and 110. Use a 50-week year. Round answers to the nearest whole number.

Country	Annual Number of Hours Worked
Britain	1731
France	1656
Japan	1889
Norway	1399
United States	1966

Source: International Labor Organization

109. What is the average number of hours worked per week by employees in Britain? 35 hours

110. On average, how many more hours per week do employees in the United States work than employees in France? 6 hours

111. **Coins** The U.S. Mint estimates that about 114,000,000,000 of the 312,000,000,000 pennies it has minted over the last 30 years are in active circulation. That works out to how many pennies in circulation for each of the 300,000,000 people living in the United States? 380 pennies

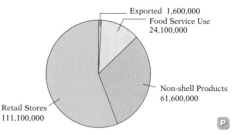

APPLYING THE CONCEPTS

112. **Payroll Deductions** Your paycheck shows deductions of $225 for savings, $98 for taxes, and $27 for insurance. Find the total of the three deductions. $350

Instructor Note
Exercises 112 to 122 are intended to provide students with practice in deciding what operation to use in order to solve an application problem.

Dairy Products The topic of the graph at the right is the eggs produced in the United States in a recent year. It shows where the eggs that were produced went or how they were used. Use this table for Exercises 113 and 114.

113. Use the graph to determine the total number of cases of eggs produced during the year. 198,400,000 cases of eggs

114. How many more cases of eggs were sold by retail stores than were used for non-shell products? 49,500,000 more cases of eggs

Exported 1,600,000
Food Service Use 24,100,000
Non-shell Products 61,600,000
Retail Stores 111,100,000

Eggs Produced in the United States (in cases)
Source: American Egg Board

115. **Fuel Consumption**   The Energy Information Administration projects that in 2020, the average U.S. household will spend $1562 annually on gasoline. Use this estimate to determine how much the average U.S. household will spend on gasoline each month in 2020. Round to the nearest dollar. $130

Salaries   The table at the right lists average starting salaries for students graduating with one of six different bachelor's degree. Use this table for Exercises 116 and 117.

116. Find the difference between the average starting salary of a computer sciences major and that of a biology major. $17,865

117. How much greater is the average starting salary of an accounting major than that of a psychology major? $13,092

Degree	Average Starting Salary
Accounting	$40,546
Biology	$29,554
Business	$37,122
Computer Sciences	$47,419
History	$32,108
Psychology	$27,454

Source: National Association of Colleges

The Military   The graph at the right shows the basic monthly pay for Army commissioned officers with 20 years of service. Use this graph for Exercises 118 and 119.

118. What is a major's annual pay? $66,336

119. What is the difference between a colonel's annual pay and a lieutenant colonel's annual pay? $10,848

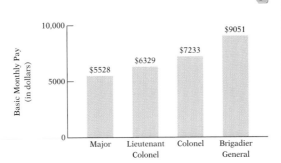

Basic Monthly Pay for Army Officers
Source: Department of Defense

120. **Farming**   A farmer harvested 48,290 pounds of avocados from one grove and 23,710 pounds of avocados from another grove. The avocados were packed in shipping boxes with 24 pounds in each box. How many boxes were needed to pack the avocados?   3000 boxes

121. **Wages**   A sales associate earns $374 for working a 40-hour week. Last week the associate worked an additional 9 hours at $13 an hour. Find the sales associate's total pay for last week's work. $491

122. **Finances**   You purchase a used car with a down payment of $2500 and monthly payments of $195 for 48 months. Find the total amount paid for the car. $11,860

123. A palindromic number is a whole number that remains unchanged when its digits are written in reverse order. For instance, 292 is a palindromic number. Find the smallest three-digit palindromic number that is divisible by 4.   212

Objective 1.6A

New Vocabulary
exponent
exponential notation
power

Discuss the Concepts
Explain what an exponent means.

Concept Check
Simplify $2^{40} \div 4^{20}$.   1

Optional Student Activity
Find the ones digit of 7^{97}.   7

Optional Student Activity
In which column will the number 1 million appear: column A, B, or C?

A	B	C
1	8	27
64	125	216
·	·	·
·	·	·
·	·	·

Column A; $1,000,000 = 100^3$

1.6 Exponential Notation and the Order of Operations Agreement

Objective A   **To simplify expressions that contain exponents**

Repeated multiplication of the same factor can be written in two ways:

$$3 \cdot 3 \cdot 3 \cdot 3 \cdot 3 \qquad \text{or} \qquad 3^5 \leftarrow \textbf{Exponent}$$

The **exponent** indicates how many times the factor occurs in the multiplication. The expression 3^5 is in **exponential notation.**

It is important to be able to read numbers written in exponential notation.

$6 = 6^1$ is read "six to the first **power**" or just "six." Usually the exponent 1 is not written.

$6 \cdot 6 = 6^2$ is read "six squared" or "six to the second power."

$6 \cdot 6 \cdot 6 = 6^3$ is read "six cubed" or "six to the third power."

$6 \cdot 6 \cdot 6 \cdot 6 = 6^4$ is read "six to the fourth power."

$6 \cdot 6 \cdot 6 \cdot 6 \cdot 6 = 6^5$ is read "six to the fifth power."

Each place value in the place-value chart can be expressed as a power of 10.

$$
\begin{aligned}
\text{Ten} &= & 10 &= & 10 &= 10^1 \\
\text{Hundred} &= & 100 &= & 10 \cdot 10 &= 10^2 \\
\text{Thousand} &= & 1000 &= & 10 \cdot 10 \cdot 10 &= 10^3 \\
\text{Ten-thousand} &= & 10,000 &= & 10 \cdot 10 \cdot 10 \cdot 10 &= 10^4 \\
\text{Hundred-thousand} &= & 100,000 &= & 10 \cdot 10 \cdot 10 \cdot 10 \cdot 10 &= 10^5 \\
\text{Million} &= & 1,000,000 &= & 10 \cdot 10 \cdot 10 \cdot 10 \cdot 10 \cdot 10 &= 10^6
\end{aligned}
$$

Integrating Technology
A calculator can be used to evaluate an exponential expression. The y^x key (or, on some calculators, a x^y or $\wedge$ key) is used to enter the exponent. For instance, for the example at the right, enter 4 y^x 3 = . The display reads 64.

To simplify a numerical expression containing exponents, write each factor as many times as indicated by the exponent and carry out the indicated multiplication.

$$4^3 = 4 \cdot 4 \cdot 4 = 64$$
$$2^2 \cdot 3^4 = (2 \cdot 2) \cdot (3 \cdot 3 \cdot 3 \cdot 3) = 4 \cdot 81 = 324$$

Example 1   Write $3 \cdot 3 \cdot 3 \cdot 5 \cdot 5$ in exponential notation.

Solution $3 \cdot 3 \cdot 3 \cdot 5 \cdot 5 = 3^3 \cdot 5^2$

You Try It 1   Write $2 \cdot 2 \cdot 2 \cdot 2 \cdot 3 \cdot 3 \cdot 3$ in exponential notation.

Your solution $2^4 \cdot 3^3$

Example 2   Write as a power of 10:
$10 \cdot 10 \cdot 10 \cdot 10$

Solution $10 \cdot 10 \cdot 10 \cdot 10 = 10^4$

You Try It 2   Write as a power of 10:
$10 \cdot 10 \cdot 10 \cdot 10 \cdot 10 \cdot 10 \cdot 10$

Your solution 10^7

Example 3   Simplify $3^2 \cdot 5^3$.

Solution $3^2 \cdot 5^3 = (3 \cdot 3) \cdot (5 \cdot 5 \cdot 5)$
$= 9 \cdot 125 = 1125$

You Try It 3   Simplify $2^3 \cdot 5^2$.

Your solution   200

Solutions on p. S4

In-Class Examples (Objective 1.6A)
Write in exponential notation.
1. $5 \cdot 5 \cdot 5 \cdot 7 \cdot 7 \cdot 7 \cdot 7$ $5^3 \cdot 7^4$
2. $2 \cdot 2 \cdot 2 \cdot 3 \cdot 3 \cdot 7 \cdot 7 \cdot 7 \cdot 7$ $2^3 \cdot 3^2 \cdot 7^4$
Simplify.
3. $2^3 \cdot 3^2$ 72
4. $5^3 \cdot 7^2$ 6125

Objective B **To use the Order of Operations Agreement to simplify expressions**

More than one operation may occur in a numerical expression. The answer may be different, depending on the order in which the operations are performed. For example, consider $3 + 4 \times 5$.

Multiply first, then add.

$$3 + \underbrace{4 \times 5}$$
$$\underbrace{3 + 20}$$
$$23$$

Add first, then multiply.

$$\underbrace{3 + 4} \times 5$$
$$\underbrace{7 \times 5}$$
$$35$$

An Order of Operations Agreement is used so that only one answer is possible.

The Order of Operations Agreement

Step 1. Do all the operations inside parentheses.

Step 2. Simplify any number expressions containing exponents.

Step 3. Do multiplication and division as they occur from left to right.

Step 4. Do addition and subtraction as they occur from left to right.

Integrating Technology

Many scientific calculators have an x^2 key. This key is used to square the displayed number. For example, after the user presses 2 x^2 = , the display reads 4.

HOW TO Simplify $3 \times (2 + 1) - 2^2 + 4 \div 2$ by using the Order of Operations Agreement.

$3 \times \underbrace{(2 + 1)} - 2^2 + 4 \div 2$ 1. Perform operations in parentheses.

$3 \times 3 - \underbrace{2^2} + 4 \div 2$ 2. Simplify expressions with exponents.

$\underbrace{3 \times 3} - 4 + \underbrace{4 \div 2}$ 3. Do multiplication and division as they occur from left to right.

$\underbrace{9 - 4} + 4 \div 2$

$\underbrace{9 - 4 + 2}$ 4. Do addition and subtraction as they occur from left to right.

$\underbrace{5 + 2}$

7

One or more of the foregoing steps may not be needed to simplify an expression. In that case, proceed to the next step in the Order of Operations Agreement.

HOW TO Simplify $5 + 8 \div 2$.   There are no parentheses or exponents. Proceed to Step 3 of the agreement.

$5 + \underbrace{8 \div 2}$ 3. Do multiplication or division.

$\underbrace{5 + 4}$ 4. Do addition or subtraction.

9

Example 4 Simplify: $64 \div (8 - 4)^2 \cdot 9 - 5^2$

Solution
$$64 \div (8 - 4)^2 \cdot 9 - 5^2$$
$$= 64 \div 4^2 \cdot 9 - 5^2 \quad \bullet \text{ Parentheses}$$
$$= 64 \div 16 \cdot 9 - 25 \quad \bullet \text{ Exponents}$$
$$= 4 \cdot 9 - 25 \quad \bullet \text{ Division and}$$
$$= 36 - 25 \quad \quad \text{ multiplicaton}$$
$$= 11 \quad \bullet \text{ Subtraction}$$

You Try It 4 Simplify: $5 \cdot (8 - 4)^2 \div 4 - 2$

Your solution 18

Solution on p. S4

Objective 1.6B

New Procedures
Order of Operations Agreement

Discuss the Concepts

1. Why do we need an Order of Operations Agreement?

2. Describe the steps in the Order of Operations Agreement.

3. What operations are in the expression $12 + (9 - 5) \cdot 3$? In what order must these operations be performed?

Concept Check
Arrange the expressions in order from the greatest value to the least value.

$$12 \div 6 + 2$$
$$8 \cdot 4 + 5$$
$$7^2 - 14$$
$$15 + 3(7)$$
$$4(6 - 3) - 10$$
$$3(6 + 2) \div 8$$

$8 \cdot 4 + 5 > 15 + 3(7) >$
$7^2 - 14 > 12 \div 6 + 2 >$
$3(6 + 2) \div 8 > 4(6 - 3) - 10;$
$37 > 36 > 35 > 4 > 3 > 2$

Optional Student Activity
How can you place twelve sugar lumps in three coffee mugs so that there is an odd number of lumps in each mug?
Not possible

Instructor Note
Have students try the *Projects and Group Activities* at the end of this chapter to determine whether their calculators use the Order of Operations Agreement.

In-Class Examples (Objective 1.6B)

Simplify.

1. $4^2 + 6(3 - 1)$ 28

2. $9 - 6 + 6 \cdot 2 \div 3$ 7

Section 1.6

Suggested Assignment
Exercises 1–75, odds
Exercises 77, 78

1.6 Exercises

Objective A   **To simplify expressions that contain exponents**

For Exercises 1 to 12, write the number in exponential notation.

1. $2 \cdot 2 \cdot 2$
2^3

2. $7 \cdot 7 \cdot 7 \cdot 7 \cdot 7$
7^5

3. $6 \cdot 6 \cdot 6 \cdot 7 \cdot 7 \cdot 7 \cdot 7$
$6^3 \cdot 7^4$

4. $6 \cdot 6 \cdot 9 \cdot 9 \cdot 9 \cdot 9$
$6^2 \cdot 9^4$

5. $2 \cdot 2 \cdot 2 \cdot 3 \cdot 3 \cdot 3$
$2^3 \cdot 3^3$

6. $3 \cdot 3 \cdot 10 \cdot 10$
$3^2 \cdot 10^2$

7. $5 \cdot 7 \cdot 7 \cdot 7 \cdot 7 \cdot 7$
$5 \cdot 7^5$

8. $4 \cdot 4 \cdot 4 \cdot 5 \cdot 5 \cdot 5$
$4^3 \cdot 5^3$

9. $3 \cdot 3 \cdot 3 \cdot 6 \cdot 6 \cdot 6 \cdot 6$
$3^3 \cdot 6^4$

10. $2 \cdot 2 \cdot 5 \cdot 5 \cdot 5 \cdot 8$
$2^2 \cdot 5^3 \cdot 8$

11. $3 \cdot 3 \cdot 3 \cdot 5 \cdot 9 \cdot 9 \cdot 9$
$3^3 \cdot 5 \cdot 9^3$

12. $2 \cdot 2 \cdot 2 \cdot 4 \cdot 7 \cdot 7 \cdot 7$
$2^3 \cdot 4 \cdot 7^3$

For Exercises 13 to 37, simplify.

13. 2^3
8

14. 2^6
64

15. $2^4 \cdot 5^2$
400

16. $2^6 \cdot 3^2$
576

17. $3^2 \cdot 10^2$
900

18. $2^3 \cdot 10^4$
80,000

19. $6^2 \cdot 3^3$
972

20. $4^3 \cdot 5^2$
1600

21. $5 \cdot 2^3 \cdot 3$
120

22. $6 \cdot 3^2 \cdot 4$
216

23. $2^2 \cdot 3^2 \cdot 10$
360

24. $3^2 \cdot 5^2 \cdot 10$
2250

25. $0^2 \cdot 4^3$
0

26. $6^2 \cdot 0^3$
0

27. $3^2 \cdot 10^4$
90,000

28. $5^3 \cdot 10^3$
125,000

29. $2^2 \cdot 3^3 \cdot 5$
540

30. $5^2 \cdot 7^3 \cdot 2$
17,150

31. $2 \cdot 3^4 \cdot 5^2$
4050

32. $6 \cdot 2^6 \cdot 7^2$
18,816

33. $5^2 \cdot 3^2 \cdot 7^2$
11,025

34. $4^2 \cdot 9^2 \cdot 6^2$
46,656

35. $3^4 \cdot 2^6 \cdot 5$
25,920

36. $4^3 \cdot 6^3 \cdot 7$
96,768

37. $4^2 \cdot 3^3 \cdot 10^4$
4,320,000

Objective B   **To use the Order of Operations Agreement to simplify expressions**

For Exercises 38 to 76, simplify by using the Order of Operations Agreement.

38. $4 - 2 + 3$
5

39. $6 - 3 + 2$
5

40. $6 \div 3 + 2$
4

41. $8 \div 4 + 8$
10

42. $6 \cdot 3 + 5$
23

43. $5 \cdot 9 + 2$
47

44. $3^2 - 4$
5

45. $5^2 - 17$
8

Quick Quiz (Objective 1.6A)
Write in exponential notation.
1. $2 \cdot 2 \cdot 3 \cdot 3 \cdot 3 \cdot 3$ $2^2 \cdot 3^4$
2. $5 \cdot 5 \cdot 5 \cdot 7 \cdot 11 \cdot 11 \cdot 11 \cdot 11$ $5^3 \cdot 7 \cdot 11^4$
Simplify.
3. $2^2 \cdot 5^3$ 500
4. $3^3 \cdot 7$ 189

46. $4 \cdot (5 - 3) + 2$
10

47. $3 + (4 + 2) \div 3$
5

48. $5 + (8 + 4) \div 6$
7

49. $8 - 2^2 + 4$
8

50. $16 \cdot (3 + 2) \div 10$
8

51. $12 \cdot (1 + 5) \div 12$
6

52. $10 - 2^3 + 4$
6

53. $5 \cdot 3^2 + 8$
53

54. $16 + 4 \cdot 3^2$
52

55. $12 + 4 \cdot 2^3$
44

56. $16 + (8 - 3) \cdot 2$
26

57. $7 + (9 - 5) \cdot 3$
19

58. $2^2 + 3 \cdot (6 - 2)^2$
52

59. $3^3 + 5 \cdot (8 - 6)^3$
67

60. $2^2 \cdot 3^2 + 2 \cdot 3$
42

61. $4 \cdot 6 + 3^2 \cdot 4^2$
168

62. $16 - 2 \cdot 4$
8

63. $12 + 3 \cdot 5$
27

64. $3 \cdot (6 - 2) + 4$
16

65. $5 \cdot (8 - 4) - 6$
14

66. $8 - (8 - 2) \div 3$
6

67. $12 - (12 - 4) \div 4$
10

68. $8 + 2 - 3 \cdot 2 \div 3$
8

69. $10 + 1 - 5 \cdot 2 \div 5$
9

70. $3 \cdot (4 + 2) \div 6$
3

71. $(7 - 3)^2 \div 2 - 4 + 8$
12

72. $20 - 4 \div 2 \cdot (3 - 1)^3$
4

73. $12 \div 3 \cdot 2^2 + (7 - 3)^2$
32

74. $(4 - 2) \cdot 6 \div 3 + (5 - 2)^2$
13

75. $18 - 2 \cdot 3 + (4 - 1)^3$
39

76. $100 \div (2 + 3)^2 - 8 \div 2$
0

APPLYING THE CONCEPTS

77. **Computers**   Memory in computers is measured in bytes. One kilobyte (1 K) is 2^{10} bytes. Write this number in standard form.   1024

78. ✎ Explain the difference that the order of operations makes between **a.** $(14 - 2) \div 2 \cdot 3$ and **b.** $(14 - 2) \div (2 \cdot 3)$. Work the two problems. What is the difference between the larger answer and the smaller answer?

Answers to Writing Exercises

78. In part a, the expression $(14 - 2)$ is first divided by 2 and then multiplied by 3. In part b, the same expression is divided by the product of 2 and 3. The difference between the two answers is $18 - 2 = 16$.

Quick Quiz (Objective 1.6B)

Simplify.

1. $3^2 - 2(12 \div 6)$ 5

2. $14 - (11 - 2) \div 3$ 11

Objective 1.7A

New Vocabulary
factors of a number

New Rules
Rules for finding factors of a
number

Discuss the Concepts
1. What does it mean for a
number to be evenly divisible
by another number?
2. How can you determine
whether 13 is a factor of 91?
Is 13 a factor of 91?

Concept Check
Find the largest factor of
111,111,111,111 that is less than
111,111,111,111.
37,037,037,037 because
111,111,111,111 is not divisible by
2 and 111,111,111,111 ÷ 3 =
37,037,037,037.

Concept Check
List all the numbers between 210
and 220 that do not have a factor
of 2, 3, 4, 5, 6, 7, 8, or 9.   211

**Optional Student
Activity**
What is the ones digit in the
product of all the factors of
100?   0

**Optional Student
Activity**
Find the difference between the
sum of all the factors of 66 and
the sum of all the factors of 70.
0. Both sums equal 144.

1.7 Prime Numbers and Factoring

Objective A   To factor numbers

Whole-number **factors of a number** divide that number evenly (there is no remainder).

1, 2, 3, and 6 are whole-number factors of 6 because they divide 6 evenly.

$$\frac{6}{1)6} \quad \frac{3}{2)6} \quad \frac{2}{3)6} \quad \frac{1}{6)6}$$

Note that both the divisor and the quotient are factors of the dividend.

To find the factors of a number, try dividing the number by 1, 2, 3, 4, 5, ... Those numbers that divide the number evenly are its factors. Continue this process until the factors start to repeat.

HOW TO Find all the factors of 42.

$42 \div 1 = 42$	1 and 42 are factors.
$42 \div 2 = 21$	2 and 21 are factors.
$42 \div 3 = 14$	3 and 14 are factors.
$42 \div 4$	Will not divide evenly
$42 \div 5$	Will not divide evenly
$42 \div 6 = 7$	6 and 7 are factors. ⎫ Factors are repeating; all the
$42 \div 7 = 6$	7 and 6 are factors. ⎬ factors of 42 have been found.

1, 2, 3, 6, 7, 14, 21, and 42 are factors of 42.

The following rules are helpful in finding the factors of a number.

2 is a factor of a number if the last digit of the number is 0, 2, 4, 6, or 8.

436 ends in 6; therefore, 2 is a factor of 436. ($436 \div 2 = 218$)

3 is a factor of a number if the sum of the digits of the number is divisible by 3.

The sum of the digits of 489 is $4 + 8 + 9 = 21$. 21 is divisible by 3. Therefore, 3 is a factor of 489. ($489 \div 3 = 163$)

5 is a factor of a number if the last digit of the number is 0 or 5.

520 ends in 0; therefore, 5 is a factor of 520. ($520 \div 5 = 104$)

Example 1 Find all the factors of 30.

Solution
$30 \div 1 = 30$
$30 \div 2 = 15$
$30 \div 3 = 10$
$30 \div 4$ Will not divide evenly
$30 \div 5 = 6$
$30 \div 6 = 5$

1 2, 3, 5, 6, 10, 15, and 30 are factors of 30.

You Try It 1 Find all the factors of 40.

Your solution 1, 2, 4, 5, 8, 10, 20, 40

Solution on p. S4

In-Class Examples (Objective 1.7A)
Find all the factors of the number.
1. 72   1, 2, 3, 4, 6, 8, 9, 12, 18, 24, 36, 72
2. 108   1, 2, 3, 4, 6, 9, 12, 18, 27, 36, 54, 108
3. 137   1, 137

Objective B   To find the prime factorization of a number

Point of Interest

Prime numbers are an important part of cryptology, the study of secret codes. To make it less likely that codes can be broken, cryptologists use prime numbers that have hundreds of digits.

A number is a **prime number** if its only whole-number factors are 1 and itself. 7 is prime because its only factors are 1 and 7. If a number is not prime, it is called a **composite number**. Because 6 has factors of 2 and 3, 6 is a composite number. The number 1 is not considered a prime number; therefore, it is not included in the following list of prime numbers less than 50.

$$2, 3, 5, 7, 11, 13, 17, 19, 23, 29, 31, 37, 41, 43, 47$$

The **prime factorization** of a number is the expression of the number as a product of its prime factors. We use a "T-diagram" to find the prime factors of 60. Begin with the smallest prime number as a trial divisor, and continue with prime numbers as trial divisors until the final quotient is 1.

$$
\begin{array}{c|c}
\multicolumn{2}{c}{60} \\
\hline
2 & 30 \\
2 & 15 \\
3 & 5 \\
5 & 1 \\
\end{array}
\qquad
\begin{aligned}
60 \div 2 &= 30 \\
30 \div 2 &= 15 \\
15 \div 3 &= 5 \\
5 \div 5 &= 1
\end{aligned}
$$

The prime factorization of 60 is $2 \cdot 2 \cdot 3 \cdot 5$.

Finding the prime factorization of larger numbers can be more difficult. Try each prime number as a trial divisor. Stop when the square of the trial divisor is greater than the number being factored.

HOW TO   Find the prime factorization of 106.

$$
\begin{array}{c|c}
\multicolumn{2}{c}{106} \\
\hline
2 & 53 \\
53 & 1 \\
\end{array}
$$

• 53 cannot be divided evenly by 2, 3, 5, 7, or 11. Prime numbers greater than 11 need not be tested because 11^2 is greater than 53.

The prime factorization of 106 is $2 \cdot 53$.

New Vocabulary
prime number
composite number
prime factorization

Discuss the Concepts
1. What is the difference between a prime number and a composite number?
2. Is $4 \cdot 5$ the prime factorization of 20? Why or why not?

Concept Check
There are three different digits such that any two of them, written in any order, serve as the digits of a two-digit prime number. What are these three digits?   1, 3, 7

Optional Student Activity
What is the smallest prime number greater than $58 - [2 + 3(4^2 - 6 \cdot 2)]$?   47

Example 2   Find the prime factorization of 315.

Solution

$$
\begin{array}{c|c}
\multicolumn{2}{c}{315} \\
\hline
3 & 105 \\
3 & 35 \\
5 & 7 \\
7 & 1 \\
\end{array}
$$

• $315 \div 3 = 105$
• $105 \div 3 = 35$
• $35 \div 5 = 7$
• $7 \div 7 = 1$

$315 = 3 \cdot 3 \cdot 5 \cdot 7$

You Try It 2   Find the prime factorization of 44.

Your solution $2 \cdot 2 \cdot 11$

Example 3   Find the prime factorization of 201.

Solution

$$
\begin{array}{c|c}
\multicolumn{2}{c}{201} \\
\hline
3 & 67 \\
67 & 1 \\
\end{array}
$$

• Try only 2, 3, 5, 7, and 11 because $11^2 > 67$.

$201 = 3 \cdot 67$

You Try It 3   Find the prime factorization of 177.

Your solution $3 \cdot 59$

Solutions on p. S4

In-Class Examples (Objective 1.7B)
Find the prime factorization.
1. 84 $2 \cdot 2 \cdot 3 \cdot 7$
2. 110 $2 \cdot 5 \cdot 11$

52

Suggested Assignment
Exercises 1–83, odds
Exercises 85, 86, 87

1.7 Exercises

Objective A   **To factor numbers**

For Exercises 1 to 40, find all the factors of the number.

1. 4
1, 2, 4

2. 6
1, 2, 3, 6

3. 10
1, 2, 5, 10

4. 20
1, 2, 4, 5, 10, 20

5. 7
1, 7

6. 12
1, 2, 3, 4, 6, 12

7. 9
1, 3, 9

8. 8
1, 2, 4, 8

9. 13
1, 13

10. 17
1, 17

11. 18
1, 2, 3, 6, 9, 18

12. 24
1, 2, 3, 4, 6,
8, 12, 24

13. 56
1, 2, 4, 7, 8,
14, 28, 56

14. 36
1, 2, 3, 4, 6, 9,
12, 18, 36

15. 45
1, 3, 5, 9, 15, 45

16. 28
1, 2, 4, 7,
14, 28

17. 29
1, 29

18. 33
1, 3, 11, 33

19. 22
1, 2, 11, 22

20. 26
1, 2, 13, 26

21. 52
1, 2, 4, 13, 26, 52

22. 49
1, 7, 49

23. 82
1, 2, 41, 82

24. 37
1, 37

25. 57
1, 3, 19, 57

26. 69
1, 3, 23, 69

27. 48
1, 2, 3, 4, 6, 8,
12, 16, 24, 48

28. 64
1, 2, 4, 8, 16,
32, 64

29. 95
1, 5, 19, 95

30. 46
1, 2, 23, 46

31. 54
1, 2, 3, 6, 9, 18,
27, 54

32. 50
1, 2, 5, 10,
25, 50

33. 66
1, 2, 3, 6, 11,
22, 33, 66

34. 77
1, 7, 11, 77

35. 80
1, 2, 4, 5, 8, 10,
16, 20, 40, 80

36. 100
1, 2, 4, 5, 10,
20, 25, 50, 100

37. 96
1, 2, 3, 4, 6, 8, 12,
16, 24, 32, 48, 96

38. 85
1, 5, 17, 85

39. 90
1, 2, 3, 5, 6, 9, 10,
15, 18, 30, 45, 90

40. 101
1, 101

Objective B   **To find the prime factorization of a number**

For Exercises 41 to 84, find the prime factorization.

41. 6
$2 \cdot 3$

42. 14
$2 \cdot 7$

43. 17
Prime

44. 83
Prime

45. 24
$2 \cdot 2 \cdot 2 \cdot 3$

46. 12
$2 \cdot 2 \cdot 3$

47. 27
$3 \cdot 3 \cdot 3$

48. 9
$3 \cdot 3$

Quick Quiz (Objective 1.7A)
Find all the factors of the number.
1. 78   1, 2, 3, 6, 13, 26, 39, 78
2. 121   1, 11, 121

Quick Quiz (Objective 1.7B)
Find the prime factorization.
1. 88 $2 \cdot 2 \cdot 2 \cdot 11$
2. 200 $2 \cdot 2 \cdot 2 \cdot 5 \cdot 5$

49. 36
2 · 2 · 3 · 3

50. 40
2 · 2 · 2 · 5

51. 19
Prime

52. 37
Prime

53. 90
2 · 3 · 3 · 5

54. 65
5 · 13

55. 115
5 · 23

56. 80
2 · 2 · 2 · 2 · 5

57. 18
2 · 3 · 3

58. 26
2 · 13

59. 28
2 · 2 · 7

60. 49
7 · 7

61. 31
Prime

62. 42
2 · 3 · 7

63. 62
2 · 31

64. 81
3 · 3 · 3 · 3

65. 22
2 · 11

66. 39
3 · 13

67. 101
Prime

68. 89
Prime

69. 66
2 · 3 · 11

70. 86
2 · 43

71. 74
2 · 37

72. 95
5 · 19

73. 67
Prime

74. 78
2 · 3 · 13

75. 55
5 · 11

76. 46
2 · 23

77. 120
2 · 2 · 2 · 3 · 5

78. 144
2 · 2 · 2 · 2 · 3 · 3

79. 160
2 · 2 · 2 · 2 · 2 · 5

80. 175
5 · 5 · 7

81. 216
2 · 2 · 2 · 3 · 3 · 3

82. 400
2 · 2 · 2 · 2 · 5 · 5

83. 625
5 · 5 · 5 · 5

84. 225
3 · 3 · 5 · 5

APPLYING THE CONCEPTS

85. Twin primes are two prime numbers that differ by 2. For instance, 17 and 19 are twin primes. Find three sets of twin primes, not including 17 and 19.   3 and 5, 5 and 7, 11 and 13   Other answers are possible.

86. In 1742, Christian Goldbach conjectured that every even number greater than 2 could be expressed as the sum of two prime numbers. Show that this conjecture is true for 8, 24, and 72. (*Note:* Mathematicians have not yet been able to determine whether Goldbach's conjecture is true or false.)
8 = 3 + 5; 24 = 11 + 13; 72 = 29 + 43. Other answers are possible.

87.   Explain why 2 is the only even prime number.

88.   If a number is divisible by 6 and 10, is the number divisible by 30? If so, explain why. If not, give an example.

Answers to Writing Exercises

87. 2 is the only even prime number because all other even numbers have 2 as a factor and thus cannot be prime.

88. Yes; 30 is a product of the factors of 6 and 10 (2 · 3 · 5).

Instructor Note

The feature entitled *Focus on Problem Solving* appears at the end of every chapter of the text. It provides optional material that can be used to enhance your students' problem-solving skills.

Focus on Problem Solving

Questions to Ask

You encounter problem-solving situations every day. Some problems are easy to solve, and you may mentally solve these problems without considering the steps you are taking in order to draw a conclusion. Others may be more challenging and may require more thought and consideration.

Suppose a friend suggests that you both take a trip over spring break. You'd like to go. What questions go through your mind? You might ask yourself some of the following questions:

How much will the trip cost? What will be the cost for travel, hotel rooms, meals, and so on?

Are some costs going to be shared by both me and my friend?

Can I afford it?

How much money do I have in the bank?

How much more money than I have now do I need?

How much time is there to earn that much money?

How much can I earn in that amount of time?

How much money must I keep in the bank in order to pay the next tuition bill (or some other expense)?

These questions require different mathematical skills. Determining the cost of the trip requires **estimation;** for example, you must use your knowledge of air fares or the cost of gasoline to arrive at an estimate of these costs. If some of the costs are going to be shared, you need to **divide** those costs by 2 in order to determine your share of the expense. The question regarding how much more money you need requires **subtraction:** the amount needed minus the amount currently in the bank. To determine how much money you can earn in the given amount of time requires **multiplication**—for example, the amount you earn per week times the number of weeks to be worked. To determine if the amount you can earn in the given amount of time is sufficient, you need to use your knowledge of **order relations** to compare the amount you can earn with the amount needed.

Facing the problem-solving situation described above may not seem difficult to you. The reason may be that you have faced similar situations before and, therefore, know how to work through this one. You may feel better prepared to deal with a circumstance such as this one because you know what questions to ask. An important aspect of learning to solve problems is learning what questions to ask. As you work through application problems in this text, try to become more conscious of the mental process you are going through. You might begin the process by asking yourself the following questions whenever you are solving an application problem.

1. Have I read the problem enough times to be able to understand the situation being described?

2. Will restating the problem in different words help me to understand the problem situation better?

3. What facts are given? (You might make a list of the information contained in the problem.)

4. What information is being asked for?

5. What relationship exists among the given facts? What relationship exists between the given facts and the solution?

6. What mathematical operations are needed in order to solve the problem?

Try to focus on the problem-solving situation, not on the computation or on getting the answer quickly. And remember, the more problems you solve, the better able you will be to solve other problems in the future, partly because you are learning what questions to ask.

Projects and Group Activities

Order of Operations

Does your calculator use the Order of Operations Agreement? To find out, try this problem:

$$2 + 4 \cdot 7$$

If your answer is 30, then the calculator uses the Order of Operations Agreement. If your answer is 42, it does not use that agreement.

Even if your calculator does not use the Order of Operations Agreement, you can still correctly evaluate numerical expressions. The parentheses keys, (and) , are used for this purpose.

Remember that $2 + 4 \cdot 7$ means $2 + (4 \cdot 7)$ because the multiplication must be completed before the addition. To evaluate this expression, enter the following:

Enter: 2   +   (4   ×   7   )   =
Display: 2   2   (4   4   7   28   30

When using your calculator to evaluate numerical expressions, insert parentheses around multiplications and around divisions. This has the effect of forcing the calculator to do the operations in the order you want.

For Exercises 1 to 10, evaluate.

1. $3 \cdot 8 - 5$

2. $6 + 8 \div 2$

3. $3 \cdot (8 - 2)^2$

4. $24 - (4 - 2)^2 \div 4$

5. $3 + (6 \div 2 + 4)^2 - 2$

6. $16 \div 2 + 4 \cdot (8 - 12 \div 4)^2 - 50$

7. $3 \cdot (15 - 2 \cdot 3) - 36 \div 3$

8. $4 \cdot 2^2 - (12 + 24 \div 6) + 5$

9. $16 \div 4 \cdot 3 + (3 \cdot 4 - 5) + 2$

10. $15 \cdot 3 \div 9 + (2 \cdot 6 - 3) + 4$

Instructor Note
Projects and Group Activities appear at the end of each chapter in the text. This feature can be used for individual assignments, such as extra credit; for cooperative learning exercises, such as small-group projects; or for class discussions.

Answers to Projects and Group Activities— Order of Operations
1. 19
2. 10
3. 108
4. 23
5. 50
6. 58
7. 15
8. 5
9. 21
10. 18

Answers to Projects and Group Activities— Patterns in Mathematics

Number of straight lines in:

A circle with 3 dots—3

A circle with 4 dots—6

A circle with 5 dots—10

A circle with 6 dots—15

The differences in the number of lines drawn in going to successive circles are

$$6 - 3 = 3$$
$$10 - 6 = 4$$
$$15 - 10 = 5$$

The difference increases by 1 each time.

For a circle with 7 dots, there will be

$$15 + 6 = 21$$

lines drawn.

For a circle with 8 dots, there will be

$$21 + 7 = 28$$

lines drawn.

For a circle with 9 dots, there will be

$$28 + 8 = 36$$

lines drawn. Therefore, 36 singles matches will be played among the 9 players.

Instructor Note

The numbers from the "Search the World Wide Web" project could lead to a discussion about population and food supply, about the need for greater technology in agriculture, and about colonizing the moon and using it as a place for producing food.

Patterns in Mathematics

For the circle at the left, use a straight line to connect each dot on the circle with every other dot on the circle. How many different straight lines are there?

Follow the same procedure for each of the circles shown below. How many different straight lines are there in each?

Find a pattern to describe the number of dots on a circle and the corresponding number of different lines drawn. Use the pattern to determine the number of different lines that would be drawn in a circle with 7 dots and in a circle with 8 dots.

Now use the pattern to answer the following question. You are arranging a tennis tournament with 9 players. How many singles matches will be played among the 9 players if each player plays each of the other players only once?

Search the World Wide Web

Go to **www.census.gov** on the Internet.

1. Find a projection for the total U.S. population 10 years from now and a projection for the total population 20 years from now. Record the two numbers.
2. Use the data from Exercise 1 to determine the expected growth in the population over the next 10 years.
3. Use the answer from Exercise 2 to find the average increase in the U.S. population per year over the next 10 years. Round to the nearest million.
4. Use data in the population table you found to write two word problems. Then state whether addition, subtraction, multiplication, or division is required to solve each of the problems.

Chapter 1 Summary

Key Words	Examples
The *whole numbers* are 0, 1, 2, 3, 4, 5, 6, 7, 8, 9, 10, ... [1.1A, p. 3]	
The *graph of a whole number* is shown by placing a heavy dot directly above that number on the number line. [1.1A, p. 3]	This is the graph of 4 on the number line.
The symbol for *is less than* is <. The symbol for *is greater than* is >. These symbols are used to show the order relation between two numbers. [1.1A, p. 3]	$3 < 7$ $9 > 2$

When a whole number is written using the digits 0, 1, 2, 3, 4, 5, 6, 7, 8, and 9, it is said to be in *standard form*. The position of each digit in the number determines the digit's *place value*. The place values are used to write the expanded form of a number. [1.1B, p. 4]

The number 598,317 is in standard form. The digit 8 is in the thousands place. The number 598,317 is written in expanded form as 500,000 + 90,000 + 8000 + 300 + 10 + 7.

Addition is the process of finding the total of two or more numbers. The numbers being added are called *addends*. The result is the *sum*. [1.2A, p. 9]

$$\begin{array}{r} \overset{1\ \ 11}{8,762} \\ +\ 1,359 \\ \hline 10,121 \end{array}$$

Subtraction is the process of finding the difference between two numbers. The *minuend* minus the *subtrahend* equals the *difference*. [1.3A, p. 17]

$$\begin{array}{r} \overset{4\ \ \ 11\ \ 11\ \ \ 6\ \ \ 13}{5\ \ 2\,,\,1\ \ 7\ \ 3} \\ -3\ \ 4\,,\,9\ \ 6\ \ 8 \\ \hline 1\ \ 7\,,\,2\ \ 0\ \ 5 \end{array}$$

Multiplication is the repeated addition of the same number. The numbers that are multiplied are called *factors*. The result is the *product*. [1.4A, p. 25]

$$\begin{array}{r} \overset{4\ \ \ 5}{3\ \ 5\ \ 8} \\ \times\ \ \ \ \ \ \ \ 7 \\ \hline 2\ \ 5\ \ 0\ \ 6 \end{array}$$

Division is used to separate objects into equal groups. The *dividend* divided by the *divisor* equals the *quotient*. [1.5A, p. 33]
For any division problem,
(*quotient · divisor*) + *remainder* = *dividend*. [1.5B, p. 36]

$$\begin{array}{r} 93\text{ r}3 \\ 7)\overline{654} \\ -63\ \ \\ \hline 24 \\ -21 \\ \hline 3 \end{array}$$

Check: $(7 \cdot 93) + 3 = 651 + 3 = 654$

The expression 4^3 is in *exponential notation*. The *exponent*, 3, indicates how many times 4 occurs as a factor in the multiplication. [1.6A, p. 46]

$5^4 = 5 \cdot 5 \cdot 5 \cdot 5 = 625$

Whole-number *factors of a number* divide that number evenly (there is no remainder). [1.7A, p. 50]

$18 \div 1 = 18$
$18 \div 2 = 9$
$18 \div 3 = 6$
$18 \div 4$ 4 does not divide 18 evenly.
$18 \div 5$ 5 does not divide 18 evenly.
$18 \div 6 = 3$ The factors are repeating.
The factors of 18 are 1, 2, 3, 6, 9, and 18.

A number greater than 1 is a *prime number* if its only whole-number factors are 1 and itself. If a number is not prime, it is a *composite number*. [1.7B, p. 51]

The prime numbers less than 20 are 2, 3, 5, 7, 11, 13, 17, and 19.
The composite numbers less than 20 are 4, 6, 8, 9, 10, 12, 14, 15, 16, and 18.

The *prime factorization* of a number is the expression of the number as a product of its prime factors. [1.7B, p. 51]

$$\begin{array}{r} 42 \\ \hline 2\,|\,21 \\ 3\,|\,7 \\ 7\,|\,1 \end{array}$$
The prime factorization of 42 is $2 \cdot 3 \cdot 7$.

Essential Rules and Procedures	**Examples**
To round a number to a given place value: If the digit to the right of the given place value is less than 5, replace that digit and all digits to the right by zeros. If the digit to the right of the given place value is greater than or equal to 5, increase the digit in the given place value by 1, and replace all other digits to the right by zeros. [1.1D, p. 6]	36,178 rounded to the nearest thousand is 36,000. 4592 rounded to the nearest thousand is 5000.

Properties of Addition [1.2A, p. 9]

Addition Property of Zero Zero added to a number does not change the number.	$7 + 0 = 7$
Commutative Property of Addition Two numbers can be added in either order; the sum will be the same.	$8 + 3 = 3 + 8$
Associative Property of Addition Numbers to be added can be grouped in any order; the sum will be the same.	$(2 + 4) + 6 = 2 + (4 + 6)$

To estimate the answer to a calculation: Round each number to the same place value. Perform the calculation using the rounded numbers. [1.2A, p. 11]	$\begin{array}{rr} 39{,}471 & 40{,}000 \\ 12{,}586 & +10{,}000 \\ \hline & 50{,}000 \end{array}$ 50,000 is an estimate of the sum of 39,471 and 12,586.

Properties of Multiplication [1.4A, p. 25]

Multiplication Property of Zero The product of a number and zero is zero.	$3 \cdot 0 = 0$
Multiplication Property of One The product of a number and one is the number.	$6 \cdot 1 = 6$
Commutative Property of Multiplication Two numbers can be multiplied in either order; the product will be the same.	$2 \cdot 8 = 8 \cdot 2$
Associative Property of Multiplication Grouping numbers to be multiplied in any order gives the same result.	$(2 \cdot 4) \cdot 6 = 2 \cdot (4 \cdot 6)$

Division Properties of Zero and One [1.5A, p. 33] Any whole number, except zero, divided by itself is 1. Any whole number divided by 1 is the whole number. Zero divided by any other whole number is zero. Division by zero is not allowed.	$3 \div 3 = 1$ $3 \div 1 = 3$ $0 \div 3 = 0$ $3 \div 0$ is not allowed.

Order of Operations Agreement [1.6B, p. 47]

Step 1 Do all the operations inside parentheses.	$5^2 - 3(2 + 4) = 5^2 - 3(6)$
Step 2 Simplify any number expressions containing exponents.	$= 25 - 3(6)$
Step 3 Do multiplications and divisions as they occur from left to right.	$= 25 - 18$
Step 4 Do addition and subtraction as they occur from left to right.	$= 7$

Chapter 1 Review Exercises

1. Simplify: $3 \cdot 2^3 \cdot 5^2$
    600   [1.6A]

2. Write 10,327 in expanded form.
    $10{,}000 + 300 + 20 + 7$ [1.1C]

3. Find all the factors of 18.
    1, 2, 3, 6, 9, 18   [1.7A]

4. Find the sum of 5894, 6301, and 298.
    12,493   [1.2A]

5. Subtract: 4926
                 $-\ 3177$
    _____
                    1749   [1.3B]

6. Divide: $7\overline{)14{,}945}$
    2135   [1.5A]

7. Place the correct symbol, $<$ or $>$, between
    the two numbers: 101   87
    $101 > 87$ [1.1A]

8. Write $5 \cdot 5 \cdot 7 \cdot 7 \cdot 7 \cdot 7 \cdot 7$ in
    exponential notation.
    $5^2 \cdot 7^5$ [1.6A]

9. What is 2019 multiplied by 307?
    619,833   [1.4B]

10. What is 10,134 decreased by 4725?
    5409   [1.3B]

11. Add: 298
            461
          $+\ 322$
    _____
          1081   [1.2A]

12. Simplify: $2^3 - 3 \cdot 2$
    2   [1.6B]

13. Round 45,672 to the nearest hundred.
    45,700   [1.1D]

14. Write 276,057 in words.
    Two hundred seventy-six thousand fifty-seven   [1.1B]

15. Find the quotient of 109,763 and 84.
    1306 r59   [1.5C]

16. Write two million eleven thousand forty-four
    in standard form.
    2,011,044   [1.1B]

17. What is 3906 divided by 8?
    488 r2   [1.5B]

18. Simplify: $3^2 + 2^2 \cdot (5 - 3)$
    17   [1.6B]

19. Simplify: $8 \cdot (6 - 2)^2 \div 4$
    32   [1.6B]

20. Find the prime factorization of 72.
    $2 \cdot 2 \cdot 2 \cdot 3 \cdot 3$ [1.7B]

Instructor Note
The notation [1.6A] following the
answer to Exercise 1 indicates
the objective that the student
should review if that question is
answered incorrectly. The
notation [1.6A] means Chapter 1,
Section 6, Objective A. This
notation is used following every
answer in all of the Prep Tests
(except Chapter 1), Chapter
Review Exercises, Chapter Tests,
and Cumulative Reviews
throughout the text.

21. What is 3895 minus 1762?
2133 [1.3A]

22. Multiply: 843
× 27
22,761 [1.4B]

23. Wages Vincent Meyers, a sales assistant, earns $480 for working a 40-hour week. Last week Vincent worked an additional 12 hours at $24 an hour. Find Vincent's total pay for last week's work.
$768 [1.4C]

24. Fuel Efficiency Louis Reyes, a sales executive, drove a car 351 miles on 13 gallons of gas. Find the number of miles driven per gallon of gasoline.
27 miles per gallon [1.5D]

25. Consumerism A car is purchased for $17,880, with a down payment of $3000. The balance is paid in 48 equal monthly payments. Find the monthly car payment.
$310 [1.5D]

26. Compensation An insurance account executive received commissions of $723, $544, $812, and $488 during a 4-week period. Find the total income from commissions for the 4 weeks.
$2567 [1.2B]

27. Banking You had a balance of $516 in your checking account before making deposits of $88 and $213. Find the total amount deposited, and determine your new account balance.
$301; $817 [1.2B]

28. Compensation You have a car payment of $246 per month. What is the total of the car payments over a 12-month period?
$2952 [1.4C]

Athletics The table at the right shows the athletic participation by males and females at U.S. colleges in 1972 and 2001. Use this information for Exercises 29 to 32.

Year	Male Athletes	Female Athletes
1972	170,384	29,977
2001	208,866	150,916

Source: U.S. Department of Education commission report

29. In which year, 1972 or 2001, were there more males involved in sports at U.S. colleges?
2001 [1.1A]

30. What is the difference between the number of males involved in sports and the number of females involved in sports at U.S. colleges in 1972?
140,407 students [1.3C]

31. Find the increase in the number of females involved in sports in U.S. colleges from 1972 to 2001.
120,939 students [1.3C]

32. How many more U.S. college students were involved in athletics in 2001 than in 1972?
159,421 more students [1.3C]

Chapter 1 Test

1. Simplify: $3^3 \cdot 4^2$
432   [1.6A]

2. Write 207,068 in words.
Two hundred seven thousand sixty-eight   [1.1B]

3. Subtract: 17,495
$- \underline{\quad 8,162}$
9333   [1.3B]

4. Find all the factors of 20.
1, 2, 4, 5, 10, 20   [1.7A]

5. Multiply: 9736
$\underline{\times \quad 704}$
6,854,144   [1.4B]

6. Simplify: $4^2 \cdot (4 - 2) \div 8 + 5$
9   [1.6B]

7. Write 906,378 in expanded form.
900,000 + 6000 + 300 + 70 + 8   [1.1C]

8. Round 74,965 to the nearest hundred.
75,000   [1.1D]

9. Divide: $97\overline{)108,764}$
1121 r27   [1.5C]

10. Write $3 \cdot 3 \cdot 3 \cdot 7 \cdot 7$ in exponential form.
$3^3 \cdot 7^2$ [1.6A]

11. Find the sum of 8756, 9094, and 37,065.
54,915   [1.2A]

12. Find the prime factorization of 84.
$2 \cdot 2 \cdot 3 \cdot 7$ [1.7B]

13. Simplify: $16 \div 4 \cdot 2 - (7 - 5)^2$
4   [1.6B]

14. Find the product of 8 and 90,763.
726,104   [1.4A]

15. Write one million two hundred four thousand six in standard form.
1,204,006   [1.1B]

16. Divide: $7\overline{)60,972}$
8710 r2   [1.5B]

17. Place the correct symbol, < or >, between the two numbers: 21 19
21 > 19 [1.1A]

18. Find the quotient of 5624 and 8.
703 [1.5A]

19. Add: 25,492
+71,306
96,798 [1.2A]

20. Find the difference between 29,736 and 9814.
19,922 [1.3B]

Education The table at the right shows the projected enrollment in public and private elementary and secondary schools in the fall of 2009 and the fall of 2012. Use this information for Exercises 21 and 22.

Year	Kindergarten through Grade 8	Grades 9 through 12
2009	37,726,000	15,812,000
2012	38,258,000	15,434,000

Source: The National Center for Education Statistics

21. Find the difference between the total enrollment in 2012 and that in 2009.
154,000 students [1.3C]

22. Find the average enrollment in each of grades 9 through 12 in 2012.
3,858,500 students [1.5D]

23. **Farming** A farmer harvested 48,290 pounds of lemons from one grove and 23,710 pounds of lemons from another grove. The lemons were packed in boxes with 24 pounds of lemons in each box. How many boxes were needed to pack the lemons?
3000 boxes [1.5D]

24. **Investments** An investor receives $237 each month from a corporate bond fund. How much will the investor receive over a 12-month period?
$2844 [1.4C]

25. **Travel** A family drives 425 miles the first day, 187 miles the second day, and 243 miles the third day of their vacation. The odometer read 47,626 miles at the start of the vacation.
a. How many miles were driven during the 3 days? 855 miles
b. What is the odometer reading at the end of the 3 days? 48,481 miles [1.2B]

2 Fractions

Members of the Los Angeles Philharmonic follow conductor Esa-Pekka Salonen's movements to guarantee synchronized playing. By paying attention to both Salonen and the time signature for each musical piece, the musicians play in time with each other. The time signature appears as a fraction at the beginning of a piece of music and tells them how many beats to play per measure. The **project on page 115** demonstrates how to interpret the time signature.

Need help? For online student resources, such as section quizzes, visit this textbook's website at **math.college.hmco.com/students**.

OBJECTIVES

Section 2.1

A To find the least common multiple (LCM)
B To find the greatest common factor (GCF)

Section 2.2

A To write a fraction that represents part of a whole
B To write an improper fraction as a mixed number or a whole number, and a mixed number as an improper fraction

Section 2.3

A To find equivalent fractions by raising to higher terms
B To write a fraction in simplest form

Section 2.4

A To add fractions with the same denominator
B To add fractions with different denominators
C To add whole numbers, mixed numbers, and fractions
D To solve application problems

Section 2.5

A To subtract fractions with the same denominator
B To subtract fractions with different denominators
C To subtract whole numbers, mixed numbers, and fractions
D To solve application problems

Section 2.6

A To multiply fractions
B To multiply whole numbers, mixed numbers, and fractions
C To solve application problems

Section 2.7

A To divide fractions
B To divide whole numbers, mixed numbers, and fractions
C To solve application problems

Section 2.8

A To identify the order relation between two fractions
B To simplify expressions containing exponents
C To use the Order of Operations Agreement to simplify expressions

Features of the IAE

For a description of the features included in the margins of this Instructor's Annotated Edition, see page 2.

PREP TEST • • •

Do these exercises to prepare for Chapter 2.

For Exercises 1 to 6, add, subtract, multiply, or divide.

1. 4×5
20   [1.4A]

2. $2 \cdot 2 \cdot 2 \cdot 3 \cdot 5$
120   [1.4A]

3. 9×1
9   [1.4A]

4. $6 + 4$
10   [1.2A]

5. $10 - 3$
7   [1.3A]

6. $63 \div 30$
2 r3   [1.5C]

7. Which of the following numbers divide evenly into 12?

1 2 3 4 5 6 7 8 9 10 11 12
1, 2, 3, 4, 6, 12   [1.5B]

8. Simplify: $8 \times 7 + 3$
59   [1.6B]

9. Complete: $8 = ? + 1$
7   [1.3A]

10. Place the correct symbol, $<$ or $>$, between the two numbers.

44   48
44 $<$ 48   [1.1A]

GO FIGURE • • •

You and a friend are swimming laps in a pool. You swim one lap every 4 minutes. Your friend swims one lap every 5 minutes. If you start at the same time from the same end of the pool, in how many minutes will both of you be at the starting point again? How many times will you have passed each other in the pool prior to that time?   20 minutes; 8 times

2.1 The Least Common Multiple and Greatest Common Factor

Objective A   To find the least common multiple (LCM)

The **multiples of a number** are the products of that number and the numbers 1, 2, 3, 4, 5, ...

$3 \times 1 = 3$
$3 \times 2 = 6$
$3 \times 3 = 9$
$3 \times 4 = 12$ The multiples of 3 are 3, 6, 9, 12, 15, ...
$3 \times 5 = 15$
  .
  .
  .

A number that is a multiple of two or more numbers is a **common multiple** of those numbers.

The multiples of 4 are 4, 8, 12, 16, 20, 24, 28, 32, 36, ...
The multiples of 6 are 6, 12, 18, 24, 30, 36, 42, ...
Some common multiples of 4 and 6 are 12, 24, and 36.

The **least common multiple (LCM)** is the smallest common multiple of two or more numbers.

The least common multiple of 4 and 6 is 12.

Listing the multiples of each number is one way to find the LCM. Another way to find the LCM uses the prime factorization of each number.

To find the LCM of 450 and 600, find the prime factorization of each number and write the factorization of each number in a table. Circle the greatest product in each column. The LCM is the product of the circled numbers.

	2	3	5
450 =	2	(3 · 3)	(5 · 5)
600 =	(2 · 2 · 2)	3	5 · 5

In the column headed by 5, the products are equal. Circle just one product.

The LCM is the product of the circled numbers.
The LCM = 2 · 2 · 2 · 3 · 3 · 5 · 5 = 1800.

Example 1   Find the LCM of 24, 36, and 50.

Solution

	2	3	5
24 =	(2 · 2 · 2)	3	
36 =	2 · 2	(3 · 3)	
50 =	2		(5 · 5)

The LCM = 2 · 2 · 2 · 3 · 3 · 5 · 5
      = 1800.

You Try It 1   Find the LCM of 12, 27, and 50.

Your solution   2700

Solution on p. S4

Objective 2.1A

New Vocabulary
multiples of a number
common multiple
least common multiple (LCM)

Discuss the Concepts
1. How can you find the multiples of 12?
2. How can you find some common multiples of 8 and 12?
3. Why is 24 the least common multiple of 8 and 12?

Concept Check
Find the LCM of 16, 20, and 40 by first listing the multiples of each number and then using the prime factorization method shown on page 65 of the textbook.   80

Optional Student Activity
The ancient Mayans used two calendars, a civil calendar of 365 days and a sacred calendar of 260 days. If a civil year and the sacred year begin on the same day, how many civil years and how many sacred years will pass before this situation occurs again?
The LCM of 365 and 260 is 18,980.
18,980 ÷ 365 = 52
18,980 ÷ 260 = 73
The situation occurs again after 52 civil years and 73 sacred years.

In-Class Examples (Objective 2.1A)
Find the LCM.
1. 14, 21   42
2. 2, 7, 14   14
3. 5, 12, 15   60

Objective 2.1B

Vocabulary to Review

factors of a number [1.7A]

New Vocabulary

common factor
greatest common factor (GCF)

Discuss the Concepts

1. How can you find the factors of 24?
2. How can you find the common factors of 12 and 24?
3. Why is 12 the greatest common factor of 12 and 24?

Concept Check

Find the product of the GCF of 225 and 444 and the LCM of 225 and 444.

GCF $= 3$; LCM $= 33,300$;
$3(33,300) = 99,900$

Concept Check

Find the GCF of $2^5 \cdot 3^9 \cdot 5^7$ and $2^7 \cdot 3^2 \cdot 5^4$.

$2^5 \cdot 3^2 \cdot 5^4 = 180,000$

Instructor Note

The following model may help some students with the LCM and GCF.

The arrow indicates "divides into."

Optional Student Activity

What number must be multiplied by 200 so that the product has exactly 15 factors?

2; Factors: 1, 2, 4, 5, 8, 10, 16, 20, 25, 40, 50, 80, 100, 200, 400

Objective B   **To find the greatest common factor (GCF)**

Recall that a number that divides another number evenly is a factor of that number. The number 64 can be evenly divided by 1, 2, 4, 8, 16, 32, and 64, so the numbers 1, 2, 4, 8, 16, 32, and 64 are factors of 64.

A number that is a factor of two or more numbers is a **common factor** of those numbers.

The factors of 30 are 1, 2, 3, 5, 6, 10, 15, and 30.
The factors of 105 are 1, 3, 5, 7, 15, 21, 35, and 105.
The common factors of 30 and 105 are 1, 3, 5, and 15.

The **greatest common factor (GCF)** is the largest common factor of two or more numbers.

The greatest common factor of 30 and 105 is 15.

Listing the factors of each number is one way of finding the GCF. Another way to find the GCF uses the prime factorization of each number.

To find the GCF of 126 and 180, find the prime factorization of each number and write the factorization of each number in a table. Circle the least product in each column that does not have a blank. The GCF is the product of the circled numbers.

	2	3	5	7
$126 =$	②	(3 · 3)		7
$180 =$	2 · 2	3 · 3	5	

In the column headed by 3, the products are equal. Circle just one product.
Columns 5 and 7 have a blank, so 5 and 7 are not common factors of 126 and 180. Do not circle any number in these columns.

The GCF is the product of the circled numbers.
The GCF $= 2 \cdot 3 \cdot 3 = 18$.

Example 2   Find the GCF of 90, 168, and 420.

Solution

	2	3	5	7
$90 =$	②	3 · 3	5	
$168 =$	2 · 2 · 2	③		7
$420 =$	2 · 2	3	5	7

The GCF $= 2 \cdot 3 = 6$.

You Try It 2   Find the GCF of 36, 60, and 72.

Your solution   12

Example 3   Find the GCF of 7, 12, and 20.

Solution

	2	3	5	7
$7 =$				7
$12 =$	2 · 2	3		
$20 =$	2 · 2		5	

Because no numbers are circled, the GCF $= 1$.

You Try It 3   Find the GCF of 11, 24, and 30.

Your solution   1

Solutions on p. S4

In-Class Examples (Objective 2.1B)

Find the GCF.

1. 12, 18   6
2. 24, 64   8
3. 41, 67   1
4. 21, 27, 33   3

2.1 Exercises

Section 2.1

Suggested Assignment
Exercises 1–67, odds
Exercises 70–72, 74
More challenging problem:
 Exercise 73

Objective A   **To find the least common multiple (LCM)**

For Exercises 1 to 34, find the LCM.

1. 5, 8
40

2. 3, 6
6

3. 3, 8
24

4. 2, 5
10

5. 5, 6
30

6. 5, 7
35

7. 4, 6
12

8. 6, 8
24

9. 8, 12
24

10. 12, 16
48

11. 5, 12
60

12. 3, 16
48

13. 8, 14
56

14. 6, 18
18

15. 3, 9
9

16. 4, 10
20

17. 8, 32
32

18. 7, 21
21

19. 9, 36
36

20. 14, 42
42

21. 44, 60
660

22. 120, 160
480

23. 102, 184
9384

24. 123, 234
9594

25. 4, 8, 12
24

26. 5, 10, 15
30

27. 3, 5, 10
30

28. 2, 5, 8
40

29. 3, 8, 12
24

30. 5, 12, 18
180

31. 9, 36, 64
576

32. 18, 54, 63
378

33. 16, 30, 84
1680

34. 9, 12, 15
180

Objective B   **To find the greatest common factor (GCF)**

For Exercises 35 to 68, find the GCF.

35. 3, 5
1

36. 5, 7
1

37. 6, 9
3

38. 18, 24
6

39. 15, 25
5

40. 14, 49
7

41. 25, 100
25

42. 16, 80
16

43. 32, 51
1

44. 21, 44
1

45. 12, 80
4

46. 8, 36
4

47. 16, 140
4

48. 12, 76
4

Quick Quiz (Objective 2.1A)
Find the LCM.
1. 10, 25   50
2. 3, 6, 7   42
3. 2, 8, 64   64

Quick Quiz (Objective 2.1B)
Find the GCF.
1. 6, 16   2
2. 4, 9   1
3. 26, 52   26
4. 12, 30, 60   6

49. 24, 30
6

50. 48, 144
48

51. 44, 96
4

52. 18, 32
2

53. 3, 5, 11
1

54. 6, 8, 10
2

55. 7, 14, 49
7

56. 6, 15, 36
3

57. 10, 15, 20
5

58. 12, 18, 20
2

59. 24, 40, 72
8

60. 3, 17, 51
1

61. 17, 31, 81
1

62. 14, 42, 84
14

63. 25, 125, 625
25

64. 12, 68, 92
4

65. 28, 35, 70
7

66. 1, 49, 153
1

67. 32, 56, 72
8

68. 24, 36, 48
12

Answers to Writing Exercises

71. The LCM of 2 and 3 is 6. The LCM of 5 and 7 is 35. The LCM of 11 and 19 is 209. The LCM of two prime numbers is the product of the two numbers. The LCM of three prime numbers is the product of the three numbers.

72. The GCF of 3 and 5 is 1. The GCF of 7 and 11 is 1. The GCF of 29 and 43 is 1. Because two prime numbers do not have a common factor other than 1, the GCF of two prime numbers is 1. Because three prime numbers do not have a common factor other than 1, the GCF of three prime numbers is 1.

73. Yes, the LCM of two numbers is always divisible by the GCF of the two numbers. Two numbers are factors of their LCM. The GCF of the two numbers is a factor of the LCM of the same numbers. That is, the LCM of two numbers always is divisible by the GCF of the two numbers. For example, the GCF of 4 and 6 is 2, and the LCM of 4 and 6 is 12. 12 is divisible by 2.

APPLYING THE CONCEPTS

69. Define the phrase *relatively prime numbers*. List three pairs of relatively prime numbers.
Relatively prime numbers are numbers with no common factors except the factor 1. Answers will vary. For example, 4 and 5, 8 and 9, 16 and 21.

70. **Work Schedules**   Joe Salvo, a lifeguard, works 3 days and then has a day off. A friend works 5 days and then has a day off. How many days after Joe and his friend have a day off together will they have another day off together?   12 days

71.   Find the LCM of each of the following pairs of numbers: 2 and 3, 5 and 7, and 11 and 19. Can you draw a conclusion about the LCM of two prime numbers? Suggest a way of finding the LCM of three distinct prime numbers.

72.   Find the GCF of each of the following pairs of numbers: 3 and 5, 7 and 11, and 29 and 43. Can you draw a conclusion about the GCF of two prime numbers? What is the GCF of three prime distinct numbers?

73.   Is the LCM of two numbers always divisible by the GCF of the two numbers? If so, explain why. If not, give an example.

74. Using the pattern for the first two triangles at the right, determine the center number of the last triangle.
4

```
       16              18              36
      /  \            /  \            /  \
     / 4  \          / 2  \          / ?  \
    /_____\        /_____\        /_____\
   20     12       20     16       60     20
```

2.2 Introduction to Fractions

Objective A   To write a fraction that represents part of a whole

Point of Interest

The fraction bar was first used in 1050 by al-Hassar. It is also called a vinculum.

A **fraction** can represent the number of equal parts of a whole. The shaded portion of the circle is represented by the fraction $\frac{4}{7}$. Four of the seven equal parts of the circle (that is, four-sevenths of it) are shaded.

Each part of a fraction has a name.

Fraction bar → $\frac{4}{7}$ ← **Numerator**
← **Denominator**

A **proper fraction** is a fraction less than 1. The numerator of a proper fraction is smaller than the denominator. The shaded portion of the circle can be represented by the proper fraction $\frac{3}{4}$.

A **mixed number** is a number greater than 1 with a whole-number part and a fractional part. The shaded portion of the circles can be represented by the mixed number $2\frac{1}{4}$.

An **improper fraction** is a fraction greater than or equal to 1. The numerator of an improper fraction is greater than or equal to the denominator. The shaded portion of the circles can be represented by the improper fraction $\frac{9}{4}$. The shaded portion of the square can be represented by $\frac{4}{4}$.

Objective 2.2A

New Vocabulary
fraction
fraction bar
numerator
denominator
proper fraction
mixed number
improper fraction

Discuss the Concepts
Explain why the shaded portion of the diagram can be described as $\frac{2}{3}$.

Concept Check
Write a proper fraction and explain why it is a proper fraction.

Write a mixed number and explain why it is a mixed number.

Write an improper fraction and explain why it is an improper fraction.

Optional Student Activity
Use the symbol $<$, $>$, or $=$ to compare each number with the number 1.

1. $\frac{3}{7}$ $\frac{3}{7} < 1$

2. $\frac{9}{9}$ $\frac{9}{9} = 1$

3. $\frac{8}{5}$ $\frac{8}{5} > 1$

Example 1 Express the shaded portion of the circles as a mixed number.

Solution $3\frac{2}{5}$

You Try It 1 Express the shaded portion of the circles as a mixed number.

Your solution $4\frac{1}{4}$

Example 2 Express the shaded portion of the circles as an improper fraction.

Solution $\frac{17}{5}$

You Try It 2 Express the shaded portion of the circles as an improper fraction.

Your solution $\frac{17}{4}$

Solutions on pp. S4–S5

In-Class Examples (Objective 2.2A)

1. Express the shaded portion of the circle as a fraction.
$\frac{3}{8}$

2. Express the shaded portion of the circles as a mixed number.
$2\frac{1}{4}$

3. Express the shaded portion of the circles as an improper fraction. $\frac{11}{6}$

Objective 2.2B

Vocabulary to Review
improper fraction [2.2A]
mixed number [2.2A]

Discuss the Concepts
1. Explain the procedure for rewriting an improper fraction as a mixed number or a whole number.
2. Explain the procedure for rewriting a mixed number as an improper fraction.

Concept Check
Match the equivalent mixed numbers and improper fractions.

1. $\dfrac{9}{2}$ **a.** $2\dfrac{7}{9}$

2. $\dfrac{25}{9}$ **b.** $2\dfrac{2}{9}$

3. $\dfrac{19}{4}$ **c.** $4\dfrac{1}{2}$

4. $\dfrac{11}{4}$ **d.** $2\dfrac{3}{4}$

5. $\dfrac{20}{9}$ **e.** $4\dfrac{3}{4}$

1. c 2. a 3. e 4. d 5. b

Instructor Note
As a classroom exercise, ask students to give real-world examples where mixed numbers are used. Some possible answers: carpentry, sewing, recipes.

Objective B **To write an improper fraction as a mixed number or a whole number, and a mixed number as an improper fraction**

Note from the diagram that the mixed number $2\dfrac{3}{5}$ and the improper fraction $\dfrac{13}{5}$ both represent the shaded portion of the circles.

$$2\dfrac{3}{5} = \dfrac{13}{5}$$

An improper fraction can be written as a mixed number or a whole number.

HOW TO Write $\dfrac{13}{5}$ as a mixed number.

Divide the numerator by the denominator.	To write the fractional part of the mixed number, write the remainder over the divisor.	Write the answer.
$\begin{array}{r} 2 \\ 5{\overline{\smash{\big)}\,13}} \\ -10 \\ \hline 3 \end{array}$	$\begin{array}{r} 2\frac{3}{5} \\ 5{\overline{\smash{\big)}\,13}} \\ -10 \\ \hline 3 \end{array}$	$\dfrac{13}{5} = 2\dfrac{3}{5}$

To write a mixed number as an improper fraction, multiply the denominator of the fractional part by the whole-number part. The sum of this product and the numerator of the fractional part is the numerator of the improper fraction. The denominator remains the same.

HOW TO Write $7\dfrac{3}{8}$ as an improper fraction.

$$7\dfrac{3}{8} = \dfrac{(8 \times 7) + 3}{8} = \dfrac{56 + 3}{8} = \dfrac{59}{8} \qquad 7\dfrac{3}{8} = \dfrac{59}{8}$$

Point of Interest
Archimedes (c. 287–212 B.C.) is the person who calculated that $\pi \approx 3\dfrac{1}{7}$. He actually showed that $3\dfrac{10}{71} < \pi < 3\dfrac{1}{7}$. The approximation $3\dfrac{10}{71}$ is more accurate but more difficult to use.

Example 3 Write $\dfrac{21}{4}$ as a mixed number.

Solution
$\begin{array}{r} 5 \\ 4{\overline{\smash{\big)}\,21}} \\ -20 \\ \hline 1 \end{array}$ $\dfrac{21}{4} = 5\dfrac{1}{4}$

You Try It 3 Write $\dfrac{22}{5}$ as a mixed number.

Your solution $4\dfrac{2}{5}$

Example 4 Write $\dfrac{18}{6}$ as a whole number.

Solution
$\begin{array}{r} 3 \\ 6{\overline{\smash{\big)}\,18}} \\ -18 \\ \hline 0 \end{array}$ $\dfrac{18}{6} = 3$

Note: The remainder is zero.

You Try It 4 Write $\dfrac{28}{7}$ as a whole number.

Your solution 4

Example 5 Write $21\dfrac{3}{4}$ as an improper fraction.

Solution $21\dfrac{3}{4} = \dfrac{84 + 3}{4} = \dfrac{87}{4}$

You Try It 5 Write $14\dfrac{5}{8}$ as an improper fraction.

Your solution $\dfrac{117}{8}$

Solutions on pp. S4–S5

In-Class Examples (Objective 2.2B)

Write the improper fraction as a mixed number or a whole number.

1. $\dfrac{10}{3}$ $3\dfrac{1}{3}$

2. $\dfrac{81}{9}$ 9

Write the mixed number as an improper fraction.

3. $3\dfrac{1}{4}$ $\dfrac{13}{4}$

4. $4\dfrac{5}{9}$ $\dfrac{41}{9}$

2.2 Exercises

Objective A To write a fraction that represents part of a whole

Section 2.2

Suggested Assignment
Exercises 1–25, odds
Exercises 27–71, every other odd

For Exercises 1 to 4, identify the fraction as a proper fraction, an improper fraction, or a mixed number.

1. $\frac{12}{7}$
Improper fraction

2. $5\frac{2}{11}$
Mixed number

3. $\frac{29}{40}$
Proper fraction

4. $\frac{19}{13}$
Improper fraction

For Exercises 5 to 8, express the shaded portion of the circle as a fraction.

5. $\frac{3}{4}$

6. $\frac{4}{7}$

7. $\frac{7}{8}$

8. $\frac{3}{5}$

For Exercises 9 to 14, express the shaded portion of the circles as a mixed number.

9. $1\frac{1}{2}$

10. $2\frac{2}{3}$

11. $2\frac{5}{8}$

12. $2\frac{3}{4}$

13. $3\frac{3}{5}$

14. $3\frac{5}{6}$

For Exercises 15 to 20, express the shaded portion of the circles as an improper fraction.

15. $\frac{5}{4}$

16. $\frac{7}{6}$

17. $\frac{8}{3}$

18. $\frac{9}{4}$

19. $\frac{28}{8}$

20. $\frac{18}{5}$

21. Shade $\frac{5}{6}$ of

22. Shade $\frac{3}{8}$ of

23. Shade $1\frac{2}{5}$ of

24. Shade $1\frac{3}{4}$ of

Quick Quiz (Objective 2.2A)

1. Express the shaded portion of the circle as a fraction.

$\frac{2}{5}$

2. Express the shaded portion of the circles as a mixed number.

$1\frac{1}{3}$

3. Express the shaded portion of the circles as an improper fraction.

$\frac{13}{8}$

25. Shade $\frac{6}{5}$ of

26. Shade $\frac{7}{3}$ of

> **Objective B**   To write an improper fraction as a mixed number or a whole number, and a mixed number as an improper fraction

For Exercises 27 to 50, write the improper fraction as a mixed number or a whole number.

27. $\frac{11}{4}$
$2\frac{3}{4}$

28. $\frac{16}{3}$
$5\frac{1}{3}$

29. $\frac{20}{4}$
5

30. $\frac{18}{9}$
2

31. $\frac{9}{8}$
$1\frac{1}{8}$

32. $\frac{13}{4}$
$3\frac{1}{4}$

33. $\frac{23}{10}$
$2\frac{3}{10}$

34. $\frac{29}{2}$
$14\frac{1}{2}$

35. $\frac{48}{16}$
3

36. $\frac{51}{3}$
17

37. $\frac{8}{7}$
$1\frac{1}{7}$

38. $\frac{16}{9}$
$1\frac{7}{9}$

39. $\frac{7}{3}$
$2\frac{1}{3}$

40. $\frac{9}{5}$
$1\frac{4}{5}$

41. $\frac{16}{1}$
16

42. $\frac{23}{1}$
23

43. $\frac{17}{8}$
$2\frac{1}{8}$

44. $\frac{31}{16}$
$1\frac{15}{16}$

45. $\frac{12}{5}$
$2\frac{2}{5}$

46. $\frac{19}{3}$
$6\frac{1}{3}$

47. $\frac{9}{9}$
1

48. $\frac{40}{8}$
5

49. $\frac{72}{8}$
9

50. $\frac{3}{3}$
1

For Exercises 51 to 74, write the mixed number as an improper fraction.

51. $2\frac{1}{3}$
$\frac{7}{3}$

52. $4\frac{2}{3}$
$\frac{14}{3}$

53. $6\frac{1}{2}$
$\frac{13}{2}$

54. $8\frac{2}{3}$
$\frac{26}{3}$

55. $6\frac{5}{6}$
$\frac{41}{6}$

56. $7\frac{3}{8}$
$\frac{59}{8}$

57. $9\frac{1}{4}$
$\frac{37}{4}$

58. $6\frac{1}{4}$
$\frac{25}{4}$

59. $10\frac{1}{2}$
$\frac{21}{2}$

60. $15\frac{1}{8}$
$\frac{121}{8}$

61. $8\frac{1}{9}$
$\frac{73}{9}$

62. $3\frac{5}{12}$
$\frac{41}{12}$

63. $5\frac{3}{11}$
$\frac{58}{11}$

64. $3\frac{7}{9}$
$\frac{34}{9}$

65. $2\frac{5}{8}$
$\frac{21}{8}$

66. $12\frac{2}{3}$
$\frac{38}{3}$

67. $1\frac{5}{8}$
$\frac{13}{8}$

68. $5\frac{3}{7}$
$\frac{38}{7}$

69. $11\frac{1}{9}$
$\frac{100}{9}$

70. $12\frac{3}{5}$
$\frac{63}{5}$

71. $3\frac{3}{8}$
$\frac{27}{8}$

72. $4\frac{5}{9}$
$\frac{41}{9}$

73. $6\frac{7}{13}$
$\frac{85}{13}$

74. $8\frac{5}{14}$
$\frac{117}{14}$

APPLYING THE CONCEPTS

75. ✏ Name three situations in which fractions are used. Provide an example of a fraction that is used in each situation.

Answers to Writing Exercises

75. Students might mention any of the following: fractional parts of an hour, as in three-quarters of an hour; lengths of nails, as in a $\frac{3}{4}$-inch nail; lengths of fabric, as in $1\frac{5}{8}$ yards of material; lengths of lumber, as in $2\frac{1}{2}$ feet of pine; ingredients in a recipe, as in $1\frac{1}{2}$ of sugar; or innings pitched, as in four and two-thirds innings.

Quick Quiz (Objective 2.2B)

Write the improper fraction as a mixed number or a whole number.

1. $\frac{15}{7}$ $2\frac{1}{7}$ **2.** $\frac{20}{5}$ 4

Write the mixed number as an improper fraction.

3. $8\frac{1}{5}$ $\frac{41}{5}$ **4.** $6\frac{2}{3}$ $\frac{20}{3}$

2.3 Writing Equivalent Fractions

Objective A **To find equivalent fractions by raising to higher terms**

Equal fractions with different denominators are called **equivalent fractions.**

$\frac{4}{6}$ is equivalent to $\frac{2}{3}$.

Remember that the Multiplication Property of One states that the product of a number and one is the number. This is true for fractions as well as whole numbers. This property can be used to write equivalent fractions.

$$\frac{2}{3} \times 1 = \frac{2}{3} \times \frac{1}{1} = \frac{2 \cdot 1}{3 \cdot 1} = \frac{2}{3}$$

$$\frac{2}{3} \times 1 = \frac{2}{3} \times \boxed{\frac{2}{2}} = \frac{2 \cdot 2}{3 \cdot 2} = \frac{4}{6} \qquad \frac{4}{6} \text{ is equivalent to } \frac{2}{3}.$$

$$\frac{2}{3} \times 1 = \frac{2}{3} \times \boxed{\frac{4}{4}} = \frac{2 \cdot 4}{3 \cdot 4} = \frac{8}{12} \qquad \frac{8}{12} \text{ is equivalent to } \frac{2}{3}.$$

$\frac{2}{3}$ was rewritten as the equivalent fractions $\frac{4}{6}$ and $\frac{8}{12}$.

HOW TO Write a fraction that is equivalent to $\frac{5}{8}$ and has a denominator of 32.

$32 \div 8 = 4$ • Divide the larger denominator by the smaller.

$\frac{5}{8} = \frac{5 \cdot 4}{8 \cdot 4} = \frac{20}{32}$ • Multiply the numerator and denominator of the given fraction by the quotient (4).

$\frac{20}{32}$ is equivalent to $\frac{5}{8}$.

Example 1 Write $\frac{2}{3}$ as an equivalent fraction that has a denominator of 42.

Solution $42 \div 3 = 14 \qquad \frac{2}{3} = \frac{2 \cdot 14}{3 \cdot 14} = \frac{28}{42}$

$\frac{28}{42}$ is equivalent to $\frac{2}{3}$.

You Try It 1 Write $\frac{3}{5}$ as an equivalent fraction that has a denominator of 45.

Your solution $\frac{27}{45}$

Example 2 Write 4 as a fraction that has a denominator of 12.

Solution Write 4 as $\frac{4}{1}$.

$12 \div 1 = 12 \qquad 4 = \frac{4 \cdot 12}{1 \cdot 12} = \frac{48}{12}$

$\frac{48}{12}$ is equivalent to 4.

You Try It 2 Write 6 as a fraction that has a denominator of 18.

Your solution $\frac{108}{18}$

Solutions on p. S5

Properties to Review
Multiplication Property of One
[1.4A]

New Vocabulary
equivalent fractions

Discuss the Concepts
Explain the procedure for finding equivalent fractions.

Optional Student Activity
1. Divide two-thirds of a circle into sixths and write the equivalent fraction.
2. Divide one-third of a circle into ninths and write the equivalent fraction.
3. Divide three-fourths of a circle into eighths and write the equivalent fraction.
4. Divide one-fourth of a circle into twelfths and write the equivalent fraction.

1. $\frac{4}{6}$ 2. $\frac{3}{9}$ 3. $\frac{6}{8}$ 4. $\frac{3}{12}$

Instructor Note
To help some students understand equivalent fractions, use a pizza. By cutting the pizza into, say, eight pieces, students are able to see that

$$\frac{1}{2} = \frac{4}{8}$$
$$\frac{1}{4} = \frac{2}{8}$$

In-Class Examples (Objective 2.3A)
Write an equivalent fraction with the given denominator.

1. $\frac{1}{2} = \frac{}{32}$ 16

2. $\frac{2}{3} = \frac{}{12}$ 8

3. $6 = \frac{}{11}$ 66

Objective 2.3B

Vocabulary to Review
common factors   [2.1B]

New Vocabulary
simplest form of a fraction

Discuss the Concepts
Explain the procedure for simplifying fractions.

Concept Check
Name the equivalent fractions in the list below. Then name the fractions that are written in simplest form.

$$\frac{6}{8} \quad \frac{12}{16} \quad \frac{9}{16} \quad \frac{3}{4} \quad \frac{18}{24}$$

The equivalent fractions are $\frac{6}{8}, \frac{12}{16}, \frac{3}{4}$, and $\frac{18}{24}$. The fractions $\frac{3}{4}$ and $\frac{9}{16}$ are in simplest form.

Instructor Note
You may prefer to explain that a fraction can be simplified by dividing the numerator and denominator by the GCF of the numerator and denominator.

Instructor Note
As mentioned earlier, one of the main pedagogical features of this text is the paired examples. Using the model of the Example, students should work the You Try It. A *complete* solution is provided in the Appendix so that students can check not only the answer but also their work.

Objective B   **To write a fraction in simplest form**

Writing the **simplest form of a fraction** means writing it so that the numerator and denominator have no common factors other than 1.

The fractions $\frac{4}{6}$ and $\frac{2}{3}$ are equivalent fractions.

$\frac{4}{6}$ has been written in simplest form as $\frac{2}{3}$.

The Multiplication Property of One can be used to write fractions in simplest form. Write the numerator and denominator of the given fraction as a product of factors. Write factors common to both the numerator and denominator as an improper fraction equivalent to 1.

$$\frac{4}{6} = \frac{2 \cdot 2}{2 \cdot 3} = \frac{2}{2} \cdot \frac{2}{3} = 1 \cdot \frac{2}{3} = \frac{2}{3}$$

The process of eliminating common factors is displayed with slashes through the common factors as shown at the right.

To write a fraction in simplest form, eliminate the common factors.

An improper fraction can be changed to a mixed number.

$$\frac{4}{6} = \frac{2 \cdot \overset{1}{\cancel{2}}}{\cancel{2} \cdot 3} = \frac{2}{3}$$

$$\frac{18}{30} = \frac{\overset{1}{\cancel{2}} \cdot \overset{1}{\cancel{3}} \cdot 3}{\cancel{2} \cdot \cancel{3} \cdot 5} = \frac{3}{5}$$

$$\frac{22}{6} = \frac{\overset{1}{\cancel{2}} \cdot 11}{\cancel{2} \cdot 3} = \frac{11}{3} = 3\frac{2}{3}$$

Example 3   Write $\frac{15}{40}$ in simplest form.

Solution
$$\frac{15}{40} = \frac{3 \cdot \overset{1}{\cancel{5}}}{2 \cdot 2 \cdot 2 \cdot \underset{1}{\cancel{5}}} = \frac{3}{8}$$

You Try It 3   Write $\frac{16}{24}$ in simplest form.

Your solution $\frac{2}{3}$

Example 4   Write $\frac{6}{42}$ in simplest form.

Solution
$$\frac{6}{42} = \frac{\overset{1}{\cancel{2}} \cdot \overset{1}{\cancel{3}}}{\underset{1}{\cancel{2}} \cdot \underset{1}{\cancel{3}} \cdot 7} = \frac{1}{7}$$

You Try It 4   Write $\frac{8}{56}$ in simplest form.

Your solution $\frac{1}{7}$

Example 5   Write $\frac{8}{9}$ in simplest form.

Solution
$$\frac{8}{9} = \frac{2 \cdot 2 \cdot 2}{3 \cdot 3} = \frac{8}{9}$$

$\frac{8}{9}$ is already in simplest form because there are no common factors in the numerator and denominator.

You Try It 5   Write $\frac{15}{32}$ in simplest form.

Your solution $\frac{15}{32}$

Example 6   Write $\frac{30}{12}$ in simplest form.

Solution
$$\frac{30}{12} = \frac{\overset{1}{\cancel{2}} \cdot \overset{1}{\cancel{3}} \cdot 5}{\underset{1}{\cancel{2}} \cdot 2 \cdot \underset{1}{\cancel{3}}} = \frac{5}{2} = 2\frac{1}{2}$$

You Try It 6   Write $\frac{48}{36}$ in simplest form.

Your solution $1\frac{1}{3}$

Solutions on p. S5

In-Class Examples (Objective 2.3B)
Write the fraction in simplest form.

1. $\frac{6}{9}$ $\frac{2}{3}$

2. $\frac{24}{64}$ $\frac{3}{8}$

3. $\frac{85}{75}$ $1\frac{2}{15}$

2.3 Exercises

Objective A   **To find equivalent fractions by raising to higher terms**

For Exercises 1 to 40, write an equivalent fraction with the given denominator.

1. $\frac{1}{2} = \frac{5}{10}$ **2.** $\frac{1}{4} = \frac{4}{16}$ **3.** $\frac{3}{16} = \frac{9}{48}$ **4.** $\frac{5}{9} = \frac{45}{81}$ **5.** $\frac{3}{8} = \frac{12}{32}$

6. $\frac{7}{11} = \frac{21}{33}$ **7.** $\frac{3}{17} = \frac{9}{51}$ **8.** $\frac{7}{10} = \frac{63}{90}$ **9.** $\frac{3}{4} = \frac{12}{16}$ **10.** $\frac{5}{8} = \frac{20}{32}$

11. $3 = \frac{27}{9}$ **12.** $5 = \frac{125}{25}$ **13.** $\frac{1}{3} = \frac{20}{60}$ **14.** $\frac{1}{16} = \frac{3}{48}$ **15.** $\frac{11}{15} = \frac{44}{60}$

16. $\frac{3}{50} = \frac{18}{300}$ **17.** $\frac{2}{3} = \frac{12}{18}$ **18.** $\frac{5}{9} = \frac{20}{36}$ **19.** $\frac{5}{7} = \frac{35}{49}$ **20.** $\frac{7}{8} = \frac{28}{32}$

21. $\frac{5}{9} = \frac{10}{18}$ **22.** $\frac{11}{12} = \frac{33}{36}$ **23.** $7 = \frac{21}{3}$ **24.** $9 = \frac{36}{4}$ **25.** $\frac{7}{9} = \frac{35}{45}$

26. $\frac{5}{6} = \frac{35}{42}$ **27.** $\frac{15}{16} = \frac{60}{64}$ **28.** $\frac{11}{18} = \frac{33}{54}$ **29.** $\frac{3}{14} = \frac{21}{98}$ **30.** $\frac{5}{6} = \frac{120}{144}$

31. $\frac{5}{8} = \frac{30}{48}$ **32.** $\frac{7}{12} = \frac{56}{96}$ **33.** $\frac{5}{14} = \frac{15}{42}$ **34.** $\frac{2}{3} = \frac{28}{42}$ **35.** $\frac{17}{24} = \frac{102}{144}$

36. $\frac{5}{13} = \frac{65}{169}$ **37.** $\frac{3}{8} = \frac{153}{408}$ **38.** $\frac{9}{16} = \frac{153}{272}$ **39.** $\frac{17}{40} = \frac{340}{800}$ **40.** $\frac{9}{25} = \frac{360}{1000}$

Objective B   **To write a fraction in simplest form**

For Exercises 41 to 75, write the fraction in simplest form.

41. $\frac{4}{12}$ **42.** $\frac{8}{22}$ **43.** $\frac{22}{44}$ **44.** $\frac{2}{14}$ **45.** $\frac{2}{12}$

$\frac{1}{3}$ $\frac{4}{11}$ $\frac{1}{2}$ $\frac{1}{7}$ $\frac{1}{6}$

Quick Quiz (Objective 2.3A)
Write an equivalent fraction with the given denominator.

1. $\frac{1}{8} = \frac{}{64}$ 8

2. $\frac{5}{6} = \frac{}{18}$ 15

3. $4 = \frac{}{15}$ 60

46. $\dfrac{50}{75}$

$\dfrac{2}{3}$

47. $\dfrac{40}{36}$

$1\dfrac{1}{9}$

48. $\dfrac{12}{8}$

$1\dfrac{1}{2}$

49. $\dfrac{0}{30}$

0

50. $\dfrac{10}{10}$

1

51. $\dfrac{9}{22}$

$\dfrac{9}{22}$

52. $\dfrac{14}{35}$

$\dfrac{2}{5}$

53. $\dfrac{75}{25}$

3

54. $\dfrac{8}{60}$

$\dfrac{2}{15}$

55. $\dfrac{16}{84}$

$\dfrac{4}{21}$

56. $\dfrac{20}{44}$

$\dfrac{5}{11}$

57. $\dfrac{12}{35}$

$\dfrac{12}{35}$

58. $\dfrac{8}{36}$

$\dfrac{2}{9}$

59. $\dfrac{28}{44}$

$\dfrac{7}{11}$

60. $\dfrac{12}{16}$

$\dfrac{3}{4}$

61. $\dfrac{16}{12}$

$1\dfrac{1}{3}$

62. $\dfrac{24}{18}$

$1\dfrac{1}{3}$

63. $\dfrac{24}{40}$

$\dfrac{3}{5}$

64. $\dfrac{44}{60}$

$\dfrac{11}{15}$

65. $\dfrac{8}{88}$

$\dfrac{1}{11}$

66. $\dfrac{9}{90}$

$\dfrac{1}{10}$

67. $\dfrac{144}{36}$

4

68. $\dfrac{140}{297}$

$\dfrac{140}{297}$

69. $\dfrac{48}{144}$

$\dfrac{1}{3}$

70. $\dfrac{32}{120}$

$\dfrac{4}{15}$

71. $\dfrac{60}{100}$

$\dfrac{3}{5}$

72. $\dfrac{33}{110}$

$\dfrac{3}{10}$

73. $\dfrac{36}{16}$

$2\dfrac{1}{4}$

74. $\dfrac{80}{45}$

$1\dfrac{7}{9}$

75. $\dfrac{32}{160}$

$\dfrac{1}{5}$

APPLYING THE CONCEPTS

76. Make a list of five different fractions that are equivalent to $\dfrac{2}{3}$.

Answers will vary. For example, $\dfrac{4}{6}, \dfrac{6}{9}, \dfrac{8}{12}, \dfrac{10}{15}, \dfrac{12}{18}$.

77. Make a list of five different fractions that are equivalent to 3.

Answers will vary. For example, $\dfrac{3}{1}, \dfrac{6}{2}, \dfrac{9}{3}, \dfrac{12}{4}, \dfrac{15}{5}$.

78. Show that $\dfrac{15}{24} = \dfrac{5}{8}$ by using a diagram.

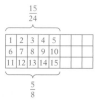

79. a. Geography What fraction of the states in the United States of America have names that begin with the letter M?
b. What fraction of the states have names that begin and end with a vowel?

a. $\dfrac{4}{25}$　**b.** $\dfrac{4}{25}$

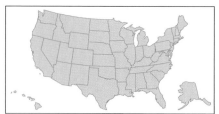

Quick Quiz (Objective 2.3B)

Write the fraction in simplest form.

1. $\dfrac{10}{25}$　$\dfrac{2}{5}$

2. $\dfrac{45}{81}$　$\dfrac{5}{9}$

3. $\dfrac{32}{24}$　$1\dfrac{1}{3}$

2.4 Addition of Fractions and Mixed Numbers

Objective A   **To add fractions with the same denominator**

Fractions with the same denominator are added by adding the numerators and placing the sum over the common denominator. After adding, write the sum in simplest form.

HOW TO   Add: $\frac{2}{7} + \frac{4}{7}$

$$\begin{array}{r} \frac{2}{7} \\ +\frac{4}{7} \\ \hline \frac{6}{7} \end{array}$$

• Add the numerators and place the sum over the common denominator.

$$\frac{2}{7} + \frac{4}{7} = \frac{2+4}{7} = \frac{6}{7}$$

Example 1   Add: $\frac{5}{12} + \frac{11}{12}$

Solution
$$\begin{array}{r} \frac{5}{12} \\ +\frac{11}{12} \\ \hline \frac{16}{12} = \frac{4}{3} = 1\frac{1}{3} \end{array}$$

• The denominators are the same. Add the numerators. Place the sum over the common denominator.

You Try It 1   Add: $\frac{3}{8} + \frac{7}{8}$

Your solution $1\frac{1}{4}$

Solution on p. S5

Objective B   **To add fractions with different denominators**

To add fractions with different denominators, first rewrite the fractions as equivalent fractions with a common denominator. The common denominator is the LCM of the denominators of the fractions.

Integrating Technology

Some scientific calculators have a fraction key, a^b/c . It is used to perform operations on fractions. To use this key to simplify the expression at the right, enter

1 a^b/c 2 + 1 a^b/c 3 =
$\frac{1}{2}$ $\frac{1}{3}$

HOW TO   Find the total of $\frac{1}{2}$ and $\frac{1}{3}$.

The common denominator is the LCM of 2 and 3. The LCM = 6. The LCM of denominators is sometimes called the **least common denominator (LCD)**.

Write equivalent fractions using the LCM.
$$\begin{array}{r} \frac{1}{2} = \frac{3}{6} \\ +\frac{1}{3} = \frac{2}{6} \\ \hline \end{array}$$

Add the fractions.
$$\begin{array}{r} \frac{1}{2} = \frac{3}{6} \\ +\frac{1}{3} = \frac{2}{6} \\ \hline \frac{5}{6} \end{array}$$

Objective 2.4A

Vocabulary to Review
numerator   [2.2A]
denominator   [2.2A]

Discuss the Concepts
Explain the procedure for adding two fractions with the same denominator.

Concept Check
Which of the following fractions, when added together, have a sum of 2?

$$\frac{1}{9}, \frac{2}{9}, \frac{4}{9}, \frac{5}{9}, \frac{7}{9}$$

$\frac{2}{9}, \frac{4}{9}, \frac{5}{9}$, and $\frac{7}{9}$

Instructor Note
We have chosen to present addition and subtraction of fractions prior to multiplication and division of fractions. If you prefer to present multiplication first, simply present the sections of this chapter in the following order:

Section 2.1
Section 2.2
Section 2.3
Section 2.6
Section 2.7
Section 2.4
Section 2.5
Section 2.8

In-Class Examples (Objective 2.4A)
Add.

1. $\frac{2}{9} + \frac{5}{9}$ $\frac{7}{9}$

2. $\frac{3}{6} + \frac{1}{6}$ $\frac{2}{3}$

3. $\frac{5}{7} + \frac{3}{7} + \frac{6}{7}$ 2

Objective 2.4B

Vocabulary to Review
least common multiple (LCM)
[2.1A]

New Vocabulary
least common denominator (LCD)

Discuss the Concepts
Explain why two fractions must have the same denominator before they can be added.

Optional Student Activity
Without calculating, decide which is greater. Explain your reasoning.

1. $\dfrac{1}{2} + \dfrac{2}{3}$ or $\dfrac{1}{4} + \dfrac{2}{5}$

2. $\dfrac{5}{6} + \dfrac{4}{9}$ or $\dfrac{3}{8} + \dfrac{3}{20}$

1. $\dfrac{1}{2} + \dfrac{2}{3}$

2. $\dfrac{5}{6} + \dfrac{4}{9}$

Explanations will vary. For example, in Exercise 1, $\dfrac{1}{2} > \dfrac{1}{4}$ and $\dfrac{2}{3} > \dfrac{2}{5}$, so the first expression must be greater than the second.

Instructor Note
Students would like to simply add numerators and denominators. They can be shown that this does not yield a correct result by using money.

$\$\dfrac{1}{4} + \$\dfrac{1}{2} = \$\dfrac{3}{4}, \; not \; \$\dfrac{2}{6}$

Example 2 Find $\dfrac{7}{12}$ more than $\dfrac{3}{8}$.

Solution

$$\dfrac{3}{8} = \dfrac{9}{24}$$

$$+\dfrac{7}{12} = \dfrac{14}{24}$$ • The LCM of 8 and 12 is 24.

$$\dfrac{23}{24}$$

You Try It 2 Find the sum of $\dfrac{5}{12}$ and $\dfrac{9}{16}$.

Your solution $\dfrac{47}{48}$

Example 3 Add: $\dfrac{5}{8} + \dfrac{7}{9}$

Solution

$$\dfrac{5}{8} = \dfrac{45}{72}$$

$$+\dfrac{7}{9} = \dfrac{56}{72}$$

$$\dfrac{101}{72} = 1\dfrac{29}{72}$$

You Try It 3 Add: $\dfrac{7}{8} + \dfrac{11}{15}$

Your solution $1\dfrac{73}{120}$

Example 4 Add: $\dfrac{2}{3} + \dfrac{3}{5} + \dfrac{5}{6}$

Solution

$$\dfrac{2}{3} = \dfrac{20}{30}$$

$$\dfrac{3}{5} = \dfrac{18}{30}$$ • The LCM of 3, 5, and 6 is 30.

$$+\dfrac{5}{6} = \dfrac{25}{30}$$

$$\dfrac{63}{30} = 2\dfrac{3}{30} = 2\dfrac{1}{10}$$

You Try It 4 Add: $\dfrac{3}{4} + \dfrac{4}{5} + \dfrac{5}{8}$

Your solution $2\dfrac{7}{40}$

Solutions on p. S5

Objective C   **To add whole numbers, mixed numbers, and fractions**

> **TAKE NOTE**
> The procedure at the right illustrates why $2 + \dfrac{2}{3} = 2\dfrac{2}{3}$. You do not need to show these steps when adding a whole number and a fraction. Here are two more examples:
>
> $7 + \dfrac{1}{5} = 7\dfrac{1}{5}$
>
> $6 + \dfrac{3}{4} = 6\dfrac{3}{4}$

The sum of a whole number and a fraction is a mixed number.

HOW TO Add: $2 + \dfrac{2}{3}$

$$\boxed{2} + \dfrac{2}{3} = \boxed{\dfrac{6}{3}} + \dfrac{2}{3} = \dfrac{8}{3} = 2\dfrac{2}{3}$$

To add a whole number and a mixed number, write the fraction and then add the whole numbers.

HOW TO Add: $7\dfrac{2}{5} + 4$

Write the fraction.

$$\begin{array}{r} 7\dfrac{2}{5} \\ +\,4 \\ \hline \dfrac{2}{5} \end{array}$$

Add the whole numbers.

$$\begin{array}{r} 7\dfrac{2}{5} \\ +\,4 \\ \hline 11\dfrac{2}{5} \end{array}$$

In-Class Examples (Objective 2.4B)
Add.

1. $\dfrac{3}{4} + \dfrac{1}{6}$ $\dfrac{11}{12}$

2. $\dfrac{7}{15} + \dfrac{2}{9}$ $\dfrac{31}{45}$

3. $\dfrac{3}{5} + \dfrac{9}{10} + \dfrac{4}{15}$ $1\dfrac{23}{30}$

To add two mixed numbers, add the fractional parts and then add the whole numbers. Remember to reduce the sum to simplest form.

Integrating Technology

Use the fraction key on a calculator to enter mixed numbers. For the example at the right, enter

5 a^b/c 4 a^b/c 9 +
$5\frac{4}{9}$

6 a^b/c 14 a^b/c 15 =
$6\frac{14}{15}$

HOW TO What is $6\frac{14}{15}$ added to $5\frac{4}{9}$?

The LCM of 9 and 15 is 45.

Add the fractional parts.

$$5\frac{4}{9} = 5\frac{20}{45}$$
$$+6\frac{14}{15} = 6\frac{42}{45}$$
$$\overline{\phantom{+6\frac{14}{15} = 6}\frac{62}{45}}$$

Add the whole numbers.

$$5\frac{4}{9} = 5\frac{20}{45}$$
$$+6\frac{14}{15} = 6\frac{42}{45}$$
$$\overline{11\frac{62}{45} = 11 + 1\frac{17}{45} = 12\frac{17}{45}}$$

Example 5 Add: $5 + \frac{3}{8}$

Solution $5 + \frac{3}{8} = 5\frac{3}{8}$

You Try It 5 What is 7 added to $\frac{6}{11}$?

Your solution $7\frac{6}{11}$

Example 6 Find 17 increased by $3\frac{3}{8}$.

Solution
$$17$$
$$+\ 3\frac{3}{8}$$
$$\overline{20\frac{3}{8}}$$

You Try It 6 Find the sum of 29 and $17\frac{5}{12}$.

Your solution $46\frac{5}{12}$

Example 7 Add: $5\frac{2}{3} + 11\frac{5}{6} + 12\frac{7}{9}$

Solution
$$5\frac{2}{3} = 5\frac{12}{18} \qquad \bullet \text{ LCM} = 18$$
$$11\frac{5}{6} = 11\frac{15}{18}$$
$$+\ 12\frac{7}{9} = 12\frac{14}{18}$$
$$\overline{\phantom{+\ 12\frac{7}{9} = }28\frac{41}{18} = 30\frac{5}{18}}$$

You Try It 7 Add: $7\frac{4}{5} + 6\frac{7}{10} + 13\frac{11}{15}$

Your solution $28\frac{7}{30}$

Example 8 Add: $11\frac{5}{8} + 7\frac{5}{9} + 8\frac{7}{15}$

Solution
$$11\frac{5}{8} = 11\frac{225}{360} \qquad \bullet \text{ LCM} = 360$$
$$7\frac{5}{9} = 7\frac{200}{360}$$
$$+\ 8\frac{7}{15} = 8\frac{168}{360}$$
$$\overline{\phantom{+\ 8\frac{7}{15} = }26\frac{593}{360} = 27\frac{233}{360}}$$

You Try It 8 Add: $9\frac{3}{8} + 17\frac{7}{12} + 10\frac{14}{15}$

Your solution $37\frac{107}{120}$

Solutions on p. S5

Objective 2.4C

Discuss the Concepts
Explain the steps involved in adding $3\frac{5}{6}$ and $4\frac{7}{12}$.

Concept Check
Find the sum of $\frac{1}{2} + 1 + 1\frac{1}{2} +$ $2 + 2\frac{1}{2} + \cdots + 19\frac{1}{2} + 20$. The three dots mean that the pattern continues. 410

Optional Student Activity
In a magic square, the sums across, down, and diagonally are the same. Determine whether these squares are magic squares. Yes

1	$\frac{3}{8}$	$\frac{1}{2}$
$\frac{1}{8}$	$\frac{5}{8}$	$1\frac{1}{8}$
$\frac{3}{4}$	$\frac{7}{8}$	$\frac{1}{4}$

2	$\frac{3}{4}$	1
$\frac{1}{4}$	$1\frac{1}{4}$	$2\frac{1}{4}$
$1\frac{1}{2}$	$1\frac{3}{4}$	$\frac{1}{2}$

In-Class Examples (Objective 2.4C)
Add.

1. $6\frac{1}{2} + 5\frac{2}{3}$ $12\frac{1}{6}$

2. $7\frac{5}{6} + 2\frac{13}{15}$ $10\frac{7}{10}$

3. $4\frac{5}{8} + 8\frac{1}{2} + 4\frac{7}{12}$ $17\frac{17}{24}$

80

80   Chapter 2 / Fractions

Objective 2.4D

Optional Student Activity

The following are the average portions of each day that a person spends for each activity.

Sleeping, $\frac{1}{3}$

Working, $\frac{1}{3}$

Personal hygiene, $\frac{1}{24}$

Eating, $\frac{1}{8}$

Rest and relaxation, $\frac{1}{12}$

Do these five activities account for an entire day? Explain your answer.

No. These activities account for only 22 hours.

Example 9

A rain gauge collected $2\frac{1}{3}$ inches of rain in October, $5\frac{1}{2}$ inches in November, and $3\frac{3}{8}$ inches in December. Find the total rainfall for the 3 months.

Strategy
To find the total rainfall for the 3 months, add the three amounts of rainfall $\left(2\frac{1}{3}, 5\frac{1}{2}, \text{ and } 3\frac{3}{8}\right)$.

Solution
$$2\frac{1}{3} = 2\frac{8}{24}$$
$$5\frac{1}{2} = 5\frac{12}{24}$$
$$+\ 3\frac{3}{8} = 3\frac{9}{24}$$
$$10\frac{29}{24} = 11\frac{5}{24}$$

The total rainfall for the 3 months was $11\frac{5}{24}$ inches.

You Try It 9

On Monday, you spent $4\frac{1}{2}$ hours in class, $3\frac{3}{4}$ hours studying, and $1\frac{1}{3}$ hours driving. Find the number of hours spent on these three activities.

Your strategy

Your solution $9\frac{7}{12}$ hours

Example 10

Barbara Walsh worked 4 hours, $2\frac{1}{3}$ hours, and $5\frac{2}{3}$ hours this week at a part-time job. Barbara is paid $9 an hour. How much did she earn this week?

Strategy
To find how much Barbara earned:
- Find the total number of hours worked.
- Multiply the total number of hours worked by the hourly wage (9).

Solution
$$
\begin{array}{ll}
4 & \quad 12 \\
2\frac{1}{3} & \quad \times\ 9 \\
+\ 5\frac{2}{3} & \quad \overline{108} \\
\overline{11\frac{3}{3}} = 12 \text{ hours worked} &
\end{array}
$$

Barbara earned $108 this week.

You Try It 10

Jeff Sapone, a carpenter, worked $1\frac{2}{3}$ hours of overtime on Monday, $3\frac{1}{3}$ hours of overtime on Tuesday, and 2 hours of overtime on Wednesday. At an overtime hourly rate of $36, find Jeff's overtime pay for these 3 days.

Your strategy

Your solution $252

Solutions on pp. S5–S6

In-Class Examples (Objective 2.4D)

1. A carpenter built a header by nailing a $1\frac{1}{4}$-inch board to a $2\frac{5}{8}$-inch beam. Find the total thickness of the header. $3\frac{7}{8}$ inches

2.4 Exercises

Objective A   To add fractions with the same denominator

For Exercises 1 to 20, add.

1. $\dfrac{2}{7} + \dfrac{1}{7}$

$\dfrac{3}{7}$

2. $\dfrac{3}{11} + \dfrac{5}{11}$

$\dfrac{8}{11}$

3. $\dfrac{1}{2} + \dfrac{1}{2}$

1

4. $\dfrac{1}{3} + \dfrac{2}{3}$

1

5. $\dfrac{8}{11} + \dfrac{7}{11}$

$1\dfrac{4}{11}$

6. $\dfrac{9}{13} + \dfrac{7}{13}$

$1\dfrac{3}{13}$

7. $\dfrac{8}{5} + \dfrac{9}{5}$

$3\dfrac{2}{5}$

8. $\dfrac{5}{3} + \dfrac{7}{3}$

4

9. $\dfrac{3}{5} + \dfrac{8}{5} + \dfrac{3}{5}$

$2\dfrac{4}{5}$

10. $\dfrac{3}{8} + \dfrac{5}{8} + \dfrac{7}{8}$

$1\dfrac{7}{8}$

11. $\dfrac{3}{4} + \dfrac{1}{4} + \dfrac{5}{4}$

$2\dfrac{1}{4}$

12. $\dfrac{2}{7} + \dfrac{4}{7} + \dfrac{5}{7}$

$1\dfrac{4}{7}$

13. $\dfrac{3}{8} + \dfrac{7}{8} + \dfrac{1}{8}$

$1\dfrac{3}{8}$

14. $\dfrac{5}{12} + \dfrac{7}{12} + \dfrac{1}{12}$

$1\dfrac{1}{12}$

15. $\dfrac{4}{15} + \dfrac{7}{15} + \dfrac{11}{15}$

$1\dfrac{7}{15}$

16. $\dfrac{3}{4} + \dfrac{3}{4} + \dfrac{1}{4}$

$1\dfrac{3}{4}$

17. $\dfrac{3}{16} + \dfrac{5}{16} + \dfrac{7}{16}$

$\dfrac{15}{16}$

18. $\dfrac{5}{18} + \dfrac{11}{18} + \dfrac{17}{18}$

$1\dfrac{5}{6}$

19. $\dfrac{3}{11} + \dfrac{5}{11} + \dfrac{7}{11}$

$1\dfrac{4}{11}$

20. $\dfrac{5}{7} + \dfrac{4}{7} + \dfrac{5}{7}$

2

21. Find the sum of $\dfrac{4}{9}$ and $\dfrac{5}{9}$.

1

22. Find the sum of $\dfrac{5}{12}$, $\dfrac{1}{12}$, and $\dfrac{11}{12}$.

$1\dfrac{5}{12}$

23. Find the total of $\dfrac{5}{8}$, $\dfrac{3}{8}$, and $\dfrac{7}{8}$.

$1\dfrac{7}{8}$

24. Find the total of $\dfrac{4}{13}$, $\dfrac{7}{13}$, and $\dfrac{11}{13}$.

$1\dfrac{9}{13}$

Objective B   To add fractions with different denominators

For Exercises 25 to 48, add.

25. $\dfrac{1}{2} + \dfrac{2}{3}$

$1\dfrac{1}{6}$

26. $\dfrac{2}{3} + \dfrac{1}{4}$

$\dfrac{11}{12}$

27. $\dfrac{3}{14} + \dfrac{5}{7}$

$\dfrac{13}{14}$

28. $\dfrac{3}{5} + \dfrac{7}{10}$

$1\dfrac{3}{10}$

29. $\dfrac{8}{15} + \dfrac{7}{20}$

$\dfrac{53}{60}$

30. $\dfrac{1}{6} + \dfrac{7}{9}$

$\dfrac{17}{18}$

31. $\dfrac{3}{8} + \dfrac{9}{14}$

$1\dfrac{1}{56}$

32. $\dfrac{5}{12} + \dfrac{5}{16}$

$\dfrac{35}{48}$

33. $\dfrac{3}{20} + \dfrac{7}{30}$

$\dfrac{23}{60}$

34. $\dfrac{5}{12} + \dfrac{7}{30}$

$\dfrac{13}{20}$

35. $\dfrac{2}{3} + \dfrac{6}{19}$

$\dfrac{56}{57}$

36. $\dfrac{1}{2} + \dfrac{3}{29}$

$\dfrac{35}{58}$

Quick Quiz (Objective 2.4A)

Add.

1. $\dfrac{7}{15} + \dfrac{4}{15}$ $\dfrac{11}{15}$

2. $\dfrac{3}{10} + \dfrac{7}{10}$ 1

3. $\dfrac{4}{9} + \dfrac{1}{9} + \dfrac{7}{9}$ $1\dfrac{1}{3}$

Quick Quiz (Objective 2.4B)

Add.

1. $\dfrac{1}{3} + \dfrac{5}{8}$ $\dfrac{23}{24}$

2. $\dfrac{3}{5} + \dfrac{11}{15}$ $1\dfrac{1}{3}$

3. $\dfrac{1}{2} + \dfrac{3}{4} + \dfrac{5}{6}$ $2\dfrac{1}{12}$

82

37. $\frac{1}{3} + \frac{5}{6} + \frac{7}{9}$

$1\frac{17}{18}$

38. $\frac{2}{3} + \frac{5}{6} + \frac{7}{12}$

$2\frac{1}{12}$

39. $\frac{5}{6} + \frac{1}{12} + \frac{5}{16}$

$1\frac{11}{48}$

40. $\frac{2}{9} + \frac{7}{15} + \frac{4}{21}$

$\frac{277}{315}$

41. $\frac{2}{3} + \frac{1}{5} + \frac{7}{12}$

$1\frac{9}{20}$

42. $\frac{3}{4} + \frac{4}{5} + \frac{7}{12}$

$2\frac{2}{15}$

43. $\frac{1}{4} + \frac{4}{5} + \frac{5}{9}$

$1\frac{109}{180}$

44. $\frac{2}{3} + \frac{3}{5} + \frac{7}{8}$

$2\frac{17}{120}$

45. $\frac{5}{16} + \frac{11}{18} + \frac{17}{24}$

$1\frac{91}{144}$

46. $\frac{3}{10} + \frac{14}{15} + \frac{9}{25}$

$1\frac{89}{150}$

47. $\frac{2}{3} + \frac{5}{8} + \frac{7}{9}$

$2\frac{5}{72}$

48. $\frac{1}{3} + \frac{2}{9} + \frac{7}{8}$

$1\frac{31}{72}$

49. What is $\frac{3}{8}$ added to $\frac{3}{5}$?

$\frac{39}{40}$

50. What is $\frac{5}{9}$ added to $\frac{7}{12}$?

$1\frac{5}{36}$

51. Find the sum of $\frac{3}{8}$, $\frac{5}{6}$, and $\frac{7}{12}$.

$1\frac{19}{24}$

52. Find the sum of $\frac{11}{12}$, $\frac{13}{24}$, and $\frac{4}{15}$.

$1\frac{29}{40}$

53. Find the total of $\frac{1}{2}$, $\frac{5}{8}$, and $\frac{7}{9}$.

$1\frac{65}{72}$

54. Find the total of $\frac{5}{14}$, $\frac{3}{7}$, and $\frac{5}{21}$.

$1\frac{1}{42}$

Objective C   **To add whole numbers, mixed numbers, and fractions**

For Exercises 55 to 82, add.

55. $\begin{array}{r} 1\frac{1}{2} \\ + 2\frac{1}{6} \\ \hline 3\frac{2}{3} \end{array}$

56. $\begin{array}{r} 2\frac{2}{5} \\ + 3\frac{3}{10} \\ \hline 5\frac{7}{10} \end{array}$

57. $\begin{array}{r} 4\frac{1}{2} \\ + 5\frac{7}{12} \\ \hline 10\frac{1}{12} \end{array}$

58. $\begin{array}{r} 3\frac{3}{8} \\ + 2\frac{5}{16} \\ \hline 5\frac{11}{16} \end{array}$

59. $\begin{array}{r} 4 \\ + 5\frac{2}{7} \\ \hline 9\frac{2}{7} \end{array}$

60. $\begin{array}{r} 6\frac{8}{9} \\ + 12 \\ \hline 18\frac{8}{9} \end{array}$

61. $\begin{array}{r} 3\frac{5}{8} \\ + 2\frac{11}{20} \\ \hline 6\frac{7}{40} \end{array}$

62. $\begin{array}{r} 4\frac{5}{12} \\ + 6\frac{11}{18} \\ \hline 11\frac{1}{36} \end{array}$

63. $7\frac{5}{12} + 2\frac{9}{16}$

$9\frac{47}{48}$

64. $9\frac{1}{2} + 3\frac{3}{11}$

$12\frac{17}{22}$

65. $6 + 2\frac{3}{13}$

$8\frac{3}{13}$

66. $8\frac{21}{40} + 6$

$14\frac{21}{40}$

67. $8\frac{29}{30} + 7\frac{11}{40}$

$16\frac{29}{120}$

68. $17\frac{5}{16} + 3\frac{11}{24}$

$20\frac{37}{48}$

69. $17\frac{3}{8} + 7\frac{7}{20}$

$24\frac{29}{40}$

70. $14\frac{7}{12} + 29\frac{13}{21}$

$44\frac{17}{84}$

Quick Quiz (Objective 2.4C)

Add.

1. $4\frac{1}{2} + 8\frac{1}{5}$ $12\frac{7}{10}$

2. $3\frac{4}{5} + 9\frac{3}{7}$ $13\frac{8}{35}$

3. $1\frac{3}{4} + 2\frac{3}{8} + 6\frac{7}{12}$ $10\frac{17}{24}$

71. $5\dfrac{7}{8} + 27\dfrac{5}{12}$

$33\dfrac{7}{24}$

72. $7\dfrac{5}{6} + 3\dfrac{5}{9}$

$11\dfrac{7}{18}$

73. $7\dfrac{5}{9} + 2\dfrac{7}{12}$

$10\dfrac{5}{36}$

74. $3\dfrac{1}{2} + 2\dfrac{3}{4} + 1\dfrac{5}{6}$

$8\dfrac{1}{12}$

75. $2\dfrac{1}{2} + 3\dfrac{2}{3} + 4\dfrac{1}{4}$

$10\dfrac{5}{12}$

76. $3\dfrac{1}{3} + 7\dfrac{1}{5} + 2\dfrac{1}{7}$

$12\dfrac{71}{105}$

77. $3\dfrac{1}{2} + 3\dfrac{1}{5} + 8\dfrac{1}{9}$

$14\dfrac{73}{90}$

78. $6\dfrac{5}{9} + 6\dfrac{5}{12} + 2\dfrac{5}{18}$

$15\dfrac{1}{4}$

79. $2\dfrac{3}{8} + 4\dfrac{7}{12} + 3\dfrac{5}{16}$

$10\dfrac{13}{48}$

80. $2\dfrac{1}{8} + 4\dfrac{2}{9} + 5\dfrac{17}{18}$

$12\dfrac{7}{24}$

81. $6\dfrac{5}{6} + 17\dfrac{2}{9} + 18\dfrac{5}{27}$

$42\dfrac{13}{54}$

82. $4\dfrac{7}{20} + \dfrac{17}{80} + 25\dfrac{23}{60}$

$29\dfrac{227}{240}$

83. Find the sum of $2\dfrac{4}{9}$ and $5\dfrac{7}{12}$.

$8\dfrac{1}{36}$

84. Find $5\dfrac{5}{6}$ more than $3\dfrac{3}{8}$.

$9\dfrac{5}{24}$

85. What is $4\dfrac{3}{4}$ added to $9\dfrac{1}{3}$?

$14\dfrac{1}{12}$

86. What is $4\dfrac{8}{9}$ added to $9\dfrac{1}{6}$?

$14\dfrac{1}{18}$

87. Find the total of 2, $4\dfrac{5}{8}$, and $2\dfrac{2}{9}$.

$8\dfrac{61}{72}$

88. Find the total of $1\dfrac{5}{8}$, 3, and $7\dfrac{7}{24}$.

$11\dfrac{11}{12}$

 Objective D   **To solve application problems**

89. **Mechanics**   Find the length of the shaft.

Length

$1\dfrac{5}{16}$ inches

90. **Mechanics**   Find the length of the shaft.

Length

$8\dfrac{9}{16}$ inches

91. **Carpentry**   A table 30 inches high has a top that is $1\dfrac{1}{8}$ inches thick. Find the total thickness of the table top after a $\dfrac{3}{16}$-inch veneer is applied.

$1\dfrac{5}{16}$ inches

92. **Wages**   Fred Thomson worked $2\dfrac{2}{3}$ hours of overtime on Monday, $1\dfrac{1}{4}$ hours on Wednesday, $1\dfrac{1}{3}$ hours on Friday, and $6\dfrac{3}{4}$ hours on Saturday.

a.   Find the total number of overtime hours worked during the week.   12 hours

b.   At an overtime hourly wage of $22 per hour, how much overtime pay does Fred receive? $264

Quick Quiz (Objective 2.4D)

1. A plumber works $1\dfrac{1}{2}$ hours of overtime on Monday, $2\dfrac{1}{4}$ hours of overtime on Tuesday, and $3\dfrac{1}{4}$ hours of overtime on Wednesday. Find the total number of hours of overtime worked during the three days.

7 hours

Answers to Writing Exercises

98. A unit fraction is a fraction in which the numerator is 1 and the denominator is any natural number greater than 1.

$$\frac{1}{2} + \frac{1}{3} + \frac{1}{4} = \frac{6}{12} + \frac{4}{12} + \frac{3}{12}$$
$$= \frac{13}{12} = 1\frac{1}{12}$$

There is no smallest unit fraction. No matter how small the unit fraction is, we can always add 1 to the denominator to make it smaller.

99. We can use 3 dimes and 1 quarter to model adding $\frac{3}{10}$ and $\frac{1}{4}$. Because the dime and the quarter are not the same unit, we change all of them to nickels and add.

$$\begin{array}{r} 3 \text{ dimes} = 6 \text{ nickels} \\ + 1 \text{ quarter} = 5 \text{ nickels} \\ \hline 11 \text{ nickels} \end{array}$$

Now consider the above problem in terms of dollars.

$$\begin{array}{r} \frac{3}{10} = \frac{6}{20} \\ + \frac{1}{4} = \frac{5}{20} \\ \hline \frac{11}{20} \end{array}$$

Note that

1 dime $= \frac{1}{10}$ of a dollar

1 quarter $= \frac{1}{4}$ of a dollar

1 nickel $= \frac{1}{20}$ of a dollar

100. $\frac{1}{3} + \frac{1}{6} + \frac{1}{8} + \frac{1}{12} + \frac{2}{5} =$
$\frac{40}{120} + \frac{20}{120} + \frac{15}{120} + \frac{10}{120} +$
$\frac{48}{120} = \frac{133}{120} = 1\frac{13}{120}$ No, this is not possible. The total should not be greater than 1, which represents all the people surveyed.

---

93. **Wages** You are working a part-time job that pays $11 an hour. You worked 5, $3\frac{3}{4}$, $2\frac{1}{3}$, $1\frac{1}{4}$, and $7\frac{2}{3}$ hours during the last five days.
   a. Find the total number of hours you worked during the last five days. 20 hours
   b. Find your total wages for the five days. $220

94. **Sports** The course of a yachting race is in the shape of a triangle with sides that measure $4\frac{3}{10}$ miles, $3\frac{7}{10}$ miles, and $2\frac{1}{2}$ miles. Find the total length of the course. $10\frac{1}{2}$ miles

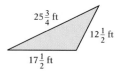

95. **Landscaping** A flower garden in the yard of a colonial home is in the shape of a triangle as shown at the right. The wooden beams lining the edge of the garden need to be replaced. Find the total length of wood beams that must be purchased in order to replace the old beams. $55\frac{3}{4}$ feet

96. 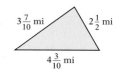 **Food Preferences** When the Heller Research Group conducted a survey to determine favorite doughnut flavors, $\frac{2}{5}$ of the respondents named glazed doughnuts, $\frac{8}{25}$ named filled doughnuts, and $\frac{3}{20}$ named frosted doughnuts. What fraction of the respondents named glazed, filled, *or* frosted as their favorite type of doughnut? $\frac{87}{100}$

97. **Mobility** During a recent year, over 42 million Americans changed homes. The graph at the right shows what fractions of the people moved within the same county, moved to a different county in the same state, and moved to a different state. (Source: Census Bureau, *Geographical Mobility*) What fractional part of those who changed homes moved outside the county they had been living in? $\frac{1}{3}$ of the people who changed homes

Where Americans Moved

Within the same county $\frac{2}{3}$

Different state $\frac{1}{7}$

$\frac{4}{21}$

Different county in same state

APPLYING THE CONCEPTS

98. What is a unit fraction? Find the sum of the three largest unit fractions. Is there a smallest unit fraction? If so, write it down. If not, explain why.

99. Use a model to illustrate and explain the addition of fractions with unlike denominators.

100. A survey was conducted to determine people's favorite color from among blue, green, red, purple, and other. The surveyor claims that $\frac{1}{3}$ of the people responded blue, $\frac{1}{6}$ responded green, $\frac{1}{8}$ responded red, $\frac{1}{12}$ responded purple, and $\frac{2}{5}$ responded some other color. Is this possible? Explain your answer.

2.5 Subtraction of Fractions and Mixed Numbers

Objective A To subtract fractions with the same denominator

Fractions with the same denominator are subtracted by subtracting the numerators and placing the difference over the common denominator. After subtracting, write the fraction in simplest form.

HOW TO Subtract: $\frac{5}{7} - \frac{3}{7}$

$$\begin{array}{r} \frac{5}{7} \\ -\frac{3}{7} \\ \hline \frac{2}{7} \end{array}$$

• Subtract the numerators and place the difference over the common denominator.

$$\frac{5}{7} - \frac{3}{7} = \frac{5-3}{7} = \frac{2}{7}$$

Example 1 Find $\frac{17}{30}$ less $\frac{11}{30}$.

Solution

$$\begin{array}{r} \frac{17}{30} \\ -\frac{11}{30} \\ \hline \frac{6}{30} = \frac{1}{5} \end{array}$$

• The denominators are the same. Subtract the numerators. Place the difference over the common denominator.

You Try It 1 Subtract: $\frac{16}{27} - \frac{7}{27}$

Your solution $\frac{1}{3}$

Solution on p. S6

Objective B To subtract fractions with different denominators

To subtract fractions with different denominators, first rewrite the fractions as equivalent fractions with a common denominator. As with adding fractions, the common denominator is the LCM of the denominators of the fractions.

HOW TO Subtract: $\frac{5}{6} - \frac{1}{4}$

The common denominator is the LCM of 6 and 4. The LCM = 12.

Write equivalent fractions using the LCM.

$$\frac{5}{6} = \frac{10}{12}$$
$$-\frac{1}{4} = \frac{3}{12}$$

Subtract the fractions.

$$\begin{array}{r} \frac{5}{6} = \frac{10}{12} \\ -\frac{1}{4} = \frac{3}{12} \\ \hline \frac{7}{12} \end{array}$$

Objective 2.5A

Discuss the Concepts
Explain the procedure for subtracting two fractions with the same denominator.

Concept Check
Which of the following fractions, when subtracted, have a difference of $\frac{2}{5}$?

$$\frac{4}{5}, \frac{3}{5}, \frac{2}{5}, \frac{1}{5}$$

$\frac{4}{5}$ and $\frac{2}{5}$, or $\frac{3}{5}$ and $\frac{1}{5}$

Optional Student Activity
Use a diagram to illustrate and explain subtraction of two fractions with the same denominator.

Discuss the Concepts
What is the difference between the procedure for adding two fractions with the same denominator and the procedure for subtracting two fractions with the same denominator?

Objective 2.5B

In-Class Examples (Objective 2.5A)
Subtract.
1. $\frac{14}{15} - \frac{1}{15}$ $\frac{13}{15}$
2. $\frac{11}{18} - \frac{5}{18}$ $\frac{1}{3}$

In-Class Examples (Objective 2.5B)
Subtract.
1. $\frac{3}{4} - \frac{2}{5}$ $\frac{7}{20}$
2. $\frac{5}{6} - \frac{4}{15}$ $\frac{17}{30}$
3. $\frac{53}{60} - \frac{7}{12}$ $\frac{3}{10}$

Concept Check

Which expression is larger?

1. $\dfrac{11}{12} - \dfrac{2}{3}$ or $\dfrac{2}{3} - \dfrac{1}{6}$

2. $\dfrac{3}{5} - \dfrac{1}{3}$ or $\dfrac{9}{10} - \dfrac{7}{8}$

1. $\dfrac{2}{3} - \dfrac{1}{6}$ **2.** $\dfrac{3}{5} - \dfrac{1}{3}$

Instructor Note

Another money example that may reinforce the common denominator concept is "Find 3 quarters minus 7 dimes." The concept of rewriting fractions as equivalent fractions with a common denominator is similar to exchanging all the coins for pennies. Three quarters equal 75 pennies, and 7 dimes equal 70 pennies.

$$\frac{3}{4} - \frac{7}{10} = \frac{75}{100} - \frac{70}{100} = \frac{5}{100}$$

$$= \frac{1}{20}$$

You might also use this example to cite that it is not necessary to find the *least* common denominator when adding and subtracting fractions with different denominators.

Objective 2.5C

Discuss the Concepts

In subtraction of mixed numbers, when is borrowing necessary?

Example 2 Subtract: $\dfrac{11}{16} - \dfrac{5}{12}$

Solution
$$\frac{11}{16} = \frac{33}{48}$$
$$-\frac{5}{12} = \frac{20}{48}$$
$$\frac{13}{48}$$

• LCM = 48

You Try It 2 Subtract: $\dfrac{13}{18} - \dfrac{7}{24}$

Your solution $\dfrac{31}{72}$

Solution on p. S6

Objective C **To subtract whole numbers, mixed numbers, and fractions**

To subtract mixed numbers without borrowing, subtract the fractional parts and then subtract the whole numbers.

HOW TO Subtract: $5\dfrac{5}{6} - 2\dfrac{3}{4}$

Subtract the fractional parts.

$$5\frac{5}{6} = 5\frac{10}{12}$$
$$-2\frac{3}{4} = 2\frac{9}{12}$$
$$\frac{1}{12}$$

• The LCM of 6 and 4 is 12.

Subtract the whole numbers.

$$5\frac{5}{6} = 5\frac{10}{12}$$
$$-2\frac{3}{4} = 2\frac{9}{12}$$
$$3\frac{1}{12}$$

Subtraction of mixed numbers sometimes involves borrowing.

HOW TO Subtract: $5 - 2\dfrac{5}{8}$

Borrow 1 from 5.

$$5 = \overset{4}{\cancel{5}}\,1$$
$$-2\frac{5}{8} = 2\frac{5}{8}$$

Write 1 as a fraction so that the fractions have the same denominators.

$$5 = 4\frac{8}{8}$$
$$-2\frac{5}{8} = 2\frac{5}{8}$$

Subtract the mixed numbers.

$$5 = 4\frac{8}{8}$$
$$-2\frac{5}{8} = 2\frac{5}{8}$$
$$2\frac{3}{8}$$

HOW TO Subtract: $7\dfrac{1}{6} - 2\dfrac{5}{8}$

Write equivalent fractions using the LCM.

$$7\frac{1}{6} = 7\frac{4}{24}$$
$$-2\frac{5}{8} = 2\frac{15}{24}$$

Borrow 1 from 7. Add the 1 to $\frac{4}{24}$. Write $1\frac{4}{24}$ as $\frac{28}{24}$.

$$7\frac{1}{6} = \overset{6}{\cancel{7}}1\frac{4}{24} = 6\frac{28}{24}$$
$$-2\frac{5}{8} = 2\frac{15}{24} = 2\frac{15}{24}$$

Subtract the mixed numbers.

$$7\frac{1}{6} = 6\frac{28}{24}$$
$$-2\frac{5}{8} = 2\frac{15}{24}$$
$$4\frac{13}{24}$$

In-Class Examples (Objective 2.5C)

Subtract.

1. $9\dfrac{19}{24} - 5\dfrac{11}{24}$ $4\dfrac{1}{3}$

2. $11 - 8\dfrac{16}{17}$ $2\dfrac{1}{17}$

3. $6\dfrac{7}{9} - 3\dfrac{5}{6}$ $2\dfrac{17}{18}$

Example 3 Subtract: $15\frac{7}{8} - 12\frac{2}{3}$

Solution
$$15\frac{7}{8} = 15\frac{21}{24} \quad \bullet \ \text{LCM} = 24$$
$$- \ 12\frac{2}{3} = 12\frac{16}{24}$$
$$\overline{\qquad\qquad 3\frac{5}{24}}$$

You Try It 3 Subtract: $17\frac{5}{9} - 11\frac{5}{12}$

Your solution $6\frac{5}{36}$

Example 4 Subtract: $9 - 4\frac{3}{11}$

Solution
$$9 \quad = 8\frac{11}{11} \quad \bullet \ \text{LCM} = 11$$
$$- \ 4\frac{3}{11} = 4\frac{3}{11}$$
$$\overline{\qquad\qquad 4\frac{8}{11}}$$

You Try It 4 Subtract: $8 - 2\frac{4}{13}$

Your solution $5\frac{9}{13}$

Example 5 Find $11\frac{5}{12}$ decreased by $2\frac{11}{16}$.

Solution
$$11\frac{5}{12} = 11\frac{20}{48} = 10\frac{68}{48} \quad \bullet \ \text{LCM} = 48$$
$$- \quad 2\frac{11}{16} = \ 2\frac{33}{48} = \ 2\frac{33}{48}$$
$$\overline{\qquad\qquad\qquad\qquad\qquad 8\frac{35}{48}}$$

You Try It 5 What is $21\frac{7}{9}$ minus $7\frac{11}{12}$?

Your solution $13\frac{31}{36}$

Solutions on p. S6

Objective D **To solve application problems**

Outside Diameter

Inside Diameter

HOW TO The outside diameter of a bushing is $3\frac{3}{8}$ inches and the wall thickness is $\frac{1}{4}$ inch. Find the inside diameter of the bushing.

$$\frac{1}{4} + \frac{1}{4} = \frac{2}{4} = \frac{1}{2}$$ • Add $\frac{1}{4}$ to $\frac{1}{4}$ to find the total thickness of the two walls.

$$3\frac{3}{8} = 3\frac{3}{8} = 2\frac{11}{8}$$
$$- \ \frac{1}{2} = \ \frac{4}{8} = \ \frac{4}{8}$$
$$\overline{\qquad\qquad\qquad 2\frac{7}{8}}$$

• Subtract the total thickness of the two walls to find the inside diameter.

The inside diameter of the bushing is $2\frac{7}{8}$ inches.

Optional Student Activity

Use the diagram below to illustrate the difference between $\frac{5}{6}$ and $\frac{1}{8}$.

Optional Student Activity

Use the diagram below to illustrate the difference between $\frac{5}{6}$ and $\frac{1}{8}$.

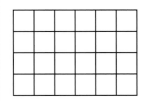

Why does the figure have 24 squares? Would it be possible to illustrate the difference between $\frac{5}{6}$ and $\frac{1}{8}$ if there were 48 squares in the figure? What if there were 16 squares? Make a list of some of the possible numbers of squares that could be used to illustrate the difference between $\frac{5}{6}$ and $\frac{1}{8}$.

Objective 2.5D

Optional Student Activity

A timed college entrance exam is $2\frac{1}{2}$ hours long. Complete the table below. Fill in the second column by calculating the test time remaining, given the time, in hours, that has already elapsed, which is shown in the first column.

Time Elapsed	Time Remaining
$\frac{1}{4}$	$2\frac{1}{4}$
$\frac{3}{4}$	$1\frac{3}{4}$
$1\frac{1}{2}$	1
$1\frac{3}{4}$	$\frac{3}{4}$
$2\frac{1}{4}$	$\frac{1}{4}$

In-Class Examples (Objective 2.5D)

1. The length of a regulation NCAA football must be no less than $10\frac{7}{8}$ inches and no more than $11\frac{7}{16}$ inches. What is the difference between the minimum and maximum lengths of an NCAA regulation football?

$\frac{9}{16}$ inch

Example 6

A $2\frac{2}{3}$-inch piece is cut from a $6\frac{5}{8}$-inch board. How much of the board is left?

Strategy
To find the length remaining, subtract the length of the piece cut from the total length of the board.

Solution

$$6\frac{5}{8} = 6\frac{15}{24} = 5\frac{39}{24}$$
$$-2\frac{2}{3} = 2\frac{16}{24} = 2\frac{16}{24}$$
$$\overline{\phantom{-2\frac{2}{3} = 2\frac{16}{24} = {}}3\frac{23}{24}}$$

$3\frac{23}{24}$ inches of the board are left.

You Try It 6

A flight from New York to Los Angeles takes $5\frac{1}{2}$ hours. After the plane has been in the air for $2\frac{3}{4}$ hours, how much flight time remains?

Your strategy

Your solution $2\frac{3}{4}$ hours

Example 7

Two painters are staining a house. In 1 day one painter stained $\frac{1}{3}$ of the house, and the other stained $\frac{1}{4}$ of the house. How much of the job remains to be done?

Strategy
To find how much of the job remains:
- Find the total amount of the house already stained $\left(\frac{1}{3} + \frac{1}{4}\right)$.
- Subtract the amount already stained from 1, which represents the complete job.

Solution

$$\frac{1}{3} = \frac{4}{12} \qquad\qquad 1 = \frac{12}{12}$$
$$+\frac{1}{4} = \frac{3}{12} \qquad\qquad -\frac{7}{12} = \frac{7}{12}$$
$$\overline{\phantom{+\frac{1}{4} = {}}\frac{7}{12}} \qquad\qquad \overline{\phantom{-\frac{7}{12} = {}}\frac{5}{12}}$$

$\frac{5}{12}$ of the house remains to be stained.

You Try It 7

A patient is put on a diet to lose 24 pounds in 3 months. The patient lost $7\frac{1}{2}$ pounds the first month and $5\frac{3}{4}$ pounds the second month. How much weight must be lost the third month to achieve the goal?

Your strategy

Your solution $10\frac{3}{4}$ pounds

Solutions on p. S6

2.5 Exercises

Objective A   To subtract fractions with the same denominator

For Exercises 1 to 10, subtract.

1. $\dfrac{9}{17}$
$-\dfrac{7}{17}$
$\dfrac{2}{17}$

2. $\dfrac{11}{15}$
$-\dfrac{3}{15}$
$\dfrac{8}{15}$

3. $\dfrac{11}{12}$
$-\dfrac{7}{12}$
$\dfrac{1}{3}$

4. $\dfrac{13}{15}$
$-\dfrac{4}{15}$
$\dfrac{3}{5}$

5. $\dfrac{9}{20}$
$-\dfrac{7}{20}$
$\dfrac{1}{10}$

6. $\dfrac{48}{55}$
$-\dfrac{13}{55}$
$\dfrac{7}{11}$

7. $\dfrac{42}{65}$
$-\dfrac{17}{65}$
$\dfrac{5}{13}$

8. $\dfrac{11}{24}$
$-\dfrac{5}{24}$
$\dfrac{1}{4}$

9. $\dfrac{23}{30}$
$-\dfrac{13}{30}$
$\dfrac{1}{3}$

10. $\dfrac{17}{42}$
$-\dfrac{5}{42}$
$\dfrac{2}{7}$

11. What is $\dfrac{5}{14}$ less than $\dfrac{13}{14}$?

$\dfrac{4}{7}$

12. What is $\dfrac{7}{19}$ less than $\dfrac{17}{19}$?

$\dfrac{10}{19}$

13. Find the difference between $\dfrac{7}{8}$ and $\dfrac{5}{8}$.

$\dfrac{1}{4}$

14. Find the difference between $\dfrac{7}{12}$ and $\dfrac{5}{12}$.

$\dfrac{1}{6}$

15. What is $\dfrac{18}{23}$ minus $\dfrac{9}{23}$?

$\dfrac{9}{23}$

16. What is $\dfrac{7}{9}$ minus $\dfrac{3}{9}$?

$\dfrac{4}{9}$

17. Find $\dfrac{17}{24}$ decreased by $\dfrac{11}{24}$.

$\dfrac{1}{4}$

18. Find $\dfrac{19}{30}$ decreased by $\dfrac{11}{30}$.

$\dfrac{4}{15}$

Objective B   To subtract fractions with different denominators

For Exercises 19 to 33, subtract.

19. $\dfrac{2}{3}$
$-\dfrac{1}{6}$
$\dfrac{1}{2}$

20. $\dfrac{7}{8}$
$-\dfrac{5}{16}$
$\dfrac{9}{16}$

21. $\dfrac{5}{8}$
$-\dfrac{2}{7}$
$\dfrac{19}{56}$

22. $\dfrac{5}{6}$
$-\dfrac{3}{7}$
$\dfrac{17}{42}$

23. $\dfrac{5}{7}$
$-\dfrac{3}{14}$
$\dfrac{1}{2}$

24. $\dfrac{5}{9}$
$-\dfrac{7}{15}$
$\dfrac{4}{45}$

25. $\dfrac{8}{15}$
$-\dfrac{7}{20}$
$\dfrac{11}{60}$

26. $\dfrac{7}{9}$
$-\dfrac{1}{6}$
$\dfrac{11}{18}$

27. $\dfrac{9}{14}$
$-\dfrac{3}{8}$
$\dfrac{15}{56}$

28. $\dfrac{5}{12}$
$-\dfrac{5}{16}$
$\dfrac{5}{48}$

Quick Quiz (Objective 2.5A)

Subtract.

1. $\dfrac{12}{17} - \dfrac{10}{17}$ $\dfrac{2}{17}$

2. $\dfrac{9}{10} - \dfrac{3}{10}$ $\dfrac{3}{5}$

29. $\dfrac{46}{51}$
$-\ \dfrac{3}{17}$
$\dfrac{37}{51}$

30. $\dfrac{9}{16}$
$-\ \dfrac{17}{32}$
$\dfrac{1}{32}$

31. $\dfrac{21}{35}$
$-\ \dfrac{5}{14}$
$\dfrac{17}{70}$

32. $\dfrac{19}{40}$
$-\ \dfrac{3}{16}$
$\dfrac{23}{80}$

33. $\dfrac{29}{60}$
$-\ \dfrac{3}{40}$
$\dfrac{49}{120}$

34. What is $\dfrac{3}{5}$ less than $\dfrac{11}{12}$?
$\dfrac{19}{60}$

35. What is $\dfrac{5}{9}$ less than $\dfrac{11}{15}$?
$\dfrac{8}{45}$

36. Find the difference between $\dfrac{11}{24}$ and $\dfrac{7}{18}$.
$\dfrac{5}{72}$

37. Find the difference between $\dfrac{9}{14}$ and $\dfrac{5}{42}$.
$\dfrac{11}{21}$

38. Find $\dfrac{11}{12}$ decreased by $\dfrac{11}{15}$.
$\dfrac{11}{60}$

39. Find $\dfrac{17}{20}$ decreased by $\dfrac{7}{15}$.
$\dfrac{23}{60}$

40. What is $\dfrac{13}{20}$ minus $\dfrac{1}{6}$?
$\dfrac{29}{60}$

41. What is $\dfrac{5}{6}$ minus $\dfrac{7}{9}$?
$\dfrac{1}{18}$

Objective C **To subtract whole numbers, mixed numbers, and fractions**

For Exercises 42 to 61, subtract.

42. $5\dfrac{7}{12}$
$-\ 2\dfrac{5}{12}$
$3\dfrac{1}{6}$

43. $16\dfrac{11}{15}$
$-\ 11\dfrac{8}{15}$
$5\dfrac{1}{5}$

44. $72\dfrac{21}{23}$
$-\ 16\dfrac{17}{23}$
$56\dfrac{4}{23}$

45. $19\dfrac{16}{17}$
$-\ 9\dfrac{7}{17}$
$10\dfrac{9}{17}$

46. $6\dfrac{1}{3}$
$-\ 2$
$4\dfrac{1}{3}$

47. $5\dfrac{7}{8}$
$-\ 1$
$4\dfrac{7}{8}$

48. 10
$-\ 6\dfrac{1}{3}$
$3\dfrac{2}{3}$

49. 3
$-\ 2\dfrac{5}{21}$
$\dfrac{16}{21}$

50. $6\dfrac{2}{5}$
$-\ 4\dfrac{4}{5}$
$1\dfrac{3}{5}$

51. $16\dfrac{3}{8}$
$-\ 10\dfrac{7}{8}$
$5\dfrac{1}{2}$

52. $25\dfrac{4}{9}$
$-\ 16\dfrac{7}{9}$
$8\dfrac{2}{3}$

53. $8\dfrac{3}{7}$
$-\ 2\dfrac{6}{7}$
$5\dfrac{4}{7}$

54. $16\dfrac{2}{5}$
$-\ 8\dfrac{4}{9}$
$7\dfrac{43}{45}$

55. $23\dfrac{7}{8}$
$-\ 16\dfrac{2}{3}$
$7\dfrac{5}{24}$

56. 6
$-\ 4\dfrac{3}{5}$
$1\dfrac{2}{5}$

57. $65\dfrac{8}{35}$
$-\ 16\dfrac{11}{14}$
$48\dfrac{31}{70}$

58. $82\dfrac{4}{33}$
$-\ 16\dfrac{5}{22}$
$65\dfrac{59}{66}$

59. $101\dfrac{2}{9}$
$-\ 16$
$85\dfrac{2}{9}$

60. $77\dfrac{5}{18}$
$-\ 61$
$16\dfrac{5}{18}$

61. 17
$-\ 7\dfrac{8}{13}$
$9\dfrac{5}{13}$

Quick Quiz (Objective 2.5B)

Subtract.

1. $\dfrac{3}{5} - \dfrac{1}{4}$ $\dfrac{7}{20}$

2. $\dfrac{22}{25} - \dfrac{43}{50}$ $\dfrac{1}{50}$

3. $\dfrac{11}{12} - \dfrac{13}{15}$ $\dfrac{1}{20}$

62. What is $5\frac{3}{8}$ less than $8\frac{1}{9}$?

$2\frac{53}{72}$

63. What is $7\frac{3}{5}$ less than $23\frac{3}{20}$?

$15\frac{11}{20}$

64. Find the difference between $9\frac{2}{7}$ and $3\frac{1}{4}$.

$6\frac{1}{28}$

65. Find the difference between $12\frac{3}{8}$ and $7\frac{5}{12}$.

$4\frac{23}{24}$

66. What is $10\frac{5}{9}$ minus $5\frac{11}{15}$?

$4\frac{37}{45}$

67. Find $6\frac{1}{3}$ decreased by $3\frac{3}{5}$.

$2\frac{11}{15}$

Objective D   **To solve application problems**

68. Mechanics   Find the missing dimension.

$8\frac{19}{24}$ feet Ⓟ

69. Mechanics   Find the missing dimension.

$9\frac{1}{2}$ inches Ⓟ

70.   **Sports**   In the Kentucky Derby the horses run $1\frac{1}{4}$ miles. In the Belmont Stakes they run $1\frac{1}{2}$ miles, and in the Preakness Stakes they run $1\frac{3}{16}$ miles. How much farther do the horses run in the Kentucky Derby than in the Preakness Stakes? How much farther do they run in the Belmont Stakes than in the Preakness Stakes?

$\frac{1}{16}$ mile; $\frac{5}{16}$ mile

71.   **Sports**   In the running high jump in the 1948 Summer Olympic Games, Alice Coachman's distance was $66\frac{1}{8}$ inches. In the same event in the 1972 Summer Olympics, Urika Meyfarth jumped $75\frac{1}{2}$ inches, and in the 1996 Olympic Games, Stefka Kostadinova jumped $80\frac{3}{4}$ inches. Find the difference between Meyfarth's distance and Coachman's distance. Find the difference between Kostadinova's distance and Meyfarth's distance.

$9\frac{3}{8}$ inches; $5\frac{1}{4}$ inches

72. Hiking   Two hikers plan a 3-day, $27\frac{1}{2}$-mile backpack trip carrying a total of 80 pounds. The hikers plan to travel $7\frac{3}{8}$ miles the first day and $10\frac{1}{3}$ miles the second day.
   a. How many miles do the hikers plan to travel the first two days?
   b. How many miles will be left to travel on the third day?

a. $17\frac{17}{24}$ miles   **b.** $9\frac{19}{24}$ miles

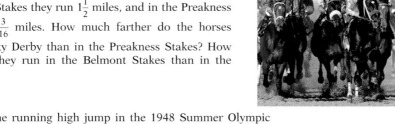

Quick Quiz (Objective 2.5C)

Subtract.

1. $23\frac{13}{16} - 12\frac{7}{16}$ $11\frac{3}{8}$

2. $14 - 5\frac{4}{7}$ $8\frac{3}{7}$

3. $8\frac{5}{12} - 5\frac{4}{9}$ $2\frac{35}{36}$

73. Fundraising A 12-mile walkathon has three checkpoints. The first is $3\frac{3}{8}$ miles from the starting point. The second checkpoint is $4\frac{1}{3}$ miles from the first.

   a. How many miles is it from the starting point to the second checkpoint?
   b. How many miles is it from the second checkpoint to the finish line?

   a. $7\frac{17}{24}$ miles   **b.** $4\frac{7}{24}$ miles

74. Health A patient with high blood pressure who weighs 225 pounds is put on a diet to lose 25 pounds in 3 months. The patient loses $8\frac{3}{4}$ pounds the first month and $11\frac{5}{8}$ pounds the second month. How much weight must be lost the third month for the goal to be achieved? $4\frac{5}{8}$ pounds

75. Sports A wrestler is entered in the 172-pound weight class in the conference finals coming up in 3 weeks. The wrestler needs to lose $12\frac{3}{4}$ pounds. The wrestler loses $5\frac{1}{4}$ pounds the first week and $4\frac{1}{4}$ pounds the second week.

   a. Without doing the calculations, determine whether the wrestler can reach his weight class by losing less in the third week than was lost in the second week.   Yes
   b. How many pounds must be lost in the third week for the desired weight to be reached? $3\frac{1}{4}$ pounds

76.   **Government** The figure at the right shows the sources of federal income. The category "Social Security, Medicare" includes unemployment and other retirement taxes. The category "Miscellaneous Taxes" includes excise, customs, estate, and gift taxes.

   a. What is the difference between the fraction of federal income derived from personal income taxes and the fraction derived from corporate income taxes?
   b. What is the difference between the fraction of federal income derived from the Social Security category and the fraction derived from miscellaneous taxes?

   a. $\frac{28}{75}$ **b.** $\frac{19}{75}$

Sources of Federal Income

Social Security, Medicare $\frac{1}{3}$

Personal Income Taxes $\frac{12}{25}$

$\frac{8}{75}$ $\frac{2}{25}$

Corporate Income Taxes   Miscellaneous Taxes

APPLYING THE CONCEPTS

77. Fill in the square to produce a true statement: $5\frac{1}{3} - \boxed{} = 2\frac{1}{2}$. $2\frac{5}{6}$

78. Fill in the square to produce a true statement: $\boxed{} - 4\frac{1}{2} = 1\frac{5}{8}$. $6\frac{1}{8}$

79. Fill in the blank squares at the right so that the sum of the numbers is the same along any row, column, or diagonal. The resulting square is called a magic square.

$\frac{3}{8}$	$\frac{3}{4}$	$\frac{3}{4}$
1	$\frac{5}{8}$	$\frac{1}{4}$
$\frac{1}{2}$	$\frac{1}{2}$	$\frac{7}{8}$

80. Finances If $\frac{4}{15}$ of an electrician's income is spent for housing, what fraction of the electrician's income is not spent for housing? $\frac{11}{15}$

Quick Quiz (Objective 2.5D)

1. A plane trip from Boston to San Francisco takes $6\frac{1}{4}$ hours. After the plane has been in the air for $3\frac{1}{2}$ hours, how much time remains before landing?

$2\frac{3}{4}$ hours

2.6 Multiplication of Fractions and Mixed Numbers

Objective A   To multiply fractions

The product of two fractions is the product of the numerators over the product of the denominators.

HOW TO   Multiply: $\frac{2}{3} \times \frac{4}{5}$

$$\frac{2}{3} \times \frac{4}{5} = \frac{2 \cdot 4}{3 \cdot 5} = \frac{8}{15}$$

- Multiply the numerators.
- Multiply the denominators.

The product $\frac{2}{3} \times \frac{4}{5}$ can be read "$\frac{2}{3}$ times $\frac{4}{5}$" or "$\frac{2}{3}$ of $\frac{4}{5}$."

Reading the times sign as "of" is useful in application problems.

$\frac{4}{5}$ of the bar is shaded.

Shade $\frac{2}{3}$ of the $\frac{4}{5}$ already shaded.

$\frac{8}{15}$ of the bar is then shaded light yellow.

$$\frac{2}{3} \text{ of } \frac{4}{5} = \frac{2}{3} \times \frac{4}{5} = \frac{8}{15}$$

After multiplying two fractions, write the product in simplest form.

HOW TO   Multiply: $\frac{3}{4} \times \frac{14}{15}$

$$\frac{3}{4} \times \frac{14}{15} = \frac{3 \cdot 14}{4 \cdot 15}$$

- Multiply the numerators.
- Multiply the denominators.

$$= \frac{3 \cdot 2 \cdot 7}{2 \cdot 2 \cdot 3 \cdot 5}$$

- Write the prime factorization of each number.

$$= \frac{\overset{1}{\cancel{3}} \cdot 2 \cdot 7}{2 \cdot 2 \cdot \underset{1}{\cancel{3}} \cdot 5} = \frac{7}{10}$$

- Eliminate the common factors. Then multiply the remaining factors in the numerator and denominator.

This example could also be worked by using the GCF.

$$\frac{3}{4} \times \frac{14}{15} = \frac{42}{60}$$

- Multiply the numerators.
- Multiply the denominators.

$$= \frac{6 \cdot 7}{6 \cdot 10}$$

- The GCF of 42 and 60 is 6. Factor 6 from 42 and 60.

$$= \frac{\overset{1}{\cancel{6}} \cdot 7}{\underset{1}{\cancel{6}} \cdot 10} = \frac{7}{10}$$

- Eliminate the GCF.

In-Class Examples (Objective 2.6A)

Multiply.

1. $\frac{3}{4} \times \frac{6}{7}$ $\frac{9}{14}$

2. $\frac{3}{5} \times \frac{7}{8}$ $\frac{21}{40}$

3. $\frac{7}{55} \times \frac{11}{35}$ $\frac{1}{25}$

Optional Student Activity

a. Shade $\frac{1}{3}$ of the diagram shown below.

b. Shade $\frac{2}{5}$ of the $\frac{1}{3}$ of the diagram you already shaded.

c. Then use the diagram to find the product of $\frac{2}{5}$ and $\frac{1}{3}$.

a. 5 of the 15 parts should be shaded.

b. 2 of the 15 parts should be shaded.

c. $\frac{2}{15}$

Example 1

Multiply $\frac{4}{15}$ and $\frac{5}{28}$.

Solution

$$\frac{4}{15} \times \frac{5}{28} = \frac{4 \cdot 5}{15 \cdot 28} = \frac{\overset{1}{2} \cdot \overset{1}{2} \cdot \overset{1}{5}}{3 \cdot \underset{1}{5} \cdot \underset{1}{2} \cdot \underset{1}{2} \cdot 7} = \frac{1}{21}$$

You Try It 1

Multiply $\frac{4}{21}$ and $\frac{7}{44}$.

Your solution $\frac{1}{33}$

Example 2

Find the product of $\frac{9}{20}$ and $\frac{33}{35}$.

Solution

$$\frac{9}{20} \times \frac{33}{35} = \frac{9 \cdot 33}{20 \cdot 35} = \frac{3 \cdot 3 \cdot 3 \cdot 11}{2 \cdot 2 \cdot 5 \cdot 5 \cdot 7} = \frac{297}{700}$$

You Try It 2

Find the product of $\frac{2}{21}$ and $\frac{10}{33}$.

Your solution $\frac{20}{693}$

Example 3

What is $\frac{14}{9}$ times $\frac{12}{7}$?

Solution

$$\frac{14}{9} \times \frac{12}{7} = \frac{14 \cdot 12}{9 \cdot 7} = \frac{2 \cdot \overset{1}{7} \cdot 2 \cdot 2 \cdot \overset{1}{3}}{3 \cdot \underset{1}{3} \cdot \underset{1}{7}} = \frac{8}{3} = 2\frac{2}{3}$$

You Try It 3

What is $\frac{16}{5}$ times $\frac{15}{24}$?

Your solution 2

Solutions on pp. S6–S7

Objective 2.6B

Discuss the Concepts

Describe the steps involved in multiplying $6\frac{1}{2}$ times 4.

Objective B **To multiply whole numbers, mixed numbers, and fractions**

To multiply a whole number by a fraction or mixed number, first write the whole number as a fraction with a denominator of 1.

HOW TO Multiply: $4 \times \frac{3}{7}$

$$4 \times \frac{3}{7} = \frac{4}{1} \times \frac{3}{7} = \frac{4 \cdot 3}{1 \cdot 7} = \frac{2 \cdot 2 \cdot 3}{7} = \frac{12}{7} = 1\frac{5}{7}$$

- Write 4 with a denominator of 1; then multiply the fractions.

When one or more of the factors in a product is a mixed number, write the mixed number as an improper fraction before multiplying.

HOW TO Multiply: $2\frac{1}{3} \times \frac{3}{14}$

$$2\frac{1}{3} \times \frac{3}{14} = \frac{7}{3} \times \frac{3}{14} = \frac{7 \cdot 3}{3 \cdot 14} = \frac{\overset{1}{7} \cdot \overset{1}{3}}{\underset{1}{3} \cdot 2 \cdot \underset{1}{7}} = \frac{1}{2}$$

- Write $2\frac{1}{3}$ as an improper fraction; then multiply the fractions.

In-Class Examples (Objective 2.6B)

Multiply.

1. $3 \times \frac{5}{6}$ $2\frac{1}{2}$

2. $5\frac{1}{4} \times \frac{2}{7}$ $1\frac{1}{2}$

3. $6 \times 2\frac{1}{3}$ 14

4. $3\frac{2}{25} \times 2\frac{1}{2}$ $7\frac{7}{10}$

Example 4

Multiply: $4\frac{5}{6} \times \frac{12}{13}$

Solution

$$4\frac{5}{6} \times \frac{12}{13} = \frac{29}{6} \times \frac{12}{13} = \frac{29 \cdot 12}{6 \cdot 13}$$

$$= \frac{29 \cdot \overset{1}{\cancel{2}} \cdot 2 \cdot \overset{1}{\cancel{3}}}{\underset{1}{\cancel{2}} \cdot \underset{1}{\cancel{3}} \cdot 13} = \frac{58}{13} = 4\frac{6}{13}$$

You Try It 4

Multiply: $5\frac{2}{5} \times \frac{5}{9}$

Your solution   3

Example 5

Find $5\frac{2}{3}$ times $4\frac{1}{2}$.

Solution

$$5\frac{2}{3} \times 4\frac{1}{2} = \frac{17}{3} \times \frac{9}{2} = \frac{17 \cdot 9}{3 \cdot 2}$$

$$= \frac{17 \cdot \overset{1}{\cancel{3}} \cdot 3}{\underset{1}{\cancel{3}} \cdot 2} = \frac{51}{2} = 25\frac{1}{2}$$

You Try It 5

Multiply: $3\frac{2}{5} \times 6\frac{1}{4}$

Your solution $21\frac{1}{4}$

Example 6

Multiply: $4\frac{2}{5} \times 7$

Solution

$$4\frac{2}{5} \times 7 = \frac{22}{5} \times \frac{7}{1} = \frac{22 \cdot 7}{5 \cdot 1}$$

$$= \frac{2 \cdot 11 \cdot 7}{5} = \frac{154}{5} = 30\frac{4}{5}$$

You Try It 6

Multiply: $3\frac{2}{7} \times 6$

Your solution $19\frac{5}{7}$

Solutions on p. S7

Concept Check

Which expression results in the largest product? Which results in the smallest product?

a. $5\frac{1}{3} \times 2\frac{5}{8}$

b. $6\frac{3}{7} \times 2\frac{4}{5}$

c. $4\frac{4}{5} \times 2\frac{1}{12}$

d. $6\frac{1}{4} \times 2\frac{6}{25}$

Part **b** is the largest product (18), and part **c** is the smallest product (10).

Optional Student Activity

Find two fractions evenly spaced between $\frac{3}{8}$ and $\frac{2}{5}$. $\frac{23}{60}$ and $\frac{47}{120}$

Objective 2.6C

Optional Student Activity

You walk at a rate of $3\frac{1}{2}$ miles per hour. Complete the table below. Fill in the second column by calculating the distance walked in the number of hours shown in the first column.

Time Walking	Distance Walked
$1\frac{1}{2}$	$5\frac{1}{4}$
2	7
$2\frac{1}{2}$	$8\frac{3}{4}$
$3\frac{1}{4}$	$11\frac{3}{8}$

Objective C   **To solve application problems**

Length (ft)	Weight (lb/ft)
$6\frac{1}{2}$	$\frac{3}{8}$
$8\frac{5}{8}$	$1\frac{1}{4}$
$10\frac{3}{4}$	$2\frac{1}{2}$
$12\frac{7}{12}$	$4\frac{1}{3}$

The table at the left lists the length of steel rods and the weight per foot. The weight per foot is measured in pounds for each foot of rod and is abbreviated as lb/ft.

HOW TO   Find the weight of the steel bar that is $10\frac{3}{4}$ feet long.

Strategy
To find the weight of the steel bar, multiply its length by the weight per foot.

Solution $10\frac{3}{4} \times 2\frac{1}{2} = \frac{43}{4} \times \frac{5}{2} = \frac{43 \cdot 5}{4 \cdot 2} = \frac{215}{8} = 26\frac{7}{8}$

The weight of the $10\frac{3}{4}$-foot rod is $26\frac{7}{8}$ pounds.

In-Class Examples (Objective 2.6C)

1. An apprentice bricklayer earns $12 an hour. What are the bricklayer's total earnings for working $7\frac{3}{4}$ hours? $93

2. A person can walk $3\frac{3}{4}$ miles in one hour. How many miles can the person walk in $1\frac{1}{4}$ hours? $4\frac{11}{16}$ miles

Example 7

An electrician earns $206 for each day worked. What are the electrician's earnings for working $4\frac{1}{2}$ days?

Strategy

To find the electrician's total earnings, multiply the daily earnings (206) by the number of days worked $\left(4\frac{1}{2}\right)$.

Solution $206 \times 4\frac{1}{2} = \frac{206}{1} \times \frac{9}{2}$

$$= \frac{206 \cdot 9}{1 \cdot 2}$$

$$= 927$$

The electrician's earnings are $927.

You Try It 7

Over the last 10 years, a house increased in value by $2\frac{1}{2}$ times. The price of the house 10 years ago was $170,000. What is the value of the house today?

Your strategy

Your solution $425,000

Example 8

The value of a small office building and the land on which it is built is $290,000. The value of the land is $\frac{1}{4}$ the total value. What is the dollar value of the building?

Strategy

To find the value of the building:

• Find the value of the land $\left(\frac{1}{4} \times 290{,}000\right)$.

• Subtract the value of the land from the total value (290,000).

Solution $\frac{1}{4} \times 290{,}000 = \frac{290{,}000}{4}$

$= 72{,}500$ • Value of the land

$290{,}000 - 72{,}500 = 217{,}500$

The value of the building is $217,500.

You Try It 8

A paint company bought a drying chamber and an air compressor for spray painting. The total cost of the two items was $160,000. The drying chamber's cost was $\frac{4}{5}$ of the total cost. What was the cost of the air compressor?

Your strategy

Your solution $32,000

Solutions on p. S7

2.6 Exercises

Objective A   To multiply fractions

For Exercises 1 to 36, multiply.

1. $\dfrac{2}{3} \times \dfrac{7}{8}$

$\dfrac{7}{12}$

2. $\dfrac{1}{2} \times \dfrac{2}{3}$

$\dfrac{1}{3}$

3. $\dfrac{5}{16} \times \dfrac{7}{15}$

$\dfrac{7}{48}$

4. $\dfrac{3}{8} \times \dfrac{6}{7}$

$\dfrac{9}{28}$

5. $\dfrac{1}{6} \times \dfrac{1}{8}$

$\dfrac{1}{48}$

6. $\dfrac{2}{5} \times \dfrac{5}{6}$

$\dfrac{1}{3}$

7. $\dfrac{11}{12} \times \dfrac{6}{7}$

$\dfrac{11}{14}$

8. $\dfrac{11}{12} \times \dfrac{3}{5}$

$\dfrac{11}{20}$

9. $\dfrac{1}{6} \times \dfrac{6}{7}$

$\dfrac{1}{7}$

10. $\dfrac{3}{5} \times \dfrac{10}{11}$

$\dfrac{6}{11}$

11. $\dfrac{1}{5} \times \dfrac{5}{8}$

$\dfrac{1}{8}$

12. $\dfrac{6}{7} \times \dfrac{14}{15}$

$\dfrac{4}{5}$

13. $\dfrac{8}{9} \times \dfrac{27}{4}$

6

14. $\dfrac{3}{5} \times \dfrac{3}{10}$

$\dfrac{9}{50}$

15. $\dfrac{5}{6} \times \dfrac{1}{2}$

$\dfrac{5}{12}$

16. $\dfrac{3}{8} \times \dfrac{5}{12}$

$\dfrac{5}{32}$

17. $\dfrac{16}{9} \times \dfrac{27}{8}$

6

18. $\dfrac{5}{8} \times \dfrac{16}{15}$

$\dfrac{2}{3}$

19. $\dfrac{3}{2} \times \dfrac{4}{9}$

$\dfrac{2}{3}$

20. $\dfrac{5}{3} \times \dfrac{3}{7}$

$\dfrac{5}{7}$

21. $\dfrac{7}{8} \times \dfrac{3}{14}$

$\dfrac{3}{16}$

22. $\dfrac{2}{9} \times \dfrac{1}{5}$

$\dfrac{2}{45}$

23. $\dfrac{1}{10} \times \dfrac{3}{8}$

$\dfrac{3}{80}$

24. $\dfrac{5}{12} \times \dfrac{6}{7}$

$\dfrac{5}{14}$

25. $\dfrac{15}{8} \times \dfrac{16}{3}$

10

26. $\dfrac{5}{6} \times \dfrac{4}{15}$

$\dfrac{2}{9}$

27. $\dfrac{1}{2} \times \dfrac{2}{15}$

$\dfrac{1}{15}$

28. $\dfrac{3}{8} \times \dfrac{5}{16}$

$\dfrac{15}{128}$

29. $\dfrac{5}{7} \times \dfrac{14}{15}$

$\dfrac{2}{3}$

30. $\dfrac{3}{8} \times \dfrac{15}{41}$

$\dfrac{45}{328}$

31. $\dfrac{5}{12} \times \dfrac{42}{65}$

$\dfrac{7}{26}$

32. $\dfrac{16}{33} \times \dfrac{55}{72}$

$\dfrac{10}{27}$

33. $\dfrac{12}{5} \times \dfrac{5}{3}$

4

34. $\dfrac{17}{9} \times \dfrac{81}{17}$

9

35. $\dfrac{16}{85} \times \dfrac{125}{84}$

$\dfrac{100}{357}$

36. $\dfrac{19}{64} \times \dfrac{48}{95}$

$\dfrac{3}{20}$

Quick Quiz (Objective 2.6A)

Multiply.

1. $\dfrac{2}{3} \times \dfrac{5}{8}$ $\dfrac{5}{12}$

2. $\dfrac{4}{5} \times \dfrac{12}{13}$ $\dfrac{48}{65}$

3. $\dfrac{2}{5} \times \dfrac{15}{16}$ $\dfrac{3}{8}$

Section 2.6

Suggested Assignment
Exercises 1–33, every other odd
Exercises 37–101, odds
Exercises 102, 104
More challenging problems:
   Exercises 105, 106

37. Multiply $\frac{7}{12}$ and $\frac{15}{42}$.

$\frac{5}{24}$

38. Multiply $\frac{32}{9}$ and $\frac{3}{8}$.

$1\frac{1}{3}$

39. Find the product of $\frac{5}{9}$ and $\frac{3}{20}$.

$\frac{1}{12}$

40. Find the product of $\frac{7}{3}$ and $\frac{15}{14}$.

$2\frac{1}{2}$

41. What is $\frac{1}{2}$ times $\frac{8}{15}$?

$\frac{4}{15}$

42. What is $\frac{3}{8}$ times $\frac{12}{17}$?

$\frac{9}{34}$

Objective B   **To multiply whole numbers, mixed numbers, and fractions**

For Exercises 43 to 82, multiply.

43. $4 \times \frac{3}{8}$

$1\frac{1}{2}$

44. $14 \times \frac{5}{7}$

10

45. $\frac{2}{3} \times 6$

4

46. $\frac{5}{12} \times 40$

$16\frac{2}{3}$

47. $\frac{1}{3} \times 1\frac{1}{3}$

$\frac{4}{9}$

48. $\frac{2}{5} \times 2\frac{1}{2}$

1

49. $1\frac{7}{8} \times \frac{4}{15}$

$\frac{1}{2}$

50. $2\frac{1}{5} \times \frac{5}{22}$

$\frac{1}{2}$

51. $55 \times \frac{3}{10}$

$16\frac{1}{2}$

52. $\frac{5}{14} \times 49$

$17\frac{1}{2}$

53. $4 \times 2\frac{1}{2}$

10

54. $9 \times 3\frac{1}{3}$

30

55. $2\frac{1}{7} \times 3$

$6\frac{3}{7}$

56. $5\frac{1}{4} \times 8$

42

57. $3\frac{2}{3} \times 5$

$18\frac{1}{3}$

58. $4\frac{2}{9} \times 3$

$12\frac{2}{3}$

59. $\frac{1}{2} \times 3\frac{3}{7}$

$1\frac{5}{7}$

60. $\frac{3}{8} \times 4\frac{4}{5}$

$1\frac{4}{5}$

61. $6\frac{1}{8} \times \frac{4}{7}$

$3\frac{1}{2}$

62. $5\frac{1}{3} \times \frac{5}{16}$

$1\frac{2}{3}$

63. $5\frac{1}{8} \times 5$

$25\frac{5}{8}$

64. $6\frac{1}{9} \times 2$

$12\frac{2}{9}$

65. $\frac{3}{8} \times 4\frac{1}{2}$

$1\frac{11}{16}$

66. $\frac{5}{7} \times 2\frac{1}{3}$

$1\frac{2}{3}$

67. $6 \times 2\frac{2}{3}$

16

68. $6\frac{1}{8} \times 0$

0

69. $1\frac{1}{3} \times 2\frac{1}{4}$

3

70. $2\frac{5}{8} \times \frac{3}{23}$

$\frac{63}{184}$

71. $2\frac{5}{8} \times 3\frac{2}{5}$

$8\frac{37}{40}$

72. $5\frac{3}{16} \times 5\frac{1}{3}$

$27\frac{2}{3}$

73. $3\frac{1}{7} \times 2\frac{1}{8}$

$6\frac{19}{28}$

74. $16\frac{5}{8} \times 1\frac{1}{16}$

$17\frac{85}{128}$

Quick Quiz (Objective 2.6B)

Multiply.

1. $\frac{4}{5} \times 30$ 24

3. $4\frac{5}{14} \times 7$ $30\frac{1}{2}$

2. $\frac{2}{3} \times 2\frac{5}{8}$ $1\frac{3}{4}$

4. $10\frac{2}{3} \times 3\frac{3}{4}$ 40

75. $2\frac{2}{5} \times 3\frac{1}{12}$ $7\frac{2}{5}$ **76.** $2\frac{2}{3} \times \frac{3}{20}$ $\frac{2}{5}$ **77.** $5\frac{1}{5} \times 3\frac{1}{13}$ 16   **78.** $3\frac{3}{4} \times 2\frac{3}{20}$ $8\frac{1}{16}$

79. $10\frac{1}{4} \times 3\frac{1}{5}$ $32\frac{4}{5}$ **80.** $12\frac{3}{5} \times 1\frac{3}{7}$ 18   **81.** $5\frac{3}{7} \times 5\frac{1}{4}$ $28\frac{1}{2}$ **82.** $6\frac{1}{2} \times 1\frac{3}{13}$ 8

83. Multiply $2\frac{1}{2}$ and $3\frac{3}{5}$. 9   **84.** Multiply $4\frac{3}{8}$ and $3\frac{3}{5}$. $15\frac{3}{4}$

85. Find the product of $2\frac{1}{8}$ and $\frac{5}{17}$. $\frac{5}{8}$ **86.** Find the product of $12\frac{2}{5}$ and $3\frac{7}{31}$. 40

87. What is $1\frac{3}{8}$ times $2\frac{1}{5}$? $3\frac{1}{40}$ **88.** What is $3\frac{1}{8}$ times $2\frac{4}{7}$? $8\frac{1}{28}$

Objective C   **To solve application problems**

89. Consumerism Salmon costs $4 per pound. Find the cost of $2\frac{3}{4}$ pounds of salmon. $11

90. Exercise Maria Rivera can walk $3\frac{1}{2}$ miles in 1 hour. At this rate, how far can Maria walk in $\frac{1}{3}$ hour? $1\frac{1}{6}$ miles

91. Carpentry A board that costs $6 is $9\frac{1}{4}$ feet long. One-third of the board is cut off.

   a. Without doing the calculation, is the piece being cut off at least 4 feet long? No

   b. What is the length of the piece cut off? $3\frac{1}{12}$ feet

$3\frac{1}{2}$ mi
1 h

?
$\frac{1}{3}$ h

92. Geometry The perimeter of a square is equal to four times the length of a side of the square. Find the perimeter of a square whose side measures $16\frac{3}{4}$ inches. 67 inches

$16\frac{3}{4}$ in.

93. Geometry To find the area of a square, multiply the length of one side of the square times itself. What is the area of a square whose side measures $5\frac{1}{4}$ feet? The area of the square will be in square feet. $27\frac{9}{16}$ square feet

94. Geometry The area of a rectangle is equal to the product of the length of the rectangle times its width. Find the area of a rectangle that has a length of $4\frac{2}{5}$ miles and a width of $3\frac{3}{10}$ miles. The area will be in square miles.
$14\frac{13}{25}$ square miles

$4\frac{2}{5}$ mi

$3\frac{3}{10}$ mi

95. Finances A family budgets $\frac{2}{5}$ of its monthly income of $4200 per month for housing and utilities.

   a. What amount is budgeted for housing and utilities? $1680

   b. What amount remains for purposes other than housing and utilities? $2520

Quick Quiz (Objective 2.6C)

1. A sports car gets 27 miles on each gallon of gasoline. How many miles can the car travel on $4\frac{2}{3}$ gallons of gasoline? 126 miles

96. Education $\frac{5}{6}$ of a chemistry class of 36 students has passing grades. $\frac{1}{5}$ of the students with passing grades received an A.
a. How many students passed the chemistry course?   30 students
b. How many students received A grades?   6 students

97. Sewing   The Booster Club is making 22 capes for the members of the high school marching band. Each cape is $1\frac{3}{8}$ yards of material at a cost of $12 per yard. Find the total cost of the material. $363

Measurement   The table at the right shows the length of steel rods and their weight per foot. Use this table for Exercises 98 to 100.

98. Find the weight of the $6\frac{1}{2}$-foot steel rod. $2\frac{7}{16}$ pounds

Length (ft)	Weight (lb/ft)
$6\frac{1}{2}$	$\frac{3}{8}$
$8\frac{5}{8}$	$1\frac{1}{4}$
$10\frac{3}{4}$	$2\frac{1}{2}$
$12\frac{7}{12}$	$4\frac{1}{3}$

99. Find the weight of the $12\frac{7}{12}$-foot steel rod. $54\frac{19}{36}$ pounds

100. Find the total weight of the $8\frac{5}{8}$-foot and the $10\frac{3}{4}$-foot steel rods.
$37\frac{21}{32}$ pounds

101. Investments   The manager of a mutual fund has $\frac{1}{2}$ of the portfolio invested in bonds. Of the amount invested in bonds, $\frac{3}{8}$ is invested in corporate bonds. What fraction of the total portfolio is invested in corporate bonds? $\frac{3}{16}$

APPLYING THE CONCEPTS

102. The product of 1 and a number is $\frac{1}{2}$. Find the number. $\frac{1}{2}$

103.   **Time**   Our calendar is based on the solar year, which is $365\frac{1}{4}$ days. Use this fact to explain leap years.

104. Is the product of two positive fractions always greater than either one of the two numbers? If so, explain why. If not, give an example.
No. $\frac{1}{4} \times \frac{1}{2} = \frac{1}{8}$ $\frac{1}{8}$ is less than either $\frac{1}{4}$ or $\frac{1}{2}$.

105. Which of the labeled points on the number line at the right could be the graph of the product of B and C?   A

106. Fill in the circles on the square at the right with the fractions $\frac{1}{6}, \frac{5}{18}, \frac{4}{9}, \frac{5}{9}, \frac{2}{3}, \frac{3}{4}, 1\frac{1}{9}, 1\frac{1}{2}$, and $2\frac{1}{4}$ so that the product of any row is equal to $\frac{5}{18}$. (*Note:* There is more than one possible answer.)

$\frac{2}{3}$ $\frac{3}{4}$ $\frac{5}{9}$
$1\frac{1}{9}$ $\frac{1}{6}$ $1\frac{1}{2}$
$2\frac{1}{4}$ $\frac{5}{18}$ $\frac{4}{9}$

2.7 Division of Fractions and Mixed Numbers

Objective A To divide fractions

The **reciprocal of a fraction** is the fraction with the numerator and denominator interchanged.

The reciprocal of $\frac{2}{3}$ is $\frac{3}{2}$.

The process of interchanging the numerator and denominator is called **inverting a fraction.**

To find the reciprocal of a whole number, first write the whole number as a fraction with a denominator of 1; then find the reciprocal of that fraction.

The reciprocal of 5 is $\frac{1}{5}$. $\left(\text{Think } 5 = \frac{5}{1}.\right)$

Reciprocals are used to rewrite division problems as related multiplication problems. Look at the following two problems:

$$8 \div 2 = 4 \qquad\qquad 8 \times \frac{1}{2} = 4$$

8 divided by 2 is 4.   8 times the reciprocal of 2 is 4.

"Divided by" means the same as "times the reciprocal of." Thus "$\div\ 2$" can be replaced with "$\times \frac{1}{2}$," and the answer will be the same. Fractions are divided by making this replacement.

HOW TO Divide: $\frac{2}{3} \div \frac{3}{4}$

$$\frac{2}{3} \div \frac{3}{4} = \frac{2}{3} \times \frac{4}{3} = \frac{2 \cdot 4}{3 \cdot 3} = \frac{2 \cdot 2 \cdot 2}{3 \cdot 3} = \frac{8}{9}$$
• Multiply the first fraction by the reciprocal of the second fraction.

Example 1 Divide: $\frac{5}{8} \div \frac{4}{9}$

Solution
$$\frac{5}{8} \div \frac{4}{9} = \frac{5}{8} \times \frac{9}{4} = \frac{5 \cdot 9}{8 \cdot 4}$$
$$= \frac{5 \cdot 3 \cdot 3}{2 \cdot 2 \cdot 2 \cdot 2 \cdot 2} = \frac{45}{32} = 1\frac{13}{32}$$

You Try It 1 Divide: $\frac{3}{7} \div \frac{2}{3}$

Your solution $\frac{9}{14}$

Example 2 Divide: $\frac{3}{5} \div \frac{12}{25}$

Solution
$$\frac{3}{5} \div \frac{12}{25} = \frac{3}{5} \times \frac{25}{12} = \frac{3 \cdot 25}{5 \cdot 12}$$
$$= \frac{\overset{1}{\cancel{3}} \cdot \overset{1}{\cancel{5}} \cdot 5}{\underset{1}{\cancel{5}} \cdot 2 \cdot 2 \cdot \underset{1}{\cancel{3}}} = \frac{5}{4} = 1\frac{1}{4}$$

You Try It 2 Divide: $\frac{3}{4} \div \frac{9}{10}$

Your solution $\frac{5}{6}$

Solutions on p. S7

Objective 2.7A

New Vocabulary
reciprocal of a fraction
inverting a fraction

Discuss the Concepts
Explain why we "invert and multiply" when dividing a fraction by a fraction.

Concept Check
Show by example that (1) the Commutative Property is not satisfied by division of fractions and (2) the Associative Property is not satisfied by division of fractions.
Answers will vary. For example:

(1) $\frac{1}{2} \div \frac{1}{4} = 2,\ \frac{1}{4} \div \frac{1}{2} = \frac{1}{2}$

(2) $\left(\frac{1}{2} \div \frac{1}{4} \right) \div \frac{1}{8} = 16$

$\frac{1}{2} \div \left(\frac{1}{4} \div \frac{1}{8} \right) = \frac{1}{4}$

Optional Student Activity
Find the sum of the reciprocals of all the whole-number factors of 24.
$2\frac{1}{2}$

Instructor Note
Here is an extra-credit problem: One quarter is the same part of one-third as one-half is of what number? One-sixth

In-Class Examples (Objective 2.7A)
Divide.
1. $\frac{2}{9} \div \frac{1}{3}$ $\frac{2}{3}$
2. $\frac{1}{6} \div \frac{4}{9}$ $\frac{3}{8}$
3. $\frac{8}{13} \div \frac{4}{7}$ $1\frac{1}{13}$

Objective 2.7B

Concept Check

(*Note:* This is a classic problem that students frequently miss.) What is 8 divided by one-half?

16

Optional Student Activity

Shown below is the net weight of four different boxes of cereal. Find the number of $\frac{3}{4}$-ounce servings in each box.

a. Kellogg Honey Crunch Corn Flakes: 24 ounces

b. Nabisco Instant Cream of Wheat: 28 ounces

c. Post Shredded Wheat: 18 ounces

d. Quaker Oats: 42 ounces

a. 32 servings

b. $37\frac{1}{3}$ servings

c. 24 servings

d. 56 servings

(*Note:* For a two-step problem, ask the students how many more $\frac{3}{4}$-ounce servings are in a box of Quaker Oats than are in a box of Nabisco Cream of Wheat. The answer is $18\frac{2}{3}$ more servings.)

Objective B   **To divide whole numbers, mixed numbers, and fractions**

To divide a fraction and a whole number, first write the whole number as a fraction with a denominator of 1.

> **HOW TO**   Divide: $\frac{3}{7} \div 5$
>
> $$\frac{3}{7} \div \boxed{5} = \frac{3}{7} \div \boxed{\frac{5}{1}} = \frac{3}{7} \times \frac{1}{5} = \frac{3 \cdot 1}{7 \cdot 5} = \frac{3}{35}$$
>
> • Write 5 with a denominator of 1. Then divide the fractions.

When a number in a quotient is a mixed number, write the mixed number as an improper fraction before dividing.

> **HOW TO**   Divide: $1\frac{13}{15} \div 4\frac{4}{5}$
>
> Write the mixed numbers as improper fractions. Then divide the fractions.
>
> $$1\frac{13}{15} \div 4\frac{4}{5} = \frac{28}{15} \div \frac{24}{5} = \frac{28}{15} \times \frac{5}{24} = \frac{28 \cdot 5}{15 \cdot 24} = \frac{\overset{1}{2} \cdot \overset{1}{2} \cdot 7 \cdot \overset{1}{5}}{3 \cdot \underset{1}{3} \cdot \underset{1}{2} \cdot \underset{1}{2} \cdot 2 \cdot 3} = \frac{7}{18}$$

Example 3   Divide $\frac{4}{9}$ by 5.

Solution

$$\frac{4}{9} \div 5 = \frac{4}{9} \div \frac{5}{1} = \frac{4}{9} \times \frac{1}{5}$$

$$= \frac{4 \cdot 1}{9 \cdot 5} = \frac{2 \cdot 2}{3 \cdot 3 \cdot 5} = \frac{4}{45}$$

• $5 = \frac{5}{1}$. The reciprocal of $\frac{5}{1}$ is $\frac{1}{5}$.

You Try It 3   Divide $\frac{5}{7}$ by 6.

Your solution $\frac{5}{42}$

Example 4   Find the quotient of $\frac{3}{8}$ and $2\frac{1}{10}$.

Solution

$$\frac{3}{8} \div 2\frac{1}{10} = \frac{3}{8} \div \frac{21}{10} = \frac{3}{8} \times \frac{10}{21}$$

$$= \frac{3 \cdot 10}{8 \cdot 21} = \frac{\overset{1}{3} \cdot \overset{1}{2} \cdot 5}{2 \cdot 2 \cdot 2 \cdot \underset{1}{3} \cdot 7} = \frac{5}{28}$$

You Try It 4   Find the quotient of $12\frac{3}{5}$ and 7.

Your solution $1\frac{4}{5}$

Example 5   Divide: $2\frac{3}{4} \div 1\frac{5}{7}$

Solution

$$2\frac{3}{4} \div 1\frac{5}{7} = \frac{11}{4} \div \frac{12}{7} = \frac{11}{4} \times \frac{7}{12} = \frac{11 \cdot 7}{4 \cdot 12}$$

$$= \frac{11 \cdot 7}{2 \cdot 2 \cdot 2 \cdot 2 \cdot 3} = \frac{77}{48} = 1\frac{29}{48}$$

You Try It 5   Divide: $3\frac{2}{3} \div 2\frac{2}{5}$

Your solution $1\frac{19}{36}$

Solutions on p. S7

In-Class Examples (Objective 2.7B)

Divide.

1. $\frac{5}{7} \div 5$ $\frac{1}{7}$

2. $\frac{5}{6} \div 3\frac{3}{4}$ $\frac{2}{9}$

3. $6\frac{2}{3} \div 2\frac{1}{2}$ $2\frac{2}{3}$

Example 6   Divide: $1\frac{13}{15} \div 4\frac{1}{5}$

Solution

$$1\frac{13}{15} \div 4\frac{1}{5} = \frac{28}{15} \div \frac{21}{5} = \frac{28}{15} \times \frac{5}{21} = \frac{28 \cdot 5}{15 \cdot 21}$$

$$= \frac{2 \cdot 2 \cdot \overset{1}{\cancel{7}} \cdot \overset{1}{\cancel{5}}}{3 \cdot \underset{1}{\cancel{5}} \cdot 3 \cdot \underset{1}{\cancel{7}}} = \frac{4}{9}$$

You Try It 6   Divide: $2\frac{5}{6} \div 8\frac{1}{2}$

Your solution $\frac{1}{3}$

Example 7   Divide: $4\frac{3}{8} \div 7$

Solution

$$4\frac{3}{8} \div 7 = \frac{35}{8} \div \frac{7}{1} = \frac{35}{8} \times \frac{1}{7}$$

$$= \frac{35 \cdot 1}{8 \cdot 7} = \frac{5 \cdot \overset{1}{\cancel{7}}}{2 \cdot 2 \cdot 2 \cdot \underset{1}{\cancel{7}}} = \frac{5}{8}$$

You Try It 7   Divide: $6\frac{2}{5} \div 4$

Your solution $1\frac{3}{5}$

Solutions on p. S7

Objective C   **To solve application problems**

Example 8

A car used $15\frac{1}{2}$ gallons of gasoline on a 310-mile trip. How many miles can this car travel on 1 gallon of gasoline?

Strategy
To find the number of miles, divide the number of miles traveled by the number of gallons of gasoline used.

Solution

$$310 \div 15\frac{1}{2} = \frac{310}{1} \div \frac{31}{2}$$

$$= \frac{310}{1} \times \frac{2}{31} = \frac{310 \cdot 2}{1 \cdot 31}$$

$$= \frac{2 \cdot 5 \cdot \overset{1}{\cancel{31}} \cdot 2}{1 \cdot \underset{1}{\cancel{31}}} = \frac{20}{1} = 20$$

The car travels 20 miles on 1 gallon of gasoline.

You Try It 8

A factory worker can assemble a product in $7\frac{1}{2}$ minutes. How many products can the worker assemble in 1 hour?

Your strategy

Your solution   Eight products

Solution on p. S7

Objective 2.7C

Optional Student Activity

A diet regime is formulated so that participants will lose $1\frac{1}{2}$ pounds per week. Complete the table below. Fill in the second column by calculating the number of weeks it will take a client to lose the amount of weight shown in the first column.

Pounds to Lose	Weeks to Lose Them
63	42
58	$38\frac{2}{3}$
32	$21\frac{1}{3}$
$21\frac{1}{2}$	$14\frac{1}{3}$

In-Class Examples (Objective 2.7C)

1. A station wagon used $15\frac{3}{10}$ gallons of gasoline on a 459-mile trip. How many miles did this car travel on 1 gallon of gasoline?   30 miles

2. A building contractor bought $8\frac{1}{4}$ acres of land for $132,000. What was the cost per acre? $16,000

Concept Check

(*Note:* Example 9 and You Try It 9 are difficult for students. Use the following problem if you have worked through these examples with your students and want to have them work another, similar problem either on their own or in small groups.)

A 14-foot piece of wood molding is cut into pieces $4\frac{1}{4}$ feet long. What is the length of the piece that remains after as many pieces as possible have been cut?

$1\frac{1}{4}$ feet

Example 9

A 12-foot board is cut into pieces $2\frac{1}{4}$ feet long for use as bookshelves. What is the length of the remaining piece after as many shelves as possible have been cut?

Strategy

To find the length of the remaining piece:

- Divide the total length of the board (12) by the length of each shelf $\left(2\frac{1}{4}\right)$. This will give you the number of shelves cut, with a certain fraction of a shelf left over.
- Multiply the fractional part of the result in step 1 by the length of one shelf to determine the length of the remaining piece.

Solution

$$12 \div 2\frac{1}{4} = \frac{12}{1} \div \frac{9}{4} = \frac{12}{1} \times \frac{4}{9}$$

$$= \frac{12 \cdot 4}{1 \cdot 9} = \frac{16}{3} = 5\frac{1}{3}$$

There are 5 pieces $2\frac{1}{4}$ feet long.

There is 1 piece that is $\frac{1}{3}$ of $2\frac{1}{4}$ feet long.

$$\frac{1}{3} \times 2\frac{1}{4} = \frac{1}{3} \times \frac{9}{4} = \frac{1 \cdot 9}{3 \cdot 4} = \frac{3}{4}$$

The length of the piece remaining is $\frac{3}{4}$ foot.

You Try It 9

A 16-foot board is cut into pieces $3\frac{1}{3}$ feet long for shelves for a bookcase. What is the length of the remaining piece after as many shelves as possible have been cut?

Your strategy

Your solution $2\frac{2}{3}$ feet

Solution on p. S8

2.7 Exercises

Objective A   To divide fractions

For Exercises 1 to 32, divide.

1. $\dfrac{1}{3} \div \dfrac{2}{5}$

$\dfrac{5}{6}$

2. $\dfrac{3}{7} \div \dfrac{3}{2}$

$\dfrac{2}{7}$

3. $\dfrac{3}{7} \div \dfrac{3}{7}$

1

4. $0 \div \dfrac{1}{2}$

0

5. $0 \div \dfrac{3}{4}$

0

6. $\dfrac{16}{33} \div \dfrac{4}{11}$

$1\dfrac{1}{3}$

7. $\dfrac{5}{24} \div \dfrac{15}{36}$

$\dfrac{1}{2}$

8. $\dfrac{11}{15} \div \dfrac{1}{12}$

$8\dfrac{4}{5}$

9. $\dfrac{15}{16} \div \dfrac{16}{39}$

$2\dfrac{73}{256}$

10. $\dfrac{2}{15} \div \dfrac{3}{5}$

$\dfrac{2}{9}$

11. $\dfrac{8}{9} \div \dfrac{4}{5}$

$1\dfrac{1}{9}$

12. $\dfrac{11}{15} \div \dfrac{5}{22}$

$3\dfrac{17}{75}$

13. $\dfrac{1}{9} \div \dfrac{2}{3}$

$\dfrac{1}{6}$

14. $\dfrac{10}{21} \div \dfrac{5}{7}$

$\dfrac{2}{3}$

15. $\dfrac{2}{5} \div \dfrac{4}{7}$

$\dfrac{7}{10}$

16. $\dfrac{3}{8} \div \dfrac{5}{12}$

$\dfrac{9}{10}$

17. $\dfrac{1}{2} \div \dfrac{1}{4}$

2

18. $\dfrac{1}{3} \div \dfrac{1}{9}$

3

19. $\dfrac{1}{5} \div \dfrac{1}{10}$

2

20. $\dfrac{4}{15} \div \dfrac{2}{5}$

$\dfrac{2}{3}$

21. $\dfrac{7}{15} \div \dfrac{14}{5}$

$\dfrac{1}{6}$

22. $\dfrac{5}{8} \div \dfrac{15}{2}$

$\dfrac{1}{12}$

23. $\dfrac{14}{3} \div \dfrac{7}{9}$

6

24. $\dfrac{7}{4} \div \dfrac{9}{2}$

$\dfrac{7}{18}$

25. $\dfrac{5}{9} \div \dfrac{25}{3}$

$\dfrac{1}{15}$

26. $\dfrac{5}{16} \div \dfrac{3}{8}$

$\dfrac{5}{6}$

27. $\dfrac{2}{3} \div \dfrac{1}{3}$

2

28. $\dfrac{4}{9} \div \dfrac{1}{9}$

4

29. $\dfrac{5}{7} \div \dfrac{2}{7}$

$2\dfrac{1}{2}$

30. $\dfrac{5}{6} \div \dfrac{1}{9}$

$7\dfrac{1}{2}$

31. $\dfrac{2}{3} \div \dfrac{2}{9}$

3

32. $\dfrac{5}{12} \div \dfrac{5}{6}$

$\dfrac{1}{2}$

33. Divide $\dfrac{7}{8}$ by $\dfrac{3}{4}$.

$1\dfrac{1}{6}$

34. Divide $\dfrac{7}{12}$ by $\dfrac{3}{4}$.

$\dfrac{7}{9}$

35. Find the quotient of $\dfrac{5}{7}$ and $\dfrac{3}{14}$.

$3\dfrac{1}{3}$

36. Find the quotient of $\dfrac{6}{11}$ and $\dfrac{9}{32}$.

$1\dfrac{31}{33}$

Section 2.7

Suggested Assignment
Exercises 1–31, every other odd
Exercises 33–107, odds

Quick Quiz (Objective 2.7A)

Divide.

1. $\dfrac{5}{12} \div \dfrac{5}{8}$ $\quad \dfrac{2}{3}$

2. $\dfrac{3}{16} \div \dfrac{9}{20}$ $\quad \dfrac{5}{12}$

3. $\dfrac{8}{15} \div \dfrac{16}{45}$ $\quad 1\dfrac{1}{2}$

106

Objective B **To divide whole numbers, mixed numbers, and fractions**

For Exercises 37 to 81, divide.

37. $4 \div \dfrac{2}{3}$

6

38. $\dfrac{2}{3} \div 4$

$\dfrac{1}{6}$

39. $\dfrac{3}{2} \div 3$

$\dfrac{1}{2}$

40. $3 \div \dfrac{3}{2}$

2

41. $\dfrac{5}{6} \div 25$

$\dfrac{1}{30}$

42. $22 \div \dfrac{3}{11}$

$80\dfrac{2}{3}$

43. $6 \div 3\dfrac{1}{3}$

$1\dfrac{4}{5}$

44. $5\dfrac{1}{2} \div 11$

$\dfrac{1}{2}$

45. $6\dfrac{1}{2} \div \dfrac{1}{2}$

13

46. $\dfrac{3}{8} \div 2\dfrac{1}{4}$

$\dfrac{1}{6}$

47. $\dfrac{5}{12} \div 4\dfrac{4}{5}$

$\dfrac{25}{288}$

48. $1\dfrac{1}{2} \div 1\dfrac{3}{8}$

$1\dfrac{1}{11}$

49. $8\dfrac{1}{4} \div 2\dfrac{3}{4}$

3

50. $3\dfrac{5}{9} \div 32$

$\dfrac{1}{9}$

51. $4\dfrac{1}{5} \div 21$

$\dfrac{1}{5}$

52. $6\dfrac{8}{9} \div \dfrac{31}{36}$

8

53. $\dfrac{11}{12} \div 2\dfrac{1}{3}$

$\dfrac{11}{28}$

54. $\dfrac{7}{8} \div 3\dfrac{1}{4}$

$\dfrac{7}{26}$

55. $\dfrac{5}{16} \div 5\dfrac{3}{8}$

$\dfrac{5}{86}$

56. $\dfrac{9}{14} \div 3\dfrac{1}{7}$

$\dfrac{9}{44}$

57. $35 \div \dfrac{7}{24}$

120

58. $\dfrac{3}{8} \div 2\dfrac{3}{4}$

$\dfrac{3}{22}$

59. $\dfrac{11}{18} \div 2\dfrac{2}{9}$

$\dfrac{11}{40}$

60. $\dfrac{21}{40} \div 3\dfrac{3}{10}$

$\dfrac{7}{44}$

61. $2\dfrac{1}{16} \div 2\dfrac{1}{2}$

$\dfrac{33}{40}$

62. $7\dfrac{3}{5} \div 1\dfrac{7}{12}$

$4\dfrac{4}{5}$

63. $1\dfrac{2}{3} \div \dfrac{3}{8}$

$4\dfrac{4}{9}$

64. $16 \div \dfrac{2}{3}$

24

65. $1\dfrac{5}{8} \div 4$

$\dfrac{13}{32}$

66. $13\dfrac{3}{8} \div \dfrac{1}{4}$

$53\dfrac{1}{2}$

67. $16 \div 1\dfrac{1}{2}$

$10\dfrac{2}{3}$

68. $9 \div \dfrac{7}{8}$

$10\dfrac{2}{7}$

69. $16\dfrac{5}{8} \div 1\dfrac{2}{3}$

$9\dfrac{39}{40}$

70. $24\dfrac{4}{5} \div 2\dfrac{3}{5}$

$9\dfrac{7}{13}$

71. $1\dfrac{1}{3} \div 5\dfrac{8}{9}$

$\dfrac{12}{53}$

72. $13\dfrac{2}{3} \div 0$

Undefined

Quick Quiz (Objective 2.7B)
Divide.

1. $8 \div \dfrac{4}{15}$ 30

2. $\dfrac{3}{8} \div 2\dfrac{5}{8}$ $\dfrac{1}{7}$

3. $3\dfrac{2}{5} \div 1\dfrac{7}{10}$ 2

73. $82\frac{3}{5} \div 19\frac{1}{10}$ $4\frac{62}{191}$

74. $45\frac{3}{5} \div 15$ $3\frac{1}{25}$

75. $102 \div 1\frac{1}{2}$ 68

76. $0 \div 3\frac{1}{2}$ 0

77. $8\frac{2}{7} \div 1$ $8\frac{2}{7}$

78. $6\frac{9}{16} \div 1\frac{3}{32}$ 6

79. $8\frac{8}{9} \div 2\frac{13}{18}$ $3\frac{13}{49}$

80. $10\frac{1}{5} \div 1\frac{7}{10}$ 6

81. $7\frac{3}{8} \div 1\frac{27}{32}$ 4

82. Divide $7\frac{7}{9}$ by $5\frac{5}{6}$. $1\frac{1}{3}$

83. Divide $2\frac{3}{4}$ by $1\frac{23}{32}$. $1\frac{3}{5}$

84. Find the quotient of $8\frac{1}{4}$ and $1\frac{5}{11}$. $5\frac{43}{64}$

85. Find the quotient of $\frac{14}{17}$ and $3\frac{1}{9}$. $\frac{9}{34}$

Objective C   **To solve application problems**

86. Consumerism   Individual cereal boxes contain $\frac{3}{4}$ ounce of cereal. How many boxes can be filled with 600 ounces of cereal?   800 boxes

87. Consumerism   A box of Post's Great Grains cereal costing $4 contains 16 ounces of cereal. How many $1\frac{1}{3}$-ounce portions can be served from this box?   12 servings

88. Gemology   A $\frac{5}{8}$-karat diamond was purchased for $1200. What would a similar diamond weighing 1 karat cost? $1920

89. Real Estate   The Inverness Investor Group bought $8\frac{1}{3}$ acres of land for $200,000. What was the cost of each acre? $24,000

90. Fuel Efficiency   A car used $12\frac{1}{2}$ gallons of gasoline on a 275-mile trip. How many miles can their car travel on 1 gallon of gasoline?   22 miles

91. Mechanics   A nut moves $\frac{5}{32}$ inch for each turn. Find the number of turns it will take for the nut to move $1\frac{7}{8}$ inches.   12 turns

92. Real Estate   The Hammond Company purchased $9\frac{3}{4}$ acres for a housing project. One and one-half acres were set aside for a park.

a. How many acres are available for housing? $8\frac{1}{4}$ acres

b. How many $\frac{1}{4}$-acre parcels of land can be sold after the land for the park is set aside?   33 parcels

Quick Quiz (Objective 2.7C)

1. A car traveled 104 miles in $3\frac{1}{4}$ hours. What was the car's average speed in miles per hour?
32 miles per hour

93. **The Food Industry**   A chef purchased a roast that weighed $10\frac{3}{4}$ pounds. After the fat was trimmed and the bone removed, the roast weighed $9\frac{1}{3}$ pounds.

a. What was the total weight of the fat and bone? $1\frac{5}{12}$ pounds

b. How many $\frac{1}{3}$-pound servings can be cut from the trimmed roast?

28 servings

94. **Carpentry**   A 15-foot board is cut into pieces $3\frac{1}{2}$ feet long for a bookcase. What is the length of the piece remaining after as many shelves as possible have been cut?   1 foot

95. **Architecture**   A scale of $\frac{1}{2}$ inch to 1 foot is used to draw the plans for a house. The scale measurements for three walls are given in the table at the right. Complete the table to determine the actual wall lengths for the three walls a, b, and c.

Wall	Scale	Actual Wall Length
a	$6\frac{1}{4}$ in.	? $12\frac{1}{2}$ ft
b	9 in.	? 18 ft
c	$7\frac{7}{8}$ in.	? $15\frac{3}{4}$ ft

APPLYING THE CONCEPTS

Loans   The figure at the right shows how the money borrowed on home equity loans is spent. Use this graph for Exercises 96 and 97.

96. What fractional part of the money borrowed on home equity loans is spent on debt consolidation and home improvement?
$\frac{31}{50}$

97. What fractional part of the money borrowed on home equity loans is spent on home improvement, cars, and tuition?
$\frac{17}{50}$

98. **Finances**   A bank recommends that the maximum monthly payment for a home be $\frac{1}{3}$ of your total monthly income. Your monthly income is $4500. What would the bank recommend as your maximum monthly house payment?

$1500

99. **Entertainment**   On Friday evening, 1200 people attended a concert at the music center. The center was $\frac{2}{3}$ full. What is the capacity of the music center?
1800 people

100. **Sports**   During the second half of the 1900s, greenskeepers mowed the grass on golf putting surfaces progressively lower. The table at the right shows the average grass height by decade. What was the difference between the average height of the grass in the 1980s and its average height in the 1950s?
$\frac{3}{32}$ inch

How Money Borrowed on Home Equity Loans Is Spent
Source: Consumer Bankers Association

Average Height of Grass on Golf Putting Surfaces	
Decade	Height (in inches)
1950s	$\frac{1}{4}$
1960s	$\frac{7}{32}$
1970s	$\frac{3}{16}$
1980s	$\frac{5}{32}$
1990s	$\frac{1}{8}$

Source: Golf Course Superintendents Association of America

Instructor Note

Exercises 96 to 107 are intended to provide students with practice in deciding what operation to use in order to solve an application problem.

101. **Puzzles**   You completed $\frac{1}{3}$ of a jigsaw puzzle yesterday and $\frac{1}{2}$ of the puzzle today. What fraction of the puzzle is left to complete?

$\frac{1}{6}$

102. **Board Games**   A wooden travel game board has hinges that allow the board to be folded in half. If the dimensions of the open board are 14 inches by 14 inches by $\frac{7}{8}$ inch, what are the dimensions of the board when it is closed?

14 inches by 7 inches by $1\frac{3}{4}$ inches

103. **Wages**   You have a part-time job that pays $9 an hour. You worked 5 hours, $3\frac{3}{4}$ hours, $1\frac{1}{4}$ hours, and $2\frac{1}{3}$ hours during the four days you worked last week. Find your total earnings for last week's work.

$111

Nutrition   According to the Center for Science in the Public Interest, the average teenage boy drinks $3\frac{1}{3}$ cans of soda per day. The average teenage girl drinks $2\frac{1}{3}$ cans of soda per day. Use this information for Exercises 104 to 106.

104. The average teenage boy drinks how many cans of soda per week?

$23\frac{1}{3}$ cans

105. If a can of soda contains 150 calories, how many calories does the average teenage boy consume each week in soda?   3500 calories

106. How many more cans of soda per week does the average teenage boy drink than the average teenage girl?   7 cans

107. **Maps**   On a map, two cites are $4\frac{5}{8}$ inches apart. If $\frac{3}{8}$ inch on the map represents 60 miles, what is the number of miles between the two cities?

740 miles

108. Is the quotient always less than the dividend in a division problem? Explain.   No. For example, $4 \div \frac{1}{2} = 8$. Answers will vary.

109. Fill in the box to make a true statement.

a. $\frac{3}{4} \cdot \boxed{} = \frac{1}{2}$ $\frac{2}{3}$ **b.** $\frac{2}{3} \cdot \boxed{} = 1\frac{3}{4}$ $2\frac{5}{8}$

110. **Publishing**   A page of type in a certain textbook is $7\frac{1}{2}$ inches wide. If the page is divided into three equal columns, with $\frac{3}{8}$ inch between columns, how wide is each column? $2\frac{1}{4}$ inches

111. A whole number is both multiplied and divided by the same proper fraction. What is greater, the product or the quotient?   The quotient

Objective 2.8A

Vocabulary to Review

number line   [1.1A]
graphing of a whole number
[1.1A]

Discuss the Concepts

1. If two fractions have the same denominator, how can you determine which fraction is larger than the other?
2. If two fractions have different denominators, how can you determine which fraction is larger than the other?

Concept Check

Put the following fractions in order from smallest to largest.

$$\frac{17}{30}, \frac{7}{12}, \frac{8}{15}, \frac{17}{25}, \frac{5}{9}, \frac{13}{24}$$

$$\frac{8}{15}, \frac{13}{24}, \frac{5}{9}, \frac{17}{30}, \frac{7}{12}, \frac{17}{25}$$

Optional Student Activity

Use a diagram to show that $\frac{2}{3}$ is greater than $\frac{5}{8}$.

Objective 2.8B

Vocabulary to Review

exponent   [1.6A]
exponential notation   [1.6A]

2.8 Order, Exponents, and the Order of Operations Agreement

Objective A To identify the order relation between two fractions

Recall that whole numbers can be graphed as points on the number line. Fractions can also be graphed as points on the number line.

The graph of $\frac{3}{4}$ on the number line

The number line can be used to determine the order relation between two fractions. A fraction that appears to the left of a given fraction is less than the given fraction. A fraction that appears to the right of a given fraction is greater than the given fraction.

$$\frac{1}{8} < \frac{3}{8} \qquad \frac{6}{8} > \frac{3}{8}$$

To find the order relation between two fractions with the same denominator, compare the numerators. The fraction that has the smaller numerator is the smaller fraction. When the denominators are different, begin by writing equivalent fractions with a common denominator; then compare numerators.

HOW TO Find the order relation between $\frac{11}{18}$ and $\frac{5}{8}$.

The LCM of 18 and 8 is 72.

$$\frac{11}{18} = \frac{44}{72} \leftarrow \text{Smaller numerator} \qquad \frac{11}{18} < \frac{5}{8} \quad \text{or} \quad \frac{5}{8} > \frac{11}{18}$$

$$\frac{5}{8} = \frac{45}{72} \leftarrow \text{Larger numerator}$$

Point of Interest

Leonardo of Pisa, who was also called Fibonacci (c. 1175–1250), is credited with bringing the Hindu-Arabic number system to the Western world and promoting its use in place of the cumbersome Roman numeral system.

He was also influential in promoting the idea of the fraction bar. His notation, however, was very different from what we have today.

For instance, he wrote $\frac{3 \quad 5}{4 \quad 7}$

to mean $\frac{5}{7} + \frac{3}{7 \cdot 4}$, which

equals $\frac{23}{28}$.

Example 1 Place the correct symbol, < or >, between the two numbers.

$$\frac{5}{12} \qquad \frac{7}{18}$$

Solution $\frac{5}{12} = \frac{15}{36}$ $\frac{7}{18} = \frac{14}{36}$

$$\frac{5}{12} > \frac{7}{18}$$

You Try It 1 Place the correct symbol, < or >, between the two numbers.

$$\frac{9}{14} \qquad \frac{13}{21}$$

Your solution $\frac{9}{14} > \frac{13}{21}$

Solution on p. S8

Objective B To simplify expressions containing exponents

Repeated multiplication of the same fraction can be written in two ways:

$$\frac{1}{2} \cdot \frac{1}{2} \cdot \frac{1}{2} \cdot \frac{1}{2} \qquad \text{or} \qquad \left(\frac{1}{2}\right)^4 \leftarrow \text{Exponent}$$

The exponent indicates how many times the fraction occurs as a factor in the multiplication. The expression $\left(\frac{1}{2}\right)^4$ is in exponential notation.

In-Class Examples (Objective 2.8A)

Place the correct symbol, < or >, between the two numbers.

1. $\frac{10}{17}$ $\frac{13}{17}$ <

2. $\frac{2}{3}$ $\frac{5}{8}$ >

3. $\frac{6}{11}$ $\frac{4}{7}$ <

In-Class Examples (Objective 2.8B)

Simplify.

1. $\left(\frac{4}{9}\right)^2$ $\frac{16}{81}$

2. $\left(\frac{2}{3}\right)^2\left(\frac{3}{4}\right)$ $\frac{1}{3}$

3. $\left(\frac{9}{16}\right)\left(\frac{2}{3}\right)^4\left(\frac{3}{5}\right)^3$ $\frac{3}{125}$

Example 2   Simplify: $\left(\frac{5}{6}\right)^3 \cdot \left(\frac{3}{5}\right)^2$

Solution $\left(\frac{5}{6}\right)^3 \cdot \left(\frac{3}{5}\right)^2 = \left(\frac{5}{6} \cdot \frac{5}{6} \cdot \frac{5}{6}\right) \cdot \left(\frac{3}{5} \cdot \frac{3}{5}\right)$

$= \dfrac{\overset{1}{\cancel{5}} \cdot \overset{1}{\cancel{5}} \cdot 5 \cdot \overset{1}{\cancel{3}} \cdot \overset{1}{\cancel{3}}}{2 \cdot \underset{1}{\cancel{3}} \cdot 2 \cdot \underset{1}{\cancel{3}} \cdot 2 \cdot 3 \cdot \underset{1}{\cancel{5}} \cdot \underset{1}{\cancel{5}}} = \dfrac{5}{24}$

You Try It 2   Simplify: $\left(\frac{7}{11}\right)^2 \cdot \left(\frac{2}{7}\right)$

Your solution $\dfrac{14}{121}$

Solution on p. S8

Objective C   **To use the Order of Operations Agreement to simplify expressions**

The Order of Operations Agreement is used for fractions as well as whole numbers.

The Order of Operations Agreement

Step 1.   Do all the operations inside parentheses.
Step 2.   Simplify any number expressions containing exponents.
Step 3.   Do multiplications and divisions as they occur from left to right.
Step 4.   Do additions and subtractions as they occur from left to right.

HOW TO   Simplify $\frac{14}{15} - \left(\frac{1}{2}\right)^2 \times \left(\frac{2}{3} + \frac{4}{5}\right)$.

$\frac{14}{15} - \left(\frac{1}{2}\right)^2 \times \underbrace{\left(\frac{2}{3} + \frac{4}{5}\right)}$ 1. Perform operations in parentheses.

$\frac{14}{15} - \left(\frac{1}{2}\right)^2 \times \frac{22}{15}$ 2. Simplify expressions with exponents.

$\frac{14}{15} - \underbrace{\frac{1}{4} \times \frac{22}{15}}$ 3. Do multiplication and division as they occur from left to right.

$\underbrace{\frac{14}{15} - \frac{11}{30}}$ 4. Do addition and subtraction as they occur from left to right.

$\frac{17}{30}$

One or more of the above steps may not be needed to simplify an expression. In that case, proceed to the next step in the Order of Operations agreement.

Example 3   Simplify: $\left(\frac{3}{4}\right)^2 \div \left(\frac{3}{8} - \frac{1}{12}\right)$

Solution $\left(\frac{3}{4}\right)^2 \div \left(\frac{3}{8} - \frac{1}{12}\right)$

$= \left(\frac{3}{4}\right)^2 \div \left(\frac{7}{24}\right) = \frac{9}{16} \div \frac{7}{24}$

$= \frac{9}{16} \cdot \frac{24}{7} = \frac{27}{14} = 1\frac{13}{14}$

You Try It 3   Simplify: $\left(\frac{1}{13}\right)^2 \cdot \left(\frac{1}{4} + \frac{1}{6}\right) \div \frac{5}{13}$

Your solution $\dfrac{1}{156}$

Solution on p. S8

Discuss the Concepts
How can you write the expression

$\dfrac{2}{3} \cdot \dfrac{2}{3} \cdot \dfrac{2}{3} \cdot \dfrac{2}{3} \cdot \dfrac{2}{3}$

as an exponential expression?

Objective 2.8C

Procedures to Review
Order of Operations Agreement
[1.6B]

Discuss the Concepts
In simplifying the expression

$\dfrac{3}{4} + \dfrac{1}{2} \cdot \dfrac{8}{9}$

why can't you begin by adding $\frac{3}{4}$ and $\frac{1}{2}$?

Optional Student Activity
Find the sum of the fraction halfway between $\frac{4}{5}$ and $\frac{2}{3}$ and the fraction halfway between $\frac{1}{2}$ and

$\dfrac{1}{3}$. $1\dfrac{3}{20}$

Optional Student Activity
Find the product of

$\left(1 - \frac{1}{2^2}\right)\left(1 - \frac{1}{3^2}\right)\left(1 - \frac{1}{4^2}\right)$

$\cdots \left(1 - \frac{1}{9^2}\right)\left(1 - \frac{1}{10^2}\right)$

The dots mean that the pattern continues. $\dfrac{11}{20}$

In-Class Examples (Objective 2.8C)
Simplify.

1. $\dfrac{7}{8} + \dfrac{1}{9} \div \dfrac{8}{9}$ 1

2. $\left(\dfrac{1}{3}\right)^2 \left(\dfrac{4}{5} - \dfrac{1}{2}\right)$ $\dfrac{1}{30}$

3. $\left(\dfrac{1}{2}\right)^2 + \left(\dfrac{3}{5} - \dfrac{1}{2}\right) \div \dfrac{4}{15}$ $\dfrac{5}{8}$

Section 2.8

Suggested Assignment
Exercises 1–47, odds
Exercise 48
More challenging problem:
   Exercise 49

2.8 Exercises

Objective A   To identify the order relation between two fractions

For Exercises 1 to 12, place the correct symbol, < or >, between the two numbers.

1. $\dfrac{11}{40} < \dfrac{19}{40}$

2. $\dfrac{92}{103} > \dfrac{19}{103}$

3. $\dfrac{2}{3} < \dfrac{5}{7}$

4. $\dfrac{2}{5} > \dfrac{3}{8}$

5. $\dfrac{5}{8} > \dfrac{7}{12}$

6. $\dfrac{11}{16} < \dfrac{17}{24}$

7. $\dfrac{7}{9} < \dfrac{11}{12}$

8. $\dfrac{5}{12} < \dfrac{7}{15}$

9. $\dfrac{13}{14} > \dfrac{19}{21}$

10. $\dfrac{13}{18} > \dfrac{7}{12}$

11. $\dfrac{7}{24} < \dfrac{11}{30}$

12. $\dfrac{13}{36} < \dfrac{19}{48}$

Objective B   To simplify expressions containing exponents

For Exercises 13 to 28, simplify.

13. $\left(\dfrac{3}{8}\right)^2$
$\dfrac{9}{64}$

14. $\left(\dfrac{5}{12}\right)^2$
$\dfrac{25}{144}$

15. $\left(\dfrac{2}{9}\right)^3$
$\dfrac{8}{729}$

16. $\left(\dfrac{1}{2}\right) \cdot \left(\dfrac{2}{3}\right)^2$
$\dfrac{2}{9}$

17. $\left(\dfrac{2}{3}\right) \cdot \left(\dfrac{1}{2}\right)^4$
$\dfrac{1}{24}$

18. $\left(\dfrac{1}{3}\right)^2 \cdot \left(\dfrac{3}{5}\right)^3$
$\dfrac{3}{125}$

19. $\left(\dfrac{2}{5}\right)^3 \cdot \left(\dfrac{5}{7}\right)^2$
$\dfrac{8}{245}$

20. $\left(\dfrac{5}{9}\right)^3 \cdot \left(\dfrac{18}{25}\right)^2$
$\dfrac{4}{45}$

21. $\left(\dfrac{1}{3}\right)^4 \cdot \left(\dfrac{9}{11}\right)^2$
$\dfrac{1}{121}$

22. $\left(\dfrac{1}{2}\right)^6 \cdot \left(\dfrac{32}{35}\right)^2$
$\dfrac{16}{1225}$

23. $\left(\dfrac{2}{3}\right)^4 \cdot \left(\dfrac{81}{100}\right)^2$
$\dfrac{81}{625}$

24. $\left(\dfrac{1}{6}\right) \cdot \left(\dfrac{6}{7}\right)^2 \cdot \left(\dfrac{2}{3}\right)$
$\dfrac{4}{49}$

25. $\left(\dfrac{2}{7}\right) \cdot \left(\dfrac{7}{8}\right)^2 \cdot \left(\dfrac{8}{9}\right)$
$\dfrac{7}{36}$

26. $3 \cdot \left(\dfrac{3}{5}\right)^3 \cdot \left(\dfrac{1}{3}\right)^2$
$\dfrac{9}{125}$

27. $4 \cdot \left(\dfrac{3}{4}\right)^3 \cdot \left(\dfrac{4}{7}\right)^2$
$\dfrac{27}{49}$

28. $11 \cdot \left(\dfrac{3}{8}\right)^3 \cdot \left(\dfrac{8}{11}\right)^2$
$\dfrac{27}{88}$

Objective C   To use the Order of Operations Agreement to simplify expressions

For Exercises 29 to 47, simplify.

29. $\dfrac{1}{2} - \dfrac{1}{3} + \dfrac{2}{3}$
$\dfrac{5}{6}$

30. $\dfrac{2}{5} + \dfrac{3}{10} - \dfrac{2}{3}$
$\dfrac{1}{30}$

31. $\dfrac{1}{3} \div \dfrac{1}{2} + \dfrac{3}{4}$
$1\dfrac{5}{12}$

32. $\dfrac{4}{5} + \dfrac{3}{7} \cdot \dfrac{14}{15}$
$1\dfrac{1}{5}$

Quick Quiz (Objective 2.8A)
Place the correct symbol, < or >, between the two numbers.

1. $\dfrac{1}{3}$ $\dfrac{5}{16}$ >

2. $\dfrac{7}{9}$ $\dfrac{5}{6}$ <

Quick Quiz (Objective 2.8B)
Simplify.

1. $\left(\dfrac{2}{5}\right)^2$ $\dfrac{4}{25}$

2. $\left(\dfrac{5}{6}\right)\left(\dfrac{3}{10}\right)^2$ $\dfrac{3}{40}$

3. $3 \cdot \left(\dfrac{5}{6}\right)^2 \left(\dfrac{1}{5}\right)^3$ $\dfrac{1}{60}$

33. $\left(\dfrac{3}{4}\right)^2 - \dfrac{5}{12}$

$\dfrac{7}{48}$

34. $\left(\dfrac{3}{5}\right)^3 - \dfrac{3}{25}$

$\dfrac{12}{125}$

35. $\dfrac{5}{6} \cdot \left(\dfrac{2}{3} - \dfrac{1}{6}\right) + \dfrac{7}{18}$

$\dfrac{29}{36}$

36. $\dfrac{3}{4} \cdot \left(\dfrac{11}{12} - \dfrac{7}{8}\right) + \dfrac{5}{16}$

$\dfrac{11}{32}$

37. $\dfrac{7}{12} - \left(\dfrac{2}{3}\right)^2 + \dfrac{5}{8}$

$\dfrac{55}{72}$

38. $\dfrac{11}{16} - \left(\dfrac{3}{4}\right)^2 + \dfrac{7}{12}$

$\dfrac{17}{24}$

39. $\dfrac{3}{4} \cdot \left(\dfrac{4}{9}\right)^2 + \dfrac{1}{2}$

$\dfrac{35}{54}$

40. $\dfrac{9}{10} \cdot \left(\dfrac{2}{3}\right)^3 + \dfrac{2}{3}$

$\dfrac{14}{15}$

41. $\left(\dfrac{1}{2} + \dfrac{3}{4}\right) \div \dfrac{5}{8}$

2

42. $\left(\dfrac{2}{3} + \dfrac{5}{6}\right) \div \dfrac{5}{9}$

$2\dfrac{7}{10}$

43. $\dfrac{3}{8} \div \left(\dfrac{5}{12} + \dfrac{3}{8}\right)$

$\dfrac{9}{19}$

44. $\dfrac{7}{12} \div \left(\dfrac{2}{3} + \dfrac{5}{9}\right)$

$\dfrac{21}{44}$

45. $\left(\dfrac{3}{8}\right)^2 \div \left(\dfrac{3}{7} + \dfrac{3}{14}\right)$

$\dfrac{7}{32}$

46. $\left(\dfrac{5}{6}\right)^2 \div \left(\dfrac{5}{12} + \dfrac{2}{3}\right)$

$\dfrac{25}{39}$

47. $\dfrac{2}{5} \div \dfrac{3}{8} \cdot \dfrac{4}{5}$

$\dfrac{64}{75}$

APPLYING THE CONCEPTS

48. **The Food Industry** The table at the right shows the results of a survey that asked fast-food patrons their criteria for choosing where to go for fast food. For example, 3 out of every 25 people surveyed said that the speed of the service was most important.
  a. According to the survey, do more people choose a fast-food restaurant on the basis of its location or the quality of the food?
  b. Which criterion was cited by the most people?
  a. Location   **b.** Location

Fast-Food Patrons' Top Criteria for Fast-Food Restaurants	
Food quality	$\dfrac{1}{4}$
Location	$\dfrac{13}{50}$
Menu	$\dfrac{4}{25}$
Price	$\dfrac{2}{25}$
Speed	$\dfrac{3}{25}$
Other	$\dfrac{3}{100}$

Source: Maritz Marketing Research, Inc.

49. A farmer died and left 17 horses to be divided among 3 children. The first child was to receive $\dfrac{1}{2}$ of the horses, the second child $\dfrac{1}{3}$ of the horses, and the third child $\dfrac{1}{9}$ of the horses. The executor for the family's estate realized that 17 horses could not be divided by halves, thirds, or ninths and so added a neighbor's horse to the farmer's. With 18 horses, the executor gave 9 horses to the first child, 6 horses to the second child, and 2 horses to the third child. This accounted for the 17 horses, so the executor returned the borrowed horse to the neighbor. Explain why this worked.

50. $\dfrac{2}{3} < \dfrac{3}{4}$. Is $\dfrac{2+3}{3+4}$ less than $\dfrac{2}{3}$, greater than $\dfrac{2}{3}$, or between $\dfrac{2}{3}$ and $\dfrac{3}{4}$?

Between $\dfrac{2}{3}$ and $\dfrac{3}{4}$

Answers to Writing Exercises

49. The "puzzle" works as it does because the sum of the fractions $\dfrac{1}{2}$, $\dfrac{1}{3}$, and $\dfrac{1}{9}$ is $\dfrac{17}{18}$, not 1. As a result, the first child actually received $\dfrac{9}{17}$ of the horses, not $\dfrac{1}{2}$; the second child received $\dfrac{6}{17}$ of the horses, not $\dfrac{1}{3}$; and the third child received $\dfrac{2}{17}$ of the horses, not $\dfrac{1}{9}$.

Quick Quiz (Objective 2.8C)

Simplify.

1. $\dfrac{1}{2} + \dfrac{3}{5} \div \dfrac{3}{10}$ $2\dfrac{1}{2}$

2. $\dfrac{1}{6} \div \left(\dfrac{1}{3}\right)^2 - \dfrac{1}{2}$ 1

Focus on Problem Solving

Common Knowledge

An application problem may not provide all the information that is needed to solve the problem. Sometimes, however, the necessary information is common knowledge.

> **HOW TO** You are traveling by bus from Boston to New York. The trip is 4 hours long. If the bus leaves Boston at 10 A.M., what time should you arrive in New York?
>
> What other information do you need to solve this problem?
>
> You need to know that, using a 12-hour clock, the hours run
>
> 10 A.M.
> 11 A.M.
> 12 P.M.
> 1 P.M.
> 2 P.M.
>
> Four hours after 10 A.M. is 2 P.M.

• You should arrive in New York at 2 P.M.

> **HOW TO** You purchase a 37¢ stamp at the Post Office and hand the clerk a one-dollar bill. How much change do you receive?
>
> What information do you need to solve this problem?
>
> You need to know that there are 100¢ in one dollar.
>
> Your change is 100¢ − 37¢.
>
> 100 − 37 = 63

• You receive 63¢ in change.

What information do you need to know to solve each of the following problems?

1. You sell a dozen tickets to a fundraiser. Each ticket costs $10. How much money do you collect?

2. The weekly lab period for your science course is 1 hour and 20 minutes long. Find the length of the science lab period in minutes.

3. An employee's monthly salary is $3750. Find the employee's annual salary.

4. A survey revealed that eighth graders spend an average of 3 hours each day watching television. Find the total time an eighth grader spends watching TV each week.

5. You want to buy a carpet for a room that is 15 feet wide and 18 feet long. Find the amount of carpet that you need.

Answers to Focus on Problem Solving— Common Knowledge

1. There are 12 in one dozen.
2. There are 60 minutes in 1 hour.
3. There are 12 months in 1 year.
4. There are 7 days in 1 week.
5. You need to know the formula for the area of a rectangle.

Projects and Group Activities

Music   In musical notation, notes are printed on a **staff,** which is a set of five horizontal lines and the spaces between them. The notes of a musical composition are grouped into **measures,** or **bars.** Vertical lines separate measures on a staff. The shape of a note indicates how long it should be held. The whole note has the longest time value of any note. Each time value is divided by 2 in order to find the next smallest time value.

The **time signature** is a fraction that appears at the beginning of a piece of music. The numerator of the fraction indicates the number of beats in a measure. The denominator indicates what kind of note receives 1 beat. For example, music written in $\frac{2}{4}$ time has 2 beats to a measure, and a quarter note receives 1 beat. One measure in $\frac{2}{4}$ time may have 1 half note, 2 quarter notes, 4 eighth notes, or any other combination of notes totaling 2 beats. Other common time signatures are $\frac{4}{4}$, $\frac{3}{4}$, and $\frac{6}{8}$.

1. Explain the meaning of the 6 and the 8 in the time signature $\frac{6}{8}$.

2. Give some possible combinations of notes in one measure of a piece written in $\frac{4}{4}$ time.

3. What does a dot at the right of a note indicate? What is the effect of a dot at the right of a half note? At the right of a quarter note? At the right of an eighth note?

4. Symbols called rests are used to indicate periods of silence in a piece of music. What symbols are used to indicate the different time values of rests?

5. Find some examples of musical compositions written in different time signatures. Use a few measures from each to show that the sum of the time values of the notes and rests in each measure equals the numerator of the time signature.

Construction   Suppose you are involved in building your own home. Design a stairway from the first floor of the house to the second floor. Some of the questions you will need to answer follow.

What is the distance from the floor of the first story to the floor of the second story?

Typically, what is the number of steps in a stairway?

What is a reasonable length for the run of each step?

What is the width of the wood being used to build the staircase?

In designing the stairway, remember that each riser should be the same height, that each run should be the same length, and that the width of the wood used for the steps will have to be incorporated into the calculation.

Answers to Projects and Group Activities— Music

1. There are 6 beats to a measure, and an eighth note receives 1 beat.

2. Answers will vary. Some examples include:
   1 whole note, 2 half notes, 4 quarter notes, 8 eighth notes, 1 half note and 2 quarter notes, 1 half note and 4 eighth notes, and 2 quarter notes and 4 eighth notes.

3. A dot at the right of a note increases its duration by one-half.
   A dotted half note equals a half note plus a quarter note.
   A dotted quarter note equals a quarter note plus an eighth note.
   A dotted eighth note equals an eighth note plus a sixteenth note.

4. Symbols used to indicate the different time values of rests are shown at the left below.

5. Examples will vary.

Answers to Projects and Group Activities— Construction

Designs will vary.

Symbols for Rests

Fractions of Diagrams

The diagram that follows has been broken up into nine areas separated by heavy lines. Eight of the areas have been labeled *A* through *H*. The ninth area is shaded. Determine which lettered areas would have to be shaded so that half of the entire diagram is shaded and half is not shaded. Write down the strategy that you or your group use to arrive at the solution. Compare your strategy with that of other individual students or groups.

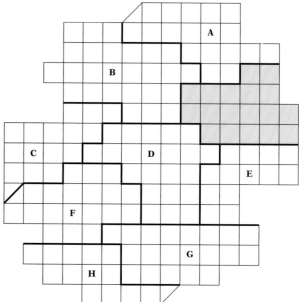

Study Tip

Three important features of this text that can be used to prepare for a test are the
• Chapter Summary
• Chapter Review Exercises
• Chapter Test
See *AIM for Success* at the front of the book.

Chapter 2 Summary

Key Words	Examples
A number that is a multiple of two or more numbers is a *common multiple* of those numbers. The *least common multiple (LCM)* is the smallest common multiple of two or more numbers. [2.1A, p. 65]	12, 24, 36, 48, . . . are common multiples of 4 and 6. The LCM of 4 and 6 is 12.
A number that is a factor of two or more numbers is a *common factor* of those numbers. The *greatest common factor (GCF)* is the largest common factor of two or more numbers. [2.1B, p. 66]	The common factors of 12 and 16 are 1, 2, and 4. The GCF of 12 and 16 is 4.
A *fraction* can represent the number of equal parts of a whole. In a fraction, the *fraction bar* separates the *numerator* and the *denominator*. [2.2A, p. 69]	In the fraction $\frac{3}{4}$, the numerator is 3 and the denominator is 4.

In a *proper fraction*, the numerator is smaller than the denominator; a proper fraction is a number less than 1. In an *improper fraction*, the numerator is greater than or equal to the denominator; an improper fraction is a number greater than or equal to 1. A *mixed number* is a number greater than 1 with a whole-number part and a fractional part. [2.2A, p. 69]

$\frac{2}{5}$ is proper fraction.

$\frac{7}{6}$ is an improper fraction.

$4\frac{1}{10}$ is a mixed number; 4 is the whole-number part and $\frac{1}{10}$ is the fractional part.

Equal fractions with different denominators are called *equivalent fractions*. [2.3A, p. 73]

$\frac{3}{4}$ and $\frac{6}{8}$ are equivalent fractions.

A fraction is in *simplest form* when the numerator and denominator have no common factors other than 1. [2.3B, p. 74]

The fraction $\frac{11}{12}$ is in simplest form.

The *reciprocal* of a fraction is the fraction with the numerator and denominator interchanged. [2.7A, p. 101]

The reciprocal of $\frac{3}{8}$ is $\frac{8}{3}$.

The reciprocal of 5 is $\frac{1}{5}$.

Essential Rules and Procedures

Examples

To find the LCM of two or more numbers, find the prime factorization of each number and write the factorization of each number in a table. Circle the greatest product in each column. The LCM is the product of the circled numbers. [2.1A, p. 65]

	2	3
12 =	⃝2·2	3
18 =	2	⃝3·3

The LCM of 12 and 18 is $2 \cdot 2 \cdot 3 \cdot 3 = 36$.

To find the GCF of two or more numbers, find the prime factorization of each number and write the factorization of each number in a table. Circle the least product in each column that does not have a blank. The GCF is the product of the circled numbers. [2.1B, p. 66]

	2	3
12 =	2·2	③
18 =	②	3·3

The GCF of 12 and 18 is $2 \cdot 3 = 6$.

To write an improper fraction as a mixed number or a whole number, divide the numerator by the denominator. [2.2B, p. 70]

$\frac{29}{6} = 29 \div 6 = 4\frac{5}{6}$

To write a mixed number as an improper fraction, multiply the denominator of the fractional part of the mixed number by the whole-number part. Add this product and the numerator of the fractional part. The sum is the numerator of the improper fraction. The denominator remains the same. [2.2B, p. 70]

$3\frac{2}{5} = \frac{5 \times 3 + 2}{5} = \frac{17}{5}$

To find equivalent fractions by raising to higher terms, multiply the numerator and denominator of the fraction by the same number. [2.3A, p. 73]

$\frac{3}{4} = \frac{3 \cdot 5}{4 \cdot 5} = \frac{15}{20}$

$\frac{3}{4}$ and $\frac{15}{20}$ are equivalent fractions.

To write a fraction in simplest form, factor the numerator and denominator of the fraction; then eliminate the common factors. [2.3B, p. 74]

$\frac{30}{45} = \frac{2 \cdot \overset{1}{\cancel{3}} \cdot \overset{1}{\cancel{5}}}{\underset{1}{\cancel{3}} \cdot 3 \cdot \underset{1}{\cancel{5}}} = \frac{2}{3}$

To add fractions with the same denominator, add the numerators and place the sum over the common denominator. [2.4A, p. 77]	$\dfrac{5}{12} + \dfrac{11}{12} = \dfrac{16}{12} = 1\dfrac{4}{12} = 1\dfrac{1}{3}$
To add fractions with different denominators, first rewrite the fractions as equivalent fractions with a common denominator. (The common denominator is the LCM of the denominators of the fractions.) Then add the fractions. [2.4B, p. 77]	$\dfrac{1}{4} + \dfrac{2}{5} = \dfrac{5}{20} + \dfrac{8}{20} = \dfrac{13}{20}$
To subtract fractions with the same denominator, subtract the numerators and place the difference over the common denominator. [2.5A, p. 85]	$\dfrac{9}{16} - \dfrac{5}{16} = \dfrac{4}{16} = \dfrac{1}{4}$
To subtract fractions with different denominators, first rewrite the fractions as equivalent fractions with a common denominator. (The common denominator is the LCM of the denominators of the fractions.) Then subtract the fractions. [2.5B, p. 85]	$\dfrac{2}{3} - \dfrac{7}{16} = \dfrac{32}{48} - \dfrac{21}{48} = \dfrac{11}{48}$
To multiply two fractions, multiply the numerators; this is the numerator of the product. Multiply the denominators; this is the denominator of the product. [2.6A, p. 93]	$\dfrac{3}{4} \cdot \dfrac{2}{9} = \dfrac{3 \cdot 2}{4 \cdot 9} = \dfrac{\overset{1}{\cancel{3}} \cdot \overset{1}{\cancel{2}}}{\underset{1}{\cancel{2}} \cdot 2 \cdot \underset{1}{\cancel{3}} \cdot 3} = \dfrac{1}{6}$
To divide two fractions, multiply the first fraction by the reciprocal of the second fraction. [2.7A, p. 101]	$\dfrac{8}{15} \div \dfrac{4}{5} = \dfrac{8}{15} \cdot \dfrac{5}{4} = \dfrac{8 \cdot 5}{15 \cdot 4}$ $= \dfrac{\overset{1}{\cancel{2}} \cdot \overset{1}{\cancel{2}} \cdot 2 \cdot \overset{1}{\cancel{5}}}{3 \cdot \underset{1}{\cancel{5}} \cdot \underset{1}{\cancel{2}} \cdot \underset{1}{\cancel{2}}} = \dfrac{2}{3}$
The find the order relation between two fractions with the same denominator, compare the numerators. The fraction that has the smaller numerator is the smaller fraction. [2.8A, p. 110]	$\dfrac{17}{25} \leftarrow$ Smaller numerator $\dfrac{19}{25} \leftarrow$ Larger numerator $\dfrac{17}{25} < \dfrac{19}{25}$
To find the order relation between two fractions with different denominators, first rewrite the fractions with a common denominator. The fraction that has the smaller numerator is the smaller fraction. [2.8A, p. 110]	$\dfrac{3}{5} = \dfrac{24}{40} \qquad \dfrac{5}{8} = \dfrac{25}{40}$ $\dfrac{24}{40} < \dfrac{25}{40}$ $\dfrac{3}{5} < \dfrac{5}{8}$

Order of Operations Agreement [2.8C, p. 111]

Step 1 Do all the operations inside parentheses.	$\left(\dfrac{1}{3}\right)^2 + \left(\dfrac{5}{6} - \dfrac{7}{12}\right) \cdot (4)$
Step 2 Simplify any numerical expressions containing exponents.	$= \left(\dfrac{1}{3}\right)^2 + \left(\dfrac{1}{4}\right) \cdot (4)$
Step 3 Do multiplication and division as they occur from left to right.	$= \dfrac{1}{9} + \left(\dfrac{1}{4}\right) \cdot (4)$
Step 4 Do addition and subtraction as they occur from left to right.	$= \dfrac{1}{9} + 1 = 1\dfrac{1}{9}$

Chapter 2 Review Exercises

1. Write $\frac{30}{45}$ in simplest form.

$\frac{2}{3}$ [2.3B]

2. Simplify: $\left(\frac{3}{4}\right)^3 \cdot \frac{20}{27}$

$\frac{5}{16}$ [2.8B]

3. Express the shaded portion of the circles as an improper fraction.

  $\frac{13}{4}$ [2.2A]

4. Find the total of $\frac{2}{3}$, $\frac{5}{6}$, and $\frac{2}{9}$.

$1\frac{13}{18}$ [2.4B]

5. Place the correct symbol, $<$ or $>$, between the two numbers.

$\frac{11}{18} < \frac{17}{24}$ [2.8A]

6. Subtract: $18\frac{1}{6}$

$-\ 3\frac{5}{7}$

$14\frac{19}{42}$ [2.5C]

7. Simplify: $\frac{2}{7}\left(\frac{5}{8} - \frac{1}{3}\right) \div \frac{3}{5}$

$\frac{5}{36}$ [2.8C]

8. Multiply: $2\frac{1}{3} \times 3\frac{7}{8}$

$9\frac{1}{24}$ [2.6B]

9. Divide: $1\frac{1}{3} \div \frac{2}{3}$

2 [2.7B]

10. Find $\frac{17}{24}$ decreased by $\frac{3}{16}$.

$\frac{25}{48}$ [2.5B]

11. Divide: $8\frac{2}{3} \div 2\frac{3}{5}$

$3\frac{1}{3}$ [2.7B]

12. Find the GCF of 20 and 48.

4 [2.1B]

13. Write an equivalent fraction with the given denominator.

$\frac{2}{3} = \frac{24}{36}$ [2.3A]

14. What is $\frac{15}{28}$ divided by $\frac{5}{7}$?

$\frac{3}{4}$ [2.7A]

15. Write an equivalent fraction with the given denominator.

$\frac{8}{11} = \frac{32}{44}$ [2.3A]

16. Multiply: $2\frac{1}{4} \times 7\frac{1}{3}$

$16\frac{1}{2}$ [2.6B]

17. Find the LCM of 18 and 12.

36 [2.1A]

18. Write $\frac{16}{44}$ in simplest form.

$\frac{4}{11}$ [2.3B]

19. Add: $\frac{3}{8} + \frac{5}{8} + \frac{1}{8}$

$1\frac{1}{8}$ [2.4A]

20. Subtract: $\begin{array}{r} 16 \\ -\ 5\frac{7}{8} \\ \hline \end{array}$

$10\frac{1}{8}$ [2.5C]

21. Add: $4\frac{4}{9} + 2\frac{1}{6} + 11\frac{17}{27}$

$18\frac{13}{54}$ [2.4C]

22. Find the GCF of 15 and 25.

5   [2.1B]

23. Write $\frac{17}{5}$ as a mixed number.

$3\frac{2}{5}$ [2.2B]

24. Simplify: $\left(\frac{4}{5} - \frac{2}{3}\right)^2 \div \frac{4}{15}$

$\frac{1}{15}$ [2.8C]

25. Add: $\frac{3}{8} + 1\frac{2}{3} + 3\frac{5}{6}$

$5\frac{7}{8}$ [2.4C]

26. Find the LCM of 18 and 27.

54   [2.1A]

27. Subtract: $\frac{11}{18} - \frac{5}{18}$

$\frac{1}{3}$ [2.5A]

28. Write $2\frac{5}{7}$ as an improper fraction.

$\frac{19}{7}$ [2.2B]

29. Divide: $\frac{5}{6} \div \frac{5}{12}$

2   [2.7A]

30. Multiply: $\frac{5}{12} \times \frac{4}{25}$

$\frac{1}{15}$ [2.6A]

31. What is $\frac{11}{50}$ multiplied by $\frac{25}{44}$?

$\frac{1}{8}$ [2.6A]

32. Express the shaded portion of the circles as a mixed number.

$1\frac{7}{8}$ [2.2A]

33. **Meteorology** During 3 months of the rainy season, $5\frac{7}{8}$, $6\frac{2}{3}$, and $8\frac{3}{4}$ inches of rain fell. Find the total rainfall for the 3 months. $21\frac{7}{24}$ inches   [2.4D]

34. **Real Estate** A home building contractor bought $4\frac{2}{3}$ acres for $168,000. What was the cost of each acre? $36,000   [2.7C]

35. **Sports** A 15-mile race has three checkpoints. The first checkpoint is $4\frac{1}{2}$ miles from the starting point. The second checkpoint is $5\frac{3}{4}$ miles from the first checkpoint. How many miles is the second checkpoint from the finish line? $4\frac{3}{4}$ miles   [2.5D]

36. **Fuel Efficiency** A compact car gets 36 miles on each gallon of gasoline. How many miles can the car travel on $6\frac{3}{4}$ gallons of gasoline?   243 miles   [2.6C]

Chapter 2 Test

1. Multiply: $\dfrac{9}{11} \times \dfrac{44}{81}$

$\dfrac{4}{9}$ [2.6A]

2. Find the GCF of 24 and 80.

8 [2.1B]

3. Divide: $\dfrac{5}{9} \div \dfrac{7}{18}$

$1\dfrac{3}{7}$ [2.7A]

4. Simplify: $\left(\dfrac{3}{4}\right)^2 \div \left(\dfrac{2}{3} + \dfrac{5}{6}\right) - \dfrac{1}{12}$

$\dfrac{7}{24}$ [2.8C]

5. Write $9\dfrac{4}{5}$ as an improper fraction.

$\dfrac{49}{5}$ [2.2B]

6. What is $5\dfrac{2}{3}$ multiplied by $1\dfrac{7}{17}$?

8 [2.6B]

7. Write $\dfrac{40}{64}$ in simplest form.

$\dfrac{5}{8}$ [2.3B]

8. Place the correct symbol, $<$ or $>$, between the two numbers.

$\dfrac{3}{8} < \dfrac{5}{12}$ [2.8A]

9. Simplify: $\left(\dfrac{1}{4}\right)^3 \div \left(\dfrac{1}{8}\right)^2 - \dfrac{1}{6}$

$\dfrac{5}{6}$ [2.8C]

10. Find the LCM of 24 and 40.

120 [2.1A]

11. Subtract: $\dfrac{17}{24} - \dfrac{11}{24}$

$\dfrac{1}{4}$ [2.5A]

12. Write $\dfrac{18}{5}$ as a mixed number.

$3\dfrac{3}{5}$ [2.2B]

13. Find the quotient of $6\dfrac{2}{3}$ and $3\dfrac{1}{6}$.

$2\dfrac{2}{19}$ [2.7B]

14. Write an equivalent fraction with the given denominator.

$\dfrac{5}{8} = \dfrac{45}{72}$ [2.3A]

15. Add: $\dfrac{5}{6}$

$\dfrac{7}{9}$

$+ \dfrac{1}{15}$

$1\dfrac{61}{90}$ [2.4B]

16. Subtract: $23\dfrac{1}{8}$

$- \ 9\dfrac{9}{44}$

$13\dfrac{81}{88}$ [2.5C]

17. What is $\dfrac{9}{16}$ minus $\dfrac{5}{12}$?

$\dfrac{7}{48}$ [2.5B]

18. Simplify: $\left(\dfrac{2}{3}\right)^4 \cdot \dfrac{27}{32}$

$\dfrac{1}{6}$ [2.8B]

19. Add: $\dfrac{7}{12} + \dfrac{11}{12} + \dfrac{5}{12}$

$1\dfrac{11}{12}$ [2.4A]

20. What is $12\dfrac{5}{12}$ more than $9\dfrac{17}{20}$?

$22\dfrac{4}{15}$ [2.4C]

21. Express the shaded portion of the circles as an improper fraction.

$\dfrac{11}{4}$ [2.2A]

22. **Compensation** An electrician earns $240 for each day worked. What is the total of the electrician's earnings for working $3\dfrac{1}{2}$ days? $840   [2.6C]

23. **Real Estate** Grant Miura bought $7\dfrac{1}{4}$ acres of land for a housing project. One and three-fourths acres were set aside for a park, and the remaining land was developed into $\dfrac{1}{2}$-acre lots. How many lots were available for sale? 11 lots   [2.7C]

24. **Real Estate** A land developer purchases $25\dfrac{1}{2}$ acres of land and plans to set aside 3 acres for an entranceway to a housing development to be built on the property. Each house will be built on a $\dfrac{3}{4}$-acre plot of land. How many houses does the developer plan to build on the property? 30 houses   [2.7C]

25. **Meterology** In 3 successive months, the rainfall measured $11\dfrac{1}{2}$ inches, $7\dfrac{5}{8}$ inches, and $2\dfrac{1}{3}$ inches. Find the total rainfall for the 3 months.

$21\dfrac{11}{24}$ inches   [2.4D]

Cumulative Review Exercises

1. Round 290,496 to the nearest thousand.
290,000   [1.1D]

2. Subtract: 390,047
$$-\ 98,769$$
291,278   [1.3B]

3. Find the product of 926 and 79.
73,154   [1.4B]

4. Divide: $57\overline{)30,792}$
540 r12   [1.5C]

5. Simplify: $4 \cdot (6 - 3) \div 6 - 1$
1   [1.6B]

6. Find the prime factorization of 44.
$2 \cdot 2 \cdot 11$ [1.7B]

7. Find the LCM of 30 and 42.
210   [2.1A]

8. Find the GCF of 60 and 80.
20   [2.1B]

9. Write $7\frac{2}{3}$ as an improper fraction.
$\frac{23}{3}$ [2.2B]

10. Write $\frac{25}{4}$ as a mixed number.
$6\frac{1}{4}$ [2.2B]

11. Write an equivalent fraction with the given denominator.
$\frac{5}{16} = \frac{15}{48}$ [2.3A]

12. Write $\frac{24}{60}$ in simplest form.
$\frac{2}{5}$ [2.3B]

13. What is $\frac{9}{16}$ more than $\frac{7}{12}$?
$1\frac{7}{48}$ [2.4B]

14. Add: $3\frac{7}{8}$
$7\frac{5}{12}$
$+\ 2\frac{15}{16}$
$14\frac{11}{48}$ [2.4C]

15. Find $\frac{3}{8}$ less than $\frac{11}{12}$.
$\frac{13}{24}$ [2.5B]

16. Subtract: $5\frac{1}{6}$
$-\ 3\frac{7}{18}$
$1\frac{7}{9}$ [2.5C]

17. Multiply: $\frac{3}{8} \times \frac{14}{15}$

$\frac{7}{20}$ [2.6A]

18. Multiply: $3\frac{1}{8} \times 2\frac{2}{5}$

$7\frac{1}{2}$ [2.6B]

19. Divide: $\frac{7}{16} \div \frac{5}{12}$

$1\frac{1}{20}$ [2.7A]

20. Find the quotient of $6\frac{1}{8}$ and $2\frac{1}{3}$.

$2\frac{5}{8}$ [2.7B]

21. Simplify: $\left(\frac{1}{2}\right)^3 \cdot \frac{8}{9}$

$\frac{1}{9}$ [2.8B]

22. Simplify: $\left(\frac{1}{2} + \frac{1}{3}\right) \div \left(\frac{2}{5}\right)^2$

$5\frac{5}{24}$ [2.8C]

23. **Banking**   Molly O'Brien had \$1359 in a checking account. During the week, Molly wrote checks for \$128, \$54, and \$315. Find the amount in the checking account at the end of the week.   \$862   [1.3C]

24. **Entertainment**   The tickets for a movie were \$10 for an adult and \$4 for a student. Find the total income from the sale of 87 adult tickets and 135 student tickets.   \$1410   [1.4C]

25. **Measurement**   Find the total weight of three packages that weigh $1\frac{1}{2}$ pounds, $7\frac{7}{8}$ pounds, and $2\frac{2}{3}$ pounds. $12\frac{1}{24}$ pounds   [2.4D]

26. **Carpentry**   A board $2\frac{5}{8}$ feet long is cut from a board $7\frac{1}{3}$ feet long. What is the length of the remaining piece? $4\frac{17}{24}$ feet   [2.5D]

27. **Fuel Efficiency**   A car travels 27 miles on each gallon of gasoline. How many miles can the car travel on $8\frac{1}{3}$ gallons of gasoline?   225 miles   [2.6C]

28. **Real Estate**   Jimmy Santos purchased $10\frac{1}{3}$ acres of land to build a housing development. Jimmy donated 2 acres for a park. How many $\frac{1}{3}$-acre parcels can be sold from the remaining land?   25 parcels   [2.7C]

This woman is the owner and operator of a wholesale groceries store. As such, she is a self-employed person. Anyone who operates a trade, business, or profession is self-employed. There are over 10 million self-employed people in the United States. Among them are authors, musicians, computer technicians, contractors, landscape designers, lawyers, psychologists, sales agents, and dentists. The table associated with **Exercises 28 to 30 on page 136** lists the numbers of self-employed persons in the United States by annual earnings, from less than $5000 per year to $50,000 or more.

OBJECTIVES

Section 3.1
A To write decimals in standard form and in words
B To round a decimal to a given place value

Section 3.2
A To add decimals
B To solve application problems

Section 3.3
A To subtract decimals
B To solve application problems

Section 3.4
A To multiply decimals
B To solve application problems

Section 3.5
A To divide decimals
B To solve application problems

Section 3.6
A To convert fractions to decimals
B To convert decimals to fractions
C To identify the order relation between two decimals or between a decimal and a fraction

Need help? For online student resources, such as section quizzes, visit this textbook's website at **math.college.hmco.com/students**.

PREP TEST ● ● ●

Do these exercises to prepare for Chapter 3.

1. Express the shaded portion of the rectangle as a fraction.

$\frac{3}{10}$ [2.2A]

2. Round 36,852 to the nearest hundred.
36,900 [1.1D]

3. Write 4791 in words.
Four thousand seven hundred ninety-one [1.1B]

4. Write six thousand eight hundred forty-two in standard form.
6842 [1.1B]

For Exercises 5 to 8, add, subtract, multiply, or divide.

5. 37 + 8892 + 465
9394 [1.2A]

6. 2403 − 765
1638 [1.3B]

7. 844 × 91
76,804 [1.4B]

8. 23)6412
278 r18 [1.5C]

GO FIGURE ● ● ●

Maria and Pedro are siblings. Pedro has as many brothers as sisters. Maria has twice as many brothers as sisters. How many children are in the family?
7 children

3.1 Introduction to Decimals

Objective A To write decimals in standard form and in words

The price tag on a sweater reads $61.88. The number 61.88 is in **decimal notation.** A number written in decimal notation is often called simply a **decimal.**

A number written in decimal notation has three parts.

61	.	88
Whole-number part	**Decimal point**	**Decimal part**

The decimal part of the number represents a number less than 1. For example, $.88 is less than $1. The decimal point (.) separates the whole-number part from the decimal part.

The position of a digit in a decimal determines the digit's place value. The place-value chart is extended to the right to show the place value of digits to the right of a decimal point.

In the decimal 458.302719, the position of the digit 7 determines that its place value is ten-thousandths.

Note the relationship between fractions and numbers written in decimal notation.

Seven tenths	Seven hundredths	Seven thousandths
$\frac{7}{10} = 0.7$	$\frac{7}{100} = 0.07$	$\frac{7}{1000} = 0.007$
1 zero in 10	2 zeros in 100	3 zeros in 1000
1 decimal place in 0.7	2 decimal places in 0.07	3 decimal places in 0.007

To write a decimal in words, write the decimal part of the number as though it were a whole number, and then name the place value of the last digit.

0.9684   Nine thousand six hundred eighty-four ten-thousandths

The decimal point in a decimal is read as "and."

372.516   Three hundred seventy-two and five hundred sixteen thousandths

Point of Interest

The decimal point did not make its appearance until the early 1600s. Stevin's notation used subscripts with circles around them after each digit: 0 for ones, 1 for tenths (which he called "primes"), 2 for hundredths (called "seconds"), 3 for thousandths ("thirds"), and so on. For example, 1.375 would have been written

To write a decimal in standard form when it is written in words, write the whole-number part, replace the word *and* with a decimal point, and write the decimal part so that the last digit is in the given place-value position.

Four and twenty-three <u>hundredths</u>

3 is in the hundredths place.   4.2<u>3</u>

When writing a decimal in standard form, you may need to insert zeros after the decimal point so that the last digit is in the given place-value position.

Ninety-one and eight <u>thousandths</u>

8 is in the thousandths place.
Insert two zeros so that the 8 is in the thousandths place.   91.00<u>8</u>

Sixty-five <u>ten-thousandths</u>

5 is in the ten-thousandths place.
Insert two zeros so that the 5 is in the ten-thousandths place.   0.006<u>5</u>

Example 1 Name the place value of the digit 8 in the number 45.687.

Solution The digit 8 is in the hundredths place.

You Try It 1 Name the place value of the digit 4 in the number 907.1342.

Your solution Thousandths

Example 2 Write $\frac{43}{100}$ as a decimal.

Solution $\frac{43}{100} = 0.43$ • Forty-three hundredths

You Try It 2 Write $\frac{501}{1000}$ as a decimal.

Your solution 0.501

Example 3 Write 0.289 as a fraction.

Solution $0.289 = \frac{289}{1000}$ • 289 thousandths

You Try It 3 Write 0.67 as a fraction.

Your solution $\frac{67}{100}$

Example 4 Write 293.50816 in words.

Solution Two hundred ninety-three and fifty thousand eight hundred sixteen hundred-thousandths

You Try It 4 Write 55.6083 in words.

Your solution Fifty-five and six thousand eighty-three ten-thousandths

Solutions on p. S8

Example 5	Write twenty-three and two hundred forty-seven millionths in standard form.	**You Try It 5**	Write eight hundred six and four hundred ninety-one hundred-thousandths in standard form.
Solution	23.000247 • **7 is in the millionths place.**	**Your solution**	806.00491

Solution on p. S8

Objective B   To round a decimal to a given place value

Study Tip

Have you considered joining a study group? Getting together regularly with other students in the class to go over material and quiz each other can be very beneficial. See *AIM for Success* at the front of the book.

In general, rounding decimals is similar to rounding whole numbers except that the digits to the right of the given place value are dropped instead of being replaced by zeros.

If the digit to the right of the given place value is less than 5, that digit and all digits to the right are dropped.

Round 6.9237 to the nearest hundredth.

Given place value (hundredths)

6.9237

3 < 5   Drop the digits 3 and 7.

6.9237 rounded to the nearest hundredth is 6.92.

If the digit to the right of the given place value is greater or equal to 5, increase the digit in the given place value by 1, and drop all digits to its right.

Round 12.385 to the nearest tenth.

Given place value (tenths)

12.385

8 > 5   Increase 3 by 1 and drop all digits to the right of 3.

12.385 rounded to the nearest tenth is 12.4.

TAKE NOTE

In the example at the right, the zero in the given place value is not dropped. This indicates that the number is rounded to the nearest thousandth. If we dropped the zero and wrote 0.47, it would indicate that the number was rounded to the nearest hundredth.

HOW TO   Round 0.46972 to the nearest thousandth.

Given place value (thousandths)

0.46972

7 > 5   Round up by adding 1 to the 9 (9 + 1 = 10). Carry the 1 to the hundredths place (6 + 1 = 7).

0.46972 rounded to the nearest thousandth is 0.470.

In-Class Examples (Objective 3.1B)

Round the decimal to the given place value.

1. 0.074 Tenths   0.1

2. 840.156 Hundredths   840.16

3. 5.60032 Nearest whole number   6

4. 0.635457 Hundred-thousandths   0.63546

5. The length of the marathon footrace in the Olympics is 42.195 kilometers. What is the length of this race to the nearest tenth of a kilometer?

42.2 kilometers

Objective 3.1B

Vocabulary to Review

rounding   [1.1D]

Discuss the Concepts

Explain how to round a decimal to the nearest hundredth.

Instructor Note

As a calculator activity, have students determine whether their calculators round or truncate. Using 2 ÷ 3 will serve as a good example.

Instructor Note

Explain to students that not all rounding is done as shown here. When sales tax is computed, the decimal is always rounded up to the nearest cent. Thus a sales tax of $.132 would be $.14.

Optional Student Activity

Bring some newspapers to class. Arrange the students in small groups. Have students find and list situations in which decimals are used. Also ask them to determine whether the decimals they find are exact or approximations. For example, large numbers (such as 3.2 billion) used for describing the balance of trade or the national debt, are approximations, whereas numbers used to describe business transactions, such as the exchange rate or stock dividends, are exact.

Example 6   Round 0.9375 to the nearest thousandth.

Solution

        ┌──── Given place value

   0.9375

       └── 5 = 5

0.9375 rounded to the nearest thousandth is 0.938.

You Try It 6   Round 3.675849 to the nearest ten-thousandth.

Your solution   3.6758

Example 7   Round 2.5963 to the nearest hundredth.

Solution

        ┌──── Given place value

   2.5963

       └── 6 > 5

2.5963 rounded to the nearest hundredth is 2.60.

You Try It 7   Round 48.907 to the nearest tenth.

Your solution   48.9

Example 8   Round 72.416 to the nearest whole number.

Solution

        ┌──── Given place value

   72.416

       └── 4 < 5

72.416 rounded to the nearest whole number is 72.

You Try It 8   Round 31.8652 to the nearest whole number.

Your solution   32

Example 9

On average, an American goes to the movies 4.56 times per year. To the nearest whole number, how many times per year does an American go to the movies?

Solution

4.56 rounded to the nearest whole number is 5. An American goes to the movies about 5 times per year.

You Try It 9

One of the driest cities in the Southwest is Yuma, Arizona, with an average annual precipitation of 2.65 inch. To the nearest inch, what is the average annual precipitation in Yuma?

Your solution

3 inches

Solutions on pp. S8–S9

3.1 Exercises

Objective A   **To write decimals in standard form and in words**

For Exercises 1 to 6, name the place value of the digit 5.

1. 76.31587
Thousandths

2. 291.508
Tenths

3. 432.09157
Ten-thousandths

4. 0.0006512
Hundred-thousandths

5. 38.2591
Hundredths

6. 0.0000853
Millionths

For Exercises 7 to 14, write the fraction as a decimal.

7. $\frac{3}{10}$
0.3

8. $\frac{9}{10}$
0.9

9. $\frac{21}{100}$
0.21

10. $\frac{87}{100}$
0.87

11. $\frac{461}{1000}$
0.461

12. $\frac{853}{1000}$
0.853

13. $\frac{93}{1000}$
0.093

14. $\frac{61}{1000}$
0.061

For Exercises 15 to 22, write the decimal as a fraction.

15. 0.1
$\frac{1}{10}$

16. 0.3
$\frac{3}{10}$

17. 0.47
$\frac{47}{100}$

18. 0.59
$\frac{59}{100}$

19. 0.289
$\frac{289}{1000}$

20. 0.601
$\frac{601}{1000}$

21. 0.09
$\frac{9}{100}$

22. 0.013
$\frac{13}{1000}$

For Exercises 23 to 31, write the number in words.

23. 0.37
Thirty-seven hundredths

24. 25.6
Twenty-five and six tenths

25. 9.4
Nine and four tenths

26. 1.004
One and four thousandths

27. 0.0053
Fifty-three ten-thousandths

28. 41.108
Forty-one and one hundred
eight thousandths

29. 0.045
Forty-five thousandths

30. 3.157
Three and one hundred
fifty-seven thousandths

31. 26.04
Twenty-six and
four hundredths

For Exercises 32 to 39, write the number in standard form.

32. Six hundred seventy-two thousandths
0.672

33. Three and eight hundred
six ten-thousandths
3.0806

34. Nine and four hundred seven
ten-thousandths
9.0407

35. Four hundred seven and three hundredths
407.03

Suggested Assignment
Exercises 1–57, odds
More challenging problems:
    Exercises 58, 59

Quick Quiz (Objective 3.1A)

1. Write $\frac{9}{1000}$ as a decimal.   0.009

Write the decimal in words.
2. 2.379
Two and three hundred seventy-nine thousandths
3. 0.00043   Forty-three hundred-thousandths

Write the decimal in standard form.
4. Nine and seven hundred four thousandths   9.704
5. Five and seventeen ten-thousandths   5.0017

36. Six hundred twelve and seven hundred four thousandths
612.704

37. Two hundred forty-six and twenty-four thousandths
246.024

38. Two thousand sixty-seven and nine thousand two ten-thousandths
2067.9002

39. Seventy-three and two thousand six hundred eighty-four hundred-thousandths
73.02684

> **Objective B**   **To round a decimal to a given place value**

For Exercises 40 to 54, round the number to the given place value.

40. 6.249   Tenths
6.2

41. 5.398   Tenths
5.4

42. 21.007   Tenths
21.0

43. 30.0092   Tenths
30.0

44. 18.40937   Hundredths
18.41

45. 413.5972   Hundredths
413.60

46. 72.4983   Hundredths
72.50

47. 6.061745   Thousandths
6.062

48. 936.2905   Thousandths
936.291

49. 96.8027   Whole number
97

50. 47.3192   Whole number
47

51. 5439.83   Whole number
5440

52. 7014.96   Whole number
7015

53. 0.023591   Ten-thousandths
0.0236

54. 2.975268   Hundred-thousandths
2.97527

55. **Measurement**   A nickel weighs about 0.1763668 ounce. Find the weight of a nickel to the nearest hundredth of an ounce.
0.18 ounce

56. **Business**   The total cost of a parka, including sales tax, is $124.1093. Round the total cost to the nearest cent to find the amount a customer pays for the parka.
$124.11

57. **Sports**   Runners in the Boston Marathon run a distance of 26.21875 miles. To the nearest tenth of a mile, find the distance that an entrant who completes the Boston Marathon runs.
26.2 miles

APPLYING THE CONCEPTS

58. Indicate which digits of the number, if any, need not be entered on a calculator.

a. 1.500   **b.** 0.908   **c.** 60.07   **d.** 0.0032
a. 1.5<u>00</u>   **b.** <u>0</u>.908   **c.** 60.07   **d.** <u>0</u>.0032

59. **a.** Find a number between 0.1 and 0.2.   **b.** Find a number between 1 and 1.1   **c.** Find a number between 0 and 0.005.
For example, **a.** 0.15   **b.** 1.05   **c.** 0.001

Quick Quiz (Objective 3.1B)
Round the decimal to the given place value.
1. 9.1384   Tenths   9.1
2. 512.677   Hundredths   512.68
3. 7.880102   Nearest whole number   8
4. 3.4978205   Hundred-thousandths   3.49782

3.2 Addition of Decimals

Objective A To add decimals

To add decimals, write the numbers so that the decimal points are on a vertical line. Add as for whole numbers, and write the decimal point in the sum directly below the decimal points in the addends.

HOW TO Add: 0.237 + 4.9 + 27.32

Note that by placing the decimal points on a vertical line, we make sure that digits of the same place value are added.

Example 1 Find the sum of 42.3, 162.903, and 65.0729.

Solution

```
      1 1 1
      42.3          • Place the decimal
     162.903          points on a vertical
  +   65.0729         line.
     270.2759
```

You Try It 1 Find the sum of 4.62, 27.9, and 0.62054.

Your solution 33.14054

Example 2 Add: 0.83 + 7.942 + 15

Solution

```
     1 1
     0.83
     7.942
  + 15.
    23.772
```

You Try It 2 Add: 6.05 + 12 + 0.374

Your solution 18.424

Solutions on p. S9

● E S T I M A T I O N ●

Estimating the Sum of Two or More Decimals

Calculate 23.037 + 16.7892. Then use estimation to determine whether the sum is reasonable.

Add to find the exact sum.   23.037 + 16.7892 = 39.8262

To estimate the sum, round each number to the same place value. Here we have rounded to the nearest whole number. Then add. The estimated answer is 40, which is very close to the exact sum, 39.8262.

```
   23.037  ≈    23
 +16.7892  ≈  +17
                40
```

Objective 3.2A

Vocabulary to Review
addend   [1.2A]
sum   [1.2A]

Discuss the Concepts
When we add decimals using a vertical format, why must the decimal points be aligned?

Concept Check
Place plus signs between three of the numbers below so that the sum is 19.24.

0.12
3.45
6.78
9.01
2.34

3.45 + 6.78 + 9.01 = 19.24

Instructor Note
You might use Example 1 to show your students that you can use zeros for placeholders by writing 42.3000 and 162.9030.

Optional Student Activity
Complete the following addition table. How is the Commutative Property of Addition represented in the table?

+	0.1	0.2	0.3	0.4	0.5
0.1					
0.2					
0.3					
0.4					
0.5					

Row 1: 0.2, 0.3, 0.4, 0.5, 0.6; Row 2: 0.3, 0.4, 0.5, 0.6, 0.7; Row 3: 0.4, 0.5, 0.6, 0.7, 0.8; Row 4: 0.5, 0.6, 0.7, 0.8, 0.9; Row 5: 0.6, 0.7, 0.8, 0.9, 1.0. Answers will vary. For example, 0.4 + 0.3 = 0.3 + 0.4.

In-Class Examples (Objective 3.2A)
Add.
1. 3.514 + 22.6981 + 145.78   171.9921
2. 7.814 + 63.109 + 2 + 0.0099   72.9329

Objective 3.2B

Optional Student Activity

The sum of six different positive numbers is 11. Four of the six numbers are 4, 3, 2, and 1. Find the other two numbers.

This problem has a number of possible solutions. For example, the answer could be 0.2 and 0.8, 0.7 and 0.3, or 0.6 and 0.4.

Objective B   **To solve application problems**

 The graph at the right shows the breakdown by age group of Americans who are hearing-impaired. Use this graph for Example 3 and You Try It 3.

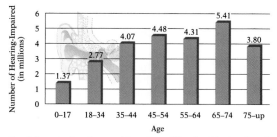

Breakdown by Age Group of Americans Who Are Hearing-Impaired

Source: American Speech-Language-Hearing Association   P

Example 3

Determine the number of Americans under the age of 45 who are hearing-imparied.

Strategy
To determine the number, add the numbers of hearing impaired ages 0 to 17, 18 to 34, and 35 to 44.

Solution

$$
\begin{array}{r}
1.37 \\
2.77 \\
+4.07 \\
\hline
8.21
\end{array}
$$

8.21 million Americans under the age of 45 are hearing-impaired.

You Try It 3

Determine the number of Americans ages 45 and older who are hearing-impaired.

Your strategy

Your solution   18 million Americans

Example 4

Dan Burhoe earned a salary of $138.50 for working 3 days this week as a food server. He also received $42.75, $35.80, and $59.25 in tips during the 3 days. Find his total income for the 3 days of work.

Strategy
To find the total income, add the tips (42.75, 35.80, and 59.25) to the salary (138.50).

Solution
138.50 + 42.75 + 35.80 + 59.25 = 276.30

Dan's total income for the 3 days of work was $276.30.

You Try It 4

Anita Khavari, an insurance executive, earns a salary of $875 every 4 weeks. During the past 4-week period, she received commissions of $985.80, $791.46, $829.75, and $635.42. Find her total income for the past 4-week period.

Your strategy

Your solution $4117.43

Solutions on p. S9

In-Class Example (Objective 3.2B)

1. A salesperson's commission checks for six months are $1649.52, $2731.18, $1711.98, $675.49, $2406.37, and $1986.06. Find the total commission income for the six months. $11,160.60

3.2 Exercises

Objective A   **To add decimals**

For Exercises 1 to 17, add.

1. 16.008 + 2.0385 + 132.06
150.1065

2. 17.32 + 1.0579 + 16.5
34.8779

3. 1.792 + 67 + 27.0526
95.8446

4. 8.772 + 1.09 + 26.5027
36.3647

5. 3.02 + 62.7 + 3.924
69.644

6. 9.06 + 4.976 + 59.6
73.636

7. 82.006 + 9.95 + 0.927
92.883

8. 0.826 + 8.76 + 79.005
88.591

9. 4.307 + 99.82 + 9.078
113.205

10.   0.3
        + 0.07
        0.37

11.   0.29
        + 0.4
        0.69

12.   1.007
        + 2.1
        3.107

13.   7.3
        + 9.005
        16.305

14.   4.9257
        27.05
        + 9.0063
        40.9820

15.   8.72
        99.073
        + 2.9736
        110.7666

16.   62.4
        9.827
        + 692.44
        764.667

17.   8
        89.43
        + 7.0659
        104.4959

For Exercises 18 to 21, use a calculator to add. Then round the numbers to the nearest whole number and use estimation to determine whether the sum you calculated is reasonable.

18.   342.42
        89.625
        + 176.2
        Cal.: 608.245
        Est.: 608

19.   219.9
        0.872
        + 13.42
        Cal.: 234.192
        Est.: 234

20.   823.9
        82.65
        + 46.923
        Cal.: 953.473
        Est.: 954

21.   678.92
        97.6
        + 5.423
        Cal.: 781.943
        Est.: 782

Objective B   **To solve application problems**

22. **Finances**   A family has a mortgage payment of $814.72, a Visa bill of $216.40, and an electric bill of $87.32.
   a. Calculate the exact amount of the three payments.
   b. Round the numbers to the nearest hundred and use estimation to determine whether the sum you calculated is reasonable.
      a. $1118.44   **b.** $1100

23. **Mechanics**   Find the length of the shaft.

5.65 inches

24. **Mechanics**   Find the length of the shaft.

4.35 feet

Quick Quiz (Objective 3.2A)
Add.
1. 18.44 + 8.3309 + 25.7   52.4709
2. 3.39 + 4.5762 + 1.8 + 0.0312   9.7974

25. Banking   You have $2143.57 in your checking account. You make deposits of $210.98, $45.32, $1236.34, and $27.99. Find the amount in your checking account after you have made the deposits if no money has been withdrawn.
$3664.20

26. Geometry   The perimeter of a triangle is the sum of the lengths of the three sides of the triangle. Find the perimeter of a triangle that has sides that measure 4.9 meters, 6.1 meters, and 7.5 meters.
18.5 meters

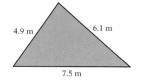

4.9 m
6.1 m
7.5 m

27.   **Demography**   The world's population in 2050 is expected to be 8.9 billion people. It is projected that in that year, Asia's population will be 5.3 billion and Africa's population will be 1.8 billion. What are the combined populations of Asia and Africa expected to be in 2050? (*Source:* United Nations Population Division, World Population Prospects)
7.1 billion people

Employment   The table at the right shows the number of self-employed people, in millions, in the United States. They are listed by annual earnings. Use this table for Exercises 28 to 30.

Annual Earnings	Number of Self-Employed (in millions)
Less than $5000	5.1
$5000 – $24,999	3.9
$25,000 – $49,999	1.6
$50,000 or more	1.0

Source: U.S. Small Business Administration

28.   How many self-employed people earn less than $50,000?
10.6 million people

29.   How many self-employed people earn $5000 or more?
6.5 million people

30.   How many people in the United States are self-employed?
11.6 million people

APPLYING THE CONCEPTS

Consumerism   The table at the right gives the prices for selected products in a grocery store. Use this table for Exercises 31 and 32.

31.   Does a customer with $10 have enough money to purchase raisin bran, bread, milk, and butter?
No

32.   Name three items that would cost more than $7 but less than $8. (There is more than one answer.)
Three possible answers are bread, butter, and mayonnaise; raisin bran, butter, and bread; and potatoes, cola, and mayonnaise.

Product	Cost
Raisin bran	$3.29
Butter	$2.79
Bread	$1.49
Popcorn	$1.19
Potatoes	$2.49
Cola (6-pack)	$1.99
Mayonnaise	$2.99
Lunch meat	$3.39
Milk	$2.59
Toothpaste	$2.69

33. Measurement   Can a piece of rope 4 feet long be wrapped around the box shown at the right?
No

1.4 ft
1.4 ft
1.4 ft

Quick Quiz (Objective 3.2B)

1. You have $655.12 in your checking account. You make deposits of $753.42, $49.90, $67.34, and $152.18. Find the amount in your checking account after you make the deposits. $1677.96

3.3 Subtraction of Decimals

Objective A To subtract decimals

To subtract decimals, write the numbers so that the decimal points are on a vertical line. Subtract as for whole numbers, and write the decimal point in the difference directly below the decimal point in the subtrahend.

HOW TO Subtract 21.532 − 9.875 and check.

	Tens	Ones		Tenths	Hundredths	Thousandths
	1 2	10 X	.	14 5	12 3	12 2
−		9	.	8	7	5
	1 1	.	6	5	7	

Placing the decimal points on a vertical line ensures that digits of the same place value are subtracted.

Check:
Subtrahend   9.875
+ Difference   + 11.657
= Minuend   21.532

HOW TO Subtract 4.3 − 1.7942 and check.

```
  3 12 9 9 10
  4 . 3 0 0 0
− 1 . 7 9 4 2
  2 . 5 0 5 8
```

If necessary, insert zeros in the minuend before subtracting.

Check: 1.7942
         + 2.5058
           4.3000

Example 1 Subtract 39.047 − 7.96 and check.

Solution
```
  8   9 14
3 9 . 0 4 7     Check:     7.96
−   7 . 9 6            + 31.087
3 1 . 0 8 7              39.047
```

You Try It 1 Subtract 72.039 − 8.47 and check.

Your solution 63.569

Example 2 Find 9.23 less than 29 and check.

Solution
```
1 18   9 10
2 9 . 0 0     Check:    9.23
−   9 . 2 3          + 19.77
1 9 . 7 7             29.00
```

You Try It 2 Subtract 35 − 9.67 and check.

Your solution 25.33

Example 3 Subtract 1.2 − 0.8235 and check.

Solution
```
0   11 9 9 10
1 . 2 0 0 0     Check:   0.8235
− 0 . 8 2 3 5         + 0.3765
  0 . 3 7 6 5           1.2000
```

You Try It 3 Subtract 3.7 − 1.9715 and check.

Your solution 1.7285

Solutions on p. S9

Objective 3.3A

Vocabulary to Review
minuend   [1.3A]
subtrahend   [1.3A]
difference   [1.3A]

Discuss the Concepts
Explain how to subtract 2.6357 from 4.79. Why is 4.79 rewritten as 4.7900 before the subtraction is performed?

Instructor Note
Inserting zeros so that each number has the same number of digits to the right of the decimal point will help some students.

Optional Student Activity
Find the pattern in each of the following lists of numbers. Then name the next three numbers in the list.
1. 0.379, 0.329, 0.279, 0.229, . . .
0.179, 0.129, 0.079
2. 1.684, 1.574, 1.464, 1.354, . . .
1.244, 1.134, 1.024
3. 9.5, 8.75, 8, 7.25, . . .
6.5, 5.75, 5

In-Class Examples (Objective 3.3A)
Subtract.
1. 18.9174 − 8.82   10.0974
2. 29.843 − 12.76   17.083

● E S T I M A T I O N ●

Estimating the Difference Between Two Decimals

Calculate $820.23 - 475.748$. Then use estimation to determine whether the difference is reasonable.

Subtract to find the exact difference. $820.23 - 475.748 = 344.482$

To estimate the difference, round each number to the same place value. Here we have rounded to the nearest ten. Then subtract. The estimated answer is 340, which is very close to the exact difference, 344.482.

$$\begin{array}{r} 820.23 \approx 820 \\ -475.748 \approx -480 \\ \hline 340 \end{array}$$

Objective 3.3B

Optional Student Activity

As of this writing, the biggest hole ever in the ozone layer was 4.63 million square miles, recorded in 1998. In 1999, the hole was 3.86 million square miles. (*Source:* World Meteorological Organization) What is the difference between these two measurements?

0.77 million square miles

Note: According to the World Meteorological Organization, we are working toward recovery of the ozone level. Ask students to find current data and then determine the difference between the 1998 measurement and the current one.

Objective B   **To solve application problems**

Example 4

You bought a book for $15.87. How much change did you receive from a $20.00 bill?

Strategy
To find the amount of change, subtract the cost of the book (15.87) from $20.00.

Solution

$$\begin{array}{r} 20.00 \\ -15.87 \\ \hline 4.13 \end{array}$$

You received $4.13 in change.

You Try It 4

Your breakfast cost $6.85. How much change did you receive from a $10.00 bill?

Your strategy

Your solution $3.15

Example 5

You had a balance of $87.93 on your student debit card. You then used the card, deducting $15.99 for a CD, $6.85 for lunch, and $18.50 for a ticket to the football game. What is your new student debit card balance?

Strategy
To find your new debit card balance:
* Add to find the total of the three deductions (15.99 + 6.85 + 18.50).
* Subtract the total of the three deductions from the old balance (87.93).

Solution

$$\begin{array}{r} 15.99 \\ 6.85 \\ +18.50 \\ \hline 41.34 \end{array} \text{ total of deductions}$$

$$\begin{array}{r} 87.93 \\ -41.34 \\ \hline 46.59 \end{array}$$

Your new debit card balance is $46.59.

You Try It 5

You had a balance of $2472.69 in your checking account. You then wrote checks for $1025.60, $79.85, and $162.47. Find the new balance in your checking account.

Your strategy

Your solution $1204.77

Solutions on p. S9

In-Class Example (Objective 3.3B)

1. A competitive swimmer beat the team's record time of 57.84 seconds in the 100-meter freestyle competition by 0.69 second. What is the new record time?   57.15 seconds

3.3 Exercises

Objective A   **To subtract decimals**

Suggested Assignment
Exercises 1–39, odds
More challenging problem:
    Exercise 41

For Exercises 1 to 24, subtract and check.

1. $24.037 - 18.41$
5.627

2. $26.029 - 19.31$
6.719

3. $123.07 - 9.4273$
113.6427

4. $214 - 7.143$
206.857

5. $16.5 - 9.7902$
6.7098

6. $13.2 - 8.6205$
4.5795

7. $235.79 - 20.093$
215.697

8. $463.27 - 40.095$
423.175

9. $63.005 - 9.1274$
53.8776

10. $23.004 - 7.2175$
15.7865

11. $92 - 19.2909$
72.7091

12. $41.2405 - 25.2709$
15.9696

13.
$$\begin{array}{r} 0.32 \\ - 0.0058 \\ \hline 0.3142 \end{array}$$

14.
$$\begin{array}{r} 0.78 \\ - 0.0073 \\ \hline 0.7727 \end{array}$$

15.
$$\begin{array}{r} 3.005 \\ - 1.982 \\ \hline 1.023 \end{array}$$

16.
$$\begin{array}{r} 6.007 \\ - 2.734 \\ \hline 3.273 \end{array}$$

17.
$$\begin{array}{r} 352.16 \\ - 90.994 \\ \hline 261.166 \end{array}$$

18.
$$\begin{array}{r} 872 \\ - 80.753 \\ \hline 791.247 \end{array}$$

19.
$$\begin{array}{r} 724.32 \\ - 69 \\ \hline 655.32 \end{array}$$

20.
$$\begin{array}{r} 625.46 \\ - 77.509 \\ \hline 547.951 \end{array}$$

21.
$$\begin{array}{r} 362.394 \\ - 19.4672 \\ \hline 342.9268 \end{array}$$

22.
$$\begin{array}{r} 421.385 \\ - 17.5293 \\ \hline 403.8557 \end{array}$$

23.
$$\begin{array}{r} 19 \\ - 10.372 \\ \hline 8.628 \end{array}$$

24.
$$\begin{array}{r} 23.4 \\ - 0.921 \\ \hline 22.479 \end{array}$$

For Exercises 25 to 27, use the relationship between addition and subtraction to complete the statement.

25. _____ $+ 2.325 = 7.01$
4.685

26. $5.392 +$ _____ $= 8.07$
2.678

27. _____ $+ 8.967 = 19.35$
10.383

For Exercises 28 to 31, use a calculator to subtract. Then round the numbers to the nearest whole number and use estimation to determine whether the difference you calculated is reasonable.

28.
$$\begin{array}{r} 93.079256 \\ - 66.09249 \\ \hline \end{array}$$
Cal.: 26.986766
Est.: 27

29.
$$\begin{array}{r} 3.7529 \\ - 1.00784 \\ \hline \end{array}$$
Cal.: 2.74506
Est.: 3

30.
$$\begin{array}{r} 76.53902 \\ - 45.73005 \\ \hline \end{array}$$
Cal.: 30.80897
Est.: 31

31.
$$\begin{array}{r} 9.07325 \\ - 1.924 \\ \hline \end{array}$$
Cal.: 7.14925
Est.: 7

Quick Quiz (Objective 3.3A)
Subtract.
1. $24.041 - 16.25$ 7.791
2. $131.13 - 90.675$ 40.455

Objective B **To solve application problems**

32. Business The manager of the Edgewater Cafe takes a reading of the cash register tape each hour. At 1:00 P.M. the tape read $967.54. At 2:00 P.M. the tape read $1437.15. Find the amount of sales between 1:00 P.M. and 2:00 P.M.
$469.61

33. Mechanics Find the missing dimension.

6.79 in.

?

14.34 in.

7.55 inches Ⓟ

34. Mechanics Find the missing dimension.

1.72 ft

?

4.31 ft

2.59 feet Ⓟ

35. Banking You had a balance of $1029.74 in your checking account. You then wrote checks for $67.92, $43.10, and $496.34.
a. Find the total amount of the checks written. $607.36
b. Find the new balance in your checking account. $422.38

36. Consumerism The price of gasoline is $1.82 per gallon after the price rose $.07 one month and $.12 the next month. Find the price of gasoline before these increases in price.
$1.63

37. Population Growth The annual number of births in the United States is projected to rise from 3.98 million in 2003 to 4.37 million in 2012. (*Sources:* U.S. Census Bureau; National Center for Health Statistics) Find the increase in the annual number of births from 2003 to 2012.
0.39 million births

38. Investments Grace Herrera owned 357.448 shares of a mutual fund on January 1. On December 31 of the same year, she had 439.917 shares. What was the increase in the number of shares Grace owned during that year?
82.469 shares

E-Commerce The table at the right shows the actual and projected number of households shopping online, in millions. Use the table for Exercises 39 and 40.

39. Calculate the growth in online shopping from 2000 to 2003.
22.6 million households

40. Calculate the growth in online shopping from 1998 to 2003.
31.6 million households

Year	Number of Households Shopping Online (in millions)
1998	8.7
1999	13.1
2000	17.7
2001	23.1
2002	30.3
2003	40.3

Source: Forrester Research Inc. Ⓟ

APPLYING THE CONCEPTS

41. Find the largest amount by which the estimate of the sum of two decimals rounded to the given place value could differ from the exact sum.
a. Tenths **b.** Hundredths **c.** Thousandths
a. 0.1 **b.** 0.01 **c.** 0.001

Quick Quiz (Objective 3.3B)

1. You buy groceries for $57.92. How much change do you receive from a $100 bill? $42.08

3.4 Multiplication of Decimals

Objective A To multiply decimals

Decimals are multiplied as though they were whole numbers. Then the decimal point is placed in the product. Writing the decimals as fractions shows where to write the decimal point in the product.

$$0.\underline{3} \times 5 = \frac{3}{10} \times \frac{5}{1} = \frac{15}{10} = 1.\underline{5}$$

1 decimal place   1 decimal place

$$0.\underline{3} \times 0.\underline{5} = \frac{3}{10} \times \frac{5}{10} = \frac{15}{100} = 0.\underline{15}$$

1 decimal place   1 decimal place   2 decimal places

$$0.\underline{3} \times 0.\underline{05} = \frac{3}{10} \times \frac{5}{100} = \frac{15}{1000} = 0.\underline{015}$$

1 decimal place   2 decimal places   3 decimal places

To multiply decimals, multiply the numbers as with whole numbers. Write the decimal point in the product so that the number of decimal places in the product is the sum of the decimal places in the factors.

HOW TO   Multiply: 21.4×0.36

21.4	1 decimal place
× 0.36	2 decimal places
1284	
642	
7.704	3 decimal places

HOW TO   Multiply: 0.037×0.08

0.037	3 decimal places	• Two zeros must be inserted between the
× 0.08	2 decimal places	2 and the decimal point so that there are
0.00296	5 decimal places	5 decimal places in the product.

To multiply a decimal by a power of 10 (10, 100, 1000, . . .), move the decimal point to the right the same number of places as there are zeros in the power of 10.

$3.8925 \times 1\underline{0} \quad = 38.925$
     1 zero   1 decimal place

$3.8925 \times 1\underline{00} \quad = 389.25$
     2 zeros   2 decimal places

$3.8925 \times 1\underline{000} \quad = 3892.5$
     3 zeros   3 decimal places

$3.8925 \times 1\underline{0,000} \quad = 38,925.$
     4 zeros   4 decimal places

$3.8925 \times 1\underline{00,000} = 389,250.$
     5 zeros   5 decimal places
            • Note that a zero must be inserted before the decimal point.

Instructor Note
Another way to practice multiplying by powers of 10 is to relate these examples to numbers given as 3.84 million, 10.4 billion, or 2.3 trillion.
    Also, multiplying or dividing (in the next section) by powers of 10 is the way one converts between various units in the metric system.

Note that if the power of 10 is written in exponential notation, the exponent indicates how many places to move the decimal point.

$$3.8925 \times 10^1 = 38.925$$
1 decimal place

$$3.8925 \times 10^2 = 389.25$$
2 decimal places

$$3.8925 \times 10^3 = 3892.5$$
3 decimal places

$$3.8925 \times 10^4 = 38,925.$$
4 decimal places

$$3.8925 \times 10^5 = 389,250.$$
5 decimal places

Example 1   Multiply: 920×3.7

Solution
$$\begin{array}{r} 920 \\ \times\ 3.7 \\ \hline 644\ 0 \\ 2760 \\ \hline 3404.0 \end{array}$$
• 1 decimal place

• 1 decimal place

You Try It 1   Multiply: 870×4.6

Your solution   4002.0

Example 2   Find 0.00079 multiplied by 0.025.

Solution
$$\begin{array}{r} 0.00079 \\ \times\ \ 0.025 \\ \hline 395 \\ 158\ \ \\ \hline 0.00001975 \end{array}$$
• 5 decimal places
• 3 decimal places

• 8 decimal places

You Try It 2   Find 0.000086 multiplied by 0.057.

Your solution   0.000004902

Example 3   Find the product of 3.69 and 2.07.

Solution
$$\begin{array}{r} 3.69 \\ \times\ 2.07 \\ \hline 2583 \\ 7380\ \ \\ \hline 7.6383 \end{array}$$
• 2 decimal places
• 2 decimal places

• 4 decimal places

You Try It 3   Find the product of 4.68 and 6.03.

Your solution   28.2204

Example 4   Multiply: $42.07 \times 10,000$

Solution $42.07 \times 10,000 = 420,700$

You Try It 4   Multiply: 6.9×1000

Your solution   6900

Example 5   Find 3.01 times 10^3.

Solution $3.01 \times 10^3 = 3010$

You Try It 5   Find 4.0273 times 10^2.

Your solution   402.73

Solutions on pp. S9–S10

● E S T I M A T I O N ●

Estimating the Product of Two Decimals

Calculate 28.259×0.029. Then use estimation to determine whether the product is reasonable.

Multiply to find the exact product. $28.259 \times 0.029 = 0.819511$

To estimate the product, round each number so there is one nonzero digit. Then multiply. The estimated answer is 0.90, which is very close to the exact product, 0.819511.

$$28.259 \approx \quad 30$$
$$\times \; 0.029 \approx \times 0.03$$
$$\overline{\qquad\qquad 0.90}$$

Objective B   **To solve application problems**

The tables that follow list water rates and meter fees for a city. These tables are used for Example 6 and You Try It 6.

Water Charges	
Commercial	$1.39/1000 gal
Comm Restaurant	$1.39/1000 gal
Industrial	$1.39/1000 gal
Institutional	$1.39/1000 gal
Res—No Sewer	
Residential—SF	
>0 <200 gal per day	$1.15/1000 gal
>200 <1500 gal per day	$1.39/1000 gal
>1500 gal per day	$1.54/1000 gal

Meter Charges	
Meter	Meter Fee
5/8" & 3/4"	$13.50
1"	$21.80
1-1/2"	$42.50
2"	$67.20
3"	$133.70
4"	$208.20
6"	$415.10
8"	$663.70

Example 6

Find the total bill for an industrial water user with a 6-inch meter that used 152,000 gallons of water for July and August.

Strategy

To find the total cost of water:

● Find the cost of water by multiplying the cost per 1000 gallons (1.39) by the number of 1000-gallon units used.
● Add the cost of the water to the meter fee (415.10).

Solution

Cost of water $= \dfrac{152,000}{1000} \cdot 1.39 = 211.28$

Total cost $= 211.28 + 415.10 = 626.38$

The total cost is $626.38.

You Try It 6

Find the total bill for a commercial user that used 5000 gallons of water per day for July and August. The user has a 3-inch meter.

Your strategy

Your solution $564.60

Solution on p. S10

Example 7

It costs $.036 an hour to operate an electric motor. How much does it cost to operate the motor for 120 hours?

Strategy

To find the cost of running the motor for 120 hours, multiply the hourly cost (0.036) by the number of hours the motor is run (120).

Solution

```
  0.036
×   120
    720
   36
  4.320
```

The cost of running the motor for 120 hours is $4.32.

You Try It 7

The cost of electricity to run a freezer for 1 hour is $.035. This month the freezer has run for 210 hours. Find the total cost of running the freezer this month.

Your strategy

Your solution $7.35

Example 8

Jason Ng earns a salary of $440 for a 40-hour workweek. This week he worked 12 hours of overtime at a rate of $16.50 for each hour of overtime worked. Find his total income for the week.

Strategy

To find Jason's total income for the week:

• Find the overtime pay by multiplying the hourly overtime rate (16.50) by the number of hours of overtime worked (12).
• Add the overtime pay to the weekly salary (440).

Solution

```
  16.50          440.00
×    12        + 198.00
  33 00          638.00
 165 0
 198.00   Overtime pay
```

Jason's total income for this week is $638.00.

You Try It 8

You make a down payment of $175 on a stereo and agree to make payments of $37.18 a month for the next 18 months to repay the remaining balance. Find the total cost of the stereo.

Your strategy

Your solution $844.24

Solutions on p. S10

In-Class Example (Objective 3.4B)

1. The cost of operating an electric saw for 1 hour is $.032. How much does it cost to operate the saw for 65 hours? $2.08

3.4 Exercises

Objective A   To multiply decimals

For Exercises 1 to 94, multiply.

1.	0.9 $\times$ 0.4	2.	0.7 $\times$ 0.9	3.	0.5 $\times$ 0.6	4.	0.3 $\times$ 0.7	5.	0.5 $\times$ 0.5
	0.36		0.63		0.30		0.21		0.25

6.	0.7 $\times$ 0.7	7.	0.9 $\times$ 0.5	8.	0.2 $\times$ 0.6	9.	7.7 $\times$ 0.9	10.	3.4 $\times$ 0.4
	0.49		0.45		0.12		6.93		1.36

11.	9.2 $\times$ 0.2	12.	2.6 $\times$ 0.7	13.	7.2 $\times$ 0.6	14.	6.8 $\times$ 0.4	15.	7.4 $\times$ 0.1
	1.84		1.82		4.32		2.72		0.74

16.	3.8 $\times$ 0.1	17.	7.9 $\times$ 5	18.	9.3 $\times$ 7	19.	0.68 $\times$ 4	20.	0.83 $\times$ 9
	0.38		39.5		65.1		2.72		7.47

21.	0.67 $\times$ 0.9	22.	0.84 $\times$ 0.3	23.	0.16 $\times$ 0.6	24.	0.47 $\times$ 0.8	25.	2.5 $\times$ 5.4
	0.603		0.252		0.096		0.376		13.50

26.	3.9 $\times$ 1.9	27.	8.4 $\times$ 9.5	28.	7.6 $\times$ 5.8	29.	0.83 $\times$ 5.2	30.	0.24 $\times$ 2.7
	7.41		79.80		44.08		4.316		0.648

31.	0.46 $\times$ 3.9	32.	0.78 $\times$ 6.8	33.	0.2 $\times$ 0.3	34.	0.3 $\times$ 0.3	35.	0.24 $\times$ 0.3
	1.794		5.304		0.06		0.09		0.072

36.	0.17 $\times$ 0.5	37.	1.47 $\times$ 0.09	38.	6.37 $\times$ 0.05	39.	8.92 $\times$ 0.004	40.	6.75 $\times$ 0.007
	0.085		0.1323		0.3185		0.03568		0.04725

Suggested Assignment
Exercises 1–117, every other odd
Exercises 121–133, odds
Exercises 134–136
More challenging problem:
    Exercise 137

Quick Quiz (Objective 3.4A)
Multiply.
1. 0.96 $\times$ 0.7   0.672
2. 5.7 $\times$ 4   22.8
3. 7.9142 $\times$ 1000   7914.2
4. 0.015 $\times$ 10^2 1.5

41.   0.49
      × 0.16
      0.0784

42.   0.38
      × 0.21
      0.0798

43.   7.6
      × 0.01
      0.076

44.   5.1
      × 0.01
      0.051

45.   8.62
      ×   4
      34.48

46.   5.83
      ×   7
      40.81

47.   64.5
      ×   9
      580.5

48.   37.8
      ×   8
      302.4

49.   2.19
      × 9.2
      20.148

50.   1.25
      × 5.6
      7.000

51.   1.85
      × 0.023
      0.04255

52.   37.8
      × 0.052
      1.9656

53.   0.478
      × 0.37
      0.17686

54.   0.526
      × 0.22
      0.11572

55.   48.3
      × 0.0041
      0.19803

56.   67.2
      × 0.0086
      0.57792

57.   2.437
      ×   6.1
      14.8657

58.   4.237
      × 0.54
      2.28798

59.   0.413
      × 0.0016
      0.0006608

60.   0.517
      × 0.0029
      0.0014993

61.   94.73
      × 0.57
      53.9961

62.   89.23
      × 0.62
      55.3226

63.   8.005
      × 0.067
      0.536335

64.   9.032
      × 0.019
      0.171608

65. 4.29 × 0.1
      0.429

66. 6.78 × 0.1
0.678

67. 5.29 × 0.4
2.116

68. 6.78 × 0.5
3.390

69. 0.68 × 0.7
0.476

70. 0.56 × 0.9
0.504

71. 1.4 × 0.73
1.022

72. 6.3 × 0.37
2.331

73. 5.2 × 7.3
37.96

74. 7.4 × 2.9
21.46

75. 3.8 × 0.61
2.318

76. 7.2 × 0.72
5.184

77. 0.32 × 10
3.2

78. 6.93 × 10
69.3

79. 0.065 × 100
6.5

80. 0.039 × 100
3.9

81. 6.2856 × 1000
6285.6

82. 3.2954 × 1000
3295.4

83. 3.2 × 1000
3200

84. 0.006 × 10,000
60

85. 3.57 × 10,000
35,700

86. 8.52×10^1
85.2

87. 0.63×10^1
6.3

88. 82.9×10^2
8290

89. 0.039×10^2
3.9

90. 6.8×10^3
6800

91. 4.9×10^4
49,000

92. 6.83×10^4
68,300

93. 0.067×10^2
6.7

94. 0.052×10^2
5.2

95. Find the product of 0.0035 and 3.45.
0.012075

96. Find the product of 237 and 0.34.
80.58

97. Multiply 3.005 by 0.00392.
0.0117796

98. Multiply 20.34 by 1.008.
20.50272

99. Multiply 1.348 by 0.23.
0.31004

100. Multiply 0.000358 by 3.56.
0.00127448

101. Find the product of 23.67 and 0.0035.
0.082845

102. Find the product of 0.00346 and 23.1.
0.079926

103. Find the product of 5, 0.45, and 2.3.
5.175

104. Find the product of 0.03, 23, and 9.45.
6.5205

For Exercises 105 to 120, use a calculator to multiply. Then use estimation to determine whether the product you calculated is reasonable.

105.   28.5
$\times$ 3.2
Cal.: 91.2
Est.: 90

106.   86.3
$\times$ 4.4
Cal.: 379.72
Est.: 360

107.   2.38
$\times$ 0.44
Cal.: 1.0472
Est.: 0.8

108.   9.82
$\times$ 0.77
Cal.: 7.5614
Est.: 8

109.   0.866
$\times$ 4.5
Cal.: 3.897
Est.: 4.5

110.   0.239
$\times$ 8.2
Cal.: 1.9598
Est.: 1.6

111.   4.34
$\times$ 2.59
Cal.: 11.2406
Est.: 12

112.   6.87
$\times$ 9.98
Cal.: 68.5626
Est.: 70

113.   8.434
$\times$ 0.044
Cal.: 0.371096
Est.: 0.32

114.   7.037
$\times$ 0.094
Cal.: 0.661478
Est.: 0.63

115.   28.44
$\times$ 1.12
Cal.: 31.8528
Est.: 30

116.   86.57
$\times$ 7.33
Cal.: 634.5581
Est.: 630

117.   49.6854
$\times$ 39.0672
Cal.: 1941.069459
Est.: 2000

118.   2.00547
$\times$ 9.672
Cal.: 19.39690584
Est.: 20

119.   0.00456
$\times$ 0.009542
Cal.: 0.00004351152
Est.: 0.00005

120.   7.00637
$\times$ 0.0128
Cal.: 0.089681536
Est.: 0.07

Objective B   **To solve application problems**

121. **Consumerism**   An electric motor costing $315.45 has an operating cost of $.027 for 1 hour of operation. Find the cost to run the motor for 56 hours. Round to the nearest cent.
$1.51

122. **Recycling**   Four hundred empty soft drink cans weigh 18.75 pounds. A recycling center pays $.75 per pound for the cans. Find the amount received for the 400 cans. Round to the nearest cent.
$14.06

123. **Recycling**   A recycling center pays $.045 per pound for newspapers.
   a. Estimate the payment for recycling 520 pounds of newspapers.
   b. Find the actual amount received from recycling the newspapers.
   a. $25.00   **b.** $23.40

124. **Geometry**   The perimeter of a square is equal to four times the length of a side of the square. Find the perimeter of a square whose side measures 2.8 meters.
11.2 meters

2.8 m

125. **Geometry**   The area of a rectangle is equal to the product of the length of the rectangle times its width. Find the area of a rectangle that has a length of 6.75 feet and a width of 3.5 feet. The area will be in square feet.
23.625 square feet

3.5 ft
6.75 ft

126. **Finance**   You bought a car for $5000 down and made payments of $399.50 each month for 36 months.
   a. Find the amount of the payments over the 36 months. $14,382
   b. Find the total cost of the car. $19,382

127. **Compensation**   A nurse earns a salary of $1156 for a 40-hour week. This week the nurse worked 15 hours of overtime at a rate of $43.35 for each hour of overtime worked.
   a. Find the nurse's overtime pay. $650.25
   b. Find the nurse's total income for the week. $1806.25

128. **Consumerism**   Bay Area Rental Cars charges $15 a day and $.15 per mile for renting a car. You rented a car for 3 days and drove 235 miles. Find the total cost of renting the car.
$80.25

129. ● **Shipping**   The graph at the right shows United States Postal Service rates for express mail. How much would it cost a company to mail 25 express mail packages, each weighing $\frac{1}{4}$ pound, post office to addressee?
$341.25

130. **Transportation**   A taxi costs $2.50 and $.20 for each $\frac{1}{8}$ mile driven. Find the cost of hiring a taxi to get from the airport to the hotel—a distance of $5\frac{1}{2}$ miles.
$11.30

$20
$15
Rate $10
$5
$0

10.40   13.65   14.60   17.85

Up to $\frac{1}{2}$ lb   Over $\frac{1}{2}$ to 1 lb

■ Post Office to Post Office
■ Post Office to Addressee   P
U.S. Postal Service Rates for Express Mail

Quick Quiz (Objective 3.4B)

1. You bought a house with payments of $1024.50 per month for 20 years. Find the total amount of the payments. $245,880

131. Business The table at the right lists three pieces of steel required for a repair project.
a. Find the total cost of grade 1.
b. Find the total cost of grade 2.
c. Find the total cost of grade 3.
d. Find the total cost of the three pieces of steel.
a. $21.12   b. $29.84   c. $201.66   d. $252.62

Grade of Steel	Weight (pounds per foot)	Required Number of Feet	Cost per Pound
1	2.2	8	$1.20
2	3.4	6.5	$1.35
3	6.75	15.4	$1.94

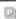

132. Business A confectioner ships holiday packs of candy and nuts anywhere in the United States. At the right is a price list for nuts and candy, and below is a table of shipping charges to zones in the United States. For any fraction of a pound, use the next higher weight. (16 oz = 1 lb)

Code	Description	Price
112	Almonds 16 oz	$4.75
116	Cashews 8 oz	$2.90
117	Cashews 16 oz	$5.50
130	Macadamias 7 oz	$5.25
131	Macadamias 16 oz	$9.95
149	Pecan halves 8 oz	$6.25
155	Mixed nuts 8 oz	$4.80
160	Cashew brittle 8 oz	$1.95
182	Pecan roll 8 oz	$3.70
199	Chocolate peanuts 8 oz	$1.90

Pounds	Zone 1	Zone 2	Zone 3	Zone 4
1–3	$6.55	$6.85	$7.25	$7.75
4–6	$7.10	$7.40	$7.80	$8.30
7–9	$7.50	$7.80	$8.20	$8.70
10–12	$7.90	$8.20	$8.60	$9.10

Find the cost of sending the following orders to the given mail zone.

a. Code	Quantity	b. Code	Quantity	c. Code	Quantity
116	2	112	1	117	3
130	1	117	4	131	1
149	3	131	2	155	2
182	4	160	3	160	4
Mail to zone 4.		182	5	182	1
		Mail to zone 3.		199	3
				Mail to zone 2.	

a. $52.90   b. $79.60   c. $61.05

133. Air Pollution An emissions test for cars requires that of the total engine exhaust, less than 1 part per thousand $\left(\frac{1}{1000} = 0.001\right)$ be hydrocarbon emissions. Using this figure, determine which of the cars in the table below would fail the emissions test.

Car	Total Engine Exhaust	Hydrocarbon Emission
1	367,921	360
2	401,346	420
3	298,773	210
4	330,045	320
5	432,989	450

Cars 2 and 5

APPLYING THE CONCEPTS

134. Automotive Repair   Chris works at B & W Garage as an auto mechanic and has just completed an engine overhaul for a customer. To determine the cost of the repair job, Chris keeps a list of times worked and parts used. A parts list and a list of the times worked are shown below.

Parts Used		Time Spent	
Item	Quantity	Day	Hours
Gasket set	1	Monday	7.0
Ring set	1	Tuesday	7.5
Valves	8	Wednesday	6.5
Wrist pins	8	Thursday	8.5
Valve springs	16	Friday	9.0
Rod bearings	8		
Main bearings	5		
Valve seals	16		
Timing chain	1		

Price List		
Item Number	Description	Unit Price
27345	Valve spring	$9.25
41257	Main bearing	$17.49
54678	Valve	$16.99
29753	Ring set	$169.99
45837	Gasket set	$174.90
23751	Timing chain	$50.49
23765	Fuel pump	$429.99
28632	Wrist pin	$13.55
34922	Rod bearing	$4.69
2871	Valve seal	$1.69

a. Organize a table of data showing the parts used, the unit price for each, and the price of the quantity used. *Hint:* Use the following headings for the table.

Quantity	Item Number	Description	Unit Price	Total
1	45837	Gasket set	$174.90	$174.90
1	29753	Ring set	$169.99	$169.99
8	54678	Valve	$ 16.99	$135.92
8	28632	Wrist pin	$ 13.55	$108.40
16	27345	Valve spring	$ 9.25	$148.00
8	34922	Rod bearing	$ 4.69	$ 37.52
5	41257	Main bearing	$ 17.49	$ 87.45
16	2871	Valve seal	$ 1.69	$ 27.04
1	23751	Timing chain	$ 50.49	$ 50.49

b. Add up the numbers in the "Total" column to find the total cost of the parts. $939.71

c. If the charge for labor is $46.75 per hour, compute the cost of labor. $1799.88

d. What is the total cost for parts and labor? $2739.59

135. Explain how the decimal point is placed when a number is multiplied by 10, 100, 1000, 10,000, etc.

136. Explain how the decimal point is placed in the product of two decimals.

137. Show how the decimal is placed in the product of 1.3×2.31 by first writing each number as a fraction and then multiplying. Then change the product back to decimal notation. $1.3 \times 2.31 = \dfrac{13}{10} \times \dfrac{231}{100} = \dfrac{3003}{1000} = 3.003$

Answers to Writing Exercises

135. When a number is multiplied by 10, 100, 1000, 10,000, etc., the decimal point is moved as many places to the right as there are zeros in the multiple of 10. For example, the decimal point in a number multiplied by 1000 would be moved three places to the right.

136. Add the number of decimal places in each of the numbers being multiplied. The number of decimal places in the product is equal to this sum.

3.5 Division of Decimals

Objective A   To divide decimals

To divide decimals, move the decimal point in the divisor to the right to make the divisor a whole number. Move the decimal point in the dividend the same number of places to the right. Place the decimal point in the quotient directly over the decimal point in the dividend, and then divide as with whole numbers.

> **HOW TO**   Divide: $3.25\overline{)15.275}$
>
> $3.25.\overline{)15.27.5}$
> • Move the decimal point 2 places to the right in the divisor and then in the dividend. Place the decimal point in the quotient.
>
> $$\begin{array}{r} 4.7 \\ 325.\overline{)\ 1527.5} \\ -1300 \\ \hline 227\ 5 \\ -227\ 5 \\ \hline 0 \end{array}$$
> • Divide as with whole numbers.

Study Tip

To learn mathematics, you must be an active participant. Listening and watching your professor do mathematics are not enough. Take notes in class, mentally think through every question your instructor asks, and try to answer it even if you are not called on to answer it verbally. Ask questions when you have them. See *AIM for Success* at the front of the book for other ways to be an active learner.

Moving the decimal point the same number of decimal places in the divisor and dividend does not change the value of the quotient, because this process is the same as multiplying the numerator and denominator of a fraction by the same number. In the example above,

$$3.25\overline{)15.275} = \frac{15.275}{3.25} = \frac{15.275 \times 100}{3.25 \times 100} = \frac{1527.5}{325} = 325\overline{)1527.5}$$

When dividing decimals, we usually round the quotient off to a specified place value, rather than writing the quotient with a remainder.

> **HOW TO**   Divide: $0.3\overline{)0.56}$
> Round to the nearest hundredth.
>
> $$\begin{array}{r} 1.866 \approx 1.87 \\ 0.3.\overline{)\ 0.5.600} \\ -\ 3 \\ \hline 2\ 6 \\ -2\ 4 \\ \hline 20 \\ -18 \\ \hline 20 \\ -18 \end{array}$$
>
> We must carry the division to the thousandths place to round the quotient to the nearest hundredth. Therefore, zeros must be inserted in the dividend so that the quotient has a digit in the thousandths place.

Vocabulary to Review
divisor [1.5A]
dividend [1.5A]
quotient [1.5A]
power of 10 [1.6A]

Discuss the Concepts
Explain how to divide a number by a power of 10.

Concept Check
Which of the following examples have the same quotient?
a. $15\overline{)180}$
b. $1.5\overline{)1.8}$
c. $1.5\overline{)18}$
d. $0.15\overline{)18}$
e. $0.15\overline{)1.8}$
a, c, and e have the same quotient.

Instructor Note
The paragraph at the left below the How To gives the justification for moving the decimal point before decimal numbers are divided.

Instructor Note
An alternative method for rounding quotients is

$$\begin{array}{r} 1.86 \\ 3\overline{)5.60} \\ -3 \\ \hline 2\ 6 \\ -2\ 4 \\ \hline 20 \\ -18 \\ \hline 2 \end{array}$$

Because the last remainder is more than one-half of the divisor, 1.86 is rounded to 1.87.

In-Class Examples (Objective 3.5A)
Divide. Round to the nearest hundredth.
1. 36.597 ÷ 5.2   7.04
2. 5.628 ÷ 19   0.30
Divide.
3. 14.08 ÷ 1000   0.01408
4. 49.65 ÷ 10^2 0.4965

152

Concept Check

Simplify:

$$\frac{2,400,000 \times 0.0000063}{0.00009 \times 480}$$

350

Optional Student Activity

Light travels at a speed of 1.86×10^5 miles per second. (See the Concept Check for Objective 3.4A.) How long would it take light to travel 1.86 million miles?   10 seconds

Integrating Technology

A calculator displays the quotient to the limit of the calculator's display. Enter

| 57 | . | 93 | ÷ | 3 | . | 24 | = |

to determine the number of places your calculator displays.

HOW TO Divide 57.93 by 3.24. Round to the nearest thousandth.

$$
\begin{array}{r}
17.8796 \approx 17.880 \\
3.24.\overline{)\,57.93.0000} \\
-32\ 4 \\
\hline
25\ 53 \\
-22\ 68 \\
\hline
2\ 85\ 0 \\
-2\ 59\ 2 \\
\hline
25\ 80 \\
-22\ 68 \\
\hline
3\ 120 \\
-2\ 916 \\
\hline
2040 \\
-1944 \\
\hline
\end{array}
$$

Zeros must be inserted in the dividend so that the quotient has a digit in the ten-thousandths place.

To divide a decimal by a power of 10 (10, 100, 1000, . . .), move the decimal point to the left the same number of places as there are zeros in the power of 10.

$34.65 \div 10 = 3.465$
1 zero   1 decimal place

$34.65 \div 100 = 0.3465$
2 zeros   2 decimal places

$34.65 \div 1000 = 0.03465$
3 zeros   3 decimal places

• Note that a zero must be inserted between the 3 and the decimal point.

$34.65 \div 10{,}000 = 0.003465$
4 zeros   4 decimal places

• Note that two zeros must be inserted between the 3 and the decimal point.

If the power of 10 is written in exponential notation, the exponent indicates how many places to move the decimal point.

$34.65 \div 10^1 = 3.465$ 1 decimal place
$34.65 \div 10^2 = 0.3465$ 2 decimal places
$34.65 \div 10^3 = 0.03465$ 3 decimal places
$34.65 \div 10^4 = 0.003465$ 4 decimal places

Example 1 Divide: $0.1344 \div 0.032$

Solution

$$
\begin{array}{r}
4.2 \\
0.032.\overline{)\,0.134.4} \\
-128 \\
\hline
6\ 4 \\
-6\ 4 \\
\hline
0
\end{array}
$$

• Move the decimal point 3 places to the right in the divisor and the dividend.

You Try It 1 Divide: $0.1404 \div 0.052$

Your solution 2.7

Solution on p. S10

Example 2 Divide: 58.092 ÷ 82
Round to the nearest
thousandth.

Solution

$$
\begin{array}{r}
0.7084 \approx 0.708 \\
82\overline{)\ 58.0920} \\
-57\ 4 \\
\hline
69 \\
-\ \ 0 \\
\hline
692 \\
-656 \\
\hline
360 \\
-328 \\
\end{array}
$$

You Try It 2 Divide: 37.042 ÷ 76
Round to the nearest
thousandth.

Your solution 0.487

Example 3 Divide: 420.9 ÷ 7.06
Round to the nearest tenth.

Solution

$$
\begin{array}{r}
59.61 \approx 59.6 \\
7.06\overline{)\ 420.90.00} \\
-353\ 0 \\
\hline
67\ 90 \\
-63\ 54 \\
\hline
4\ 36\ 0 \\
-4\ 23\ 6 \\
\hline
12\ 40 \\
-\ 7\ 06 \\
\end{array}
$$

You Try It 3 Divide: 370.2 ÷ 5.09
Round to the nearest tenth.

Your solution 72.7

Example 4 Divide: 402.75 ÷ 1000

Solution 402.75 ÷ 1000 = 0.40275

You Try It 4 Divide: 309.21 ÷ 10,000

Your solution 0.030921

Example 5 What is 0.625 divided by 10^2?

Solution $0.625 \div 10^2 = 0.00625$

You Try It 5 What is 42.93 divided by 10^4?

Your solution 0.004293

Solutions on p. S10

● E S T I M A T I O N ●

Estimating the Quotient of Two Decimals

Calculate 282.18 ÷ 0.48. Then use estimation to determine whether the
quotient is reasonable.

Divide to find the exact quotient.   282.18 ÷ 0.48 = 587.875

To estimate the quotient, round each $282.18 \div 0.48 \approx 300 \div 0.5$
number so there is one nonzero digit.   = 600
Then divide. The estimated answer is 600,
which is very close to the exact quotient,
587.875.

Objective 3.5B

Objective B   **To solve application problems**

The graph at the right shows average hourly earnings in the United States. Use this table for Example 6 and You Try It 6.

Average Hourly Earnings in the United States

Sources: Statistical Abstract of the United States; Bureau of Labor Statistics

---

Example 6

How many times greater were the average hourly earnings in 2000 than in 1970? Round to the nearest whole number.

Strategy

To find how many times greater the average hourly earnings were, divide the 2000 average hourly earnings (14.38) by the 1970 average hourly earnings (3.23).

Solution

$14.38 \div 3.23 \approx 4$

The average hourly earnings in 2000 were about 4 times greater than in 1970.

You Try It 6

How many times greater were the average hourly earnings in 1990 than in 1970? Round to the nearest tenth.

Your strategy

Your solution

3.1 times greater

Example 7

A 1-year subscription to a monthly magazine costs $90. The price of each issue at the newsstand is $9.80. How much would you save per issue by buying a year's subscription rather than buying each issue at the newsstand?

Strategy

To find the amount saved:

• Find the subscription price per issue by dividing the cost of the subscription (90) by the number of issues (12).
• Subtract the subscription price per issue from the newsstand price (9.80).

Solution

$90 \div 12 = 7.50$
$9.80 - 7.50 = 2.30$

The savings would be $2.30 per issue.

You Try It 7

A Nielsen survey of the number of people (in millions) who watch television each day of the week is given in the table below.

Mon.	Tues.	Wed.	Thu.	Fri.	Sat.	Sun.
91.9	89.8	90.6	93.9	78.0	77.1	87.7

Find the average number of people who watch television per day.

Your strategy

Your solution

87 million people

Solutions on pp. S10–S11

In-Class Examples (Objective 3.5B)

1. A jogger ran 7.4 miles in 45.88 minutes. What was the jogger's average time per mile?   6.2 minutes

3.5 Exercises

Objective A   **To divide decimals**

For Exercises 1 to 20, divide.

1. $3\overline{)2.46}$ — 0.82

2. $7\overline{)3.71}$ — 0.53

3. $0.8\overline{)3.84}$ — 4.8

4. $0.9\overline{)6.93}$ — 7.7

5. $0.7\overline{)62.3}$ — 89

6. $0.4\overline{)52.8}$ — 132

7. $0.4\overline{)24}$ — 60

8. $0.5\overline{)65}$ — 130

9. $0.7\overline{)59.01}$ — 84.3

10. $0.9\overline{)8.721}$ — 9.69

11. $0.5\overline{)16.15}$ — 32.3

12. $0.8\overline{)77.6}$ — 97

13. $0.7\overline{)3.542}$ — 5.06

14. $0.6\overline{)2.436}$ — 4.06

15. $6.3\overline{)8.19}$ — 1.3

16. $3.2\overline{)7.04}$ — 2.2

17. $3.6\overline{)0.396}$ — 0.11

18. $2.7\overline{)0.648}$ — 0.24

19. $6.9\overline{)26.22}$ — 3.8

20. $1.7\overline{)84.66}$ — 49.8

For Exercises 21 to 29, divide. Round to the nearest tenth.

21. $55.62 \div 8.8$
6.3

22. $25.43 \div 5.4$
4.7

23. $5.427 \div 9.5$
0.6

24. $1.837 \div 1.4$
1.3

25. $18.4 \div 7.3$
2.5

26. $52.9 \div 8.1$
6.5

27. $0.183 \div 0.17$
1.1

28. $0.381 \div 0.47$
0.8

29. $6.924 \div 0.053$
130.6

For Exercises 30 to 38, divide. Round to the nearest hundredth.

30. $4.817 \div 16$
0.30

31. $6.467 \div 8$
0.81

32. $0.0418 \div 0.53$
0.08

33. $0.0647 \div 0.72$
0.09

34. $7 \div 0.55$
12.73

35. $38.665 \div 0.95$
40.70

36. $13.97 \div 25.4$
0.55

37. $27.738 \div 60.8$
0.46

38. $3.171 \div 45.6$
0.07

Suggested Assignment
Exercises 1–93, every other odd
Exercises 97–117, odds

Quick Quiz (Objective 3.5A)
Divide. Round to the nearest hundredth.
1. $8.48 \div 0.3$ 28.27
2. $2.178 \div 0.039$ 55.85
Divide.
3. $257.93 \div 100$ 2.5793
4. $3.761 \div 10^3$ 0.003761

For Exercises 39 to 47, divide. Round to the nearest thousandth.

39. $1.028 \div 54$
0.019

40. $6.729 \div 27$
0.249

41. $0.0437 \div 0.5$
0.087

42. $75.469 \div 77.8$
0.970

43. $34.31 \div 95.3$
0.360

44. $0.2695 \div 2.67$
0.101

45. $0.4871 \div 4.72$
0.103

46. $0.1142 \div 17.2$
0.007

47. $0.2307 \div 26.7$
0.009

For Exercises 48 to 56, divide. Round to the nearest whole number.

48. $16.5 \div 4$
4

49. $89.76 \div 90$
1

50. $1.94 \div 0.3$
6

51. $1.0478 \div 0.413$
3

52. $2.148 \div 0.519$
4

53. $0.79 \div 0.778$
1

54. $3.092 \div 0.075$
41

55. $392 \div 6.9$
57

56. $8.729 \div 0.075$
116

For Exercises 57 to 74, divide.

57. $4.07 \div 10$
0.407

58. $0.039 \div 10$
0.0039

59. $42.67 \div 10$
4.267

60. $389.7 \div 100$
3.897

61. $1.037 \div 100$
0.01037

62. $237.835 \div 100$
2.37835

63. $8.295 \div 1000$
0.008295

64. $82,547 \div 1000$
82.547

65. $825.37 \div 1000$
0.82537

66. $8.35 \div 10^1$
0.835

67. $0.32 \div 10^1$
0.032

68. $87.65 \div 10^1$
8.765

69. $23.627 \div 10^2$
0.23627

70. $2.954 \div 10^2$
0.02954

71. $0.0053 \div 10^2$
0.000053

72. $289.32 \div 10^3$
0.28932

73. $1.8932 \div 10^3$
0.0018932

74. $0.139 \div 10^3$
0.000139

75. Divide 44.208 by 2.4.
18.42

76. Divide 0.04664 by 0.44.
0.106

77. Find the quotient of 723.15 and 45.
16.07

78. Find the quotient of 3.3463 and 3.07.
1.09

79. Divide 13.5 by 10^3.
0.0135

80. Divide 0.045 by 10^5.
0.00000045

81. Find the quotient of 23.678 and 1000.
0.023678

82. Find the quotient of 7.005 and 10,000.
0.0007005

83. What is 0.0056 divided by 0.05?
0.112

84. What is 123.8 divided by 0.02?
6190

For Exercises 85 to 96, use a calculator to divide. Round to the nearest ten-thousandth. Then use estimation to determine whether the quotient you calculated is reasonable.

85. $42.42 \div 3.8$
Cal.: 11.1632
Est.: 10

86. $69.8 \div 7.2$
Cal.: 9.6944
Est.: 10

87. $389 \div 0.44$
Cal.: 884.0909
Est.: 1000

88. $642 \div 0.83$
Cal.: 773.4940
Est.: 750

89. $6.394 \div 3.5$
Cal.: 1.8269
Est.: 1.5

90. $8.429 \div 4.2$
Cal.: 2.0069
Est.: 2

91. $1.235 \div 0.021$
Cal.: 58.8095
Est.: 50

92. $7.456 \div 0.072$
Cal.: 103.5556
Est.: 100

93. $95.443 \div 1.32$
Cal.: 72.3053
Est.: 100

94. $423.0925 \div 4.0927$
Cal.: 103.3774
Est.: 100

95. $1.000523 \div 429.07$
Cal.: 0.0023
Est.: 0.0025

96. $0.03629 \div 0.00054$
Cal.: 67.2037
Est.: 80

Objective B   **To solve application problems**

97. **Sports**   Ramon, a high school football player, gained 162 yards on 26 carries in a high school football game. Find the average number of yards gained per carry. Round to the nearest hundredth.
6.23 yards

98. **Fuel Efficiency**   A car with an odometer reading of 17,814.2 is filled with 9.4 gallons of gas. At an odometer reading of 18,130.4, the tank is empty and the car is filled with 12.4 gallons of gas. How many miles does the car travel on 1 gallon of gasoline?
25.5 miles

99. **Consumerism**   A case of diet cola costs $6.79. If there are 24 cans in a case, find the cost per can. Round to the nearest cent.
$.28

100. **Carpentry**   Anne is building bookcases that are 3.4 feet long. How many complete shelves can be cut from a 12-foot board?
3 shelves

101. **Travel**   When the Massachusetts Turnpike opened, the toll for a passenger car that traveled the entire 136 miles of it was $5.60. Calculate the cost per mile. Round to the nearest cent.
$.04

Quick Quiz (Objective 3.5B)

1. You pay $1284.72 per year in life insurance premiums. You pay the premiums in 12 equal monthly payments. Find the amount of each monthly payment. $107.06

102. Investments An oil company has issued 3,541,221,500 shares of stock. The company paid $6,090,990,120 in dividends. Find the dividend for each share of stock. Round to the nearest cent.
$1.72

103. Insurance Earl is 52 years old and is buying $70,000 of life insurance for an annual premium of $703.80. If he pays each annual premium in 12 equal installments, how much is each monthly payment?
$58.65

104. ● **Email** The graph at the right shows the growth in the number of spam messages sent daily worldwide. Figures given are in billions. How many times greater is the number of spam messages sent daily in 2004 than the number sent in 2000? Round to the nearest tenth.
3.8 times greater

Spam Messages Sent World Wide
Source: IDC

APPLYING THE CONCEPTS

105. ● **Education** According to the National Center for Education Statistics, 10.2 million women and 7.3 million men will be enrolled at institutions of higher learning in 2010. How many more women than men are expected to be attending institutions of higher learning in 2010?
2.9 million more women

● **The Military** The table at the right shows the advertising budgets of four branches of the U.S. armed services in a recent year. Use this table for Exercises 106 to 108.

Service	Advertising Budget
Army	$85.3 million
Air Force	$41.1 million
Navy	$20.5 million
Marines	$15.9 million

Source: CMR/TNS Media Intelligence

106. Find the difference between the Army's advertising budget and the Marines' advertising budget.
$69.4 million

107. How many times greater was the Army's advertising budget than the Navy's advertising budget? Round to the nearest tenth.
4.2 times greater

108. What was the total of the advertising budgets for the four branches of the service?
$162.8 million

109. ● **Population Growth** The U.S. population of people ages 85 and over is expected to grow from 4.2 million in 2000 to 8.9 million in 2030. How many times greater is the population of this segment expected to be in 2030 than in 2000? Round to the nearest tenth.
2.1 times greater

Health   The graph at the right shows that the worldwide consumption of cigarettes has been increasing. Use this table for Exercises 110 to 112.

Consumption of Cigarettes (in trillions)

1.7 2.2 3.1 4.4 5.4 5.5

1950 1960 1970 1980 1990 2000

World Wide Consumption of Cigarettes

Source: The Tobacco Atlas; U.S. Department of Agriculture

110. Find the increase in cigarette consumption from 1950 to 1990.
3.7 trillion cigarettes

111. How many times greater was the cigarette consumption in 2000 than in 1960?
2.5 times greater

112. **a.** During which 10-year period was the increase in cigarette consumption greatest?   1970 to 1980
b. During which 10-year period was the increase in cigarette consumption the least?   1990 to 2000

Forestry   The table at the right shows the number of acres burned in wildfires through July 24 of each year listed. Use this table for Exercises 113 and 114.

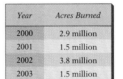

Year	Acres Burned
2000	2.9 million
2001	1.5 million
2002	3.8 million
2003	1.5 million

Source: National Interagency Fire Center

113. Find the total number of acres burned in the 4 years listed.
9.7 million acres

114. How many more acres were burned in 2002 and 2003 than in 2000 and 2001?
0.9 million acres

115. ✎ Explain how the decimal point is moved when a number is divided by 10, 100, 1000, 10,000, etc.

116. ✎ **Sports**   Explain how baseball batting averages are determined. Then find Chicago Cubs shortstop Nomar Garciaparra's batting average with 190 hits out of 532 at bats. Round to the nearest thousandth.

117. ✎ Explain how the decimal point is placed in the quotient when a number is divided by a decimal.

For Exercises 118 to 123, insert +, −, ×, or ÷ into the square so that the statement is true.

118. 3.45 ☐ 0.5 = 6.9
÷

119. 3.46 ☐ 0.24 = 0.8304
×

120. 6.009 ☐ 4.68 = 1.329
−

121. 0.064 ☐ 1.6 = 0.1024
×

122. 9.876 ☐ 23.12 = 32.996
+

123. 3.0381 ☐ 1.23 = 2.47
÷

For Exercises 124 to 126, fill in the square to make a true statement.

124. 6.47 − ☐ = 1.253
5.217

125. 6.47 + ☐ = 9
2.53

126. 0.009 ÷ ☐ = 0.36
0.025

Answers to Writing Exercises

115. When a number is divided by 10, 100, 1000, 10,000, etc., the decimal point is moved as many places to the left as there are zeros in the power of 10. For example, the decimal point in a number divided by 1000 would be moved three places to the left.

116. To calculate a batting average, divide the number of hits by the number of times at bat. Round to the nearest thousandth. Nomar Garciaparra's batting average = 190 ÷ 532 ≈ 0.357.

117. To determine where a decimal point is placed in a quotient, first move the decimal point in the divisor to make it a whole number. Then move the decimal point in the dividend the same number of places to the right. The decimal point in the quotient lines up vertically with the decimal point in the dividend.

Objective 3.6A

Vocabulary to Review
numerator [2.2A]
denominator [2.2A]

Discuss the Concepts
1. Explain how to write a proper fraction as a decimal.
2. Explain how to write a mixed number as a decimal.

Concept Check

Round $\frac{49}{50}$ to the nearest tenth.

Explain why it is incorrect to write the answer as 1.

1.0 indicates that the decimal has been rounded to the nearest tenth; 1 does not.

Optional Student Activity

Stock prices used to be written as mixed numbers rather than as decimals. The table below shows the stock prices on August 28, 2003, of four companies on the New York Stock Exchange. Convert each price to a decimal. Discuss the advantage or disadvantage of switching from using fractions to using decimals.

Company	Stock Price
Choice Hotels	$32\frac{3}{10}$
Fairmont Hotels	$26\frac{7}{20}$
Four Seasons	$47\frac{3}{4}$
Trump Hotels	$1\frac{4}{5}$

$32.30, $26.35, $47.75, $1.80

Objective 3.6B

Vocabulary to Review
place value [1.1B]

Discuss the Concepts
Explain how to write a decimal as a fraction.

3.6 Comparing and Converting Fractions and Decimals

Objective A To convert fractions to decimals

Every fraction can be written as a decimal. To write a fraction as a decimal, divide the numerator of the fraction by the denominator. The quotient can be rounded to the desired place value.

> **HOW TO** Convert $\frac{3}{7}$ to a decimal.
>
> $$7\overline{)3.00000} \quad \frac{0.42857}{}$$
>
> $\frac{3}{7}$ rounded to the nearest hundredth is 0.43.
>
> $\frac{3}{7}$ rounded to the nearest thousandth is 0.429.
>
> $\frac{3}{7}$ rounded to the nearest ten-thousandth is 0.4286.

> **HOW TO** Convert $3\frac{2}{9}$ to a decimal. Round to the nearest thousandth.
>
> $$3\frac{2}{9} = \frac{29}{9} \qquad 9\overline{)29.0000} \quad \frac{3.2222}{}$$
>
> $3\frac{2}{9}$ rounded to the nearest thousandth is 3.222.

Example 1 Convert $\frac{3}{8}$ to a decimal. Round to the nearest hundredth.

Solution $\dfrac{0.375}{8\overline{)3.000}} \approx 0.38$

You Try It 1 Convert $\frac{9}{16}$ to a decimal. Round to the nearest tenth.

Your solution 0.6

Example 2 Convert $2\frac{3}{4}$ to a decimal. Round to the nearest tenth.

Solution $2\frac{3}{4} = \frac{11}{4} \qquad \dfrac{2.75}{4\overline{)11.00}} \approx 2.8$

You Try It 2 Convert $4\frac{1}{6}$ to a decimal. Round to the nearest hundredth.

Your solution 4.17

Solutions on p. S11

Objective B To convert decimals to fractions

To convert a decimal to a fraction, remove the decimal point and place the decimal part over a denominator equal to the place value of the last digit in the decimal.

$$0.47 = \overset{\text{hundredths}}{\frac{47}{100}}$$

$$7.45 = 7\overset{\text{hundredths}}{\frac{45}{100}} = 7\frac{9}{20}$$

$$0.275 = \overset{\text{thousandths}}{\frac{275}{1000}} = \frac{11}{40}$$

$$0.16\frac{2}{3} = \overset{\text{hundredths}}{\frac{16\frac{2}{3}}{100}} = 16\frac{2}{3} \div 100 = \frac{50}{3} \times \frac{1}{100} = \frac{1}{6}$$

In-Class Examples (Objective 3.6A)

Convert the fraction to a decimal. Round to the nearest thousandth.

1. $\frac{4}{15}$ 0.267   **3.** $7\frac{1}{4}$ 7.25

2. $\frac{25}{6}$ 4.167

Example 3 Convert 0.82 and 4.75 to fractions.

Solution $0.82 = \dfrac{82}{100} = \dfrac{41}{50}$

$4.75 = 4\dfrac{75}{100} = 4\dfrac{3}{4}$

You Try It 3 Convert 0.56 and 5.35 to fractions.

Your solution $\dfrac{14}{25}, 5\dfrac{7}{20}$

Example 4 Convert $0.15\dfrac{2}{3}$ to a fraction.

Solution $0.15\dfrac{2}{3} = \dfrac{15\frac{2}{3}}{100} = 15\dfrac{2}{3} \div 100$

$= \dfrac{47}{3} \times \dfrac{1}{100} = \dfrac{47}{300}$

You Try It 4 Convert $0.12\dfrac{7}{8}$ to a fraction.

Your solution $\dfrac{103}{800}$

Solutions on p. S11

Discuss the Concepts
Given both a fraction and a decimal, how can you determine which is the larger number?

Concept Check
Which of the following is true?

a. $\dfrac{137}{300} = 0.456666667$ False

b. $\dfrac{137}{300} < 0.45666666$ False

c. $\dfrac{137}{300} > 0.45666666$ True

Instructor Note
Have students, when comparing decimals, write each number with the same number of place values by adding zeros if necessary. For example, it's easier to see that $0.375 < 0.38$ when you write it as $0.375 < 0.380$.

Objective C To identify the order relation between two decimals or between a decimal and a fraction

Decimals, like whole numbers and fractions, can be graphed as points on the number line. The number line can be used to show the order of decimals. A decimal that appears to the right of a given number is greater than the given number. A decimal that appears to the left of a given number is less than the given number.

```
 ◄──┼────┼────┼────┼────┼────┼────┼────┼────┼──►
   3.00 3.05 3.10 3.15 3.20 3.25 3.30 3.35 3.40
```

Note that 3, 3.0, and 3.00 represent the same number.

HOW TO Find the order relation between $\dfrac{3}{8}$ and 0.38.

$\dfrac{3}{8} = 0.375$ $0.38 = 0.380$ • Convert the fraction $\dfrac{3}{8}$ to a decimal.

$0.375 < 0.380$ • Compare the two decimals.

$\dfrac{3}{8} < 0.38$ • Convert 0.375 back to a fraction.

Example 5 Place the correct symbol, $<$ or $>$, between the numbers.

$\dfrac{5}{16}$ 0.32

Solution $\dfrac{5}{16} \approx 0.313$ • Convert $\dfrac{5}{16}$ to a decimal.

$0.313 < 0.32$ • Compare the two decimals.

$\dfrac{5}{16} < 0.32$ • Convert 0.313 back to a fraction.

You Try It 5 Place the correct symbol, $<$ or $>$, between the numbers.

0.63 $\dfrac{5}{8}$

Your solution $0.63 > \dfrac{5}{8}$

Solution on p. S11

Optional Student Activity

Convert $\dfrac{1}{9}, \dfrac{2}{9}, \dfrac{3}{9}$, and $\dfrac{4}{9}$ to decimals. Describe the pattern. Use the pattern to convert $\dfrac{5}{9}, \dfrac{7}{9}$, and $\dfrac{8}{9}$ to decimals.

$\dfrac{1}{9} = 0.111\ldots, \dfrac{2}{9} = 0.222\ldots,$

$\dfrac{3}{9} = 0.333\ldots,$ and

$\dfrac{4}{9} = 0.444\ldots.$

The pattern is a decimal with a repeating digit that is the same as the numerator of the original fraction.

$\dfrac{5}{9} = 0.555\ldots, \dfrac{7}{9} = 0.777\ldots,$

and $\dfrac{8}{9} = 0.888\ldots$

In-Class Examples (Objective 3.6B)
Convert the decimal to a fraction.

1. 0.6 $\dfrac{3}{5}$

2. 0.42 $\dfrac{21}{50}$

3. 8.808 $8\dfrac{101}{125}$

In-Class Examples (Objective 3.6C)
Place the correct symbol, $<$ or $>$, between the numbers.

1. 0.39   0.7 $<$

2. $\dfrac{7}{9}$ 0.76 $>$

3. 0.42 $\dfrac{5}{12}$ $>$

162 Chapter 3 / Decimals

Section 3.6

Suggested Assignment
Exercises 1–69, odds
Exercises 70–72
More challenging problems:
    Exercises 73, 74

3.6 Exercises

Objective A   **To convert fractions to decimals**

For Exercises 1 to 24, convert the fraction to a decimal. Round to the nearest thousandth.

1. $\dfrac{5}{8}$
0.625

2. $\dfrac{7}{12}$
0.583

3. $\dfrac{2}{3}$
0.667

4. $\dfrac{5}{6}$
0.833

5. $\dfrac{1}{6}$
0.167

6. $\dfrac{7}{8}$
0.875

7. $\dfrac{5}{12}$
0.417

8. $\dfrac{9}{16}$
0.563

9. $\dfrac{7}{4}$
1.750

10. $\dfrac{5}{3}$
1.667

11. $1\dfrac{1}{2}$
1.500

12. $2\dfrac{1}{3}$
2.333

13. $\dfrac{16}{4}$
4.000

14. $\dfrac{36}{9}$
4.000

15. $\dfrac{3}{1000}$
0.003

16. $\dfrac{5}{10}$
0.500

17. $7\dfrac{2}{25}$
7.080

18. $16\dfrac{7}{9}$
16.778

19. $37\dfrac{1}{2}$
37.500

20. $\dfrac{5}{24}$
0.208

21. $\dfrac{4}{25}$
0.160

22. $3\dfrac{1}{3}$
3.333

23. $8\dfrac{2}{5}$
8.400

24. $5\dfrac{4}{9}$
5.444

Objective B   **To convert decimals to fractions**

For Exercises 25 to 49, convert the decimal to a fraction.

25. 0.8
$\dfrac{4}{5}$

26. 0.4
$\dfrac{2}{5}$

27. 0.32
$\dfrac{8}{25}$

28. 0.48
$\dfrac{12}{25}$

29. 0.125
$\dfrac{1}{8}$

30. 0.485
$\dfrac{97}{200}$

31. 1.25
$1\dfrac{1}{4}$

32. 3.75
$3\dfrac{3}{4}$

33. 16.9
$16\dfrac{9}{10}$

34. 17.5
$17\dfrac{1}{2}$

35. 8.4
$8\dfrac{2}{5}$

36. 10.7
$10\dfrac{7}{10}$

37. 8.437
$8\dfrac{437}{1000}$

38. 9.279
$9\dfrac{279}{1000}$

39. 2.25
$2\dfrac{1}{4}$

40. 7.75
$7\dfrac{3}{4}$

41. $0.15\dfrac{1}{3}$
$\dfrac{23}{150}$

42. $0.17\dfrac{2}{3}$
$\dfrac{53}{300}$

43. $0.87\dfrac{7}{8}$
$\dfrac{703}{800}$

44. $0.12\dfrac{5}{9}$
$\dfrac{113}{900}$

45. 7.38
$7\dfrac{19}{50}$

46. 0.33
$\dfrac{33}{100}$

47. 0.57
$\dfrac{57}{100}$

48. $0.33\dfrac{1}{3}$
$\dfrac{1}{3}$

49. $0.66\dfrac{2}{3}$
$\dfrac{2}{3}$

Quick Quiz (Objective 3.6A)
Convert the fraction to a decimal. Round to the nearest thousandth.

1. $\dfrac{1}{12}$ 0.083

2. $\dfrac{53}{7}$ 7.571

3. $12\dfrac{1}{6}$ 12.167

Quick Quiz (Objective 3.6B)
Convert the decimal to a fraction.

1. 0.5 $\dfrac{1}{2}$

2. 0.78 $\dfrac{39}{50}$

3. 5.146 $5\dfrac{73}{500}$

Objective C **To identify the order relation between two decimals or between a decimal and a fraction**

For Exercises 50 to 69, place the correct symbol, < or >, between the numbers.

50. 0.15 < 0.5

51. 0.6 > 0.45

52. 6.65 > 6.56

53. 3.89 < 3.98

54. 2.504 > 2.054

55. 0.025 < 0.105

56. $\frac{3}{8}$ > 0.365

57. $\frac{4}{5}$ < 0.802

58. $\frac{2}{3}$ > 0.65

59. 0.85 < $\frac{7}{8}$

60. $\frac{5}{9}$ > 0.55

61. $\frac{7}{12}$ > 0.58

62. 0.62 > $\frac{7}{15}$

63. $\frac{11}{12}$ < 0.92

64. 0.161 > $\frac{1}{7}$

65. 0.623 > 0.6023

66. 0.86 > 0.855

67. 0.87 > 0.087

68. 1.005 > 0.5

69. 0.033 < 0.3

APPLYING THE CONCEPTS

Demography The graph at the right shows the U.S. population according to the 2000 census. Use this table for Exercises 70 to 72.

70. Are there more individuals ages 0 to 39 or ages 40 and over in the United States?
Ages 0 to 39

71. Is the population ages 0 to 19 more than or less than $\frac{1}{4}$ of the total population?
More than

72. Is the population ages 0 to 39 more than or less than $\frac{1}{2}$ of the total population?
More than

U.S. Population
Source: U.S. Census Bureau

73. Is $\frac{7}{13}$ in decimal form a repeating decimal? Why or why not? (*Hint:* See Projects and Group Activities, page 165.)
Yes, the digits 538461 repeat.

74. If a number is rounded to the nearest thousandth, is it always greater than if it were rounded to the nearest hundredth? Give examples to support your answer.
No. 0.0402 rounded to hundredths is 0.04; to thousandths it is 0.040.

75. Explain how terminating, repeating, and nonrepeating decimals differ. Give an example of each kind of decimal.

Answers to Writing Exercises

75. A terminating decimal ends, or stops. For example, 3.25 and 9.762104 are terminating decimals.

A repeating decimal never ends. One or more digits to the right of the decimal point repeat without end. The decimals in the Optional Student Activity for Objective 3.6C are examples of repeating decimals: 0.111..., 0.222..., 0.333..., 0.444..., etc.

A nonrepeating decimal never ends, nor does it have any digits to the right of the decimal point that repeat. For example, 1.20200200020000200000... is a nonrepeating decimal.

Quick Quiz (Objective 3.6C)

Place the correct symbol, < or >, between the numbers.

1. 0.25   0.3   <

2. 0.313 $\frac{5}{16}$ >

3. $\frac{6}{7}$ 0.84   >

Focus on Problem Solving

Relevant Information

Problems in mathematics or real life involve a question or a need and information or circumstances related to that question or need. Solving problems in the sciences usually involves a question, an observation, and measurements of some kind.

One of the challenges of problem solving in the sciences is to separate the information that is relevant to the problem from other information. Following is an example from the physical sciences in which some relevant information was omitted.

Hooke's Law states that the distance that a weight will stretch a spring is directly proportional to the weight on the spring. That is, $d = kF$, where d is the distance the spring is stretched and F is the force. In an experiment to verify this law, some physics students were continually getting inconsistent results. Finally, the instructor discovered that the heat produced when the lights were turned on was affecting the experiment. In this case, relevant information was omitted—namely, that the temperature of the spring can affect the distance it will stretch.

A lawyer drove 8 miles to the train station. After a 35-minute ride of 18 miles, the lawyer walked 10 minutes to the office. Find the total time it took the lawyer to get to work.

From this situation, answer the following before reading on.

a. What is asked for?

b. Is there enough information to answer the question?

c. Is information given that is not needed?

Here are the answers.

a. We want the total time for the lawyer to get to work.

b. No. We do not know the time it takes the lawyer to get to the train station.

c. Yes. Neither the distance to the train station nor the distance of the train ride is necessary to answer the question.

For each of the following problems, answer the questions printed in red above.

1. A customer bought 6 boxes of strawberries and paid with a $20 bill. What was the change?

2. A board is cut into two pieces. One piece is 3 feet longer than the other piece. What is the length of the original board?

3. A family rented a car for their vacation and drove 680 miles. The cost of the rental car was $21 per day with 150 free miles per day and $.15 for each mile driven above the number of free miles allowed. How many miles did the family drive per day?

4. An investor bought 8 acres of land for $80,000. One and one-half acres were set aside for a park, and the remainder was developed into one-half-acre lots. How many lots were available for sale?

5. You wrote checks of $43.67, $122.88, and $432.22 after making a deposit of $768.55. How much do you have left in your checking account?

Answers to Focus on Problem Solving—Relevant Information

1a. The amount of change is asked for.

b. No. We need to know the cost of the strawberries.

c. No.

2a. The length of the original board is asked for.

b. No. We need to know the length of the other piece.

c. No.

3a. The number of miles driven per day is asked for.

b. No. We are given only the total number of miles driven. We need the number of days the car is rented.

c. Yes. Only the total number of miles driven can be used to answer the question.

4a. We are asked for the number of lots available for sale.

b. Yes.

c. Yes. We do not need the price of the land.

5a. We are asked for the checking account balance.

b. No. We don't know the balance before writing the checks and making the deposit.

c. No.

Projects and Group Activities

Fractions as Terminating or Repeating Decimals

The fraction $\frac{3}{4}$ is equivalent to 0.75. The decimal 0.75 is a terminating decimal because there is a remainder of zero when 3 is divided by 4. The fraction $\frac{1}{3}$ is equivalent to 0.333 . . . . The three dots mean the pattern continues on and on. 0.333 . . . is a repeating decimal. To determine whether a fraction can be written as a terminating decimal, first write the fraction in simplest form. Then look at the denominator of the fraction. If it contains prime factors of only 2s and/or 5s, then it can be expressed as a terminating decimal. If it contains prime factors other than 2s or 5s, it represents a repeating decimal.

TAKE NOTE

If the denominator of a fraction in simplest form is 20, then it can be written as a terminating decimal because $20 = 2 \cdot 2 \cdot 5$ (only prime factors of 2 and 5). If the denominator of a fraction in simplest form is 6, it represents a repeating decimal because it contains the prime factor 3 (a number other than 2 or 5).

1. Assume that each of the following numbers is the denominator of a fraction written in simplest form. Does the fraction represent a terminating or repeating decimal?

   a. 4   **b.** 5   **c.** 7   **d.** 9   **e.** 10   **f.** 12   **g.** 15
   h. 16   **i.** 18   **j.** 21   **k.** 24   **l.** 25   **m.** 28   **n.** 40

2. Write two other numbers that, as denominators of fractions in simplest form, represent terminating decimals, and write two other numbers that, as denominators of fractions in simplest form, represent repeating decimals.

Answers to Projects and Group Activities— Fractions as Terminating or Repeating Decimals

1a. Terminating
 b. Terminating
 c. Repeating
 d. Repeating
 e. Terminating
 f. Repeating
 g. Repeating
 h. Terminating
 i. Repeating
 j. Repeating
 k. Repeating
 l. Terminating
 m. Repeating
 n. Terminating

2. Answers will vary.

Chapter 3 Summary

Key Words	**Examples**
A number written in *decimal notation* has three parts: a *whole-number part*, a *decimal point*, and a *decimal part*. The decimal part of a number represents a number less than 1. A number written in decimal notation is often simply called a *decimal*. [3.1A, p. 127]	For the decimal 31.25, 31 is the whole-number part and 25 is the decimal part.

Essential Rules and Procedures	**Examples**
To write a decimal in words, write the decimal part as if it were a whole number. Then name the place value of the last digit. The decimal point is read as "and." [3.1A, p. 127]	The decimal 12.875 is written in words as twelve and eight hundred seventy-five thousandths.
To write a decimal in standard form when it is written in words, write the whole-number part, replace the word *and* with a decimal point, and write the decimal part so that the last digit is in the given place-value position. [3.1A, p. 128]	The decimal forty-nine and sixty-three thousandths is written in standard form as 49.063.
To round a decimal to a given place value, use the same rules used with whole numbers, except drop the digits to the right of the given place value instead of replacing them with zeros. [3.1B, p. 129]	2.7134 rounded to the nearest tenth is 2.7. 0.4687 rounded to the nearest hundredth is 0.47.

To add decimals, write the decimals so that the decimal points are on a vertical line. Add as you would with whole numbers. Then write the decimal point in the sum directly below the decimal points in the addends. [3.2A, p. 133]	$\begin{array}{r} 1\ 1 \\ 1.35 \\ 20.8 \\ +\ \ 0.76 \\ \hline 22.91 \end{array}$

To subtract decimals, write the decimals so that the decimal points are on a vertical line. Subtract as you would with whole numbers. Then write the decimal point in the difference directly below the decimal point in the subtrahend. [3.3A, p. 137]	$\begin{array}{r} {\scriptstyle 2\ 15\ \ \ \ 6\ 10} \\ 3\,5\,.\,8\,7\,0 \\ -\ \ 9\,.\,6\,4\,1 \\ \hline 2\,6\,.\,2\,2\,9 \end{array}$

To multiply decimals, multiply the numbers as you would whole numbers. Then write the decimal point in the product so that the number of decimal places in the product is the sum of the decimal places in the factors. [3.4A, p. 141]	$\begin{array}{rl} 26.83 & \text{2 decimal places} \\ \times\ \ \ \ 0.45 & \text{2 decimal places} \\ \hline 13415 & \\ 10732\ \ & \\ \hline 12.0735 & \text{4 decimal places} \end{array}$

To multiply a decimal by a power of 10, move the decimal point to the right the same number of places as there are zeros in the power of 10. If the power of 10 is written in exponential notation, the exponent indicates how many places to move the decimal point. [3.4A, pp. 141, 142]	$3.97 \cdot 10{,}000 = 39{,}700$ $0.641 \cdot 10^5 = 64{,}100$

To divide decimals, move the decimal point in the divisor to the right so that it is a whole number. Move the decimal point in the dividend the same number of places to the right. Place the decimal point in the quotient directly above the decimal point in the dividend. Then divide as you would with whole numbers. [3.5A, p. 151]	$\begin{array}{r} 6.2 \\ 0.39.\overline{)2.41.8} \\ -2\,34 \\ \hline 7\,8 \\ -7\,8 \\ \hline 0 \end{array}$

To divide a decimal by a power of 10, move the decimal point to the left the same number of places as there are zeros in the power of 10. If the power of 10 is written in exponential notation, the exponent indicates how many places to move the deicmal point. [3.5A, p. 152]	$972.8 \div 1000 = 0.9728$ $61.305 \div 10^4 = 0.0061305$

To convert a fraction to a decimal, divide the numerator of the fraction by the denominator. [3.6A, p. 160]	$\dfrac{7}{8} = 7 \div 8 = 0.875$

To convert a decimal to a fraction, remove the decimal point and place the decimal part over a denominator equal to the place value of the last digit in the decimal. [3.6B, p. 160]	0.85 is eighty-five <u>hundredths</u>. $0.85 = \dfrac{85}{100} = \dfrac{17}{20}$

To find the order relation between a decimal and a fraction, first rewrite the fraction as a decimal. Then compare the two decimals. [3.6C, p. 161]	Because $\dfrac{3}{11} \approx 0.273$, and $0.273 > 0.26$, $\dfrac{3}{11} > 0.26$.

Chapter 3 Review Exercises

1. Find the quotient of 3.6515 and 0.067.
54.5 [3.5A]

2. Find the sum of 369.41, 88.3, 9.774, and 366.474.
833.958 [3.2A]

3. Place the correct symbol, < or >, between the two numbers.
0.055 < 0.1 [3.6C]

4. Write 22.0092 in words.
Twenty-two and ninety-two ten-thousandths [3.1A]

5. Round 0.05678235 to the nearest hundred-thousandth.
0.05678 [3.1B]

6. Convert $2\frac{1}{3}$ to a decimal. Round to the nearest hundredth.
2.33 [3.6A]

7. Convert 0.375 to a fraction.
$\frac{3}{8}$ [3.6B]

8. Add: 3.42 + 0.794 + 32.5
36.714 [3.2A]

9. Write thirty-four and twenty-five thousandths in standard form.
34.025 [3.1A]

10. Place the correct symbol, < or >, between the two numbers.
$\frac{5}{8}$ > 0.62 [3.6C]

11. Convert $\frac{7}{9}$ to a decimal. Round to the nearest thousandth.
0.778 [3.6A]

12. Convert 0.66 to a fraction.
$\frac{2}{3}$ [3.6B]

13. Subtract: 27.31 − 4.4465
22.8635 [3.3A]

14. Round 7.93704 to the nearest hundredth.
7.94 [3.1B]

15. Find the product of 3.08 and 2.9.
8.932 [3.4A]

16. Write 342.37 in words.
Three hundred forty-two and thirty-seven hundredths
[3.1A]

17. Write three and six thousand seven hundred fifty-three hundred-thousandths in standard form.
3.06753 [3.1A]

18. Multiply: 34.79
× 0.74
25.7446 [3.4A]

19. Divide: $0.053\overline{)0.349482}$
6.594 [3.5A]

20. What is 7.796 decreased by 2.9175?
4.8785 [3.3A]

21. Banking You had a balance of $895.68 in your checking account. You then wrote checks of $145.72 and $88.45. Find the new balance in your checking account.
$661.51 [3.3B]

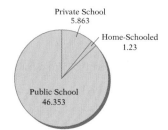

Education The graph at the right shows where American children in grades K–12 are being educated. Figures are in millions of children. Use this graph for Exercises 22 and 23.

Private School
5.863

Home-Schooled
1.23

Public School
46.353

Where Children in Grades K–12 Are
Being Educated in America
(in millions)

Sources: U.S. Department of Education;
Home School Legal Defense Association

22. Find the total number of American children in grades K–12.
53.446 million children [3.2B]

23. How many more children are being educated in public school than in private school?
40.49 million children [3.3B]

24. **Nutrition** According to the American School Food Service Association, 1.9 million gallons of milk are served in school cafeterias every day. How many gallons of milk are served in school cafeterias during a 5-day school week?
9.5 million gallons [3.4B]

25. **Travel** In a recent year, 30.6 million Americans drove to their destinations over Thanksgiving, and 4.8 million Americans traveled by plane. (*Source:* AAA) How many times greater is the number who drove than the number who flew? Round to the nearest tenth.
6.4 times greater [3.5B]

Chapter 3 Test

1. Place the correct symbol, $<$ or $>$, between the two numbers.
0.66 $<$ 0.666 [3.6C]

2. Subtract: 13.027
                  $-$ 8.94
                       4.087   [3.3A]

3. Write 45.0302 in words.
Forty-five and three hundred two ten-thousandths
[3.1A]

4. Convert $\frac{9}{13}$ to a decimal. Round to the nearest thousandth.
0.692 [3.6A]

5. Convert 0.825 to a fraction.
$\frac{33}{40}$ [3.6B]

6. Round 0.07395 to the nearest ten-thousandth.
0.0740 [3.1B]

7. Find 0.0569 divided by 0.037. Round to the nearest thousandth.
1.538 [3.5A]

8. Find 9.23674 less than 37.003.
27.76626 [3.3A]

9. Round 7.0954625 to the nearest thousandth.
7.095 [3.1B]

10. Divide: $0.006\overline{)1.392}$
        232 [3.5A]

11. Add: 270.93
               97.
               1.976
           $+$ 88.675
               458.581 [3.2A]

12. **Mechanics**   Find the missing dimension.

4.86 in.

?

6.23 in.

1.37 inches   [3.3B]

13. Multiply: 1.37
                   × 0.004
                   ─────────
                   0.00548   [3.4A]

14. What is the total of 62.3, 4.007, and 189.65?
255.957   [3.2A]

15. Write two hundred nine and seven thousand eighty-six hundred-thousandths in standard form.
209.07086   [3.1A]

16. Finances A car was bought for $16,734.40, with a down payment of $2500. The balance was paid in 36 monthly payments. Find the amount of each monthly payment.
$395.40   [3.5B]

17. Compensation You received a salary of $727.50, a commission of $1909.64, and a bonus of $450. Find your total income.
$3087.14   [3.2B]

18. Consumerism A long-distance telephone call costs $.85 for the first 3 minutes and $.42 for each additional minute. Find the cost of a 12-minute long-distance telephone call.
$4.63   [3.4B]

Computers The table at the right shows the average number of hours per week that students use a computer. Use this table for Exercises 19 and 20.

19. On average, how many hours per year does a 10th-grade student use a computer? Use a 52-week year.
348.4 hours   [3.4B]

Grade Level	Average Number of Hours of Computer Use per Week
Prekindergarten – kindergarten	3.9
1st – 3rd	4.9
4th – 6th	4.2
7th – 8th	6.9
9th – 12th	6.7

Source: Find/SVP American Learning Household Survey

20. On average, how many more hours per year does a 2nd-grade student use a computer than a 5th-grade student? Use a 52-week year.
36.4 hours   [3.4B]

Cumulative Review Exercises

1. Divide: $89\overline{)20{,}932}$
   235 r17   [1.5C]

2. Simplify: $2^3 \cdot 4^2$
   128   [1.6A]

3. Simplify: $2^2 - (7 - 3) \div 2 + 1$
   3   [1.6B]

4. Find the LCM of 9, 12, and 24.
   72   [2.1A]

5. Write $\frac{22}{5}$ as a mixed number.
   $4\frac{2}{5}$ [2.2B]

6. Write $4\frac{5}{8}$ as an improper fraction.
   $\frac{37}{8}$ [2.2B]

7. Write an equivalent fraction with the given denominator.
   $$\frac{5}{12} = \frac{}{60}$$
   $\frac{25}{60}$ [2.3A]

8. Add: $\frac{3}{8} + \frac{5}{12} + \frac{9}{16}$
   $1\frac{17}{48}$ [2.4B]

9. What is $5\frac{7}{12}$ increased by $3\frac{7}{18}$?
   $8\frac{35}{36}$ [2.4C]

10. Subtract: $9\frac{5}{9} - 3\frac{11}{12}$
    $5\frac{23}{36}$ [2.5C]

11. Multiply: $\frac{9}{16} \times \frac{4}{27}$
    $\frac{1}{12}$ [2.6A]

12. Find the product of $2\frac{1}{8}$ and $4\frac{5}{17}$.
    $9\frac{1}{8}$ [2.6B]

13. Divide: $\frac{11}{12} \div \frac{3}{4}$
    $1\frac{2}{9}$ [2.7A]

14. What is $2\frac{3}{8}$ divided by $2\frac{1}{2}$?
    $\frac{19}{20}$ [2.7B]

15. Simplify: $\left(\frac{2}{3}\right)^2 \cdot \left(\frac{3}{4}\right)^3$
    $\frac{3}{16}$ [2.8B]

16. Simplify: $\left(\frac{2}{3}\right)^2 - \left(\frac{2}{3} - \frac{1}{2}\right) + 2$
    $2\frac{5}{18}$ [2.8C]

17. Write 65.0309 in words.
    Sixty-five and three hundred nine
    ten-thousandths   [3.1A]

18. Add: 379.006
             27.523
             9.8707
    $+$ 88.2994
             504.6991   [3.2A]

19. What is 29.005 decreased by 7.9286?
21.0764 [3.3A]

20. Multiply: 9.074
             × 6.09
            55.26066   [3.4A]

21. Divide: 8.09)‾17.42963.
Round to the nearest thousandth.
2.154 [3.5A]

22. Convert $\frac{11}{15}$ to a decimal. Round to the nearest thousandth.
0.733 [3.6A]

23. Convert $0.16\frac{2}{3}$ to a fraction.
$\frac{1}{6}$ [3.6B]

24. Place the correct symbol, < or >, between the two numbers.
$\frac{8}{9}$ < 0.98   [3.6C]

25. Vacation   The graph at the right shows the number of vacation days per year that are legally mandated in several countries. How many more vacation days does Sweden mandate than Germany?
14 days   [1.3C]

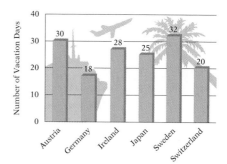

Number of Legally Mandated Vacation Days
Sources: Economic Policy Institute; *World Almanac*

26. Health   A patient is put on a diet to lose 24 pounds in 3 months. The patient loses $9\frac{1}{2}$ pounds the first month and $6\frac{3}{4}$ pounds the second month. How much weight must this patient lose the third month to achieve the goal?
$7\frac{3}{4}$ [2.5D]

27. Banking   You have a checking account balance of $814.35. You then write checks for $42.98, $16.43, and $137.56. Find your checking account balance after you write the checks.
$617.38   [3.3B]

28. Mechanics   A machine lathe takes 0.017 inch from a brass bushing that is 1.412 inches thick. Find the resulting thickness of the bushing.
1.395 inches   [3.3B]

29. Taxes   The state income tax on your business is $820 plus 0.08 times your profit. You made a profit of $64,860 last year. Find the amount of income tax you paid last year.
$6008.80   [3.4B]

30. Finances   You bought a camera costing $210.96. The down payment was $20, and the balance is to be paid in 8 equal monthly payments. Find the monthly payment.
$23.87   [3.5B]

4 Ratio and Proportion

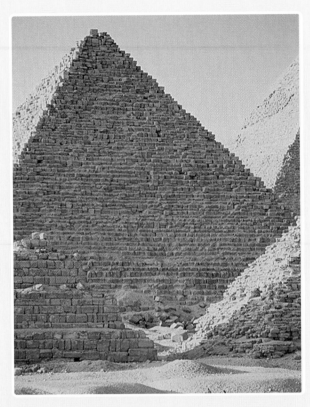

Egypt is known around the world for its enormous pyramids, built thousands of years ago by the pharaohs. This photo shows the pyramids at Giza. The largest pyramid at Giza, called the Great Pyramid, dates to approximately 2600 B.C. and is the oldest of the Seven Ancient Wonders of the World. It is also the only ancient wonder that still survives today. Some historians believe that some of the pyramids of Egypt incorporate a special ratio called the golden ratio. This ratio, which is described in the **project on page 192,** has been used extensively in art and architecture for centuries. The ratio of the slant height to a side of the base of the pyramid reflects the golden ratio.

OBJECTIVES

Section 4.1
A To write the ratio of two quantities in simplest form
B To solve application problems

Section 4.2
A To write rates
B To write unit rates
C To solve application problems

Section 4.3
A To determine whether a proportion is true
B To solve proportions
C To solve application problems

Need help? For online student resources, such as section quizzes, visit this textbook's website at **math.college.hmco.com/students.**

Do these exercises to prepare for Chapter 4.

1. Simplify: $\dfrac{8}{10}$

   $\dfrac{4}{5}$ [2.3B]

2. Simplify: $\dfrac{450}{650 + 250}$

   $\dfrac{1}{2}$ [2.3B]

3. Write as a decimal: $\dfrac{372}{15}$

   24.8 [3.6A]

4. Which is greater, 4×33 or 62×2?

   4×33 [1.4A]

5. Complete: $? \times 5 = 20$

   4 [1.5A]

GO FIGURE • • •

Luis, Kim, Reggie, and Dave are standing in line. Dave is not first. Kim is between Luis and Reggie. Luis is between Dave and Kim. Give the order in which the men are standing.

Reggie, Kim, Luis, Dave

4.1 Ratio

Objective A — **To write the ratio of two quantities in simplest form**

Quantities such as 3 feet, 12 cents, and 9 cars are number quantities written with units.

$$\left.\begin{array}{l} 3 \text{ feet} \\ 12 \text{ cents} \\ 9 \text{ cars} \end{array}\right\} \uparrow \text{ units}$$ These are some examples of units. Shirts, dollars, trees, miles, and gallons are further examples.

A **ratio** is a comparison of two quantities that have the *same* units. This comparison can be written three different ways:

1. As a fraction
2. As two numbers separated by a colon (:)
3. As two numbers separated by the word *to*

The ratio of the lengths of two boards, one 8 feet long and the other 10 feet long, can be written as

1. $\dfrac{8 \text{ feet}}{10 \text{ feet}} = \dfrac{8}{10} = \dfrac{4}{5}$

2. 8 feet : 10 feet = 8 : 10 = 4 : 5

3. 8 feet to 10 feet = 8 to 10 = 4 to 5

Writing the **simplest form of a ratio** means writing it so that the two numbers have no common factor other than 1.

This ratio means that the smaller board is $\dfrac{4}{5}$ the length of the longer board.

Example 1

Write the comparison $6 to $8 as a ratio in simplest form using a fraction, a colon, and the word *to*.

Solution $\dfrac{\$6}{\$8} = \dfrac{6}{8} = \dfrac{3}{4}$

$6 : $8 = 6 : 8 = 3 : 4

$6 to $8 = 6 to 8 = 3 to 4

You Try It 1

Write the comparison 20 pounds to 24 pounds as a ratio in simplest form using a fraction, a colon, and the word *to*.

Your solution $\dfrac{5}{6}$ 5 : 6   5 to 6

Example 2

Write the comparison 18 quarts to 6 quarts as a ratio in simplest form using a fraction, a colon, and the word *to*.

Solution $\dfrac{18 \text{ quarts}}{6 \text{ quarts}} = \dfrac{18}{6} = \dfrac{3}{1}$

18 quarts : 6 quarts = 18 : 6 = 3 : 1

18 quarts to 6 quarts = 18 to 6 = 3 to 1

You Try It 2

Write the comparison 64 miles to 8 miles as a ratio in simplest form using a fraction, a colon, and the word *to*.

Your solution $\dfrac{8}{1}$ 8 : 1   8 to 1

Solutions on p. S11

Objective B   To solve application problems

Use the table below for Example 3 and You Try It 3.

Board Feet of Wood at a Lumber Store			
Pine	Ash	Oak	Cedar
20,000	18,000	10,000	12,000

Example 3

Find, as a fraction in simplest form, the ratio of the number of board feet of pine to the number of board feet of oak.

Strategy
To find the ratio, write the ratio of board feet of pine (20,000) to board feet of oak (10,000) in simplest form.

Solution
$$\frac{20,000}{10,000} = \frac{2}{1}$$

The ratio is $\frac{2}{1}$.

You Try It 3

Find, as a fraction in simplest form, the ratio of the number of board feet of cedar to the number of board feet of ash.

Your strategy

Your solution
$\frac{2}{3}$

Example 4

The cost of building a patio cover was $500 for labor and $700 for materials. What, as a fraction in simplest form, is the ratio of the cost of materials to the total cost for labor and materials?

Strategy
To find the ratio, write the ratio of the cost of materials ($700) to the total cost ($500 + $700) in simplest form.

Solution
$$\frac{\$700}{\$500 + \$700} = \frac{700}{1200} = \frac{7}{12}$$

The ratio is $\frac{7}{12}$.

You Try It 4

A company spends $60,000 a month for television advertising and $45,000 a month for radio advertising. What, as a fraction in simplest form, is the ratio of the cost of radio advertising to the total cost of radio and television advertising?

Your strategy

Your solution
$\frac{3}{7}$

Solutions on p. S11

In-Class Examples (Objective 4.1B)

1. You sleep 8 hours per day. Find the ratio of the number of hours you sleep to the number of hours in one day. Write the ratio as a fraction in simplest form.

$\dfrac{1}{3}$

2. A house with an original value of $144,000 had increased in value to $187,200 five years later. What is the ratio, as a fraction in simplest form, of the increase in value to the original value?

$\dfrac{3}{10}$

4.1 Exercises

Objective A   To write the ratio of two quantities in simplest form

Suggested Assignment
Exercises 1–33, odds
More challenging problems:
    Exercises 35–38

For Exercises 1 to 24, write the comparison as a ratio in simplest form using a fraction, a colon (:), and the word *to*.

1. 3 pints to 15 pints
$\frac{1}{5}$ 1:5   1 to 5

2. 6 pounds to 8 pounds
$\frac{3}{4}$ 3:4   3 to 4

3. $40 to $20
$\frac{2}{1}$ 2:1   2 to 1

4. 10 feet to 2 feet
$\frac{5}{1}$ 5:1   5 to 1

5. 3 miles to 8 miles
$\frac{3}{8}$ 3:8   3 to 8

6. 2 hours to 3 hours
$\frac{2}{3}$ 2:3   2 to 3

7. 37 hours to 24 hours
$\frac{37}{24}$ 37:24   37 to 24

8. 29 inches to 12 inches
$\frac{29}{12}$ 29:12   29 to 12

9. 6 minutes to 6 minutes
$\frac{1}{1}$ 1:1   1 to 1

10. 8 days to 12 days
$\frac{2}{3}$ 2:3   2 to 3

11. 35 cents to 50 cents
$\frac{7}{10}$ 7:10   7 to 10

12. 28 inches to 36 inches
$\frac{7}{9}$ 7:9   7 to 9

13. 30 minutes to 60 minutes
$\frac{1}{2}$ 1:2   1 to 2

14. 25 cents to 100 cents
$\frac{1}{4}$ 1:4   1 to 4

15. 32 ounces to 16 ounces
$\frac{2}{1}$ 2:1   2 to 1

16. 12 quarts to 4 quarts
$\frac{3}{1}$ 3:1   3 to 1

17. 3 cups to 4 cups
$\frac{3}{4}$ 3:4   3 to 4

18. 6 years to 7 years
$\frac{6}{7}$ 6:7   6 to 7

19. $5 to $3
$\frac{5}{3}$ 5:3   5 to 3

20. 30 yards to 12 yards
$\frac{5}{2}$ 5:2   5 to 2

21. 12 quarts to 18 quarts
$\frac{2}{3}$ 2:3   2 to 3

22. 20 gallons to 28 gallons
$\frac{5}{7}$ 5:7   5 to 7

23. 14 days to 7 days
$\frac{2}{1}$ 2:1   2 to 1

24. 9 feet to 3 feet
$\frac{3}{1}$ 3:1   3 to 1

Objective B   To solve application problems

For Exercises 25 to 28, write ratios in simplest form using a fraction.

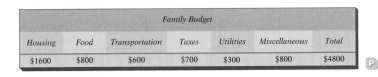

Family Budget						
Housing	Food	Transportation	Taxes	Utilities	Miscellaneous	Total
$1600	$800	$600	$700	$300	$800	$4800

25. Budgets   Use the table to find the ratio of housing cost to total expenses.
$\frac{1}{3}$

26. Budgets   Use the table to find the ratio of food cost to total expenses.
$\frac{1}{6}$

27. Budgets   Use the table to find the ratio of utilities cost to food cost.
$\frac{3}{8}$

28. Budgets   Use the table to find the ratio of transportation cost to housing cost.
$\frac{3}{8}$

Quick Quiz (Objective 4.1A)
Write the comparison as a ratio in simplest form using a fraction, a colon (:), and the word *to*.

1. 2 cups to 8 cups $\frac{1}{4}$ 1:4   1 to 4

2. 18 hours to 3 hours $\frac{6}{1}$ 6:1   6 to 1

3. 7 ounces to 13 ounces $\frac{7}{13}$ 7:13   7 to 13

29. **Sports** National Collegiate Athletic Association (NCAA) statistics show that for every 154,000 high school seniors playing basketball, only 4000 will play college basketball as first-year students. Write the ratio of the number of first-year students playing college basketball to the number of high school seniors playing basketball. $\frac{2}{77}$

30. **Sports** NCAA statistics show that for every 2800 college seniors playing college basketball, only 50 will play as rookies in the National Basketball Association. Write the ratio of the number of National Basketball Association rookies to the number of college seniors playing basketball. $\frac{1}{56}$

31. **Electricity** A transformer has 40 turns in the primary coil and 480 turns in the secondary coil. State the ratio of the number of turns in the primary coil to the number of turns in the secondary coil. $\frac{1}{12}$

Primary coil   Secondary coil

32. **Consumerism** Rita Sterling bought a computer system for $2400. Five years later she sold the computer for $900. Find the ratio of the amount she received for the computer to the cost of the computer. $\frac{3}{8}$

33. **Real Estate** A house with an original value of $90,000 increased in value to $110,000 in 5 years.
  a. Find the increase in the value of the house. $20,000
  b. What is the ratio of the increase in value to the original value of the house? $\frac{2}{9}$

34. **Energy Prices** The price of gasoline jumped from $1.35 to $1.62 in 1 year.
  a. What was the increase in the price per gallon? $.27
  b. What is the ratio of the increase in price to the original price? $\frac{1}{5}$

APPLYING THE CONCEPTS

Banking A bank uses the ratio of a borrower's total monthly debts to the borrower's total monthly income to determine the maximum monthly payment for a potential homeowner. This ratio is called the debt–income ratio. Use the homeowner's debt–income table at the right for Exercises 35 and 36.

Income	Debt
$5500	$1200
$450	$300
$250	$450
	$250

35. Compute the debt–income ratio for the potential homeowner. $\frac{11}{31}$

36. Central Trust Bank will make a loan to a customer whose debt–income ratio is less than $\frac{1}{3}$. Will the potential homeowner qualify? Explain your answer. No, $\frac{11}{31}$ is greater than $\frac{1}{3}$.

37. **Banking** To make a home loan, First National Bank requires a debt–income ratio that is less than $\frac{2}{5}$. Would the home-owner whose debt–income table is given at the right qualify for a loan using these standards? Explain. No, $\frac{830}{1389}$ is greater than $\frac{2}{5}$.

Income		Debt	
Salary	$3400	Mortgage	$1800
Interest	$83	Property tax	$104
Rent	$650	Insurance	$35
Dividends	$34	Liabilities	$120
		Credit card	$234
		Car loan	$197

38. Is the value of a ratio always less than 1? Explain.

Quick Quiz (Objective 4.1B)

1. You study 4 hours per day. Find the ratio of the number of hours you study to the number of hours in one day. Write the ratio as a fraction in simplest form.
$\frac{1}{6}$

2. A computer that sold for $2400 two years ago is now selling for $1650. What is the ratio, as a fraction in simplest form, of the decrease in price to the original price?
$\frac{5}{16}$

4.2 Rates

Objective 4.2A

Objective A   To write rates

New Vocabulary
rate
simplest form of a rate

Point of Interest

Listed below are rates at which some crimes are committed in our nation.

Crime	Every
Larceny	4 seconds
Burglary	14 seconds
Robbery	60 seconds
Rape	6 minutes
Murder	31 minutes

A **rate** is a comparison of two quantities that have *different* units. A rate is written as a fraction.

A distance runner ran 26 miles in 4 hours. The distance-to-time rate is written

$$\frac{26 \text{ miles}}{4 \text{ hours}} = \frac{13 \text{ miles}}{2 \text{ hours}}$$

Writing the **simplest form of a rate** means writing it so that the two numbers that form the rate have no common factor other than 1.

Instructor Note

Emphasize the difference between *ratio*, a comparison of two quantities with the same units, and *rate*, a comparison of two quantities with different units.

Discuss the Concepts

1. Explain what a rate is. Provide three examples of rates.
2. What is the difference between a ratio and a rate?
3. What does it mean for a rate to be in simplest form?

Example 1   Write "6 roof supports for every 9 feet" as a rate in simplest form.

Solution $\dfrac{6 \text{ supports}}{9 \text{ feet}} = \dfrac{2 \text{ supports}}{3 \text{ feet}}$

You Try It 1   Write "15 pounds of fertilizer for 12 trees" as a rate in simplest form.

Your solution $\dfrac{5 \text{ pounds}}{4 \text{ trees}}$

Solution on p. S11

Concept Check

State whether each of the following is a rate or a ratio. Explain your answer.

1. $\dfrac{3}{7}$ Ratio
2. $\dfrac{121 \text{ words}}{2 \text{ minutes}}$ Rate
3. $\dfrac{13 \text{ houses}}{20 \text{ acres}}$ Rate
4. $\dfrac{14}{5}$ Ratio

Objective B   To write unit rates

Point of Interest

According to a Gallup Poll, women see doctors more often than men do. On average, men visit the doctor 3.8 times per year, whereas women go to the doctor 5.8 times per year.

A **unit rate** is a rate in which the number in the denominator is 1.

$$\frac{\$3.25}{1 \text{ pound}}$$ or \$3.25/pound is read "\$3.25 per pound."

To find unit rates, divide the number in the numerator of the rate by the number in the denominator of the rate.

A car traveled 344 miles on 16 gallons of gasoline. To find the miles per gallon (unit rate), divide the numerator of the rate by the denominator of the rate.

$$\frac{344 \text{ miles}}{16 \text{ gallons}}$$ is the rate.

$$16)\overline{344.0} \quad \frac{21.5}{}$$ 21.5 miles/gallon is the unit rate.

Objective 4.2B

New Vocabulary
unit rate

Example 2   Write "300 feet in 8 seconds" as a unit rate.

Solution $\dfrac{300 \text{ feet}}{8 \text{ seconds}}$ $8)\overline{300.0} \quad \frac{37.5}{}$

37.5 feet/second

You Try It 2   Write "260 miles in 8 hours" as a unit rate.

Your solution   32.5 miles/hour

Solution on p. S11

In-Class Examples (Objective 4.2A)

Write as a rate in simplest form.

1. 102 miles in 4 hours $\dfrac{51 \text{ miles}}{2 \text{ hours}}$

2. 84 feet in 9 seconds $\dfrac{28 \text{ feet}}{3 \text{ seconds}}$

In-Class Examples (Objective 4.2B)

Write as a unit rate.

1. 297 miles on 9 gallons   33 miles/gallon
2. 365 pounds on 20 square inches
   18.25 pounds/square inch

Objective 4.2C

Instructor Note

Unit rates are given in many situations. The EPA evaluates cars on the basis of their miles per gallon. A more difficult unit rate for students is the one used in the airline industry: cubic feet of fresh air per minute per person. Typical rates are economy class: 7 ft³/min/person; first class: 50 ft³/min/person; cockpit: 150 ft³/min/person.

Optional Student Activity

Bring the business section of a newspaper to class. Have students find the price-earnings (PE) ratio of different companies. (See page 182, Exercise 35.)

Companies that are considered to be good prospects for increasing profits in the future tend to have higher PE ratios than companies whose profits are not growing or are perceived to be high-risk investments.

Ask students to make comparisons between companies on the basis of their PE ratios.

(*Note:* Investors are using the PE ratio, which is based on past performance, to predict future performance. Therefore, the PE ratio is generally used only in conjunction with other indicators.)

Objective C   **To solve application problems**

Denver Airport

HOW TO The table at the right shows air fares for some routes in the continental United States. Find the cost per mile for the four routes in order to determine the most expensive route and the least expensive route on the basis of mileage flown.

Long Routes	Miles	Fare
New York–Los Angeles	2475	$683
San Francisco–Dallas	1464	$536
Denver–Pittsburgh	1302	$525
Minneapolis–Hartford	1050	$483

Integrating Technology

To calculate the costs per mile using a calculator, perform four divisions:

683 ÷ 2475 =
536 ÷ 1464 =
525 ÷ 1302 =
483 ÷ 1050 =

In each case, round the number in the display to the nearest hundredth.

Strategy

To find the cost per mile, divide the fare by the miles flown for each route. Compare the costs per mile to determine the most expensive and least expensive routes per mile.

Solution

New York–Los Angeles $\dfrac{683}{2475} \approx 0.28$

San Francisco–Dallas $\dfrac{536}{1464} \approx 0.37$

Denver–Pittsburgh $\dfrac{525}{1302} \approx 0.40$

Minneapolis–Hartford $\dfrac{483}{1050} = 0.46$

$0.28 < 0.37 < 0.40 < 0.46$

The Minneapolis–Hartford route is the most expensive per mile, and the New York–Los Angeles route is the least expensive per mile.

Example 3

As an investor, Jung Ho purchased 100 shares of stock for $1500. One year later, Jung sold the 100 shares for $1800. What was his profit per share?

Strategy

To find Jung's profit per share:

- Find the total profit by subtracting the original cost ($1500) from the selling price ($1800).
- Find the profit per share (unit rate) by dividing the total profit by the number of shares of stock (100).

Solution

$1800 - 1500 = 300$

$300 \div 100 = 3$

Jung Ho's profit was $3/share.

You Try It 3

Erik Peltier, a jeweler, purchased 5 ounces of gold for $1625. Later, he sold the 5 ounces for $1720. What was Erik's profit per ounce?

Your strategy

Your solution
$19/ounce

Solution on p. S11

In-Class Examples (Objective 4.2C)

1. An investor purchased 475 shares of stock for $21,375. What was the cost per share? $45

2. The total cost of making 5000 video tapes was $12,054. One hundred of the tapes made did not meet company standards. What was the cost per tape for those tapes that did meet company standards? $2.46

4.2 Exercises

Objective A   To write rates

Suggested Assignment
Exercises 1–33, odds
Exercise 34

For Exercises 1 to 10, write each phrase as a rate in simplest form.

1. 3 pounds of meat for 4 people
    $\dfrac{3 \text{ pounds}}{4 \text{ people}}$

2. 30 ounces in 24 glasses
    $\dfrac{5 \text{ ounces}}{4 \text{ glasses}}$

3. $80 for 12 boards
    $\dfrac{\$20}{3 \text{ boards}}$

4. 84 cents for 3 bars of soap
    $\dfrac{28 \text{ cents}}{1 \text{ bar}}$

5. 300 miles on 15 gallons
    $\dfrac{20 \text{ miles}}{1 \text{ gallon}}$

6. 88 feet in 8 seconds
    $\dfrac{11 \text{ feet}}{1 \text{ second}}$

7. 20 children in 8 families
    $\dfrac{5 \text{ children}}{2 \text{ families}}$

8. 48 leaves on 9 plants
    $\dfrac{16 \text{ leaves}}{3 \text{ plants}}$

9. 16 gallons in 2 hours
    $\dfrac{8 \text{ gallons}}{1 \text{ hour}}$

10. 25 ounces in 5 minutes
    $\dfrac{5 \text{ ounces}}{1 \text{ minute}}$

Objective B   To write unit rates

For Exercises 11 to 24, write each phrase as a unit rate.

11. 10 feet in 4 seconds
    2.5 feet/second

12. 816 miles in 6 days
    136 miles/day

13. $3900 earned in 4 weeks
    $975/week

14. $51,000 earned in 12 months
    $4250/month

15. 1100 trees planted on 10 acres
    110 trees/acre

16. 3750 words on 15 pages
    250 words/page

17. $131.88 earned in 7 hours
    $18.84/hour

18. $315.70 earned in 22 hours
    $14.35/hour

19. 628.8 miles in 12 hours
    52.4 miles/hour

20. 388.8 miles in 8 hours
    48.6 miles/hour

21. 344.4 miles on 12.3 gallons of gasoline
    28 miles/gallon

22. 409.4 miles on 11.5 gallons of gasoline
    35.6 miles/gallon

23. $349.80 for 212 pounds
    $1.65/pound

24. $11.05 for 3.4 pounds
    $3.25/pound

Objective C   To solve application problems

25. **Fuel Efficiency**   An automobile was driven 326.6 miles on 11.5 gallons of gas. Find the number of miles driven per gallon of gas.
    28.4 miles/gallon

26. **Travel**   You drive 246.6 miles in 4.5 hours. Find the number of miles you drove per hour.
    54.8 miles/hour

27. **Fuel Efficiency**   The Saturn-5 rocket uses 534,000 gallons of fuel in 2.5 minutes. How much fuel does the rocket use per minute?
    213,600 gallons/minute

Quick Quiz (Objective 4.2A)

Write as a rate in simplest form.

1. 6 tablets in 24 hours $\dfrac{1 \text{ tablet}}{4 \text{ hours}}$

2. $324 earned in 40 hours $\dfrac{\$81}{10 \text{ hours}}$

Quick Quiz (Objective 4.2B)

Write as a unit rate.

1. $27 for 30 pounds $.90/pound

2. 198 words in 4.5 minutes   44 words/minute

28.   Manufacturing   Assume that Regency Computer produced 5000 zip disks for $26,536.32. Of the disks made, 122 did not meet company standards.
a. How many disks did meet company standards?   4878 disks
b. What was the cost per disk for those disks that met company standards? $5.44/disk

29.   Consumerism   The Pierre family purchased a 250-pound side of beef for $365.75 and had it packaged. During the packaging, 75 pounds of beef were discarded as waste.
a. How many pounds of beef were packaged?   175 pounds
b. What was the cost per pound for the packaged beef?
$2.09/pound

30.   **Advertising**   In 2003, the average price of a 30-second commercial during the sitcom *Friends* was $455,700. (*Source: Time*) Find the price per second.
$15,190/second

31.   **The Film Industry**   During filming, an IMAX camera uses 65-mm film at a rate of 5.6 feet per second.
a. At what rate per minute does the camera go through film?   336 feet/minute
b. At what rate does it use a 500-foot roll of 65-mm film? Round to the nearest second.   89 seconds/roll

32.   **Demography**   The table at the right shows the population and area of three countries. The population density of a country is the number of people per square mile.
a. Which country has the least population density?   Australia
b. How many more people per square mile are there in India than in the United States? Round to the nearest whole number.
743 more people

Country	Population	Area (in square miles)
Australia	19,547,000	2,968,000
India	1,045,845,000	1,269,000
United States	291,929,000	3,619,000

Another application of rates is in the area of international trade. Suppose a company in Canada purchases a shipment of sneakers from an American company. The Canadian company must exchange Canadian dollars for U.S. dollars in order to pay for the order. The number of Canadian dollars that are equivalent to 1 U.S. dollar is called the **exchange rate.**

33.   **Exchange Rates**   The table at the right shows the exchange rates per U.S. dollar for three foreign countries and for the euro at the time of this writing.
a. How many euros would be paid for an order of American computer hardware costing $120,000?   109,236 euros
b. Calculate the cost, in Japanese yen, of an American car costing $34,000.   3,978,000 yen

Exchange Rates per U.S. Dollar	
Australian Dollar	1.545
Canadian Dollar	1.386
Japanese Yen	117.000
The Euro	0.9103

APPLYING THE CONCEPTS

34.   Compensation   You have a choice of receiving a wage of $34,000 per year, $2840 per month, $650 per week, or $18 per hour. Which pay choice would you take? Assume a 40-hour week with 52 weeks per year.
$18/hour

35.   The price–earnings ratio of a company's stock is one measure used by stock market analysts to assess the financial well-being of the company. Explain the meaning of the price–earnings ratio.

Answers to Writing Exercises

35. The price-earnings ratio of a company's stock is computed by dividing the current price per share of the stock by the annual earnings per share. For example, if the price-earnings ratio of a company's stock is 8.5, the price of the stock is 8.5 times the annual earnings per share of the stock.

Quick Quiz (Objective 4.2C)

1. A grocery store sells 3 pounds of tomatoes for $4.00. What is the cost per pound? Round to the nearest cent. $1.33

2. A store bought 175 ice scrapers wholesale for $304.50 and sold them for $486.50. What was the store's profit per ice scraper? $1.04

Proportions

Objective A   To determine whether a proportion is true

A **proportion** is an expression of the equality of two ratios or rates.

$$\frac{50 \text{ miles}}{4 \text{ gallons}} = \frac{25 \text{ miles}}{2 \text{ gallons}}$$ Note that the units of the numerators are the same and the units of the denominators are the same.

$$\frac{3}{6} = \frac{1}{2}$$ This is the equality of two ratios.

A proportion is **true** if the fractions are equal when written in lowest terms.

In any true proportion, the **cross products** are equal.

> **HOW TO**   Is $\frac{2}{3} = \frac{8}{12}$ a true proportion?
>
> $$\frac{2}{3} \quad \frac{8}{12} \quad 3 \times 8 = 24$$ The cross products *are* equal.
> $$2 \times 12 = 24 \qquad \frac{2}{3} = \frac{8}{12} \text{ is a true proportion.}$$

A proportion is **not true** if the fractions are not equal when reduced to lowest terms.

If the cross products are not equal, then the proportion is not true.

> **HOW TO**   Is $\frac{4}{5} = \frac{8}{9}$ a true proportion?
>
> $$\frac{4}{5} \quad \frac{8}{9} \quad 5 \times 8 = 40$$ The cross products *are not* equal.
> $$4 \times 9 = 36 \qquad \frac{4}{5} = \frac{8}{9} \text{ is not a true proportion.}$$

Example 1   Is $\frac{5}{8} = \frac{10}{16}$ a true proportion?

Solution
$$\frac{5}{8} \quad \frac{10}{16} \quad 8 \times 10 = 80$$
$$5 \times 16 = 80$$

The cross products are equal.
The proportion is true.

Example 2   Is $\frac{62 \text{ miles}}{4 \text{ gallons}} = \frac{33 \text{ miles}}{2 \text{ gallons}}$ a true proportion?

Solution
$$\frac{62}{4} \quad \frac{33}{2} \quad 4 \times 33 = 132$$
$$62 \times 2 = 124$$

The cross products are not equal.
The proportion is not true.

You Try It 1   Is $\frac{6}{10} = \frac{9}{15}$ a true proportion?

Your solution   True

You Try It 2   Is $\frac{\$32}{6 \text{ hours}} = \frac{\$90}{8 \text{ hours}}$ a true proportion?

Your solution   Not true

Solutions on p. S11

Point of Interest
Proportions were studied by the earliest mathematicians. Clay tablets uncovered by archaeologists show evidence of proportions in Egyptian and Babylonian cultures dating from 1800 B.C.

Objective 4.3A

Vocabulary to Review
ratio   [4.1A]
rate   [4.2A]

New Vocabulary
proportion
cross products

Discuss the Concepts
1. What is a proportion?
2. What does the phrase "the cross products are equal" mean?

Optional Student Activity
On a calendar for any month, box in a few two-by-two number squares.

Su	Mo	Tu
	1	2
7	8	9
14	15	16
21	22	23
28	29	30

1. Compare the cross products in the number squares. What pattern do you notice?
   The products always differ by 7.
2. What pattern do you observe in the "cross sums"?
   The sums are always the same.

In-Class Examples (Objective 4.3A)
Determine whether the proportion is true or not true.

1. $\frac{3}{7} = \frac{6}{14}$ True

2. $\frac{111 \text{ miles}}{3 \text{ hours}} = \frac{812 \text{ miles}}{22 \text{ hours}}$ Not true

Objective 4.3B

New Vocabulary

solve a proportion

Instructor Note

The solution of these equations is based on the relationship between multiplication and division. You may want to show the solution by dividing each side by 9. For instance,

$$\frac{9}{6} = \frac{3}{n}$$
$$9 \times n = 6 \times 3$$
$$9 \times n = 18$$
$$\frac{9 \times n}{9} = \frac{18}{9}$$
$$n = 2$$

Discuss the Concepts

In preparation for solving proportions, have students explain how to solve each of the following equations.

1. $2 \times n = 24$ $n = 24 \div 2$
2. $n \times 7 = 49$ $n = 49 \div 7$
3. $40 = n \times 8$ $40 \div 8 = n$
4. $63 = 9 \times n$ $63 \div 9 = n$

Optional Student Activity

(*Note:* This is a challenging problem for students.) If 5 bats eat 6 mosquitoes in 6 seconds, how many mosquitoes would be eaten by 30 bats in 30 seconds?
180 mosquitoes

Point of Interest

In late summer, the Congress Avenue Bridge in Austin, Texas, is home to about 1.5 million Mexican freetail bats, who eat approximately 10,000 to 30,000 pounds of insects per night.

Objective B   **To solve proportions**

Study Tip

An important element of success is practice. We cannot do anything well if we do not practice it repeatedly. Practice is crucial to success in mathematics. In this objective you are learning a new skill: how to solve a proportion. You will need to practice this skill over and over again in order to be successful at it.

Sometimes one of the numbers in a proportion is unknown. In this case, it is necessary to *solve* the proportion.

To **solve a proportion**, find a number to replace the unknown so that the proportion is true.

HOW TO   Solve: $\frac{9}{6} = \frac{3}{n}$

$$\frac{9}{6} = \frac{3}{n}$$

$9 \times n = 6 \times 3$ • Find the cross products.

$9 \times n = 18$

$n = 18 \div 9$ • Think of $9 \times n = 18$ as $9\overline{)18}$.

$n = 2$

Check:

$$\frac{9}{6} \bowtie \frac{3}{2} \quad \begin{array}{l} 6 \times 3 = 18 \\ 9 \times 2 = 18 \end{array}$$

Example 3   Solve $\frac{n}{12} = \frac{25}{60}$ and check.

Solution
$n \times 60 = 12 \times 25$ • Find the cross
$n \times 60 = 300$ products. Then
$n = 300 \div 60$ solve for n.
$n = 5$

Check:

$$\frac{5}{12} \bowtie \frac{25}{60} \quad \begin{array}{l} 12 \times 25 = 300 \\ 5 \times 60 = 300 \end{array}$$

You Try It 3   Solve $\frac{n}{14} = \frac{3}{7}$ and check.

Your solution   6

Example 4   Solve $\frac{4}{9} = \frac{n}{16}$. Round to the nearest tenth.

Solution
$4 \times 16 = 9 \times n$ • Find the cross
$64 = 9 \times n$ products. Then
$64 \div 9 = n$ solve for n.
$7.1 \approx n$

Note: A rounded answer is an approximation. Therefore, the answer to a check will not be exact.

You Try It 4   Solve $\frac{5}{7} = \frac{n}{20}$. Round to the nearest tenth.

Your solution   14.3

Solutions on p. S12

In-Class Examples (Objective 4.3B)

Solve. Round to the nearest hundredth, if necessary.

1. $\frac{60}{n} = \frac{24}{7}$ 17.5

2. $\frac{18}{20} = \frac{15}{n}$ 16.67

Example 5 Solve $\frac{28}{52} = \frac{7}{n}$ and check.

Solution $28 \times n = 52 \times 7$ • Find the cross
$\quad\quad 28 \times n = 364$ products. Then
$\quad\quad\quad\quad n = 364 \div 28$ solve for n.
$\quad\quad\quad\quad n = 13$

Check:

$\frac{28}{52} \bowtie \frac{7}{13}$ → $52 \times 7 = 364$
$\quad\quad\quad\quad$ → $28 \times 13 = 364$

You Try It 5 Solve $\frac{15}{20} = \frac{12}{n}$ and check.

Your solution 16

Example 6 Solve $\frac{15}{n} = \frac{8}{3}$. Round to the
nearest hundredth.

Solution $15 \times 3 = n \times 8$
$\quad\quad\quad 45 = n \times 8$
$\quad 45 \div 8 = n$
$\quad\quad 5.63 \approx n$

You Try It 6 Solve $\frac{12}{n} = \frac{7}{4}$. Round to the
nearest hundredth.

Your solution 6.86

Example 7 Solve $\frac{n}{9} = \frac{3}{1}$ and check.

Solution $n \times 1 = 9 \times 3$
$\quad\quad n \times 1 = 27$
$\quad\quad\quad\quad n = 27 \div 1$
$\quad\quad\quad\quad n = 27$

Check:

$\frac{27}{9} \bowtie \frac{3}{1}$ → $9 \times 3 = 27$
$\quad\quad\quad\quad$ → $27 \times 1 = 27$

You Try It 7 Solve $\frac{n}{12} = \frac{4}{1}$ and check.

Your solution 48

Solutions on p. S12

Objective C   **To solve application problems**

The application problems in this objective require you to write and solve a proportion. When setting up a proportion, remember to keep the same units in the numerator and the same units in the denominator.

Objective 4.3C

Instructor Note

Students will have some difficulty setting up the proportions in this objective. Although there are a number of ways to set up a proportion correctly, you might tell them to write a proportion so that the units in the numerators are the same and the units in the denominators are the same.

Concept Check

Use the student-faculty ratio at your school and the number of students enrolled to estimate the number of faculty at the school.

Optional Student Activity

(*Note:* In this Information Age, we are bombarded by statistics. Rates are frequently quoted for everything from students per computer in public schools to state divorce rates per 1000 couples. Using real data, such as in the problem below, might help students see meaning in the numbers they encounter.)

In the United States, the average annual number of deaths per million people age 5 to 34 from asthma is 3.5. (*Source:* National Center for Health Statistics) Approximately how many people age 5 to 34 die from asthma each year in this country? Use a figure of 150,000,000 for the number of residents 5 to 34 years old.

525 people

Example 8

The dosage of a certain medication is 2 ounces for every 50 pounds of body weight. How many ounces of this medication are required for a person who weighs 175 pounds?

Strategy

To find the number of ounces of medication for a person weighing 175 pounds, write and solve a proportion using n to represent the number of ounces of medication for a 175-pound person.

Solution

$$\frac{2 \text{ ounces}}{50 \text{ pounds}} = \frac{n \text{ ounces}}{175 \text{ pounds}}$$

• The unit "ounces" is in the numerator. The unit "pounds" is in the denominator.

$$2 \times 175 = 50 \times n$$
$$350 = 50 \times n$$
$$350 \div 50 = n$$
$$7 = n$$

A 175-pound person requires 7 ounces of medication.

You Try It 8

Three tablespoons of a liquid plant fertilizer are to be added to every 4 gallons of water. How many tablespoons of fertilizer are required for 10 gallons of water?

Your strategy

Your solution

7.5 tablespoons

Example 9

A mason determines that 9 cement blocks are required for a retaining wall 2 feet long. At this rate, how many cement blocks are required for a retaining wall that is 24 feet long?

Strategy

To find the number of cement blocks for a retaining wall 24 feet long, write and solve a proportion using n to represent the number of blocks required.

Solution

$$\frac{9 \text{ cement blocks}}{2 \text{ feet}} = \frac{n \text{ cement blocks}}{24 \text{ feet}}$$

$$9 \times 24 = 2 \times n$$
$$216 = 2 \times n$$
$$216 \div 2 = n$$
$$108 = n$$

A 24-foot retaining wall requires 108 cement blocks.

You Try It 9

Twenty-four jars can be packed in 6 identical boxes. At this rate, how many jars can be packed in 15 boxes?

Your strategy

Your solution

60 jars

Solutions on p. S12

In-Class Examples (Objective 4.3C)

1. A stock investment of 150 shares paid a dividend of $555. At this rate, what dividend would be paid on 180 shares of stock? $666

2. A life insurance policy costs $8.52 for every $1000 of insurance. At this rate, what is the cost for $20,000 worth of life insurance? $170.40

4.3 Exercises

Section 4.3

Section 4.3

Objective A   **To determine whether a proportion is true**

Suggested Assignment
Exercises 1–21, every other odd
Exercises 25–71, odds
More challenging problems:
    Exercises 72–75

For Exercises 1 to 24, determine whether the proportion is true or not true.

1. $\dfrac{4}{8} = \dfrac{10}{20}$
True

2. $\dfrac{39}{48} = \dfrac{13}{16}$
True

3. $\dfrac{7}{8} = \dfrac{11}{12}$
Not true

4. $\dfrac{15}{7} = \dfrac{17}{8}$
Not true

5. $\dfrac{27}{8} = \dfrac{9}{4}$
Not true

6. $\dfrac{3}{18} = \dfrac{4}{19}$
Not true

7. $\dfrac{45}{135} = \dfrac{3}{9}$
True

8. $\dfrac{3}{4} = \dfrac{54}{72}$
True

9. $\dfrac{16}{3} = \dfrac{48}{9}$
True

10. $\dfrac{15}{5} = \dfrac{3}{1}$
True

11. $\dfrac{7}{40} = \dfrac{7}{8}$
Not true

12. $\dfrac{9}{7} = \dfrac{6}{5}$
Not true

13. $\dfrac{50 \text{ miles}}{2 \text{ gallons}} = \dfrac{25 \text{ miles}}{1 \text{ gallon}}$
True

14. $\dfrac{16 \text{ feet}}{10 \text{ seconds}} = \dfrac{24 \text{ feet}}{15 \text{ seconds}}$
True

15. $\dfrac{6 \text{ minutes}}{5 \text{ cents}} = \dfrac{30 \text{ minutes}}{25 \text{ cents}}$
True

16. $\dfrac{16 \text{ pounds}}{12 \text{ days}} = \dfrac{20 \text{ pounds}}{14 \text{ days}}$
Not true

17. $\dfrac{\$15}{4 \text{ pounds}} = \dfrac{\$45}{12 \text{ pounds}}$
True

18. $\dfrac{270 \text{ trees}}{6 \text{ acres}} = \dfrac{90 \text{ trees}}{2 \text{ acres}}$
True

19. $\dfrac{300 \text{ feet}}{4 \text{ rolls}} = \dfrac{450 \text{ feet}}{7 \text{ rolls}}$
Not true

20. $\dfrac{1 \text{ gallon}}{4 \text{ quarts}} = \dfrac{7 \text{ gallons}}{28 \text{ quarts}}$
True

21. $\dfrac{\$65}{5 \text{ days}} = \dfrac{\$26}{2 \text{ days}}$
True

22. $\dfrac{80 \text{ miles}}{2 \text{ hours}} = \dfrac{110 \text{ miles}}{3 \text{ hours}}$
Not true

23. $\dfrac{7 \text{ tiles}}{4 \text{ feet}} = \dfrac{42 \text{ tiles}}{20 \text{ feet}}$
Not true

24. $\dfrac{15 \text{ feet}}{3 \text{ yards}} = \dfrac{90 \text{ feet}}{18 \text{ yards}}$
True

Quick Quiz (Objective 4.3A)

Determine whether the proportion is true or not true.

1. $\dfrac{4}{5} = \dfrac{13}{16}$ Not true

2. $\dfrac{\$300}{36 \text{ hours}} = \dfrac{\$200}{24 \text{ hours}}$ True

Objective B **To solve proportions**

For Exercises 25 to 52, solve. Round to the nearest hundredth, if necessary.

25. $\dfrac{n}{4} = \dfrac{6}{8}$
3

26. $\dfrac{n}{7} = \dfrac{9}{21}$
3

27. $\dfrac{12}{18} = \dfrac{n}{9}$
6

28. $\dfrac{7}{21} = \dfrac{35}{n}$
105

29. $\dfrac{6}{n} = \dfrac{24}{36}$
9

30. $\dfrac{3}{n} = \dfrac{15}{10}$
2

31. $\dfrac{n}{45} = \dfrac{17}{135}$
5.67

32. $\dfrac{9}{4} = \dfrac{18}{n}$
8

33. $\dfrac{n}{6} = \dfrac{2}{3}$
4

34. $\dfrac{5}{12} = \dfrac{n}{144}$
60

35. $\dfrac{n}{5} = \dfrac{7}{8}$
4.38

36. $\dfrac{4}{n} = \dfrac{9}{5}$
2.22

37. $\dfrac{n}{11} = \dfrac{32}{4}$
88

38. $\dfrac{3}{4} = \dfrac{8}{n}$
10.67

39. $\dfrac{5}{12} = \dfrac{n}{8}$
3.33

40. $\dfrac{36}{20} = \dfrac{12}{n}$
6.67

41. $\dfrac{n}{15} = \dfrac{21}{12}$
26.25

42. $\dfrac{40}{n} = \dfrac{15}{8}$
21.33

43. $\dfrac{32}{n} = \dfrac{1}{3}$
96

44. $\dfrac{5}{8} = \dfrac{42}{n}$
67.2

45. $\dfrac{18}{11} = \dfrac{16}{n}$
9.78

46. $\dfrac{25}{4} = \dfrac{n}{12}$
75

47. $\dfrac{28}{8} = \dfrac{12}{n}$
3.43

48. $\dfrac{n}{30} = \dfrac{65}{120}$
16.25

49. $\dfrac{0.3}{5.6} = \dfrac{n}{25}$
1.34

50. $\dfrac{1.3}{16} = \dfrac{n}{30}$
2.44

51. $\dfrac{0.7}{9.8} = \dfrac{3.6}{n}$
50.4

52. $\dfrac{1.9}{7} = \dfrac{13}{n}$
47.89

Objective C **To solve application problems**

For Exercises 53 to 71, solve. Round to the nearest hundredth.

53. **Nutrition** A 6-ounce package of Puffed Wheat contains 600 calories. How many calories are in a 0.5-ounce serving of the cereal?
50 calories

54. **Fuel Efficiency** A car travels 70.5 miles on 3 gallons of gas. Find the distance that the car can travel on 14 gallons of gas.
329 miles

Quick Quiz (Objective 4.3B)

Solve. Round to the nearest hundredth, if necessary.

1. $\dfrac{n}{14} = \dfrac{3}{7}$ 6

2. $\dfrac{4}{9} = \dfrac{n}{7}$ 3.11

55. **Landscaping**   Ron Stokes uses 2 pounds of fertilizer for every 100 square feet of lawn for landscape maintenance. At this rate, how many pounds of fertilizer did he use on a lawn that measures 3500 square feet?
70 pounds

56. **Gardening**   A nursery prepares a liquid plant food by adding 1 gallon of water for each 2 ounces of plant food. At this rate, how many gallons of water are required for 25 ounces of plant food?
12.5 gallons

57. **Manufacturing**   A manufacturer of baseball equipment makes 4 aluminum bats for every 15 bats made from wood. On a day when 100 aluminum bats are made, how many wooden bats are produced?
375 wooden bats

58. **Masonry**   A brick wall 20 feet in length contains 1040 bricks. At the same rate, how many bricks would it take to build a wall 48 feet in length?
2496 bricks

59. **Cartography**   The scale on the map at the right is "1.25 inches equals 10 miles." Find the distance between Carlsbad and Del Mar, which are 2 inches apart on the map.
16 miles

60. **Architecture**   The scale on the plans for a new house is "1 inch equals 3 feet." Find the width and the length of a room that measures 5 inches by 8 inches on the drawing.
15 feet by 24 feet

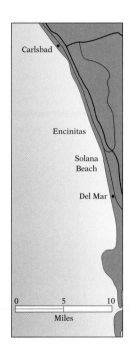

61. **Medicine**   The dosage for a medication is $\frac{1}{3}$ ounce for every 40 pounds of body weight. At this rate, how many ounces of medication should a physician prescribe for a patient who weighs 150 pounds? Write the answer as a decimal.
1.25 ounces

62. **Banking**   A bank requires a monthly payment of $33.45 on a $2500 loan. At the same rate, find the monthly payment on a $10,000 loan.
$133.80 per month

63. **Elections**   A pre-election survey showed that 2 out of every 3 eligible voters would cast ballots in the county election. At this rate, how many people in a county of 240,000 eligible voters would vote in the election?
160,000 people

64. **Interior Design**   A paint manufacturer suggests using 1 gallon of paint for every 400 square feet of a wall. At this rate, how many gallons of paint would be required for a room that has 1400 square feet of wall?
3.5 gallons

65. **Insurance**   A 60-year-old male can obtain $10,000 of life insurance for $35.35 per month. At this rate, what is the monthly cost of $50,000 of life insurance?
$176.75

66. **Manufacturing**   Suppose a computer chip manufacturer knows from experience that in an average production run of 2000 circuit boards, 60 will be defective. How many defective circuit boards can be expected from a run of 25,000 circuit boards?
750 defective boards

Quick Quiz (Objective 4.3C)

1. A liquid plant food is prepared by using 1 gallon of water for each 1.5 teaspoons of plant food. At this rate, how many teaspoons of plant food are required for 5 gallons of water?   7.5 teaspoons

2. For every 10 people who work in a city, 3 of them do not commute by public transportation. If 34,600 people work in the city, how many of them do not take public transportation?   10,380 people

Answers to Writing Exercises

72. From the given information, we can write the ratio of the number of Atkins books sold to the number of Grisham books sold: $\frac{100}{7.3}$. Let the unknown number of copies of Atkins book sold be n. We'll use 20,000 for the number of Grisham books sold. The ratio $\frac{100}{7.3}$ is equal to $\frac{n}{20,000}$, and we can write the proportion $\frac{100}{7.3} = \frac{n}{20,000}$. Solve this proportion for n to determine the number of copies of Atkins book sold.

73. The fact that the number of workers per retiree is decreasing means that for each retiree drawing money out of Social Security, there are fewer and fewer workers paying into the Social Security system. In other words, fewer workers are supporting each retiree. Therefore, unless the amount paid into the system by each worker is increased or other radical changes are made, the funds to pay the Social Security benefits will be depleted.

74. To determine the average pay of a CEO in 2000, you would need to know the pay of the average factory worker in 2000. Suppose the average factory worker's pay in 2000 was $50,000 and we let S be the average pay of a CEO in 2000. Then we can use the ratio $\frac{531}{1}$ and write the proportion $\frac{531}{1} = \frac{S}{50,000}$. Solve this proportion for S to determine the average pay of a CEO in 2000.

75. No, it is not possible. The sum of the fractions is $\frac{2}{5} + \frac{3}{4} = \frac{23}{20} = 1\frac{3}{20}$, which is greater than 1. In order for the responses to be possible, the sum of the fractions must be 1.

---

67. **Investments** You own 240 shares of stock in a computer company. The company declares a stock split of 5 shares for every 3 owned. How many shares of stock will you own after the stock split?
400 shares

68. **Computers** The director of data processing at a college estimates that the ratio of student time to administrative time on a certain computer is 3:2. During a month in which the computer was used 200 hours for administration, how many hours was it used by students?
300 hours

69. **Physics** The ratio of weight on the moon to weight on Earth is 1:6. If a bowling ball weighs 16 pounds on Earth, what would it weigh on the moon?
2.67 pounds

70. **Automobiles** When engineers designed a new car, they first built a model of the car. The ratio of the size of a part on the model to the actual size of the part is 2:5. If a door is 1.3 feet long on the model, what is the length of the door on the car?
3.25 feet

71. **Investments** Carlos Capasso owns 50 shares of Texas Utilities that pay dividends of $153. At this rate, what dividend would Carlos receive after buying 300 additional shares of Texas Utilities?
$1071

APPLYING THE CONCEPTS

72. **Publishing** In January 2003, *USA Today* reported that for every 100 copies of *Dr. Atkins' New Diet Revolution* sold, John Grisham's *The Summons* sold 7.3 copies. Explain how a proportion can be used to determine the number of copies of *Dr. Atkins' New Diet Revolution* sold given the number of copies of *The Summons* sold.

73. **Social Security** According to the Social Security Administration, the numbers of workers per retiree in the future are expected to be as given in the table below.

Year	2010	2020	2030	2040
Number of workers per retiree	3.1	2.5	2.1	2.0

Why is the shrinking number of workers per retiree of importance to the Social Security Administration?

74. **Compensation** In June 2002, *Time* magazine reported, "In 1980 the average CEO made 40 times the pay of the average factory worker; by 2000 the ratio had climbed to 531 to 1." What information would you need to know in order to determine the average pay of a CEO in 2000? With that information, how would you calculate the average pay of a CEO in 2000?

75. **Elections** A survey of voters in a city claimed that 2 people of every 5 who voted cast a ballot in favor of city amendment A and that 3 people of every 4 who voted cast a ballot against amendment A. Is this possible? Explain your answer.

76. Write a word problem that requires solving a proportion to find the answer.

---

76. Answers will vary. Here's an example: One 200-pound bag of fertilizer will cover 5000 square feet of lawn. How many pounds of fertilizer are required for 18,000 square feet of lawn?

Focus on Problem Solving

Looking for a Pattern

A very useful problem-solving strategy is looking for a pattern.

Problem   A legend says that a peasant invented the game of chess and gave it to a very rich king as a present. The king so enjoyed the game that he gave the peasant the choice of anything in the kingdom. The peasant's request was simple: "Place one grain of wheat on the first square, 2 grains on the second square, 4 grains on the third square, 8 on the fourth square, and continue doubling the number of grains until the last square of the chessboard is reached." How many grains of wheat must the king give the peasant?

Solution   A chessboard consists of 64 squares. To find the total number of grains of wheat on the 64 squares, we begin by looking at the amount of wheat on the first few squares.

Square 1	Square 2	Square 3	Square 4	Square 5	Square 6	Square 7	Square 8
1	2	4	8	16	32	64	128
1	3	7	15	31	63	127	255

The bottom row of numbers represents the sum of the number of grains of wheat up to and including that square. For instance, the number of grains of wheat on the first 7 squares is $1 + 2 + 4 + 8 + 16 + 32 + 64 = 127$.

One pattern to observe is that the number of grains of wheat on a square can be expressed as a power of 2.

The number of grains on square $n = 2^{n-1}$.

For example, the number of grains on square $7 = 2^{7-1} = 2^6 = 64$.

A second pattern of interest is that **the number *below* a square** (the total number of grains up to and including that square) **is 1 less than the number of grains of wheat *on the next* square.** For example, the number *below* square 7 is 1 less than the number *on* square 8 ($128 - 1 = 127$). From this observation, the number of grains of wheat on the first 8 squares is the number on square 8 (128) plus 1 less than the number on square 8 (127): The total number of grains of wheat on the first 8 squares is $128 + 127 = 255$.

From this observation,

$$\begin{array}{ll} \text{Number of grains of} \\ \text{wheat on the chessboard} \end{array} = \begin{array}{l} \text{number of grains} \\ \text{on square 64} \end{array} + \begin{array}{l} \text{1 less than the number} \\ \text{of grains on square 64} \end{array}$$

$$= 2^{64-1} + (2^{64-1} - 1)$$

$$= 2^{63} + 2^{63} - 1 \approx 18{,}000{,}000{,}000{,}000{,}000{,}000$$

To give you an idea of the magnitude of this number, this is more wheat than has been produced in the world since chess was invented.

The same king decided to have a banquet in the long banquet room of the palace to celebrate the invention of chess. The king had 50 square tables, and each table could seat only one person on each side. The king pushed the tables together to form one long banquet table. How many people could sit at this table? *Hint:* Try constructing a pattern by using 2 tables, 3 tables, and 4 tables.

Projects and Group Activities

The Golden Ratio

There are certain designs that have been repeated over and over in both art and architecture. One of these involves the **golden rectangle.**

A golden rectangle is drawn at the right. Begin with a square that measures, say, 2 inches on a side. Let *A* be the midpoint of a side (halfway between two corners). Now measure the distance from *A* to *B*. Place this length along the bottom of the square, starting at *A*. The resulting rectangle is a golden rectangle.

The **golden ratio** is the ratio of the length of the golden rectangle to its width. If you have drawn the rectangle following the procedure above, you will find that the golden ratio is approximately 1.6 to 1.

The golden ratio appears in many different situations. Some historians claim that some of the great pyramids of Egypt are based on the golden ratio. The drawing at the right shows the Pyramid of Giza, which dates from approximately 2600 B.C. The ratio of the height to a side of the base is approximately 1.6 to 1.

1. There are instances of the golden rectangle in the Mona Lisa painted by Leonardo da Vinci. Do some research on this painting and write a few paragraphs summarizing your findings.
2. What do 3 × 5 and 5 × 8 index cards have to do with the golden rectangle?
3. What does the United Nations Building in New York City have to do with the golden rectangle?
4. When was the Parthenon in Athens, Greece, built? What does the front of that building have to do with the golden rectangle?

Answers to Focus on Problem Solving— Looking for a Pattern

102 people could sit at this table.

Answers to Projects and Group Activities— The Golden Ratio

1. Instances of the golden rectangle include the Mona Lisa's face; the upper portion of her face, bounded by the eyes; the region from the neck to just below the hands; the region from the neckline on the dress to just below the arms.

2. The ratio $\frac{5}{3} = 1.666...$, which is close to the golden ratio. The ratio $\frac{8}{5} = 1.6$, which is close to the golden ratio.

3. The upright portion of the United Nations building is in the shape of the golden rectangle.

4. Construction of the Parthenon began in 447 B.C. The face of the building is in the shape of the golden rectangle.

Drawing the Floor Plans for a Building

BED

BED

BEDROOM

BATH

UNDER CABINET LIGHT

FRIDGE UNDER CABINET

LIVING ROOM

DECK

9" LOG SUPPORT POSTS

1/2 POST FOR LOG HANDRAIL

The drawing at the left is a sketch of the floor plan for a cabin at a resort in the mountains of Utah. The measurements are missing. Assume that you are the architect and will finish the drawing. You will have to decide the size of the rooms and put in the measurements to scale.

Design a cabin that you would like to own. Select a scale and draw all the rooms to scale.

If you are interested in architecture, visit an architect who is using CAD (computer-aided design) to create a floor plan. Computer technology has revolutionized the field of architectural design.

The U.S. House of Representatives

The framers of the Constitution decided to use a ratio to determine the number of representatives from each state. It was determined that each state would have one representative for every 30,000 citizens, with a minimum of one representative. Congress has changed this ratio over the years, and we now have 435 representatives.

Find the number of representatives from your state. Determine the ratio of citizens to representatives. Also do this for the most populous state and for the least populous state.

 You might consider getting information on the number of representatives for each state and the populations of different states via the Internet.

Chapter 4 Summary

Key Words	**Examples**
A *ratio* is the comparison of two quantities with the same units. A ratio can be written in three ways: as a fraction, as two numbers separated by a colon (:), or as two numbers separated by the word *to*. A ratio is in *simplest form* when the two numbers do not have a common factor. [4.1A, p. 175]	The comparison 16 to 24 ounces can be written as a ratio in simplest form as $\frac{2}{3}$, 2:3, or 2 to 3.
A *rate* is the comparison of two quantities with different units. A rate is written as a fraction. A rate is in *simplest form* when the numbers that form the ratio do not have a common factor. [4.2A, p. 179]	You earned $63 for working 6 hours. The rate is written in simplest form as $\frac{\$21}{2 \text{ hours}}$.
A *unit rate* is a rate in which the number in the denominator is 1. [4.2B, p. 179]	You traveled 144 miles in 3 hours. The unit rate is 48 miles per hour.
A *proportion* is an expression of the equality of two ratios or rates. A proportion is true if the fractions are equal when written in lowest terms; in any true proportion, the *cross products* are equal. A proportion is not true if the fractions are not equal when written in lowest terms; if the cross products are not equal, the proportion is not true. [4.3A, p. 183]	The proportion $\frac{3}{5} = \frac{12}{20}$ is true because the cross products are equal: $3 \times 20 = 5 \times 12$. The proportion $\frac{3}{4} = \frac{12}{20}$ is not true because the cross products are not equal: $3 \times 20 \neq 4 \times 12$.

Essential Rules and Procedures	**Examples**
To find a unit rate, divide the number in the numerator of the rate by the number in the denominator of the rate. [4.2B, p. 179]	You earned $41 for working 4 hours. $41 \div 4 = 10.25$ The unit rate is $10.25/hour.
To solve a proportion, find a number to replace the unknown so that the proportion is true. [4.3B, p. 184]	$\frac{6}{24} = \frac{9}{n}$ $6 \times n = 24 \times 9$ • Find the cross products. $6 \times n = 216$ $n = 216 \div 6$ $n = 36$
To set up a proportion, keep the same units in the numerator and the same units in the denominator. [4.3C, p. 185]	Three machines fill 5 cereal boxes per minute. How many boxes can 8 machines fill per minute? $\frac{3 \text{ machines}}{5 \text{ cereal boxes}} = \frac{8 \text{ machines}}{n \text{ cereal boxes}}$

Chapter 4 Review Exercises

1. Determine whether the proportion is true or not true.
    $$\frac{2}{9} = \frac{10}{45}$$
    True   [4.3A]

2. Write the comparison 32 dollars to 80 dollars as a ratio in simplest form using a fraction, a colon (:), and the word *to*.
    $\frac{2}{5}$ 2:5   2 to 5   [4.1A]

3. Write "250 miles in 4 hours" as a unit rate.
    62.5 miles/hour   [4.2B]

4. Determine whether the proportion is true or not true.
    $$\frac{8}{15} = \frac{32}{60}$$
    True   [4.3A]

5. Solve the proportion.
    $$\frac{16}{n} = \frac{4}{17}$$
    68   [4.3B]

6. Write "$300 earned in 40 hours" as a unit rate.
    $7.50/hour   [4.2B]

7. Write "$8.75 for 5 pounds" as a unit rate.
    $1.75/pound   [4.2B]

8. Write the comparison 8 feet to 28 feet as a ratio in simplest form using a fraction, a colon (:), and the word *to*.
    $\frac{2}{7}$ 2:7   2 to 7   [4.1A]

9. Solve the proportion.
    $$\frac{n}{8} = \frac{9}{2}$$
    36   [4.3B]

10. Solve the proportion. Round to the nearest hundredth.
    $$\frac{18}{35} = \frac{10}{n}$$
    19.44   [4.3B]

11. Write the comparison 6 inches to 15 inches as a ratio in simplest form using a fraction, a colon (:), and the word *to*.
    $\frac{2}{5}$ 2:5   2 to 5   [4.1A]

12. Determine whether the proportion is true or not true.
    $$\frac{3}{8} = \frac{10}{24}$$
    Not true   [4.3A]

13. Write "$15 in 4 hours" as a rate in simplest form.
    $\dfrac{\$15}{4 \text{ hours}}$ [4.2A]

14. Write "326.4 miles on 12 gallons" as a unit rate.
    27.2 miles/gallon   [4.2B]

15. Write the comparison 12 days to 12 days as a ratio in simplest form using a fraction, a colon (:), and the word *to*.
    $\frac{1}{1}$ 1:1   1 to 1   [4.1A]

16. Determine whether the proportion is true or not true.
    $$\frac{5}{7} = \frac{25}{35}$$
    True   [4.3A]

17. Solve the proportion. Round to the nearest hundredth.

$$\frac{24}{11} = \frac{n}{30}$$

65.45 [4.3B]

18. Write "100 miles in 3 hours" as a rate in simplest form.

$\frac{100 \text{ miles}}{3 \text{ hours}}$ [4.2A]

19. **Business**   In 5 years, the price of a calculator went from $40 to $24. What is the ratio, as a fraction in simplest form, of the decrease in price to the original price?

$\frac{2}{5}$ [4.1B]

20. **Taxes**   The property tax on a $245,000 home is $4900. At the same rate, what is the property tax on a home valued at $320,000?

$6400 [4.3C]

21. **Meteorology**   The high temperature during a 24-hour period was 84 degrees, and the low temperature was 42 degrees. Write the ratio, as a fraction in simplest form, of the high temperature to the low temperature for the 24-hour period.

$\frac{2}{1}$ [4.1B]

22. **Manufacturing**   The total cost of manufacturing 1000 cordless phones was $36,600. Of the phones made, 24 did not pass inspection. What is the cost per phone of the phones that *did* pass inspection?

$37.50 [4.2C]

23. **Masonry**   A brick wall 40 feet in length contains 448 concrete blocks. At the same rate, how many blocks would it take to build a wall that is 120 feet in length?

1344 blocks [4.3C]

24. **Advertising**   A retail computer store spends $30,000 a year on radio advertising and $12,000 on newspaper advertising. Find the ratio, as a fraction in simplest form, of radio advertising to newspaper advertising.

$\frac{5}{2}$ [4.1B]

25. **Consumerism**   A 15-pound turkey costs $13.95. What is the cost per pound?

$.93/pound [4.2C]

26. **Travel**   Mahesh drove 198.8 miles in 3.5 hours. Find the average number of miles he drove per hour.

56.8 miles/hour [4.2C]

27. **Insurance**   An insurance policy costs $9.87 for every $1000 of insurance. At this rate, what is the cost of $50,000 of insurance?

$493.50 [4.3C]

28. **Investments**   Pascal Hollis purchased 80 shares of stock for $3580. What was the cost per share?

$44.75/share [4.2C]

29. **Landscaping**   Monique uses 1.5 pounds of fertilizer for every 200 square feet of lawn. How many pounds of fertilizer will she have to use on a lawn that measures 3000 square feet?

22.5 pounds [4.3C]

30. **Real Estate**   A house had an original value of $80,000, but its value increased to $120,000 in 2 years. Find the ratio, as a fraction in simplest form, of the increase to the original value.

$\frac{1}{2}$ [4.1B]

Chapter 4 Test

1. Write "$46,036.80 earned in 12 months" as a unit rate.
   $3836.40/month   [4.2B]

2. Write the comparison 40 miles to 240 miles as a ratio in simplest form using a fraction, a colon (:), and the word *to*.
   $\frac{1}{6}$ 1:6   1 to 6   [4.1A]

3. Write "18 supports for every 8 feet" as a rate in simplest form.
   $\frac{9 \text{ supports}}{4 \text{ feet}}$ [4.2A]

4. Determine whether the proportion is true or not true.
   $\frac{40}{125} = \frac{5}{25}$
   Not true   [4.3A]

5. Write the comparison 12 days to 8 days as a ratio in simplest form using a fraction, a colon (:), and the word *to*.
   $\frac{3}{2}$ 3:2   3 to 2   [4.1A]

6. Solve the proportion.
   $\frac{5}{12} = \frac{60}{n}$
   144   [4.3B]

7. Write "256.2 miles on 8.4 gallons of gas" as a unit rate.
   30.5 miles/gallon   [4.2B]

8. Write the comparison 27 dollars to 81 dollars as a ratio in simplest form using a fraction, a colon (:), and the word *to*.
   $\frac{1}{3}$ 1:3   1 to 3   [4.1A]

9. Determine whether the proportion is true or not true.
   $\frac{5}{14} = \frac{25}{70}$
   True   [4.3A]

10. Solve the proportion.
    $\frac{n}{18} = \frac{9}{4}$
    40.5   [4.3B]

11. Write "$81 for 12 boards" as a rate in simplest form.
    $\frac{\$27}{4 \text{ boards}}$ [4.2A]

12. Write the comparison 18 feet to 30 feet as a ratio in simplest form using a fraction, a colon (:), and the word *to*.
    $\frac{3}{5}$ 3:5   3 to 5   [4.1A]

13. **Investments** Fifty shares of a utility stock pay a dividend of $62.50. At the same rate, what is the dividend paid on 500 shares of the utility stock?
$625   [4.3C]

14. **Meteorology** The average summer temperature in a California desert is 112 degrees. In a city 100 miles away, the average summer temperature is 86 degrees. Write the ratio, as a fraction in simplest form, of the average city temperature to the average desert temperature.
$\frac{43}{56}$ [4.1B]

15. **Travel** A plane travels 2421 miles in 4.5 hours. Find the plane's speed in miles per hour.
538 miles/hour   [4.2C]

16. **Physiology** A research scientist estimates that the human body contains 88 pounds of water for every 100 pounds of body weight. At this rate, estimate the number of pounds of water in a college student who weighs 150 pounds.
132 pounds   [4.3C]

17. **Business** If 40 feet of lumber costs $69.20, what is the per-foot cost of the lumber?
$1.73/foot   [4.2C]

18. **Medicine** The dosage of a certain medication is $\frac{1}{4}$ ounce for every 50 pounds of body weight. How many ounces of this medication are required for a person who weighs 175 pounds? Write the answer as a decimal.
0.875 ounce   [4.3C]

19. **Sports** A basketball team won 20 games and lost 5 games during the season. Write, as a fraction in simplest form, the ratio of the number of games won to the total number of games played.
$\frac{4}{5}$ [4.1B]

20. **Manufacturing** A computer manufacturer discovers through experience that an average of 3 defective hard drives are found in every 100 hard drives manufactured. How many defective hard drives are expected to be found in the production of 1200 hard drives?
36 defective hard drives   [4.3C]

Cumulative Review Exercises

1. Subtract: 20,095
         − 10,937
             9158 [1.3B]

2. Write $2 \cdot 2 \cdot 2 \cdot 2 \cdot 3 \cdot 3 \cdot 3$ in exponential notation.
   $2^4 \cdot 3^3$ [1.6A]

3. Simplify: $4 - (5 - 2)^2 \div 3 + 2$
   3 [1.6B]

4. Find the prime factorization of 160.
   $2 \cdot 2 \cdot 2 \cdot 2 \cdot 2 \cdot 5$ [1.7B]

5. Find the LCM of 9, 12, and 18.
   36 [2.1A]

6. Find the GCF of 28 and 42.
   14 [2.1B]

7. Write $\frac{40}{64}$ in simplest form.
   $\frac{5}{8}$ [2.3B]

8. Find $4\frac{7}{15}$ more than $3\frac{5}{6}$.
   $8\frac{3}{10}$ [2.4C]

9. What is $4\frac{5}{9}$ less than $10\frac{1}{6}$?
   $5\frac{11}{18}$ [2.5C]

10. Multiply: $\frac{11}{12} \times 3\frac{1}{11}$
    $2\frac{5}{6}$ [2.6B]

11. Find the quotient of $3\frac{1}{3}$ and $\frac{5}{7}$.
    $4\frac{2}{3}$ [2.7B]

12. Simplify: $\left(\frac{2}{5} + \frac{3}{4}\right) \div \frac{3}{2}$
    $\frac{23}{30}$ [2.8C]

13. Write 4.0709 in words.
    Four and seven hundred nine ten-thousandths [3.1A]

14. Round 2.09762 to the nearest hundredth.
    2.10 [3.1B]

15. Divide: $8.09\overline{)16.0976}$
    Round to the nearest thousandth.
    1.990 [3.5A]

16. Convert $0.06\frac{2}{3}$ to a fraction.
    $\frac{1}{15}$ [3.6B]

17. Write the comparison 25 miles to 200 miles as a ratio in simplest form using a fraction.
$\frac{1}{8}$ [4.1A]

18. Write "87 cents for 6 pencils" as a rate in simplest form.
$\frac{29¢}{2 \text{ pencils}}$ [4.2A]

19. Write "250.5 miles on 7.5 gallons of gas" as a unit rate.
33.4 miles/gallon [4.2B]

20. Solve $\frac{40}{n} = \frac{160}{17}$.
4.25 [4.3B]

21. **Travel** A car traveled 457.6 miles in 8 hours. Find the car's speed in miles per hour.
57.2 miles/hour [4.2C]

22. Solve the proportion.
$\frac{12}{5} = \frac{n}{15}$ 36 [4.3B]

23. **Banking** You had $1024 in your checking account. You then wrote checks for $192 and $88. What is your new checking account balance?
$744 [1.3C]

24. **Finance** Malek Khatri buys a tractor for $32,360. A down payment of $5000 is required. The balance remaining is paid in 48 equal monthly installments. What is the monthly payment?
$570 [1.5D]

25. **Homework Assignments** Yuko is assigned to read a book containing 175 pages. She reads $\frac{2}{5}$ of the book during Thanksgiving vacation. How many pages of the assignment remain to be read?
105 pages [2.6C]

26. **Real Estate** A building contractor bought $2\frac{1}{3}$ acres of land for $84,000. What was the cost of each acre?
$36,000 [2.7C]

27. **Consumerism** Benjamin Eli bought a shirt for $22.79 and a tie for $9.59. He used a $50 bill to pay for the purchases. Find the amount of change.
$17.62 [3.3B]

28. **Compensation** If you earn an annual salary of $41,619, what is your monthly salary?
$3468.25 [3.5B]

29. **Erosion** A soil conservationist estimates that a river bank is eroding at the rate of 3 inches every 6 months. At this rate, how many inches will be eroded in 50 months?
25 inches [4.3C]

30. **Medicine** The dosage of a certain medication is $\frac{1}{2}$ ounce for every 50 pounds of body weight. How many ounces of this medication are required for a person who weighs 160 pounds? Write the answer as a decimal.
1.6 ounces [4.3C]

5 Percents

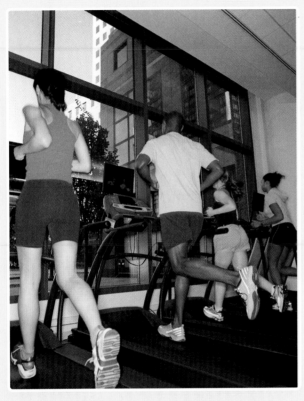

Everyone knows that good health depends on eating right, watching your weight, not smoking, and exercising regularly. In order to get the most out of your workout, the American College of Sports Medicine (ACSM) recommends that you know how to determine your target heart rate. Your target heart rate is the rate at which your heart should beat during any aerobic exercise, such as running, fast walking, or bicycling. Your target heart rate depends on how fit you are, so athletes have higher target heart rates than people who are more sedentary. According to the ACSM, you should reach and then maintain your target heart rate for 20 minutes or more during a workout to achieve cardiovascular fitness. The **Projects and Group Activities on page 224** explain how you can calculate your target heart rate.

OBJECTIVES

Section 5.1

A To write a percent as a fraction or a decimal

B To write a fraction or a decimal as a percent

Section 5.2

A To find the amount when the percent and the base are given

B To solve application problems

Section 5.3

A To find the percent when the base and amount are given

B To solve application problems

Section 5.4

A To find the base when the percent and amount are given

B To solve application problems

Section 5.5

A To solve percent problems using proportions

B To solve application problems

Need help? For online student resources, such as section quizzes, visit this textbook's website at **math.college.hmco.com/students.**

PREP TEST ● ● ●

Do these exercises to prepare for Chapter 5.

For Exercises 1 to 6, multiply or divide.

1. $19 \times \dfrac{1}{100}$

$\dfrac{19}{100}$ [2.6B]

2. 23×0.01

0.23 [3.4A]

3. 0.47×100

47 [3.4A]

4. $0.06 \times 47,500$

2850 [3.4A]

5. $60 \div 0.015$

4000 [3.5A]

6. $8 \div \dfrac{1}{4}$

32 [2.7B]

7. Multiply $\dfrac{5}{8} \times 100$. Write the answer as a decimal.

62.5 [3.6A]

8. Write $\dfrac{200}{3}$ as a mixed number.

$66\dfrac{2}{3}$ [2.2B]

9. Divide $28 \div 16$. Write the answer as a decimal.

1.75 [3.5A]

GO FIGURE ● ● ●

A whole number that remains unchanged when its digits are written in reverse order is a palindrome. For example, 818 is a palindrome.
a. Find the smallest three-digit multiple of 6 that is a palindrome. 222
b. Many nonpalindrome numbers can be converted into palindromes: reverse the digits of the number and add the result to the original; continue until a palindrome is achieved. Using this process, what is the palindrome created from 874? 3883

5.1 Introduction to Percents

Objective A To write a percent as a fraction or a decimal

Percent means "parts of 100." In the figure at the right, there are 100 parts. Because 13 of the 100 parts are shaded, 13% of the figure is shaded. The symbol % is the **percent sign**.

In most applied problems involving percents, it is necessary either to rewrite a percent as a fraction or a decimal or to rewrite a fraction or a decimal as a percent.

> **TAKE NOTE**
> Recall that division is defined as multiplication by the reciprocal. Therefore, multiplying by $\frac{1}{100}$ is equivalent to dividing by 100.

To write a percent as a fraction, remove the percent sign and multiply by $\frac{1}{100}$.

$$13\% = 13 \times \frac{1}{100} = \frac{13}{100}$$

To write a percent as a decimal, remove the percent sign and multiply by 0.01.

$$13\% = 13 \times 0.01 = 0.13$$

> Move the decimal point two places to the left. Then remove the percent sign.

Example 1 **a.** Write 120% as a fraction.
b. Write 120% as a decimal.

Solution **a.** $120\% = 120 \times \frac{1}{100} = \frac{120}{100}$

$$= 1\frac{1}{5}$$

b. $120\% = 120 \times 0.01 = 1.2$

Note that percents larger than 100 are greater than 1.

You Try It 1 **a.** Write 125% as a fraction.
b. Write 125% as a decimal.

Your solution a. $1\frac{1}{4}$

b. 1.25

Example 2 Write $16\frac{2}{3}\%$ as a fraction.

Solution $16\frac{2}{3}\% = 16\frac{2}{3} \times \frac{1}{100}$

$$= \frac{50}{3} \times \frac{1}{100} = \frac{50}{300} = \frac{1}{6}$$

You Try It 2 Write $33\frac{1}{3}\%$ as a fraction.

Your solution $\frac{1}{3}$

Example 3 Write 0.5% as a decimal.

Solution $0.5\% = 0.5 \times 0.01 = 0.005$

You Try It 3 Write 0.25% as a decimal.

Your solution 0.0025

Solutions on pp. S12–S13

New Vocabulary
percent
percent sign

Discuss the Concepts
Why do we multiply a percent by $\frac{1}{100}$ in order to rewrite it as a fraction and by 0.01 in order to rewrite it as a decimal?

Concept Check

1. Is 9.4% the same as $9\frac{2}{5}\%$? If not, what is the difference between the decimal equivalents of 9.4% and $9\frac{2}{5}\%$?
  Yes

2. Is $\frac{1}{2}\%$ the same as 0.5?   No
  If not, what is the difference between 0.5 and the decimal equivalent of $\frac{1}{2}\%$?   0.495

Instructor Note
Example 2 and You Try It 2 are difficult for students. Here is an additional in-class example to use.

Write $12\frac{1}{2}\%$ as a fraction.

Solution:

$$12\frac{1}{2}\% = 12\frac{1}{2} \times \frac{1}{100}$$

$$= \frac{25}{2} \times \frac{1}{100}$$

$$= \frac{1}{8}$$

In-Class Examples (Objective 5.1A)

1a. Write 72% as a fraction. $\frac{18}{25}$

b. Write 72% as a decimal.   0.72

2. Write $15\frac{2}{3}\%$ as a fraction. $\frac{47}{300}$

3. Write 82.9% as a decimal.   0.829

Objective 5.1B

Discuss the Concepts

Why is it that we can multiply a number by 100% and not change the value of the number?

Concept Check

Have students complete a table of equivalent fractions, decimals, and percents. For example:

Fraction	Decimal	Percent
$\frac{1}{2}$	0.5	50%
$\frac{3}{4}$	0.75	75%
$\frac{2}{5}$	0.4	40%
$\frac{3}{8}$	0.375	37.5%
$\frac{13}{25}$	0.52	52%
$\frac{9}{16}$	0.5625	56.25%
$\frac{11}{20}$	0.55	55%
$\frac{3}{5}$	0.6	60%
$\frac{1}{3}$	$0.33\frac{1}{3}$	$33\frac{1}{3}\%$

Instructor Note

Students will ask whether to write their answers as fractions or decimals. As a general rule, if the fraction can be written as a terminating decimal, the answer is written in decimal form. If the answer is a repeating decimal, the answer is written as a fraction.

Objective B To write a fraction or a decimal as a percent

A fraction or a decimal can be written as a percent by multiplying by 100%.

HOW TO Write $\frac{3}{8}$ as a percent.

$$\frac{3}{8} = \frac{3}{8} \times 100\% = \frac{3}{8} \times \frac{100}{1}\% = \frac{300}{8}\% = 37\frac{1}{2}\% \text{ or } 37.5\%$$

HOW TO Write 0.37 as a percent.

$$0.37 \quad = \quad 0.37 \times 100\% \quad = \quad 37\%$$

Move the decimal point two places to the right. Then write the percent sign.

Example 4 Write 0.015 as a percent.

Solution
$$0.015 = 0.015 \times 100\%$$
$$= 1.5\%$$

You Try It 4 Write 0.048 as a percent.

Your solution 4.8%

Example 5 Write 2.15 as a percent.

Solution $2.15 = 2.15 \times 100\% = 215\%$

You Try It 5 Write 3.67 as a percent.

Your solution 367%

Example 6 Write $0.33\frac{1}{3}$ as a percent.

Solution
$$0.33\frac{1}{3} = 0.33\frac{1}{3} \times 100\%$$
$$= 33\frac{1}{3}\%$$

You Try It 6 Write $0.62\frac{1}{2}$ as a percent.

Your solution $62\frac{1}{2}\%$

Example 7 Write $\frac{2}{3}$ as a percent.
Write the remainder in fractional form.

Solution
$$\frac{2}{3} = \frac{2}{3} \times 100\% = \frac{200}{3}\%$$
$$= 66\frac{2}{3}\%$$

You Try It 7 Write $\frac{5}{6}$ as a percent.
Write the remainder in fractional form.

Your solution $83\frac{1}{3}\%$

Example 8 Write $2\frac{2}{7}$ as a percent.
Round to the nearest tenth.

Solution
$$2\frac{2}{7} = \frac{16}{7} = \frac{16}{7} \times 100\%$$
$$= \frac{1600}{7}\% \approx 228.6\%$$

You Try It 8 Write $1\frac{4}{9}$ as a percent.
Round to the nearest tenth.

Your solution 144.4%

Solutions on p. S13

In-Class Examples (Objective 5.1B)

1. Write 0.16 as a percent.   16%

2. Write $\frac{5}{12}$ as a percent. Round to the nearest tenth of a percent.   41.7%

3. Write $\frac{4}{9}$ as a percent. Write the remainder in fractional form. $44\frac{4}{9}\%$

5.1 Exercises

Objective A   To write a percent as a fraction or a decimal

For Exercises 1 to 16, write as a fraction and as a decimal.

1. 25%
$\frac{1}{4}$, 0.25

2. 40%
$\frac{2}{5}$, 0.40

3. 130%
$1\frac{3}{10}$, 1.30

4. 150%
$1\frac{1}{2}$, 1.50

5. 100%
1, 1.00

6. 87%
$\frac{87}{100}$, 0.87

7. 73%
$\frac{73}{100}$, 0.73

8. 45%
$\frac{9}{20}$, 0.45

9. 383%
$3\frac{83}{100}$, 3.83

10. 425%
$4\frac{1}{4}$, 4.25

11. 70%
$\frac{7}{10}$, 0.70

12. 55%
$\frac{11}{20}$, 0.55

13. 88%
$\frac{22}{25}$, 0.88

14. 64%
$\frac{16}{25}$, 0.64

15. 32%
$\frac{8}{25}$, 0.32

16. 18%
$\frac{9}{50}$, 0.18

For Exercises 17 to 28, write as a fraction.

17. $66\frac{2}{3}$%
$\frac{2}{3}$

18. $12\frac{1}{2}$%
$\frac{1}{8}$

19. $83\frac{1}{3}$%
$\frac{5}{6}$

20. $3\frac{1}{8}$%
$\frac{1}{32}$

21. $11\frac{1}{9}$%
$\frac{1}{9}$

22. $\frac{3}{8}$%
$\frac{3}{800}$

23. $45\frac{5}{11}$%
$\frac{5}{11}$

24. $15\frac{3}{8}$%
$\frac{123}{800}$

25. $4\frac{2}{7}$%
$\frac{3}{70}$

26. $5\frac{3}{4}$%
$\frac{23}{400}$

27. $6\frac{2}{3}$%
$\frac{1}{15}$

28. $8\frac{2}{3}$%
$\frac{13}{150}$

For Exercises 29 to 43, write as a decimal.

29. 6.5%
0.065

30. 9.4%
0.094

31. 12.3%
0.123

32. 16.7%
0.167

33. 0.55%
0.0055

34. 0.45%
0.0045

35. 8.25%
0.0825

36. 6.75%
0.0675

37. 5.05%
0.0505

38. 3.08%
0.0308

39. 2%
0.02

40. 7%
0.07

41. 80.4%
0.804

42. 36.2%
0.362

43. 4.9%
0.049

Objective B   To write a fraction or a decimal as a percent

For Exercises 44 to 55, write as a percent.

44. 0.16
16%

45. 0.73
73%

46. 0.05
5%

47. 0.01
1%

48. 1.07
107%

49. 2.94
294%

50. 0.004
0.4%

51. 0.006
0.6%

52. 1.012
101.2%

53. 3.106
310.6%

54. 0.8
80%

55. 0.7
70%

Quick Quiz (Objective 5.1A)

1. a. Write 65% as a fraction. $\frac{13}{20}$

  b. Write 65% as a decimal. 0.65

2. Write $45\frac{1}{3}$% as a fraction. $\frac{34}{75}$

3. Write 34.27% as a decimal. 0.3427

For Exercises 56 to 67, write as a percent. Round to the nearest tenth of a percent.

56. $\frac{27}{50}$ 54%   **57.** $\frac{37}{100}$ 37%   **58.** $\frac{1}{3}$ 33.3%   **59.** $\frac{2}{5}$ 40%

60. $\frac{5}{8}$ 62.5%   **61.** $\frac{1}{8}$ 12.5%   **62.** $\frac{1}{6}$ 16.7%   **63.** $1\frac{1}{2}$ 150%

64. $\frac{7}{40}$ 17.5%   **65.** $1\frac{2}{3}$ 166.7%   **66.** $1\frac{7}{9}$ 177.8%   **67.** $\frac{7}{8}$ 87.5%

For Exercises 68 to 75, write as a percent. Write the remainder in fractional form.

68. $\frac{15}{50}$ 30%   **69.** $\frac{12}{25}$ 48%   **70.** $\frac{7}{30}$ $23\frac{1}{3}$%   **71.** $\frac{1}{3}$ $33\frac{1}{3}$%

72. $2\frac{3}{8}$ $237\frac{1}{2}$%   **73.** $1\frac{2}{3}$ $166\frac{2}{3}$%   **74.** $2\frac{1}{6}$ $216\frac{2}{3}$%   **75.** $\frac{7}{8}$ $87\frac{1}{2}$%

76. Write the part of the square that is shaded as a fraction, as a decimal, and as a percent. Write the part of the square that is not shaded as a fraction, as a decimal, and as a percent.
$\frac{1}{4}$,   0.25,   25%; $\frac{3}{4}$,   0.75,   75%

APPLYING THE CONCEPTS

77. **The Food Industry**   In a survey conducted by Opinion Research Corp. for Lloyd's Barbeque Co., people were asked to name their favorite barbeque side dishes. 38% named corn on the cob, 35% named cole slaw, 11% named corn bread, and 10% named fries. What percent of those surveyed named something other than corn on the cob, cole slaw, corn bread, or fries?
6%

78. **Consumerism**   A sale on computers advertised $\frac{1}{3}$ off the regular price. What percent of the regular price does this represent?
$33\frac{1}{3}$%

79. **Consumerism**   A suit was priced at 50% off the regular price. What fraction of the regular price does this represent?
$\frac{1}{2}$

80. **Elections**   If $\frac{2}{5}$ of the population voted in an election, what percent of the population did not vote?
60%

81. a. Is the statement "Multiplying a number by a percent always decreases the number" true or false?   False

b. If it is false, given an example to show that the statement is false.
For example, 200% × 4 = 2 × 4 = 8

Quick Quiz (Objective 5.1B)

1. Write 0.56 as a percent.   56%

2. Write $\frac{7}{45}$ as a percent. Round to the nearest tenth of a percent.   15.6%

3. Write $\frac{5}{6}$ as a percent. Write the remainder in fractional form. $83\frac{1}{3}$%

5.2 Percent Equations: Part I

Objective A   **To find the amount when the percent and the base are given**

A real estate broker receives a payment that is 4% of a $285,000 sale. To find the amount the broker receives requires answering the question "4% of $285,000 is what?"

This sentence can be written using mathematical symbols and then solved for the unknown number.

4%	of	$285,000	is	what?
↓	↓	↓	↓	↓

of is written as × (times)
is is written as = (equals)
what is written as *n* (the unknown number)

$$\boxed{\text{Percent } 4\%} \times \boxed{\text{base } 285{,}000} = \boxed{\text{amount } n}$$

$$0.04 \times 285{,}000 = n$$
$$11{,}400 = n$$

Note that the percent is written as a decimal.

The broker receives a payment of $11,400.

The solution was found by solving the **basic percent equation** for amount.

> **The Basic Percent Equation**
> $$\boxed{\text{Percent}} \times \boxed{\text{base}} = \boxed{\text{amount}}$$

In most cases, the percent is written as a decimal before the basic percent equation is solved. However, some percents are more easily written as a fraction than as a decimal. For example,

$$33\tfrac{1}{3}\% = \frac{1}{3} \qquad 66\tfrac{2}{3}\% = \frac{2}{3} \qquad 16\tfrac{2}{3}\% = \frac{1}{6} \qquad 83\tfrac{1}{3}\% = \frac{5}{6}$$

Example 1   Find 5.7% of 160.

Solution

Percent × base = amount
$0.057 \times 160 = n$
$9.12 = n$

• The word *Find* is used instead of the words *what is*.

Example 2   What is $33\tfrac{1}{3}\%$ of 90?

Solution   Percent × base = amount
$$\frac{1}{3} \times 90 = n$$
$$30 = n$$

• $33\tfrac{1}{3}\% = \tfrac{1}{3}$

You Try It 1   Find 6.3% of 150.

Your solution   9.45

You Try It 2   What is $16\tfrac{2}{3}\%$ of 66?

Your solution   11

Solutions on p. S13

Objective 5.2A

New Vocabulary
the basic percent equation
base
amount

Concept Check
Create grids and ask the students to shade a given percent of the region. For example, shade 40% of the region below.

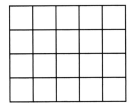

There are 20 squares.

40% × 20 = 0.40 × 20 = 8

8 squares should be shaded.

Shade different percents of the same grid (for example, for the grid above, shade 25% and 90%). Also create different grids; for example, shade 75% of a 5 × 8 grid or 37.5% of a 4 × 8 grid.

Instructor Note
Effective use of the percent equation is one of the most important skills a student can acquire. This section and the next two sections are devoted to solving this equation.
The last section in the chapter, Section 5.5, gives you the option of teaching the percent equation using proportions.

In-Class Examples (Objective 5.2A)

1. 7% of 50 is what?   3.5

2. What is 45% of 80?   36

3. Find 12% of 425.   51

Objective 5.2B

Discuss the Concepts

1. Multiplying a number by $33\frac{1}{3}\%$ is the same as multiplying it by what fraction?

2. Multiplying a number by 200% is the same as multiplying it by what whole number?

3. Multiplying a number by 150% is the same as multiplying it by what mixed number?

Concept Check

The figure below shows the results of a survey of 1500 people who were asked how much they worry about privacy on the Internet.

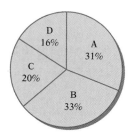

A—Worry a great deal
B—Worry somewhat
C—Don't worry very much
D—Don't worry at all

1. How many of the people said they worried somewhat about privacy on the Internet?
   495 people

2. How many of the people said they did not worry at all about privacy on the Internet?
   240 people

Objective B **To solve application problems**

Solving percent problems requires identifying the three elements of the basic percent equation. Recall that these three parts are the *percent*, the *base*, and the *amount*. Usually the base follows the phrase "percent of."

During a recent year, Americans gave $212 billion to charities. The circle graph at the right shows where that money came from. Use these data for Example 3 and You Try It 3.

Charitable Giving

Sources: American Association of Fundraising Counsel; AP

Example 3

How much of the amount given to charities came from individuals?

Strategy
To determine the amount that came from individuals, write and solve the basic percent equation using n to represent the amount. The percent is 76%. The base is $212 billion.

Solution
$$\text{Percent} \times \text{base} = \text{amount}$$
$$76\% \times 212 = n$$
$$0.76 \times 212 = n$$
$$161.12 = n$$

Individuals gave $161.12 billion to charities.

You Try It 3

How much of the amount given to charities was given by corporations?

Your strategy

Your solution $8.48 billion

Example 4

A quality control inspector found that 1.2% of 2500 telephones inspected were defective. How many telephones inspected were not defective?

Strategy
To find the number of nondefective phones:

• Find the number of defective phones. Write and solve the basic percent equation using n to represent the number of defective phones (amount). The percent is 1.2% and the base is 2500.

• Subtract the number of defective phones from the number of phones inspected (2500).

Solution
$$1.2\% \times 2500 = n$$
$$0.012 \times 2500 = n$$
$$30 = n \text{ defective phones}$$

$$2500 - 30 = 2470$$

2470 telephones were not defective.

You Try It 4

An electrician's hourly wage was $33.50 before an 8% raise. What is the new hourly wage?

Your strategy

Your solution $36.18

Solutions on p. S13

In-Class Examples (Objective 5.2B)

1. A truck retail sales company made a 4.5% profit on sales of $360,000. Find the company's profit.
   $16,200

5.2 Exercises

Objective A   To find the amount when the percent and the base are given

1. 8% of 100 is what?
8

2. 16% of 50 is what?
8

3. 27% of 40 is what?
10.8

4. 52% of 95 is what?
49.4

5. 0.05% of 150 is what?
0.075

6. 0.075% of 625 is what?
0.46875

7. 125% of 64 is what?
80

8. 210% of 12 is what?
25.2

9. Find 10.7% of 485.
51.895

10. Find 12.8% of 625.
80

11. What is 0.25% of 3000?
7.5

12. What is 0.06% of 250?
0.15

13. 80% of 16.25 is what?
13

14. 26% of 19.5 is what?
5.07

15. What is $1\frac{1}{2}$% of 250?
3.75

16. What is $5\frac{3}{4}$% of 65?
3.7375

17. $16\frac{2}{3}$% of 120 is what?
20

18. $83\frac{1}{3}$% of 246 is what?
205

19. What is $33\frac{1}{3}$% of 630?
210

20. What is $66\frac{2}{3}$% of 891?
594

21. Which is larger: 5% of 95, or 75% of 6?
5% of 95

22. Which is larger: 112% of 5, or 0.45% of 800?
112% of 5

23. Which is smaller: 79% of 16, or 20% of 65?
79% of 16

24. Which is smaller: 15% of 80, or 95% of 15?
15% of 80

25. Which is smaller: 2% of 1500, or 72% of 40?
72% of 40

26. Which is larger: 22% of 120, or 84% of 32?
84% of 32

27. Find 31.294% of 82,460.
25,805.0324

28. Find 123.94% of 275,976.
342,044.6544

Objective B   To solve application problems

29. **Health Insurance** Approximately 30% of the 44 million people in the United States who do not have health insurance are between the ages of 18 and 24. (*Source:* U.S. Census Bureau) About how many people in the United States aged 18 to 24 do not have health insurance?
13.2 million people

30. **Aviation** The Federal Aviation Administration reported that 55,422 new student pilots were flying single-engine planes last year. The number of new student pilots flying single-engine planes this year is 106% of the number flying single-engine planes last year. How many new student pilots are flying single-engine planes this year?
58,747 new student pilots

Quick Quiz (Objective 5.2A)

1. 29% of 60 is what?   17.4

2. What is 35% of 73?   25.55

3. Find 25% of 112.   28

Politics The results of a survey in which 32,840 full-time college and university faculty members were asked to describe their political views is shown at the right. Use these data for Exercises 31 and 32.

Political View	Percent of Faculty Members Responding
Far Left	5.3%
Liberal	42.3%
Middle of the road	34.3%
Conservative	17.7%
Far right	0.3%

Source: Higher Education Research Institute, UCLA

31. How many more faculty members described their political views as liberal than described their views as far left?
12,151 more faculty members

32. How many fewer faculty members described their political views as conservative than described their views as middle of the road?
5451 fewer faculty members

33. Taxes A sales tax of 6% of the cost of a car was added to the purchase price of $29,500.
   a. How much was the sales tax? $1770
   b. What is the total cost of the car, including sales tax? $31,270

34. Business During the packaging process for oranges, spoiled oranges are discarded by an inspector. In one day an inspector found that 4.8% of the 20,000 pounds of oranges inspected were spoiled.
   a. How many pounds of oranges were spoiled? 960 pounds
   b. How many pounds of oranges were not spoiled? 19,040 pounds

35. Entertainment A USA TODAY.com online poll asked 8878 Internet users, "Would you use software to cut out objectionable parts of movies?" 29.8% of the respondents answered yes. How many respondents did not answer yes to the question? Round to the nearest whole number.
6232 respondents

36. Employment Funtimes Amusement Park has 550 employees and must hire an additional 22% for the vacation season. What is the total number of employees needed for the vacation season?
671 employees

APPLYING THE CONCEPTS

Sociology The two circle graphs at the right show how surveyed employees actually spend their time and the way they would prefer to spend their time. Assuming that employees have 112 hours a week of time that is not spent sleeping, answer Exercises 37 to 39. Round to the nearest tenth of an hour.

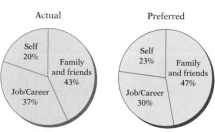

Actual   Preferred

Self 20%, Family and friends 43%, Job/Career 37%

Self 23%, Family and friends 47%, Job/Career 30%

Source: WSJ Supplement, Work & Family, from *Families and Work Institute*

37. What is the actual number of hours per week that employees spend with family and friends?
48.2 hours

38. What is the number of hours per week that employees would prefer to spend on job/career?
33.6 hours

39. What is the difference between the number of hours an employee preferred to spend on self and the actual amount of time the employee spent on self?
3.4 hours

Quick Quiz (Objective 5.2B)

1. An office building has an appraised value of $5,000,000. The real estate taxes are 1.85% of the appraised value of the building. Find the real estate taxes. $92,500

5.3 Percent Equations: Part II

Objective A   **To find the percent when the base and amount are given**

A recent promotional game at a grocery store listed the probability of winning a prize as "1 chance in 2." A percent can be used to describe the chance of winning. This requires answering the question "What percent of 2 is 1?"

The chance of winning can be found by solving the basic percent equation for *percent*.

Integrating Technology

The percent key % on a scientific calculator moves the decimal point to the left two places when pressed after a multiplication or division computation. For the example at the right, enter

1 ÷ 2 % =

The display reads 50.

What percent of   2   is   1?

$$\begin{array}{ccccc} \boxed{\begin{array}{c}\text{Percent}\\ n\end{array}} & \times & \boxed{\begin{array}{c}\text{base}\\ 2\end{array}} & = & \boxed{\begin{array}{c}\text{amount}\\ 1\end{array}} \end{array}$$

$$\begin{aligned} n \times 2 &= 1 \\ n &= 1 \div 2 \\ n &= 0.5 \\ n &= 50\% \end{aligned}$$

• The solution must be written as a percent to answer the question.

There is a 50% chance of winning a prize.

Vocabulary to Review

the basic percent equation [5.2A]
base   [5.2A]
amount   [5.2A]

Concept Check

Create grids and ask the students what percent of the entire region is shaded. For example, ask students what percent of the region below is shaded.

There are 20 squares, of which 4 are shaded, so students are answering the question "What percent of 20 is 4?"   20%

Shade different numbers of squares within the same grid (for example, for the grid above, shade 6 squares, 10 squares, and 13 squares). Also create different grids. For example, shade 8 squares in a 4 × 6 grid, 9 squares in a 3 × 5 grid, or 7 squares in a 5 × 8 grid.

Instructor Note

Although this is not foolproof, the base in the basic percent equation will generally follow the phrase "percent of" in application problems. Having students find this phrase will help them identify all the components of the percent equation.

Example 1   What percent of 40 is 30?

Solution   Percent × base = amount
$$\begin{aligned} n \times 40 &= 30 \\ n &= 30 \div 40 \\ n &= 0.75 \\ n &= 75\% \end{aligned}$$

Example 2   What percent of 12 is 27?

Solution   Percent × base = amount
$$\begin{aligned} n \times 12 &= 27 \\ n &= 27 \div 12 \\ n &= 2.25 \\ n &= 225\% \end{aligned}$$

Example 3   25 is what percent of 75?

Solution   Percent × base = amount
$$\begin{aligned} n \times 75 &= 25 \\ n &= 25 \div 75 \\ n &= \frac{1}{3} = 33\frac{1}{3}\% \end{aligned}$$

You Try It 1   What percent of 32 is 16?

Your solution   50%

You Try It 2   What percent of 15 is 48?

Your solution   320%

You Try It 3   30 is what percent of 45?

Your solution $66\frac{2}{3}\%$

Solutions on p. S13

In-Class Examples (Objective 5.3A)

1. What percent of 80 is 25?   31.25%

2. 19 is what percent of 95?   20%

Objective 5.3B

Concept Check

1. Employee A had an annual salary of $22,000, Employee B had an annual salary of $28,000, and Employee C had an annual salary of $26,000 before each employee was given a 5% raise. Which of the three employee's annual salary is now the largest? Explain how you arrived at your answer.

   Employee B's salary will be the largest after the raise. You want students to explain that 5% of the largest original salary will be the largest raise.

2. Each of three employees earned an annual salary of $25,000 before Employee A was given a 3% raise, Employee B was given a 6% raise, and Employee C was given a 4.5% raise. Which of the three employee's annual salary is now the largest? Explain how you arrived at your answer.

   Employee B's salary will be the largest after the raise. You want students to explain that because all the original salaries are the same, the largest resulting salary will go to the employee with the largest percent raise.

Objective B   **To solve application problems**

To solve percent problems, remember that it is necessary to identify the percent, base, and amount. Usually the base follows the phrase "percent of."

Example 4

The monthly house payment for the Kaminski family is $787.50. What percent of the Kaminskis' monthly income of $3750 is the house payment?

Strategy

To find what percent of the income the house payment is, write and solve the basic percent equation using n to represent the percent. The base is $3750 and the amount is $787.50.

Solution

$n \times 3750 = 787.50$
$n = 787.50 \div 3750$
$n = 0.21 = 21\%$

The house payment is 21% of the monthly income.

You Try It 4

Tomo Nagata had an income of $33,500 and paid $5025 in income tax. What percent of the income is the income tax?

Your strategy

Your solution
15%

Example 5

On one Monday night, 31.39 million of the approximately 40.76 million households watching television were not watching David Letterman. What percent of these households were watching David Letterman? Round to the nearest percent.

Strategy

To find the percent of households watching David Letterman:

- Subtract to find the number of households that were watching David Letterman (40.76 million − 31.39 million).
- Write and solve the basic percent equation using n to represent the percent. The base is 40.76, and the amount is the number of households watching David Letterman.

Solution

40.76 million − 31.39 million = 9.37 million

9.37 million households were watching David Letterman.

$n \times 40.76 = 9.37$
$n = 9.37 \div 40.76$
$n \approx 0.23$

Approximately 23% of the households were watching David Letterman.

You Try It 5

According to the U.S. Department of Defense, of the 518,921 enlisted personnel in the U.S. Army in 1950, 512,370 people were men. What percent of the enlisted personnel in the U.S. Army in 1950 were women? Round to the nearest tenth of a percent.

Your strategy

Your solution
1.3%

Solutions on p. S13

In-Class Examples (Objective 5.3B)

1. An investor received a dividend of $360 on an investment of $4500. What percent of the investment is the dividend?   8%

5.3 Exercises

Objective A   To find the percent when the base and amount are given

1. What percent of 75 is 24?
32%

2. What percent of 80 is 20?
25%

3. 15 is what percent of 90?
$16\frac{2}{3}$%

4. 24 is what percent of 60?
40%

5. What percent of 12 is 24?
200%

6. What percent of 6 is 9?
150%

7. What percent of 16 is 6?
37.5%

8. What percent of 24 is 18?
75%

9. 18 is what percent of 100?
18%

10. 54 is what percent of 100?
54%

11. 5 is what percent of 2000?
0.25%

12. 8 is what percent of 2500?
0.32%

13. What percent of 6 is 1.2?
20%

14. What percent of 2.4 is 0.6?
25%

15. 16.4 is what percent of 4.1?
400%

16. 5.3 is what percent of 50?
10.6%

17. 1 is what percent of 40?
2.5%

18. 0.3 is what percent of 20?
1.5%

19. What percent of 48 is 18?
37.5%

20. What percent of 11 is 88?
800%

21. What percent of 2800 is 7?
0.25%

22. What percent of 400 is 12?
3%

23. 4.2 is what percent of 175?
2.4%

24. 41.79 is what percent of 99.5?
42%

25. What percent of 86.5 is 8.304?
9.6%

26. What percent of 1282.5 is 2.565?
0.2%

Objective B   To solve application problems

27. **Sociology**   Seven in ten couples disagree about financial issues. (*Source:* Yankelovich Partners for Lutheran Brotherhood) What percent of couples disagree about financial matters?
70%

28. **Sociology**   In a survey, 1236 adults nationwide were asked, "What irks you most about the actions of other motorists?" The response "tailgaters" was given by 293 people. (*Source:* Reuters/Zogby) What percent of those surveyed were most irked by tailgaters? Round to the nearest tenth of a percent.
23.7%

Quick Quiz (Objective 5.3A)

1. What percent of 35 is 21?   60%

2. 33 is what percent of 60?   55%

29. **Agriculture** According to the U.S. Department of Agriculture, of the 63 billion pounds of vegetables produced in the United States in 1 year, 16 billion pounds were wasted. What percent of the vegetables produced were wasted? Round to the nearest tenth of a percent.
25.4%

30. **Agriculture** In a recent year, Wisconsin growers produced 281.72 million pounds of the 572 million pounds of cranberries grown in the United States. What percent of the total cranberry crop was produced in Wisconsin? Round to the nearest percent.
49%

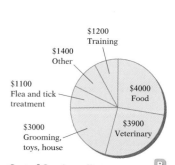

31. **Energy** The typical American household spends $1355 a year on energy utilities. Of this amount, approximately $81.30 is spent on lighting. (*Source:* Department of Energy, Owens Corning) What percent of the total amount spent on energy utilities is spent on lighting?
6%

32. **Education** To receive a license to sell insurance, an insurance account executive must answer correctly 70% of the 250 questions on a test. Nicholas Mosley answered 177 questions correctly. Did he pass the test?
Yes

33. **Agriculture** According to the U.S. Department of Agriculture, of the 356 billion pounds of food produced in the United States annually, 260 billion pounds are not wasted. What percent of the food produced in the United States is wasted? Round to the nearest percent.
27%

34. **Construction** In a test of the breaking strength of concrete slabs for freeway construction, 3 of the 200 slabs tested did not meet safety requirements. What percent of the slabs did meet safety requirements?
98.5%

APPLYING THE CONCEPTS

Pets The graph at the right shows several categories of average lifetime costs of dog ownership. Use this graph for Exercises 35 to 37. Round answers to the nearest tenth of a percent.

35. What percent of the total amount is spent on food?
27.4%

36. What percent of the total is spent on veterinary care?
26.7%

37. What percent of the total is spent on all categories except training?
91.8%

$1200 Training
$1400 Other
$1100 Flea and tick treatment
$4000 Food
$3000 Grooming, toys, house
$3900 Veterinary

Cost of Owning a Dog
Source: Based on data from the American Kennel Club, *USA Today* research

Answers to Writing Exercises

38. The sum of the percents in the percent column is 113%. In order for the responses to be possible, the sum of the percents must be 100%.

38. **Sports** The Fun in the Sun organization claims to have taken a survey of 350 people, asking them to give their favorite outdoor temperature for hiking. The results are given in the table at the right. Explain why these results are not possible.

Favorite Temperature	Percent
Greater than 90	5%
80–89	28%
70–79	35%
60–69	32%
Below 60	13%

Quick Quiz (Objective 5.3B)

1. A survey of 1650 people showed that 462 people favored the incumbent mayor. What percent of the people surveyed favored the incumbent mayor?
28%

5.4 Percent Equations: Part III

Objective A   **To find the base when the percent and amount are given**

In 1780, the population of Virginia was 538,000. This was 19% of the total population of the United States at that time. To find the total population at that time, you must answer the question "19% of what number is 538,000?"

Study Tip

After completing this objective, you will have learned to solve the basic percent equation for each of the three elements: percent, base, and amount. You will need to be able to recognize these three different types of problems. To test yourself, you might do the Chapter 5 Review Exercises.

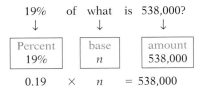

19%	of	what	is	538,000?
↓		↓		↓
Percent 19%		base n		amount 538,000

$$0.19 \times n = 538,000$$
$$n = 538,000 \div 0.19$$
$$n \approx 2,832,000$$

• The population of the United States in 1780 can be found by solving the basic percent equation for the base.

The population of the United States in 1780 was approximately 2,832,000.

Example 1   18% of what is 900?

Solution
Percent × base = amount
$$0.18 \times n = 900$$
$$n = 900 \div 0.18$$
$$n = 5000$$

You Try It 1   86% of what is 215?

Your solution   250

Example 2   30 is 1.5% of what?

Solution
Percent × base = amount
$$0.015 \times n = 30$$
$$n = 30 \div 0.015$$
$$n = 2000$$

You Try It 2   15 is 2.5% of what?

Your solution   600

Example 3 $33\frac{1}{3}$% of what is 7?

Solution
Percent × base = amount
$$\frac{1}{3} \times n = 7$$
• Note that the percent is written as a fraction.
$$n = 7 \div \frac{1}{3}$$
$$n = 21$$

You Try It 3 $16\frac{2}{3}$% of what is 5?

Your solution   30

Solutions on p. S14

Objective B   **To solve application problems**

To solve percent problems, it is necessary to identify the percent, base, and amount. Usually the base follows the phrase "percent of."

Discuss the Concepts

1. What is the basic percent equation?
2. Describe each component in the basic percent equation.

Concept Check

For each of the following problems, do not do any calculation. Determine the answer by thinking about what the given percent does to the base in order to be equal to the given amount.

1. 25% of what is 800?
   a. 200
   b. 1600
   c. 3200   c
2. 20% of what is 16?
   a. 3.2
   b. 80
   c. 8   b
3. 90 is 150% of what?
   a. 60
   b. 135
   c. 100   a
4. 300 is $33\frac{1}{3}$% of what?
   a. 100
   b. 330
   c. 900   c
5. 200% of what is 50?
   a. 200
   b. 100
   c. 25   c

Vocabulary to Review

the basic percent equation [5.2A]
base   [5.2A]
amount   [5.2A]

In-Class Examples (Objective 5.4A)
1. 10% of what is 20?   200
2. 7 is 14% of what?   50

Objective 5.4B

Concept Check

A survey by the *Boston Globe* involved questioning elementary and middle-school students about television. 68 students, or 42.5% of those surveyed, said that they had a television in their bedroom at home. How many students were included in the *Globe* survey?   160 students

Optional Student Activity

(*Note :* This is a challenging problem.)

Five percent of the students enrolled at a state university were asked to name their favorite TV show. Of those surveyed, $\frac{1}{4}$ were first-year students, $\frac{1}{6}$ were sophomores, $\frac{1}{3}$ were juniors, $\frac{1}{5}$ were seniors, and 21 were graduate students. How many students are enrolled in the university?   8400 students

---

Example 4

A business office bought a used copy machine for $900, which was 75% of the original cost. What was the original cost of the copier?

Strategy

To find the original cost of the copier, write and solve the basic percent equation using n to represent the original cost (base). The percent is 75% and the amount is $900.

Solution

$75\% \times n = 900$
$0.75 \times n = 900$
$\qquad n = 900 \div 0.75$
$\qquad n = 1200$

The original cost of the copier was $1200.

Example 5

A carpenter's wage this year is $26.40 per hour, which is 110% of last year's wage. What was the increase in the hourly wage over last year?

Strategy

To find the increase in the hourly wage over last year:

- Find last year's wage. Write and solve the basic percent equation using n to represent last year's wage (base). The percent is 110% and the amount is $26.40.
- Subtract last year's wage from this year's wage (26.40).

Solution

$110\% \times n = 26.40$
$1.10 \times n = 26.40$
$\qquad n = 26.40 \div 1.10$
$\qquad n = 24.00$ • Last year's wage

$26.40 - 24.00 = 2.40$

The increase in the hourly wage was $2.40.

You Try It 4

A used car has a value of $10,458, which is 42% of the car's original value. What was the car's original value?

Your strategy

Your solution
$24,900

You Try It 5

Chang's Sporting Goods has a tennis racket on sale for $89.60, which is 80% of the original price. What is the difference between the original price and the sale price?

Your strategy

Your solution
$22.40

Solutions on p. S14

In-Class Examples (Objective 5.4B)

1. A student answered 16 of the questions on a 2-hour exam incorrectly. This was 25% of the total number of exam questions. How many questions were on the exam?   64 questions

5.4 Exercises

Objective A   To find the base when the percent and amount are given

For Exercises 1 to 26, solve. Round to the nearest hundredth.

1. 12% of what is 9?
   75

2. 38% of what is 171?
   450

3. 8 is 16% of what?
   50

4. 54 is 90% of what?
   60

5. 10 is 10% of what?
   100

6. 37 is 37% of what?
   100

7. 30% of what is 25.5?
   85

8. 25% of what is 21.5?
   86

9. 2.5% of what is 30?
   1200

10. 10.4% of what is 52?
    500

11. 125% of what is 24?
    19.2

12. 180% of what is 21.6?
    12

13. 18 is 240% of what?
    7.5

14. 24 is 320% of what?
    7.5

15. 4.8 is 15% of what?
    32

16. 87.5 is 50% of what?
    175

17. 25.6 is 12.8% of what?
    200

18. 45.014 is 63.4% of what?
    71

19. 0.7% of what is 0.56?
    80

20. 0.25% of what is 1?
    400

21. 30% of what is 2.7?
    9

22. 78% of what is 3.9?
    5

23. 84 is $16\frac{2}{3}$% of what?
    504

24. 120 is $33\frac{1}{3}$% of what?
    360

25. $66\frac{2}{3}$% of what is 72?
    108

26. $83\frac{1}{3}$% of what is 13.5?
    16.2

Objective B   To solve application problems

27. **Travel** Of the travelers who, during a recent year, allowed their children to miss school to go along on a trip, approximately 1.738 million allowed their children to miss school for more than a week. This represented 11% of the travelers who allowed their children to miss school. (*Source:* Travel Industry Association) About how many travelers allowed their children to miss school to go along on a trip?
    15.8 million travelers

Quick Quiz (Objective 5.4A)
1. 42% of what is 105?   250
2. 56 is 70% of what?   80

28. **Education** In the United States today, 23.1% of the women and 27.5% of the men have earned a bachelor's or graduate degree. (*Source:* Census Bureau) How many women in the United States have earned a bachelor's or graduate degree?
Insufficient information

29. **Taxes** A TurboTax online survey asked people how they planned to use their tax refunds. 740 people, or 22% of the respondents, said they would save the money. How many people responded to the survey?
3364 people

30. **Taxes** The Internal Revenue Service says that the average deduction for medical expenses for taxpayers in the $40,000–$50,000 bracket is $4500. This is 26% of the medical expenses claimed by taxpayers in the over-$200,000 bracket. How much is the average deduction for medical expenses claimed by taxpayers in the over-$200,000 bracket? Round to the nearest thousand.
$17,000

31. **Manufacturing** During a quality control test, Micronics found that 24 computer boards were defective. This amount was 0.8% of the computer boards tested.
    a. How many computer boards were tested? 3000 boards
    b. How many computer boards tested were not defective? 2976 boards

32. **Directory Assistance** Of the calls a directory assistance operator received, 441 were requests for telephone numbers listed in the current directory. This accounted for 98% of the calls for assistance that the operator received.
    a. How many calls did the operator receive? 450 calls
    b. How many telephone numbers requested were not listed in the current directory? 9 numbers

APPLYING THE CONCEPTS

33. **Demography** At the last census, of the 281,422,000 people in the United States, 28.6% were under the age of 20. (*Source:* U.S. Census 2000) How many people in the United States were age 20 or older in 2000?
200,935,308 people

Nutrition The table at the right contains nutrition information about a breakfast cereal. Solve Exercises 34 and 35 using information from this table.

34. The amount of thiamin in one serving of cereal with skim milk is 0.45 milligram. Find the recommended daily allowance of thiamin for an adult.
1.5 milligrams

35. The amount of copper in one serving of cereal with skim milk is 0.08 milligram. Find the recommended daily allowance of copper for an adult.
2 milligrams

36. Increase a number by 10%. Now decrease the number by 10%. Is the result the original number? Explain.

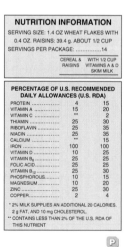

NUTRITION INFORMATION

SERVING SIZE: 1.4 OZ WHEAT FLAKES WITH 0.4 OZ. RAISINS: 39.4 g. ABOUT 1/2 CUP
SERVINGS PER PACKAGE: 14

	CEREAL & RAISINS	WITH 1/2 CUP VITAMINS A & D SKIM MILK

PERCENTAGE OF U.S. RECOMMENDED DAILY ALLOWANCES (U.S. RDA)

	CEREAL & RAISINS	WITH 1/2 CUP SKIM MILK
PROTEIN	4	15
VITAMIN A	15	20
VITAMIN C	**	2
THIAMIN	25	30
RIBOFLAVIN	25	35
NIACIN	25	35
CALCIUM	**	15
IRON	100	100
VITAMIN D	10	25
VITAMIN B₆	25	25
FOLIC ACID	25	25
VITAMIN B₁₂	25	30
PHOSPHOROUS	10	15
MAGNESIUM	10	20
ZINC	25	30
COPPER	2	4

* 2% MILK SUPPLIES AN ADDITIONAL 20 CALORIES. 2 g FAT, AND 10 mg CHOLESTEROL.
** CONTAINS LESS THAN 2% OF THE U.S. RDA OF THIS NUTRIENT

Answer to Writing Exercise

36. Suppose we increase 100 by 10%.

$$100 + 0.10(100) = 100 + 10$$
$$= 110$$

Now we decrease 110 by 10%.

$$110 - 0.10(110) = 110 - 11$$
$$= 99$$

No, the new number is not the original number. The 10% increase applied to the number 100, but the 10% decrease applied to 110. Therefore, the decrease was greater than the increase.

Quick Quiz (Objective 5.4B)

1. A company spent $128,000 on advertising in one year. This was 16% of the company's annual budget. What was the company's annual budget?
$800,000

5.5 Percent Problems: Proportion Method

Objective A To solve percent problems using proportions

Vocabulary to Review
ratio   [4.1A]
proportion   [4.3A]
base   [5.2A]
amount   [5.2A]

Problems that can be solved using the basic percent equation can also be solved using proportions.

The proportion method is based on writing two ratios. One ratio is the percent ratio, written as $\frac{percent}{100}$. The second ratio is the amount-to-base ratio, written as $\frac{amount}{base}$. These two ratios form the proportion

$$\frac{\text{percent}}{100} = \frac{\text{amount}}{\text{base}}$$

To use the proportion method, first identify the percent, the amount, and the base (the base usually follows the phrase "percent of").

Instructor Note

This section explains the proportion method of solving the basic percent equation. If you choose not to use this method, you can nonetheless use the exercises as practice in solving problems that involve percent. This will help students review the topic and recognize the different types of percent problems.

Integrating Technology

To use a calculator to solve the proportions at the right for *n*, enter

23 × 45 ÷ 100 =

100 × 4 ÷ 25 =

100 × 12 ÷ 60 =

What is 23% of 45?

$$\frac{23}{100} = \frac{n}{45}$$
$$23 \times 45 = 100 \times n$$
$$1035 = 100 \times n$$
$$1035 \div 100 = n$$
$$10.35 = n$$

What percent of 25 is 4?

$$\frac{n}{100} = \frac{4}{25}$$
$$n \times 25 = 100 \times 4$$
$$n \times 25 = 400$$
$$n = 400 \div 25$$
$$n = 16$$
16% of 25 is 4.

12 is 60% of what number?

$$\frac{60}{100} = \frac{12}{n}$$
$$60 \times n = 100 \times 12$$
$$60 \times n = 1200$$
$$n = 1200 \div 60$$
$$n = 20$$

Discuss the Concepts

Do you prefer to solve percent problems using the basic percent equation or using proportions? Why?

Instructor Note

Some students find it easier to remember the proportion method by using the equation

$$\frac{\text{is}}{\text{of}} = \frac{n}{100}$$

Example 1 15% of what is 7? Round to the nearest hundredth.

Solution
$$\frac{15}{100} = \frac{7}{n}$$
$$15 \times n = 100 \times 7$$
$$15 \times n = 700$$
$$n = 700 \div 15$$
$$n \approx 46.67$$

You Try It 1 26% of what is 22? Round to the nearest hundredth.

Your solution 84.62

Example 2 30% of 63 is what?

Solution
$$\frac{30}{100} = \frac{n}{63}$$
$$30 \times 63 = 100 \times n$$
$$1890 = 100 \times n$$
$$1890 \div 100 = n$$
$$18.90 = n$$

You Try It 2 16% of 132 is what?

Your solution 21.12

Solutions on p. S14

In-Class Examples (Objective 5.5A)

1. What is 28% of 950?   266
2. 48 is what percent of 160?   30%
3. 90% of what is 63?   70

Objective 5.5B

Optional Student Activity

1. At a department store, you purchase two items, one for $129 and the other for $99. The cash register automatically adds the sales tax and displays the total, $239.40. What is the sales tax rate expressed as a percent?   5%

2. In an election for president of the student government, one candidate received more than 94% of the votes cast but less than 100% of the votes cast. What is the least possible number of votes cast in the election?   17 votes

3. A soccer team has won 25 of its first 30 games this season. The team has 20 more games to play. How many of the remaining games must the team win in order to win 80% of its games this season?

   15 games

Objective B   To solve application problems

Example 3

An antiques dealer found that 86% of the 250 items that were sold during one month sold for under $1000. How many items sold for under $1000?

Strategy
To find the number of items that sold for under $1000, write and solve a proportion using n to represent the number of items sold (amount) for less than $1000. The percent is 86% and the base is 250.

Solution
$$\frac{86}{100} = \frac{n}{250}$$
$$86 \times 250 = 100 \times n$$
$$21{,}500 = 100 \times n$$
$$21{,}500 \div 100 = n$$
$$215 = n$$

215 items sold for under $1000.

Example 4

In a test of the strength of nylon rope, 5 pieces of the 25 pieces tested did not meet the test standards. What percent of the nylon ropes tested did meet the standards?

Strategy
To find the percent of ropes tested that met the standards:

• Find the number of ropes that met the test standards (25 − 5).
• Write and solve a proportion using n to represent the percent of ropes that met the test standards. The base is 25 and the amount is the number of ropes that met the standards.

Solution
$$25 - 5 = 20 \quad \text{ropes met test standards}$$

$$\frac{n}{100} = \frac{20}{25}$$
$$n \times 25 = 100 \times 20$$
$$n \times 25 = 2000$$
$$n = 2000 \div 25$$
$$n = 80$$

80% of the ropes tested did meet the test standards.

You Try It 3

Last year it snowed 64% of the 150 days of the ski season at a resort. How many days did it snow?

Your strategy

Your solution
96 days

You Try It 4

Five ballpoint pens in a box of 200 were found to be defective. What percent of the pens were not defective?

Your strategy

Your solution
97.5%

Solutions on p. S14

In-Class Examples (Objective 5.5B)

(*Note:* Solve for percent in example 1, base in 2, and amount in 3.)

1. A soccer team won 42 out of the 56 games they played this season. What percent of the games played did the team win?   75%

2. A down payment of $4110 was paid on a new car. The down payment is 15% of the cost of the car. Find the cost of the car. $27,400

3. A growing company "plowed back" 54% of the $80,000 it earned into research and development. How much of the money earned was reinvested in research and development? $43,200

5.5 Exercises

Objective A   **To solve percent problems using proportions**

1. 26% of 250 is what?
   65

2. What is 18% of 150?
   27

3. 37 is what percent of 148?
   25%

4. What percent of 150 is 33?
   22%

5. 68% of what is 51?
   75

6. 126 is 84% of what?
   150

7. What percent of 344 is 43?
   12.5%

8. 750 is what percent of 50?
   1500%

9. 82 is 20.5% of what?
   400

10. 2.4% of what is 21?
    875

11. What is 6.5% of 300?
    19.5

12. 96% of 75 is what?
    72

13. 7.4 is what percent of 50?
    14.8%

14. What percent of 1500 is 693?
    46.2%

15. 50.5% of 124 is what?
    62.62

16. What is 87.4% of 255?
    222.87

17. 120% of what is 6?
    5

18. 14 is 175% of what?
    8

19. What is 250% of 18?
    45

20. 325% of 4.4 is what?
    14.3

21. 33 is 220% of what?
    15

22. 160% of what is 40?
    25

Objective B   **To solve application problems**

23. **Charities**   A charitable organization spent $2940 for administrative expenses. This amount is 12% of the money it collected. What is the total amount of money that the organization collected?
    $24,500

24. **Medicine**   A manufacturer of an anti-inflammatory drug claims that the drug will be effective for 6 hours. An independent testing service determined that the drug was effective for only 80% of the length of time claimed by the manufacturer. Find the length of time the drug will be effective as determined by the testing service.
    4.8 hours

25. **Geography**   The land area of North America is approximately 9,400,000 square miles. This represents approximately 16% of the total land area of the world. What is the approximate total land area of the world?
    58,750,000 square miles

26. **Fire Departments**   The Rincon Fire Department received 24 false alarms out of a total of 200 alarms received. What percent of the alarms received were false alarms?
    12%

Section 5.5

Suggested Assignment
Exercises 1–31, odds
More challenging problem:
  Exercise 33

Quick Quiz (Objective 5.5A)

1. What is 14% of 250?   35

2. What percent of 140 is 49?   35%

3. 166 is 83% of what?   200

27. Lodging   The graph at the right shows the break-down of the locations of the 53,500 hotels throughout the United States. How many hotels in the United States are located along highways?
22,577 hotels

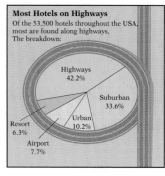

Most Hotels on Highways
Of the 53,500 hotels throughout the USA,
most are found along highways,
The breakdown:

Highways 42.2%
Suburban 33.6%
Urban 10.2%
Airport 7.7%
Resort 6.3%

Source: American Hotel and Lodging Association

28. Poultry   In a recent year, North Carolina produced 1,300,000,000 pounds of turkey. This was 18.6% of the U.S. total in that year. Calculate the U.S. total turkey production for that year. Round to the nearest billion.
7 billion pounds

29. Mining   During 1 year, approximately 2,240,000 ounces of gold went into the manufacturing of electronic equipment in the United States. This is 16% of all the gold mined in the United States that year. How many ounces of gold were mined in the United States that year?
14,000,000 ounces

30. Demography   The table at the right shows the predicted increase in population from 2000 to 2040 for each of four counties in the Central Valley of California.
a. What percent of the 2000 population of Sacramento County is the increase in population?   75%
b. What percent of the 2000 population of Kern County is the increase in population? Round to the nearest tenth of a percent.   145.5%

County	2000 Population	Projected Increase
Sacramento	1,200,000	900,000
Kern	651,700	948,300
Fresno	794,200	705,800
San Joaquin	562,000	737,400

Source: California Department of Finance

31. Police Officers   The graph at the right shows the causes of death for all police officers killed in the line of duty during a recent year. What percent of the deaths were due to traffic accidents? Round to the nearest tenth of a percent.
46.8%

Job-related illness 19
Other 6
Violent attacks 58
Traffic accidents 73

Causes of Death for Police Officers Killed in the Line of Duty

Source: International Union of Police Associations

32. Demography   According to a 25-city survey of the status of hunger and homelessness by the U.S. Conference of Mayors, 41% of the homeless in the United States are single men, 41% are families with children, 13% are single women, and 5% are unaccompanied minors. How many homeless people in the United States are single men?
Insufficient information

APPLYING THE CONCEPTS

33. The Federal Government   In the 108th Senate, there were 51 Republicans, 48 Democrats, and 1 Independent. In the 108th House of Representatives, there were 229 Republicans, 205 Democrats, and 1 Independent. Which had the larger percent of Republicans, the 108th Senate or the 108th House of Representatives?
The 108th House of Representatives

Quick Quiz (Objective 5.5B)

(*Note:* Solve for percent in question 1, base in 2, and amount in 3.)

1. A down payment of $31,200 was paid on a new house costing $156,000. What percent of the purchase price is the down payment?   20%

2. A supermarket reduced the price of melon to $2.24 per pound, which is 80% of the original price. What was the original price? $2.80

3. You bought a boat 1 year ago for $49,475. Since then it has depreciated 12% of the price you paid for it. What is the value of the boat today?
$43,538

Focus on Problem Solving

Using a Calculator as a Problem-Solving Tool

A calculator is an important tool for problem solving. Here are a few problems to solve with a calculator. You may need to research some of the questions to find information you do not know.

1. Choose any single-digit positive number. Multiply the number by 1507 and 7373. What is the answer? Choose another positive single-digit number and again multiply by 1507 and 7373. What is the answer? What pattern do you see? Why does this work?

2. The gross domestic product in 2002 was $10,446,200,000. Is this more or less than the amount of money that would be placed on the last square of a standard checkerboard if 1 cent were placed on the first square, 2 cents were placed on the second square, 4 cents were placed on the third square, 8 cents were placed on the fourth square, and so on, until the 64th square was reached?

3. Which of the reciprocals of the first 16 natural numbers have a terminating-decimal representation and which have a repeating-decimal representation?

4. What is the largest natural number n for which $4^n > 1 \cdot 2 \cdot 3 \cdot 4 \cdot 5 \cdot \cdots \cdot n$?

5. If $1000 bills are stacked one on top of another, is the height of $1 billion less than or greater than the height of the Washington Monument?

6. What is the value of $1 + \dfrac{1}{1 + \dfrac{1}{1 + \dfrac{1}{1 + \dfrac{1}{1 + 1}}}}$?

7. Calculate 15^2, 35^2, 65^2, and 85^2. Study the results. Make a conjecture about a relationship between a number ending in 5 and its square. Use your conjecture to find 75^2 and 95^2. Does your conjecture work for 125^2?

8. Find the sum of the first 1000 natural numbers. (*Hint:* You could just start adding $1 + 2 + 3 + \cdots$, but even if you performed one operation every 3 seconds, it would take you an hour to find the sum. Instead, try pairing the numbers and then adding the pairs. Pair 1 and 1000, 2 and 999, 3 and 998, and so on. What is the sum of each pair? How many pairs are there? Use this information to answer the original question.)

9. For a borrower to qualify for a home loan, a bank requires that the monthly mortgage payment be less than 25% of a borrower's monthly take-home income. A laboratory technician has deductions for taxes, insurance, and retirement that amount to 25% of the technician's monthly gross income. What minimum monthly income must this technician earn to receive a bank loan that has a mortgage payment of $1200 per month?

Answers to Focus on Problem Solving— Using a Calculator as a Problem-Solving Tool

1. $1507 \times 7373 = 11,111,111$
   By the Multiplication Property of One, a number times 1 equals the number. Because the product of 1507 and 7373 is 11,111,111, the product of 11,111,111 and a number between 1 and 9 will be an eight-digit number in which all digits equal the number between 1 and 9.

2. It is less; see page 191.

3. The reciprocals of 1, 2, 4, 5, 8, 10, and 16 have terminating-decimal representations. The reciprocals of 3, 6, 7, 9, 11, 12, 13, 14, and 15 have repeating-decimal representations.

4. $n = 8$

5. The height of the stack of bills is less than the height of the Washington Monument. (*Note:* The Washington Monument is about 550 feet tall. A bill is 0.0043 inch thick. The stack of bills is $358\frac{1}{3}$ feet high.)

6. $1\frac{5}{8}$

7. $15^2 = 225$
   $35^2 = 1225$
   $65^2 = 4225$
   $85^2 = 7225$
   Multiply the first digit of the number being squared by 1 more than that number. Attach 25 to the product.
   $75^2 = 5625$
   $95^2 = 9025$
   Yes; $125^2 = 15,625$

8. 500,500

9. The technician must earn more than $6400 per month.

Using Estimation as a Problem- Solving Tool

You can use your knowledge of rounding, your understanding of percent, and your experience with the basic percent equation to quickly estimate the answer to a percent problem. Here is an example.

HOW TO   What is 11.2% of 978?

Round the given numbers. $11.2\% \approx 10\%$
$978 \approx 1000$

Mentally calculate with the rounded numbers. 10% of $1000 = \frac{1}{10}$ of $1000 = 100$

11.2% of 978 is approximately 100.

TAKE NOTE

The exact answer is $0.112 \times 978 = 109.536$. The exact answer 109.536 is close to the approximation of 100.

For Exercises 1 to 8, state which quantity is greater.

1. 49% of 51, or 201% of 15   2. 99% of 19, or 22% of 55
3. 8% of 31, or 78% of 10   4. 24% of 402, or 76% of 205
5. 10.2% of 51, or 20.9% of 41   6. 51.8% of 804, or 25.3% of 1223
7. 26% of 39.217, or 9% of 85.601 8. 66% of 31.807, or 33% of 58.203

For Exercises 9 to 12, use estimation to provide an approximate number.

9. A company survey found that 24% of its 2096 employees favored a new dental plan. How many employees favored the new dental plan?

10. A local newspaper reported that 52.3% of the 29,875 eligible voters in the town voted in the last election. How many people voted in the last election?

11. 19.8% of the 2135 first-year students at a community college have part-time jobs. How many of the first-year students at the college have part-time jobs?

12. A couple made a down payment of 33% of the $310,000 cost of the home. Find the down payment.

Projects and Group Activities

Health

The American College of Sports Medicine (ACSM) recommends that you know how to determine your target heart rate in order to get the full benefit of exercise. Your **target heart rate** is the rate at which your heart should beat during any aerobic exercise such as running, cycling, fast walking, or participating in an aerobics class. According to the ACSM, you should reach your target rate and then maintain it for 20 minutes or more to achieve cardiovascular fitness. The intensity level varies for different individuals. A sedentary person might begin at the 60% level and gradually work up to 70%, whereas athletes and very fit individuals might work at the 85% level. The ACSM suggests that you calculate both 50% and 85% of your maximum heart rate. This will give you the low and high ends of the range within which your heart rate should stay.

To calculate your target heart rate:

	Example
Subtract your age from 220. This is your maximum heart rate.	$220 - 20 = 200$
Multiply your maximum heart rate by 50%. This is the low end of your range.	$200(0.50) = 100$
Divide the low end by 6. This is your low 10-second heart rate.	$100 \div 6 \approx 17$
Multiply your maximum heart rate by 85%. This is the high end of your range.	$200(0.85) = 170$
Divide the high end by 6. This is your high 10-second heart rate.	$170 \div 6 \approx 28$

1. Why are the low end and high end divided by 6 in order to determine the low and high 10-second heart rates?

2. Calculate your target heart rate, both the low and high end of your range.

Consumer Price Index The consumer price index (CPI) is a percent that is written without the percent sign. For instance, a CPI of 160.1 means 160.1%. This number means that an item that cost $100 between 1982 and 1984 (the base years) would cost $160.10 today. Determining the cost is an application of the basic percent equation.

$$\text{Percent} \times \text{base} = \text{amount}$$
$$\text{CPI} \times \text{cost in base year} = \text{cost today}$$
$$1.601 \times 100 = 160.1 \qquad \bullet \quad 160.1\% = 1.601$$

The table below gives the CPI for various products in July of 2003. If you have Internet access, you can obtain current data for the items below, as well as other items not on this list, by visiting the website of the Bureau of Labor Statistics.

Product	*CPI*
All items	183.9
Food and beverages	180.3
Housing	185.9
Clothes	116.2
Transportation	156.8
Medical care	297.6
Entertainment	107.7
Education	108.9

1. Of the items listed, are there any items that in 2003 cost more than twice as much as they cost during the base year? If so, which items?

2. Of the items listed, are there any items that in 2003 cost more than one-and-one-half times as much as they cost during the base years but less than twice as much as they cost during the base years? If so, which items?

3. If the cost for textbooks for one semester was $120 in the base years, how much did similar textbooks cost in 2003? Use the "Education" category.

4. If a new car cost $20,000 in 2003, what would a comparable new car have cost during the base years? Use the "Transportation" category.

5. If a movie ticket cost $8 in 2003, what would a comparable movie ticket have cost during the base years? Use the "Entertainment" category.

6. The base year for the CPI was 1967 before the change to 1982–1984. If 1967 were still used as the base year, the CPI for all items in 2003 (not just those listed above) would be 550.9.
   a. Using the base year of 1967, explain the meaning of a CPI of 550.9.
   b. Using the base year of 1967 and a CPI of 550.9, if textbooks cost $75 for one semester in 1967, how much did similar textbooks cost in 2003?
   c. Using the base year of 1967 and a CPI of 550.9, if a family's food budget in 2003 is $800 per month, what would a comparable family budget have been in 1967?

Chapter 5 Summary

Key Words	**Examples**
Percent means "parts of 100." [5.1A, p. 203]	23% means 23 of 100 equal parts.

Essential Rules and Procedures	**Examples**
To write a percent as a fraction, drop the percent sign and multiply by $\frac{1}{100}$. [5.1A, p. 203]	$56\% = 56\left(\dfrac{1}{100}\right) = \dfrac{56}{100} = \dfrac{14}{25}$
To write a percent as a decimal, drop the percent sign and multiply by 0.01. [5.1A, p. 203]	$87\% = 87(0.01) = 0.87$
To write a fraction as a percent, multiply by 100%. [5.1B, p. 204]	$\dfrac{7}{20} = \dfrac{7}{20}(100\%) = \dfrac{700}{20}\% = 35\%$
To write a decimal as a percent, multiply by 100%. [5.1B, p. 204]	$0.325 = 0.325(100\%) = 32.5\%$
The Basic Percent Equation [5.2A, p. 207] The basic percent equation is $$\text{Percent} \times \text{base} = \text{amount}$$ Solving percent problems requires identifying the three elements of this equation. Usually the base follows the phrase "percent of."	8% of 250 is what number? Percent × base = amount $$0.08 \times 250 = n$$ $$20 = n$$
Proportion Method of Solving a Percent Problem [5.5A, p. 219] The following proportion can be used to solve percent problems. $$\frac{\text{percent}}{100} = \frac{\text{amount}}{\text{base}}$$ To use the proportion method, first identify the percent, the amount, and the base. The base usually follows the phrase "percent of."	8% of 250 is what number? $$\frac{\text{percent}}{100} = \frac{\text{amount}}{\text{base}}$$ $$\frac{8}{100} = \frac{n}{250}$$ $$8 \times 250 = 100 \times n$$ $$2000 = 100 \times n$$ $$2000 \div 100 = n$$ $$20 = n$$

Chapter 5 Review Exercises

1. What is 30% of 200?
60   [5.2A]

2. 16 is what percent of 80?
20%   [5.3A]

3. Write $1\frac{3}{4}$ as a percent.
175%   [5.1B]

4. 20% of what is 15?
75   [5.4A]

5. Write 12% as a fraction.
$\frac{3}{25}$ [5.1A]

6. Find 22% of 88.
19.36   [5.2A]

7. What percent of 20 is 30?
150%   [5.3A]

8. $16\frac{2}{3}$% of what is 84?
504   [5.4A]

9. Write 42% as a decimal.
0.42   [5.1A]

10. What is 7.5% of 72?
5.4   [5.2A]

11. $66\frac{2}{3}$% of what is 105?
157.5   [5.4A]

12. Write 7.6% as a decimal.
0.076   [5.1A]

13. Find 125% of 62.
77.5   [5.2A]

14. Write $16\frac{2}{3}$% as a fraction.
$\frac{1}{6}$ [5.1A]

15. Use the proportion method to find what percent of 25 is 40.
160%   [5.5A]

16. 20% of what number is 15? Use the proportion method.
75   [5.5A]

17. Write 0.38 as a percent.
38%   [5.1B]

18. 78% of what is 8.5? Round to the nearest tenth.
10.9   [5.4A]

19. What percent of 30 is 2.2? Round to the nearest tenth of a percent.
7.3% [5.3A]

20. What percent of 15 is 92? Round to the nearest tenth of a percent.
613.3% [5.3A]

21. **Education** Trent missed 9 out of 60 questions on a history exam. What percent of the questions did he answer correctly? Use the proportion method.
85% [5.5B]

22. **Advertising** A company used 7.5% of its $60,000 advertising budget for newspaper advertising. How much of the advertising budget was spent for newspaper advertising?
$4500 [5.2B]

23. **Energy** The graph at the right shows the amounts that the average U.S. household spends for energy use. What percent of these costs is for electricity? Round to the nearest tenth of a percent.
31.7% [5.3B]

Where Your Energy Dollar Goes
The average U.S. household spent $2868 on energy use in a recent year. How it was spent:

Motor gasoline $1492
Fuel oil, kerosene $83
Natural gas $383
Electricity $910

Source: Energy Information Administration

24. **Consumerism** Joshua purchased a video camera for $980 and paid a sales tax of 6.25% of the cost. What was the total cost of the video camera?
$1041.25 [5.2B]

25. **Health** In a survey of 350 women and 420 men, 275 of the women and 300 of the men reported that they wore sunscreen often. To the nearest tenth of a percent, what percent of the women wore sunscreen often?
78.6% [5.3B]

26. **Demography** It is estimated that the world's population will be 9,100,000,000 by the year 2050. This is 149% of the population in 2000. (*Source:* U.S. Census Bureau). What was the world's population in 2000? Round to the nearest hundred million.
6,100,000,000 people [5.4B]

27. **Computers** A computer system can be purchased for $1800. This is 60% of what the computer cost 4 years ago. What was the cost of the computer 4 years ago? Use the proportion method.
$3000 [5.5B]

28. **Online Transactions** In a recent year, $25.96 billion worth of online transactions were paid for using a Visa card. This represented 50.4% of the online transactions made during the year. (*Source:* The Nilson Report) Find the dollar value of all the online transactions made that year. Round to the nearest billion.
$52 billion [5.4B]

Chapter 5 Test

1. Write 97.3% as a decimal.

0.973 [5.1A]

2. Write $83\frac{1}{3}$% as a fraction.

$\frac{5}{6}$ [5.1A]

3. Write 0.3 as a percent.

30% [5.1B]

4. Write 1.63 as a percent.

163% [5.1B]

5. Write $\frac{3}{2}$ as a percent.

150% [5.1B]

6. Write $\frac{2}{3}$ as a percent.

$66\frac{2}{3}$% [5.1B]

7. What is 77% of 65?

50.05 [5.2A]

8. 47.2% of 130 is what?

61.36 [5.2A]

9. Which is larger:
7% of 120, or 76% of 13?

76% of 13 [5.2A]

10. Which is smaller:
13% of 200, or 212% of 12?

212% of 12 [5.2A]

11. Advertising A travel agency uses 6% of its $75,000 budget for advertising. What amount of the budget is spent on advertising?

$4500 [5.2B]

12. Agriculture During the packaging process for vegetables, spoiled vegetables are discarded by an inspector. In one day an inspector found that 6.4% of the 1250 pounds of vegetables were spoiled. How many pounds of vegetables were not spoiled?

1170 pounds [5.2B]

Nutrition The table at the right contains nutrition information about a breakfast cereal. Solve Exercises 13 and 14 with information taken from this table.

13. The recommended amount of potassium per day for an adult is 3000 milligrams (mg). What percent, to the nearest tenth of a percent, of the daily recommended amount of potassium is provided by one serving of this cereal with skim milk?

14.7% [5.3B]

14. The daily recommended number of calories for a 190-pound man is 2200 calories. What percent, to the nearest tenth of a percent, of the daily recommended number of calories is provided by one serving of this cereal with 2% milk?

9.1% [5.3B]

NUTRITION INFORMATION

SERVING SIZE: 1.4 OZ WHEAT FLAKES WITH 0.4 OZ. RAISINS: 39.4 g. ABOUT 1/2 CUP
SERVINGS PER PACKAGE: 14

	CEREAL & RAISINS	WITH 1/2 CUP VITAMINS A & D SKIM MILK
CALORIES................	120	180
PROTEIN, g...............	3	7
CARBOHYDRATE, g	28	34
FAT, TOTAL, g	1	1*
UNSATURATED, g	1	
SATURATED, g	0	
CHOLESTEROL, mg	0	0*
SODIUM, mg	125	190
POTASSIUM, mg	240	440

* 2% MILK SUPPLIES AN ADDITIONAL 20 CALORIES, 2 g FAT, AND 10 mg CHOLESTEROL.
** CONTAINS LESS THAN 2% OF THE U.S. RDA OF THIS NUTRIENT

15. **Employment** The Urban Center Department Store has 125 permanent employees and must hire an additional 20 temporary employees for the holiday season. What percent of the permanent employees is the number hired as temporary employees for the holiday season?
16% [5.3B]

16. **Education** Conchita missed 7 out of 80 questions on a math exam. What percent of the questions did she answer correctly? Round to the nearest tenth of a percent.
91.3% [5.3B]

17. 12 is 15% of what?
80 [5.4A]

18. 42.5 is 150% of what? Round to the nearest tenth.
28.3 [5.4A]

19. **Manufacturing** A manufacturer of PDAs found 384 defective PDAs during a quality control study. This amount was 1.2% of the PDAs tested. Find the number of PDAs tested.
32,000 PDAs [5.4B]

20. **Real Estate** A new house was bought for $95,000. Five years later the house sold for $152,000. The increase was what percent of the original price?
60% [5.3B]

21. 123 is 86% of what number? Use the proportion method. Round to the nearest tenth.
143.0 [5.5A]

22. What percent of 12 is 120? Use the proportion method.
1000% [5.5A]

23. **Wages** A secretary receives a wage of $16.24 per hour. This amount is 112% of last year's wage. What is the dollar increase in the hourly wage over last year? Use the proportion method.
$1.74 [5.5B]

24. **Demography** A city has a population of 71,500. Ten years ago the population was 32,500. The population now is what percent of what the population was 10 years ago? Use the proportion method.
220% [5.5B]

25. **Fees** The annual license fee on a car is 1.4% of the value of the car. If the license fee during a year is $175.00, what is the value of the car? Use the proportion method.
$12,500 [5.5B]

Cumulative Review Exercises

1. Simplify: $18 \div (7 - 4)^2 + 2$.
4 [1.6B]

2. Find the LCM of 16, 24, and 30.
240 [2.1A]

3. Find the sum of $2\frac{1}{3}$, $3\frac{1}{2}$, and $4\frac{5}{8}$.
$10\frac{11}{24}$ [2.4C]

4. Subtract: $27\frac{5}{12} - 14\frac{9}{16}$
$12\frac{41}{48}$ [2.5C]

5. Multiply: $7\frac{1}{3} \times 1\frac{5}{7}$
$12\frac{4}{7}$ [2.6B]

6. What is $\frac{14}{27}$ divided by $1\frac{7}{9}$?
$\frac{7}{24}$ [2.7B]

7. Simplify: $\left(\frac{3}{4}\right)^3 \cdot \left(\frac{8}{9}\right)^2$
$\frac{1}{3}$ [2.8B]

8. Simplify: $\left(\frac{2}{3}\right)^2 - \left(\frac{3}{8} - \frac{1}{3}\right) \div \frac{1}{2}$
$\frac{13}{36}$ [2.8C]

9. Round 3.07973 to the nearest hundredth.
3.08 [3.1B]

10. Subtract: $\begin{array}{r} 3.0902 \\ -\ 1.9706 \\ \hline \end{array}$
1.1196 [3.3A]

11. Divide: $0.032\overline{)1.097}$
Round to the nearest ten-thousandth.
34.2813 [3.5A]

12. Convert $3\frac{5}{8}$ to a decimal.
3.625 [3.6A]

13. Convert 1.75 to a fraction.
$1\frac{3}{4}$ [3.6B]

14. Place the correct symbol, < or >, between the two numbers.
$\frac{3}{8} < 0.87$ [3.6C]

15. Solve the proportion $\frac{3}{8} = \frac{20}{n}$. Round to the nearest tenth.
53.3 [4.3B]

16. Write "$76.80 earned in 8 hours" as a unit rate.
$9.60/hour [4.2B]

17. Write $18\frac{1}{3}\%$ as a fraction.

$\dfrac{11}{60}$ [5.1A]

18. Write $\frac{5}{6}$ as a percent.

$83\frac{1}{3}\%$ [5.1B]

19. 16.3% of 120 is what? Round to the nearest hundredth.
19.56   [5.2A/5.5A]

20. 24 is what percent of 18?
$133\frac{1}{3}\%$ [5.3A/5.5A]

21. 12.4 is 125% of what?
9.92   [5.4A/5.5A]

22. What percent of 35 is 120? Round to the nearest tenth.
342.9%   [5.3A/5.5A]

23. **Taxes**   Sergio has an income of $740 per week. One-fifth of his income is deducted for income tax payments. Find his take-home pay.
$592   [2.6C]

24. **Finance**   Eunice bought a used car for $8353, with a down payment of $1000. The balance was paid in 36 equal monthly payments. Find the monthly payment.
$204.25   [3.5B]

25. **Taxes**   The gasoline tax is $.19 a gallon. Find the number of gallons of gasoline used during a month in which $79.80 was paid in gasoline taxes.
420 gallons   [3.5B]

26. **Taxes**   The real estate tax on a $172,000 home is $3440. At the same rate, find the real estate tax on a home valued at $250,000.
$5000   [4.3C]

27. **Consumerism**   Ken purchased a camera for $490 and paid $29.40 in sales tax. What percent of the purchase price was the sales tax?
6%   [5.3B/5.5B]

28. **Elections**   A survey of 300 people showed that 165 people favored a certain candidate for mayor. What percent of the people surveyed did not favor this candidate?
45%   [5.3B/5.5B]

29. **Television**   According to the Cabletelevision Advertising Bureau, cable households watch television 36.5% of the time. On average, how many hours per week do cable households spend watching TV? Round to the nearest tenth.
61.3 hours   [5.2B/5.5B]

30. **Health**   The Environmental Protection Agency found that 990 out of 5500 children tested had levels of lead in their blood exceeding federal guidelines. What percent of the children tested had levels of lead in the blood that exceeded federal standards?
18%   [5.3B/5.5B]

6

Applications for Business and Consumers

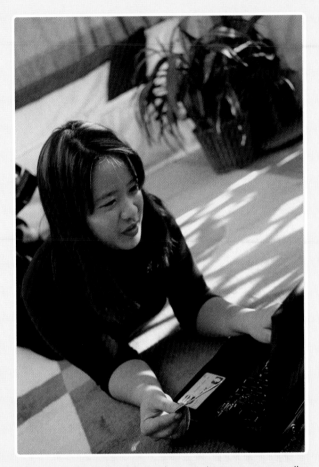

When you use a credit card to make a purchase, you are actually receiving a loan. This service allows you to defer payment on your purchase until a predetermined date in the future. In exchange for this convenient service, credit card companies will frequently charge you money for using their credit card. This added cost may be in the form of an annual fee, or it may be in the form of interest charges on balances that are not paid off after a certain deadline. These interest charges on unpaid balances are called finance charges and can be calculated using the simple interest formula. **Exercises 27 to 32 on page 257** ask you to use this formula to calculate the finance charges on unpaid balances.

Need help? For online student resources, such as section quizzes, visit this textbook's website at **math.college.hmco.com/students.**

OBJECTIVES

Section 6.1

A To find unit cost
B To find the most economical purchase
C To find total cost

Section 6.2

A To find percent increase
B To apply percent increase to business—markup
C To find percent decrease
D To apply percent decrease to business—discount

Section 6.3

A To calculate simple interest
B To calculate finance charges on a credit card bill
C To calculate compound interest

Section 6.4

A To calculate the initial expenses of buying a home
B To calculate the ongoing expenses of owning a home

Section 6.5

A To calculate the initial expenses of buying a car
B To calculate the ongoing expenses of owning a car

Section 6.6

A To calculate commissions, total hourly wages, and salaries

Section 6.7

A To calculate checkbook balances
B To balance a checkbook

PREP TEST • • •

Do these exercises to prepare for Chapter 6.

For Exercises 1 to 6, add, subtract, multiply, or divide.

1. Divide: 3.75 ÷ 5
0.75 [3.5A]

2. Multiply: 3.47 × 15
52.05 [3.4A]

3. Subtract: 874.50 − 369.99
504.51 [3.3A]

4. Multiply: 0.065 × 150,000
9750 [3.4A]

5. Multiply: 1500 × 0.06 × 0.5
45 [3.4A]

6. Add: 1372.47 + 36.91 + 5.00 + 2.86
1417.24 [3.2A]

7. Divide 10 ÷ 3. Round to the nearest hundredth.
3.33 [3.5A]

8. Divide 345 ÷ 570. Round to the nearest thousandth.
0.605 [3.5A]

9. Place the correct symbol, < or >, between the two numbers.
0.379 0.397
0.379 < 0.397 [3.6C]

GO FIGURE • • •

A store manager priced a pair of earrings in dollars and cents such that when 4% sales tax was added, the result was a whole number of dollars and 0 cents. Find the smallest possible number of dollars the items sold for, including the sales tax. $13

6.1 Applications to Purchasing

Objective A To find unit cost

Frequently, stores promote items for purchase by advertising, say, 2 Red Baron Bake to Rise Pizzas for $10.50 or 5 cans of StarKist tuna for $4.25.

The **unit cost** is the cost of *one* Red Baron Pizza or of *one* can of StarKist tuna. To find the unit cost, divide the total cost by the number of units.

2 pizzas for $10.50	5 cans for $4.25
$10.50 \div 2 = 5.25$	$4.25 \div 5 = 0.85$
$5.25 is the cost of one pizza.	$.85 is the cost of one can.
Unit cost: $5.25 per pizza	Unit cost: $.85 per can

Example 1

Find the unit cost. Round to the nearest tenth of a cent.
a. 3 gallons of mint chip ice cream for $17
b. 4 ounces of Crest toothpaste for $2.29

Strategy
To find the unit cost, divide the total cost by the number of units.

Solution
a. $17 \div 3 \approx 5.667$
   $5.667 per gallon
b. $2.29 \div 4 = 0.5725$
   $.573 per ounce

You Try It 1

Find the unit cost. Round to the nearest tenth of a cent.
a. 8 size-AA Energizer batteries for $7.67
b. 15 ounces of Revlon shampoo for $2.29

Your strategy

Your solution
a. $.959 per battery
b. $.153 per ounce

Solution on p. S15

Objective B To find the most economical purchase

Comparison shoppers often find the most economical buy by comparing unit costs.

One store is selling 6 twelve-ounce cans of ginger ale for $2.99, and a second store is selling 24 twelve-ounce cans of ginger ale for $11.79. To find the better buy, compare the unit costs.

$2.99 \div 6 \approx 0.498$	$11.79 \div 24 \approx 0.491$
Unit cost: $.498 per can	Unit cost: $.491 per can

Because $.491 < $.498, the better buy is 24 cans for $11.79.

Objective 6.1A

New Vocabulary
unit cost

Concept Check
State the division necessary to calculate the unit cost. Do not perform the division.
1. Juice, 32 ounces for $2.78
   $2.78 \div 32$
2. Shampoo, 15 ounces for $5.75
   $5.75 \div 15$
3. Photographic processing, 24 photos for $6.28
   $6.28 \div 24$

Instructor Note
One way to help students calculate unit cost is to tell them to divide by the number associated with the unit following the word *per*. For instance, to find the cost *per* gallon, divide cost by the number of gallons.
   Alternatively, you may wish to provide students with the following formula for unit price.

$$\text{Unit price} = \frac{\text{price per package}}{\text{measure or count}}$$

Objective 6.1B

Discuss the Concepts
1. When we compare unit prices, is the size with the lower unit cost or the higher unit cost the better buy?
2. Assume the larger size of a product is the more economical purchase. Is it the better buy if you will not be able to use up the product before it spoils?

In-Class Examples (Objective 6.1A)
Find the unit cost. Round to the nearest tenth of a cent.
1. Salad dressing, 8 ounces for $2.39 $.299
2. Spaghetti, 12 ounces for $1.15 $.096

In-Class Examples (Objective 6.1B)
Find the more economical purchase.
1. Syrup, 15 ounces for $1.84 or 24 ounces for $3.05
   15 ounces for $1.84
2. Catsup, 32 ounces for $1.98 or 18 ounces for $1.15
   32 ounces for $1.98

236 Chapter 6 / Applications for Business and Consumers

Instructor Note

You might explain that using unit prices to determine which item to buy is basing the better buy solely on price; quality is not a factor. The issue of quality does not arise, however, when comparing the unit prices of the same product in different-size packages.

Example 2

Find the more economical purchase:
5 pounds of nails for $4.80, or 4 pounds of nails for $3.78.

Strategy
To find the more economical purchase, compare the unit costs.

Solution
$4.80 \div 5 = 0.96$
$3.78 \div 4 = 0.945$
$\$.945 < \$.96$

The more economical purchase is 4 pounds for $3.78.

You Try It 2

Find the more economical purchase: 6 cans of fruit for $5.70, or 4 cans of fruit for $3.96.

Your strategy

Your solution
6 cans for $5.70

Solution on p. S15

Objective 6.1C

New Vocabulary

total cost

Optional Student Activity

(*Note:* This activity extends the objective of finding total cost to include adding sales tax to the total price.)
   Find the total purchase price. Round to the nearest cent.

1. 3 CDs at $17.99 each, plus 7% sales tax $57.75

2. 4 candles at $2.25 each, plus 6% sales tax $9.54

3. 8 batteries at $1.96 each, plus 5.5% sales tax $16.54

4. 6 quarts of motor oil at $1.99 per quart, plus 6.5% sales tax $12.72

Objective C To find total cost

An installer of floor tile found the unit cost of identical floor tiles at three stores.

Store 1	Store 2	Store 3
$1.22 per tile	$1.18 per tile	$1.28 per tile

By comparing the unit costs, the installer determined that store 2 would provide the most economical purchase.

The installer also uses the unit cost to find the total cost of purchasing 300 floor tiles at store 2. The **total cost** is found by multiplying the unit cost by the number of units purchased.

Unit cost	×	number of units	=	total cost
1.18	×	300	=	354

The total cost is $354.

Example 3

Clear redwood lumber costs $5.43 per foot. How much would 25 feet of clear redwood cost?

Strategy
To find the total cost, multiply the unit cost (5.43) by the number of units (25).

Solution

Unit cost	×	number of units	=	total cost
5.43	×	25	=	135.75

The total cost is $135.75.

You Try It 3

Pine saplings cost $9.96 each. How much would 7 pine saplings cost?

Your strategy

Your solution
$69.72

Solution on p. S15

In-Class Examples (Objective 6.1C)

1. Decorative stepping stones cost $3.30 per stone. Find the cost of 24 stones. $79.20

2. Grapes cost $1.39 per pound. Find the cost of 2.8 pounds. Round to the nearest cent. $3.89

6.1 Exercises

Objective A   To find unit cost

Suggested Assignment
Exercises 1–33, odds

For Exercises 1 to 12, find the unit cost. Round to the nearest tenth of a cent.

1. Heinz B·B·Q sauce, 18 ounces for $.99
$.055 per ounce

2. Birds-eye maple, 6 feet for $18.75
$3.125 per foot

3. Diamond walnuts, $2.99 for 8 ounces
$.374 per ounce

4. A&W root beer, 6 cans for $2.99
$.498 per can

5. Ibuprofen, 50 tablets for $3.99
$.080 per tablet

6. Visine eye drops, 0.5 ounce for $3.89
$7.78 per ounce

7. Adjustable wood clamps, 2 for $13.95
$6.975 per clamp

8. Corn, 6 ears for $1.85
$.308 per ear

9. Cheerios cereal, 15 ounces for $2.99
$.199 per ounce

10. Doritos Cool Ranch chips, 14.5 ounces for $2.99
$.206 per ounce

11. Sheet metal screws, 8 for $.95
$.119 per screw

12. Folgers coffee, 11.5 ounces for $4.79
$.417 per ounce

Objective B   To find the most economical purchase

For Exercises 13 to 24, suppose your local supermarket offers the following products at the given prices. Find the more economical purchase.

13. Sutter Home pasta sauce, 25.5 ounces for $3.29, or Muir Glen Organic pasta sauce, 26 ounces for $3.79
Sutter Home

14. Kraft mayonnaise, 40 ounces for $2.98, or Springfield mayonnaise, 32 ounces for $2.39
Kraft

15. Ortega salsa, 20 ounces for $3.29 or 12 ounces for $1.99
20 ounces for $3.29

16. L'Oréal shampoo, 13 ounces for $4.69, or Cortexx shampoo, 12 ounces for $3.99
Cortexx

17. Golden Sun vitamin E, 200 tablets for $7.39 or 400 tablets for $12.99
400 tablets for $12.99

18. Ultra Mr. Clean, 20 ounces for $2.67, or Ultra Spic and Span, 14 ounces for $2.19
Ultra Mr. Clean

19. 16 ounces of Kraft cheddar cheese, $4.37, or 9 ounces of Land O' Lakes cheddar cheese, $2.29
Land O' Lakes

20. Bertolli olive oil, 34 ounces for $9.49, or Pompeian olive oil, 8 ounces for $2.39
Bertolli

Quick Quiz (Objective 6.1A)
Find the unit cost. Round to the nearest tenth of a cent.
1. Potatoes, 5 pounds for $2.29 $.458
2. Corn chips, 8 ounces for $3.25 $.406

Quick Quiz (Objective 6.1B)
Find the more economical purchase.
1. Aspirin tablets, 50 for $3.78 or 75 for $6.00
50 for $3.78
2. Tuna, 6.5 ounces for $.79 or 12 ounces for $1.32
12 ounces for $1.32

21. Maxwell House coffee, 4 ounces for $3.99, or Sanka coffee, 2 ounces for $2.39
Maxwell House

22. Wagner's vanilla extract, $3.29 for 1.5 ounces, or Durkee vanilla extract, 1 ounce for $2.74
Wagner's

23. Purina Cat Chow, $4.19 for 56 ounces, or Friskies Chef's Blend, $3.37 for 50.4 ounces
Friskies Chef's Blend

24. Kleenex tissues, $1.73 for 250 tissues, or Puffs tissues, $1.23 for 175 tissues
Kleenex

Objective C   **To find total cost**

25. If sliced bacon costs $4.59 per pound, find the total cost of 3 pounds.
$13.77

26. Used red brick costs $.98 per brick. Find the total cost of 75 bricks.
$73.50

27. Kiwi fruit cost $.23 each. Find the total cost of 8 kiwi.
$1.84

28. Boneless chicken filets cost $4.69 per pound. Find the cost of 3.6 pounds. Round to the nearest cent.
$16.88

29. Herbal tea costs $.98 per ounce. Find the total cost of 6.5 ounces.
$6.37

30. If Stella Swiss Lorraine cheese costs $5.99 per pound, find the total cost of 0.65 pound. Round to the nearest cent.
$3.89

31. Red Delicious apples cost $1.29 per pound. Find the total cost of 2.1 pounds. Round to the nearest cent.
$2.71

32. Choice rib eye steak costs $8.49 per pound. Find the total cost of 2.8 pounds. Round to the nearest cent.
$23.77

33. If Godiva chocolate costs $7.95 per pound, find the total cost of $\frac{3}{4}$ pound. Round to the nearest cent.
$5.96

34. Color photocopying costs $.89 per page. Find the total cost for photocopying 120 pages.
$106.80

APPLYING THE CONCEPTS

35. ✏ Explain in your own words the meaning of unit pricing.

36. ✏ What is the UPC (Universal Product Code) and how is it used?

Answers to Writing Exercises

35. Students might explain that unit pricing is used in grocery stores. The unit price of a product is the total price divided by the number of units the product contains. Consumers can use this information to determine which size of a product is the more economical purchase.

36. The Universal Product Code, or UPC, is a series of lines, bars, and numbers found on the packages of consumer products. The UPC is used with an optical scanning device that "reads" the UPC and signals the computer to search its memory for the price of the item.

ISBN 0−395−75524−7

0 4644 27701 8 7   65)

Quick Quiz (Objective 6.1C)

1. Honeydew melons cost $2.99 each. Find the total cost of 4 honeydew melons. $11.96

2. Raisins cost $3.29 per pound. Find the total cost of 1.25 pounds. Round to the nearest cent. $4.11

Objective 6.2A

6.2 Percent Increase and Percent Decrease

Objective A To find percent increase

New Vocabulary
percent increase

Percent increase is used to show how much a quantity has increased over its original value. The statements "Food prices increased by 2.3% last year" and "City council members received a 4% pay increase" are examples of percent increase.

HOW TO According to the Energy Information Administration, the number of alternative-fuel vehicles increased from approximately 277,000 to 352,000 in four years. Find the percent increase in alternative-fuel vehicles. Round to the nearest percent.

Point of Interest

According to the U.S. Census Bureau, the number of persons aged 65 and over in the United States will increase to about 82.0 million by 2050, a 136% increase from 2000.

New value	−	original value	=	amount of increase

$$352{,}000 - 277{,}000 = 75{,}000$$

Now solve the basic percent equation for percent.

Percent	×	base	=	amount

Percent increase	×	original value	=	amount of increase

$$n \times 277{,}000 = 75{,}000$$
$$n = 75{,}000 \div 277{,}000$$
$$n \approx 0.27$$

Amount of increase (75,000)
Original value (277,000)
New value (352,000)

The number of alternative-fuel vehicles increased by approximately 27%.

Instructor Note

For students to be successful in calculating percent increase and percent decrease, they must remember that the base (in the basic percent equation) is always the quantity *before* the increase or decrease. That is, it is the *original value*. For the problem at the left, the base is 277,000—the quantity before the increase.

Discuss the Concepts

1. What is the difference between amount of increase and percent increase?
2. Explain how to find the new value when you are given the original value and the percent increase.

Example 1

The average wholesale price of coffee increased from $2 per pound to $3 per pound in one year. What was the percent increase in the price of 1 pound of coffee?

Strategy
To find the percent increase:

• Find the amount of the increase.
• Solve the basic percent equation for *percent*.

Solution

New value	−	original value	=	amount of increase
3	−	2	=	1

Percent × base = amount
$$n \times 2 = 1$$
$$n = 1 \div 2$$
$$n = 0.5 = 50\%$$

The percent increase was 50%.

You Try It 1

The average price of gasoline rose from $1.46 to $1.83 in 5 months. What was the percent increase in the price of gasoline? Round to the nearest percent.

Your strategy

Your solution
25%

Solution on p. S15

In-Class Examples (Objective 6.2A)

1. The amount of gasoline used by a fleet of cars increased from 200 to 230 gallons per day. What percent increase does this represent?   15%
2. A manufacturer of ceiling fans increased its monthly output of 1500 fans by 10%. Find the company's monthly output of fans now.   1650 fans

---

Optional Student Activity

In a recent quarter, Wal-Mart reported the following percent increases in sales shown in the table to the right. Use the information to calculate the sales for the previous quarter. Round to the nearest million.

Store	Sales (billions of dollars)	Percent Increase	Sales for Previous Quarter
Wal-Mart	27.5	15.1%	$23,892 million
Sam's Club	6.1	8.9%	$5,601 million
International	7.2	118.7%	$3,292 million
McLane	2.2	13%	$1,947 million

Example 2

Chris Carley was earning $13.50 an hour as a nursing assistant before receiving a 10% increase in pay. What is Chris's new hourly pay?

Strategy

To find the new hourly wage:

- Solve the basic percent equation for *amount*.
- Add the amount of the increase to the original wage.

Solution

Percent × base = amount
0.10 × 13.50 = n
1.35 = n

The amount of the increase was $1.35.

13.50 + 1.35 = 14.85

The new hourly wage is $14.85.

You Try It 2

Yolanda Liyama was making a wage of $12.50 an hour as a baker before receiving a 14% increase in hourly pay. What is Yolanda's new hourly wage?

Your strategy

Your solution
$14.25

Solution on p. S15

New Vocabulary
cost
selling price
markup
markup rate

Discuss the Concepts

1. Explain the difference between cost and selling price.

2. What is the difference between markup and markup rate?

3. How can you determine the markup when you know the cost and the selling price?

4. How can you determine the selling price when you know the markup and the cost?

5. How can you determine the markup when you know the cost and the markup rate?

Instructor Note

It will help students to know that markup is an application of percent increase. In business situations, markup can be based on *cost* or on *selling price*. We have chosen cost, which is the more common practice. This means that the base in the basic percent equation is cost.

Having students focus on definitions of words in this objective will help them solve these problems.

Objective B To apply percent increase to business—markup

Some of the expenses involved in operating a business are salaries, rent, equipment, and utilities. To pay these expenses and earn a profit, a business must sell a product at a higher price than it paid for the product.

Cost is the price a business pays for a product, and **selling price** is the price at which a business sells a product to a customer. The difference between selling price and cost is called **markup.**

Point of Interest

According to *Managing a Small Business*, from Liraz Publishing Company, goods in a store are often marked up 50% to 100% of the cost. This allows a business to make a profit of 5% to 10%.

| Selling price | − | cost | = | markup |

or

| Cost | + | markup | = | selling price |

Markup is frequently expressed as a percent of a product's cost. This percent is called the **markup rate.**

| Markup rate | × | cost | = | markup |

HOW TO Suppose Bicycles Galore purchases an AMP Research B-5 bicycle for $2119.20 and sells it for $2649. What markup rate does Bicycles Galore use?

| Selling price | − | cost | = | markup |
| 2649.00 | − 2119.20 = | 529.80 |

• First find the markup.

Percent	×	base	=	amount
Markup rate	×	cost	=	markup
n	× 2119.20 =	529.80

• Then solve the basic percent equation for *percent*.

n = 529.80 ÷ 2119.20 = 0.25

The markup rate is 25%.

In-Class Examples (Objective 6.2B)

1. An automobile tire dealer uses a markup rate of 32%. What is the markup on tires that cost the dealer $84? $26.88

2. The markup on an appliance that costs a store $210 is $84. What markup rate does this represent? 40%

3. A garden shop uses a markup rate of 35% on a rose trellis that costs the store $52. What is the selling price? $70.20

Example 3

The manager of a sporting goods store determines that a markup rate of 36% is necessary to make a profit. What is the markup on a pair of skis that costs the store $225?

Strategy

To find the markup, solve the basic percent equation for *amount*.

Solution

Percent × base = amount

$$\boxed{\begin{array}{c} \text{Markup} \\ \text{rate} \end{array}} \times \boxed{\text{cost}} = \boxed{\text{markup}}$$

$$0.36 \quad \times \quad 225 \quad = \quad n$$
$$81 = n$$

The markup is $81.

You Try It 3

A bookstore manager determines that a markup rate of 20% is necessary to make a profit. What is the markup on a book that costs the bookstore $8?

Your strategy

Your solution

$1.60

Example 4

A plant nursery bought a yellow twig dogwood for $4.50 and used a markup rate of 46%. What is the selling price?

Strategy

To find the selling price:

- Find the markup by solving the basic percent equation for *amount*.
- Add the markup to the cost.

Solution

Percent × base = amount

$$\boxed{\begin{array}{c} \text{Markup} \\ \text{rate} \end{array}} \times \boxed{\text{cost}} = \boxed{\text{markup}}$$

$$0.46 \quad \times \quad 4.50 \quad = \quad n$$
$$2.07 = n$$

$$\boxed{\text{Cost}} + \boxed{\text{markup}} = \boxed{\text{selling price}}$$

$$4.50 \quad + \quad 2.07 \quad = \quad 6.57$$

The selling price is $6.57.

You Try It 4

A clothing store bought a leather suit for $72 and used a markup rate of 55%. What is the selling price?

Your strategy

Your solution

$111.60

Solutions on p. S15

Objective 6.2C

New Vocabulary
percent decrease

Instructor Note
Remind students that the base (in the basic percent equation) is always the original value—that is, the quantity *before* the decrease. For the problem at the right, the base is $60.60, which is the quantity before the decrease.

Also in connection with the problem at the right, explain to the students why we do not need the unit "billions" in the calculations.

Discuss the Concepts
1. What is the difference between amount of decrease and percent decrease?
2. Explain how to find the amount of decrease when you are given the original value and the percent decrease.
3. Explain how to find the new value when you are given the original value and the amount of decrease.
4. Explain how to find the new value when you are given the original value and the percent decrease.

Objective C **To find percent decrease**

Percent decrease is used to show how much a quantity has decreased from its original value. The statements "The number of family farms decreased by 2% last year" and "There has been a 50% decrease in the cost of a Pentium chip" are examples of percent decrease.

HOW TO During a 2-year period, the value of U.S. agricultural products exported decreased from approximately $60.6 billion to $52.0 billion. Find the percent decrease in the value of U.S. agricultural exports. Round to the nearest tenth of a percent.

$$\boxed{\text{Original value}} - \boxed{\text{new value}} = \boxed{\text{amount of decrease}}$$
$$60.6 \quad - \quad 52.0 \quad = \quad 8.6$$

Now solve the basic percent equation for percent.

$$\text{Percent} \times \text{base} = \text{amount}$$

$$\boxed{\text{Percent decrease}} \times \boxed{\text{original value}} = \boxed{\text{amount of decrease}}$$

$$n \quad \times \quad 60.6 \quad = \quad 8.6$$
$$n = 8.6 \div 60.6$$
$$n \approx 0.142$$

Amount of decrease (8.6)

New value (52.0)

Original value (60.6)

The value of agricultural exports decreased approximately 14.2%.

Study Tip
Note in the example below that solving a word problem includes stating a strategy and using the strategy to find a solution. If you have difficulty with a word problem, write down the known information. Be very specific. Write out a phrase or sentence that states what you are trying to find. See *AIM for Success* at the front of the book.

Example 5

During an 8-year period, the population of Baltimore, Maryland, decreased from approximately 736,000 to 646,000. Find the percent decrease in Baltimore's population. Round to the nearest tenth of a percent.

Strategy
To find the percent decrease:
- Find the amount of the decrease.
- Solve the basic percent equation for *percent*.

Solution

$$\boxed{\text{Original value}} - \boxed{\text{new value}} = \boxed{\text{amount of decrease}}$$
$$736,000 \quad - \quad 646,000 \quad = \quad 90,000$$

$$\text{Percent} \times \text{base} = \text{amount}$$
$$n \quad \times 736,000 = 90,000$$
$$n = 90,000 \div 736,000$$
$$n \approx 0.122$$

Baltimore's population decreased approximately 12.2%.

You Try It 5

During an 8-year period, the population of Norfolk, Virginia, decreased from approximately 261,000 to 215,000. Find the percent decrease in Norfolk's population. Round to the nearest tenth of a percent.

Your strategy

Your solution
17.6%

Solution on p. S15

In-Class Examples (Objective 6.2C)
1. A new bypass around a small town reduced the normal 30-minute driving time between two cities by 9 minutes. What percent decrease does this represent? 30%
2. Last year a company earned a profit of $175,000. This year the company's profits were 6% less than last year's. What was the profit this year? $164,500

Example 6

The total sales for December for a stationery store were $96,000. For January, total sales showed an 8% decrease from December's sales. What were the total sales for January?

Strategy

To find the total sales for January:

- Find the amount of decrease by solving the basic percent equation for *amount.*
- Subtract the amount of decrease from the December sales.

Solution

Percent × base = amount

0.08 × 96,000 = n

7680 = n

The decrease in sales was $7680.

96,000 − 7680 = 88,320

The total sales for January were $88,320.

You Try It 6

Fog decreased the normal 5-mile visibility at an airport by 40%. What was the visibility in the fog?

Your strategy

Your solution

3 miles

Solution on p. S16

Objective D **To apply percent decrease to business—discount**

To promote sales, a store may reduce the regular price of some of its products temporarily. The reduced price is called the **sale price.** The difference between the regular price and the sale price is called the **discount.**

| Regular price | − | sale price | = | discount |

or

| Regular price | − | discount | = | sale price |

Discount is frequently stated as a percent of a product's regular price. This percent is called the **discount rate.**

| Discount rate | × | regular price | = | discount |

In-Class Examples (Objective 6.2D)

1. A department store is giving a discount of $3 on an ice chest that normally sells for $20. What is the discount rate? 15%

2. A jewelry store is selling $150 quartz watches at 30% off the regular price. What is the discount? $45

3. A store is offering 35% off its stock of art supplies. What is the sale price of a set of paint brushes that regularly sells for $90? $58.50

Concept Check

A family reduced its normal monthly food bill from $960 to $840, a decrease of $120, or 12.5%.

a. Name the original amount. $960

b. Name the amount of decrease. $120

c. Name the new amount. $840

d. Name the percent decrease. 12.5%

Objective 6.2D

New Vocabulary
sale price
discount
discount rate

Instructor Note

Remind students that discount is an application of percent decrease. The base in the basic percent equation is the regular price.

Discuss the Concepts

1. What is the difference between discount and discount rate?

2. How can you determine the discount when you know the sale price and the regular price?

3. How can you determine the sale price when you know the discount and the regular price?

4. How can you determine the discount when you know the regular price and the discount rate?

Concept Check

(*Note:* This problem requires understanding of both markup and discount.)

The manager of a store that sells sports equipment uses a markup rate of 40% on tennis rackets. After the summer season, the manager reduces the price of a tennis racket that cost the store $150 by offering it for 20% off the regular price. Find the difference between the sale price and the cost of the tennis racket. $18

You may prefer to ask intermediate questions:
What is the regular price? $210
What is the sale price? $168
What is the discount? $42

Example 7

A GE 25-inch stereo television that regularly sells for $299 is on sale for $250. Find the discount rate. Round to the nearest tenth of a percent.

Strategy
To find the discount rate:

• Find the discount.
• Solve the basic percent equation for *percent*.

Solution

Regular price	−	sale price	=	discount
299	−	250	=	49

Percent	×	base	=	amount
Discount rate	×	regular price	=	discount
n	×	299	=	49

$$n = 49 \div 299$$
$$n \approx 0.164$$

The discount rate is 16.4%.

You Try It 7

A white azalea that regularly sells for $12.50 is on sale for $10.99. Find the discount rate. Round to the nearest tenth of a percent.

Your strategy

Your solution
12.1%

Example 8

A 20-horsepower lawn mower is on sale for 25% off the regular price of $1125. Find the sale price.

Strategy
To find the sale price:

• Find the discount by solving the basic percent equation for *amount*.
• Subtract to find the sale price.

Solution

Percent	×	base	=	amount
Discount rate	×	regular price	=	discount
0.25	×	1125	=	n
		281.25	=	n

Regular price	−	discount	=	sale price
1125	−	281.25	=	843.75

The sale price is $843.75.

You Try It 8

A hardware store is selling a Newport security door for 15% off the regular price of $110. Find the sale price.

Your strategy

Your solution
$93.50

Solutions on p. S16

6.2 Exercises

Objective A   **To find percent increase**

For Exercises 1 to 5, solve. Round percents to the nearest tenth of a percent.

1.   **Education**   The figure at the right shows the actual and projected enrollments in grades K–12 in the United States. What percent increase is expected from 1988 to 2008?
19.6%

2.   **Fuel Efficiency**   An automobile manufacturer increased the average mileage on a car from 17.5 miles per gallon to 18.2 miles per gallon. Find the percent increase in mileage.
4%

3.   **Business**   In the 1990s, the number of Target stores increased from 420 stores to 914 stores. (*Source:* Target) What was the percent increase in the number of Target stores in the 1990s?
117.6%

4.   **Demography**   The graph at the right shows the number of unmarried American couples living together. (*Source:* U.S. Census Bureau) Find the percent increase in the number of unmarried couples living together from 1980 to 2000. Round to the nearest tenth.
193.8%

5.   **Sports**   In 1924, the number of events in the Winter Olympics was 14. The 2002 Winter Olympics in Salt Lake City included 78 medal events. (*Source:* David Wallenchinsky's *The Complete Book of the Winter Olympics*) Find the percent increase in the number of events in the Winter Olympics from 1924 to 2002.
457.1%

6.   **Wealth**   The table at the right shows the estimated number of millionaire households in the United States. Find the percent increase in the estimated number of households containing millionaires from 1975 to 2005.
1500%

7.   **Demography**   The population of Boise City, Idaho, increased 24.3% over an 8-year period. The population initially was approximately 127,000. (*Source:* U.S. Census Bureau) Find the population of Boise City 8 years later.
157,861 people

8.   **Television**   During 1 year, the number of people subscribing to direct broadcasting satellite systems increased 87%. If the number of subscribers at the beginning of the year was 2.3 million, how many subscribers were there at the end of the year?
4.301 million subscribers

9.   **Demography**   From 1970 to 2000, the average age of American mothers giving birth to their first child rose 16.4%. (*Source:* Centers for Disease Control and Prevention) If the average age in 1970 was 21.4 years, what was the average age in 2000? Round to the nearest tenth.
24.9 years

Elementary and Secondary School Enrollments
Source: National Center for Education

Unmarried U.S. Couples Living Together

Year	Number of Households Containing Millionaires
1975	350,000
1997	3,500,000
2005	5,600,000

Source: Affluent Market Institute

Section 6.2

Suggested Assignment
Exercises 1–39, odds
Exercise 41
More challenging problems:
   Exercises 42, 43

Quick Quiz (Objective 6.2A)

1. The value of a $3000 investment increased $750. What percent increase does this represent?
25%

2. A supervisor's salary this year is $48,000. This salary will increase by 8% next year. What will the salary be next year? $51,840

Objective B **To apply percent increase to business—markup**

10. A window air conditioner cost AirRite Air Conditioning Systems $285. Find the markup on the air conditioner if the markup rate is 25% of the cost.
$71.25

11. The owner of Kerr Electronics purchased 300 Craig portable CD players at a cost of $85 each. If the owner uses a markup rate of 42%, what is the markup on each of the Craig CD players?
$35.70

12. The manager of Brass Antiques has determined that a markup rate of 38% is necessary for a profit to be made. What is the markup on a brass doorknob that costs $45?
$17.10

13. Computer Inc. uses a markup of $975 on a computer system that costs $3250. What is the markup rate on this system?
30%

14. Saizon Pen & Office Supply uses a markup of $12 on a calculator that costs $20. What markup rate does this amount represent?
60%

15. Giant Photo Service uses a markup rate of 48% on its Model ZA cameras, which cost the shop $162.
a. What is the markup? $77.76
b. What is the selling price? $239.76

16. The Circle R golf pro shop uses a markup rate of 45% on a set of Tour Pro golf clubs that costs the shop $210.
a. What is the markup? $94.50
b. What is the selling price? $304.50

17. Harvest Time Produce Inc. uses a 55% markup rate and pays $2 for a box of strawberries.
a. What is the markup? $1.10
b. What is the selling price of a box of strawberries? $3.10

18. Resner Builders' Hardware uses a markup rate of 42% for a table saw that costs $160. What is the selling price of the table saw?
$227.20

19. Brad Burt's Magic Shop uses a markup rate of 48%. What is the selling price of a telescoping sword that costs $50?
$74

Objective C **To find percent decrease**

20. **Travel**   A new bridge reduced the normal 45-minute travel time between two cities by 18 minutes. What percent decrease does this represent?
40%

21. **Energy**   By installing energy-saving equipment, the Pala Rey Youth Camp reduced its normal $800-per-month utility bill by $320. What percent decrease does this amount represent?
40%

22. **Depreciation**   It is estimated that the value of a new car is reduced 30% after 1 year of ownership. Using this estimate, find how much value a $18,200 new car loses after 1 year.
$5460

Quick Quiz (Objective 6.2B)

1. If a business uses a markup rate of 38% on video games, what is the markup on a video game that costs the business $28? $10.64

2. The markup on a necklace that cost a jeweler $120 is $72. What markup rate does this represent? 60%

Quick Quiz (Objective 6.2C)

1. The price of a new model camera dropped from $450 to $396 in 10 months. What percent decrease does this represent? 12%

2. A golf resort employs 240 people during the golfing season. At the end of the season, the resort reduces the number of employees by 55%. How many employees are employed by the resort in the off-season? 108 employees

23. **Employment**   A department store employs 1200 people during the holi-day. At the end of the holiday season, the store reduces the number of employees by 45%. What is the decrease in the number of employees?
    540 employees

24. **Business**   Because of a decrease in demand for super-8 video cameras, Kit's Cameras reduced the orders for these models from 20 per month to 8 per month.
    a. What is the amount of the decrease?   12 cameras
    b. What percent decrease does this amount represent?   60%

25. **Business**   A new computer system reduced the time for printing the payroll from 52 minutes to 39 minutes.
    a. What is the amount of the decrease?   13 minutes
    b. What percent decrease does this amount represent?   25%

26. **Finance**   Juanita's average monthly expense for gasoline was $76. After joining a car pool, she was able to reduce the expense by 20%.
    a. What was the amount of the decrease? $15.20
    b. What is the average monthly gasoline bill now? $60.80

27. **Investments**   An oil company paid a dividend of $1.60 per share. After a reorganization, the company reduced the dividend by 37.5%.
    a. What was the amount of the decrease? $.60
    b. What is the new dividend? $1.00

28. **Traffic Patterns**   Because of an improved traffic pattern at a sports stadium, the average amount of time a fan waits to park decreased from 3.5 minutes to 2.8 minutes. What percent decrease does this amount represent?   20%

29. **Airplanes**   One configuration of the Boeing 777-300 has a seating capacity of 394. This is 26 more than that of the corresponding 777-200 jet. What is the percent decrease in capacity from the 777-300 to the 777-200 model? Round to nearest tenth of a percent.   6.6%

30. **The Military**   In 2000, the Pentagon revised its account of the num-ber of Americans killed in the Korean War from 54,246 to 36,940. (*Source: Time*, June 12, 2000) What is the percent decrease in the reported number of military personnel killed in the Korean War? Round to nearest tenth of a percent.   31.9%

Objective D   **To apply percent decrease to business—discount**

31. The Austin College Bookstore is giving a discount of $8 on calculators that normally sell for $24. What is the discount rate? $33\frac{1}{3}$%

32. A discount clothing store is selling a $72 sport jacket for $24 off the regu-lar price. What is the discount rate? $33\frac{1}{3}$%

33. A disc player that regularly sells for $340 is selling for 20% off the regu-lar price. What is the discount? $68

34. Dacor Appliances is selling its $450 washing machine for 15% off the regular price. What is the discount? $67.50

35. An electric grill that regularly sells for $140 is selling for $42 off the regular price. What is the discount rate?   30%

Quick Quiz (Objective 6.2D)

1. An auto body shop has regularly priced $500 paint jobs on sale for $410. What is the discount rate?
18%

2. A hardware store is selling its $64 lock set for 15% off the regular price. What is the discount? $9.60

3. A lawn mower with a regular price of $460 is on sale for 40% off the regular price. Find the sale price. $276

36. Quick Service Gas Station has its regularly priced $45 tune-up on sale for 16% off the regular price.
 a. What is the discount? $7.20
 b. What is the sale price? $37.80

37. Tomatoes that regularly sell for $1.25 per pound are on sale for 20% off the regular price.
 a. What is the discount? $.25 per pound
 b. What is the sale price? $1.00 per pound

38. An outdoor supply store has its regularly priced $160 sleeping bags on sale for $120.
 a. What is the discount? $40
 b. What is the discount rate? 25%

39. Standard Brands paint that regularly sells for $20 per gallon is on sale for $16 per gallon.
 a. What is the discount? $4
 b. What is the discount rate? 20%

40. **The Military** The graph at the right shows the number of active-duty U.S. military personnel in 1990 and in 2000.
 a. Which branch of the military had the greatest percent decrease in personnel from 1990 to 2000? Navy
 b. What was the percent decrease for this branch of the service? 36.2%

Number of Active-Duty U.S. Military Personnel Ⓟ
Source: Fiscal 2000 Annual Report to the President and Congress by the Secretary of Defense

APPLYING THE CONCEPTS

41. **Compensation** A welder earning $12 per hour is given a 10% raise. To find the new wage, we can multiply $12 by 0.10 and add the product to $12. Can the new wage be found by multiplying $12 by 1.10? Try both methods and compare your answers.
Yes

42. **Business** Grocers, florists, bakers, and other businesses must consider spoilage when deciding the markup of a product. For instance, suppose a florist purchased 200 roses at a cost of $.86 per rose. The florist wants a markup rate of 50% of the total cost of all the roses and expects 7% of the roses to wilt and therefore not be salable. Find the selling price per rose by answering each of the following questions.
 a. What is the florist's total cost for the 200 roses? $172
 b. Find the total selling price without spoilage. $258
 c. Find the number of roses the florist expects to sell. *Hint:* The number of roses the florist expects to sell is

  % of salable roses × number of roses purchased 186 roses

 d. To find the selling price per rose, divide the total selling price without spoilage by the number of roses the florist expects to sell. Round to the nearest cent. $1.39

43. **Business** A promotional sale at a department store offers 25% off the sale price. The sale price itself is 25% off the regular price. Is this the same as a sale that offers 50% off the regular price? If not, which sale gives the better price? Explain your answer.

Answers to Writing Exercises

43. No. Suppose the regular price is $100. Then the sale price is $75 because $100 - 0.25(100) = 100 - 25 = 75$. The promotional sale offers 25% off the sale price of $75, so the price is lowered to $56.25 because $75 - 0.25(75) = 75 - 18.75 = 56.25$. A sale that offers 50% off the regular price of $100 offers the product for $50. Because $50 < $56.25, the better price is the one that is 50% off the regular price.

6.3 Interest

Objective A   To calculate simple interest

When you deposit money in a bank—for example, in a savings account—you are permitting the bank to use your money. The bank may use the deposited money to lend customers the money to buy cars or make renovations on their homes. The bank pays you for the privilege of using your money. The amount paid to you is called **interest.** If you are the one borrowing money from the bank, the amount you pay for the privilege of using that money is also called interest.

The original amount deposited or borrowed is called the **principal.** The amount of interest paid is usually given as a percent of the principal. The percent used to determine the amount of interest is the **interest rate.**

Interest paid on the original principal is called **simple interest.** To calculate simple interest, multiply the principal by the interest rate per period by the number of time periods. In this objective, we are working with annual interest rates, so the time periods are years. The simple interest formula for an annual interest rate is given below.

> **Simple Interest Formula for Annual Interest Rates**
> Principal × annual interest rate × time (in years) = interest

Interest rates are generally given as percents. Before performing calculations involving an interest rate, write the interest rate as a decimal.

HOW TO   Calculate the simple interest due on a 2-year loan of $1500 that has an annual interest rate of 7.5%.

Principal	×	annual interest rate	×	time (in years)	=	interest
1500	×	0.075	×	2	=	225

The simple interest due is $225.

When we borrow money, the total amount to be repaid to the lender is the sum of the principal and the interest. This amount is called the **maturity value of a loan.**

> **Maturity Value Formula for Simple Interest Loans**
> Principal + interest = maturity value

In the example above, the simple interest due on the loan of $1500 was $225. The maturity value of the loan is therefore $1500 + $225 = $1725.

TAKE NOTE
If you deposit $1000 in a savings account paying 5% interest, the $1000 is the principal and 5% is the interest rate.

In-Class Examples (Objective 6.3A)

1. A rancher borrowed $120,000 for 180 days at an annual interest rate of 8.75%. What is the simple interest due on the loan? $5178.08

2. To finance the purchase of four new taxicabs, the owner of the fleet borrowed $84,000 for 8 months at an annual interest rate of 6.5%. Find the maturity value of the loan. $87,640

3. A software company borrowed $75,000 for 6 months at an annual interest rate of 7.25%. Find the monthly payment on the loan. $12,953.13

Objective 6.3A

New Vocabulary
interest
principal
interest rate
simple interest
maturity value of a loan

New Formulas
Simple Interest Formula for Annual Interest Rates: Principal × annual interest rate × time (in years) = interest
Maturity Value Formula for Simple Interest Loans: Principal + interest = maturity value
Monthly Payment on a Simple Interest Loan: Maturity value ÷ length of the loan in months = monthly payment

Discuss the Concepts
1. Explain the difference between interest and interest rate.
2. Explain how to convert a number of months to a fractional part of a year.
3. Explain how to convert a number of days to a fractional part of a year.

Concept Check
1. Find the difference between the maturity value of a 6-month $10,000 loan at an annual interest rate of 8.75% and that of the same loan at an annual interest rate of 6.25%. $125
2. Find the difference between the monthly payments for the two loans in Exercise 1. $20.83

Instructor Note
In this objective, we have presented the *maturity value* of a loan. For an investment, such as a deposit in a bank savings account, the sum of the principal and the interest is called the *future value* of the investment. The formula is the same, but the name applied to the sum is different. You may want to introduce this term in class. (*Continued on the next page*)

(*Continued*)

Those students who take subsequent courses in which finance is discussed will be introduced to the term *present value*. The present value of an investment is the original principal invested, or the value of the investment before it earns any interest. A present-value formula is used to find how much money must be invested today in order for the investment to have a specific value at a future date. In other words, it determines the present value given the future value.

Optional Student Activity

Have students complete the table below by calculating the simple interest due on the loan. Then have them use the pattern of their answers in the table to find the simple interest due on a 1-month loan that has an annual simple interest rate of 6% when the principal is:

a. $6000 $30
b. $7000 $35
c. $8000 $40
d. $9000 $45

You may want to create another table, substituting $10,000, $20,000, $30,000, $40,000, and $50,000 for the loan amounts and 9.3% for the interest rate.

The answers in the interest column will be $77.50, $155, $232.50, $310, and $387.50.

Then have students answer the following questions:

e. If you know the simple interest due on a $1000 loan, explain how you can use that figure to calculate the simple interest due on an $8000 loan for the same time period and the same interest rate.

Multiply the interest due by 8.

(*Continued on the next page*)

TAKE NOTE

The time of the loan must be in years. Eight months is $\frac{8}{12}$ of a year.

See Example 1. The time of the loan must be in years. 180 days is $\frac{180}{365}$ of a year.

HOW TO Calculate the maturity value of a simple interest, 8-month loan of $8000 if the annual interest rate is 9.75%.

First find the interest due on the loan.

Principal	×	annual interest rate	×	time (in years)	=	interest
8000	×	0.0975	×	$\frac{8}{12}$	=	520

Find the maturity value.

Principal	+	interest	=	maturity value
8000	+	520	=	8520

The maturity value of the loan is $8520.

The monthly payment on a loan can be calculated by dividing the maturity value by the length of the loan in months.

> **Monthly Payment on a Simple Interest Loan**
>
> Maturity value ÷ length of the loan in months = monthly payment

In the example above, the maturity value of the loan is $8520. To find the monthly payment on the 8-month loan, divide 8520 by 8.

Maturity value	÷	length of the loan in months	=	monthly payment
8520	÷	8	=	1065

The monthly payment on the loan is $1065.

Example 1

Kamal borrowed $500 from a savings and loan association for 180 days at an annual interest rate of 7%. What is the simple interest due on the loan?

Strategy
To find the simple interest due, multiply the principal (500) times the annual interest rate (7% = 0.07) times the time, in years (180 days = $\frac{180}{365}$ year).

Solution

Principal	×	annual interest rate	×	time (in years)	=	interest
500	×	0.07	×	$\frac{180}{365}$	≈	17.26

The simple interest due is $17.26.

You Try It 1

A company borrowed $15,000 from a bank for 18 months at an annual interest rate of 8%. What is the simple interest due on the loan?

Your strategy

Your solution
$1800

Solution on p. S16

Amount	Interest	Period	Interest
$1000	6%	1 month	$ 5
$2000	6%	1 month	$10
$3000	6%	1 month	$15
$4000	6%	1 month	$20
$5000	6%	1 month	$25

Example 2

Calculate the maturity value of a simple interest, 9-month loan of $4000 if the annual interest rate is 8.75%.

Strategy
To find the maturity value:

• Use the simple interest formula to find the simple interest due.
• Find the maturity value by adding the principal and the interest.

Solution

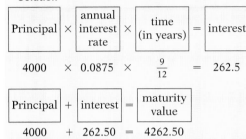

| Principal | × | annual interest rate | × | time (in years) | = | interest |

$$4000 \quad \times \quad 0.0875 \quad \times \quad \frac{9}{12} \quad = \quad 262.5$$

| Principal | + | interest | = | maturity value |

$$4000 \quad + \quad 262.50 \quad = \quad 4262.50$$

The maturity value is $4262.50.

Example 3

The simple interest due on a 3-month loan of $1400 is $26.25. Find the monthly payment on the loan.

Strategy
To find the monthly payment:

• Find the maturity value by adding the principal and the interest.
• Divide the maturity value by the length of the loan in months (3).

Solution

Principal + interest = maturity value
  1400 + 26.25 = 1426.25

Maturity value ÷ length of the loan = payment
  1426.25 ÷ 3 ≈ 475.42

The monthly payment is $475.42.

You Try It 2

Calculate the maturity value of a simple interest, 90-day loan of $3800. The annual interest rate is 6%.

Your strategy

Your solution
$3856.22

You Try It 3

The simple interest due on a 1-year loan of $1900 is $152. Find the monthly payment on the loan.

Your strategy

Your solution
$171

Solutions on p. S16

Objective B   **To calculate finance charges on a credit card bill**

When a customer uses a credit card to make a purchase, the customer is actually receiving a loan. Therefore, there is frequently an added cost to the consumer who purchases on credit. This may be in the form of an annual fee and interest charges on purchases. The interest charges on purchases are called **finance charges.**

In-Class Examples (Objective 6.3B)

1. A credit card company charges a customer 1.5% per month on the customer's unpaid balance. Find the interest owed to the credit card company when the customer's unpaid balance for the month is $1400. $21

(Continued)

f. If the principal of a loan is doubled but the time period and interest rate remain the same, how many times larger is the simple interest due on the loan?   2 times larger

---

Objective 6.3B

New Vocabulary
finance charges

Instructor Note
Emphasize that the simple interest formula requires that the interest rate and the time have comparable units. If the *annual* interest rate is given, then the time must be in *years*. If a *monthly* interest rate is given (as on most credit cards), then the time must be in *months*.

Optional Student Activity
Many traditional college students have little knowledge of finances but are very interested in learning about them. This objective may lend itself to some instruction on the use of credit.

(*Continued on the next page*)

(*Continued*)

Although the calculation of finance charges, given periodic payments and added purchases, is beyond the scope of this course, you may want to present a simplified version for the sake of illustration. For example, you might present a case in which the unpaid balance (say, $500) and the minimum payment due (perhaps $10) both remain constant. You might use a charge of 1.5% per month on the unpaid balance. Illustrate the total amount paid if only the minimum payment is paid each month. Calculate the total amount paid in finance charges. Using the same figures, illustrate the difference in these totals when the customer pays more than the minimum payment each month.

The finance charge on a credit card bill is calculated using the simple interest formula. In the last objective, the interest rates were annual interest rates. However, credit card companies generally issue *monthly* bills and express interest rates on credit card purchases as *monthly* interest rates. Therefore, when using the simple interest formula to calculate finance charges on credit card purchases, use a monthly interest rate and express the time in months.

Note: In the simple interest formula, the time must be expressed in the same period as the rate. For an *annual* interest rate, the time must be expressed in years. For a *monthly* interest rate, the time must be in months.

Example 4

A credit card company charges a customer 1.5% per month on the unpaid balance of charges on the credit card. What is the finance charge in a month when the customer has an unpaid balance of $254?

Strategy
To find the finance charge, multiply the principal, or unpaid balance (254), times the monthly interest rate (1.5%) times the number of months (1).

Solution

$$\text{Principal} \times \begin{array}{c}\text{monthly}\\\text{interest}\\\text{rate}\end{array} \times \begin{array}{c}\text{time}\\\text{(in months)}\end{array}$$

254 × 0.015 × 1 = 3.81

The simple interest due is $3.81.

You Try It 4

The credit card that Francesca uses charges her 1.6% per month on her unpaid balance. Find the finance charge when her unpaid balance one month is $1250.

Your strategy

Your solution
$20

Solution on pp. S16–S17

Objective 6.3C

New Vocabulary
compound interest

Discuss the Concepts
Explain the fundamental difference between simple interest and compound interest.

Objective C   To calculate compound interest

Usually, the interest paid on money deposited or borrowed is compound interest. **Compound interest** is computed not only on the original principal but also on interest already earned. Here is an illustration.

Suppose $1000 is invested for 3 years at an annual interest rate of 9% compounded annually. Because this is an *annual* interest rate, we will calculate the interest earned each year.

During the first year, the interest earned is calculated as follows:

Principal		annual interest rate		time (in years)		interest
1000	×	0.09	×	1	=	90

In-Class Examples (Objective 6.3C)
Note: You will need the Compound Interest Table in the Appendix.

1. An investment of $1500 pays 10% annual interest, compounded quarterly. What is the value of the investment after 10 years? $4027.59

2. A business invested $9000 in an account that paid 9% annual interest, compounded monthly. How much interest was earned in 5 years? $5091.13

At the end of the first year, the total amount in the account is

$$1000 + 90 = 1090$$

During the second year, the interest earned is calculated on the amount in the account at the end of the first year.

Principal	×	annual interest rate	×	time (in years)	=	interest
1090	×	0.09	×	1	=	98.10

Note that the interest earned during the second year ($98.10) is greater than the interest earned during the first year ($90). This is because the interest earned during the first year was added to the original principal, and the interest for the second year was calculated using this sum. If the account earned simple interest, the interest earned would be the same every year ($90).

At the end of the second year, the total amount in the account is the sum of the amount in the account at the end of the first year and the interest earned during the second year.

$$1090 + 98.10 = 1188.10$$

The interest earned during the third year is calculated using the amount in the account at the end of the second year ($1188.10).

Principal	×	annual interest rate	×	time (in years)	=	interest
1188.10	×	0.09	×	1	≈	106.93

The amount in the account at the end of the third year is

$$1188.10 + 106.93 = 1295.03$$

TAKE NOTE

The interest earned each year keeps increasing. This is the effect of compound interest.

To find the interest earned for the three years, subtract the original principal from the new principal.

New principal	−	original principal	=	interest earned
1295.03	−	1000	=	295.03

Note that the compound interest earned is $295.03. The simple interest earned on the investment would have been only $1000 × 0.09 × 3 = $270.

In this example, the interest was compounded annually. However, compound interest can be compounded

Compounding periods:	annually (once a year)
	semiannually (twice a year)
	quarterly (four times a year)
	monthly (12 times a year)
	daily (365 times a year)

The more frequent the compounding periods, the more interest the account earns. For example, if, in the above example, the interest had been compounded quarterly rather than annually, the interest earned would have been greater.

Concept Check

1. You are considering investing some money. One investment pays 7.25% annual interest compounded quarterly. A second investment pays 6.75% annual interest compounded quarterly. On which investment will you earn more interest?
   The 7.25% investment

2. You need to borrow some money. One bank offers a loan at 8.6% annual interest compounded monthly. A second institution offers a loan at 9.8% annual interest compounded monthly. On which loan will you pay less interest?   The 8.6% loan

3. You are considering investing some money. One investment pays 7.5% annual interest compounded daily. A second investment pays 7.5% annual interest compounded annually. On which investment will you earn more interest?
   Compounded daily

4. How much more interest is earned in 5 years on a $10,000 investment paying 8% compounded daily than on the same investment paying 8% compounded semiannually?
   $115.20

5. How much more interest is earned in 5 years on a $10,000 investment paying 9% compounded daily than on the same investment paying 7% compounded semiannually?
   $1576.30

Optional Student Activity

Have students assume that they invest $2000 in an IRA today and that they don't add any additional money to the account before they retire. Have them find the value of the account in 40 or 45 years. Here are a few examples:

For 40 years, compounded monthly, at a 7% annual interest rate, use a compounding factor of 16.311411. The value will be $32,622.82. (*Note:* For 40 years, you can find the future value after 20 years and then use that future value as the principal for another 20 years. The result is the same.)

For 40 years, compounded monthly, at an 8% annual interest rate, use a compounding factor of 24.273386. The value will be $48,546.77.

For 45 years, compounded monthly, at an 8% annual interest rate, use a compounding factor of 36.163599. The value will be $72,327.20.

Instructor Note

If students have a scientific calculator, you might show them the compound interest formula

$$A = P\left(1 + \frac{i}{m}\right)^{mt}$$

where P is the amount invested, i is the annual interest rate written as a decimal, m is the number of compounding periods per year, and t is the number of years.

The calculator sequence for Example 5 is

650 ×

(1 + .08 ÷ 2) $\boxed{y^x}$ (2 × 5) =

Calculating compound interest can be very tedious, so there are tables that can be used to simplify these calculations. A portion of a Compound Interest Table is given in the Appendix.

HOW TO What is the value after 5 years of $1000 invested at 7% annual interest compounded quarterly?

To find the interest earned, multiply the original principal (1000) by the factor found in the Compound Interest Table. To find the factor, first find the table headed "Compounded Quarterly" in the Compound Interest Table in the Appendix. Then look at the number where the 7% column and the 5-year row meet.

| | Compounded Quarterly | | | | | | |
	4%	5%	6%	7%	8%	9%	10%
1 year	1.04060	1.05094	1.06136	1.07186	1.08243	1.09308	1.10381
5 years	1.22019	1.28204	1.34686	**1.41478**	1.48595	1.56051	1.63862
10 years	1.48886	1.64362	1.81402	2.00160	2.20804	2.43519	2.68506
15 years	1.81670	2.10718	2.44322	2.83182	3.28103	3.80013	4.39979
20 years	2.21672	2.70148	3.29066	4.00639	4.87544	5.93015	7.20957

The factor is 1.41478.

$$1000 \times 1.41478 = 1414.78$$

The value of the investment after 5 years is $1414.78.

Example 5

An investment of $650 pays 8% annual interest compounded semiannually. What is the interest earned in 5 years?

Strategy
To find the interest earned:

• Find the new principal by multiplying the original principal (650) by the factor found in the Compound Interest Table (1.48024).
• Subtract the original principal from the new principal.

Solution
$650 \times 1.48024 \approx 962.16$

The new principal is $962.16.

$962.16 - 650 = 312.16$

The interest earned is $312.16.

You Try It 5

An investment of $1000 pays 6% annual interest compounded quarterly. What is the interest earned in 20 years?

Your strategy

Your solution
$2290.66

Solution on p. S17

Instructor Note

The Compound Interest Table in the Appendix has an accompanying Ⓟ indicating that Microsoft PowerPoint® slides of this table are available.

6.3 Exercises

Objective A   **To calculate simple interest**

1. A 2-year student loan of $10,000 is made at an annual simple interest rate of 4.25%. The simple interest on the loan is $850. Identify **a.** the principal, **b.** the interest, **c.** the interest rate, and **d.** the time period of the loan.
   a. $10,000   b. $850   c. 4.25%   d. 2 years

2. A contractor obtained a 9-month loan for $80,000 at an annual simple interest rate of 9.75%. The simple interest on the loan is $5850. Identify **a.** the principal, **b.** the interest, **c.** the interest rate, and **d.** the time period of the loan.
   a. $80,000   b. $5850   c. 9.75%   d. 9 months

3. Find the simple interest Jacob Zucker owes on a 2-year student loan of $8000 at an annual interest rate of 6%.
   $960

4. Find the simple interest Kara Tanamachi owes on a $1\frac{1}{2}$-year loan of $1500 at an annual interest rate of 7.5%.
   $168.75

5. To finance the purchase of 15 new cars, the Tropical Car Rental Agency borrowed $100,000 for 9 months at an annual interest rate of 4.5%. What is the simple interest due on the loan?
   $3375

6. A home builder obtained a preconstruction loan of $50,000 for 8 months at an annual interest rate of 9.5%. What is the simple interest due on the loan?
   $3166.67

7. A bank lent Gloria Masters $20,000 at an annual interest rate of 8.8%. The period of the loan was 9 months. Find the simple interest due on the loan.
   $1320

8. Eugene Madison obtained an 8-month loan of $4500 at an annual interest rate of 6.2%. Find the simple interest Eugene owes on the loan.
   $186

9. Shannon O'Hara borrowed $5000 for 90 days at an annual simple interest rate of 7.5%. Find the simple interest due on the loan.
   $92.47

10. Jon McCloud borrowed $8500 for 180 days at an annual simple interest rate of 9.25%. Find the simple interest due on the loan.
    $387.74

11. Jorge Elizondo took out a 75-day loan of $7500 at an annual interest rate of 5.5%. Find the simple interest due on the loan.
    $84.76

12. Kristi Yang borrowed $15,000. The term of the loan was 90 days, and the annual simple interest rate was 7.4%. Find the simple interest due on the loan.
    $273.70

Quick Quiz (Objective 6.3A)

1. A mechanic borrowed $15,000 for 90 days at an annual interest rate of 7.2%. What is the simple interest due on the loan? $266.30

2. The owner of a convenience store borrowed $60,000 for 9 months at an annual interest rate of 8.6%. Find the maturity value of the loan. $63,870

3. A company borrowed $175,000 for 10 months at an annual interest rate of 9.9%. The simple interest on the loan was $14,437.50. Find the monthly payment on the loan. $18,943.75

13. The simple interest due on a 4-month loan of $4800 is $320. What is the maturity value of the loan?

$5120

14. The simple interest due on a 60-day loan of $6500 is $80.14. Find the maturity value of the loan.

$6580.14

15. An auto parts dealer borrowed $150,000 at a 9.5% annual simple interest rate for 1 year. Find the maturity value of the loan.

$164,250

16. A corporate executive took out a $25,000 loan at an 8.2% annual simple interest rate for 1 year. Find the maturity value of the loan.

$27,050

17. William Carey borrowed $12,500 for 8 months at an annual simple interest rate of 4.5%. Find the total amount due on the loan.

$12,875

18. You arrange for a 9-month bank loan of $9000 at an annual simple interest rate of 8.5%. Find the total amount you must repay to the bank.

$9573.75

19. Capital City Bank approves a home-improvement loan application for $14,000 at an annual simple interest rate of 5.25% for 270 days. What is the maturity value of the loan?

$14,543.70

20. A credit union lends a member $5000 for college tuition. The loan is made for 18 months at an annual simple interest rate of 6.9%. What is the maturity value of this loan?

$5517.50

21. Action Machining Company purchased a robot-controlled lathe for $225,000 and financed the full amount at 8% annual simple interest for 4 years. The simple interest on the loan is $72,000. Find the monthly payment.

$6187.50

22. For the purchase of an entertainment center, a $1900 loan is obtained for 2 years at an annual simple interest rate of 9.4%. The simple interest due on the loan is $357.20. What is the monthly payment on the loan?

$94.05

23. To attract new customers, Heller Ford is offering car loans at an annual simple interest rate of 4.5%.
    a. Find the interest charged to a customer who finances a car loan of $12,000 for 2 years. $1080
    b. Find the monthly payment. $545

24. Cimarron Homes Inc. purchased a snow plow for $57,000 and financed the full amount for 5 years at an annual simple interest rate of 9%.
    a. Find the interest due on the loan. $25,650
    b. Find the monthly payment. $1377.50

Quick Quiz (Objective 6.3B)

1. Suppose you have an unpaid balance of $879.40 on a credit card that charges 1.2% per month on any unpaid balance. What finance charge do you owe the company? $10.55

25. Dennis Pappas decided to build onto his present home instead of buying a new, larger house. He borrowed $42,000 for $3\frac{1}{2}$ years at an annual simple interest rate of 9.5%. Find the monthly payment.
    $1332.50

26. Rosalinda Johnson took out a 6-month, $12,000 loan. The annual simple interest rate on the loan was 8.5%. Find the monthly payment.
    $2085

Objective B   **To calculate finance charges on a credit card bill**

27. A credit card company charges a customer 1.25% per month on the unpaid balance of charges on the credit card. What is the finance charge in a month when the customer has an unpaid balance of $118.72?
    $1.48

28. The credit card that Dee Brown uses charges her 1.75% per month on her unpaid balance. Find the finance charge when her unpaid balance one month is $391.64.
    $6.85

29. What is the finance charge on an unpaid balance of $12,368.92 on a credit card that charges 1.5% per month on any unpaid balance?
    $185.53

30. Suppose you have an unpaid balance of $995.04 on a credit card that charges 1.2% per month on any unpaid balance. What finance charge do you owe the company?
    $11.94

31. A credit card customer has an unpaid balance of $1438.20. What is the difference between monthly finance charges of 1.15% per month on the unpaid balance and monthly finance charges of 1.85% per month?
    $10.07

32. One credit card company charges 1.25% per month on any unpaid balance, and a second company charges 1.75%. What is the difference between the finance charges that these two companies assess on an unpaid balance of $687.45?
    $3.44

Objective C   **To calculate compound interest**

33. North Island Federal Credit Union pays 4% annual interest, compounded daily, on time savings deposits. Find the value of $750 deposited in this account after 1 year.
    $780.60

34. Tanya invested $2500 in a tax-sheltered annuity that pays 8% annual interest compounded daily. Find the value of her investment after 20 years.
    $12,380.43

35. Sal Travato invested $3000 in a corporate retirement account that pays 6% annual interest compounded semiannually. Find the value of his investment after 15 years.
    $7281.78

Quick Quiz (Objective 6.3C)

Note: Students will need the Compound Interest Table in the Appendix.

1. An investment group invests $40,000 in a certificate of deposit that pays 7% annual interest compounded quarterly. Find the value of this investment after 20 years. $160,255.60

2. An interior decorator deposited $4000 in an account that paid 8% annual interest compounded monthly. How much interest was earned in 15 years? $9227.68

36. To replace equipment, a farmer invested $20,000 in an account that pays 7% annual interest compounded monthly. What is the value of the investment after 5 years?
$28,352.50

37. Green River Lodge invests $75,000 in a trust account that pays 8% interest compounded quarterly.
a. What will the value of the investment be in 5 years? $111,446.25
b. How much interest will be earned in the 5 years? $36,446.25

38. To save for retirement, a couple deposited $3000 in an account that pays 7% annual interest compounded daily.
a. What will the value of the investment be in 10 years? $6040.86
b. How much interest will be earned in the 10 years? $3040.86

39. To save for a child's education, the Petersens deposited $2500 into an account that pays 6% annual interest compounded daily. Find the amount of interest earned on this account over a 20-year period.
$5799.48

40. How much interest is earned in 2 years on $4000 deposited in an account that pays 6% interest, compounded quarterly?
$505.94

APPLYING THE CONCEPTS

41. Banking   The Mission Valley Credit Union charges its customers an interest rate of 2% per month on money that is transferred into an account that is overdrawn. Find the interest owed to the credit union for 1 month when $800 is transferred into an overdrawn account.
$16

42. Banking   Suppose you have a savings account that earns interest at the rate of 6% per year compounded monthly. On January 1, you open this account with a deposit of $100.
a. On February 1, you deposit an additional $100 into the account. What is the value of the account after the deposit? $200.50
b. On March 1, you deposit an additional $100 into the account. What is the value of the account after the deposit? $301.50
Note: This type of savings plan, wherein equal amounts ($100) are saved at equal time intervals (every month), is called an annuity.

43. Banking   At 4 P.M. on July 31, you open a savings account that pays 5% annual interest and you deposit $500 in the account. Your deposit is credited as of August 1. At the beginning of September, you receive a statement from the bank that shows that during the month of August, you received $2.12 in interest. The interest has been added to your account, bringing the total on deposit to $502.12. At the beginning of October, you receive a statement from the bank that shows that during the month of September, you received $2.06 in interest on the $502.12 on deposit. Explain why you received less interest during the second month when there was more money on deposit.

Answers to Writing Exercises

43. You received less interest during the second month because there are fewer days in the month of September (30 days) than in the month of August (31 days). Using the simple interest formula:

$$500 \times 0.05 \times \frac{31}{365} \approx 2.12$$

$$502.12 \times 0.05 \times \frac{30}{365} \approx 2.06$$

Even though the principal is greater during the second month, the interest earned is less because there are fewer days in the month.

6.4 Real Estate Expenses

Objective A   **To calculate the initial expenses of buying a home**

One of the largest investments most people ever make is the purchase of a home. The major initial expense in the purchase is the **down payment,** which is normally a percent of the purchase price. This percent varies among banks, but it usually ranges from 5% to 25%.

The **mortgage** is the amount that is borrowed to buy real estate. The mortgage amount is the difference between the purchase price and the down payment.

> **HOW TO**   A home is purchased for $140,000, and a down payment of $21,000 is made. Find the mortgage.

Purchase price	−	down payment	=	mortgage
140,000	−	21,000	=	119,000

The mortgage is $119,000.

TAKE NOTE

Because *points* means percent, a loan origination fee of
$2\frac{1}{2}$ points = $2\frac{1}{2}$% =
2.5% = 0.025.

Another initial expense in buying a home is the **loan origination fee,** which is a fee that the bank charges for processing the mortgage papers. The loan origination fee is usually a percent of the mortgage and is expressed in **points,** which is the term banks use to mean percent. For example, "5 points" means "5 percent."

Points	×	mortgage	=	loan origination fee

Example 1

A house is purchased for $125,000, and a down payment, which is 20% of the purchase price, is made. Find the mortgage.

Strategy

To find the mortgage:

- Find the down payment by solving the basic percent equation for *amount.*
- Subtract the down payment from the purchase price.

Solution

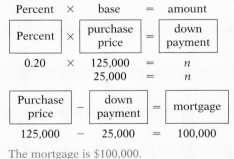

Percent	×	base	=	amount
Percent	×	purchase price	=	down payment
0.20	×	125,000	=	n
		25,000	=	n

Purchase price	−	down payment	=	mortgage
125,000	−	25,000	=	100,000

The mortgage is $100,000.

You Try It 1

An office building is purchased for $216,000, and a down payment, which is 25% of the purchase price, is made. Find the mortgage.

Your strategy

Your solution

$162,000

Solution on p. S17

Discuss the Concepts

True or False:

1. A mortgage is simply a loan, or money that is borrowed, to buy real estate.

2. The word *points* simply means "percent."

Optional Student Activity

1. A couple has saved $28,000 for a down payment on a home. Their bank requires a minimum down payment of 20%. What is the maximum price they can offer for a house in order to have enough money for the down payment? $140,000

2. You have saved $24,000 for a down payment on a house. Your bank requires a minimum down payment of 15%. What is the maximum price you can offer for a home in order to have enough money for the down payment? $160,000

In-Class Examples (Objective 6.4A)

1. A delicatessen is purchased for $120,000 and a down payment of $25,000 is made. Find the mortgage. $95,000

2. A savings and loan association requires a borrower to pay $2\frac{1}{2}$ points for a loan. Find the loan origination fee for a loan of $90,000. $2250

3. A mortgage lender requires a down payment of 8% of the $70,000 purchase price of a house. How much is the mortgage? $64,400

Instructor Note

Many students will have heard of equity loans but will not understand what they are. You may want to discuss the following with them.

Home equity is the difference between the fair market value of your home (the price at which it would probably sell if you were to put it on the market today) and the amount you still owe on the mortgage. For example, if the market value of your home is $125,000 and you still owe $75,000 on the mortgage, your home equity is $125,000 − $75,000 = $50,000.

Many banks allow homeowners to borrow on their equity. The bank charges fees to originate the loan and charges interest. The borrower pays back the loan just as she or he would any other loan, normally with monthly payments. In a home equity loan, the house serves as collateral, or security, for the loan.

Objective 6.4B

New Vocabulary

monthly mortgage payment
property tax
fixed-rate mortgage

Example 2

A home is purchased with a mortgage of $65,000. The buyer pays a loan origination fee of $3\frac{1}{2}$ points. How much is the loan origination fee?

Strategy

To find the loan origination fee, solve the basic percent equation for *amount*.

Solution

Percent	×	base	= amount
Points	×	mortgage	= fee
0.035	×	65,000	= n
		2275	= n

The loan origination fee is $2275.

You Try It 2

The mortgage on a real estate investment is $80,000. The buyer paid a loan origination fee of $4\frac{1}{2}$ points. How much was the loan origination fee?

Your strategy

Your solution
$3600

Solution on p. S17

Objective B   To calculate the ongoing expenses of owning a home

Point of Interest

The number-one response of adults asked what they would spend money on first if they suddenly became wealthy (for example, by winning the lottery) was a house; 31% gave this response. (*Source:* Yankelovich Partners for Lutheran Brotherhood)

Besides the initial expenses of buying a house, there are continuing monthly expenses involved in owning a home. The **monthly mortgage payment** (one of 12 payments due each year to the lender of money to buy real estate), utilities, insurance, and **property tax** (a tax based on the value of real estate) are some of these ongoing expenses. Of these expenses, the largest one is normally the monthly mortgage payment.

For a **fixed-rate mortgage,** the monthly mortgage payment remains the same throughout the life of the loan. The calculation of the monthly mortgage payment is based on the amount of the loan, the interest rate on the loan, and the number of years required to pay back the loan. Calculating the monthly mortgage payment is fairly difficult, so tables such as the one in the Appendix are used to simplify these calculations.

Integrating Technology

In general, when a problem requests a monetary payment, the answer is rounded to the nearest cent. For the example at the right, enter

60000 × 0.0080462 =

The display reads 482.772. Round this number to the nearest hundredth: 482.77. The answer is $482.77.

HOW TO   Find the monthly mortgage payment on a 30-year $60,000 mortgage at an interest rate of 9%. Use the Monthly Payment Table in the Appendix.

$$60{,}000 \times \underset{\substack{\uparrow \\ \text{From the} \\ \text{table}}}{0.0080462} \approx 482.77$$

The monthly mortgage payment is $482.77.

The monthly mortgage payment includes the payment of both principal and interest on the mortgage. The interest charged during any one month is charged on the unpaid balance of the loan. Therefore, during the early years of the mortgage, when the unpaid balance is high, most of the monthly mortgage payment is interest charged on the loan. During the last few years of a mortgage, when the unpaid balance is low, most of the monthly mortgage payment goes toward paying off the loan.

In Class Examples (Objective 6.4B)

Note: Students will need the Monthly Payment Table.

1. A home has a mortgage of $80,000 for 30 years at an annual interest rate of 9%.
   a. Find the monthly mortgage payment. $643.70
   b. During a month when $240.54 of the monthly mortgage payment is principal, how much of the payment is interest? $403.16

2. The monthly mortgage on a home is $1100.10. The homeowner must pay an annual property tax of $768. Find the total monthly payment for the mortgage and property tax. $1164.10

Point of Interest

Home buyers rated the following characteristics "extremely important" in their purchase decision.

Natural, open space: 77%
Walking and biking paths: 74%
Gardens with native plants: 56%
Clustered retail stores: 55%
Wilderness area: 52%
Outdoor pool: 52%
Community recreation center: 52%
Interesting little parks: 50%

(*Sources:* American Lives, Inc; Intercommunications, Inc.)

HOW TO Find the interest paid on a mortgage during a month when the monthly mortgage payment is $186.26 and $58.08 of that amount goes toward paying off the principal.

Monthly mortgage payment	−	principal	=	interest
186.26	−	58.08	=	128.18

The interest paid on the mortgage is $128.18.

Property tax is another ongoing expense of owning a house. Property tax is normally an annual expense that may be paid on a monthly basis. The monthly property tax, which is determined by dividing the annual property tax by 12, is usually added to the monthly mortgage payment.

HOW TO A homeowner must pay $534 in property tax annually. Find the property tax that must be added each month to the homeowner's monthly mortgage payment.

$534 \div 12 = 44.5$

Each month, $44.50 must be added to the monthly mortgage payment for property tax.

Example 3

Serge purchased some land for $120,000 and made a down payment of $25,000. The savings and loan association charges an annual interest rate of 8% on Serge's 25-year mortgage. Find the monthly mortgage payment.

Strategy

To find the monthly mortgage payment:

- Subtract the down payment from the purchase price to find the mortgage.
- Multiply the mortgage by the factor found in the Monthly Payment Table in the Appendix.

Solution

Purchase price	−	down payment	=	mortgage
120,000	−	25,000	=	95,000

$95,000 \times 0.0077182 \approx 733.23$
        ↑
   From the table

The monthly mortgage payment is $733.23.

You Try It 3

A new condominium project is selling townhouses for $75,000. A down payment of $15,000 is required, and a 20-year mortgage at an annual interest rate of 9% is available. Find the monthly mortgage payment.

Your strategy

Your solution
$539.84

Solution on p. S17

Instructor Note

The Monthly Payment Table in the Appendix has an accompanying Ⓟ indicating that Microsoft PowerPoint® slides of this table are available.

Concept Check

(*Note:* Students will need the Monthly Payment Table.)

1. How much greater is the monthly mortgage payment on a $90,000 mortgage at 5% if the term is 15 years rather than 30 years? $228.57

2. For a $112,000, 30-year, fixed-rate mortgage, calculate the difference in the monthly mortgage payment for a 9% interest rate and a 7% interest rate. $156.04

Instructor Note

If you are discussing fixed-rate mortgages, you might want to describe the bimonthly payment option that many institutions are offering now.

With bimonthly payments, you pay half of the monthly mortgage payment every two weeks, rather than the total monthly payment once a month. This is comparable to 13 monthly payments a year, which results in a shorter life of the loan and thus reduces the amount of interest paid on the loan. For example, using the biweekly mortgage payment process, the borrower can pay off a 30-year loan in about 19 years.

Optional Student Activity

(*Note:* Students will need the Monthly Payment Table. Also, students are generally shocked at the amount of interest paid throughout the life of a mortgage.)

A person obtains a 30-year mortgage of $125,000 at an interest rate of 9%.

a. Find the monthly payment.
$1005.78

b. Find the total of the mortgage payments over the life of the loan.
$1005.78 × 30 × 12 = $362,080.80

c. Find the total amount of interest paid on the loan.
$237,080.80

Instructor Note

As an optional exercise for students with a scientific calculator, you can give them the following keystrokes to calculate a monthly payment:

$$\text{Payment} = B \times i \div (1 - 1 \div (1 + i)\boxed{y^x}n) =$$

where B is the amount borrowed, i is the annual interest rate as a decimal divided by the number of payments per year, and n is the number of months of the loan.

Here is an example you can use:
Find the monthly payment on $100,000 borrowed at an annual interest rate of 9.6% for 15 years.
$1050.27

Example 4

A home has a mortgage of $134,000 for 25 years at an annual interest rate of 7%. During a month when $375.88 of the monthly mortgage payment is principal, how much of the payment is interest?

Strategy

To find the interest:

- Multiply the mortgage by the factor found in the Monthly Payment Table in the Appendix to find the monthly mortgage payment.
- Subtract the principal from the monthly mortgage payment.

Solution

$$134{,}000 \times 0.0070678 \approx 947.09$$

From the table ↑   ↑ Monthly mortgage payment

| Monthly mortgage payment | − | principal | = | interest |

947.09   −   375.88   =   571.21

$571.21 of the payment is interest on the mortgage.

You Try It 4

An office building has a mortgage of $125,000 for 25 years at an annual interest rate of 9%. During a month when $492.65 of the monthly mortgage payment is principal, how much of the payment is interest?

Your strategy

Your solution
$556.35

Example 5

The monthly mortgage payment for a home is $598.75. The annual property tax is $900. Find the total monthly payment for the mortgage and property tax.

Strategy

To find the monthly payment:

- Divide the annual property tax by 12 to find the monthly property tax.
- Add the monthly property tax to the monthly mortgage payment.

Solution

900 ÷ 12 = 75   • Monthly property tax

598.75 + 75 = 673.75

The total monthly payment is $673.75.

You Try It 5

The monthly mortgage payment for a home is $415.20. The annual property tax is $744. Find the total monthly payment for the mortgage and property tax.

Your strategy

Your solution
$477.20

Solutions on pp. S17–S18

6.4 Exercises

Objective A   To calculate the initial expenses of buying a home

Suggested Assignment
Exercises 1–23, odds
More challenging problem:
   Exercise 25

1. A condominium at Mt. Baldy Ski Resort was purchased for $97,000, and a down payment of $14,550 was made. Find the mortgage.
   $82,450

2. An insurance business was purchased for $173,000, and a down payment of $34,600 was made. Find the mortgage.
   $138,400

3. A building lot was purchased for $25,000. The lender requires a down payment of 30% of the purchase price. Find the down payment.
   $7500

4. Ian Goldman purchased a new home for $188,500. The lender requires a down payment of 20% of the purchase price. Find the down payment.
   $37,700

5. Brian Stedman made a down payment of 25% of the $850,000 purchase price of an apartment building. How much was the down payment?
   $212,500

6. A clothing store was purchased for $125,000, and a down payment that was 25% of the purchase price was made. How much was the down payment?
   $31,250

7. A loan of $150,000 is obtained to purchase a home. The loan origination fee is $2\frac{1}{2}$ points. Find the amount of the loan origination fee.
   $3750

8. Security Savings & Loan requires a borrower to pay $3\frac{1}{2}$ points for a loan. Find the amount of the loan origination fee for a loan of $90,000.
   $3150

9. Baja Construction Inc. is selling homes for $150,000. A down payment of 5% is required.
   a. Find the down payment. $7500
   b. Find the mortgage. $142,500

10. A cattle rancher purchased some land for $240,000. The bank requires a down payment of 15% of the purchase price.
    a. Find the down payment. $36,000
    b. Find the mortgage. $204,000

11. Vivian Tom purchased a home for $210,000. Find the mortgage if the down payment Vivian made is 10% of the purchase price.
    $189,000

12. A mortgage lender requires a down payment of 5% of the $80,000 purchase price of a condominium. How much is the mortgage?
    $76,000

Objective B   To calculate the ongoing expenses of owning a home

For Exercises 13 to 24, solve. Use the Monthly Payment Table in the Appendix. Round to the nearest cent.

13. An investor obtained a loan of $150,000 to buy a car wash business. The monthly mortgage payment was based on 25 years at 8%. Find the monthly mortgage payment.
    $1157.73

14. A beautician obtained a 20-year mortgage of $90,000 to expand the business. The credit union charges an annual interest rate of 6%. Find the monthly mortgage payment.
    $644.79

Quick Quiz (Objective 6.4A)

1. A beach house is purchased for $52,500, and a down payment of $12,600 is made. Find the mortgage. $39,900

2. The loan origination fee on a $180,000 mortgage is $4\frac{1}{2}$ points. Find the loan origination fee. $8100

3. An architect purchases a home for $125,000. Find the mortgage if the down payment is 6% of the purchase price. $117,500

15. A couple interested in buying a home determines that they can afford a monthly mortgage payment of $800. Can they afford to buy a home with a 30-year $110,000 mortgage at 8% interest?
No

16. A lawyer is considering purchasing a new office building with a 15-year $400,000 mortgage at 6% interest. The lawyer can afford a monthly mortgage payment of $3500. Can the lawyer afford the monthly mortgage payment on the new office building?
Yes

17. The county tax assessor has determined that the annual property tax on a $125,000 house is $1348.20. Find the monthly property tax.
$112.35

18. The annual property tax on a $155,000 home is $1992. Find the monthly property tax.
$166

19. Abacus Imports Inc. has a warehouse with a 25-year mortgage of $200,000 at an annual interest rate of 9%.
    a. Find the monthly mortgage payment. $1678.40
    b. During a month when $941.72 of the monthly mortgage payment is principal, how much of the payment is interest? $736.68

20. A vacation home has a mortgage of $135,000 for 30 years at an annual interest rate of 7%.
    a. Find the monthly mortgage payment. $898.16
    b. During a month when $392.47 of the monthly mortgage payment is principal, how much of the payment is interest? $505.69

21. The annual mortgage payment on a duplex is $10,844.40. The owner must pay an annual property tax of $948. Find the total monthly payment for the mortgage and property tax.
$982.70

22. The monthly mortgage payment on a home is $716.40, and the homeowner pays an annual property tax of $792. Find the total monthly payment for the mortgage and property tax.
$782.40

23. Maria Hernandez purchased a home for $210,000 and made a down payment of $15,000. The balance was financed for 15 years at an annual interest rate of 6%. Find the monthly mortgage payment.
$1645.53

24. A customer of a savings and loan purchased a $185,000 home and made a down payment of $20,000. The savings and loan charges its customers an annual interest rate of 7% for 30 years for a home mortgage. Find the monthly mortgage payment.
$1097.75

APPLYING THE CONCEPTS

25. **Mortgages**   A couple considering a mortgage of $100,000 have a choice of loans. One loan is an 8% loan for 20 years, and the other loan is at 8% for 30 years. Find the amount of interest that the couple can save by choosing the 20-year loan.
$63,408

26.   **Mortgages**   Find out what an adjustable-rate mortgage is. What is the difference between this type of loan and a fixed-rate mortgage? List some of the advantages and disadvantages of each.

Answers to Writing Exercises

26. In a fixed-rate mortgage, a fixed interest rate is charged, and fixed payments are made on the loan throughout the term of the loan. This means that the interest rate of the loan never changes and the amount of the monthly payment remains the same throughout the life of the loan.

An adjustable-rate mortgage (ARM) does not have a fixed interest rate throughout the term of the mortgage. Instead, the interest rate charged is adjusted periodically to more closely reflect current interest rates. In return for taking a risk that interest rates may rise, borrowers usually get a lower interest rate at the beginning of the ARM than they would have if they had taken out a fixed-rate mortgage for the same term.

Quick Quiz (Objective 6.4B)

Note: Students will need the Monthly Payment Table.

1. A homeowner has a 15-year mortgage of $150,000 at an annual interest rate of 8%.
   a. Find the monthly mortgage payment. $1433.48
   b. During a month when $268.88 of the monthly mortgage payment is principal, how much of the payment is interest? $1164.60

2. The monthly mortgage on a home is $1244.30. The owner must pay an annual property tax of $984. Find the total monthly payment for the mortgage and the property tax. $1326.30

6.5 Car Expenses

Objective A   **To calculate the initial expenses of buying a car**

The initial expenses in the purchase of a car usually include the down payment, the **license fees** (fees charged for authorization to operate a vehicle), and the **sales tax** (a tax levied by a state or municipality on purchases). The down payment may be very small or as much as 25% or 30% of the purchase price of the car, depending on the lending institution. License fees and sales tax are regulated by each state, so these expenses vary from state to state.

New Vocabulary
license fees
sales tax

Concept Check
Suppose you are purchasing a car for $23,500.

a. How much more money must you put down if the down payment is 25% of the purchase price rather than 15% of the purchase price?
$2350

b. How much more do you have to pay in sales tax if the tax rate is 7.8% rather than 4.2%?
$846

c. How much more do you have to pay for the license fee if it is 3% of the purchase price rather than 1.25% of the purchase price? $411.25

Example 1

A car is purchased for $18,500, and the lender requires a down payment of 15% of the purchase price. Find the amount financed.

Strategy
To find the amount financed:

• Find the down payment by solving the basic percent equation for *amount*.
• Subtract the down payment from the purchase price.

Solution
Percent × base = amount

Percent	×	purchase price	=	down payment

$0.15 \times 18{,}500 = n$
$2775 = n$

$18{,}500 - 2775 = 15{,}725$

The amount financed is $15,725.

You Try It 1

A down payment of 20% of the $19,200 purchase price of a new car is made. Find the amount financed.

Your strategy

Your solution
$15,360

Example 2

A sales clerk purchases a car for $16,500 and pays a sales tax that is 5% of the purchase price. How much is the sales tax?

Strategy
To find the sales tax, solve the basic percent equation for *amount*.

Solution
Percent × base = amount

Percent	×	purchase price	=	sales tax

$0.05 \times 16{,}500 = n$
$825 = n$

The sales tax is $825.

You Try It 2

A car is purchased for $17,350. The car license fee is 1.5% of the purchase price. How much is the license fee?

Your strategy

Your solution
$260.25

Solutions on p. S18

Optional Student Activity
Connie is ordering a new car. The base price is $18,750, the options she is ordering total $5420, and there is a destination charge of $385. Connie must pay a 5.4% sales tax on these charges. The state license fee is 2.5% of the total of the base price and the cost of the options. Find the total amount Connie is paying for the car, the sales tax, and the license fee. $26,485.22

In-Class Examples (Objective 6.5A)

1. A carpenter purchases a truck for $25,300 and pays a sales tax of 4% of the purchase price. Find the sales tax. $1012

2. A state charges a car license fee of 2% of the purchase price of a car. How much is the license fee for a car that costs $17,595? $351.90

3. An airline employee buys a sports car for $34,000 and makes a down payment of 20% of the purchase price. Find the amount financed.
$27,200

Objective 6.5B

Instructor Note

See page 284 for a project that involves determining the cost of owning and operating a car.

Concept Check

(*Note:* Students will need the Monthly Payment Table.)

1. How much greater is the monthly car payment on a $20,000 car loan financed at 9% if the term is 3 years rather than 5 years? $220.82

2. Suppose you obtain a car loan for $15,000. You agree to pay off the loan in 4 years. Calculate the difference in the monthly car payment for a 9% interest rate and a 4% interest rate. $34.59

Optional Student Activity

(*Note:* Students will need the Monthly Payment Table.)

A person obtains a 4-year car loan of $17,500 at an interest rate of 8%.

a. Find the monthly car payment. $427.23

b. Find the total of the car payments over the life of the loan.
$427.23 × 4 × 12 = $20,507.04

c. Find the total amount of interest paid on the loan. $3007.04

Objective B **To calculate the ongoing expenses of owning a car**

> **TAKE NOTE**
> The same formula that is used to calculate a monthly mortgage payment is used to calculate a monthly car payment.

Besides the initial expenses of buying a car, there are continuing expenses involved in owning a car. These ongoing expenses include car insurance, gas and oil, general maintenance, and the monthly car payment. The monthly car payment is calculated in the same manner as the monthly mortgage payment on a home loan. A monthly payment table, such as the one in the Appendix, is used to simplify the calculation of monthly car payments.

Example 3

At a cost of $.33 per mile, how much does it cost to operate a car during a year in which the car is driven 15,000 miles?

Strategy
To find the cost, multiply the cost per mile (0.33) by the number of miles driven (15,000).

Solution
15,000 × 0.33 = 4950

The cost is $4950.

You Try It 3

At a cost of $.31 per mile, how much does it cost to operate a car during a year in which the car is driven 23,000 miles?

Your strategy

Your solution
$7130

Example 4

During one month, your total gasoline bill was $84 and the car was driven 1200 miles. What was the cost per mile for gasoline?

Strategy
To find the cost per mile, divide the cost for gasoline (84) by the number of miles driven (1200).

Solution
84 ÷ 1200 = 0.07

The cost per mile was $.07.

You Try It 4

In a year in which your total car insurance bill was $360 and the car was driven 15,000 miles, what was the cost per mile for car insurance?

Your strategy

Your solution
$.024

Example 5

A car is purchased for $18,500 with a down payment of $3700. The balance is financed for 3 years at an annual interest rate of 6%. Find the monthly car payment.

Strategy
To find the monthly payment:

- Subtract the down payment from the purchase price to find the amount financed.
- Multiply the amount financed by the factor found in the Monthly Payment Table in the Appendix.

Solution
18,500 − 3700 = 14,800
14,800 × 0.0304219 ≈ 450.24

The monthly payment is $450.24.

You Try It 5

A truck is purchased for $25,900 with a down payment of $6475. The balance is financed for 4 years at an annual interest rate of 8%. Find the monthly car payment.

Your strategy

Your solution
$474.22

Solutions on p. S18

In-Class Examples (Objective 6.5B)

Note: Students will need the Monthly Payment Table.

1. A car loan of $15,400 is financed through a credit union at an annual interest rate of 6% for 3 years. Find the monthly car payment. $468.50

2. In a year when a car owner's total gasoline bill was $1920, the car was driven 12,000 miles. What was the cost per mile for gasoline? $.16

3. A used car is purchased for $16,275, and a down payment of $1275 is made. The balance is financed for 3 years at an interest rate of 7%. Find the monthly car payment. $463.16

6.5 Exercises

Objective A   **To calculate the initial expenses of buying a car**

1. Amanda has saved $780 to make a down payment on a used minivan that costs $7100. The car dealer requires a down payment of 12% of the purchase price. Has she saved enough money to make the down payment?
No

2. A sedan was purchased for $23,500. A down payment of 15% of the purchase price was required. How much was the down payment?
$3525

3. A drapery installer bought a van to carry drapery samples. The purchase price of the van was $26,500, and a 4.5% sales tax was paid. How much was the sales tax?
$1192.50

4. A & L Lumber Company purchased a delivery truck for $28,500. A sales tax of 4% of the purchase price was paid. Find the sales tax.
$1140

5. A license fee of 2% of the purchase price is paid on a pickup truck costing $22,500. Find the license fee for the truck.
$450

6. Your state charges a license fee of 1.5% on the purchase price of a car. How much is the license fee for a car that costs $16,998?
$254.97

7. An electrician bought a $32,000 flatbed truck. A state license fee of $275 and a sales tax of 3.5% of the purchase price are required.
   a. Find the sales tax. $1120
   b. Find the total cost of the sales tax and the license fee. $1395

8. A physical therapist bought a used car for $9375 and made a down payment of $1875. The sales tax is 5% of the purchase price.
   a. Find the sales tax. $468.75
   b. Find the total cost of the sales tax and the down payment. $2343.75

9. Martin bought a motorcycle for $16,200 and made a down payment of 25% of the purchase price.
   a. Find the down payment. $4050
   b. Find the amount financed. $12,150

10. A carpenter bought a utility van for $24,900 and made a down payment of 15% of the purchase price.
    a. Find the down payment. $3735
    b. Find the amount financed. $21,165

11. An author bought a sports car for $35,000 and made a down payment of 20% of the purchase price. Find the amount financed.
$28,000

12. Tania purchased a used car for $13,500 and made a down payment of 25% of the cost. Find the amount financed.
$10,125

Objective B   **To calculate the ongoing expenses of owning a car**

For Exercises 13 to 24, solve. Use the Monthly Payment Table in the Appendix. Round to the nearest cent.

13. A rancher financed $24,000 for the purchase of a truck through a credit union at 5% interest for 4 years. Find the monthly truck payment.
$552.70

Quick Quiz (Objective 6.5A)

1. A couple buy a station wagon. The purchase price is $21,050, and a 4% sales tax is paid. How much is the sales tax? $842

2. A license fee of 1.5% of the purchase price of a car is paid on a convertible costing $23,400. How much is the license fee for the car? $351

3. An SUV is purchased for $24,500. A down payment of 15% is required. Find the amount that is financed. $20,825

14. A car loan of $18,000 is financed for 3 years at an annual interest rate of 4%. Find the monthly car payment.
$531.43

15. An estimate of the cost of owning a compact car is $.32 per mile. Using this estimate, find how much it costs to operate a car during a year in which the car is driven 16,000 miles.
$5120

16. An estimate of the cost of care and maintenance of automobile tires is $.015 per mile. Using this estimate, find how much it costs for care and maintenance of tires during a year in which the car is driven 14,000 miles.
$210

17. A family spent $1600 on gas, oil, and car insurance during a period in which the car was driven 14,000 miles. Find the cost per mile for gas, oil, and car insurance.
$.11

18. Last year you spent $2100 for gasoline for your car. The car was driven 15,000 miles. What was your cost per mile for gasoline?
$.14

19. Elena's monthly car payment is $143.50. During a month in which $68.75 of the monthly payment is principal, how much of the payment is interest?
$74.75

20. The cost for a pizza delivery truck for the year included $2868 in truck payments, $2400 for gasoline, and $675 for insurance. Find the total cost for truck payments, gasoline, and insurance for the year.
$5943

21. The city of Colton purchased a fire truck for $164,000 and made a down payment of $10,800. The balance is financed for 5 years at an annual interest rate of 6%.
   a. Find the amount financed. $153,200
   b. Find the monthly truck payment. $2961.78

22. A used car is purchased for $14,999, and a down payment of $2999 is made. The balance is financed for 3 years at an annual interest rate of 5%.
   a. Find the amount financed. $12,000
   b. Find the monthly car payment. $359.65

23. An artist purchased a new car costing $27,500 and made a down payment of $5500. The balance is financed for 3 years at an annual interest rate of 4%. Find the monthly car payment.
$649.53

24. A camper is purchased for $39,500, and a down payment of $5000 is made. The balance is financed for 4 years at an annual interest rate of 6%. Find the monthly payment.
$810.23

APPLYING THE CONCEPTS

25. **Car Loans**   One bank offers a 4-year car loan at an annual interest rate of 7% plus a loan application fee of $45. A second bank offers 4-year car loans at an annual interest rate of 8% but charges no loan application fee. If you need to borrow $5800 to purchase a car, which of the two bank loans has the lesser loan costs? Assume you keep the car for 4 years.
The 7% loan with the application fee

26. **Car Loans**   How much interest is paid on a 5-year car loan of $9000 if the interest rate is 9%?
$2209.54

Quick Quiz (Objective 6.5B)

Note: Students will need the Monthly Payment Table.

1. A truck purchased for $21,900 is financed through a bank at 7% interest for 4 years. Find the monthly car payment. $524.42

2. A car owner spent $3300 on gas, oil, and car insurance during a period when the car was driven 15,000 miles. Find the cost per mile for gasoline, oil, and car insurance. $.22

3. A car is purchased for $22,000, and a down payment of $5500 is made. The balance is financed for 4 years at an interest rate of 8%. Find the monthly car payment. $402.81

6.6 Wages

Objective A To calculate commissions, total hourly wages, and salaries

Commissions, hourly wage, and salary are three ways to receive payment for doing work.

Commissions are usually paid to salespersons and are calculated as a percent of total sales.

HOW TO As a real estate broker, Emma Smith receives a commission of 4.5% of the selling price of a house. Find the commission she earned for selling a home for $175,000.

To find the commission Emma earned, solve the basic percent equation for *amount*.

Percent $\times$ base = amount

Commission rate	$\times$	total sales	=	commission

0.045 $\times$ 175,000 = 7875

The commission is $7875.

An employee who receives an **hourly wage** is paid a certain amount for each hour worked.

HOW TO A plumber receives an hourly wage of $18.25. Find the plumber's total wages for working 37 hours.

To find the plumber's total wages, multiply the hourly wage by the number of hours worked.

Hourly wage	$\times$	number of hours worked	=	total wages

18.25 $\times$ 37 = 675.25

The plumber's total wages for working 37 hours are $675.25.

An employee who is paid a **salary** receives payment based on a weekly, biweekly (every other week), monthly, or annual time schedule. Unlike the employee who receives an hourly wage, the salaried worker does not receive additional pay for working more than the regularly scheduled workday.

HOW TO Ravi Basar is a computer operator who receives a weekly salary of $695. Find his salary for 1 month (4 weeks).

To find Ravi's salary for 1 month, multiply the salary per pay period by the number of pay periods.

Salary per pay period	$\times$	number of pay periods	=	total salary

695 $\times$ 4 = 2780

Ravi's total salary for 1 month is $2780.

New Vocabulary
commission
hourly wage
salary

Concept Check

1. Melanie is currently earning an annual salary of $28,600 at Delta Electronics. She has been offered a job at Alpha Tech that pays $34,580 per year. How much more will Melanie earn per week if she accepts the job offer from Alpha Tech? $115
(*Note:* You might ask students to describe how they arrived at the answer so that you can compare the two possible approaches: (1) dividing the difference between the salaries by 52 and (2) subtracting the weekly salaries for the two jobs.)

2. Ron earns minimum wage. If the minimum wage increases $.75 per hour, what will be Ron's increase in pay for a 40-hour week? $30

In-Class Examples (Objective 6.6A)

1. A part-time sales clerk earns an hourly wage of $8.85. How much does the sales clerk earn during a 24-hour workweek? $212.40

2. A golf pro receives a commission of 20% for selling a set of golf clubs. Find the commission earned by the golf pro for selling a set of golf clubs that cost $320. $64

3. A junior executive for a marketing firm receives an annual salary of $41,700. How much does the executive receive per month? $3475

Optional Student Activity

Sometimes commission rates increase as sales increase. A **graduated commission** assigns a different commission rate for each of several levels of sales. The following problem is an application of graduated commissions.

Each employee at Tri-City Sales receives a graduated commission of 10% on the first $4000 of sales, 12% on the next $6000 of sales, and 15% on sales over $10,000.

a. Pat's sales for the past month totaled $18,312. Find Pat's commission for the month.
$2366.80

b. Chris has sales totaling $9880 this month. Find Chris's commission for the month.
$1105.60

Example 1

A pharmacist's hourly wage is $28. On Saturday, the pharmacist earns time and a half (1.5 times the regular hourly wage). How much does the pharmacist earn for working 6 hours on Saturday?

Strategy
To find the pharmacist's earnings:
- Find the hourly wage for working on Saturday by multiplying the hourly wage by 1.5.
- Multiply the hourly wage by the number of hours worked.

Solution
$28 \times 1.5 = 42 \qquad 42 \times 6 = 252$

The pharmacist earns $252.

Example 2

An efficiency expert received a contract for $3000. The expert spent 75 hours on the project. Find the consultant's hourly wage.

Strategy
To find the hourly wage, divide the total earnings by the number of hours worked.

Solution
$3000 \div 75 = 40$

The hourly wage was $40.

Example 3

Dani Greene earns $28,500 per year plus a 5.5% commission on sales over $100,000. During one year, Dani sold $150,000 worth of computers. Find Dani's total earnings for the year.

Strategy
To find the total earnings:
- Find the sales over $100,000.
- Multiply the commission rate by sales over $100,000.
- Add the commission to the annual pay.

Solution
$150,000 - 100,000 = 50,000$
$50,000 \times 0.055 = 2750$ • Commission
$28,500 + 2750 = 31,250$

Dani earned $31,250.

You Try It 1

A construction worker, whose hourly wage is $18.50, earns double time (2 times the regular hourly wage) for working overtime. Find the worker's wages for working 8 hours of overtime.

Your strategy

Your solution
$296

You Try It 2

A contractor for a bridge project receives an annual salary of $48,228. What is the contractor's salary per month?

Your strategy

Your solution
$4019

You Try It 3

An insurance agent earns $27,000 per year plus a 9.5% commission on sales over $50,000. During one year, the agent's sales totaled $175,000. Find the agent's total earnings for the year.

Your strategy

Your solution
$38,875

Solutions on pp. S18–S19

6.6 Exercises

Objective A **To calculate commissions, total hourly wages, and salaries**

1. Lewis works in a clothing store and earns $9.50 per hour. How much does he earn in a 40-hour week?
$380

2. Sasha pays a gardener an hourly wage of $11. How much does she pay the gardener for working 25 hours?
$275

3. A real estate agent receives a 3% commission for selling a house. Find the commission that the agent earned for selling a house for $131,000.
$3930

4. Ron Caruso works as an insurance agent and receives a commission of 40% of the first year's premium. Find Ron's commission for selling a life insurance policy with a first-year premium of $1050.
$420

5. A stockbroker receives a commission of 1.5% of the price of stock that is bought or sold. Find the commission on 100 shares of stock that were bought for $5600.
$84

6. The owner of the Carousel Art Gallery receives a commission of 20% on paintings that are sold on consignment. Find the commission on a painting that sold for $22,500.
$4500

7. Keisha Brown receives an annual salary of $38,928 as a teacher of Italian. How much does Keisha receive each month?
$3244

8. An apprentice plumber receives an annual salary of $27,900. How much does the plumber receive per month?
$2325

9. An electrician's hourly wage is $25.80. For working overtime, the electrician earns double time. What is the electrician's hourly wage for working overtime?
$51.60

10. Carlos receives a commission of 12% of his weekly sales as a sales representative for a medical supply company. Find the commission he earned during a week in which sales were $4500.
$540

11. A golf pro receives a commission of 25% for selling a golf set. Find the commission the pro earned for selling a golf set costing $450.
$112.50

12. Steven receives $3.75 per square yard to install carpet. How much does he receive for installing 160 square yards of carpet?
$600

13. A typist charges $2.75 per page for typing technical material. How much does the typist earn for typing a 225-page book?
$618.75

Quick Quiz (Objective 6.6A)

1. A food service worker earns $8.70 per hour. How much does the worker earn in a 40-hour work week? $348

2. A sales representative receives a commission of 5.5% on weekly sales. Find the commission earned during a week when sales were $8100.
$445.50

3. A legal assistant receives $35,700 annually. How much does the assistant earn each month?
$2975

14. A nuclear chemist received $15,000 in consulting fees while working on a nuclear power plant. The chemist worked 120 hours on the project. Find the chemist's hourly wage.
$125

15. Maxine received $3400 for working on a project as a computer consultant for 40 hours. Find her hourly wage.
$85

16. Gil Stratton's hourly wage is $10.78. For working overtime, he receives double time.
   a. What is Gil's hourly wage for working overtime? $21.56
   b. How much does he earn for working 16 hours of overtime? $344.96

17. Mark is a lathe operator and receives an hourly wage of $15.90. When working on Saturday, he receives time and a half.
   a. What is Mark's hourly wage on Saturday? $23.85
   b. How much does he earn for working 8 hours on Saturday? $190.80

18. A stock clerk at a supermarket earns $8.20 an hour. For working the night shift, the clerk's wage increases by 15%.
   a. What is the increase in hourly pay for working the night shift? $1.23
   b. What is the clerk's hourly wage for working the night shift? $9.43

19. A nurse earns $21.50 an hour. For working the night shift, the nurse receives a 10% increase in pay.
   a. What is the increase in hourly pay for working the night shift? $2.15
   b. What is the hourly pay for working the night shift? $23.65

20. Tony's hourly wage as a service station attendant is $9.40. For working the night shift, his wage is increased 25%. What is Tony's hourly wage for working the night shift?
$11.75

21. Nicole Tobin, a salesperson, receives a salary of $250 per week plus a commission of 15% on all sales over $1500. Find her earnings during a week in which sales totaled $3000.
$475

APPLYING THE CONCEPTS

Compensation The table at the right shows the average starting salaries for recent college graduates. Use this table for Exercises 22 to 25. Round to the nearest dollar.

Average Starting Salaries		
Bachelor's Degree	Average Starting Salary	Change from Previous Year
Chemical Engineering	$52,169	1.8% increase
Electrical Engineering	$50,566	0.4% increase
Computer Science	$46,536	7.6% decrease
Accounting	$41,360	2.6% increase
Business	$36,515	3.7% increase
Biology	$29,554	1.0% decrease
Political Science	$28,546	12.6% decrease
Psychology	$26,738	10.7% decrease

Source: National Association of Colleges

22. What was the starting salary in the previous year for an accountant?
$40,312

23. How much did the starting salary for a chemical engineer increase over that of the previous year?
$922

24. What was the starting salary in the previous year for a computer science major?
$50,364

25. How much did the starting salary for a political science major decrease from that of the previous year?
$4115

6.7 Bank Statements

Objective A To calculate checkbook balances

New Vocabulary
checking account
check
deposit slip

Discuss the Concepts

1. What is the purpose of a check register?
2. Explain how to record the writing of a check in a check register.
3. Explain how to record the making of a deposit in a check register.

TAKE NOTE

A **checking account** is a bank account that enables you to withdraw money or make payments to other people using checks. A **check** is a printed form that, when filled out and signed, instructs a bank to pay a specified sum of money to the person named on it.
A **deposit slip** is a form for depositing money in a checking account.

A checking account can be opened at most banks and savings and loan associations by depositing an amount of money in the bank. A checkbook contains checks and deposit slips and a checkbook register in which to record checks written and amounts deposited in the checking account. A sample check is shown below.

```
                    Payee                    Date Check is Written

                                                              NO. 2023    ← Check Number
East Phoenix Rental Equipment
3011 N.W. Ventura Street                                      68 - 461
Phoenix, Arizona 85280                                        1052

                          Date  October 11, 2005
PAY TO THE
ORDER OF  Tellas Manufacturing Co.          $ 827 00/100     ← Amount of Check

Eight Hundred Twenty-Seven and 00/100 — —      DOLLARS

MEYERS' NATIONAL BANK
11 N.W. Nova Street
Phoenix, Arizona 85215

Memo
I: 1052'''0461 I: 5008  2023''        Eugene L. Madison

        Amount of Check in Words          Depositor's Signature
```

Each time a check is written, the amount of the check is subtracted from the amount in the account. When a deposit is made, the amount deposited is added to the amount in the account.

A portion of a checkbook register is shown below. The account holder had a balance of $587.93 before writing two checks, one for $286.87 and the other for $202.38, and making one deposit of $345.00.

Point of Interest

There are a number of computer programs that serve as "electronic" checkbooks. With these programs, you can pay your bills by using a computer to write the check and then transmit the check over telephone lines using a modem.

RECORD ALL CHARGES OR CREDITS THAT AFFECT YOUR ACCOUNT

NUMBER	DATE	DESCRIPTION OF TRANSACTION	PAYMENT/DEBIT (−)	√T	FEE (IF ANY) (−)	DEPOSIT/CREDIT (+)	BALANCE $ 587 93	
108	8/4	Plumber	$286 87		$	$	301	06
109	8/10	Car Payment	202 38				98	68
	8/14	Deposit				345 00	443	68

To find the current checking account balance, subtract the amount of each check from the previous balance. Then add the amount of the deposit.

The current checking account balance is $443.68.

In-Class Examples (Objective 6.7A)

1. A credit manager had a checking account balance of $535.25 before making a deposit of $216.18. The manager then wrote two checks, one for $52.63 and another for $260.17. Find the current checkbook balance. $438.63

2. An inventory clerk's checkbook balance is $1434.51. The clerk wants to purchase a TV for $695 and a sofa for $675. Is there enough money in the account to make the two purchases? Yes

Optional Student Activity

Make copies of the checkbook register shown at the bottom of the page, and then have students complete it by recording the following:

1. Check number 347 is written on 3/17 to the college bookstore. The amount is $73.50.
2. Check number 348 is written on 3/19 to the warehouse club for $83.94.
3. A deposit of $468.25 is made on 3/22.
4. Check number 349 is written on 3/24 to the phone company for $47.92.

The balances are:

1. $1546.92
2. $1462.98
3. $1931.23
4. $1883.31

Example 1

A mail carrier had a checking account balance of $485.93 before writing two checks, one for $18.98 and another for $35.72, and making a deposit of $250. Find the current checking account balance.

Strategy

To find the current balance:

- Subtract the amount of each check from the old balance.
- Add the amount of the deposit.

Solution

$$
\begin{array}{rl}
485.93 & \\
-\;\;18.98 & \text{first check} \\
\hline
466.95 & \\
-\;\;35.72 & \text{second check} \\
\hline
431.23 & \\
+\,250.00 & \text{deposit} \\
\hline
681.23 &
\end{array}
$$

The current checking account balance is $681.23.

You Try It 1

A cement mason had a checking account balance of $302.46 before writing a check for $20.59 and making two deposits, one in the amount of $176.86 and another in the amount of $94.73. Find the current checking account balance.

Your strategy

Your solution

$553.46

Solution on p. S19

New Vocabulary

bank statement
balancing a checkbook
service charges

Objective B To balance a checkbook

Each month a bank statement is sent to the account holder. A **bank statement** is a document showing all the transactions in a bank account during the month. It shows the checks that the bank has paid, the deposits received, and the current bank balance.

A bank statement and checkbook register are shown on the next page.

Balancing a checkbook, or determining whether the checking account balance is accurate, requires a number of steps.

1. In the checkbook register, put a check mark (√) by each check paid by the bank and by each deposit recorded by the bank.

NUMBER	DATE	DESCRIPTION OF TRANSACTION	PAYMENT/DEBIT (−)	√ T	FEE (IF ANY) (−)	DEPOSIT/CREDIT (+)	BALANCE $	1620	42
			$		$	$			

RECORD ALL CHARGES OR CREDITS THAT AFFECT YOUR ACCOUNT

NUMBER	DATE	DESCRIPTION OF TRANSACTION	PAYMENT/DEBIT (−)		√ T	FEE (IF ANY) (−)	DEPOSIT/CREDIT (+)		BALANCE $ 840	27
263	5/20	Dentist	$ 75	00	√	$	$		765	27
264	5/22	Post Office	33	61	√				731	66
265	5/22	Gas Company	67	14					664	52
	5/29	Deposit			√		192	00	856	52
266	5/29	Pharmacy	38	95	√				817	57
267	5/30	Telephone	63	85					753	72
268	6/2	Groceries	73	19	√				680	53
	6/3	Deposit			√		215	00	895	53
269	6/7	Insurance	103	00	√				792	53
	6/10	Deposit					225	00	1017	53
270	6/15	Photo Shop	16	63	√				1000	90
271	6/18	Newspaper	27	00					973	90

Ⓟ

CHECKING ACCOUNT Monthly Statement   Account Number: 924-297-8

Date	Transaction	Amount	Balance
5/20	OPENING BALANCE		840.27
5/21	CHECK	75.00	765.27
5/23	CHECK	33.61	731.66
5/29	DEPOSIT	192.00	923.66
6/1	CHECK	38.95	884.71
6/1	INTEREST	4.47	889.18
6/3	CHECK	73.19	815.99
6/3	DEPOSIT	215.00	1030.99
6/9	CHECK	103.00	927.99
6/16	CHECK	16.63	911.36
6/20	SERVICE CHARGE	3.00	908.36
6/20	CLOSING BALANCE		908.36

Ⓟ

2. Add to the current checkbook balance all checks that have been written but have not yet been paid by the bank and any interest paid on the account.

Current checkbook balance:	973.90
Checks: 265	67.14
267	63.85
271	27.00
Interest:	+ 4.47
	1136.36

3. Subtract any service charges and any deposits not yet recorded by the bank. This is the checkbook balance.

Service charge:	− 3.00
	1133.36
Deposit:	− 225.00
Checkbook balance:	908.36

4. Compare the balance with the bank balance listed on the bank statement. If the two numbers are equal, the bank statement and the checkbook "balance."

Closing bank balance from bank statement	Checkbook balance
$908.36	= $908.36

The bank statement and checkbook balance.

In-Class Examples (Objective 6.7B)

1. Your checkbook shows a balance of $375.85. The bank statement does not show a deposit of $126.32, and checks for $56.19 and $275.05 have not been cashed. What balance does the bank statement show? $580.77

2. Your checkbook shows a balance of $1300.95. The bank statement does not show a deposit of $750, and checks for $105.49, $315, and $88.76 have not been cashed. What balance does the bank statement show? $1060.20

Discuss the Concepts
Why are check marks placed in the checkbook register by each check paid by the bank and by each deposit recorded by the bank?

Optional Student Activity

Match each of the following words with the correct definition.

1. Deposit   b

2. Statement   f

3. Balance a checkbook   a

4. Check   d

5. Balance   c

a. To obtain agreement between the bank statement and the current checkbook balance by accounting for outstanding items

b. An amount of money that you put into a checking account

c. The amount of money in your checking account

d. A written order directing the bank to deduct money from your checking account and use it to make a payment to another party

e. An item that did not reach the bank in time to be processed and listed in your bank statement

f. A record prepared by the bank listing all transactions that the bank has recorded

HOW TO

RECORD ALL CHARGES OR CREDITS THAT AFFECT YOUR ACCOUNT

NUMBER	DATE	DESCRIPTION OF TRANSACTION	PAYMENT/DEBIT (−)		√ T	FEE (IF ANY) (−)	DEPOSIT/CREDIT (+)		BALANCE $ 1620	42
413	3/2	Car Payment	$232	15	√	$	$		1388	27
414	3/2	Utilities	67	14	√				1321	13
415	3/5	Restaurant	78	14					1242	99
	3/8	Deposit			√		1842	66	3085	65
416	3/10	House Payment	672	14	√				2413	51
417	3/14	Insurance	177	10					2236	41

CHECKING ACCOUNT   Monthly Statement   Account Number: 924-297-8

Date	Transaction	Amount	Balance
3/1	OPENING BALANCE		1620.42
3/4	CHECK	232.15	1388.27
3/5	CHECK	67.14	1321.13
3/8	DEPOSIT	1842.66	3163.79
3/10	INTEREST	6.77	3170.56
3/12	CHECK	672.14	2498.42
3/25	SERVICE CHARGE	2.00	2496.42
3/30	CLOSING BALANCE		2496.42

Balance the bank statement shown above.

1. In the checkbook register, put a check mark (√) by each check paid by the bank and by each deposit recorded by the bank.

2. Add to the current checkbook balance all checks that have been written but have not yet been paid by the bank and any interest paid on the account.

3. Subtract any service charges and any deposits not yet recorded by the bank. This is the checkbook balance.

4. Compare the balance with the bank balance listed on the bank statement. If the two numbers are equal, the bank statement and the checkbook "balance."

Current checkbook balance:	2236.41
Checks: 415	78.14
417	177.10
Interest:	+ 6.77
	2498.42
Service charge:	− 2.00
Checkbook balance:	2496.42

Closing bank balance from bank statement	Checkbook balance
$2496.42	= $2496.42

The bank statement and checkbook balance.

Example 2   Balance the bank statement shown below.

RECORD ALL CHARGES OR CREDITS THAT AFFECT YOUR ACCOUNT

NUMBER	DATE	DESCRIPTION OF TRANSACTION	PAYMENT/DEBIT (−)		√ T	FEE (IF ANY) (−)	DEPOSIT/CREDIT (+)		BALANCE $ 412 64	
345	1/14	Phone Bill	$ 54	75	√	$	$		357	89
346	1/19	News Shop	18	98	√				338	91
347	1/23	Theater Tickets	95	00					243	91
	1/31	Deposit			√		947	00	1190	91
348	2/5	Cash	250	00	√				940	91
349	2/12	Rent	840	00					100	91

(P)

CHECKING ACCOUNT Monthly Statement		Account Number: 924-297-8	
Date	Transaction	Amount	Balance
1/10	OPENING BALANCE		412.64
1/18	CHECK	54.75	357.89
1/23	CHECK	18.98	338.91
1/31	DEPOSIT	947.00	1285.91
2/1	INTEREST	4.52	1290.43
2/10	CHECK	250.00	1040.43
2/10	CLOSING BALANCE		1040.43

(P)

Solution

Current checkbook
balance: 100.91

Checks: 347   95.00

         349   840.00

Interest: +   4.52

                                1040.43

Service charge: −   0.00

                                1040.43

Deposit: −   0.00

Checkbook balance: 1040.43

Closing bank balance from
bank statement: $1040.43

Checkbook balance:
$1040.43

The bank statement and
the checkbook balance.

You Try It 2

Balance the bank statement shown below.

		RECORD ALL CHARGES OR CREDITS THAT AFFECT YOUR ACCOUNT							BALANCE	
NUMBER	DATE	DESCRIPTION OF TRANSACTION	PAYMENT/DEBIT (−)		√ T	FEE (IF ANY) (−)	DEPOSIT/CREDIT (+)	$	903	17
	2/15	Deposit	$			$	$ 523	84	1427	01
234	2/20	Mortgage	773	21					653	80
235	2/27	Cash	200	00					453	80
	3/1	Deposit					523	84	977	64
236	3/12	Insurance	275	50					702	14
237	3/12	Telephone	78	73					623	41

CHECKING ACCOUNT Monthly Statement			Account Number: 314-271-4	
Date	Transaction		Amount	Balance
2/14	OPENING BALANCE			903.17
2/15	DEPOSIT		523.84	1427.01
2/21	CHECK		773.21	653.80
2/28	CHECK		200.00	453.80
3/1	INTEREST		2.11	455.91
3/14	CHECK		275.50	180.41
3/14	CLOSING BALANCE			180.41

Your solution

The bank statement and the checkbook balance.

Solution on p. S19

6.7 Exercises

Objective A   To calculate checkbook balances

Suggested Assignment
Exercises 1–15, odds
Exercises 16–19

1. You had a checking account balance of $342.51 before making a deposit of $143.81. What is your new checking account balance?
    $486.32

2. Carmen had a checking account balance of $493.26 before writing a check for $48.39. What is the current checking account balance?
    $444.87

3. A real estate firm had a balance of $2431.76 in its rental-property checking account. What is the balance in this account after a check for $1209.29 has been written?
    $1222.47

4. The business checking account for R and R Tires showed a balance of $1536.97. What is the balance in this account after a deposit of $439.21 has been made?
    $1976.18

5. A nutritionist had a checking account balance of $1204.63 before writing one check for $119.27 and another check for $260.09. Find the current checkbook balance.
    $825.27

6. Sam had a checking account balance of $3046.93 before writing a check for $1027.33 and making a deposit of $150.00. Find the current checkbook balance.
    $2169.60

7. The business checking account for Rachael's Dry Cleaning had a balance of $3476.85 before a deposit of $1048.53 was made. The store manager then wrote two checks, one for $848.37 and another for $676.19. Find the current checkbook balance.
    $3000.82

8. Joel had a checking account balance of $427.38 before a deposit of $127.29 was made. Joel then wrote two checks, one for $43.52 and one for $249.78. Find the current checkbook balance.
    $261.37

9. A carpenter had a checkbook balance of $404.96 before making a deposit of $350 and writing a check for $71.29. Is there enough money in the account to purchase a refrigerator for $675?
    Yes

10. A taxi driver had a checkbook balance of $149.85 before making a deposit of $245 and writing a check for $387.68. Is there enough money in the account for the bank to pay the check?
    Yes

11. A sporting goods store has the opportunity to buy downhill skis and cross-country skis at a manufacturer's closeout sale. The downhill skis will cost $3500, and the cross-country skis will cost $2050. There is currently $5625.42 in the sporting goods store's checking account. Is there enough money in the account to make both purchases by check?
    Yes

12. A lathe operator's current checkbook balance is $1143.42. The operator wants to purchase a utility trailer for $525 and a used piano for $650. Is there enough money in the account to make the two purchases?
    No

Quick Quiz (Objective 6.7A)

1. The business checking account for a toy store had a balance of $4385.94 before making a deposit of $918.62. The store manager then wrote two checks, one for $747.56 and another for $785.23. Find the current checkbook balance. $3771.77

2. A dental hygienist's checkbook balance is $1909.70. The hygienist wants to purchase a laser printer for $775 and patio furniture for $1180. Is there enough money in the account to make the two purchases?   No

Objective B **To balance a checkbook**

13. Balance the checkbook.

RECORD ALL CHARGES OR CREDITS THAT AFFECT YOUR ACCOUNT									
NUMBER	DATE	DESCRIPTION OF TRANSACTION	PAYMENT/DEBIT (−)		√ T	FEE (IF ANY) (−)	DEPOSIT/CREDIT (+)		BALANCE $ 2466 79
223	3/2	Groceries	$ 167	32		$	$		2299 47
	3/5	Deposit					960	70	3260 17
224	3/5	Rent	860	00					2400 17
225	3/7	Gas & Electric	142	35					2257 82
226	3/7	Cash	300	00					1957 82
227	3/7	Insurance	218	44					1739 38
228	3/7	Credit Card	419	32					1320 06
229	3/12	Dentist	92	00					1228 06
230	3/13	Drug Store	47	03					1181 03
	3/19	Deposit					960	70	2141 73
231	3/22	Car Payment	241	35					1900 38
232	3/25	Cash	300	00					1600 38
233	3/25	Oil Company	166	40					1433 98
234	3/28	Plumber	155	73					1278 25
235	3/29	Department Store	288	39					989 86

P

CHECKING ACCOUNT Monthly Statement   Account Number: 122-345-1

Date	Transaction	Amount	Balance
3/1	OPENING BALANCE		2466.79
3/5	DEPOSIT	960.70	3427.49
3/7	CHECK	167.32	3260.17
3/8	CHECK	860.00	2400.17
3/8	CHECK	300.00	2100.17
3/9	CHECK	142.35	1957.82
3/12	CHECK	218.44	1739.38
3/14	CHECK	92.00	1647.38
3/18	CHECK	47.03	1600.35
3/19	DEPOSIT	960.70	2561.05
3/25	CHECK	241.35	2319.70
3/27	CHECK	300.00	2019.70
3/29	CHECK	155.73	1863.97
3/30	INTEREST	13.22	1877.19
4/1	CLOSING BALANCE		1877.19

P

The bank statement and the checkbook balance.

Quick Quiz (Objective 6.7B)

1. Your checkbook shows a balance of $1505.29. The bank statement does not show a deposit of $810.70, and checks for $298.65, $169.47, and $79.40 have not been cashed. What balance does the bank statement show? $1242.11

14. Balance the checkbook.

RECORD ALL CHARGES OR CREDITS THAT AFFECT YOUR ACCOUNT

NUMBER	DATE	DESCRIPTION OF TRANSACTION	PAYMENT/DEBIT (–)		√T	FEE (IF ANY) (–)	DEPOSIT/CREDIT (+)		BALANCE $ 1219 43	
	5/1	Deposit	$			$	$ 619	14	1838	57
515	5/2	Electric Bill	42	35					1796	22
516	5/2	Groceries	95	14					1701	08
517	5/4	Insurance	122	17					1578	91
518	5/5	Theatre Tickets	84	50					1494	41
	5/8	Deposit					619	14	2113	55
519	5/10	Telephone	37	39					2076	16
520	5/12	Newspaper	22	50					2053	66
	5/15	Deposit					619	14	2672	80
521	5/20	Computer Store	172	90					2499	90
522	5/21	Credit Card	313	44					2186	46
523	5/22	Eye Exam	82	00					2104	46
524	5/24	Groceries	107	14					1997	32
525	5/24	Deposit					619	14	2616	46
526	5/25	Oil Company	144	16					2472	30
527	5/30	Car Payment	288	62					2183	68
528	5/30	Mortgage Payment	877	42					1306	26

CHECKING ACCOUNT Monthly Statement			Account Number: 122-345-1	
Date	Transaction		Amount	Balance
5/1	OPENING BALANCE			1219.43
5/1	DEPOSIT		619.14	1838.57
5/3	CHECK		95.14	1743.43
5/4	CHECK		42.35	1701.08
5/6	CHECK		84.50	1616.58
5/8	CHECK		122.17	1494.41
5/8	DEPOSIT		619.14	2113.55
5/15	INTEREST		7.82	2121.37
5/15	CHECK		37.39	2083.98
5/15	DEPOSIT		619.14	2703.12
5/23	CHECK		82.00	2621.12
5/23	CHECK		172.90	2448.22
5/24	CHECK		107.14	2341.08
5/24	DEPOSIT		619.14	2960.22
5/30	CHECK		288.62	2671.60
6/1	CLOSING BALANCE			2671.60

The bank statement and the checkbook balance.

15. Balance the checkbook.

		RECORD ALL CHARGES OR CREDITS THAT AFFECT YOUR ACCOUNT						
NUMBER	DATE	DESCRIPTION OF TRANSACTION	PAYMENT/DEBIT (–)	√T	FEE (IF ANY) (–)	DEPOSIT/CREDIT (+)	BALANCE $ 2035	18
218	7/2	Mortgage	$ 984	60	$	$	1050	58
219	7/4	Telephone	63	36			987	22
220	7/7	Cash	200	00			787	22
	7/12	Deposit				792	60	1579	82
221	7/15	Insurance	292	30			1287	52
222	7/18	Investment	500	00			787	52
223	7/20	Credit Card	414	83			372	69
	7/26	Deposit				792	60	1165	29
224	7/27	Department Store	113	37			1051	92

CHECKING ACCOUNT Monthly Statement		Account Number: 122-345-1	
Date	Transaction	Amount	Balance
7/1	OPENING BALANCE		2035.18
7/1	INTEREST	5.15	2040.33
7/4	CHECK	984.60	1055.73
7/6	CHECK	63.36	992.37
7/12	DEPOSIT	792.60	1784.97
7/20	CHECK	292.30	1492.67
7/24	CHECK	500.00	992.67
7/26	DEPOSIT	792.60	1785.27
7/28	CHECK	200.00	1585.27
7/30	CLOSING BALANCE		1585.27

The bank statement and the checkbook balance.

APPLYING THE CONCEPTS

16. When a check is written, the amount is ___subtracted___ from the balance.

17. When a deposit is made, the amount is ___added___ to the balance.

18. In checking the bank statement, ___add___ to the checkbook balance all checks that have been written but not processed.

19. In checking the bank balance, ___subtract___ any service charge and any deposits not yet recorded.

20. ✎ Define the words *credit* and *debit* as they apply to checkbooks.

Answers to Writing Exercises

20. When applied to a checking account, a credit is a deposit into the account. A debit is a payment or withdrawal from the account.

Focus on Problem Solving

Counterexamples

An example that is given to show that a statement is not true is called a **counterexample.** For instance, suppose someone makes the statement "All colors are red." A counterexample to that statement would be to show someone the color blue or some other color.

If a statement is *always* true, there are no counterexamples. The statement "All even numbers are divisible by 2" is always true. It is not possible to give an example of an even number that is not divisible by 2.

In mathematics, statements that are always true are called *theorems,* and mathematicians are always searching for theorems. Sometimes a conjecture by a mathematician appears to be a theorem. That is, the statement appears to be always true, but later on someone finds a counterexample.

One example of this occurred when the French mathematician Pierre de Fermat (1601–1665) conjectured that $2^{(2^n)} + 1$ is always a prime number for any natural number n. For instance, when $n = 3$, we have $2^{(2^3)} + 1 = 2^8 + 1 = 257$, and 257 is a prime number. However, in 1732 Leonhard Euler (1707–1783) showed that when $n = 5$, $2^{(2^5)} + 1 = 4,294,967,297$ and that $4,294,967,297 = 641 \cdot 6,700,417$—without a calculator! Because 4,294,967,297 is the product of two numbers (other than itself and 1), it is not a prime number. This counterexample showed that Fermat's conjecture is not a theorem.

For Exercises 1 and 5, find at least one counterexample.

1. All composite numbers are divisible by 2.

2. All prime numbers are odd numbers.

3. The square of any number is always bigger than the number.

4. The reciprocal of a number is always less than 1.

5. A number ending in 9 is always larger than a number ending in 3.

When a problem is posed, it may not be known whether the problem statement is true or false. For instance, Christian Goldbach (1690–1764) stated that every even number greater than 2 can be written as the sum of two prime numbers. For example,

$$12 = 5 + 7 \qquad 32 = 3 + 29$$

Although this problem is approximately 250 years old, mathematicians have not been able to prove it is a theorem, nor have they been able to find a counterexample.

For Exercises 6 to 9, answer true if the statement is always true. If there is an instance in which the statement is false, give a counterexample.

6. The sum of two positive numbers is always larger than either of the two numbers.

7. The product of two positive numbers is always larger than either of the two numbers.

8. Percents always represent a number less than or equal to 1.

9. It is never possible to divide by zero.

Answers to Focus on Problem Solving— Counterexamples

1. 9 is a composite number that is not divisible by 2.

2. 2 is a prime number that is not an odd number.

3. $\left(\frac{1}{2}\right)^2 = \frac{1}{4}$ and $\frac{1}{4} < \frac{1}{2}$

4. The reciprocal of $\frac{1}{5}$ is 5, and $5 > 1$.

5. $49 < 63$

6. True

7. $\frac{1}{2} \times \frac{1}{4} = \frac{1}{8}, \frac{1}{8} < \frac{1}{2}$, and $\frac{1}{8} < \frac{1}{4}$

8. $200\% = 2$, and $2 > 1$.

9. True

Projects and Group Activities

Buying a Car Suppose a student has an after-school job to earn money to buy and maintain a car. We will make assumptions about the monthly costs in several categories in order to determine how many hours per week the student must work to support the car. Assume the student earns $8.50 per hour.

1. Monthly payment

   Assume that the car cost $8500 with a down payment of $1020. The remainder is financed for 3 years at an annual simple interest rate of 9%.

   Monthly payment = _____

2. Insurance

   Assume that insurance costs $1500 per year.

   Monthly insurance payment = _____

3. Gasoline

   Assume that the student travels 750 miles per month, that the car travels 25 miles per gallon of gasoline, and that gasoline costs $1.50 per gallon.

   Number of gallons of gasoline purchased per month = _____

   Monthly cost for gasoline = _____

4. Miscellaneous

   Assume $.33 per mile for upkeep.

   Monthly expense for upkeep = _____

5. Total monthly expenses for the monthly payment, insurance, gasoline, and miscellaneous = _____

6. To find the number of hours per month that the student must work to finance the car, divide the total monthly expenses by the hourly rate.

   Number of hours per month = _____

7. To find the number of hours per week that the student must work, divide the number of hours per month by 4.

   Number of hours per week = _____

   The student has to work _____ hours per week to pay the monthly car expenses.

If you own a car, make out your own expense record. If you do not own a car, make assumptions about the kind of car that you would like to purchase, and calculate the total monthly expenses that you would have. An insurance company will give you rates on different kinds of insurance. An automobile club can give you approximations of miscellaneous expenses.

Chapter 6 Summary

Key Words	Examples
The *unit cost* is the cost of one item. [6.1A, p. 235]	Three paperback books cost $36. The unit cost is the cost of one paperback book, $12.
Percent increase is used to show how much a quantity has increased over its original value. [6.2A, p. 239]	The city's population increased 5%, from 10,000 people to 10,500 people.
Cost is the price a business pays for a product. *Selling price* is the price at which a business sells a product to a customer. *Markup* is the difference between selling price and cost. *Markup rate* is the markup expressed as a percent of a product's cost. [6.2B, p. 240]	A business pays $90 for a pair of cross trainers; the cost is $90. The business sells the cross trainers for $135; the selling price is $135. The markup is $135 − $90 = $45.
Percent decrease is used to show how much a quantity has decreased from its original value. [6.2C, p. 242]	Sales decreased 10%, from 10,000 units in the third quarter to 9000 units in the fourth quarter.
Sale price is the price after a reduction from the regular price. *Discount* is the difference between the regular price and the sale price. *Discount rate* is the discount as a percent of a product's regular price. [6.2D, p. 243]	A movie video that regularly sells for $50 is on sale for $40. The regular price is $50. The sale price is $40. The discount is $50 − $40 = $10.
Interest is the amount paid for the privilege of using someone else's money. *Principal* is the amount of money originally deposited or borrowed. The percent used to determine the amount of interest is the *interest rate*. Interest computed on the original amount is called *simple interest*. The principal plus the interest owed on a loan is called the *maturity value*. [6.3A, p. 249]	Consider a 1-year loan of $5000 at an annual simple interest rate of 8%. The principal is $5000. The interest rate is 8%. The interest paid on the loan is $5000 × 0.08 = $400. The maturity value is $5000 + $400 = $5400.
The interest charged on purchases made with a credit card are called *finance charges*. [6.3B, p. 251]	A credit card company charges 1.5% per month on any unpaid balance. The finance charge on an unpaid balance of $1000 is $1000 × 0.015 × 1 = $15.
Compound interest is computed not only on the original principal but also on the interest already earned. [6.3C, p. 252]	$10,000 is invested at 5% annual interest compounded monthly. The value of the investment after 5 years can be found by multiplying 10,000 by the factor found in the Compound Interest Table in the Appendix. $10,000 × 1.283359 = $12,833.59
A *mortgage* is an amount that is borrowed to buy real estate. The *loan origination fee* is usually a percent of the mortgage and is expressed as *points*. [6.4A, p. 259]	The loan origination fee of 3 points paid on a mortgage of $200,000 is 0.03 × $200,000 = $6000.
A *commission* is usually paid to a salesperson and is calculated as a percent of sales. [6.6A, p. 269]	A commission of 5% on sales of $50,000 is 0.05 × $50,000 = $2500.

An employee who receives an *hourly wage* is paid a certain amount for each hour worked. [6.6A, p. 269]	An employee is paid an hourly wage of $15. The employee's wages for working 10 hours are $15 × 10 = $150.
An employee who is paid a *salary* receives payment based on a weekly, biweekly, monthly, or annual time schedule. [6.6A, p. 269]	An employee paid an annual salary of $60,000 is paid $60,000 ÷ 12 = $5000 per month.
Balancing a checkbook is determining whether the checkbook balance is accurate. [6.7B, pp. 274–275]	To balance a checkbook: (1) Put a checkmark in the checkbook register by each check paid by the bank and by each deposit recorded by the bank. (2) Add to the current checkbook balance all checks that have been written but have not yet been paid by the bank and any interest paid on the account. (3) Subtract any charges and any deposits not yet recorded by the bank. This is the checkbook balance. (4) Compare the balance with the bank balance listed on the bank statement. If the two numbers are equal, the bank statement and the checkbook "balance."

Essential Rules and Procedures	**Examples**
To find unit cost, divide the total cost by the number of units. [6.1A, p. 235]	Three paperback books cost $36. The unit cost is $36 ÷ 3 = $12 per book.
To find total cost, multiply the unit cost by the number of units purchased. [6.1C, p. 236]	One melon costs $3. The total cost for 5 melons is $3 × 5 = $15.
Basic Markup Equations [6.2B, p. 240] Selling price − cost = markup Cost + markup = selling price Markup rate × cost = markup	A pair of cross trainers that cost a business $90 has a 50% markup rate. The markup is 0.50 × $90 = $45. The selling price is $90 + $45 = $135.
Basic Discount Equations [6.2D, p. 243] Regular price − sale price = discount Regular price − discount = sale price Discount rate × regular price = discount	A movie video is on sale for 20% off the regular price of $50. The discount is 0.20 × $50 = $10. The sale price is $50 − $10 = $40.
Simple Interest Formula for Annual Interest Rates [6.3A, p. 249]	Principal × annual interest rate × time (in years) = interest The simple interest due on a 2-year
Maturity Value Formula for Simple Interest Loan [6.3A, p. 249]	Principal + interest = maturity value The interest to be paid on a 2-year loan
Monthly Payment on a Simple Interest Loan [6.3A, p. 250] Maturity value ÷ length of the loan in months = monthly payment	The maturity value of a simple interest 8-month loan is $8000. The monthly payment is $8000 ÷ 8 = $1000.

Chapter 6 Review Exercises

1. **Consumerism**   A 20-ounce box of cereal costs $3.90. Find the unit cost.
    19.5¢/ounce   [6.1A]

2. **Car Expenses**   An account executive had car expenses of $1025.58 for insurance, $605.82 for gas, $37.92 for oil, and $188.27 for maintenance during a year in which 11,320 miles were driven. Find the cost per mile for these four items taken as a group. Round to the nearest tenth of a cent.
    16.4¢/mile   [6.5B]

3. **Investments**   An oil stock was bought for $42.375 per share. Six months later, the stock was selling for $55.25 per share. Find the percent increase in the price of the stock for the 6 months. Round to the nearest tenth of a percent.
    30.4%   [6.2A]

4. **Markup**   A sporting goods store uses a markup rate of 40%. What is the markup on a ski suit that costs the store $180?
    $72   [6.2B]

5. **Simple Interest**   A contractor borrowed $100,000 from a credit union for 9 months at an annual interest rate of 4%. What is the simple interest due on the loan?
    $3000   [6.3A]

6. **Compound Interest**   A computer programmer invested $25,000 in a retirement account that pays 6% interest, compounded daily. What is the value of the investment in 10 years? Use the Compound Interest Table in the Appendix. Round to the nearest cent.
    $45,550.75   [6.3C]

7. **Investments**   Last year an oil company had earnings of $4.12 per share. This year the earnings are $4.73 per share. What is the percent increase in earnings per share? Round to the nearest percent.
    15%   [6.2A]

8. **Real Estate**   The monthly mortgage payment for a condominium is $523.67. The owner must pay an annual property tax of $658.32. Find the total monthly payment for the mortgage and property tax.
    $578.53   [6.4B]

9. **Car Expenses**   A used pickup truck is purchased for $24,450. A down payment of 8% is made, and the remaining cost is financed for 4 years at an annual interest rate of 5%. Find the monthly payment. Use the Monthly Payment Table in the Appendix. Round to the nearest cent.
    $518.02   [6.5B]

10. **Compound Interest**   A fast-food restaurant invested $50,000 in an account that pays 7% annual interest compounded quarterly. What is the value of the investment in 1 year? Use the Compound Interest Table in the Appendix.
    $53,593   [6.3C]

11. **Real Estate**   Paula Mason purchased a home for $125,000. The lender requires a down payment of 15%. Find the amount of the down payment.
    $18,750   [6.4A]

12. **Car Expenses**   A plumber bought a truck for $18,500. A state license fee of $315 and a sales tax of 6.25% of the purchase price are required. Find the total cost of the sales tax and the license fee.
    $1471.25   [6.5A]

13. **Markup**   Techno-Center uses a markup rate of 35% on all computer systems. Find the selling price of a computer system that costs the store $1540.
$2079   [6.2B]

14. **Car Expenses**   Mien pays a monthly car payment of $122.78. During a month in which $25.45 is principal, how much of the payment is interest?
$97.33   [6.5B]

15. **Compensation**   The manager of the retail store at a ski resort receives a commission of 3% on all sales at the alpine shop. Find the total commission received during a month in which the shop had $108,000 in sales.
$3240   [6.6A]

16. **Discount**   A suit that regularly costs $235 is on sale for 40% off the regular price. Find the sale price.
$141   [6.2D]

17. **Banking**   Luke had a checking account balance of $1568.45 before writing checks for $123.76, $756.45, and $88.77. He then deposited a check for $344.21. Find Luke's current checkbook balance.
$943.68   [6.7A]

18. **Simple Interest**   Pros' Sporting Goods borrowed $30,000 at an annual interest rate of 8% for 6 months. Find the maturity value of the loan.
$31,200   [6.3A]

19. **Real Estate**   A credit union requires a borrower to pay $2\frac{1}{2}$ points for a loan. Find the origination fee for a $75,000 loan.
$1875   [6.4A]

20. **Consumerism**   Sixteen ounces of mouthwash cost $3.49. A 33-ounce container of the same brand of mouthwash costs $6.99. Which is the better buy?
33 ounces for $6.99   [6.1B]

21. **Real Estate**   The Sweeneys bought a home for $156,000. The family made a 10% down payment and financed the remainder with a 30-year loan at an annual interest rate of 7%. Find the monthly mortgage payment. Use the Monthly Payment Table in the Appendix. Round to the nearest cent.
$934.08   [6.4B]

22. **Compensation**   Richard Valdez receives $12.60 per hour for working 40 hours a week and time and a half for working over 40 hours. Find his total income during a week in which he worked 48 hours.
$655.20   [6.6A]

23. **Banking**   The business checking account of a donut shop showed a balance of $9567.44 before checks of $1023.55, $345.44, and $23.67 were written and checks of $555.89 and $135.91 were deposited. Find the current checkbook balance.
$8866.58   [6.7A]

24. **Simple Interest**   The simple interest due on a 4-month loan of $55,000 is $1375. Find the monthly payment on the loan.
$14,093.75   [6.3A]

25. **Simple Interest**   A credit card company charges a customer 1.25% per month on the unpaid balance of charges on the card. What is the finance charge in a month in which the customer has an unpaid balance of $576?
$7.20   [6.3B]

Chapter 6 Test

1. **Consumerism**   Twenty feet of lumber cost $138.40. What is the cost per foot?
   $6.92   [6.1A]

2. **Consumerism**   Which is the more economical purchase: 3 pounds of tomatoes for $7.49 or 5 pounds of tomatoes for $12.59?
   3 pounds for $7.49   [6.1B]

3. **Consumerism**   Red snapper costs $4.15 per pound. Find the cost of $3\frac{1}{2}$ pounds. Round to the nearest cent.
   $14.53   [6.1C]

4. **Business**   An exercise bicycle increased in price from $415 to $498. Find the percent increase in the cost of the exercise bicycle.
   20%   [6.2A]

5. **Markup**   A department store uses a 40% markup rate. Find the selling price of a compact disc player that the store purchased for $215.
   $301   [6.2B]

6. **Investments**   The price of gold dropped from $390 per ounce to $360 per ounce. What percent decrease does this amount represent? Round to the nearest tenth of a percent.
   7.7%   [6.2C]

7. **Consumerism**   The price of a video camera dropped from $1120 to $896. What percent decrease does this price drop represent?
   20%   [6.2C]

8. **Discount**   A corner hutch with a regular price of $299 is on sale for 30% off the regular price. Find the sale price.
   $209.30   [6.2D]

9. **Discount**   A box of stationery that regularly sells for $9.50 is on sale for $5.70. Find the discount rate.
   40%   [6.2D]

10. **Simple Interest**   A construction company borrowed $75,000 for 4 months at an annual interest rate of 8%. Find the simple interest due on the loan.
    $2000   [6.3A]

11. **Simple Interest**   Craig Allen borrowed $25,000 at an annual interest rate of 9.2% for 9 months. Find the maturity value of the loan.
    $26,725   [6.3A]

12. **Simple Interest**   A credit card company charges a customer 1.2% per month on the unpaid balance of charges on the card. What is the finance charge in a month in which the customer has an unpaid balance of $374.95?
    $4.50   [6.3B]

13. **Compound Interest**   Jorge, who is self-employed, placed $30,000 in an account that pays 6% annual interest compounded quarterly. How much interest was earned in 10 years? Use the Compound Interest Table in the Appendix.
    $24,420.60   [6.3C]

14. **Real Estate**   A savings and loan institution is offering mortgage loans that have a loan origination fee of $2\frac{1}{2}$ points. Find the loan origination fee when a home is purchased with a loan of $134,000.
    $3350   [6.4A]

15. **Real Estate** A new housing development offers homes with a mortgage of $222,000 for 25 years at an annual interest rate of 8%. Find the monthly mortgage payment. Use the Monthly Payment Table in the Appendix.
$1713.44 [6.4B]

16. **Car Expenses** A Chevrolet was purchased for $23,750, and a 20% down payment was made. Find the amount financed.
$19,000 [6.5A]

17. **Car Expenses** A rancher purchased an SUV for $23,714 and made a down payment of 15% of the cost. The balance was financed for 4 years at an annual interest rate of 7%. Find the monthly truck payment. Use the Monthly Payment Table in the Appendix.
$482.68 [6.5B]

18. **Compensation** Shaney receives an hourly wage of $20.40 an hour as an emergency room nurse. When called in at night, she receives time and a half. How much does Shaney earn in a week when she works 30 hours at normal rates and 15 hours during the night?
$1071 [6.6A]

19. **Banking** The business checking account for a pottery store had a balance of $7349.44 before checks for $1349.67 and $344.12 were written. The store manager then made a deposit of $956.60. Find the current checkbook balance.
$6612.25 [6.7A]

20. **Banking** Balance the checkbook shown.

RECORD ALL CHARGES OR CREDITS THAT AFFECT YOUR ACCOUNT

NUMBER	DATE	DESCRIPTION OF TRANSACTION	PAYMENT/DEBIT (−)	√ T	FEE (IF ANY) (−)	DEPOSIT/CREDIT (+)	BALANCE $ 1422 13	
843	8/1	House Payment	$ 713 72		$	$	708	41
	8/4	Deposit				852 60	1561	01
844	8/5	Loan Payment	162 40				1398	61
845	8/6	Groceries	166 44				1232	17
846	8/10	Car Payment	322 37				909	80
	8/15	Deposit				852 60	1762	40
847	8/16	Credit Card	413 45				1348	95
848	8/18	Pharmacy	92 14				1256	81
849	8/22	Utilities	72 30				1184	51
850	8/28	Telephone	78 20				1106	31

CHECKING ACCOUNT Monthly Statement			Account Number: 122-345-1	
Date	Transaction		Amount	Balance
8/1	OPENING BALANCE			1422.13
8/3	CHECK		713.72	708.41
8/4	DEPOSIT		852.60	1561.01
8/8	CHECK		166.44	1394.57
8/8	CHECK		162.40	1232.17
8/15	DEPOSIT		852.60	2084.77
8/23	CHECK		72.30	2012.47
8/24	CHECK		92.14	1920.33
9/1	CLOSING BALANCE			1920.33

The bank statement and the checkbook balance. [6.7B]

Cumulative Review Exercises

1. Simplify: $12 - (10 - 8)^2 \div 2 + 3$
13   [1.6B]

2. Add: $3\frac{1}{3} + 4\frac{1}{8} + 1\frac{1}{12}$
$8\frac{13}{24}$ [2.4C]

3. Find the difference between $12\frac{3}{16}$ and $9\frac{5}{12}$.
$2\frac{37}{48}$ [2.5C]

4. Find the product of $5\frac{5}{8}$ and $1\frac{9}{15}$.
9   [2.6B]

5. Divide: $3\frac{1}{2} \div 1\frac{3}{4}$
2   [2.7B]

6. Simplify $\left(\frac{3}{4}\right)^2 \div \left(\frac{3}{8} - \frac{1}{4}\right) + \frac{1}{2}$.
5   [2.8C]

7. Divide: $0.059\overline{)3.0792}$
Round to the nearest tenth.
52.2   [3.5A]

8. Convert $\frac{17}{12}$ to a decimal. Round to the nearest thousandth.
1.417   [3.6A]

9. Write "$410 in 8 hours" as a unit rate.
$51.25/hour   [4.2B]

10. Solve the proportion $\frac{5}{n} = \frac{16}{35}$.
Round to the nearest hundredth.
10.94   [4.3B]

11. Write $\frac{5}{8}$ as a percent.
62.5%   [5.1B]

12. Find 6.5% of 420.
27.3   [5.2A]

13. Write 18.2% as a decimal.
0.182   [5.1A]

14. What percent of 20 is 8.4?
42%   [5.3A]

15. 30 is 12% of what?
250   [5.4A]

16. 65 is 42% of what? Round to the nearest hundredth.
154.76   [5.4A/5.5A]

17. **Meteorology** A series of late-summer storms produced rainfall of $3\frac{3}{4}$, $8\frac{1}{2}$, and $1\frac{2}{3}$ inches during a 3-week period. Find the total rainfall during the 3 weeks.
$13\frac{11}{12}$ inches [2.4D]

18. **Taxes** The Homer family pays $\frac{1}{5}$ of its total monthly income for taxes. The family has a total monthly income of $4850. Find the amount of their monthly income that the Homers pay in taxes.
$970 [2.6C]

19. **Consumerism** In 5 years, the cost of a scientific calculator went from $75 to $30. What is the ratio of the decrease in price to the original price?
$\frac{3}{5}$ [4.1B]

20. **Fuel Efficiencies** A compact car was driven 417.5 miles on 12.5 gallons of gasoline. Find the number of miles driven per gallon of gasoline.
33.4 miles per gallon [4.2C]

21. **Consumerism** A 14-pound turkey costs $12.96. Find the unit cost. Round to the nearest cent.
$.93 per pound [4.2C]

22. **Investments** Eighty shares of a stock paid a dividend of $112. At the same rate, find the dividend on 200 shares of the stock.
$280 [4.3C]

23. **Discount** A video camera that regularly sells for $900 is on sale for 20% off the regular price. What is the sale price?
$720 [6.2D]

24. **Markup** A department store bought a portable disc player for $85 and used a markup rate of 40%. Find the selling price of the disc player.
$119 [6.2B]

25. **Compensation** Sook Kim, an elementary school teacher, received an increase in salary from $2800 per month to $3024 per month. Find the percent increase in her salary.
8% [6.2A]

26. **Simple Interest** A contractor borrowed $120,000 for 6 months at an annual interest rate of 4.5%. How much simple interest is due on the loan?
$2700 [6.3A]

27. **Car Expenses** A red Ford Mustang was purchased for $26,900, and a down payment of $2000 was made. The balance is financed for 3 years at an annual interest rate of 9%. Find the monthly payment. Use the Monthly Payment Table in the Appendix. Round to the nearest cent.
$791.81 [6.5B]

28. **Banking** A family had a checking account balance of $1846.78. A check of $568.30 was deposited into the account, and checks of $123.98 and $47.33 were written. Find the new checking account balance.
$2243.77 [6.7A]

29. **Car Expenses** During 1 year, Anna Gonzalez spent $840 on gasoline and oil, $520 on insurance, $185 on tires, and $432 on repairs. Find the cost per mile to drive the car 10,000 miles during the year. Round to the nearest cent.
$.20 [6.5B]

30. **Real Estate** A house has a mortgage of $172,000 for 20 years at an annual interest rate of 6%. Find the monthly mortgage payment. Use the Monthly Payment Table in the Appendix. Round to the nearest cent.
$1232.26 [6.4B]

chapter

7 Statistics and Probability

Every 10 years, the U.S. Census Bureau collects data about the country's population. It then issues statistical reports that indicate changes and trends in the U.S. population. For example, according to the 2000 Census, there were approximately 105 men for every 100 women in the category of people under the age of 20. Statistics are found in many different sources, such as the radio, television, newspapers, and the Internet. Statistics can also be about anything, from the habits of mall shoppers like those in the photo above, to unusual biological data like these two statistics from www.geocities.com: (1) 70% of the dust in your home consists of shed human skin and hair. (2) Every human spent about one-half hour as a single cell. **Exercises 9 to 12 on page 309** present data gathered from a poll of mall shoppers. You will be asked to interpret the data in order to determine certain statistics.

OBJECTIVES

Section 7.1

A To read a pictograph
B To read a circle graph

Section 7.2

A To read a bar graph
B To read a broken-line graph

Section 7.3

A To read a histogram
B To read a frequency polygon

Section 7.4

A To find the mean, median, and mode of a distribution
B To draw a box-and-whiskers plot

Section 7.5

A To calculate the probability of simple events

Need help? For online student resources, such as section quizzes, visit this textbook's website at **math.college.hmco.com/students**.

Do these exercises to prepare for Chapter 7.

1. **Mail** Bill-related mail accounted for 49 billion of the 102 billion pieces of first-class mail handled by the U.S. Postal Service during a recent year. (*Source:* US Postal Service) What percent of the pieces of first-class mail handled by the U.S. Postal Service was bill-related mail? Round to the nearest tenth of a percent.
48.0% [5.3B]

2. **Education** The table at the right shows the estimated costs of funding an education at a public college.

   a. Between which two enrollment years is the increase in cost greatest? Between 2009 and 2010

   b. What is the increase between these two years? $5318 [1.3C]

Enrollment Year	Cost of Public College
2005	$70,206
2006	$74,418
2007	$78,883
2008	$83,616
2009	$88,633
2010	$93,951

Source: The College Board's Annual Survey of Colleges

3. **Sports** During the 1924 Summer Olympics in Paris, France, the United States won 45 gold medals, 27 silver medals, and 27 bronze medals. (*Source: The Ultimate Book of Sports Lists*)

   a. Find the ratio of gold medals won by the United States to silver medals won by the United States during the 1924 Summer Olympics. Write the ratio as a fraction in simplest form. $\frac{5}{3}$

   b. Find the ratio of silver medals won by the United States to bronze medals won by the United States during the 1924 Summer Olympics. Write the ratio using a colon.
   1:1 [4.1B]

4. **Television** The table below shows the number of television viewers, in millions, who watch pay-cable channels, such as HBO and Showtime, each night of the week. (*Source:* Neilsen Media Research analyzed by Initiative Media North America)

Mon	Tue	Wed	Thu	Fri	Sat	Sun
3.9	4.5	4.2	3.9	5.2	7.1	5.5

   a. Arrange the numbers in the table from least to greatest. 3.9, 3.9, 4.2, 4.5, 5.2, 5.5, 7.1 [3.6C]

   b. Find the average number of viewers per night. 4.9 million [3.5B]

5. **The Military** Approximately 90,000 women serve in the U.S. military. Five percent of these women serve in the Marine Corps. (*Source:* U.S. Department of Defense)

   a. Approximately how many women are in the Marine Corps? 4500 women [5.2B]

   b. What fractional amount of women in the military are in the Marine Corps? $\frac{1}{20}$ [5.1A]

I have 2 brothers and 1 sister. My father's parents have 10 grandchildren. My mother's parents have 11 grandchildren. If no divorces or remarriages occurred, how many first cousins do I have?
13 first cousins

7.1 Pictographs and Circle Graphs

Objective A   To read a pictograph

Statistics is the branch of mathematics concerned with **data,** or numerical information. **Graphs** are displays that provide a pictorial representation of data. The advantage of graphs is that they present information in a way that is easily read. The disadvantage of graphs is that they can be misleading. (See "Projects and Group Activities" at the end of this chapter.)

Bill Gates

A **pictograph** uses symbols to represent information. The pictograph in Figure 1 represents the net worth of America's richest billionaires. Each symbol represents 10 billion dollars.

Net Worth (in tens of billions of dollars)

Bill Gates
Warren Buffet
Paul Allen
Larry Ellison
Jim C. Walton

Figure 1 Net worth of America's richest billionaires
Source: **www.Forbes.com**

From the pictograph, we can determine that Bill Gates has the greatest net worth. Warren Buffet's net worth is $10 billion more than Paul Allen's net worth.

HOW TO 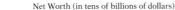 The pictograph in Figure 2 represents the responses of 600 young Americans when asked what they would like to have with them on a desert island. "Books" was the response of what percent of the respondents?

Strategy
Use the basic percent equation. The base is 600 (the total number of responses), and the amount is 90 (the number responding "Books").

Solution

Percent × base = amount

$$n \times 600 = 90$$

$$n = 90 \div 600$$

$$n = 0.15$$

15% of the respondents wanted books on a desert island.

Music
Parents
Computer
Books
TV

= 30 responses

Figure 2 What 600 young Americans want on a desert island
Source: Time Magazine

Objective 7.1A

New Vocabulary
statistics
data
graphs
pictograph

Discuss the Concepts
1. Why must a pictograph display what the symbol used in the graph represents?
2. Provide some examples of data that would display well in a pictograph.

Concept Check
How could we use a pictograph to display the students in this class and their majors?

Optional Student Activity
A good source of pictographs is *USA Today.* As an assignment, and to bring writing into the course, have students select a pictograph from that or another newspaper and write a paragraph that describes the data.

In-Class Examples (Objective 7.1A)
1. According to the data in Figure 2, how many more young people responded "Parents" than responded "Computer"?   60 more people
2. According to the data in Figure 2, what percent of the respondents answered "TV"?   10%

The pictograph in Figure 3 shows the number of new cellular phones purchased in a particular city during a 4-month period.

The ratio of the number of cellular phones purchased in March to the number purchased in January is

$$\frac{3000}{4500} = \frac{2}{3}$$

Figure 3 Monthly cellular phone purchases

Example 1

Use Figure 3 to find the total number of cellular phones purchased during the 4-month period.

Strategy

To find the total number of cellular phones purchased in the 4-month period:

• Read the pictograph to determine the number of cellular phones purchased each month.
• Add the four numbers.

Solution

Purchases for January: 4500
Purchases for February: 3500
Purchases for March: 3000
Purchases for April: 1500

Total purchases for the 4-month period:

```
   4,500
   3,500
   3,000
+  1,500
  12,500
```

There were 12,500 cellular phones purchased in the 4-month period.

You Try It 1

According to Figure 3, the number of cellular phones purchased in March represents what percent of the total number of cellular phones purchased in that city during that 4-month period?

Your strategy

Your solution
24%

Solution on p. S19

Objective B   **To read a circle graph**

Objective 7.1B

TAKE NOTE
One quadrillion is
1,000,000,000,000,000.

A **circle graph** represents data by the size of the sectors. A **sector of a circle** is one of the "pieces of the pie" into which a circle graph is divided.

The circle graph in Figure 4 shows the consumption of energy sources in the United States during a recent year. The complete circle graph represents the total amount of energy consumed, 96.6 quadrillion Btu. Each sector of the circle represents the consumption of energy from a different source.

Point of Interest

Fossil fuels include coal, natural gas, and petroleum. Renewable energy includes hydroelectric power, solar energy, wood burning, and wind energy.

To find the percent of the total energy consumed that originated from nuclear power, solve the basic percent equation for percent. The base is 96.6 quadrillion Btu, and the amount is 8.2 quadrillion Btu.

Percent × base = amount

$$n \times 96.6 = 8.2$$
$$n = 8.2 \div 96.6$$
$$n \approx 0.085$$

To the nearest tenth of a percent, 8.5% of the energy consumed originated from nuclear power.

Figure 4 Annual energy consumption in quadrillion Btu in the United States
Source: The World Almanac and Book of Facts 2003

In a recent year, the top 25 companies in the United States spent a total of $17.8 billion for national advertising. The circle graph in Figure 5 shows what percents of the $17.8 billion went to the various advertising media. The complete circle represents 100% of all the money spent by these companies. Each sector of the graph represents the percent of the total spent for a particular medium.

HOW TO   According to Figure 5, how much money was spent for magazine advertising? Round to the nearest hundred million dollars.

Strategy
Use the basic percent equation. The base is $17.8 billion, and the percent is 16%.

Solution

Percent × base = amount

$$0.16 \times 17.8 = n$$
$$2.848 = n$$

2.848 billion = 2,848,000,000

To the nearest hundred million, the amount spent for magazine advertising was $2,800,000,000.

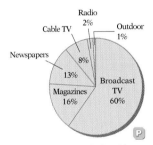

Figure 5 Distribution of advertising dollars for 25 companies
Source: Interep research

New Vocabulary
circle graph
sector of a circle

Discuss the Concepts
1. Can the data in Figure 2 on page 295 be displayed in a circle graph?
2. Can the data in Figure 1 on page 295 be displayed in a circle graph?

Concept Check
1. Why would the population of New York City in the years 1850, 1900, 1950, and 2000 be inappropriate data to represent in a circle graph?
2. Some circle graphs show data values and some show percents. (See Figures 4 and 5.) Can we create a circle graph showing percents if we are given one showing the data values? Can we create a circle graph showing data values if we are given one showing the percents?

In-Class Examples (Objective 7.1B)

1. According to the data in Figure 5, how many times greater was the amount spent on broadcast TV advertising than the amount spent on radio advertising?   30 times greater

2. According to the data in Figure 5, how much more money was spent on newspaper advertising than on cable TV advertising? $890 million

Optional Student Activity

Have small groups of students prepare three questions and answers based on the data in the circle graph shown below. Have the groups exchange questions and, after answering the other groups' questions, compare their answers with the other groups' answers.

Responses of 1000 men to the question "What type of footwear do you wear at home?"

Source: Based on data from L. B. Evans

The circle graph in Figure 6 shows typical annual expenses of owning, operating, and financing a new car. Use this figure for Example 2 and You Try It 2.

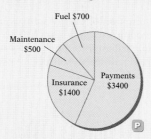

Figure 6 Annual expenses of $6000 for owning, operating, and financing a car
Source: Based on data from IntelliChoice

Example 2

Use Figure 6 to find the ratio of the annual insurance expense to the total annual cost of the car.

Strategy
To find the ratio:

- Locate the annual insurance expense in the circle graph.
- Write in simplest form the ratio of the annual insurance expense to the total annual cost of operating the car.

Solution
Annual insurance expense: $1400

$$\frac{1400}{6000} = \frac{7}{30}$$

The ratio is $\frac{7}{30}$.

Example 3

Use Figure 7 to find the employee's take-home pay.

Strategy
To find the take-home pay:

- Locate the percent of the distribution that is take-home pay.
- Solve the basic percent equation for amount.

Solution
Take-home pay: 54%

Percent × base = amount
0.54   × 2900 = n
             1566 = n

The employee's take-home pay is $1566.

The circle graph in Figure 7 shows the distribution of an employee's gross monthly income. Use this figure for Example 3 and You Try It 3.

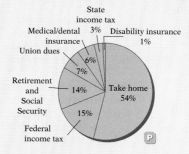

Figure 7 Distribution of gross monthly income of $2900

You Try It 2

Use Figure 6 to find the ratio of the annual cost of fuel to the annual cost of maintenance.

Your strategy

Your solution
$\frac{7}{5}$

You Try It 3

Use Figure 7 to find the amount paid for medical/dental insurance.

Your strategy

Your solution
$174

Solutions on pp. S19–S20

7.1 Exercises

Objective A   **To read a pictograph**

The Film Industry   The pictograph in Figure 8 shows the approximate gross revenues in the United States from four Walt Disney animated movies. Use this graph for Exercises 1 to 3.

1. Find the total gross revenues from the four movies.
$750 million

2. Find the ratio of the gross revenue of *Beauty and the Beast* to the gross revenue of *The Hunchback of Notre Dame*. $\frac{3}{2}$

3. Find the percent of the total gross revenue that was earned by *The Lion King*.
40%

Beauty and the Beast
The Lion King
Aladdin
The Hunchback of Notre Dame

= $50 million

Figure 8 Gross revenues of four Walt Disney animated movies
Source: **www.worldboxoffice.com**

Space Exploration   The pictograph in Figure 9 is based on a survey of adults who were asked whether they agreed with each statement. Use this graph for Exercises 4 to 6.

4. Find the ratio of the number of people who agreed that space exploration impacts daily life to the number of people who agreed that space will be colonized in their lifetime. $\frac{3}{1}$

5. How many more people agreed that humanity should explore planets than agreed that space exploration impacts daily life?
50 people

6. Is the number of people who said they would travel in space more than twice the number of people who agreed that space would be colonized in their lifetime?
No

Humanity should explore planets

Space exploration impacts daily life

Given a chance, I'd travel in space

Space will be colonized in my lifetime

= 100 people

Figure 9 Number of adults who agree with the statement
Source: Opinion Research for Space Day Partners

Children's Behavior   The pictograph in Figure 10 is based on a survey of children aged 7 through 12. The percent of children's responses to the survey are shown. Assume that 500 children were surveyed. Use this graph for Exercises 7 to 9.

7. Find the number of children who said they hid vegetables under a napkin.
150 children

8. What is the difference between the number of children who fed vegetables to the dog and the number who dropped them on the floor?
75 children

9. Were the responses given in the graph the only responses given by the children? Explain your answer.

Hide vegetables under napkin

Feed them to the dog

Hide vegetables under something else on the plate

Drop vegetables on the floor

= 10%

Figure 10 How children try to hide vegetables
Source: Strategic Consulting and Research for Del Monte

Section 7.1

Suggested Assignment
Exercises 1–35, odds

Answers to Writing Exercises

9. No, the sum of the percents given in the graph is only 80%, not 100%.

Quick Quiz (Objective 7.1A)

1. According to the data in Figure 8, what is the difference between the gross revenue of *The Lion King* and the gross revenue of *Beauty and the Beast*? $150 million

2. According to the data in Figure 8, how many times greater was the gross revenue of *The Lion King* than the gross revenue of *The Hunchback of Notre Dame*?   3 times greater

> **Objective B**   **To read a circle graph**

Education   An accounting major recorded the number of units required in each discipline to graduate with a degree in accounting. The results are shown in the circle graph in Figure 11. Use this graph for Exercises 10 to 13.

10. How many units are required to graduate with a degree in accounting?
    128 units

11. What is the ratio of the number of units in finance to the number of units in accounting?
    $\frac{1}{3}$

12. What percent of the units required to graduate are taken in accounting? Round to the nearest tenth of a percent.
    35.2%

13. What percent of the units required to graduate are taken in mathematics? Round to the nearest tenth of a percent.
    9.4%

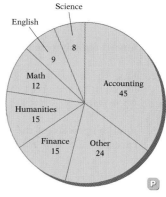

Figure 11 Number of units required to graduate with an accounting degree

Theaters   The circle graph in Figure 12 shows the result of a survey in which people were asked, "What bothers you most about movie theaters?" Use this graph for Exercises 14 to 17.

14. **a.** What complaint was mentioned the most often?   People talking
    b. What complaint was mentioned the least often?
    Uncomfortable seats

15. How many people were surveyed?
    150 people

16. What is the ratio of the number of people responding "Dirty floors" to the number responding "High ticket prices"?
    $\frac{9}{11}$

17. What percent of the respondents said that people talking bothered them most?
    28%

Figure 12 Distribution of responses in a survey

Quick Quiz (Objective 7.1B)

1. According to the data in Figure 11, how many more units are required in accounting than in English?   36 units

2. According to the data in Figure 13, how much more money was spent on portable game machines than on accessories? $31,000,000

Video Games The circle graph in Figure 13 shows the break-down of the approximately $3,100,000,000 that Americans spent on home video game equipment in one year. Use this graph for Exercises 18 to 21.

18. Find the amount of money spent on TV game machines.
    $1,085,000,000

19. Find the amount of money spent on portable game machines.
    $279,000,000

20. What fractional amount of the total money spent was spent on accessories?
    $\frac{2}{25}$

21. Is the amount spent for TV game machines more than three times the amount spent for portable game machines?
    Yes

Figure 13 Percents of $3,100,000,000 spent annually on home video games

Source: The NPD Group, Toy Manufacturers of America

Demographics The circle graph in Figure 14 shows a break-down, according to age, of the homeless in America. Use this graph for Exercises 22 to 25.

22. What age group represents the largest segment of the homeless population?
    Ages 35 to 44

23. Is the number of homeless who are aged 25 to 34 more or less than twice the number who are under the age of 25?
    More than twice

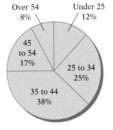

Figure 14 Ages of the homeless in America

Source: The Department of Housing and Urban Development

24. What percent of the homeless population is under the age of 35?
    37%

25. On average, how many of every 100,000 homeless people in America are over the age of 54?
    8000 people

Answers to Writing Exercises

34a. You might have students supply examples of pictographs with their explanations. Students can use the reference section at the library to find copies of *USA Today* or some other source for pictographs. In that case, it will be necessary for them to photocopy any examples they submit to you with their explanations.

Students may state that pictographs are more dramatic or visually appealing than some other type of graph. Presenting data in the form of a pictograph gives an overall picture of the relationship of each category to the whole and illustrates the relationship between categories.

b. A disadvantage is that the pictograph usually sacrifices some degree of accuracy.

35. Answers will vary. For example:
The couple's largest single expense was rent.
Food represents approximately one-quarter of the month's expenditures.
Rent represents approximately one-third of the month's expenditures.
The expense for food is approximately the same as the expense for entertainment and transportation.
The couple spent more for transportation than for entertainment.

Geography The circle graph in Figure 15 shows the land area of each of the seven continents in square miles. Use this graph for Exercises 26 to 29.

26. Find the total land area of the seven continents.
57,240,000 square miles

27. How much larger is North America than South America?
2,550,000 square miles

28. What percent of the total land area is the land area of Asia? Round to the nearest tenth of a percent.
30.0%

29. What percent of the total land area is the land area of Australia? Round to the nearest tenth of a percent.
5.2%

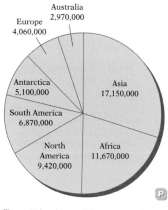

Figure 15 Land area of the seven continents (in square miles)

Cost of Living A typical household in the United States has an average after-tax income of $40,550. The circle graph in Figure 16 represents how this annual income is spent. (*Note:* It is because of rounding that the percents do not add up to 100%.) Use this graph for Exercises 30 to 33.

30. What amount is spent on food?
$5677

31. What amount is spent on health care?
$2027.50

32. How much more is spent on clothing than on entertainment?
$405.50

33. Is the amount spent on housing more than twice the amount spent on transportation?
No

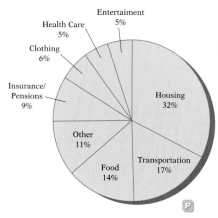

Figure 16 Average annual expenses in a U.S. household
Source: American Demographics

APPLYING THE CONCEPTS

34. a. What are the advantages of presenting data in the form of a pictograph?
b. What are the disadvantages?

35. The circle graph at the right shows a couple's expenditures last month. Write two observations about this couple's expenses.

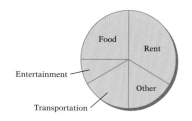

7.2 Bar Graphs and Broken-Line Graphs

Objective A To read a bar graph

New Vocabulary
bar graph
double-bar graph

Point of Interest

The first bar graph appeared in 1786 in the book *The Commercial and Political Atlas*. The author, William Playfair (1759–1823), was a pioneer in the use of graphical displays.

A **bar graph** represents data by the height of the bars. The bar graph in Figure 17 shows temperature data recorded for Cincinnati, Ohio, for the months March through November. For each month, the height of the bar indicates the average daily high temperature during that month. The jagged line near the bottom of the graph indicates that the vertical scale is missing the numbers between 0 and 50.

The daily high temperature in September was 78°F. Because the bar for July is the tallest, the daily high temperature was highest in July.

Figure 17 Daily high temperatures in Cincinnati, Ohio
Source: U.S. Weather Bureau

Instructor Note

It is important that students note the jagged portion of the vertical axis in Figure 17. One of the ways to distort the impact of a graph is to choose the vertical axis to show data in its best (or worst) light. A project at the end of this chapter discusses this problem.

You may want to discuss the characteristics of a well-prepared bar graph. One axis labels categories or variables, and the other is a numerical scale. The categories being described must be identified. The numerical scale must be labeled and presented in equal intervals. Bars should be of equal width.

TAKE NOTE

The bar for athletic females is halfway between the marks for 50 and 60. Therefore, we estimate that the lung capacity is halfway between these two numbers, at 55.

A **double-bar graph** is used to display data for purposes of comparison. The double-bar graph in Figure 18 shows the lung capacity of inactive, versus that of athletic, 45-year-olds.

The lung capacity of an athletic female is 55 milliliters of oxygen per kilogram of body weight per minute.

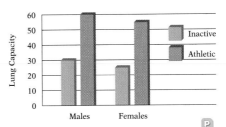

Figure 18 Lung capacity (in milliliters of oxygen per kilogram of body weight per minute)

Discuss the Concepts

What characteristics of a bar graph make it easy to compare the data displayed in one?

Concept Check

Can the numerical scale in a bar graph be in percents? Could the data in Figure 17 be recreated to show percents along the vertical axis?

Example 1

What is the ratio of the lung capacity of an inactive male to that of an athletic male?

Strategy
To write the ratio:
• Read the graph to find the lung capacity of an inactive male and of an athletic male.
• Write the ratio in simplest form.

Solution
Lung capacity of inactive male: 30
Lung capacity of athletic male: 60

$$\frac{30}{60} = \frac{1}{2}$$

The ratio is $\frac{1}{2}$.

You Try It 1

What is the ratio of the lung capacity of an inactive female to that of an athletic female?

Your strategy

Your solution
$\frac{5}{11}$

Solution on p. S20

In-Class Examples (Objective 7.2A)

1. According to the data in Figure 17, during which month is the normal daily high temperature in Cincinnati 54°F? During which month is it lowest?
   November; March

2. According to the data in Figure 17, what is the normal daily high temperature in April in Cincinnati?   64°F

Objective 7.2B

New Vocabulary
broken-line graph

Instructor Note
William Playfair, who is mentioned in the Point of Interest on page 303, was the first to display changes in a variable over time. These data, which are called time-series data, are often best represented in a line graph.

Because time-series data involve two variables, they are displayed in a coordinate grid. If the points plotted are not connected, the graph is called a scatter diagram. If the points plotted are connected with line segments, the graph is called a line graph.

Concept Check
In Objective 7.1B, we asked, "Why would the population of New York City in the years 1850, 1900, 1950, and 2000 be inappropriate data to represent in a circle graph?" Why would such data be appropriately represented in a line graph? (*Note:* You might discuss with your students the preference for presenting this data in a line graph over a bar graph by describing the population as continuous rather than discrete.)

Objective B   **To read a broken-line graph**

A **broken-line graph** represents data by the position of the lines. It is used to show trends.

The broken-line graph in Figure 19 shows the effect of inflation on the value of a $100,000 life insurance policy. The height of each dot indicates the value of the policy.

After 10 years, the purchasing power of the $100,000 has decreased to approximately $60,000.

Figure 19 Effect of inflation on the value of a $100,000 life insurance policy

Two broken-line graphs are often shown in the same figure for comparison. Figure 20 shows the net incomes of two software companies, Math Associates and MatheMentors, before their merger.

Several things can be determined from the graph:

The net income for Math Associates in 2004 was $12 million.

The net income for MatheMentors declined from 2000 to 2001.

The net income for Math Associates increased for each year shown.

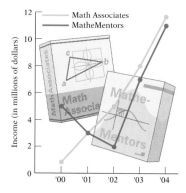

Figure 20 Net incomes of Math Associates and MatheMentors

Example 2

Use Figure 20 to approximate the difference between the net income of Math Associates and that of MatheMentors in 2002.

Strategy
To write the difference:
- Read the line graph to determine the net income of Math Associates and that of MatheMentors in 2002.
- Subtract to find the difference.

Solution
Net income for Math Associates: $5 million
Net income for MatheMentors: $2 million

$5 - 2 = 3$

The difference between the net incomes in 2002 was $3 million.

You Try It 2

Use Figure 20 to determine between which two years the net income of Math Associates increased the most.

Your strategy

Your solution
2003 and 2004

Solution on p. S20

In-Class Examples (Objective 7.2B)

1. According to the data in Figure 20, during which year was the net income of Mathementors lowest?   2002

2. According to the data in Figure 20, what was the difference between the net income of Mathementors and that of Math Associates in 2000? $4 million

7.2 Exercises

Objective A   **To read a bar graph**

Automobile Production   The bar graph in Figure 21 shows the regions in which all the passenger cars were produced during a recent year. Use this graph for Exercises 1 to 3.

1. How many passenger cars were produced worldwide?
39 million passenger cars

2. What is the difference between the number of passenger cars produced in Western Europe and the number produced in North America?
12 million passenger cars

3. What percent of the passenger cars were produced in Asia? Round to the nearest percent.
28%

Figure 21 Number of passenger cars produced (in a recent year)
Source: Copyright © 2000 by the *Los Angeles Times.* Reprinted with permission.

Fuel Efficiency   The double-bar graph in Figure 22 shows the fuel efficiency of four vehicles, as rated by the Environmental Protection Agency. They are among the most fuel-efficient 2003 model-year cars for city and highway mileage. Use this graph for Exercises 4 to 6.

4. Is the fuel efficiency of the Toyota Prius greater on the highway or in city driving?
In city driving

5. Approximately how many more miles per gallon does the Mini Cooper get while traveling on the highway than in the city?
Approximately 10 miles per gallon

6. Estimate the difference between the fuel efficiency of the Honda Insight in city driving and the fuel efficiency of the VW New Beetle in city driving.
Approximately 20 miles per gallon

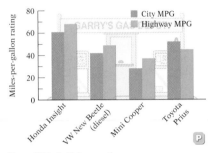

Figure 22 Fuel efficiency ratings
Source: Environmental Protection Agency

Compensation   The double-bar graph in Figure 23 shows maximum salaries for police officers in selected cities and the corresponding maximum salaries for officers in the suburbs of that city. Use this graph for Exercises 7 to 10.

7. Estimate the difference between the maximum salaries of police officers in the suburbs of New York City and in the city of New York.
$16,000

8. Is there a city for which the maximum salary of a police officer in the city is greater than the salary in the suburbs?
No

9. For which city is the difference between the maximum salary in the suburbs and that in the city the greatest?
Philadelphia

10. Of the cities shown on the graph, which city has the lowest maximum salary for police officers in the suburbs?
Detroit

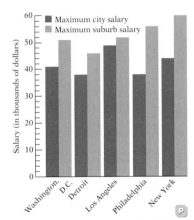

Figure 23 Maximum salaries of police officers in the city and the suburbs
Source: USA Today

Quick Quiz (Objective 7.2A)

1. According to the data in Figure 21, what is the ratio of the number of cars produced in Latin America to the number produced in North America? Write the answer using a colon.   2 : 3

2. Of the cities shown in Figure 23, which city has the highest maximum salary for police officers in the suburbs?   New York

Objective B **To read a broken-line graph**

Meteorology The broken-line graph in Figure 24 shows the average monthly snowfall during ski season around Aspen, Colorado. Use this graph for Exercises 11 to 14.

11. What is the average snowfall during January?
20 inches

12. During which month is the snowfall the greatest?
January

13. What is the total average snowfall during March and April?
25 inches

14. Find the ratio of the average snowfall in November to the average snowfall in December.
$\frac{2}{3}$

Figure 24 Average snowfall in Aspen, Colorado

Source: Weather America, by Alfred Garwood

Health The double-broken-line graph in Figure 25 shows the number of Calories per day that should be consumed by women and men in various age groups. Use this graph for Exercises 15 to 17.

15. What is the difference between the number of Calories recommended for men and the number recommended for women 19–22 years of age?
800 Calories

16. People of what age and gender have the lowest recommended number of Calories?
75+ women

17. Find the ratio of the number of Calories recommended for women 15 to 18 years old to the number recommended for women 51 to 74 years old.
$\frac{7}{6}$

Figure 25 Recommended number of Calories per day for women and men

Source: Numbers, by Andrea Sutcliffe (HarperCollins)

APPLYING THE CONCEPTS

Drug Prevention The graph in Figure 26 shows the amount of money, in billions of dollars, spent by the U.S. government for drug prevention in the 1990s. Use this graph for Exercises 18 and 19.

18. Create a table that shows the total amount spent for foreign and domestic aid for each year from 1991 to 1999.
See the table at the left.

19. Create a table that shows the difference between the amount spent for foreign and for domestic aid in each year from 1991 to 1999.
See the table at the left.

Figure 26 Aid provided by the U.S. government for drug prevention

Source: Reprinted by permission of the *San Diego Union-Tribune.*

18.

Year	Total
1991	$2.9 billion
1992	$2.7 billion
1993	$2.4 billion
1994	$2.1 billion
1995	$1.8 billion
1996	$1.8 billion
1997	$2.5 billion
1998	$2.5 billion
1999	$3.6 billion

19.

Year	Difference
1991	$1.3 billion
1992	$1.1 billion
1993	$1.0 billion
1994	$0.9 billion
1995	$0.8 billion
1996	$0.8 billion
1997	$1.1 billion
1998	$0.9 billion
1999	$1.0 billion

Quick Quiz (Objective 7.2B)

1. According to the data in Figure 24, what is the difference between the average monthly snowfall in November and the average monthly snowfall in October?   10 inches

2. According to the data in Figure 25, what is the difference between the number of Calories recommended for males and the number recommended for females aged 11 to 14?
500 Calories

7.3 Histograms and Frequency Polygons

Objective A   To read a histogram

A research group measured the fuel usage of 92 cars. The results are recorded in the histogram in Figure 27. A **histogram** is a special type of bar graph. The width of each bar corresponds to a range of numbers called a **class interval.** The height of each bar corresponds to the number of occurrences of data in each class interval and is called the **class frequency.**

Class Intervals (miles per gallon)	Class Frequencies (number of cars)
18–20	12
20–22	19
22–24	24
24–26	17
26–28	15
28–30	5

Figure 27

Twenty-four cars get between 22 and 24 miles per gallon.

A precision tool company has 85 employees. Their hourly wages are recorded in the histogram in Figure 28.

The ratio of the number of employees whose hourly wage is between $14 and $16 to the total number of employees is $\frac{17 \text{ employees}}{85 \text{ employees}} = \frac{1}{5}$.

Figure 28

Example 1

Use Figure 28 to find the number of employees whose hourly wage is between $16 and $20.

Strategy
To find the number of employees:

• Read the histogram to find the number of employees whose hourly wage is between $16 and $18 and the number whose hourly wage is between $18 and $20.
• Add the two numbers.

Solution
Number with wages between $16 and $18: 20
Number with wages between $18 and $20: 14

20 + 14 = 34

34 employees have an hourly wage between $16 and $20.

You Try It 1

Use Figure 28 to find the number of employees whose hourly wage is between $10 and $14.

Your strategy

Your solution
22 employees

Solution on p. S20

Objective 7.3B

New Vocabulary
frequency polygon
class midpoint

Discuss the Concepts
1. What does the height of a dot in a frequency polygon represent?
2. How is the class midpoint in a frequency polygon determined?

Optional Student Activity
1. From the frequency polygon in Figure 30, create a histogram.
2. From the histogram in Figure 28 on page 307, create a frequency polygon.

Objective B **To read a frequency polygon**

The speeds of 70 cars on a highway were measured by radar. The results are recorded in the frequency polygon in Figure 29. A **frequency polygon** is a graph that displays information in a manner similar to a histogram. A dot is placed above the center of each class interval at a height corresponding to that class's frequency. The dots are then connected to form a broken-line graph. The center of a class interval is called the **class midpoint.**

> **TAKE NOTE**
> The blue portion of the graph at the right is a histogram. The red portion of the graph is a frequency polygon.

Class Interval (miles per hour)	Class Midpoint	Class Frequency
30–40	35	7
40–50	45	13
50–60	55	25
60–70	65	21
70–80	75	4

Twenty-five cars were traveling between 50 and 60 miles per hour.

The per capita incomes in a recent year for the 50 states are recorded in the frequency polygon in Figure 30.

The number of states with a per capita income between $24,000 and $28,000 is 14.

Figure 29

Figure 30
Source: Bureau of Economic Analysis

Example 2

According to Figure 30, what percent of the states have a per capita income between $24,000 and $28,000?

Strategy
To find the percent, solve the basic percent equation for percent. The base is 50. The amount is 14.

Solution
Percent × base = amount
n × 50 = 14
$n = 14 \div 50$
$n = 0.28$

28% of the states have a per capita income between $24,000 and $28,000.

You Try It 2

Use Figure 30 to find the ratio of the number of states with a per capita income between $28,000 and $32,000 to the number with a per capita income between $32,000 and $36,000.

Your strategy

Your solution
$\frac{2}{1}$

Solution on p. S20

In-Class Examples (Objective 7.3B)
1. According to the data in Figure 30, how many states have a per capita income between $0 and $24,000? 4 states
2a. According to the data in Figure 30, how many states have a per capita income of $36,000 or more? 5 states
 b. What percent of the states is this? 10%

7.3 Exercises

Objective A   **To read a histogram**

🔵 **Education**   The annual tuition for undergraduate college students attending 4-year institutions varies depending on the college. The histogram in Figure 31 shows the tuition amounts for a representative sample of 120 students from various parts of the United States. Use this figure for Exercises 1 to 4.

1. How many students have a tuition that is between $3000 and $6000 per year?
   44 students

2. What is the ratio of the number of students whose tuition is between $9000 per year and $12,000 per year to the total number of students represented? $\frac{1}{10}$

3. How many students pay more than $12,000 annually for tuition?
   18 students

4. What percent of the total number of students spend less than $6000 annually?
   70%

🔵 **Automobiles**   The histogram in Figure 32 is based on data from the American Automobile Manufacturers Association. It shows the ages of a sample of 1000 cars in a typical city in the United States. Use this figure for Exercises 5 to 8.

5. How many cars are between 6 and 12 years old?
   410 cars

6. Find the ratio of the number of cars between 12 and 15 years old to the total number of cars. $\frac{9}{100}$

7. Find the number of cars more than 12 years old.
   230 cars

8. Find the percent of cars that are less than 9 years old.
   58%

🔵 **Malls**   According to a Maritz AmeriPoll, the average U.S. adult goes to a shopping mall about two times a month. The histogram in Figure 33 shows the average time 100 adults spend in the mall per trip. Use this figure for Exercises 9 to 12.

9. Find the number of adults who spend between 1 and 2 hours at the mall.
   54 adults

10. Find the number of adults who spend between 3 and 4 hours at the mall.
    18 adults

11. What percent of the adults spend less than 1 hour at the mall?
    22%

12. What percent of the adults spend 5 or more hours at the mall?
    6%

Figure 31
Source: Educational Testing Service

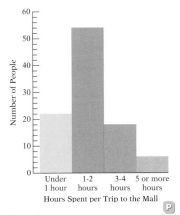

Figure 32
Source: American Automobile Manufacturers Association

Figure 33
Source: Maritz AmeriPoll

Suggested Assignment
Exercises 1–23, odds
More challenging problem:
   Exercise 25

Quick Quiz (Objective 7.3A)

1. According to the data in Figure 33, which class has the lowest frequency?   5 or more hours

2. According to the data in Figure 33, how many people spend 3 or more hours at the mall?
   24 people

Objective B   **To read a frequency polygon**

● **Sports**   The frequency polygon in Figure 34 shows the distances thrown by the entrants in the University and College Discus Finals at the 2003 Drake Relays. Use this figure for Exercises 13 to 15.

13. Determine the number of entrants in the discus finals.
    24 entrants

14. Find the number of entrants with distances of more than 170 feet.
    7 entrants

15. What percent of the entrants had distances between 160 feet and 170 feet?
    37.5%

Figure 34

● **The Lottery**   The frequency polygon in Figure 35 is based on data from a Gallup poll survey of 74 people who purchased lottery tickets. Use this figure for Exercises 16 to 19.

16. How many people purchased between 0 and 10 tickets?
    44 people

17. What percent of the people purchased between 20 and 30 tickets each month? Round to the nearest tenth of a percent.
    10.8%

18. What percent of the people purchased more than 10 tickets each month? Round to the nearest tenth of a percent.
    40.5%

19. ✎ Is it possible to determine from the graph how many people purchased 15 lottery tickets? Explain.

Figure 35

● **Education**   The frequency polygon in Figure 36 shows the distribution of scores of the approximately 1,080,000 students who took an SAT exam. Use this figure for Exercises 20 to 23.

20. How many students scored between 1200 and 1400 on the exam?
    170,000 students

21. What percent of the number of students who took the exam scored between 800 and 1000? Round to the nearest tenth of a percent.
    32.4%

22. How many students scored below 1000?
    530,000 students

23. How many students scored above 800?
    900,000 students

APPLYING THE CONCEPTS

24. ✎ Write a paragraph explaining the difference between a histogram and a bar graph.

25. ✎ In your own words, describe a frequency table.

Answers to Writing Exercises

19. No, a frequency polygon shows only the number of occurrences in a class. It does not show the number of occurrences for a particular value.

24. A bar graph represents categorical data, whereas a histogram breaks the range of values of a set of data into intervals of equal width and displays the number of values that fall into each interval.

25. A frequency table is another method of organizing data. In a frequency table, or a frequency distribution, data are combined into categories called classes, and the frequency, which is the number of pieces of data in each class, is shown.

Figure 36

Source: Educational Testing Service

Quick Quiz (Objective 7.3B)

1. According to the data in Figure 35, what is the ratio of the number of people who purchased between 30 and 40 tickets each month to the number who purchased between 20 and 30 tickets each month? Write the answer using a colon.
   3:4

2. According to the data in Figure 35, how many times more people purchased between 10 and 20 tickets each month than purchased between 20 and 30 tickets each month?   2 times more

7.4 Statistical Measures

Objective A   **To find the mean, median, and mode of a distribution**

The average score on the math portion of the SAT was 432. The EPA estimates that a 2005 Ford Focus averages 35 miles per gallon on the highway. The average rainfall for portions of Kauai is 350 inches per year. Each of these statements uses one number to describe an entire collection of numbers. Such a number is called an *average*.

In statistics there are various ways to calculate an average. Three of the most common—*mean, median,* and *mode*—are discussed here.

An automotive engineer tests the miles-per-gallon ratings of 15 cars and records the results as follows:

Miles-per-Gallon Ratings of 15 Cars														
25	22	21	27	25	35	29	31	25	26	21	39	34	32	28

The **mean** of the data is the sum of the measurements divided by the number of measurements. The symbol for the mean is $\bar{x}$.

> **Formula for the Mean**
>
> $$\bar{x} = \frac{\text{sum of the data values}}{\text{number of data values}}$$

To find the mean for the data above, add the numbers and then divide by 15.

$$\bar{x} = \frac{25 + 22 + 21 + 27 + 25 + 35 + 29 + 31 + 25 + 26 + 21 + 39 + 34 + 32 + 28}{15}$$

$$= \frac{420}{15} = 28$$

The mean number of miles per gallon for the 15 cars tested was 28 miles per gallon.

The mean is one of the most frequently computed averages. It is the one that is commonly used to calculate a student's performance in a class.

Integrating Technology

When using a calculator to calculate the mean, use parentheses to group the sum in the numerator.

(78 + 82 + 91 + 87 + 93) ÷ 5 =

HOW TO   The test scores for a student taking American history were 78, 82, 91, 87, and 93. What was the mean score for this student?

Strategy
To find the mean, divide the sum of the test scores by 5, the number of scores.

Solution
$$\bar{x} = \frac{78 + 82 + 91 + 87 + 93}{5} = \frac{431}{5} = 86.2$$

The mean score for the history student was 86.2.

Objective 7.4A

New Vocabulary
mean
median
mode

New Formulas
$$\bar{x} = \frac{\text{sum of the data values}}{\text{number of data values}}$$

Instructor Note
You might explain to your students that the mean is an appropriate measure when all the data values are relatively close. However, when the range of values is large compared to the values themselves, the mean may give an unrealistic picture of the data.

Discuss the Concepts
You may find it beneficial to have students work on Exercise 1 on page 317 in small groups and then discuss their answers as a class.

Concept Check
1. In a typical college classroom, do you expect the mean age or the median age to be larger?
   The mean

2. Provide a reason for the Bureau of Labor Statistics to cite median per capita income rather than mean per capita income.
   Only a few very large salaries would increase the mean to the point of distorting what the "average" income is.

(Continued on the next page)

In-Class Examples (Objective 7.4A)

1. A truck driver's records show the number of gallons of diesel fuel purchased each day of a 5-day trip.

Wed	Thu	Fri	Sat	Sun
74	86	93	79	88

   a. Find the mean number of gallons purchased.
      84 gallons

   b. Find the median numbers of gallons purchased.
      86 gallons

2. The ages of the six children in a small day-care center are 3, 4, 4, 5, 4, and 2 years. What is the mode of these data?   4 years

(*Continued*)

3. Suppose that a real estate agent is interested in the cost of homes in the town of Wilmington. The realtor goes to city hall and learns that the selling prices of the last 10 homes sold in the community were

$ 100,000
$ 100,000
$ 100,000
$ 100,000
$ 100,000
$ 100,000
$ 100,000
$ 100,000
$ 100,000
$1,100,000

Would a couple interested in buying a home in Wilmington be more interested in the mean or the median of these data? Why?
The mean is $200,000. The median is $100,000. The couple is more interested in the median because it is more representative of the data. The couple should expect to pay $100,000 for a home in Wilmington. The $1,100,000 selling price was not typical of the prices of homes in that town, and it raised the mean by $100,000.

Optional Student Activity

1. On five tests, on which scores could range from 0 to 100, you had an average of exactly 88. Find the lowest score you could have received on one of these tests.   40

2. If 51 is the mean of 50 consecutive even numbers, what is the largest of the numbers?   100

Study Tip

Word problems are difficult because we must read the problem, determine the quantity we must find, think of a method to find it, actually solve the problem, and then check the answer. In short, we must devise a *strategy* and then use that strategy to find the *solution*. See *AIM for Success* at the front of the book.

The **median** of the data is the number that separates the data into two equal parts when the numbers are arranged from smallest to largest (or from largest to smallest). There is an equal number of values above the median and below the median.

To find the median of a set of numbers, first arrange the numbers from smallest to largest. The median is the number in the middle.

The result of arranging the miles-per-gallon ratings given on the previous page from smallest to largest is shown below.

21 21 22 25 25 25 26   27   28 29 31 32 34 35 39

7 values below the median   Middle number **Median**   7 values above the median

The median is 27 miles per gallon.

If data contain an *even* number of values, the median is the mean of the two middle numbers.

HOW TO   The selling prices of the last six homes sold by a real estate agent were $175,000, $150,000, $250,000, $130,000, $245,000, and $190,000. Find the median selling price of these homes.

Strategy
To find the median, arrange the numbers from smallest to largest. Because there is an even number of values, the median is the mean of the two middle numbers.

Solution
130,000 150,000 175,000 190,000 245,000 250,000

Middle 2 numbers

$$\text{Median} = \frac{175,000 + 190,000}{2} = 182,500$$

The median selling price was $182,500.

The **mode** of a set of numbers is the value that occurs most frequently. If a set of numbers has no number occurring more than once, then the data have no mode.

Here again are the data for the gasoline mileage ratings of 15 cars.

Miles-per-Gallon Ratings of 15 Cars														
25	22	21	27	25	35	29	31	25	26	21	39	34	32	28

25 is the number that occurs most frequently.

The mode is 25 miles per gallon.

Note from the miles-per-gallon example that the mean, median, and mode may be different.

Example 1

Twenty students were asked the number of units in which they were enrolled. The responses were as follows:

| 15 | 12 | 13 | 15 | 17 | 18 | 13 | 20 | 9 | 16 |
| 14 | 10 | 15 | 12 | 17 | 16 | 6 | 14 | 15 | 12 |

Find the mean number of units taken by these students.

Strategy

To find the mean number of units:

- Find the sum of the 20 numbers.
- Divide the sum by 20.

Solution

$15 + 12 + 13 + 15 + 17 + 18 + 13 + 20 + 9 +$
$\quad 16 + 14 + 10 + 15 + 12 + 17 + 16 + 6 +$
$\quad 14 + 15 + 12 = 279$

$$\bar{x} = \frac{279}{20} = 13.95$$

The mean is 13.95 units.

You Try It 1

The amounts spent by 12 customers at a McDonald's restaurant were as follows:

| 6.26 | 8.23 | 5.09 | 8.11 | 7.50 | 6.69 |
| 5.66 | 4.89 | 5.25 | 9.36 | 6.75 | 7.05 |

Find the mean amount spent by these customers. Round to the nearest cent.

Your strategy

Your solution

$6.74

Example 2

The starting hourly wages for an apprentice electrician for six different work locations are $10.90, $11.25, $10.10, $11.08, $11.56, and $10.55. Find the median starting hourly wage.

Strategy

To find the median starting hourly wage:

- Arrange the numbers from smallest to largest.
- Because there is an even number of values, the median is the mean of the two middle numbers.

Solution

10.10, 10.55, 10.90, 11.08, 11.25, 11.56

$$\text{Median} = \frac{10.90 + 11.08}{2} = 10.99$$

The median starting hourly wage is $10.99.

You Try It 2

The amounts of weight lost, in pounds, by 10 participants in a 6-month weight-reduction program were 22, 16, 31, 14, 27, 16, 29, 31, 40, and 10. Find the median weight loss for these participants.

Your strategy

Your solution

24.5 pounds

Solutions on pp. S20–S21

Objective 7.4B

Objective 7.4B

Vocabulary to Review
median [7.4A]

New Vocabulary
first quartile
third quartile
range
interquartile range
box-and-whiskers plot

Instructor Note
Questions from statistics are included on many teacher's state competency exams. Hence the inclusion in this text of topics such as box-and-whiskers plots.

Discuss the Concepts
Suppose that the interquartile range on an exam was 10 points. What does that tell you about the test scores?

Instructor Note
You may want to explain to your students that the word *quartile* is used because it divides the set of data into four sets of approximately equal size.
   The second quartile, symbolized by Q_2, is the number that one-half of the data lie below and one-half of the data lie above. Therefore, it is the median of the data.

Concept Check
1. For each of the following, state whether the whiskers will be short or long relative to the box in a box-and-whiskers plot.
   a. The range is 50 and the interquartile range is 40.
   Short
   b. The range is 98 and the interquartile range is 31.
   Long
   c. The range is 60, Q_1 is 10, and Q_3 is 25.   Long
   d. The range is 100, Q_1 is 250, and Q_3 is 325.   Short
(*Continued on the next page*)

Objective B   **To draw a box-and-whiskers plot**

Recall from the last objective that an average is one number that helps to describe all the numbers in a set of data. For example, we know from the following statement that Erie gets a lot of snow each winter.

> The average annual snowfall in Erie, Pennsylvania, is 85 inches.

Now look at these two statements.

> The average annual temperature in San Francisco, California, is 57°F.
> The average annual temperature in St. Louis, Missouri, is 57°F.

San Francisco

The average annual temperature in both cities is the same. However, we do not expect the climate in St. Louis to be like San Francisco's climate. Although both cities have the same average annual temperature, their temperature ranges differ. In fact, the difference between the average monthly high temperatures in July and January in San Francisco is 14°F, whereas the difference between the average monthly high temperatures in July and January in St. Louis is 50°F.

Note that for this example, a single number (the average annual temperature) does not provide us with a very comprehensive picture of the climate of either of these two cities.

St. Louis

One method used to picture an entire set of data is a box-and-whiskers plot. To prepare a box-and-whiskers plot, we begin by separating a set of data into four parts, called quartiles. We will illustrate this by using the average monthly high temperatures for St. Louis, in degrees Fahrenheit. These are listed below from January through December.

Source: The Weather Channel

First list the numbers in order from smallest to largest and determine the median.

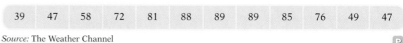

Median = 74

Now find the median of the data values below the median. The median of the data values below the median is called the **first quartile,** symbolized by Q_1. Also find the median of the data values above the median. The median of the data values above the median is called the **third quartile,** symbolized by Q_3.

$Q_1 = 48$ Median $Q_3 = 86.5$

The first quartile, Q_1, is the number that one-quarter of the data lie below. This means that 25% of the data lie below the first quartile. The third quartile, Q_3, is the number that one-quarter of the data lie above. This means that 25% of the data lie above the third quartile.

In-Class Examples (Objective 7.4B)

1. The list below is the numbers of years that U.S. presidents survived after leaving office.

   (*Note:* There are 28 values in this list. You may prefer to present your students with these numbers out of sequential order. Then the first step in organizing the data is to write the numbers from smallest to largest.) Find the lowest value, Q_1, the median, Q_3, and the highest value.
   (*Continued on the next page*)

The **range** of a set of numbers is the difference between the largest number and the smallest number in the set. The range describes the spread of the data. For the data above,

$$\text{Range} = \text{largest value} - \text{smallest value} = 89 - 39 = 50$$

The **interquartile range** is the difference between the third quartile, Q_3, and the first quartile, Q_1. For the data above,

$$\text{Interquartile range} = Q_3 - Q_1 = 86.5 - 48 = 38.5$$

The interquartile range is the distance that spans the "middle" 50% of the data values. Because it excludes the bottom fourth of the data values and the top fourth of the data values, it excludes any extremes in the numbers in the set.

A **box-and-whiskers plot,** or **boxplot,** is a graph that shows five numbers: the smallest value, the first quartile, the median, the third quartile, and the greatest value. Here are these five values for the data on St. Louis temperatures.

The smallest number	39
The first quartile, Q_1	48
The median	74
The third quartile, Q_3	86.5
The largest number	89

Think of a number line that includes the five values listed above. With this in mind, mark off the five values. Draw a box that spans the distance from Q_1 to Q_3. Draw a vertical line the height of the box at the median.

Listed below are the average monthly high temperatures for San Francisco.

57	60	61	64	68	71	71	73	74	73	60	59

Source: The Weather Channel

We can perform the same calculations on these data to determine the five values needed for the box-and-whiskers plot.

The smallest number	57
The first quartile, Q_1	60
The median	66
The third quartile, Q_3	72
The largest number	74

The box-and-whiskers plot is shown at the right with the same scale used for the data on the St. Louis temperatures.

Note that by comparing the two boxplots, we can see that the range of temperatures in St. Louis is greater than the range of temperatures in San Francisco. For the St. Louis temperatures, there is a greater spread of the data below the median than above the median, whereas the spreads of the data above and below the median of the San Francisco boxplot are nearly equal.

(*Continued*)
2. In the Concept Check in the last objective, we listed the selling prices of the last 10 homes sold in the town of Wilmington.

$ 100,000
$ 100,000
$ 100,000
$ 100,000
$ 100,000
$ 100,000
$ 100,000
$ 100,000
$ 100,000
$1,100,000

If data such as these are represented in a box-and-whiskers plot, does it result in a long left whisker, a long box, or a long right whisker?
A long right whisker

Instructor Note
Emphasize that the boxplot at the left shows that there are as many months with average high temperatures between 48°F and 74°F as there are months with average high temperatures between 74°F and 86.5°F.

Optional Student Activity
Listed on the next page are the numbers of electoral college votes allotted to each of the 50 states and the District of Columbia. (*Source:* National Archives and Records Administration)

a. Draw a box-and-whiskers plot of the data.

b. What data value is responsible for the long whisker at the right? 55
c. What state does this correspond to? California

(*Continued*)
a. Draw a box-and-whiskers plot of the data.
Lowest value: 0 years, $Q_1 = 6$ years, median = 10 years, $Q_3 = 19$ years, highest value = 31 years

b. Determine the range. 31 years

c. Determine the interquartile range. 13 years

States and Electoral College Votes	
Alabama	9
Alaska	3
Arizona	10
Arkansas	6
California	55
Colorado	9
Connecticut	7
Delaware	3
D.C.	3
Florida	27
Georgia	15
Hawaii	4
Idaho	4
Illinois	21
Indiana	11
Iowa	7
Kansas	6
Kentucky	8
Louisiana	9
Maine	4
Maryland	10
Massachusetts	12
Michigan	17
Minnesota	10
Mississippi	6
Missouri	11
Montana	3
Nebraska	5
Nevada	5
New Hampshire	4
New Jersey	15
New Mexico	5
New York	31
North Carolina	15
North Dakota	3
Ohio	20
Oklahoma	7
Oregon	7
Pennsylvania	21
Rhode Island	4
South Carolina	8
South Dakota	3
Tennessee	11
Texas	34
Utah	5
Vermont	3
Virginia	13
Washington	11
West Virginia	5
Wisconsin	10
Wyoming	3

HOW TO The numbers of avalanche deaths in the United States during each of nine consecutive winters were 8, 24, 29, 13, 28, 30, 22, 26, and 32. (*Source:* Colorado Avalanche Information Center) Draw a box-and-whiskers plot of the data and determine the interquartile range.

Strategy

To draw the box-and-whiskers plot, arrange the data from smallest to largest. Then find the median, Q_1, and Q_3. Use the smallest value, Q_1, the median, Q_3, and the largest value to draw the box-and-whiskers plot.

To find the interquartile range, find the difference between Q_3 and Q_1.

Solution

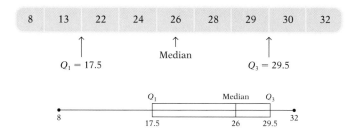

Interquartile range $= Q_3 - Q_1 = 29.5 - 17.5 = 12$

The interquartile range is 12 deaths.

TAKE NOTE

Note that the left whisker in this box-and-whiskers plot is quite long. This indicates a set of data in which the median is closer to the largest data value. If a boxplot has a long whisker on the right, the median is closer to the smallest data value. If the two whiskers are approximately the same length, then the smallest and largest values are about the same distance from the median. See Example 3.

Example 3

The average monthly snowfall, in inches, in Buffalo, New York, from October through April is 1, 12, 24, 25, 18, 12, and 3. (*Source:* The Weather Channel) Draw a box-and-whiskers plot of the data.

Strategy

To draw the box-and-whiskers plot:

• Arrange the data from smallest to largest.
• Find the median, Q_1, and Q_3.
• Use the smallest value, Q_1, the median, Q_3, and the largest value to draw the box-and-whiskers plot.

Solution

You Try It 3

The average monthly snowfall, in inches, in Denver, Colorado, from October through April is 4, 7, 7, 8, 8, 9, and 13. (*Source:* The Weather Channel)
a. Draw a box-and-whiskers plot of the data.
b. How does the spread of the data within the interquartile range compare with that in Example 3?

Your strategy

Your solution

a.

b. Answers about the spread of the data will vary.

Solution on p. S21

7.4 Exercises

Suggested Assignment
Exercises 1–21, odds
Exercises 24, 25
More challenging problems:
 Exercises 23, 27, 28

Objective A   **To find the mean, median, and mode of a distribution**

1. State whether the mean, median, or mode is being used.
   a. Half of the houses in the new development are priced under $125,000.   Median
   b. The average bill for lunch at the college union is $7.95.   Mean
   c. The college bookstore sells more green college sweatshirts than any other color.   Mode
   d. In a recent year, there were as many people age 26 and younger in the world as there were people age 26 and older.   Median
   e. The majority of full-time students carry a load of 12 credit hours per semester.   Mode
   f. The average annual return on this investment is 6.5%.   Mean

2. **Consumerism**   The number of big-screen televisions sold each month for one year was recorded by an electronics store. The results were 15, 12, 20, 20, 19, 17, 22, 24, 17, 20, 15, and 27. Calculate the mean, the median, and the mode of the number of televisions sold per month.
   Mean: 19 TVs; median: 19.5 TVs; mode: 20 TVs

3. **The Airline Industry**   The number of seats occupied on a jet for 16 trans–Atlantic flights was recorded. The numbers were 309, 422, 389, 412, 401, 352, 367, 319, 410, 391, 330, 408, 399, 387, 411, and 398. Calculate the mean, the median, and the mode of the number of seats occupied per flight.
   Mean: 381.5625 seats; median: 394.5 seats; mode: no mode

4. **Sports**   The times, in seconds, for a 100-meter dash at a college track meet were 10.45, 10.23, 10.57, 11.01, 10.26, 10.90, 10.74, 10.64, 10.52, and 10.78.
   a. Calculate the mean time for the 100-meter dash.   10.61 seconds
   b. Calculate the median time for the 100-meter dash.   10.605 seconds

5. **Consumerism**   A consumer research group purchased identical items in eight grocery stores. The costs for the purchased items were $45.89, $52.12, $41.43, $40.67, $48.73, $42.45, $47.81, and $45.82. Calculate the mean and the median costs of the purchased items.
   Mean: $45.615; median: $45.855

6. **Computers**   One measure of a computer's hard-drive speed is called access time; this is measured in milliseconds (thousandths of a second). Find the mean and median for 11 hard drives whose access times were 5, 4.5, 4, 4.5, 5, 5.5, 6, 5.5, 3, 4.5, and 4.5. Round to the nearest tenth.
   Mean: 4.7 milliseconds; median: 4.5 milliseconds

7. **Health Plans**   Eight health maintenance organizations (HMOs) presented group health insurance plans to a company. The monthly rates per employee were $423, $390, $405, $396, $426, $355, $404, and $430. Calculate the mean and the median monthly rates for these eight companies.
   Mean: $403.625; median: $404.50

8.   **Government**   The lengths of the terms, in years, of all the former Supreme Court chief justices are given in the table below. Find the mean and median length of term for the chief justices.
   Mean: 12.4 years; median: 10 years

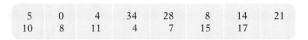

5	0	4	34	28	8	14	21
10	8	11	4	7	15	17	

Quick Quiz (Objective 7.4A)

1. A tourist center recorded the number of requests for information for a five-day period.

Mon	Tue	Wed	Thu	Fri
124	130	127	126	148

a. Find the mean number of requests.   131 requests
b. Find the median numbers of requests.
127 requests

2. The numbers of bags of flour used at a bakery during each of six days were 22, 24, 23, 25, 23, and 21. What is the mode of the data?

23 bags of flour

9. Life Expectancy The life expectancies, in years, in ten selected Central and South American countries are given at the right.
   a. Find the mean life expectancy in this group of countries.
   b. Find the median life expectancy in this group of countries.
   Mean: 70.8 years; median: 72 years

Country	Life Expectancy
Brazil	62
Chile	75
Costa Rica	78
Ecuador	70
Guatemala	64
Panama	75
Peru	66
Trinidad and Tobago	71
Uruguay	74
Venezuela	73

10. Education Your scores on six history tests were 78, 92, 95, 77, 94, and 88. If an "average score" of 90 receives an A for the course, which average, the mean or the median, would you prefer that the instructor use?
    Median

11. Education One student received scores of 85, 92, 86, and 89. A second student received scores of 90, 97, 91, and 94 (exactly 5 points more on each exam). Are the means of the two students the same? If not, what is the relationship between the means of the two students?

Answers to Writing Exercises

11. No, the mean scores of the two students are not the same. The mean score of the second student is 5 points higher than the mean score of the first student.

12c. If the year 1965 were eliminated from the data, the mean would increase and the median would increase, because the expenditures for 1965 are the lowest in the set of data.

12. Defense Spending The table below shows the defense expenditures, in billions of dollars, by the federal government for 1965 through 1973, years during which the United States was actively involved in the Vietnam War.
    a. Calculate the mean annual defense expenditure. Round to the nearest tenth of a billion. $72.3 billion
    b. Find the median annual defense expenditure. $77.7 billion
    c. If the year 1965 were eliminated from the data, how would that affect the mean? The median?

Year	1965	1966	1967	1968	1969	1970	1971	1972	1973
Expenditures	$49.6	$56.8	$70.1	$80.5	$81.2	$80.3	$77.7	$78.3	$76.0

Source: Statistical Abstract of the United States

Objective B To draw a box-and-whiskers plot

13. **a.** What percent of the data in a set of numbers lie above Q_3? 25%
    b. What percent of the data in a set of numbers lie above Q_1? 75%
    c. What percent of the data in a set of numbers lie below Q_3? 75%
    d. What percent of the data in a set of numbers lie below Q_1? 25%

14. U.S. Presidents The box-and-whiskers plot below shows the distribution of the ages of presidents of the United States at the time of their inauguration.
    a. What is the youngest age in the set of data? 42 years
    b. What is the oldest age? 69 years
    c. What is the first quartile? 51 years
    d. What is the third quartile? 58 years
    e. What is the median? 55 years
    f. Find the range. 27 years
    g. Find the interquartile range. 7 years

42   51 55 58   69

Quick Quiz (Objective 7.4B)

1. The numbers of federal, state, and local law enforcement officers killed in the line of duty during each of seven consecutive years were 157, 153, 169, 170, 132, 160, 155
   (*Source:* National Law Enforcement Officers Memorial Fund)
   a. Draw a box-and-whiskers plot of the data.

a.

Q_1 median Q_3

132   153 157   169 170

b. Determine the range and the interquartile range.
b. Range = 38 officers; interquartile range = 16 officers

15. **Compensation** The box-and-whiskers plot below shows the distribution of median incomes for 50 states and the District of Columbia. What is the lowest value in the set of data? The highest value? The first quartile? The third quartile? The median? Find the range and the interquartile range.

Lowest = $46,596;
Highest = $82,879;
Q_1 = $56,067;
Q_3 = $66,507;
Median = $61,036;
Range = $36,283;
Interquartile range = $10,440

46,596   56,067   61,036   66,507   82,879

16. **Education** An aptitude test was taken by 200 students at the Fairfield Middle School. The box-and-whiskers plot at the right shows the distribution of their scores.

43   54   72   88   98

   a. How many students scored over 88? 50 students
   b. How many students scored below 72?   100 students
   c. How many scores are represented in each quartile? 50 scores
   d. What percent of the students had scores of at least 54? 75%

17. **Health** The cholesterol levels for 80 adults were recorded and then displayed in the box-and-whiskers plot shown at the right.

172   198 217 254   345

   a. How many adults had a cholesterol level above 217? 40 adults
   b. How many adults had a cholesterol level below 254? 60 adults
   c. How many cholesterol levels are represented in each quartile? 20 cholesterol levels
   d. What percent of the adults had a cholesterol level of not more than 198? 25%

18. **Fuel Efficiency** The gasoline consumption of 19 cars was tested, and the results were recorded in the table below.

   a. Find the range, the first quartile, the third quartile, and the interquartile range. Range = 17 mpg; Q_1 = 20 mpg; Q_3 = 30 mpg; interquartile range = 10 mpg
   b. Draw a box-and-whiskers plot of the data.
   c. Is the data value 21 in the interquartile range? Yes

16   20   25   30   33

Miles per Gallon for 19 Cars									
33	21	30	32	20	31	25	20	16	24
22	31	30	28	26	19	21	17	26	

19. **Environment** Carbon dioxide is among the gases that contribute to global warming. The world's biggest emitters of carbon dioxide are listed below. The figures are emissions in millions of metric tons per year.

   a. Find the range, the first quartile, the third quartile, and the interquartile range. In millions of metric tons per year, the range is 4.39 emissions; Q_1 = 0.56 emissions; Q_3 = 2.10 emissions; the interquartile range is 1.54 emissions.
   b. Draw a box-and-whiskers plot of the data.
   c. What data value is responsible for the long "whisker" at the right? 4.80

0.41   2.10   4.80
0.56 0.815

Carbon Dioxide Emissions (in millions of metric tons per year)			
Canada	0.41	Japan	1.06
China	2.60	Russian Federation	2.10
Germany	0.87	Ukraine	0.61
India	0.76	United Kingdom	0.56
Italy	0.41	United States	4.80

Source: U.S. State Department

Answers to Writing Exercises

(These exercises appear on page 320.)

20d. Answers will vary. For example, Seattle has months that average less rainfall than any month in Houston, and it has months that average more rainfall than any month in Houston. The range of the Seattle data is wider than the range of the Houston data. Almost all of the Houston data lie between the median and the third quartile of the Seattle data. The average amount of rain per month in Houston varies relatively little compared to the average amount of rain per month in Seattle.

21d. Answers will vary. For example, the distribution of the data is relatively similar for the two cities. However, the value of each of the 5 points on the boxplot for the Portland data is less than the corresponding value on the boxplot for the Orlando data. The average monthly rainfall in Portland is less than the average monthly rainfall in Orlando.

Copyright © Houghton Mifflin Company. All rights reserved.

Answers to Writing Exercises

22a. To determine the mean, find the sum of the numbers in the set of data. Divide by the number of numbers in the set.

b. To find the median, arrange the numbers from smallest to largest. If there is an *odd* number of values in the set of data, the median is the middle number. If there is an *even* number of values in the set of data, the median is the mean of the two middle numbers.

c. To find the mode, find the number in the set of data that occurs more often than any other number in the set.

24. The mode must be a value in the data because it is the number that occurs most often in a set of data.

25a. Q_1 is the number that one-quarter of the data lie below.

b. Q_3 is the number that one-quarter of the data lie above.

c. $\overline{x}$ is the symbol for the mean of a set of data.

26. A box-and-whiskers plot displays the following values in a set of data: the lowest value, Q_1, the median, Q_3, and the highest value.
   The lowest value is shown at the far left of the plot (at the end of the left whisker). The highest value is shown at the far right of the plot (at the end of the right whisker). Q_1 is shown at the left end of the box, and Q_3 is shown at the right end of the box. The median is represented by the vertical line within the box.

27. The box does represent 50% of the data, but it provides a picture of the spread of the data. The box is not one-half the entire length of the box-and-whiskers plot because there is a greater spread of data in the interquartile range than in the first or fourth quarters of the data.

---

20. **Meteorology** The average monthly amounts of rainfall, in inches, from January through December for Seattle, Washington, and Houston, Texas, are listed below.

a. Is the difference between the means greater than 1 inch?   No
b. What is the difference between the medians?   0.8 inch
c. Draw a box-and-whiskers plot of each set of data. Use the same scale.
d. Describe the difference between the distributions of the data for Seattle and Houston.

| Seattle | 6.0 | 4.2 | 3.6 | 2.4 | 1.6 | 1.4 | 0.7 | 1.3 | 2.0 | 3.4 | 5.6 | 6.3 |
| Houston | 3.2 | 3.3 | 2.7 | 4.2 | 4.7 | 4.1 | 3.3 | 3.7 | 4.9 | 3.7 | 3.4 | 3.7 |

Source: The Weather Channel

21. **Meteorology** The average monthly amounts of rainfall, in inches, from January through December for Orlando, Florida, and Portland, Oregon, are listed below.

a. Is the difference between the means greater than 1 inch?   No
b. What is the difference between the medians?   0.3 inch
c. Draw a box-and-whiskers plot of each set of data. Use the same scale.
d. Describe the difference between the distributions of the data for Orlando and Portland.

| Orlando | 2.1 | 2.8 | 3.2 | 2.2 | 4.0 | 7.4 | 7.8 | 6.3 | 5.6 | 2.8 | 1.8 | 1.8 |
| Portland | 6.2 | 3.9 | 3.6 | 2.3 | 2.1 | 1.5 | 0.5 | 1.1 | 1.6 | 3.1 | 5.2 | 6.4 |

Source: The Weather Channel

APPLYING THE CONCEPTS

22. **a.** Explain how to determine the mean of a set of data.
   b. Explain how to determine the median of a set of data.
   c. Explain how to determine the mode of a set of data.

23. Write a set of data with five data values for which the mean, median, and mode are all 55.
   Answers will vary. For example, 55, 55, 55, 55, 55, or 50, 55, 55, 55, 60

24. A set of data has a mean of 16, a median of 15, and a mode of 14. Which of these numbers must be a value in the data? Explain your answer.

25. Explain each notation.
   a. Q_1 **b.** Q_3 **c.** $\overline{x}$

26. What values are shown on a box-and-whiskers plot? Explain how each is displayed.

27. The box in a box-and-whiskers plot represents 50%, or one-half, of the data in a set. Why is the box in Example 3 in this section not one-half of the entire length of the box-and-whiskers plot?

28. Create a set of data containing 25 numbers that would correspond to the box-and-whiskers plot shown at the right.
   Answers will vary. For example, 20, 21, 22, 24, 26, 27, 29, 31, 31, 32, 32, 33, 33, 36, 37, 37, 39, 40, 41, 43, 45, 46, 50, 54, 57

(box-and-whiskers plot: 20 28 33 42 57)

7.5 Introduction to Probability

Objective A To calculate the probability of simple events

Point of Interest

It was dice playing that led Antoine Gombaud, Chevalier de Mere, to ask Blaise Pascal, a French mathematician, to figure out the probability of throwing two sixes. Pascal and Pierre Fermat solved the problem, and their explorations led to the birth of probability theory.

A weather forecaster estimates that there is a 75% chance of rain. A state lottery director claims that there is a $\frac{1}{9}$ chance of winning a prize in a new game offered by the lottery. Each of these statements involves uncertainty to some extent. The degree of uncertainty is called **probability.** For the statements above, the probability of rain is 75%, and the probability of winning a prize in the new lottery game is $\frac{1}{9}$.

A probability is determined from an **experiment,** which is any activity that has an observable outcome. Examples of experiments include

> Tossing a coin and observing whether it lands heads or tails
> Interviewing voters to determine their preference for a political candidate
> Drawing a card from a standard deck of 52 cards

All the possible outcomes of an experiment are called the **sample space** of the experiment. The outcomes are listed between braces. For example:

The number cube shown at the left is rolled once. Any of the numbers from 1 to 6 could show on the top of the cube. The sample space is

$$\{1, 2, 3, 4, 5, 6\}$$

A fair coin is tossed once. (A fair coin is one for which heads and tails have an equal chance of landing face up.) If H represents "heads up" and T represents "tails up," then the sample space is

$$\{H, T\}$$

An **event** is one or more outcomes of an experiment. For the experiment of rolling the six-sided cube described above, some possible events are

> The number is even: $\{2, 4, 6\}$
> The number is a multiple of 3: $\{3, 6\}$
> The number is less than 10: $\{1, 2, 3, 4, 5, 6\}$
> Note that in this case, the event is the entire sample space.

HOW TO The spinner at the left is spun once. Assume that the spinner does not come to rest on a line.

a. What is the sample space?

The arrow could come to rest on any one of the four sectors. The sample space is $\{1, 2, 3, 4\}$.

b. List the outcomes in the event that the spinner points to an odd number.

$\{1, 3\}$

In discussing experiments and events, it is convenient to refer to the **favorable outcomes** of an experiment. These are the outcomes of an experiment that satisfy the requirements of a particular event.

Objective 7.5A

New Vocabulary

probability
experiment
sample space
event
favorable outcomes
at random
theoretical probability
empirical probability

New Formulas

Theoretical Probability Formula
Empirical Probability Formula

Concept Check

1. For the situation of throwing one die, name an event that has a probability of $\frac{1}{2}$.

   Examples will vary. For example, throwing a 1, 2, or 3; throwing a 1, 3, or 5; throwing a 2, 4, or 6; throwing a 4, 5, or 6

2. For the experiment of throwing one die, what is the sum of the probabilities of throwing a 1, a 2, a 3, a 4, a 5, and a 6?   1

3. Find the probability of drawing an ace or a spade in one draw from a 52-card deck.

   $\frac{4}{13}$

4. A married couple wants to have three children. If the probability of having a boy is $\frac{1}{2}$ and the probability of having a girl is $\frac{1}{2}$, what is the probability that the couple will have exactly two girls?

   $\frac{3}{8}$

In-Class Examples (Objective 7.5A)

1. A coin is tossed three times. What is the probability that the outcomes of the tosses are exactly TTH?

   $\frac{1}{8}$

2. Two dice are rolled. What is the probability that the sum of the dots on the upward faces is 4?

   $\frac{1}{12}$

Optional Student Activity

1. Have students work with the following: Each of three coins has a 1 on one side and a 2 on the other. All three coins are tossed. What is the probability of throwing a sum of 3? 4? 5? 6? 7?

   $\frac{1}{8}, \frac{3}{8}, \frac{3}{8}, \frac{1}{8}, 0$

2. In how many different ways can a panel of four on-off switches be set if no two adjacent switches can be off?

   8

3. What is the probability that a one-digit natural number selected at random is relatively prime to 12? ("Relatively prime to 12" means that the only common factor is 1.)

   $\frac{1}{3}$

4. The integers 1 through 300 are written on 300 identical pieces of paper. One of the pieces of paper is drawn randomly. What is the probability that the number drawn is divisible by 3? (When a number is divisible by 3, the sum of its digits is divisible by 3.)

   $\frac{1}{3}$

For instance, consider the experiment of rolling a fair die once. The sample space is

$$\{1, 2, 3, 4, 5, 6\}$$

and one possible event would be rolling a number that is divisible by 3. The outcomes of the experiment that are favorable to the event are 3 and 6:

$$\{3, 6\}$$

The outcomes of the experiment of tossing a fair coin are *equally likely*. Either one of the outcomes is just as likely as the other. If a fair coin is tossed once, the probability of a head is $\frac{1}{2}$, and the probability of a tail is $\frac{1}{2}$. Both events are equally likely. The theoretical probability formula, given below, applies to experiments for which the outcomes are equally likely.

Theoretical Probability Formula

The theoretical probability of an event is a fraction with the number of favorable outcomes of the experiment in the numerator and the total number of possible outcomes in the denominator.

$$\text{Probability of an event} = \frac{\text{number of favorable outcomes}}{\text{number of possible outcomes}}$$

A probability of an event is a number from 0 to 1 that tells us how likely it is that this outcome will happen.

A probability of 0 means that the event is impossible.
The probability of getting a heads when rolling the die shown at the left is 0.

A probability of 1 means that the event must happen.
The probability of getting either a heads or tails when tossing a coin is 1.

A probability of $\frac{1}{4}$ means that it is expected that the outcome will happen 1 in every 4 times the experiment is performed.

TAKE NOTE

The phrase **at random** means that each card has an equal chance of being drawn.

HOW TO Each of the letters of the word *TENNESSEE* is written on a card, and the cards are placed in a hat. If one card is drawn at random from the hat, what is the probability that the card has the letter *E* on it?

Count the possible outcomes of the experiment.

There are 9 letters in *TENNESSEE*.
There are 9 possible outcomes of the experiment.

Count the number of outcomes of the experiment that are favorable to the event that a card with the letter *E* on it is drawn.

There are 4 cards with an *E* on them.

Use the probability formula.

$$\text{Probability of the event} = \frac{\text{number of favorable outcomes}}{\text{number of possible outcomes}} = \frac{4}{9}$$

The probability of drawing an *E* is $\frac{4}{9}$.

As shown above, calculating the probability of an event requires counting the number of possible outcomes of an experiment and the number of outcomes that are favorable to the event. One way to do this is to list the outcomes of the experiment in a systematic way. Using a table is often helpful.

When two dice are rolled, the sample space for the experiment can be recorded systematically as in the following table.

Point of Interest

Romans called a die that was marked on four faces a *talus*, which meant "anklebone." The anklebone was considered an ideal die because it is roughly a rectangular solid and it has no marrow, so loose anklebones from sheep were more likely than other bones to be lying around after the wolves had left their prey.

Possible Outcomes from Rolling Two Dice

HOW TO   Two dice are rolled once. Calculate the probability that the sum of the numbers on the two dice is 7.

Use the table above to count the number of possible outcomes of the experiment.

There are 36 possible outcomes.

Count the number of outcomes of the experiment that are favorable to the event that a sum of 7 is rolled.

There are 6 favorable outcomes: (1, 6), (2, 5), (3, 4), (4, 3), (5, 2), and (6, 1).

Use the probability formula.

$$\text{Probability of the event} = \frac{\text{number of favorable outcomes}}{\text{number of possible outcomes}} = \frac{6}{36} = \frac{1}{6}$$

The probability of a sum of 7 is $\frac{1}{6}$.

The probabilities calculated above are theoretical probabilities. The calculation of a **theoretical probability** is based on theory—for example, that either side of a coin is equally likely to land face up or that each of the six sides of a fair die is equally likely to land face up. Not all probabilities arise from such assumptions.

An **empirical probability** is based on observations of certain events. For instance, a weather forecast of a 75% chance of rain is an empirical probability. From historical records kept by the weather bureau, when a similar weather pattern existed, rain occurred 75% of the time. It is theoretically impossible to predict the weather, and only observations of past weather patterns can be used to predict future weather conditions.

> **Empirical Probability Formula**
>
> The empirical probability of an event is a ratio of the number of observations of the event to the total number of observations.
>
> $$\text{Probability of an event} = \frac{\text{number of observations of the event}}{\text{total number of observations}}$$

For example, suppose the records of an insurance company show that of 2549 claims for theft filed by policy holders, 927 were claims for more than $5000. The empirical probability that the next claim for theft that this company receives will be a claim for more than $5000 is the ratio of the number of claims for over $5000 to the total number of claims.

$$\frac{927}{2549} \approx 0.36$$

The probability is approximately 0.36.

Example 1	**You Try It 1**
There are three choices, *a*, *b*, or *c*, for each of the two questions on a multiple-choice quiz. If the instructor randomly chooses which questions will have an answer of *a*, *b*, or *c*, what is the probability that the two correct answers on this quiz will be the same letter?	A professor writes three true/false questions for a quiz. If the professor randomly chooses which questions will have a true answer and which will have a false answer, what is the probability that the test will have 2 true questions and 1 false question?

Strategy

To find the probability:

- List the outcomes of the experiment in a systematic way.
- Count the number of possible outcomes of the experiment.
- Count the number of outcomes of the experiment that are favorable to the event that the two correct answers on the quiz will be the same letter.
- Use the probability formula.

Your strategy

Solution

Possible outcomes: (a, a) (b, a) (c, a)
                     (a, b) (b, b) (c, b)
                     (a, c) (b, c) (c, c)

There are 9 possible outcomes.

There are 3 favorable outcomes:
(a, a), (b, b), (c, c)

$$\text{Probability} = \frac{\text{number of favorable outcomes}}{\text{number of possible outcomes}}$$

$$= \frac{3}{9} = \frac{1}{3}$$

The probability that the two correct answers will be the same letter is $\frac{1}{3}$.

Your solution

$\frac{3}{8}$

Solution on p. S21

7.5 Exercises

Suggested Assignment

Exercises 1–23, odds
More challenging problem:
  Exercise 22

Objective A   **To calculate the probability of simple events**

1. A coin is tossed four times. List all the possible outcomes of the experiment as a sample space.
{(HHHH), (HHHT), (HHTT), (HHTH), (HTTT), (HTHH), (HTTH), (HTHT), (TTTT), (TTTH), (TTHH), (THHH), (TTHT), (THHT), (THTT), (THTH)}

2. Three cards—one red, one green, and one blue—are to be arranged in a stack. Using R for red, G for green, and B for blue, list all the different stacks that can be formed. (Some computer monitors are called RGB monitors for the colors red, green, and blue.)
RGB, RBG, GRB, GBR, BRG, BGR

Red	Green	Blue

3. A tetrahedral die is one with four triangular sides. The sides show the numbers from 1 to 4. Say two tetrahedral dice are rolled. List all the possible outcomes of the experiment as a sample space.
{(1, 1), (1, 2), (1, 3), (1, 4), (2, 1), (2, 2), (2, 3), (2, 4), (3, 1), (3, 2), (3, 3), (3, 4), (4, 1), (4, 2), (4, 3), (4, 4)}

Tetrahedral die

4. A coin is tossed and then a die is rolled. List all the possible outcomes of the experiment as a sample space. [To get you started, (H, 1) is one of the possible outcomes.]
{(H, 1), (H, 2), (H, 3), (H, 4), (H, 5), (H, 6), (T, 1), (T, 2), (T, 3), (T, 4), (T, 5), (T, 6)}

5. The spinner at the right is spun once. Assume that the spinner does not come to rest on a line.
  a. What is the sample space?   {1, 2, 3, 4, 5, 6, 7, 8}
  b. List the outcomes in the event that the number is less than 4.   {1, 2, 3}

6. A coin is tossed four times.
  a. What is the probability that the outcomes of the tosses are exactly in the order HHTT? (See Exercise 1.) $\frac{1}{16}$
  b. What is the probability that the outcomes of the tosses consist of two heads and two tails? $\frac{3}{8}$
  c. What is the probability that the outcomes of the tosses consist of one head and three tails? $\frac{1}{4}$

7. Two dice are rolled.
  a. What is the probability that the sum of the dots on the upward faces is 5? $\frac{1}{9}$
  b. What is the probability that the sum of the dots on the upward faces is 15?   0
  c. What is the probability that the sum of the dots on the upward faces is less than 15?   1
  d. What is the probability that the sum of the dots on the upward faces is 2? $\frac{1}{36}$

8. A dodecahedral die has 12 sides numbered from 1 to 12. The die is rolled once.
  a. What is the probability that the upward face shows an 11? $\frac{1}{12}$
  b. What is the probability that the upward face shows a 5? $\frac{1}{12}$

9. A dodecahedral die has 12 sides numbered from 1 to 12. The die is rolled once.
  a. What is the probability that the upward face shows a number divisible by 4? $\frac{1}{4}$
  b. What is the probability that the upward face shows a number that is a multiple of 3? $\frac{1}{3}$

Dodecahedral die

10. Two tetrahedral dice are rolled (see Exercise 3).
  a. What is the probability that the sum on the upward faces is 4? $\frac{3}{16}$
  b. What is the probability that the sum on the upward faces is 6? $\frac{3}{16}$

Quick Quiz (Objective 7.5A)

1. A coin is tossed three times. What is the probability that the outcomes of the tosses consist of one tail and two heads?
$\frac{3}{8}$

2. Two dice are rolled. What is the probability that the sum of the dots on the upward faces is less than 4?
$\frac{1}{12}$

11. Two dice are rolled. Which has the greater probability, throwing a sum of 10 or throwing a sum of 5?
    Throwing a sum of 5

12. Two dice are rolled once. Calculate the probability that the two numbers on the dice are equal. $\dfrac{1}{6}$

13. Each of the letters of the word *MISSISSIPPI* is written on a card, and the cards are placed in a hat. One card is drawn at random from the hat.
    a. What is the probability that the card has the letter *I* on it? $\dfrac{4}{11}$
    b. Which is greater, the probability of choosing an *S* or that of choosing a *P*? Choosing an *S*

14. There are five choices—choices *a* through *e*—for each question on a multiple-choice test. What is the probability of choosing the correct answer for a certain question by just guessing? $\dfrac{1}{5}$

15. Three blue marbles, four green marbles, and five red marbles are placed in a bag. One marble is chosen at random.
    a. What is the probability that the marble chosen is green? $\dfrac{1}{3}$
    b. Which is greater, the probability of choosing a blue marble or that of choosing a red marble? Choosing a red marble

16. Which has the greater probability, drawing a jack, queen, or king from a deck of cards or drawing a spade?
    Drawing a spade

17. In a history class, a set of exams earned the following grades: 4 A's, 8 B's, 22 C's, 10 D's, and 3 F's. If a single student's exam is chosen from this class, what is the probability that it received a B? $\dfrac{8}{47}$

18. A survey of 95 people showed that 37 preferred (to using a credit card) a cash discount of 2% if an item was purchased using cash or a check. Judging on the basis of this survey, what is the empirical probability that a person prefers a cash discount? Write the answer as a decimal rounded to the nearest hundredth.
    0.39

19. A survey of 725 people showed that 587 had a group health insurance plan where they worked. On the basis of this survey, what is the empirical probability that an employee has a group health insurance plan? Write the answer as a decimal rounded to the nearest hundredth.
    0.81

20. A television cable company surveyed some of its customers and asked them to rate the cable service as excellent, satisfactory, average, unsatisfactory, or poor. The results are recorded in the table at the right. What is the probability that a customer who was surveyed rated the service as satisfactory or excellent? $\dfrac{185}{377}$

Quality of Service	Number Who Voted
Excellent	98
Satisfactory	87
Average	129
Unsatisfactory	42
Poor	21

APPLYING THE CONCEPTS

21. If the spinner at the right is spun once, is each of the numbers 1 through 5 equally likely? Why or why not?

22. The probability of tossing a coin and having it land heads up is $\dfrac{1}{2}$.
    Does this mean that if that coin is tossed 100 times, it will land heads up 50 times? Explain your answer.

23. Why can the probability of an event not be $\dfrac{5}{3}$?

Answers to Writing Exercises

21. No, the numbers 1 through 5 are not equally likely because the sizes of the sectors are different.

22. The probability of tossing a fair coin and having it land heads up is $\dfrac{1}{2}$. This does not mean that if the coin is tossed 100 times, it will land heads up 50 times. Theoretical probabilities are obtained from logical reasoning, not from experimental data.

23. The mathematical probability of an event is a number from 0 to 1. Because $\dfrac{5}{3} = 1\dfrac{2}{3}$, $\dfrac{5}{3} > 1$. The probability of an event cannot be $\dfrac{5}{3}$ because the probability cannot be greater than 1.

Focus on Problem Solving

Inductive Reasoning   Suppose that, beginning in January, you save $25 each month. The total amount you have saved at the end of each month can be described by a list of numbers.

25	50	75	100	125	150	175	
Jan	Feb	Mar	Apr	May	June	July	. . .

The list of numbers that indicates your total savings is an *ordered* list of numbers called a **sequence.** Each of the numbers in a sequence is called a **term** of the sequence. The list is ordered because the position of a number in the list indicates the month in which that total amount has been saved. For example, the 7th term of the sequence (indicating July) is 175. This number means that a total of $175 has been saved by the end of the 7th month.

Now consider a person who has a different savings plan. The total amount saved by this person for the first seven months is given by the sequence

$$20, 35, 50, 65, 80, 95, 110, \ldots$$

The process you use to discover the next number in the above sequence is *inductive reasoning.* **Inductive reasoning** involves making generalizations from specific examples; in other words, we reach a conclusion by making observations about particular facts or cases. In the case of the above sequence, the person saved $15 per month after the first month.

Here is another example of inductive reasoning. Find the next two letters of the sequence A, B, E, F, I, J, ...

By trying different patterns, we can determine that a pattern for this sequence is

$$\underline{A, B}, C, D, \underline{E, F}, G, H, \underline{I, J}, \ldots$$

That is, write two letters, skip two letters, write two letters, skip two letters, and so on. The next two letters are M, N.

Use inductive reasoning to solve the following problems.

1. What is the next term of the sequence, ban, ben, bin, bon, . . . ?

2. ▦ Using a calculator, determine the decimal representation of several proper fractions that have a denominator of 99. For instance, you may use $\frac{8}{99}$, $\frac{23}{99}$, and $\frac{75}{99}$. Now use inductive reasoning to explain the pattern, and use your reasoning to find the decimal representation of $\frac{53}{99}$ without a calculator.

3. Find the next number in the sequence 1, 1, 2, 3, 5, 8, 13, 21, ...

4. The decimal representation of a number begins 0.10100100010000100000 . . . . What are the next 10 digits in this number?

5. The first seven rows of a triangle of numbers called Pascal's triangle are given below. Find the next row.

```
              1
            1   1
          1   2   1
        1   3   3   1
      1   4   6   4   1
    1   5  10  10   5   1
  1   6  15  20  15   6   1
```

Answers to Focus on Problem Solving— Inductive Reasoning

1. bun

2. Each decimal representation is a repeating decimal in which two digits repeat. If the numerator of the original fraction is one digit, then the repeating digits are 0 and the numerator. If the numerator of the original fraction is a two-digit number, then the repeating digits are the numerator.

$$\frac{53}{99} = 0.535353\ldots$$

3. 34

4. 1000000100

5. The next row is

1   7   21   35   35   21   7   1

Projects and Group Activities

Deceptive Graphs   A graphical representation of data can sometimes be misleading. Consider the graphs shown below. A financial advisor with an investment firm claims that an investment with the firm grew as shown in the graph on the left, whereas an investment with a competitor grew as shown in the graph on the right. Apparently, one would have accumulated more money by choosing the investment on the left.

Growth of Investment
with Broker A

Growth of Investment
with Broker B

However, these graphs have a serious flaw. There are no labels on the horizontal and vertical axes. Therefore, it is impossible to tell which investment increased more or over what time interval. When labels are not placed on the axes of a graph, the data represented are meaningless. This is one way in which advertisers use a visual impact to distort the true meaning of data.

The graphs below are the same as those drawn above except that scales have been drawn along each axis. Now it is possible to tell how each investment performed. Note that both turned in exactly the same performance!

Drawing a circle graph as an oval is another way of distorting data. This is especially true if a three-dimensional representation is given. From the appearance of the circle graph at the left, one would think that region *B* is larger than region *A*. However, that isn't true. Measure the angle of each sector to see this for yourself. As you read newspapers and magazines, find examples of graphs that may distort the actual data. Discuss how these graphs should be drawn to be more accurate.

Collecting, Organizing, Displaying, and Analyzing Data

Before standardized units of measurement, measurements were made in terms of the human body. For example, the cubit was the distance from the end of the elbow to the tips of the fingers. The yard was the distance from the tip of the nose to the tip of the fingers on an outstretched arm.

For each student in the class, find the measure from the tip of the nose to the tip of the fingers on an outstretched arm. Round each measure to the nearest centimeter. Record all the measurements on the board.

1. From the data collected, determine each of the following.

   Mean _____

   Median _____

   Mode _____

   Range _____

   First quartile, Q_1 _____

   Third quartile, Q_3 _____

   Interquartile range _____

2. Prepare a box-and-whiskers plot of the data.

3. Write a description of the spread of the data.

4. Explain why we need standardized units of measurement.

Projects and Group Activities—Collecting, Organizing, Displaying, and Analyzing Data

You will need centimeter tape measures for this project.

   If time permits, you might have the students prepare a histogram and/or a frequency polygon in addition to the box-and-whiskers plot. The class intervals you instruct them to use will depend on the number of values in the data collected.

Chapter 7 Summary

Key Words

Statistics is the branch of mathematics concerned with *data*, or numerical information. A *graph* is a pictorial representation of data. A *pictograph* represents data by using a symbol that is characteristic of the data. [7.1A, p. 295]

Examples

The pictograph shows the annual per capita turkey consumption in different counties.

Per Capita Turkey Consumption
Source: National Turkey Federation

A *circle graph* represents data by the size of the sectors. [7.1B, p. 297]

The circle graph shows the result of a survey of 300 people who were asked to name their favorite sport.

Distribution of Responses in a Survey

A *bar graph* represents data by the height of the bars. [7.2A, p. 303]

The bar graph shows the expected U.S. population aged 100 and over.

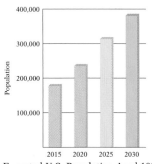

Expected U.S. Population Aged 100 and Over
Source: Census Bureau

A *broken-line graph* represents data by the position of the lines and shows trends or comparisons. [7.2B, p. 304]

The line graph shows a recent graduate's cumulative debt in college loans at the end of each of the four years of college.

Cumulative Debt in College Loans

A *histogram* is a special kind of bar graph. In a histogram, the width of each bar corresponds to a range of numbers called a *class interval*. The height of each bar corresponds to the number of occurrences of data in each class interval and is called the *class frequency*. [7.3A, p. 307]

An Internet service provider (ISP) surveyed 1000 of its subscribers to determine the time required for each subscriber to download a particular file. The results of the survey are shown in the histogram below.

A *frequency polygon* is a graph that displays information in a manner similar to a histogram. A dot is placed above the center of each class interval at a height corresponding to that class's frequency. The dots are connected to form a broken-line graph. The center of a class interval is called the *class midpoint*. [7.3B, p. 308]

Below is a frequency polygon for the data in the histogram above.

The *mean, median,* and *mode* are three types of averages used in statistics. The *mean* of the set of data is the sum of the data values divided by the number of values in the set. The *median* of a set of data is the number that separates the data into two equal parts when the data have been arranged from least to greatest (or greatest to least). There is an equal number of values above the median and below the median. The *mode* of a set of numbers is the value that occurs most frequently. [7.4A, pp. 311, 312]

Consider the following set of data.

24, 28, 33, 45, 45

The median is 33.
The mode is 45.

A *box-and-whiskers plot,* or *boxplot,* is a graph that shows five numbers: the least value, the first quartile, the median, the third quartile, and the greatest value. The *first quartile, Q_1,* is the number below which one-fourth of the data lie. The *third quartile, Q_3,* is a number above which one-fourth of the data lie. The box is placed around the values between the first quartile and the third quartile. The *range* is the difference between the largest number and the smallest number in the set. The range describes the spread of the data. The *interquartile range* is the difference between Q_3 and Q_1. [7.4B, pp. 314–315]

The box-and-whiskers plot for a set of test scores is shown below.

The range $= 96 - 45 = 51$
$Q_1 = 65$
$Q_3 = 86$
The interquartile range
$$= Q_3 - Q_1 = 86 - 65 = 21$$

Probability is a number from 0 to 1 that tells us how likely it is that a certain outcome of an experiment will happen. An *experiment* is an activity with an observable outcome. All the possible outcomes of an experiment are called the *sample space* of the experiment. An *event* is one or more outcomes of an experiment. The *favorable outcomes* of an experiment are the outcomes that satisfy the requirements of a particular event. [7.5A, p. 321]

Tossing a single die is an example of an experiment. The sample space for this experiment is the set of possible outcomes:

$$\{1, 2, 3, 4, 5, 6\}$$

The event that the number landing face up is an odd number is represented by

$$\{1, 3, 5\}$$

Essential Rules and Procedures

Examples

To Find the Mean of a Set of Data [7.4A, p. 311]
Divide the sum of the numbers by the number of values in the set.

$$\bar{x} = \frac{\text{sum of the data values}}{\text{number of data values}}$$

Consider the following set of data.

$$24, 28, 33, 45, 45$$

$$\bar{x} = \frac{24 + 28 + 33 + 45 + 45}{5} = 35$$

To Find the Median [7.4A, p. 312]
1. Arrange the numbers from least to greatest.
2. If there is an *odd* number of values in the set of data, the median is the middle number. If there is an *even* number of values in the set of data, the median is the mean of the two middle numbers.

Consider the following set of data.

$$24, 28, 33, 35, 45, 45$$

The median is $\frac{33 + 35}{2} = 34$.

To Find Q_1 [7.4B, p. 314]
Arrange the numbers from least to greatest and locate the median. Q_1 is the median of the lower half of the data.

Consider the following data.

$$8 \quad 10 \quad 12 \quad 14 \quad 16 \quad 19 \quad 22$$
$$\qquad \uparrow \qquad\qquad \uparrow$$
$$\qquad Q_1 \qquad\quad \text{Median}$$

To Find Q_3 [7.4B, p. 314]
Arrange the numbers from least to greatest and locate the median, Q_3 is the median of the upper half of the data.

Consider the following data.

$$8 \quad 10 \quad 12 \quad 14 \quad 16 \quad 19 \quad 22$$
$$\qquad\qquad\qquad \uparrow \qquad\qquad \uparrow$$
$$\qquad\qquad\quad \text{Median} \qquad Q_3$$

Theoretical Probability Formula [7.5A, p. 322]

$$\text{Probability of an event} = \frac{\text{number of favorable outcomes}}{\text{number of possible outcomes}}$$

A die is rolled. The probability of rolling a 2 or a 4 is $\frac{2}{6} = \frac{1}{3}$.

Empirical Probability Formula [7.5A, p. 324]

$$\text{Probability of an event} = \frac{\text{number of observations of the event}}{\text{total number of observations}}$$

A thumbtack is tossed 100 times. It lands point up 15 times and lands on its side 85 times. From this experiment, the empirical probability of "point up" is $\frac{15}{100} = \frac{3}{20}$.

Chapter 7 Review Exercises

Internet   The circle graph in Figure 37 shows the approximate amount of money that government agencies spent on maintaining Internet websites for a 3-year period. Use this graph for Exercises 1 to 3.

1. Find the total amount of money that these agencies spent on maintaining websites.
   $349 million   [7.1B]

2. What is the ratio of the amount spent by the Department of Commerce to the amount spent by the EPA?
   $\frac{9}{8}$ [7.1B]

3. What percent of the total money spent did NASA spend? Round to the nearest tenth of a percent.
   8.9%   [7.1B]

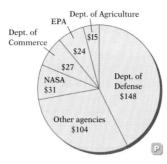

Figure 37 Millions of dollars that federal agencies spent on websites
Source: General Accounting Office

Demographics   The double-line graph in Figure 38 shows the populations of California and Texas. Use this graph for Exercises 4 to 6.

4. In 1900, which state had the larger population?
   Texas   [7.2B]

5. In 2000, approximately how much greater was the population of California than the population of Texas?
   12.5 million people   [7.2B]

6. During which 25-year period did the population of Texas increase the least?
   1925 to 1950   [7.2B]

Figure 38 Populations of California and Texas

Sports   The frequency polygon in Figure 39 shows the range of scores for the first 80 games of a season for the New York Knicks basketball team. Use this figure for Exercises 7 to 9.

7. Find the number of games in which fewer than 100 points were scored by the Knicks.
   54 games   [7.3B]

8. What is the ratio of the number of games in which 90 to 100 points were scored to the number of games in which 110 to 120 points were scored?
   $\frac{31}{8}$ [7.3B]

9. In what percent of the games were 110 points or more scored? Round to the nearest tenth of a percent.
   11.3%   [7.3B]

Figure 39
Source: Sports Illustrated website at
http://CNNSI.com

Airports The pictograph in Figure 40 shows the numbers of passengers annually passing through the five busiest airports in the United States. Use this graph for Exercises 10 and 11.

10. How many more passengers pass through O'Hare each year than pass through the Los Angeles airport each year?
10 million more passengers   [7.1A]

11. What is the ratio of the number of passengers passing through the San Francisco airport to the number of passengers passing through the Dallas/Ft. Worth airport each year? Write your answer using a colon.
2:3   [7.1A]

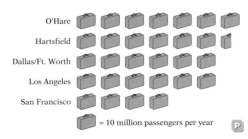

Figure 40 The busiest U.S. airports
Source: Airports Association Council International–North America; Air Transport Association of America

Sports The double-bar graph in Figure 41 shows the total days open and the days of full operation of ski resorts in different regions of the country. Use this graph for Exercises 12 to 14.

12. Find the difference between the total days open and the days of full operation for Midwest ski areas.
50 days   [7.2A]

13. What percent of the total days open were the days of full operation for the Rocky Mountain ski areas?
50%   [7.2A]

14. Which region had the lowest number of days of full operation? How many days of full operation did this region have?
Southeast; 30 days   [7.2A]

15. A coin is tossed four times. What is the probability that the outcomes of the tosses consist of one tail and three heads?
$\frac{1}{4}$ [7.5A]

Health Based on a Gallup poll, the numbers of hours that the 46 people surveyed slept during a typical weekday night are shown in the histogram in Figure 42. Use this figure for Exercises 16 and 17.

16. How many people slept 8 hours or more?
15 people   [7.3A]

17. What percent of the people surveyed slept 7 hours? Round to the nearest tenth of a percent.
28.3%   [7.3A]

18. **Sports** The heart rates of 24 women tennis players were measured after each of them had run one-quarter of a mile. The results are listed in the table below.

| 80 | 82 | 99 | 91 | 93 | 87 | 103 | 94 | 73 | 96 | 86 | 80 |
| 97 | 94 | 108 | 81 | 100 | 109 | 91 | 84 | 78 | 96 | 96 | 100 |

  a. Find the mean, median, and mode for the data. Round to the nearest tenth.
  b. Find the range and the interquartile range for the data.
  a. Mean: 91.6 heartbeats per minute; median: 93.5 heartbeats per minute; mode: 96 heartbeats per minute   [7.4A]
  b. Range: 36 heartbeats per minute; interquartile range: 15 heartbeats per minute   [7.4B]

Figure 41
Source: Economic Analysis of United States Ski Areas

Hours of Sleep per Night

Figure 42

Chapter 7 Test

Consumerism   Forty college students were surveyed to see how much money they spent each week on dining out in restaurants. The results are recorded in the frequency polygon shown in Figure 43. Use this figure for Exercises 1 to 3.

1. How many students spent between $15 and $25 per week?
   19 students   [7.3B]

2. Find the ratio of the number of students who spent between $10 and $15 to the number who spent between $15 and $20.
   $\frac{2}{3}$ [7.3B]

3. What percent of the students surveyed spent less than $15 per week?
   45%   [7.3B]

Figure 43

Marriage   The pictograph in Figure 44 is based on the results of a Gallup poll survey of married couples. Each individual was asked to give a letter grade to the marriage. Use this graph for Exercises 4 to 6.

4. Find the total number of people who were surveyed.
   36 people   [7.1A]

5. Find the ratio of the number of people who gave their marriage a B to the number who gave it a C.
   $\frac{5}{2}$ [7.1A]

6. What percent of the total number of people surveyed gave their marriage an A? Round to the nearest tenth of a percent.
   58.3%   [7.1A]

Figure 44 Survey of married couples rating their marriage

Amusement Rides   The bar graph in Figure 45 shows the number of fatalities that occurred in accidents on amusement rides in the 1990s in the United States. Use this graph for Exercises 7 to 9.

7. During which two consecutive years were the numbers of fatalities the same?
   1995 and 1996   [7.2A]

8. Find the total number of fatalities on amusement rides during 1991 through 1999.
   32 fatal accidents   [7.2A]

9. How many more such fatalities occurred during the years 1995 through 1998 than occurred during the years 1991 through 1994?
   4 more fatalities   [7.2A]

Figure 45 Number of fatalities in accidents on amusement rides
Source: USA Today, April 7, 2000

336

The Film Industry The circle graph in Figure 46 categorizes the 655 films released during a recent year by their ratings. Use this graph for Exercises 10 to 12.

10. How many more R-rated films were released than PG films?
    355 more   [7.1B]

11. How many times more PG-13 films were released than NC-17?
    16 times more   [7.1B]

12. What percent of the films released were rated G? Round to the nearest tenth of a percent.
    5.6%   [7.1B]

Compensation The histogram in Figure 47 gives information about median incomes, by state, in the United States. Use this figure for Exercises 13 to 15.

13. How many states have median incomes between $40,000 and $60,000?
    24 states   [7.3A]

14. What percent of the states have a median income that is between $50,000 and $70,000?
    72%   [7.3A]

15. What percent of the states have a median income that is $70,000 or more?
    18%   [7.3A]

16. **Probability** A box contains 50 balls, of which 15 are red. If 1 ball is randomly selected from the box, what is the probability of the ball's being red? $\dfrac{3}{10}$ [7.5A]

Education The broken-line graph in Figure 48 shows the number of students enrolled in colleges. Use this figure for Exercises 17 and 18.

17. During which decade did the student population increase the least?
    The 1990s   [7.2B]

18. Approximate the increase in the college enrollment from 1960 to 2000.
    11 million students   [7.2B]

19. **Quality Control** The length of time (in days) that various batteries operated a portable CD player continuously are given in the table below.

| 2.9 | 2.4 | 3.1 | 2.5 | 2.6 | 2.0 | 3.0 | 2.3 | 2.4 | 2.7 |
| 2.0 | 2.4 | 2.6 | 2.7 | 2.1 | 2.9 | 2.8 | 2.4 | 2.0 | 2.8 |

    a. Find the mean for the data.   2.53 days
    b. Find the median for the data.   2.55 days   [7.4A]
    c. Draw a box-and-whiskers plot for the data.

2.0   2.35   2.55   2.8   3.1   [7.4B]

Figure 46 Ratings of films released
Source: MPA Worldwide Market Research

Figure 47
Source: U.S. Census Bureau

Figure 48 Student enrollment in public and private colleges
Source: National Center for Educational Statistics

Cumulative Review Exercises

1. Simplify: $2^2 \cdot 3^3 \cdot 5$
   540 [1.6A]

2. Simplify: $3^2 \cdot (5 - 2) \div 3 + 5$
   14 [1.6B]

3. Find the LCM of 24 and 40.
   120 [2.1A]

4. Write $\frac{60}{144}$ in simplest form.
   $\frac{5}{12}$ [2.3B]

5. Find the total of $4\frac{1}{2}$, $2\frac{3}{8}$, and $5\frac{1}{5}$.
   $12\frac{3}{40}$ [2.4C]

6. Subtract: $12\frac{5}{8} - 7\frac{11}{12}$
   $4\frac{17}{24}$ [2.5C]

7. Multiply: $\frac{5}{8} \times 3\frac{1}{5}$
   2 [2.6B]

8. Find the quotient of $3\frac{1}{5}$ and $4\frac{1}{4}$.
   $\frac{64}{85}$ [2.7B]

9. Simplify: $\frac{5}{8} \div \left(\frac{3}{4} - \frac{2}{3}\right) + \frac{3}{4}$
   $8\frac{1}{4}$ [2.8C]

10. Write two hundred nine and three hundred five thousandths in standard form.
    209.305 [3.1A]

11. Find the product of 4.092 and 0.69.
    2.82348 [3.4A]

12. Convert $16\frac{2}{3}$ to a decimal. Round to the nearest hundredth.
    16.67 [3.6A]

13. Write "330 miles on 12.5 gallons of gas" as a unit rate.
    26.4 miles/gallon [4.2B]

14. Solve the proportion: $\frac{n}{5} = \frac{16}{25}$
    3.2 [4.3B]

15. Write $\frac{4}{5}$ as a percent.
    80% [5.1B]

16. 8 is 10% of what?
    80 [5.4A]

17. What is 38% of 43?
    16.34 [5.2A]

18. What percent of 75 is 30?
    40% [5.3A]

19. **Compensation** Tanim Kamal, a salesperson at a department store, receives $100 per week plus 2% commission on sales. Find the income for a week in which Tanim had $27,500 in sales.
$650   [6.6A]

20. **Insurance** A life insurance policy costs $8.15 for every $1000 of insurance. At this rate, what is the cost for $50,000 of life insurance?
$407.50   [4.3C]

21. **Simple Interest** A contractor borrowed $125,000 for 6 months at an annual simple interest rate of 6%. Find the interest due on the loan.
$3750   [6.3A]

22. **Markup** A compact disc player with a cost of $180 is sold for $279. Find the markup rate.
55%   [6.2B]

23. **Finance** The circle graph in Figure 49 shows how a family's monthly income of $3000 is budgeted. How much is budgeted for food?
$570   [7.1B]

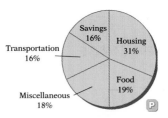

Figure 49 Budget for a monthly income of $3000

24. **Education** The double-broken-line graph in Figure 50 shows two students' scores on 5 math tests of 30 problems each. Find the difference between the numbers of problems that the two students answered correctly on Test 1.
12 problems   [7.2B]

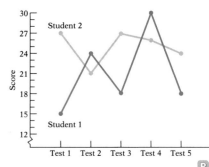

Figure 50

25. **Meteorology** The average daily high temperatures, in degrees Fahrenheit, for a week in Newtown were 56°, 72°, 80°, 75°, 68°, 62°, and 74°. Find the mean high temperature for the week. Round to the nearest tenth of a degree.
69.6°F   [7.4A]

26. **Probability** Two dice are rolled. What is the probability that the sum of the dots on the upward faces is 8?
$\frac{5}{36}$ [7.5A]

8 U.S. Customary Units of Measurement

In this chapter, you will be studying the U.S. Customary Units of length, weight, and capacity. As consumers, we are accustomed to these measurements. We purchase fabric by the yard, lumber by the foot, and tomatoes by the pound. We also buy perfume by the ounce, juice by the fluid ounce, and gasoline by the gallon. It is important to know how to convert between units and to perform operations that include different units. For example, in **Exercise 44 on page 350,** you are asked to determine how many 6-ounce bags of candy a storeowner can package from 12 pounds of candy.

OBJECTIVES

Section 8.1

A To convert measurements of length in the U.S. Customary System

B To perform arithmetic operations with measurements of length

C To solve application problems

Section 8.2

A To convert measurements of weight in the U.S. Customary System

B To perform arithmetic operations with measurements of weight

C To solve application problems

Section 8.3

A To convert measurements of capacity in the U.S. Customary System

B To perform arithmetic operations with measurements of capacity

C To solve application problems

Section 8.4

A To convert units of time

Section 8.5

A To use units of energy in the U.S. Customary System

B To use units of power in the U.S. Customary System

Need help? For online student resources, such as section quizzes, visit this textbook's website at **math.college.hmco.com/students.**

PREP TEST • • •

Do these exercises to prepare for Chapter 8.

For Exercises 1 to 8, add, subtract, multiply, or divide.

1.
$$\begin{array}{r} 485 \\ + 217 \\ \hline \end{array}$$
702 [1.2A]

2.
$$\begin{array}{r} 145 \\ - 87 \\ \hline \end{array}$$
58 [1.3B]

3. $36 \times \dfrac{1}{9}$
4 [2.6B]

4. $\dfrac{5}{3} \times 6$
10 [2.6B]

5. $400 \times \dfrac{1}{8} \times \dfrac{1}{2}$
25 [2.6B]

6. $5\dfrac{3}{4} \times 8$
46 [2.6B]

7. $3\overline{)714}$
238 [1.5A]

8. $12\overline{)18}$
1.5 [3.5A]

GO FIGURE • • •

Mandy walks to work at a constant rate. One-third of the way to work, she passes a bank. Three-fourths of the way to work, she passes a book store. At the bank her watch reads 7:52 A.M., and at the book store it reads 8:02 A.M. At what time is Mandy one-half of the way to work? 7:56 A.M.

8.1 Length

Objective A

To convert measurements of length in the U.S. Customary System

A **measurement** includes a number and a unit.

$$
\begin{array}{cc}
3 & \text{feet} \\
7 & \text{miles} \\
12 & \text{yards} \\
\underbrace{} & \underbrace{} \\
\text{Number} & \text{Unit}
\end{array}
$$

Standard units of measurement have been established to simplify trade and commerce.

The unit of **length,** or distance, that is called the yard was originally defined as the length of a specified bronze bar located in London.

The standard U.S. Customary System units of length are **inch, foot, yard,** and **mile.**

> **Equivalences Between Units of Length in the U.S. Customary System**
>
> 12 inches (in.) = 1 foot (ft)
> 3 ft = 1 yard (yd)
> 36 in. = 1 yard (yd)
> 5280 ft = 1 mile (mi)

These equivalences can be used to form conversion rates; a **conversion rate** is a relationship used to change one unit of measurement to another. For example, because 3 ft = 1 yd, the conversion rates $\frac{3\,\text{ft}}{1\,\text{yd}}$ and $\frac{1\,\text{yd}}{3\,\text{ft}}$ are both equivalent to 1.

HOW TO Convert 27 ft to yards.

$$27\,\text{ft} = 27\,\text{ft} \times \boxed{\frac{1\,\text{yd}}{3\,\text{ft}}}$$

$$= 27\,\cancel{\text{ft}} \times \frac{1\,\text{yd}}{3\,\cancel{\text{ft}}}$$

$$= \frac{27\,\text{yd}}{3}$$

$$= 9\,\text{yd}$$

HOW TO Convert 5 yd to feet.

$$5\,\text{yd} = 5\,\text{yd} \times \boxed{\frac{3\,\text{ft}}{1\,\text{yd}}}$$

$$= 5\,\cancel{\text{yd}} \times \frac{3\,\text{ft}}{1\,\cancel{\text{yd}}}$$

$$= \frac{15\,\text{ft}}{1}$$

$$= 15\,\text{ft}$$

Note that in the conversion rate chosen, the unit in the numerator is the same as the unit desired in the answer. The unit in the denominator is the same as the unit in the given measurement.

Point of Interest

The ancient Greeks devised the foot measurement, which they usually divided into 16 fingers. It was the Romans who subdivided the foot into 12 units called inches. The word *inch* is derived from the Latin word *uncia*, which means "a twelfth part."

The Romans also used a unit called *pace*, which equaled two steps. One thousand paces equaled 1 mile. The word *mile* is derived from the Latin word *mille*, which means "1000."

New Vocabulary
measurement
length
inch
foot
yard
mile
conversion rate

Discuss the Concepts

1. Refer to the "How To" examples. Why can we multiply 27 ft by $\frac{1\,\text{yd}}{3\,\text{ft}}$ without changing the measurement 27 ft? Why can we multiply 5 yd by $\frac{3\,\text{ft}}{1\,\text{yd}}$ without changing the measurement 5 yd?

2. Explain how to decide what conversion rate to use when converting (a) feet to miles and (b) miles to feet.

Optional Student Activity

1. How many inches are in $66\frac{2}{3}\%$ of a foot?   8 in.

2. How many inches are in $16\frac{2}{3}\%$ of a yard?   6 in.

3. How many feet are in $83\frac{1}{3}\%$ of a mile?   4400 ft

---

In-Class Examples (Objective 8.1A)

1. 11 yd = ____ ft   33

2. $4\frac{1}{3}$ yd = ____ in.   156

3. $1\frac{1}{4}$ mi = ____ ft   6600

Objective 8.1B

Discuss the Concepts
Explain why 2 ft 18 in. is not written in simplest form.

Concept Check
Eight people adopt 2 mi of highway. If each person cleans up an equal length of the highway, how many feet does each person clean up?   1320 ft

Optional Student Activity
A milling machine will bore into a metal plate, in accordance with an operator's instructions, starting at the center of the plate. Given below are dimensions of plates inserted in a milling machine. At what distances from the edges will the milling machine start to bore into the plate?

1. 12 in. × 8 in.
   4 in. from the 12-inch edge
   and 6 in. from the 8-inch edge

2. $1\frac{1}{2}$ ft × 9 in.
   $4\frac{1}{2}$ in. from the $1\frac{1}{2}$-foot edge
   and $\frac{3}{4}$ ft. from the 9-inch edge

3. $2\frac{1}{4}$ ft × $1\frac{3}{4}$ ft
   $\frac{7}{8}$ ft from the $2\frac{1}{4}$-foot edge
   and $1\frac{1}{8}$ ft from the $1\frac{3}{4}$-foot edge

Example 1 Convert 40 in. to feet.

Solution
$$40 \text{ in.} = 40 \text{ in.} \times \frac{1 \text{ ft}}{12 \text{ in.}}$$
$$= \frac{40 \text{ ft}}{12} = 3\frac{1}{3} \text{ ft}$$

You Try It 1 Convert 14 ft to yards.

Your solution $4\frac{2}{3}$ yd

Example 2 Convert $3\frac{1}{4}$ yd to feet.

Solution
$$3\frac{1}{4} \text{ yd} = \frac{13}{4} \text{ yd} = \frac{13}{4} \text{ yd} \times \frac{3 \text{ ft}}{1 \text{ yd}}$$
$$= \frac{39 \text{ ft}}{4} = 9\frac{3}{4} \text{ ft}$$

You Try It 2 Convert 9240 ft to miles.

Your solution $1\frac{3}{4}$ mi

Solutions on p. S22

Objective B   **To perform arithmetic operations with measurements of length**

When performing arithmetic operations with measurements of length, write the answer in simplest form. For example, 1 ft 14 in. should be written as 2 ft 2 in.

HOW TO Convert: 50 in. = ____ ft ____ in.

$$
\begin{array}{r}
4 \text{ ft 2 in.} \\
12\overline{)\,50} \\
-48 \\
\hline
2
\end{array}
$$

- Because 12 in. = 1 ft, divide 50 in. by 12. The whole-number part of the quotient is the number of feet. The remainder is the number of inches.

50 in. = 4 ft 2 in.

Example 3 Convert: 17 in. = ____ ft ____ in.

Solution
$$
\begin{array}{r}
1 \text{ ft 5 in.} \\
12\overline{)\,17} \\
-12 \\
\hline
5
\end{array}
$$
• 12 in. = 1 ft

17 in. = 1 ft 5 in.

You Try It 3 Convert: 42 in. = ____ ft ____ in.

Your solution 3 ft 6 in.

Example 4 Convert: 31 ft = ____ yd ____ ft

Solution
$$
\begin{array}{r}
10 \text{ yd 1 ft} \\
3\overline{)\,31} \\
-30 \\
\hline
1
\end{array}
$$
• 3 ft = 1 yd

31 ft = 10 yd 1 ft

You Try It 4 Convert: 14 ft = ____ yd ____ ft

Your solution 4 yd 2 ft

Solutions on p. S22

In-Class Examples (Objective 8.1B)

1. 80 in. = ____ ft ___ in.   6 ft 8 in.
2. 5 ft 9 in. + 4 ft 5 in.   10 ft 2 in.
3. 10 yd 1 ft − 2 yd 2 ft   7 yd 2 ft
4. 2 ft 7 in. × 5   12 ft 11 in.
5. 5 ft 9 in. ÷ 3   1 ft 11 in.

343

Example 5   Find the sum of 4 ft 4 in. and 1 ft 11 in.

Solution

$$\begin{array}{r} 4\text{ ft }\ 4\text{ in.} \\ +\ 1\text{ ft }11\text{ in.} \\ \hline 5\text{ ft }15\text{ in.} \end{array}$$ • 15 in. = 1 ft 3 in.

5 ft 15 in. = 6 ft 3 in.

You Try It 5   Find the sum of 3 ft 5 in. and 4 ft 9 in.

Your solution   8 ft 2 in.

Example 6   Subtract: 9 ft 6 in. − 3 ft 8 in.

Solution

$$\begin{array}{r} \overset{8\text{ ft}}{\cancel{9}}\text{ ft }\ \overset{18\text{ in.}}{\cancel{6}}\text{ in.} \\ -\ 3\text{ ft }\ 8\text{ in.} \\ \hline 5\text{ ft }10\text{ in.} \end{array}$$

• **Borrow 1 ft (12 in.) from 9 ft and add to 6 in.**

You Try It 6   Subtract: 4 ft 2 in. − 1 ft 8 in.

Your solution   2 ft 6 in.

Example 7   Multiply: 3 yd 2 ft × 4

Solution

$$\begin{array}{r} 3\text{ yd }2\text{ ft} \\ \times\qquad 4 \\ \hline 12\text{ yd }8\text{ ft} \end{array}$$ • 8 ft = 2 yd 2 ft

12 yd 8 ft = 14 yd 2 ft

You Try It 7   Multiply: 4 yd 1 ft × 8

Your solution   34 yd 2 ft

Example 8   Find the quotient of 4 ft 3 in. and 3.

Solution

$$\begin{array}{r} 1\text{ ft }\quad\ 5\text{ in.} \\ 3\overline{)\ 4\text{ ft }\quad 3\text{ in.}} \\ -\ 3\text{ ft} \\ \hline 1\text{ ft }=\ 12\text{ in.} \\ 15\text{ in.} \\ -15\text{ in.} \\ \hline 0 \end{array}$$

You Try It 8   Find the quotient of 7 yd 1 ft and 2.

Your solution   3 yd 2 ft

Example 9   Multiply: $2\frac{3}{4}$ ft × 3

Solution

$$2\frac{3}{4}\text{ ft} \times 3 = \frac{11}{4}\text{ ft} \times 3$$

$$= \frac{33}{4}\text{ ft}$$

$$= 8\frac{1}{4}\text{ ft}$$

You Try It 9   Subtract: $6\frac{1}{4}$ ft − $3\frac{2}{3}$ ft

Your solution $2\frac{7}{12}$ ft

Solutions on p. S22

Objective 8.1C

 Objective C **To solve application problems**

Optional Student Activity

1. The javelin is an Olympic event. The men's javelin is 8 ft 10 $\frac{1}{4}$ in. long. The women's javelin is 7 ft 6 $\frac{1}{2}$ in. long. Find the difference between the lengths of the men's and women's javelins.

   1 ft 3 $\frac{3}{4}$ in.

2. Rick is standing in a hole that is 3 ft deep. He plans to dig the hole 5 ft 6 in. deeper. At that point, the top of his head will be the same distance below ground level that it is now above ground level. How tall is Rick?   5 ft 9 in.

3. A *palindromic number* is a number that remains unchanged when its digits are written in reverse order.

   The odometer on Meagan's car read 14941 miles, a palindromic number. After 2 hours of continuous driving, the odometer showed another palindromic number. How fast was the car traveling in those 2 hours?   55 miles per hour

---

Example 10

A concrete block is 9 in. high. How many rows of blocks are required for a retaining wall that is 6 ft high?

Strategy

To find the number of rows of blocks, convert 9 in. to feet. Then divide the height of the wall (6 ft) by the height of each block.

Solution

$9 \text{ in.} = \frac{9 \text{ in.}}{1} \cdot \frac{1 \text{ ft}}{12 \text{ in.}} = \frac{9 \text{ ft}}{12} = 0.75 \text{ ft}$

$\frac{6 \text{ ft}}{0.75 \text{ ft}} = 8$

The wall will have 8 rows of blocks.

You Try It 10

The floor of a storage room is being tiled. Eight tiles, each a 9-inch square, fit across the width of the floor. Find the width, in feet, of the storage room.

Your strategy

Your solution

6 ft

Example 11

A plumber used 3 ft 9 in., 2 ft 6 in., and 11 in. of copper tubing to install a sink. Find the total length of copper tubing used.

Strategy

To find the total length of copper tubing used, add the three lengths of copper tubing (3 ft 9 in., 2 ft 6 in., and 11 in.).

Solution

$$\begin{array}{r} 3 \text{ ft } \ \ 9 \text{ in.} \\ 2 \text{ ft } \ \ 6 \text{ in.} \\ + \ \ \ \ \ 11 \text{ in.} \\ \hline 5 \text{ ft } 26 \text{ in.} \end{array}$$ • 26 in. = 2 ft 2 in.

5 ft 26 in. = 7 ft 2 in.

The plumber used 7 ft 2 in. of copper tubing.

You Try It 11

A board 9 ft 8 in. is cut into four pieces of equal length. How long is each piece?

Your strategy

Your solution

2 ft 5 in.

Solutions on p. S22

In-Class Examples (Objective 8.1C)

1. How long must a board be if five pieces, each 3 ft 6 in. long, are to be cut from the board?   17 ft 6 in.

2. A roof is to be constructed with 12 rafters, each 8 ft 6 in. long. Find the total number of feet of material needed to build the rafters.   102 ft

8.1 Exercises

Objective A To convert measurements of length in the U.S. Customary System

Suggested Assignment
Exercises 1–41, odds
More challenging problem:
Exercise 43

For Exercises 1 to 15, convert.

1. 6 ft = __72__ in.

2. 9 ft = __108__ in.

3. 30 in. = __$2\frac{1}{2}$__ ft

4. 64 in. = __$5\frac{1}{3}$__ ft

5. 13 yd = __39__ ft

6. $4\frac{1}{2}$ yd = __$13\frac{1}{2}$__ ft

7. 16 ft = __$5\frac{1}{3}$__ yd

8. $4\frac{1}{2}$ ft = __$1\frac{1}{2}$__ yd

9. $2\frac{1}{3}$ yd = __84__ in.

10. 5 yd = __180__ in.

11. 120 in. = __$3\frac{1}{3}$__ yd

12. 66 in. = __$1\frac{5}{6}$__ yd

13. 2 mi = __10,560__ ft

14. $1\frac{1}{2}$ mi = __7920__ ft

15. $7\frac{1}{2}$ in. = __$\frac{5}{8}$__ ft

Objective B To perform arithmetic operations with measurements of length

For Exercises 16 to 30, perform the arithmetic operation.

16. 100 in. = __8__ ft __4__ in.

17. 6400 ft = __1__ mi __1120__ ft

18. 15 in. = __1__ ft __3__ in.

19. 6 ft 7 in.
    + 3 ft 4 in.
    9 ft 11 in.

20. 9 ft 11 in.
    + 3 ft 6 in.
    13 ft 5 in.

21. 5 ft 3 in.
    − 2 ft 6 in.
    2 ft 9 in.

22. 9 yd 1 ft
    − 3 yd 2 ft
    5 yd 2 ft

23. 2 ft 5 in.
    × 6
    14 ft 6 in.

24. $3\frac{2}{3}$ ft × 4
    $14\frac{2}{3}$ ft

25. 2)5 ft 4 in.
    2 ft 8 in.

26. $12\frac{1}{2}$ in. ÷ 3
    $4\frac{1}{6}$ in.

27. $4\frac{2}{3}$ ft + $6\frac{1}{2}$ ft
    $11\frac{1}{6}$ ft

28. 3 yd 2 ft
    + 6 yd 2 ft
    10 yd 1 ft

29. 1 mi 4200 ft
    + 2 mi 3600 ft
    4 mi 2520 ft

30. 5 yd 1 ft
    − 2 yd 2 ft
    2 yd 2 ft

Quick Quiz (Objective 8.1A)

1. $10\frac{1}{4}$ ft = ___ in.   123

2. 16 yd = ___ ft   48

3. 7920 ft = ___ mi $1\frac{1}{2}$

Quick Quiz (Objective 8.1B)

1. 8 ft 9 in. + 2 ft 7 in.   11 ft 4 in.
2. 16 yd − 9 yd 1 ft   6 yd 2 ft
3. 4 ft 6 in. × 8   36 ft
4. 7 ft 4 in. ÷ 4   1 ft 10 in.

Objective C   **To solve application problems**

31. **Interior Decorating**   A kitchen counter is to be covered with tile that is 4 in. square. How many tiles can be placed along one row of a counter top that is 4 ft 8 in. long?   14 tiles

32. **Interior Decorating**   Thirty-two yards of material were used for making pleated draperies. How many feet of material were used?   96 ft

33. **Measurement**   Find the missing dimension.

$1\frac{5}{6}$ ft

34. **Measurement**   Find the total length of the shaft.   3 ft 1 in.

35. **Measurement**   What length of material is needed to drill two holes 3 in. in diameter and leave $\frac{1}{2}$ in. between the holes and on either side as shown in the diagram?

$7\frac{1}{2}$ in.

36. **Measurement**   Find the missing dimension in the figure.

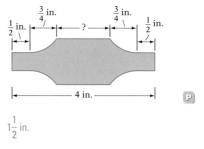

$1\frac{1}{2}$ in.

37. **Carpentry**   A board $6\frac{2}{3}$ ft long is cut into four equal pieces. How long is each piece? $1\frac{2}{3}$ ft

38. **Carpentry**   How long must a board be if four pieces, each 3 ft 4 in. long, are to be cut from it?   13 ft 4 in.

39. **Interior Decorating**   A picture is 1 ft 9 in. high and 1 ft 6 in. wide. Find the length of framing needed to frame the picture.   6 ft 6 in.

40. **Interior Decorating**   You bought 32 ft of baseboard to install in the kitchen of your house. How many inches of baseboard did you purchase?   384 in.

41. **Masonry**   Forty-five bricks, each 9 in. long, are laid end to end to make the base for a wall. Find the length of the wall in feet. $33\frac{3}{4}$ ft

42. **Construction**   A roof is constructed with nine rafters, each 8 ft 4 in. long. Find the total number of feet of material needed to build the rafters.   75 ft

APPLYING THE CONCEPTS

43. **Measurement**   There are approximately 200,000,000 adults living in the United States. Assume that the average adult is 19 in. wide from shoulder to shoulder. If all the adults in the United States are standing shoulder to shoulder, could they reach around Earth at the equator, a distance of approximately 25,000 mi?   Yes (59,975 > 25,000)

44. How good are you at estimating lengths or distances? Estimate the length of a pencil, the width of your room, the length of a block, and the distance to the grocery store. Then measure these lengths and compare the results with your estimates.   Answers will vary.

Instructor Note

In Exercise 44, student estimates will vary, as will the accuracy of their estimates. You might suggest that students then estimate the lengths of other objects, such as the width of the textbook, the height of the ceiling, and the width of the road, to see whether their estimates have become closer to the actual measurements as a consequence of prior experience.

Quick Quiz (Objective 8.1C)

1. Twenty-eight concrete blocks, each 10 in. long, are laid end to end to make the foundation for a wall. Find the length of the wall in feet.

$23\frac{1}{3}$ ft

2. You use 38 ft of a roll of copper tubing containing 50 yd of tubing. How many feet of tubing are left on the roll?   112 ft

8.2 Weight

Objective A

To convert measurements of weight in the U.S. Customary System

Point of Interest

The Romans used two different systems of weights. In both systems, the smallest unit was the *uncia*, abbreviated to "oz," from which the term *ounce* is derived. In one system, there were 16 ounces to 1 pound. In the second system, a pound, which was called the libra, equaled 12 unciae. The abbreviation "lb" for pound comes from the word *libra*.

The avoirdupois system of measurement and the troy system of measurement have their heritage in the two Roman systems.

Weight is a measure of how strongly Earth is pulling on an object. The unit of weight called the pound is defined as the weight of a standard solid kept at the Bureau of Standards in Washington, D.C. The U.S. Customary System units of weight are **ounce, pound,** and **ton.**

Equivalences Between Units of Weight in the U.S. Customary System

16 ounces (oz) = 1 pound (lb)

2000 lb = 1 ton

These equivalences can be used to form conversion rates to change one unit of measurement to another. For example, because 16 oz = 1 lb, the conversion rates $\frac{16\ oz}{1\ lb}$ and $\frac{1\ lb}{16\ oz}$ are both equivalent to 1.

HOW TO Convert 62 oz to pounds.

$$62\ oz = 62\ oz \times \frac{1\ lb}{16\ oz}$$

• The conversion rate must contain lb (the unit desired in the answer) in the numerator and must contain oz (the original unit) in the denominator.

$$= \frac{62\ \cancel{oz}}{1} \times \frac{1\ lb}{16\ \cancel{oz}}$$

$$= \frac{62\ lb}{16}$$

$$= 3\frac{7}{8}\ lb$$

Example 1 Convert $3\frac{1}{2}$ tons to pounds.

Solution

$$3\frac{1}{2}\ tons = \frac{7}{2}\ \cancel{tons} \times \frac{2000\ lb}{1\ \cancel{ton}}$$

$$= \frac{14,000\ lb}{2} = 7000\ lb$$

You Try It 1 Convert 3 lb to ounces.

Your solution 48 oz

Example 2 Convert 42 oz to pounds.

Solution

$$42\ oz = 42\ \cancel{oz} \times \frac{1\ lb}{16\ \cancel{oz}}$$

$$= \frac{42\ lb}{16} = 2\frac{5}{8}\ lb$$

You Try It 2 Convert 4200 lb to tons.

Your solution $2\frac{1}{10}$ tons

Solutions on p. S22

Objective 8.2A

Vocabulary to Review
conversion rate [8.1A]

New Vocabulary
weight
ounce
pound
ton

Discuss the Concepts

1. Refer to the example at the left. Why can we multiply 62 oz by $\frac{1\ lb}{16\ oz}$ without changing the measurement 62 oz?

2. Explain how to decide what conversion rate to use when converting (a) pounds to tons and (b) tons to pounds.

Optional Student Activity

1. How many ounces are in $12\frac{1}{2}\%$ of a pound?   2 oz

2. How many ounces are in $8\frac{1}{3}\%$ of a pound?
$1\frac{1}{3}$ oz

3. How many pounds are in $33\frac{1}{3}\%$ of a ton?
$666\frac{2}{3}$ lb

In-Class Examples (Objective 8.2A)

1. $6\frac{1}{2}$ lb = ___ oz   104

2. 72 oz = ___ lb $4\frac{1}{2}$

3. 5000 lb = ___ tons $2\frac{1}{2}$

Objective 8.2B

Optional Student Activity

A milling machine bores into a metal plate and cuts the material according to the operator's instructions. Given below are weights of plates inserted in a milling machine and their subsequent weights when the job was completed. In each case, determine the weight of the material removed by the milling machine.

1. Before: 20 lb
   After: 17 lb 2 oz   2 lb 14 oz
2. Before: 8 lb 8 oz
   After: 6 lb 10 oz   1 lb 14 oz
3. Before: 12 lb 9 oz
   After: 9 lb 12 oz   2 lb 13 oz

Objective B **To perform arithmetic operations with measurements of weight**

When performing arithmetic operations with measurements of weight, write the answer in simplest form. For example, 1 lb 22 oz should be written 2 lb 6 oz.

Example 3 Find the difference between 14 lb 5 oz and 8 lb 14 oz.

Solution

$$
\begin{array}{r}
{}^{13\text{ lb}} \quad {}^{21\text{ oz}} \\
\cancel{14\text{ lb}} \quad \cancel{5\text{ oz}} \\
-\ 8\text{ lb}\ 14\text{ oz} \\
\hline
5\text{ lb}\ 7\text{ oz}
\end{array}
$$

- Borrow 1 lb (16 oz) from 14 lb and add it to 5 oz.

You Try It 3 Find the difference between 7 lb 1 oz and 3 lb 4 oz.

Your solution 3 lb 13 oz

Example 4 Divide: 7 lb 14 oz ÷ 3

Solution

$$
\begin{array}{r}
2\text{ lb}\quad 10\text{ oz} \\
3\overline{)\ 7\text{ lb}\quad 14\text{ oz}} \\
-6\text{ lb} \\
\hline
1\text{ lb} = 16\text{ oz} \\
30\text{ oz} \\
-30\text{ oz} \\
\hline
0
\end{array}
$$

You Try It 4 Multiply: 3 lb 6 oz × 4

Your solution 13 lb 8 oz

Solutions on pp. S22–S23

Objective 8.2C

Optional Student Activity

Body mass index, or *BMI,* is a measurement for gauging a person's weight-related level of risk for high blood pressure, heart disease, and diabetes. A BMI of 25 or less is desirable. The formula is

$$
\text{BMI} = \frac{705 \times \text{weight}}{(\text{height})^2}
$$

where weight is in pounds and height is in inches.

1. Calculate the BMI of a man 6 ft 1 in. who weighs 225 lb. Round to the nearest tenth.
   29.8
2. Calculate your BMI.

Objective C **To solve application problems**

Example 5

Sirina Jasper purchased 4 lb 8 oz of oat bran and 2 lb 11 oz of wheat bran. She plans to blend the two brans and then repackage the mixture in 3-ounce packages for a diet supplement. How many 3-ounce packages can she make?

Strategy

To find the number of 3-ounce packages:

- Add the amount of oat bran (4 lb 8 oz) to the amount of wheat bran (2 lb 11 oz).
- Convert the sum to ounces.
- Divide the total ounces by the weight of each package (3 oz).

Solution

$$
\begin{array}{r}
4\text{ lb}\ \ 8\text{ oz} \\
+\ 2\text{ lb}\ 11\text{ oz} \\
\hline
6\text{ lb}\ 19\text{ oz} = 7\text{ lb}\ 3\text{ oz} = 115\text{ oz}
\end{array}
$$

$$
\frac{115\text{ oz}}{3\text{ oz}} \approx 38.3
$$

She can make 38 packages.

You Try It 5

Find the weight in pounds of 12 bars of soap. Each bar weighs 7 oz.

Your strategy

Your solution
$5\frac{1}{4}$ lb

Solution on p. S23

In-Class Examples (Objective 8.2B)

1. 5 lb 8 oz + 4 lb 10 oz   10 lb 2 oz
2. 6 lb 6 oz − 2 lb 9 oz   3 lb 13 oz
3. 2 lb 6 oz × 3   7 lb 2 oz
4. 10 lb 2 oz ÷ 6   1 lb 11 oz

In-Class Examples (Objective 8.2C)

1. Maple syrup weighing 3 lb 8 oz is divided equally and poured into four containers. How much syrup is in each container?   14 oz
2. Two books are mailed at the postal rate of $.24 per ounce. The books weigh 1 lb 2 oz and 1 lb 15 oz. Find the total cost of mailing the books. $11.76

8.2 Exercises

Objective A   To convert measurements of weight in the U.S. Customary System

Suggested Assignment
Exercises 1–45, odds

For Exercises 1 to 18, convert.

1. 64 oz = _____4_____ lb

2. 36 oz = _____$2\frac{1}{4}$_____ lb

3. 8 lb = _____128_____ oz

4. 7 lb = _____112_____ oz

5. 3200 lb = _____$1\frac{3}{5}$_____ tons

6. 9000 lb = _____$4\frac{1}{2}$_____ tons

7. 6 tons = _____12,000_____ lb

8. $1\frac{1}{4}$ tons = _____2500_____ lb

9. 66 oz = _____$4\frac{1}{8}$_____ lb

10. 90 oz = _____$5\frac{5}{8}$_____ lb

11. $1\frac{1}{2}$ lb = _____24_____ oz

12. $2\frac{5}{8}$ lb = _____42_____ oz

13. $1\frac{3}{10}$ tons = _____2600_____ lb

14. $\frac{4}{5}$ ton = _____1600_____ lb

15. 500 lb = _____$\frac{1}{4}$_____ ton

16. 5000 lb = _____$2\frac{1}{2}$_____ tons

17. 180 oz = _____$11\frac{1}{4}$_____ lb

18. 12 oz = _____$\frac{3}{4}$_____ lb

Objective B   To perform arithmetic operations with measurements of weight

For Exercises 19 to 33, perform the arithmetic operation.

19. 9000 lb =
_____4_____ tons _____1000_____ lb

20. 85 oz =
_____5_____ lb _____5_____ oz

21. 40 oz =
_____2_____ lb _____8_____ oz

22.   4 lb 7 oz
    + 3 lb 12 oz
    _____
      8 lb 3 oz

23.   1 ton   800 lb
    + 3 tons 1600 lb
    _____
      5 tons 400 lb

24.   7 lb 5 oz
     − 3 lb 8 oz
    _____
      3 lb 13 oz

25.   3 tons 500 lb
     − 1 ton 800 lb
    _____
      1 ton 1700 lb

26.   3 lb 6 oz
    ×   4
    _____
      13 lb 8 oz

27. $5\frac{1}{2}$ lb × 6
    33 lb

28. $2\overline{)3\text{ lb }8\text{ oz}}$
    1 lb 12 oz

29. $4\frac{2}{3}$ lb × 3
    14 lb

30.   7 lb 7 oz
    + 6 lb 9 oz
    _____
      14 lb

31. $6\frac{1}{2}$ oz
    + $2\frac{1}{2}$ oz
    _____
      9 oz

32. $6\frac{3}{8}$ lb
    − $2\frac{5}{6}$ lb
    _____
      $3\frac{13}{24}$ lb

33. 5 lb 12 oz ÷ 4
    1 lb 7 oz

Quick Quiz (Objective 8.2A)

1. $5\frac{1}{2}$ lb = ___ oz   88

2. 60 oz = ___ lb $3\frac{3}{4}$

3. $3\frac{1}{2}$ tons = ___ lb   7000

Quick Quiz (Objective 8.2B)

1. 8 lb 10 oz + 7 lb 9 oz   16 lb 3 oz
2. 10 lb 5 oz − 6 lb 8 oz   3 lb 13 oz
3. 5 lb 4 oz × 8   42 lb
4. 15 tons 400 lb ÷ 8   1 ton 1800 lb

Objective C   **To solve application problems**

34. **Iron Works**   A machinist has 25 iron rods to mill. Each rod weighs 20 oz. Find the total weight of the rods in pounds.

$31\frac{1}{4}$ lb

35. **Masonry**   A fireplace brick weighs $2\frac{1}{2}$ lb. What is the weight of a load of 800 bricks?

2000 lb

36. **Weights**   A college bookstore received 1200 textbooks, each weighing 9 oz. Find the total weight of the 1200 textbooks in pounds.

675 lb

37. **Weights**   A 4- × 4-inch tile weighs 7 oz. Find the weight, in pounds, of a package of 144 tiles.

63 lb

38. **Ranching**   A farmer ordered 20 tons of feed for 100 cattle. After 15 days, the farmer has 5 tons of feed left. On average, how many pounds of food has each cow eaten per day?

20 lb/day

39. **Weights**   A case of soft drinks contains 24 cans, each weighing 6 oz. Find the weight, in pounds, of the case of soft drinks.

9 lb

40. **Child Development**   A baby weighed 7 lb 8 oz at birth. At 6 months of age, the baby weighed 15 lb 13 oz. Find the baby's increase in weight during the 6 months.

8 lb 5 oz

41. **Packaging**   Shampoo weighing 5 lb 4 oz is divided equally and poured into four containers. How much shampoo is in each container?

1 lb 5 oz

42. **Weights**   A steel rod weighing 16 lb 11 oz is cut into three pieces. Find the weight of each piece of steel rod.

5 lb 9 oz

43. **Consumerism**   Find the cost of a ham roast that weighs 5 lb 10 oz if the price per pound is $4.80.

$27

44. **Markup**   A candy store buys candy weighing 12 lb for $14.40. The candy is repackaged and sold in 6-ounce packages for $1.15 each. Find the markup on the 12 lb of candy.

$22.40

45. **Shipping**   A manuscript weighing 2 lb 3 oz is mailed at the parcel post rate of $.25 per ounce. Find the cost of mailing the manuscript.

$8.75

APPLYING THE CONCEPTS

46.   Write a paragraph describing the growing need for precision in our measurements as civilization progressed. Include a discussion of the need for precision in the space industry.

47.   Estimate the weight of a nickel, a textbook, a friend, and a car. Then find the actual weights and compare them with your estimates.

Answers will vary.

Answers to Writing Exercises

46. Student essays should describe the increase in technology in society as civilization has progressed. Some might discuss the increased dependence on and sophistication of trade in the world.

With respect to the need for precision in space, students might describe lunar landings, landings on Mars, or orbiting satellites (the position of which is crucial for everything from cellular phones to television broadcasts).

Instructor Note on Exercise 47

As noted in Exercise 44 in Section 8.1, student estimates will vary, as will the accuracy of their estimates. You might suggest that students then estimate the weights of other objects, such as a calculator, a chair, and a stack of books, to see whether their estimates have become closer to the actual measurements.

Quick Quiz (Objective 8.2C)

1. A shoe store receives a shipment of 200 pairs of shoes. Each pair weighs 20 oz. Find the total weight in pounds of the 200 pairs of shoes.

250 lb

2. Hamburger meat weighing 15 lb is equally divided and placed into 6 containers. How many pounds of hamburger meat is in each container?

$2\frac{1}{2}$ lb

8.3 Capacity

Objective A

To convert measurements of capacity in the U.S. Customary System

Liquid substances are measured in units of **capacity**. The standard U.S. Customary units of capacity are the **fluid ounce, cup, pint, quart,** and **gallon.**

Equivalences Between Units of Capacity in the U.S. Customary System

8 fluid ounces (fl oz) = 1 cup (c)
2 c = 1 pint (pt)
2 pt = 1 quart (qt)
4 qt = 1 gallon (gal)

These equivalences can be used to form conversion rates to change one unit of measurement to another. For example, because 8 fl oz = 1 c, the conversion rates $\frac{8 \text{ fl oz}}{1 \text{ c}}$ and $\frac{1 \text{ c}}{8 \text{ fl oz}}$ are both equivalent to 1.

HOW TO Convert 36 fl oz to cups.

$$36 \text{ fl oz} = 36 \text{ fl oz} \times \boxed{\frac{1 \text{ c}}{8 \text{ fl oz}}}$$

$$= \frac{36 \text{ fl oz}}{1} \times \frac{1 \text{ c}}{8 \text{ fl oz}}$$

$$= \frac{36 \text{ c}}{8}$$

$$= 4\frac{1}{2} \text{ c}$$

• The conversion rate must contain c in the numerator and fl oz in the denominator.

HOW TO Convert 3 qt to cups.

$$3 \text{ qt} = 3 \text{ qt} \times \boxed{\frac{2 \text{ pt}}{1 \text{ qt}}} \times \boxed{\frac{2 \text{ c}}{1 \text{ pt}}}$$

$$= \frac{3 \text{ qt}}{1} \times \frac{2 \text{ pt}}{1 \text{ qt}} \times \frac{2 \text{ c}}{1 \text{ pt}}$$

$$= \frac{12 \text{ c}}{1}$$

$$= 12 \text{ c}$$

• The direct equivalence is not given above. Use two conversion rates. First convert quarts to pints, and then convert pints to cups. The unit in the denominator of the second conversion rate and the unit in the numerator of the first conversion rate must be the same in order to cancel.

Example 1 Convert 42 c to quarts.

Solution

$$42 \text{ c} = 42 \text{ c} \times \frac{1 \text{ pt}}{2 \text{ c}} \times \frac{1 \text{ qt}}{2 \text{ pt}}$$

$$= \frac{42 \text{ qt}}{4} = 10\frac{1}{2} \text{ qt}$$

You Try It 1 Convert 18 pt to gallons.

Your solution $2\frac{1}{4}$ gal

Solution on p. S23

Vocabulary to Review
conversion rate   [8.1A]

New Vocabulary
capacity
fluid ounce
cup
pint
quart
gallon

Optional Student Activity

Have students determine how many "sips" are in 1 c of water. Ask them to take sips of water, as uniformly as possible, until the cup of water is gone. Record the number of sips. Have the students write conversion rates for converting cups to sips and sips to cups. Have them convert sips to fluid ounces.

If time permits, you might connect this lesson with concepts covered in Chapter 7. Have students determine the mean, median, and modal numbers of sips per cup. Ask them to prepare a histogram and a frequency polygon of the data.

Instructor Note

You might want to introduce the equivalence

$$4 \text{ c} = 1 \text{ qt}$$

and the associated conversion rates. Students can then perform the conversion from 3 qt to cups, shown in the second "How To" using only one conversion rate, $\frac{4 \text{ c}}{1 \text{ qt}}$, rather than two.

In-Class Examples (Objective 8.3A)

1. 30 fl oz = ____ c $3\frac{3}{4}$

2. 3 gal = ____ qt   12

3. 8 pt = ____ fl oz   128

Objective 8.3B

Optional Student Activity

It is recommended that antifreeze and water be put into the cooling system of a car engine in the ratio 3:2. A mechanic is putting 16 qt of antifreeze and water in the cooling system of a V-8 engine. How many gallons of antifreeze and how many gallons of water should the mechanic put in the cooling system?

9.6, or $9\frac{3}{5}$, qt of antifreeze; 6.4,

or $6\frac{2}{5}$, qt of water

Objective 8.3C

Optional Student Activity

1. According to the Petroleum Equipment Institute of America, an average of 10 gal of gasoline flows from the nozzle of a gasoline pump every 60 seconds. Given that Americans use 150 billion gallons of gasoline each year, how many hours do Americans spend pumping gas every year?

250,000,000 hours

2. When 12 gal of gasoline is put into a gas tank, the indicator goes from $\frac{1}{8}$ to $\frac{7}{8}$ full. What is the total capacity of the gas tank? 16 gal

---

Objective B **To perform arithmetic operations with measurements of capacity**

When performing arithmetic operations with measurements of capacity, write the answer in simplest form. For example, 1 c 12 fl oz should be written as 2 c 4 fl oz.

Example 2 What is 4 gal 1 qt decreased by 2 gal 3 qt?

Solution

$$\begin{array}{r} \overset{3\ gal}{\cancel{4\ gal}}\ \overset{5\ qt}{\cancel{1\ qt}} \\ -\ 2\ gal\ 3\ qt \\ \hline 1\ gal\ 2\ qt \end{array}$$

• Borrow 1 gal (4 qt) from 4 gal and add to 1 qt.

You Try It 2 Find the quotient of 4 gal 2 qt and 3.

Your solution 1 gal 2 qt

Solution on p. S23

Objective C **To solve application problems**

Example 3 A can of apple juice contains 25 fl oz. Find the number of quarts of apple juice in a case of 24 cans.

Strategy To find the number of quarts of apple juice in one case:

• Multiply the number of cans (24) by the number of fluid ounces per can (25) to find the total number of fluid ounces in the case.
• Convert the number of fluid ounces in the case to quarts.

Solution
24×25 fl oz $= 600$ fl oz

$$600 \text{ fl oz} = \frac{600 \cancel{\text{fl oz}}}{1} \cdot \frac{1\ \cancel{c}}{8\ \cancel{\text{fl oz}}} \cdot \frac{1\ \cancel{\text{pt}}}{2\ \cancel{c}} \cdot \frac{1\ \text{qt}}{2\ \cancel{\text{pt}}}$$

$$= \frac{600\ \text{qt}}{32} = 18\frac{3}{4}\ \text{qt}$$

One case of apple juice contains $18\frac{3}{4}$ qt.

You Try It 3 Five students are going backpacking in the desert. Each student requires 5 qt of water per day. How many gallons of water should they take for a 3-day trip?

Your strategy

Your solution

$18\frac{3}{4}$ gal

Solution on p. S23

In-Class Examples (Objective 8.3B)

1. 10 qt = ___ gal ___ qt   2 gal 2 qt
2. 1 gal 3 qt + 5 gal 1 qt   7 gal
3. 4 c 4 fl oz − 1 c 6 fl oz   2 c 6 fl oz
4. 3 qt 1 pt × 4   14 qt

In-Class Examples (Objective 8.3C)

1. If a serving contains 1 c, how many servings can be made from 8 gal of punch?   128 servings
2. An outdoor supply store buys kerosene in 50-gallon containers for refilling the tanks on camping stoves. After 48 sales of 3 qt each, how much kerosene is left in the 50-gallon container?   14 gal

8.3 Exercises

Objective A   To convert measurements of capacity in the U.S.
Customary System

For Exercises 1 to 18, convert.

1. 60 fl oz = $7\frac{1}{2}$ c

2. 48 fl oz = 6 c

3. 3 c = 24 fl oz

4. $2\frac{1}{2}$ c = 20 fl oz

5. 8 c = 4 pt

6. 5 c = $2\frac{1}{2}$ pt

7. $3\frac{1}{2}$ pt = 7 c

8. 12 pt = 6 qt

9. 22 qt = $5\frac{1}{2}$ gal

10. 10 qt = $2\frac{1}{2}$ gal

11. $2\frac{1}{4}$ gal = 9 qt

12. 7 gal = 28 qt

13. $7\frac{1}{2}$ pt = $3\frac{3}{4}$ qt

14. $3\frac{1}{2}$ qt = 7 pt

15. 20 fl oz = $1\frac{1}{4}$ pt

16. $1\frac{1}{2}$ pt = 24 fl oz

17. 17 c = $4\frac{1}{4}$ qt

18. $1\frac{1}{2}$ qt = 6 c

Objective B   To perform arithmetic operations with measurements
of capacity

For Exercises 19 to 36, perform the arithmetic operation.

19. 14 qt =
3 gal 2 qt

20. 9 pt =
4 qt 1 pt

21. 5 pt =
2 qt 1 pt

22.   3 gal 2 qt
  + 4 gal 3 qt
  8 gal 1 qt

23.   4 qt 1 pt
  + 2 qt 1 pt
  7 qt

24.   3 gal 1 qt
  − 1 gal 2 qt
  1 gal 3 qt

25.   3 c 3 fl oz
  − 2 c 5 fl oz
  6 fl oz

26.   2 qt 1 pt
  ×   5
  12 qt 1 pt

27. $3\frac{1}{2}$ pt × 5
$17\frac{1}{2}$ pt

28. 5)6 gal 1 qt
1 gal 1 qt

29. $3\frac{1}{2}$ gal ÷ 4
$\frac{7}{8}$ gal

30.   5 c 3 fl oz
  + 3 c 6 fl oz
  9 c 1 fl oz

31.   3 gal 3 qt
  + 1 gal 2 qt
  5 gal 1 qt

32.   4 c 6 fl oz
  − 2 c 7 fl oz
  1 c 7 fl oz

33.   3 gal
  − 1 gal 2 qt
  1 gal 2 qt

Section 8.3

Suggested Assignment
Exercises 1–45, odds
More challenging problem:
  Exercise 48

Quick Quiz (Objective 8.3A)

1. $3\frac{1}{2}$ c = ___ fl oz   28

2. 30 qt = ___ gal $7\frac{1}{2}$

3. 6 pt = ___ fl oz   96

Quick Quiz (Objective 8.3B)

1. 17 qt = ___ gal ___ qt   4 gal 1 qt
2. 6 c 2 fl oz + 4 c 7 fl oz   11 c 1 fl oz
3. 8 gal 2 qt − 4 gal 3 qt   3 gal 3 qt
4. 1 gal 2 qt × 3   4 gal 2 qt

34. $1\frac{1}{2}$ pt $+ 2\frac{2}{3}$ pt $4\frac{1}{6}$ pt

35. $4\frac{1}{2}$ gal $- 1\frac{3}{4}$ gal $2\frac{3}{4}$ gal

36. $2\overline{)3\text{ gal }2\text{ qt}}$

1 gal 3 qt

> *Objective C*   **To solve application problems**

37. Catering   Sixty adults are expected to attend a book signing. Each adult will drink 2 c of coffee. How many gallons of coffee should be prepared? $7\frac{1}{2}$ gal

38. Catering   The Bayside Playhouse serves punch during intermission. Assume that 200 people will each drink 1 c of punch. How many gallons of punch should be ordered?
12.5 gal

39. Chemistry   A solution needed for a chemistry class required 72 fl oz of water, 16 fl oz of one solution, and 48 fl oz of another solution. Find the number of quarts of the final solution. $4\frac{1}{4}$ qt

40. Food Service   A cafeteria sold 124 cartons of milk in 1 day. Each carton contained 1 c of milk. How many quarts of milk were sold that day?
31 qt

41. Vehicle Maintenance   A farmer changed the oil in a tractor seven times during the year. Each oil change required 5 qt of oil. How many gallons of oil did the farmer use in the seven oil changes? $8\frac{3}{4}$ gal

42. Capacity   There are 24 cans in a case of tomato juice. Each can contains 10 fl oz of tomato juice. Find the number of 1-cup servings in the case of tomato juice.
30 servings

43. Consumerism   One brand of tomato juice costs $1.59 for 1 qt. Another brand costs $1.25 for 24 fl oz. Which is the more economical purchase?
$1.59 for 1 qt

44. Camping   Mandy carried 12 qt of water for 3 days of desert camping. Water weighs $8\frac{1}{3}$ lb per gallon. Find the weight of water that she carried.
25 lb

45. Business   A department store bought hand lotion in 5-quart containers and then repackaged the lotion in 8-fluid-ounce bottles. The lotion and bottles cost $81.50, and each 8-fluid-ounce bottle was sold for $8.25. How much profit was made on each 5-quart package of lotion?
$83.50

46. Business   Orlando bought oil in 50-gallon containers for changing the oil in his customers' cars. He paid $240 for the 50 gal of oil and charged customers $2.10 per quart. Find the profit Orlando made on one 50-gallon container of oil.
$180

APPLYING THE CONCEPTS

47.   Define the following units: grain, dram, furlong, and rod. Give an example where each would be used.
1 grain = 0.002286 oz; 1 dram = 0.0625 oz; 1 furlong = $\frac{1}{8}$ mi; 1 rod = 16.5 ft   Examples will vary.

48.   ✎   Assume that you wanted to invent a new measuring system. Discuss some of the features that would have to be incorporated into the system.

Answers to Writing Exercises

48. Students may enjoy exercising the creativity required to answer this question. However, they should note that a standard unit of measurement must be developed first. Related units can then be developed using equivalences that allow for conversion of units.

Quick Quiz (Objective 8.3C)

1. One hundred twenty-eight people attended the opening of an art exhibit. Assume that each person drank a cup of punch. How many gallons of punch were served?   8 gal

2. There are 24 bottles in a case of soda. Each bottle contains 16 oz of soda. Find the number of 1-cup servings in the case of soda.   48 servings

8.4 Time

Objective A   To convert units of time

The units in which time is generally measured are the **second, minute, hour, day,** and **week**.

> **Equivalences Between Units of Time**
>
> 60 seconds (s) = 1 minute (min)
> 60 min = 1 hour (h)
> 24 h = 1 day
> 7 days = 1 week

These equivalences can be used to form conversion rates to change one unit of time to another. For example, because 24 h = 1 day, the conversion rates $\frac{24\ h}{1\ day}$ and $\frac{1\ day}{24\ h}$ are both equivalent to 1. An example using each of these two rates is shown below.

HOW TO   Convert $5\frac{1}{2}$ days to hours.

$$5\frac{1}{2}\ \text{days} = 5\frac{1}{2}\ \text{days} \times \boxed{\frac{24\ h}{1\ day}}$$

$$= \frac{11\ \cancel{days}}{2} \times \frac{24\ h}{1\ \cancel{day}}$$

$$= \frac{264\ h}{2}$$

$$= 132\ h$$

• The conversion rate must contain h (the unit desired in the answer) in the numerator and must contain day (the original unit) in the denominator.

HOW TO   Convert 156 h to days.

$$156\ h = 156\ h \times \boxed{\frac{1\ day}{24\ h}}$$

$$= \frac{156\ \cancel{h}}{1} \times \frac{1\ day}{24\ \cancel{h}}$$

$$= \frac{156\ days}{24}$$

$$= 6\frac{1}{2}\ \text{days}$$

• The conversion rate must contain day (the unit desired in the answer) in the numerator and must contain h (the original unit) in the denominator.

Example 1   Convert 2880 min to days.

Solution

$$2880\ \text{min} = 2880\ \cancel{min} \times \frac{1\ \cancel{h}}{60\ \cancel{min}} \times \frac{1\ day}{24\ \cancel{h}}$$

$$= \frac{2880\ days}{1440} = 2\ \text{days}$$

You Try It 1   Convert 18,000 s to hours.

Your solution   5 h

Solution on p. S23

Objective 8.4A

Vocabulary to Review
conversion rate   [8.1A]

New Vocabulary
second
minute
hour
day
week

Discuss the Concepts
1. Name two conversion rates that can be formed from the equivalence 60 min = 1 h. For each rate, what units would you be converting from and to?
2. Name a situation in which we need to convert units of time.

Concept Check
1. When we convert from hours to days, what unit must be in the numerator of the rate?
   days
2. What conversion rates would you use in converting from seconds to days?
   $\frac{1\ min}{60\ s}, \frac{1\ h}{60\ min}, \frac{1\ day}{24\ h}$

Optional Student Activity
Find the charge for leaving your car in the airport garage for $2\frac{1}{2}$ days. The charge is $1.50 per hour. $90

In-Class Examples (Objective 8.4A)

1. Convert 930 s to minutes.   15.5 min
2. Convert $7\frac{1}{4}$ h to seconds.   26,100 s
3. Convert 3960 min to days. $2\frac{3}{4}$ days

Section 8.4

Suggested Assignment
Exercises 1–23, odds

8.4 Exercises

Objective A   **To convert units of time**

For Exercises 1 to 24, convert.

1. 98 days = _____14_____ weeks   **2.** 12 weeks = _____84_____ days   **3.** $6\frac{1}{4}$ days = _____150_____ h

4. 114 h = _____$4\frac{3}{4}$_____ days   **5.** 555 min = _____$9\frac{1}{4}$_____ h   **6.** $7\frac{3}{4}$ h = _____465_____ min

7. $18\frac{1}{2}$ min = _____1110_____ s   **8.** 750 s = _____$12\frac{1}{2}$_____ min   **9.** 12,600 s = _____$3\frac{1}{2}$_____ h

10. 15,300 s = _____$4\frac{1}{4}$_____ h   **11.** $6\frac{1}{2}$ h = _____23,400_____ s   **12.** $5\frac{3}{4}$ h = _____20,700_____ s

13. 5040 min = _____$3\frac{1}{2}$_____ days   **14.** 6840 min = _____$4\frac{3}{4}$_____ days   **15.** $2\frac{1}{2}$ days = _____3600_____ min

16. $6\frac{1}{4}$ days = _____9000_____ min   **17.** 672 h = _____4_____ weeks   **18.** 588 h = _____$3\frac{1}{2}$_____ weeks

19. 3 weeks = _____504_____ h   **20.** $5\frac{1}{2}$ weeks = _____924_____ h   **21.** 172,800 s = _____2_____ days

22. 20,160 min = _____2_____ weeks   **23.** 3 days = _259,200_ s   **24.** 3 weeks = _30,240_ min

APPLYING THE CONCEPTS

Another unit of time is the year. One year is equivalent to $365\frac{1}{4}$ days. However, our calendar does not include quarter days. Instead, we say that a year is 365 days, and every fourth year is a leap year of 366 days. If a year is divisible by 4, it is a leap year, unless it is a year at the beginning of a century not divisible by 400. 1600, 2000, 2004, 2008, and 2012 are leap years. 1700, 1800, and 1900 are not leap years. For Exercises 25 to 27, state whether the given year is a leap year.

25. 1984
Yes

26. 1994
No

27. 2144
Yes

For Exercises 28 to 33, convert. Use a 365-day year.

28. 1825 days = _____5_____ years   **29.** 2555 days = _____7_____ years   **30.** 4 years = _____1460_____ days

31. 6 years = _____2190_____ days   **32.** 2 years = _____17,520_____ h   **33.** $3\frac{1}{2}$ years = _____30,660_____ h

Quick Quiz (Objective 8.4A)

1. Convert 465 min to hours. $7\frac{3}{4}$ h

2. Convert 34,200 s to hours. $9\frac{1}{2}$ h

3. Convert $5\frac{1}{2}$ days to minutes.   7920 min

8.5 Energy and Power

Objective A   **To use units of energy in the U.S. Customary System**

Energy can be defined as the ability to do work. Energy is stored in coal, in gasoline, in water behind a dam, and in one's own body.

One **foot-pound** (ft · lb) of energy is the amount of energy necessary to lift 1 pound a distance of 1 foot.

To lift 50 lb a distance of 5 ft requires
$50 \times 5 = 250$ ft · lb of energy.

Consumer items that use energy, such as furnaces, stoves, and air conditioners, are rated in terms of the **British thermal unit** (Btu). For example, a furnace might have a rating of 35,000 Btu per hour, which means that it releases 35,000 Btu of energy in one hour (1 h).

Because 1 Btu = 778 ft · lb, the following conversion rate, equivalent to 1, can be written:

$$\frac{778 \text{ ft} \cdot \text{lb}}{1 \text{ Btu}} = 1$$

Example 1

Convert 250 Btu to foot-pounds.

Solution

$250 \text{ Btu} = 250 \text{ Btu} \times \dfrac{778 \text{ ft} \cdot \text{lb}}{1 \text{ Btu}}$

$\qquad = 194,500$ ft · lb

You Try It 1

Convert 4.5 Btu to foot-pounds.

Your solution
3501 ft · lb

Example 2

Find the energy required for a 125-pound person to climb a mile-high mountain.

Solution

In climbing the mountain, the person is lifting 125 lb a distance of 5280 ft.

Energy = 125 lb × 5280 ft
$\qquad = 660,000$ ft · lb

You Try It 2

Find the energy required for a motor to lift 800 lb a distance of 16 ft.

Your solution
12,800 ft · lb

Solutions on p. S23

New Vocabulary
energy
foot-pound
British thermal unit

Concept Check
How much more energy is required for a 180-pound person to climb Mt. Washington, a height of 6288 ft, than for a 120-pound person to climb the same mountain? Mount Washington is the highest mountain in the Northeast.
377,280 more foot-pounds of energy

Optional Student Activity
Torque is a turning or a twisting force.
   A 1965 Mercury V-8 with a 352-cubic-inch engine has an advertised torque of 352 ft · lb at 2800 rpm. (This is the point at which it reaches its peak torque.)
   A 1965 GTO V-8 with a 389-cubic-inch engine has an advertised torque of 431 ft · lb at 3200 rpm.
   A 1965 Mustang V-8 with a 289-cubic-inch engine has an advertised torque of 282 ft · lb at 2400 rpm.
   A 1965 Pontiac Grand Prix with a 389-cubic-inch engine has an advertised torque of 433 ft · lb at 3600 rpm.
   Rank these cars, from least to greatest, by their engines' foot-pounds of energy.
Mustang, Mercury, GTO, Pontiac

In-Class Examples (Objective 8.5A)

1. Convert 32 Btu to foot-pounds.   24,896 ft · lb

2. Find the energy required to lift 180 lb a distance of 6 ft.   1080 ft · lb

3. A furnace is rated at 30,000 Btu per hour. How many foot-pounds of energy are released by the furnace in 1 h?   23,340,000 ft · lb

Solution on p. S23

Objective 8.5B

New Vocabulary

power
foot-pounds per second
horsepower

Discuss the Concepts

What is the difference between energy and work?

Optional Student Activity

A 1965 Mercury V-8 with a 352-cubic-inch engine has an advertised horsepower of 250 at 4400 rpm. (This is the point at which it reaches its peak horsepower.)

A 1965 GTO V-8 with a 389-cubic-inch engine has an advertised horsepower of 335 at 5000 rpm.

A 1965 Mustang V-8 with a 289-cubic-inch engine has an advertised horsepower of 200 at 4400 rpm.

A 1965 Pontiac Grand Prix with a 389-cubic-inch engine has an advertised horsepower of 338 at 4800 rpm.

Convert the horsepower of each model to foot-pounds per second.

Mercury: 137,500 ft · lb/s;
GTO: 184,250 ft · lb/s;
Mustang: 110,000 ft · lb/s;
Pontiac: 185,900 ft · lb/s

Rank these cars, from least to greatest, by their horsepower. Is the ranking the same as that in the Optional Student Activity for Objective 8.5A?

Mustang, Mercury, GTO, Pontiac. Yes.

Example 3

A furnace is rated at 80,000 Btu per hour. How many foot-pounds of energy are released in 1 h?

Solution

$$80,000 \text{ Btu} = 80,000 \text{ Btu} \times \frac{778 \text{ ft} \cdot \text{lb}}{1 \text{ Btu}}$$
$$= 62,240,000 \text{ ft} \cdot \text{lb}$$

You Try It 3

A furnace is rated at 56,000 Btu per hour. How many foot-pounds of energy are released in 1 h?

Your solution
43,568,000 ft · lb

Objective B — **To use units of power in the U.S. Customary System**

Power is the rate at which work is done or the rate at which energy is released. Power is measured in **foot-pounds per second** $\left(\frac{\text{ft} \cdot \text{lb}}{\text{s}}\right)$. In each of the following examples, the amount of energy released is the same, but the time taken to release the energy is different; thus the power is different.

100 lb is lifted 10 ft in 10 s.

$$\text{Power} = \frac{10 \text{ ft} \times 100 \text{ lb}}{10 \text{ s}} = 100 \, \frac{\text{ft} \cdot \text{lb}}{\text{s}}$$

100 lb is lifted 10 ft in 5 s.

$$\text{Power} = \frac{10 \text{ ft} \times 100 \text{ lb}}{5 \text{ s}} = 200 \, \frac{\text{ft} \cdot \text{lb}}{\text{s}}$$

The U.S. Customary unit of power is the **horsepower.** A horse doing average work can pull 550 lb a distance of 1 ft in 1 s and can continue this work all day.

$$1 \text{ horsepower (hp)} = 550 \, \frac{\text{ft} \cdot \text{lb}}{\text{s}}$$

Example 4

Find the power needed to raise 300 lb a distance of 30 ft in 15 s.

Solution $\text{Power} = \dfrac{30 \text{ ft} \times 300 \text{ lb}}{15 \text{ s}}$

$$= 600 \, \frac{\text{ft} \cdot \text{lb}}{\text{s}}$$

You Try It 4

Find the power needed to raise 1200 lb a distance of 90 ft in 24 s.

Your solution $4500 \, \dfrac{\text{ft} \cdot \text{lb}}{\text{s}}$

Example 5

A motor has a power of $2750 \, \dfrac{\text{ft} \cdot \text{lb}}{\text{s}}$. Find the horsepower of the motor.

Solution $\dfrac{2750}{550} = 5 \text{ hp}$

You Try It 5

A motor has a power of $3300 \, \dfrac{\text{ft} \cdot \text{lb}}{\text{s}}$. Find the horsepower of the motor.

Your solution 6 hp

Solutions on p. S23

In-Class Examples (Objective 8.5B)

1. Convert $2200 \, \frac{\text{ft} \cdot \text{lb}}{\text{s}}$ to horsepower.   4 hp

2. Convert 8 hp to foot-pounds per second.
4400 ft · lb/s

3. Find the power, in foot-pounds per second, needed to raise 150 lb a distance of 8 ft in 2 s.
600 ft · lb/s

4. A motor has a power of $7700 \, \frac{\text{ft} \cdot \text{lb}}{\text{s}}$. Find the horsepower of the motor.   14 hp

8.5 Exercises

Objective A   To use units of energy in the U.S. Customary System

Suggested Assignment
Exercises 1–33, odds
More challenging problem:
    Exercise 34

1. Convert 25 Btu to foot-pounds.
19,450 ft · lb

2. Convert 6000 Btu to foot-pounds.
4,668,000 ft · lb

3. Convert 25,000 Btu to foot-pounds.
19,450,000 ft · lb

4. Convert 40,000 Btu to foot-pounds.
31,120,000 ft · lb

5. Find the energy required to lift 150 lb a distance of 10 ft.
1500 ft · lb

6. Find the energy required to lift 300 lb a distance of 16 ft.
4800 ft · lb

7. Find the energy required to lift a 3300-pound car a distance of 9 ft.
29,700 ft · lb

8. Find the energy required to lift a 3680-pound elevator a distance of 325 ft.
1,196,000 ft · lb

9. Three tons are lifted 5 ft. Find the energy required in foot-pounds.
30,000 ft · lb

10. Seven tons are lifted 12 ft. Find the energy required in foot-pounds.
168,000 ft · lb

11. A construction worker carries 3-pound blocks up a 10-foot flight of stairs. How many foot-pounds of energy are required to carry 850 blocks up the stairs?
25,500 ft · lb

12. A crane lifts an 1800-pound steel beam to the roof of a building 36 ft high. Find the amount of energy the crane requires in lifting the beam.
64,800 ft · lb

13. A furnace is rated at 45,000 Btu per hour. How many foot-pounds of energy are released by the furnace in 1 h?
35,010,000 ft · lb

14. A furnace is rated at 22,500 Btu per hour. How many foot-pounds of energy does the furnace release in 1 h?
17,505,000 ft · lb

15. Find the amount of energy in foot-pounds given off when 1 lb of coal is burned. One pound of coal gives off 12,000 Btu of energy when burned.
9,336,000 ft · lb

16. Find the amount of energy in foot-pounds given off when 1 lb of gasoline is burned. One pound of gasoline gives off 21,000 Btu of energy when burned.
16,338,000 ft · lb

Quick Quiz (Objective 8.5A)

1. Convert 2000 Btu to foot-pounds.   1,556,000 ft · lb

2. Find the energy required to lift 400 lb a distance of 15 ft.   6000 ft · lb

3. A furnace is rated at 65,000 Btu per hour. How many foot-pounds of energy are released by the furnace in 1 h?   50,570,000 ft · lb

Objective B   **To use units of power in the U.S. Customary System**

17. Convert $1100 \frac{\text{ft} \cdot \text{lb}}{\text{s}}$ to horsepower.
2 hp

18. Convert $6050 \frac{\text{ft} \cdot \text{lb}}{\text{s}}$ to horsepower.
11 hp

19. Convert $4400 \frac{\text{ft} \cdot \text{lb}}{\text{s}}$ to horsepower.
8 hp

20. Convert $1650 \frac{\text{ft} \cdot \text{lb}}{\text{s}}$ to horsepower.
3 hp

21. Convert 9 hp to foot-pounds per second.
$4950 \frac{\text{ft} \cdot \text{lb}}{\text{s}}$

22. Convert 4 hp to foot-pounds per second.
$2200 \frac{\text{ft} \cdot \text{lb}}{\text{s}}$

23. Convert 7 hp to foot-pounds per second.
$3850 \frac{\text{ft} \cdot \text{lb}}{\text{s}}$

24. Convert 8 hp to foot-pounds per second.
$4400 \frac{\text{ft} \cdot \text{lb}}{\text{s}}$

25. Find the power in foot-pounds per second needed to raise 125 lb a distance of 12 ft in 3 s.
$500 \frac{\text{ft} \cdot \text{lb}}{\text{s}}$

26. Find the power in foot-pounds per second needed to raise 500 lb a distance of 60 ft in 8 s.
$3750 \frac{\text{ft} \cdot \text{lb}}{\text{s}}$

27. Find the power in foot-pounds per second needed to raise 3000 lb a distance of 40 ft in 25 s.
$4800 \frac{\text{ft} \cdot \text{lb}}{\text{s}}$

28. Find the power in foot-pounds per second of an engine that can raise 12,000 lb to a height of 40 ft in 60 s.
$8000 \frac{\text{ft} \cdot \text{lb}}{\text{s}}$

29. Find the power in foot-pounds per second of an engine that can raise 180 lb to a height of 40 ft in 5 s.
$1440 \frac{\text{ft} \cdot \text{lb}}{\text{s}}$

30. Find the power in foot-pounds per second of an engine that can raise 1200 lb to a height of 18 ft in 30 s.
$720 \frac{\text{ft} \cdot \text{lb}}{\text{s}}$

31. A motor has a power of $4950 \frac{\text{ft} \cdot \text{lb}}{\text{s}}$. Find the horsepower of the motor.
9 hp

32. A motor has a power of $16,500 \frac{\text{ft} \cdot \text{lb}}{\text{s}}$. Find the horsepower of the motor.
30 hp

33. A motor has a power of $6600 \frac{\text{ft} \cdot \text{lb}}{\text{s}}$. Find the horsepower of the motor.
12 hp

APPLYING THE CONCEPTS

34.   Pick a source of energy and write an article about it. Include the source, possible pollution problems, and future prospects associated with this form of energy.

Answers to Writing Exercises

34. Students can select from a number of energy sources: natural gas, coal, wood, oil, nuclear power, hydroelectric power, or solar energy. Research will be required to answer the questions concerning sources, possible pollution problems, and future prospects associated with the form of energy.

Quick Quiz (Objective 8.5B)

1. Convert $3850 \frac{\text{ft} \cdot \text{lb}}{\text{s}}$ to horsepower.   7 hp

2. Convert 12 hp to foot-pounds per second.
6600 ft · lb/s

3. Find the power, in foot-pounds per second, needed to raise 1800 lb a distance of 20 ft in 15 s.
2400 ft · lb/s

4. A motor has a power of $13,200 \frac{\text{ft} \cdot \text{lb}}{\text{s}}$. Find the horsepower of the motor.   24 hp

Focus on Problem Solving

Applying Solutions to Other Problems

Problem solving in the previous chapters concentrated on solving specific problems. After a problem is solved, however, there is an important question to be asked: "Does the solution to this problem apply to other types of problems?"

To illustrate this extension of problem solving, we will consider *triangular numbers,* which were studied by ancient Greek mathematicians. The numbers 1, 3, 6, 10, 15, and 21 are the first six triangular numbers. What is the next triangular number?

To answer this question, note in the diagram below that a triangle can be formed using the number of dots that correspond to a triangular number.

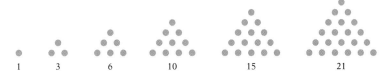

| 1 | 3 | 6 | 10 | 15 | 21 |

Observe that the number of dots in each row is one more than the number of dots in the row above. The total number of dots can be found by addition.

The pattern suggests that the next triangular number (the seventh one) is the sum of the first 7 natural numbers. The seventh triangular number is 28. The diagram at the right shows the seventh triangular number.

Using the pattern for triangular numbers, it is easy to determine that the tenth triangular number is

$$1 + 2 + 3 + 4 + 5 + 6 + 7 + 8 + 9 + 10 = 55$$

Now consider a situation that may seem to be totally unrelated to triangular numbers. Suppose you are in charge of scheduling softball games for a league. There are seven teams in the league, and each team must play every other team once. How many games must be scheduled?

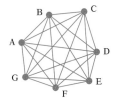

We label the teams A, B, C, D, E, F, and G. (See the figure at the left.) A line between two teams indicates that the two teams play each other. Beginning with A, there are 6 lines for the 6 teams that A must play. There are 6 teams that B must play, but the line between A and B has already been drawn, so there are only 5 remaining games to schedule for B. Now move on to C. The lines between C and A and between C and B have already been drawn, so there are only 4 additional lines to be drawn to represent the teams C will play. Moving on to D, we see that the lines between D and A, D and B, and D and C have already been drawn, so there are 3 more lines to be drawn to represent the teams D will play.

Note that each time we move from team to team, one fewer line needs to be drawn. When we reach F, there is only one line to be drawn, the one between F and G. The total number of lines drawn is 6 + 5 + 4 + 3 + 2 + 1 = 21, the sixth triangular number. For a league with 7 teams, the number of games that must be scheduled so that each team plays every other team once is the sixth triangular number. If there were ten teams in the league, the number of games that must be scheduled would be the ninth triangular number, which is 45.

A college chess team wants to schedule a match so that each of its 15 members plays each other member of the team once. How many matches must be scheduled?

Answers to Focus on Problem Solving— Applying Solutions to Other Problems

The number of matches that must be scheduled is the 14th triangular number, which is 105.

Projects and Group Activities

Nomographs

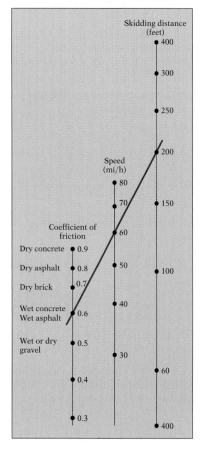

A chart is another tool that is used in problem solving. The chart at the left is a nomograph. A **nomograph** is a chart that represents numerical relationships among variables.

One of the details a traffic accident investigator checks when looking into a car accident is the length of the skid marks of the car. This length can help the investigator determine the speed of the car when the brakes were applied.

The nomograph at the left can be used to determine the speed of a car under given conditions. It shows the relationship among the speed of the car, the skidding distance, and the *coefficient of friction.*

The coefficient of friction is an experimentally obtained value that reflects how easy or hard it is to drag one object over another. For instance, it is easier to drag a box across ice than to drag it across a carpet. The coefficient of friction is smaller for the box and ice than it is for the box and carpet.

To use the nomograph at the left, an investigator would draw a line from the coefficient of friction to the skidding distance. The point at which the line crosses the speed line shows how fast the car was going when the brakes were applied. The line from 0.6 to 200 intersects the speed line at 60. This indicates that a car that skidded 200 ft on wet asphalt or wet concrete was traveling at 60 mi/h.

Answers to Projects and Group Activities— Nomographs

1. approximately 40 mi/h

2. About 62 mi/h

3. 150 ft

1. Use the nomograph to determine the speed of a car when the brakes were applied for a car traveling on gravel and for skid marks of 100 ft.

2. Use the nomograph to determine the speed of a car when the brakes were applied for a car traveling on dry concrete and for skid marks of 150 ft.

3. Suppose a car is traveling 80 mi/h when the brakes are applied. Find the difference in skidding distance if the car is traveling on wet concrete rather than dry concrete.

Averages

If two towns are 150 mi apart and you drive between the two towns in 3 h, then your

$$\text{Average speed} = \frac{\text{total distance}}{\text{total time}} = \frac{150 \text{ mi}}{3 \text{ h}} = 50 \text{ mi/h}$$

It is highly unlikely that your speed was *exactly* 50 mi/h the entire time of the trip. Sometimes you will have traveled faster than 50 mi/h, and other times you will have traveled slower than 50 mi/h. Dividing the total distance you traveled by the total time it took to go that distance is an example of calculating an average.

There are many other averages that may be calculated. For instance, the Environmental Protection Agency calculates an estimated miles per gallon (mpg) for new cars. Miles per gallon is an average calculated from the formula

$$\frac{\text{Miles traveled}}{\text{Gallons of gasoline consumed}}$$

For instance, the miles per gallon for a car that travels 308 mi on 11 gal of gas is $\frac{308 \text{ mi}}{11 \text{ gal}} = 28$ mpg.

A pilot would not use miles per gallon as a measure of fuel efficiency. Rather, pilots use gallons per hour. A plane that travels 5 h and uses 400 gal of fuel has an average that is calculated as

$$\frac{\text{Gallons of fuel}}{\text{Hours flown}} = \frac{400 \text{ gal}}{5 \text{ h}} = 80 \text{ gal/h}$$

Using the examples above, calculate the following averages.

1. Determine the average speed of a car that travels 355 mi in 6 h. Round to the nearest tenth.

2. Determine the miles per gallon of a car that can travel 405 mi on 12 gal of gasoline. Round to the nearest tenth.

3. If a plane flew 2000 mi in 5 h and used 1000 gal of fuel, determine the average number of gallons per hour that the plane used.

Another type of average is grade-point average (GPA). It is calculated by multiplying the units for each class by the grade point for that class, adding the results, and dividing by the total number of units taken. Here is an example using the grading scale A = 4, B = 3, C = 2, D = 1, and F = 0.

Class	Units	Grade
Math	4	B (= 3)
English	3	A (= 4)
French	5	C (= 2)
Biology	3	B (= 3)

$$\text{GPA} = \frac{4 \cdot 3 + 3 \cdot 4 + 5 \cdot 2 + 3 \cdot 3}{4 + 3 + 5 + 3} = \frac{43}{15} \approx 2.87$$

4. A grading scale that provides for plus or minus grades uses A = 4, A− = 3.7, B+ = 3.3, B = 3, B− = 2.7, C+ = 2.3, C = 2, C− = 1.7, D+ = 1.3, D = 1, D− = 0.7, and F = 0. Calculate the GPA of the student whose grades are given below.

Class	Units	Grade
Math	5	B +
English	3	C +
Spanish	5	A −
Physical science	3	B −

Chapter 8 Summary

Key Words	Examples
A *measurement* includes a number and a unit. [8.1A, p. 341]	9 inches, 6 feet, 3 yards, and 50 miles are measurements.
Equivalent measures are used to form *conversion rates* to change one unit in the U.S. Customary System of measurement to another. In the conversion rate chosen, the unit in the numerator is the same as the unit desired in the answer. The unit in the denominator is the same as the unit in the given measurement. [8.1A, p. 341]	Because 12 in. = 1 ft, the conversion rate $\frac{12 \text{ in.}}{1 \text{ ft}}$ is used to convert feet to inches. The conversion rate $\frac{1 \text{ ft}}{12 \text{ in.}}$ is used to convert inches to feet.

Energy is the ability to do work. One *foot-pound* (ft · lb) of energy is the amount of energy necessary to lift 1 pound a distance of 1 foot. Consumer items that use energy are rated in *British thermal units* (Btu). [8.5A, p. 357]

Find the energy required for a 110-pound person to climb a set of stairs 12 ft high.

Energy = 110 lb × 12 ft = 1320 ft · lb

Power is the rate at which work is done or energy is released.

Power is measured in *foot-pounds per second* $\left(\frac{\text{ft} \cdot \text{lb}}{\text{s}}\right)$ and *horsepower* (hp). [8.5B, p. 358]

Find the power needed to raise 250 lb a distance of 20 ft in 10 s.

$$\text{Power} = \frac{20 \text{ ft} \times 250 \text{ lb}}{10 \text{ s}} = 500\frac{\text{ft} \cdot \text{lb}}{\text{s}}$$

Essential Rules and Procedures

Examples

Equivalences Between Units of Length [8.1A, p. 341]
The U.S. Customary units of length are inch (in.), foot (ft), yard (yd), and mile (mi).
12 in. = 1 ft
3 ft = 1 yd
36 in. = 1 yd
5280 ft = 1 mi

Convert 52 in. to ft.

$$52 \text{ in.} = 52 \text{ in.} \times \frac{1 \text{ ft}}{12 \text{ in.}}$$
$$= \frac{52 \text{ ft}}{12} = 4\frac{1}{3} \text{ ft}$$

Equivalences Between Units of Weight [8.2A, p. 347]
Weight is a measure of how strongly Earth is pulling on an object. The U.S. Customary units of weight are ounce (oz), pound (lb), and ton.
16 oz = 1 lb
2000 lb = 1 ton

Convert 9 lb to ounces.

$$9 \text{ lb} = 9 \text{ lb} \times \frac{16 \text{ oz}}{1 \text{ lb}} = 144 \text{ oz}$$

Equivalences Between Units of Capacity [8.3A, p. 351]
Liquid substances are measured in units of *capacity*. The U.S. Customary units of capacity are fluid ounce (fl oz), cup (c), pint (pt), quart (qt), and gallon (gal).
8 fl oz = 1 c
2 c = 1 pt
2 pt = 1 qt
4 qt = 1 gal

Convert 14 qt to gallons.

$$14 \text{ qt} = 14 \text{ qt} \times \frac{1 \text{ gal}}{4 \text{ qt}}$$
$$= \frac{14 \text{ gal}}{4} = 3\frac{1}{2} \text{ gal}$$

Equivalences Between Units of Time [8.4A, p. 355]
Units of time are seconds (s), minutes (min), hours (h), days, and weeks.
60 s = 1 min
60 min = 1 h
24 h = 1 day
7 days = 1 week

Convert 8 days to hours.

$$8 \text{ days} = 8 \text{ days} \times \frac{24 \text{ h}}{1 \text{ day}} = 192 \text{ h}$$

Equivalences Between Units of Energy [8.5A, p. 357]
1 Btu = 778 ft · lb

Convert 70 Btu to foot-pounds.

$$70 \text{ Btu} = 70 \text{ Btu} \times \frac{778 \text{ ft} \cdot \text{lb}}{1 \text{ Btu}}$$
$$= 54,460 \text{ ft} \cdot \text{lb}$$

Equivalences Between Units of Power [8.5B, p. 358]
The U.S. Customary unit of power is the horsepower (hp).

$$1 \text{ hp} = 550 \frac{\text{ft} \cdot \text{lb}}{\text{s}}$$

Convert 5 hp to foot-pounds per second.

$$5 \times 550 = 2750 \frac{\text{ft} \cdot \text{lb}}{\text{s}}$$

Chapter 8 Review Exercises

1. Convert 4 ft to inches.
   48 in. [8.1A]

2. What is 7 ft 6 in. divided by 3?
   2 ft 6 in. [8.1B]

3. Find the energy needed to lift 200 lb a distance of 8 ft.
   1600 ft · lb [8.5A]

4. Convert $2\frac{1}{2}$ pt to fluid ounces.
   40 fl oz [8.3A]

5. Convert 14 ft to yards.
   $4\frac{2}{3}$ yd [8.1A]

6. Convert 2400 lb to tons.
   $1\frac{1}{5}$ tons [8.2A]

7. Find the quotient of 7 lb 5 oz and 3.
   2 lb 7 oz [8.2B]

8. Convert $3\frac{3}{8}$ lb to ounces.
   54 oz [8.2A]

9. Add: 3 ft 9 in.
         + 5 ft 6 in.
         9 ft 3 in. [8.1B]

10. Subtract: 3 tons 500 lb
              − 1 ton 1500 lb
              1 ton 1000 lb [8.2B]

11. Add: 4 c 7 fl oz
         + 2 c 3 fl oz
         7 c 2 fl oz [8.3B]

12. Subtract: 5 yd 1 ft
              − 3 yd 2 ft
              1 yd 2 ft [8.1B]

13. Convert 12 c to quarts.
    3 qt [8.3A]

14. Convert 375 min to hours.
    $6\frac{1}{4}$ h [8.4A]

15. Convert 2.5 hp to foot-pounds per second. $\left(1 \text{ hp} = 550 \frac{\text{ft} \cdot \text{lb}}{\text{s}}.\right)$

$1375 \frac{\text{ft} \cdot \text{lb}}{\text{s}}$ [8.5B]

16. Multiply:
$$
\begin{array}{r}
5 \text{ lb } 8 \text{ oz} \\
\times \quad\quad 8 \\
\hline
44 \text{ lb}
\end{array}
$$
[8.2B]

17. Convert 50 Btu to foot-pounds. (1 Btu = 778 ft · lb.)

38,900 ft · lb [8.5A]

18. Convert $3850 \frac{\text{ft} \cdot \text{lb}}{\text{s}}$ to horsepower. $\left(1 \text{ hp} = 550 \frac{\text{ft} \cdot \text{lb}}{\text{s}}.\right)$

7 hp [8.5B]

19. **Carpentry** A board 6 ft 11 in. long is cut from a board 10 ft 5 in. long. Find the length of the remaining piece of board.

3 ft 6 in. [8.1C]

20. **Shipping** A book weighing 2 lb 3 oz is mailed at the parcel post rate of $.24 per ounce. Find the cost of mailing the book.

$8.40 [8.2C]

21. **Capacity** A can of pineapple juice contains 18 fl oz. Find the number of quarts in a case of 24 cans.

$13\frac{1}{2}$ qt [8.3C]

22. **Food Service** A cafeteria sold 256 cartons of milk in one school day. Each carton contains 1 c of milk. How many gallons of milk were sold that day?

16 gal [8.3C]

23. **Energy** A furnace is rated at 35,000 Btu per hour. How many foot-pounds of energy does the furnace release in 1 h? (1 Btu = 778 ft · lb)

27,230,000 ft · lb [8.5A]

24. **Power** Find the power in foot-pounds per second of an engine that can raise 800 lb to a height of 15 ft in 25 s.

$480 \frac{\text{ft} \cdot \text{lb}}{\text{s}}$ [8.5B]

Chapter 8 Test

1. Convert $2\frac{1}{2}$ ft to inches.

30 in. [8.1A]

2. Subtract: 4 ft 2 in. − 1 ft 9 in.

2 ft 5 in. [8.1B]

3. **Carpentry**   A board $6\frac{2}{3}$ ft long is cut into five equal pieces. How long is each piece?

$1\frac{1}{3}$ ft [8.1C]

4. **Masonry**   Seventy-two bricks, each 8 in. long, are laid end to end to make the base for a wall. Find the length of the wall in feet.

48 ft [8.1C]

5. Convert $2\frac{7}{8}$ lb to ounces.

46 oz [8.2A]

6. Convert: 40 oz = __2__ lb __8__ oz [8.2B]

7. Find the sum of 9 lb 6 oz and 7 lb 11 oz.

17 lb 1 oz [8.2B]

8. Divide: 6 lb 12 oz ÷ 4

1 lb 11 oz [8.2B]

9. **Weights**   A college bookstore received 1000 workbooks, each weighing 12 oz. Find the total weight of the 1000 workbooks in pounds.

750 lb [8.2C]

10. **Recycling**   An elementary school class gathered 800 aluminum cans for recycling. Four aluminum cans weigh 3 oz. Find the amount the class received if the rate of pay was $.75 per pound for the aluminum cans. Round to the nearest cent.

$28.13 [8.2C]

11. Convert 13 qt to gallons.

$3\frac{1}{4}$ gal [8.3A]

12. Convert $3\frac{1}{2}$ gal to pints.

28 pt [8.3A]

13. What is $1\frac{3}{4}$ gal times 7?

$12\frac{1}{4}$ gal   [8.3B]

14. Add: 5 gal 2 qt + 2 gal 3 qt

8 gal 1 qt   [8.3B]

15. Convert 756 h to weeks.

$4\frac{1}{2}$ weeks   [8.4A]

16. Convert $3\frac{1}{4}$ days to minutes.

4680 min   [8.4A]

17. Capacity   A can of grapefruit juice contains 20 fl oz. Find the number of cups of grapefruit juice in a case of 24 cans.

60 c   [8.3C]

18. Business   Nick, a mechanic, bought oil in 40-gallon containers for changing the oil in customers' cars. He paid $200 for the 40 gal of oil and charged customers $2.15 per quart. Find the profit Nick made on one 40-gallon container of oil.

$144   [8.3C]

19. Energy   Find the energy required to lift 250 lb a distance of 15 ft.

3750 ft · lb   [8.5A]

20. Energy   A furnace is rated at 40,000 Btu per hour. How many foot-pounds of energy are released by the furnace in 1 h? (1 Btu = 778 ft · lb)

31,120,000 ft · lb   [8.5A]

21. Power   Find the power needed to lift 200 lb a distance of 20 ft in 25 s.

$160\dfrac{\text{ft} \cdot \text{lb}}{\text{s}}$ [8.5B]

22. Power   A motor has a power of 2200 $\dfrac{\text{ft} \cdot \text{lb}}{\text{s}}$. Find the motor's horsepower. $\left(1 \text{ hp} = 550 \dfrac{\text{ft} \cdot \text{lb}}{\text{s}}\right)$

4 hp   [8.5B]

Cumulative Review Exercises

1. Find the LCM of 9, 12, and 15.
   180   [2.1A]

2. Write $\frac{43}{8}$ as a mixed number.
   $5\frac{3}{8}$ [2.2B]

3. Subtract: $5\frac{7}{8} - 2\frac{7}{12}$
   $3\frac{7}{24}$ [2.5C]

4. What is $5\frac{1}{3}$ divided by $2\frac{2}{3}$?
   2   [2.7B]

5. Simplify: $\frac{5}{8} \div \left(\frac{3}{8} - \frac{1}{4}\right) - \frac{5}{8}$
   $4\frac{3}{8}$ [2.8C]

6. Round 2.0972 to the nearest hundredth.
   2.10   [3.1B]

7. Multiply: 0.0792
           $\times$ 0.49
             0.038808   [3.4A]

8. Solve the proportion: $\frac{n}{12} = \frac{44}{60}$
   8.8   [4.3B]

9. Find $2\frac{1}{2}\%$ of 50.
   1.25   [5.2A]

10. 18 is 42% of what? Round to the nearest hundredth.
    42.86   [5.4A]

11. **Consumerism** A 7.2-pound roast costs $37.08. Find the unit cost.
    $5.15/lb   [6.1A]

12. Add: $3\frac{2}{5}$ in. + $5\frac{1}{3}$ in.
    $8\frac{11}{15}$ in.   [8.1B]

13. Convert: 24 oz = __1__ lb __8__ oz
    [8.2B]

14. Multiply: 3 lb 8 oz $\times$ 9
    31 lb 8 oz   [8.2B]

15. Subtract: $4\frac{1}{3}$ qt − $1\frac{5}{6}$ qt
    $2\frac{1}{2}$ qt   [8.3B]

16. Find 2 lb 10 oz less than 4 lb 6 oz.
    1 lb 12 oz   [8.2B]

17. **Investments** An investor receives a dividend of $56 from 40 shares of stock. At the same rate, find the dividend that 200 shares of stock would yield.
$280 [4.3C]

18. **Banking** Anna had a balance of $578.56 in her checkbook. She wrote checks for $216.98 and $34.12 and made a deposit of $315.33. What is her new checking account balance?
$642.79 [6.7A]

19. **Compensation** An account executive receives a salary of $1800 per month plus a commission of 2% on all sales over $25,000. Find the total monthly income of an account executive who has monthly sales of $140,000.
$4100 [6.6A]

20. **Produce** A health inspector found that 3% of a 2500-pound shipment of carrots were spoiled and could not be sold. Find the amount of carrots that could be sold.
2425 lb [5.2B]

21. **Education** The scores on the final exam of a trigonometry class are recorded in the histogram in the figure to the right. What percent of the class received a score between 80% and 90%? Round to the nearest percent.
18% [7.3A]

22. **Markup** Hayes Department Store uses a markup rate of 40% on all merchandise. What is the selling price of a compact disc player that cost the store $220?
$308 [6.2B]

23. **Simple Interest** A construction firm received a loan of $200,000 for 8 months at a simple interest rate of 6%. Find the interest paid on the loan.
$8000 [6.3A]

24. **Income** Six college students spent several weeks panning for gold during their summer vacation. The students obtained 1 lb 3 oz of gold, which they sold for $200 per ounce. How much money did each student receive if they shared the money equally? Round to the nearest dollar.
$633 [8.2C]

25. **Shipping** Four books were mailed at the postal rate of $.28 per ounce. The books weighed 1 lb 3 oz, 13 oz, 1 lb 8 oz, and 1 lb. Find the cost of mailing the books.
$20.16 [8.2C]

26. **Consumerism** One brand of yogurt costs $.79 for 8 oz, and 36 oz of another brand can be bought for $2.98. Which purchase is the better buy?
36 oz for $2.98 [6.1B]

27. **Probability** Two dice are rolled. What is the probability that the sum of the dots on the upward faces is 9? $\frac{1}{9}$ [7.5A]

28. **Energy** Find the energy required to lift 400 lb a distance of 8 ft.
3200 ft · lb [8.5A]

29. **Power** Find the power, in foot-pounds per second, needed to raise 600 lb a distance of 8 ft in 12 s.
$400\dfrac{\text{ft} \cdot \text{lb}}{\text{s}}$ [8.5B]

9 The Metric System of Measurement

We use electrical energy in countless ways each and every day. When you turn on a light, use a microwave, start up the air conditioner, or turn on the television, you are using electrical energy. The watt-hour is used for measuring the amount of electrical energy an appliance uses. For example, light bulbs usually use 60 watts, and microwaves can use 500 watts. **Exercises 11 to 19 on page 388** ask you to determine quantities such as the number of kilowatt-hours of energy used to watch television, or how much it costs to listen to a CD player over the course of two weeks when you know how much each kilowatt-hour costs.

OBJECTIVES

Section 9.1

A To convert units of length in the metric system of measurement

B To solve application problems

Section 9.2

A To convert units of mass in the metric system of measurement

B To solve application problems

Section 9.3

A To convert units of capacity in the metric system of measurement

B To solve application problems

Section 9.4

A To use units of energy in the metric system of measurement

Section 9.5

A To convert U.S. Customary units to metric units

B To convert metric units to U.S. Customary units

Need help? For online student resources, such as section quizzes, visit this textbook's website at **math.college.hmco.com/students.**

PREP TEST • • •

Do these exercises to prepare for Chapter 9.

For Exercises 1 to 10, add, subtract, multiply, or divide.

1. $3.732 \times 10{,}000$
37,320   [3.4A]

2. 65.9×10^4
659,000   [3.4A]

3. $41.07 \div 1000$
0.04107   [3.5A]

4. $28{,}496 \div 10^3$
28.496   [3.5A]

5. $6 - 0.875$
5.125   [3.3A]

6. $5 + 0.96$
5.96   [3.2A]

7. 3.25×0.04
0.13   [3.4A]

8. $35 \times \dfrac{1.61}{1}$
56.35   [2.6A, 3.4A]

9. $1.67 \times \dfrac{1}{3.34}$
0.5   [2.6B, 3.5A]

10. $4\dfrac{1}{2} \times 150$
675   [2.6B]

GO FIGURE • • •

Suppose you threw six darts and all six hit the target shown. Which of the following could be your score?

4, 15, 58, 28, 29, 31
28

9.1 Length

Objective A

To convert units of length in the metric system of measurement

In 1789, an attempt was made to standardize units of measurement internationally in order to simplify trade and commerce between nations. A commission in France developed a system of measurement known as the **metric system.**

The basic unit of length in the metric system is the **meter.** One meter is approximately the distance from a doorknob to the floor. All units of length in the metric system are derived from the meter. Prefixes to the basic unit denote the length of each unit. For example, the prefix "centi-" means one-hundredth, so 1 centimeter is 1 one-hundredth of a meter.

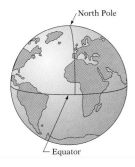

≈1 meter

North Pole

Equator

Prefixes and Units of Length in the Metric System

kilo- = 1000	1 kilometer (km) = 1000 meters (m)
hecto- = 100	1 hectometer (hm) = 100 m
deca- = 10	1 decameter (dam) = 10 m
	1 meter (m) = 1 m
deci- = 0.1	1 decimeter (dm) = 0.1 m
centi- = 0.01	1 centimeter (cm) = 0.01 m
milli- = 0.001	1 millimeter (mm) = 0.001 m

Conversion between units of length in the metric system involves moving the decimal point to the right or to the left. Listing the units in order from largest to smallest will indicate how many places to move the decimal point and in which direction.

To convert 4200 cm to meters, write the units in order from largest to smallest.

km hm dam m dm cm mm

2 positions

4200 cm = 42.00 m

2 places

* Converting cm to m requires moving 2 positions to the left.

* Move the decimal point the same number of places and in the same direction.

A metric measurement that involves two units is customarily written in terms of one unit. Convert the smaller unit to the larger unit and then add.

To convert 8 km 32 m to kilometers, first convert 32 m to kilometers.

km hm dam m dm cm mm

32 m = 0.032 km

8 km 32 m = 8 km + 0.032 km
= 8.032 km

* Converting m to km requires moving 3 positions to the left.

* Move the decimal point the same number of places and in the same direction.

* Add the result to 8 km.

Copyright © Houghton Mifflin Company. All rights reserved.

Point of Interest

Originally the meter (spelled *metre* in some countries) was defined as $\frac{1}{10,000,000}$ of the distance from the equator to the North Pole. Modern scientists have redefined the meter as 1,650,763.73 wavelengths of the orange-red light given off by the element krypton.

Study Tip

The prefixes introduced here are used throughout the chapter. As you study the material in the remaining sections, use the table above for a reference or refer to the Chapter Summary at the end of this chapter.

Objective 9.1A

New Vocabulary
metric system
meter
kilo-
hecto-
deca-
deci-
centi-
milli-

Discuss the Concepts
1. Explain how to convert meters to centimeters.
2. Explain how to convert meters to kilometers.

Concept Check
1. One of the events in the summer Olympics is the 50,000-meter walk. How many kilometers do the entrants in this event walk?   50 km
2. One of the events in the winter Olympic Games is the 10,000-meter speed skating. How many kilometers do the entrants in this event skate?
   10 km

Optional Student Activity
A column assembly is being constructed in a building. The components are shown in the diagram below. What length of column must be cut?   215.5 cm

2.4 m   Column B

A: 22-centimeter girder
B: 1.25-centimeter plate
C: 20-centimeter concrete footing

In-Class Examples (Objective 9.1A)
Convert.
1. 51 cm = ___ mm   510 mm
2. 1.725 km = ___ m   1725 m
3. 6860 m = ___ km   6.860 km
4. 38 m 4 cm = ___ m   38.04 m

Example 1 Convert 0.38 m to millimeters.

Solution 0.38 m = 380 mm

You Try It 1 Convert 3.07 m to centimeters.

Your solution 307 cm

Example 2 Convert 4 m 62 cm to meters.

Solution 62 cm = 0.62 m

4 m 62 cm = 4 m + 0.62 m
= 4.62 m

You Try It 2 Convert 3 km 750 m to kilometers.

Your solution 3.750 km

Solutions on p. S23

Objective 9.1B

Concept Check

Scientists have observed that a glacier on the coast of Greenland has been flowing at the rate of 20 m/day. At this rate, in how many days does the glacier move a distance of 1 km?

50 days

Optional Student Activity

1. The perimeter of a triangle is the sum of the lengths of the three sides of the triangle. Find the perimeter of a triangle in which the first side measures 18 cm, the second side measures twice the first side, and the third side measures twice the first side.

90 cm

2. To find the area of a square, multiply the length of one side of the square times itself. If the area of a square is more than 16 m² (square meters) but less than 64 m², what are the possible values for the length of one side of the square?

Between 4 m and 8 m

3. The area of a rectangle is equal to the product of the length of the rectangle times its width. If a rectangle is 5 cm longer than it is wide, and the area is less than 36 cm² (square centimeters), what is the largest possible value for the width of the rectangle?

4 cm

Objective B **To solve application problems**

> **TAKE NOTE**
> Although in this text we will always change units to the larger unit, it is possible to perform the calculation by changing to the smaller unit.
>
> 2 m − 85 cm
> = 200 cm − 85 cm
> = 115 cm
>
> Note that 115 cm = 1.15 m.

In the application problems in this section, we perform arithmetic operations with the measurements of length in the metric system. It is important to remember that before measurements can be added or subtracted, they must be expressed in terms of the same unit. In this textbook, unless otherwise stated, the units should be changed to the larger unit before the arithmetic operation is performed.

To subtract 85 cm from 2 m, convert 85 cm to meters.

$$2 \text{ m} - 85 \text{ cm} = 2 \text{ m} - 0.85 \text{ m}$$
$$= 1.15 \text{ m}$$

Example 3

A piece measuring 142 cm is cut from a board 4.20 m long. Find the length of the remaining piece.

Strategy

To find the length of the remaining piece:

• Convert the length of the piece cut (142 cm) to meters.
• Subtract the length of the piece cut from the original length.

Solution

142 cm = 1.42 m

4.20 m − 142 cm = 4.20 m − 1.42 m
= 2.78 m

The length of the remaining piece is 2.78 m.

You Try It 3

A bookcase 175 cm long has four shelves. Find the cost of the shelves when the price of the lumber is $15.75 per meter.

Your strategy

Your solution
$110.25

Solution on pp. S23–S24

In-Class Examples (Objective 9.1B)

1. A bicycle rider rode 8 km 500 m in 1 h. At this rate, how far can the biker ride in 2 h? 17 km
2. A living room is 3 m 90 cm wide and 4 m 80 cm long. Find the length of molding needed to put around the top edge of the four walls. 17.4 m

9.1 Exercises

Objective A   **To convert units of length in the metric system of measurement**

Suggested Assignment
Exercises 1–37, odds

For Exercises 1 to 27, convert.

1. 42 cm = _420_ mm

2. 62 cm = _620_ mm

3. 81 mm = _8.1_ cm

4. 68.2 mm = _6.82_ cm

5. 6804 m = _6.804_ km

6. 3750 m = _3.750_ km

7. 2.109 km = _2109_ m

8. 32.5 km = _32,500_ m

9. 432 cm = _4.32_ m

10. 61.7 cm = _0.617_ m

11. 0.88 m = _88_ cm

12. 3.21 m = _321_ cm

13. 7038 m = _7.038_ km

14. 2589 m = _2.589_ km

15. 3.5 km = _3500_ m

16. 9.75 km = _9750_ m

17. 260 cm = _2.60_ m

18. 705 cm = _7.05_ m

19. 1.685 m = _168.5_ cm

20. 0.975 m = _97.5_ cm

21. 14.8 cm = _148_ mm

22. 6 m 42 cm = _6.42_ m

23. 62 m 7 cm = _62.07_ m

24. 42 cm 6 mm = _42.6_ cm

25. 31 cm 9 mm = _31.9_ cm

26. 62 km 482 m = _62.482_ km

27. 8 km 75 m = _8.075_ km

Objective B   **To solve application problems**

28. **Carpentry** How many shelves, each 140 cm long, can be cut from a board that is 4.20 m in length? Find the length of the board remaining after the shelves are cut.
3 shelves; no length remaining

Quick Quiz (Objective 9.1A)
Convert.
1. 79 mm = ___ cm   7.9 cm
2. 5370 m = ___ km   5.370 km
3. 311 cm = ___ m   3.11 m
4. 51 km 376 m = ___ km   51.376 km

29. Measurements Find the missing dimension, in centimeters, in the diagram at the right.
7.8 cm

30. Sports A walk-a-thon had two checkpoints. One checkpoint was 1400 m from the starting point. The second checkpoint was 1200 m from the first checkpoint. The second checkpoint was 1800 m from the finish line. How long was the walk? Express the answer in kilometers.
4.4 km

31. Metal Works Twenty rivets are used to fasten two steel plates together. The plates are 3.4 m long, and the rivets are equally spaced, with a rivet at each end. Find the distance between the rivets. Round to the nearest tenth of a centimeter.
17.9 cm

32. Measurements Find the total length, in centimeters, of the shaft in the diagram at the right.
181 cm

33. Fencing You purchase a 50-meter roll of fencing, at a cost of $14.95 per meter, in order to build a dog run that is 340 cm wide and 1380 cm long. After you cut the four pieces of fencing from the roll, how much of the fencing is left on the roll?
15.6 m

34. Highway Cleanup Carmine is a member of a group that has adopted 10 km of highway. During this week, Carmine cleaned up litter along the highway for 2500 m, 1500 m, 1200 m, 1300 m, and 1400 m. Find the average number of kilometers Carmine cleaned up on each of the 5 days this week.
1.58 km

35. ⬤ **Astronomy** The distance between Earth and the sun is 150,000,000 km. Light travels 300,000,000 m in 1 s. How long does it take for light to reach Earth from the sun?
500 s

36. ▦ ⬤ **Earth Science** The circumference of Earth is 40,000 km. How long would it take to travel the circumference of Earth at a speed of 85 km per hour? Round to the nearest tenth.
470.6 h

37. ▦ ⬤ **Physics** Light travels 300,000 km in 1 s. How far does light travel in 1 day?
25,920,000,000 km

APPLYING THE CONCEPTS

38. Other prefixes in the metric system are becoming more commonly used as a result of technological advances. Find the meaning of the following prefixes: tera-, giga-, mega-, micro-, nano-, and pico.
tera: trillion, giga: billion, mega: million, micro: millionth, nano: billionth, pico: trillionth

39. ✎ Write a short history of the metric system.

Quick Quiz (Objective 9.1B)

1. A contractor needs 30 rafters, each 5 m 60 cm long. Find the total length in meters of the rafters needed.   168 m

2. An aluminum frame 6 m 72 cm long is cut into four equal pieces. Find the length of each piece.
1.68 m

9.2 Mass

Objective A To convert units of mass in the metric system of measurement

1 gram = the mass of water in the box

Mass and weight are closely related. Weight is a measure of how strongly Earth is pulling on an object. Therefore, an object's weight is less in space than on Earth's surface. However, the amount of material in the object, its **mass,** remains the same. On the surface of Earth, mass and weight can be used interchangeably.

The basic unit of mass in the metric system is the **gram.** If a box that is 1 cm long on each side is filled with water, then the mass of that water is 1 gram.

The gram is a very small unit of mass. A paper clip weighs about 1 gram. The kilogram (1000 grams) is a more useful unit of mass in consumer applications. This textbook weighs about 1 kilogram.

The units of mass in the metric system have the same prefixes as the units of length.

Point of Interest

An average snowflake weighs about $\frac{1}{300}$ g and contains approximately 100,000,000,000,000,000,000 water molecules. You may have heard the expression "No two snowflakes are the same." It was this large number of water molecules, and the number of their possible arrangements, that led to that statement.

> **Units of Mass in The Metric System**
>
> 1 kilogram (kg) = 1000 grams (g)
> 1 hectogram (hg) = 100 g
> 1 decagram (dag) = 10 g
> 1 gram (g) = 1 g
> 1 decigram (dg) = 0.1 g
> 1 centigram (cg) = 0.01 g
> 1 milligram (mg) = 0.001 g

Weight ≈ 1 gram

Conversion between units of mass in the metric system involves moving the decimal point to the right or to the left. Listing the units in order from largest to smallest will indicate how many places to move the decimal point and in which direction.

To convert 324 g to kilograms, first write the units in order from largest to smallest.

kg hg dag g dg cg mg
        3 positions

• Converting g to kg requires moving 3 positions to the left.

324 g = 0.324 kg
        3 places

• Move the decimal point the same number of places and in the same direction.

Example 1	Convert 4.23 g to milligrams.	**You Try It 1**	Convert 42.3 mg to grams.
Solution	4.23 g = 4230 mg	**Your solution**	0.0423 g

Solution on p. S24

Vocabulary to Review
kilo-   [9.1A]
hecto-   [9.1A]
deca-   [9.1A]
deci-   [9.1A]
centi-   [9.1A]
milli-   [9.1A]

New Vocabulary
mass
gram

Instructor Note
The difference between weight and mass is difficult for students to understand. As another illustration, mention that the mass of an astronaut is the same on Earth and on the moon; the weight of the astronaut, however, is different.

For students unfamiliar with the metric system, further examples of weights may be helpful. For instance, $4\frac{1}{2}$ sticks of butter weigh about 1 kg, a large thumbtack weighs about 1 g, and a nickel weighs about 5 g.

Discuss the Concepts
Why is it necessary to have internationally standardized units of measurement?

Instructor Note
Here is another example of the use of the prefix giga-, which means "billion":
The mass of Earth gains 40 gigagrams (Gg) each year from captured meteorites and cosmic dust.

In-Class Examples (Objective 9.2A)
Convert.
1. 360 g = ___ kg   0.360 kg
2. 354 mg = ___ g   0.354 g
3. 3.1 kg = ___ g   3100 g
4. 8 kg 713 g = ___ kg   8.713 kg

Example 2 Convert 2 kg 564 g to kilograms.

Solution   564 g = 0.564 kg

2 kg 564 g = 2 kg + 0.564 kg
= 2.564 kg

You Try It 2 Convert 3 g 54 mg to grams.

Your solution 3.054 g

Solution on p. S24

Concept Check

The size of muffins sold in the United States has been increasing. A 28-gram muffin is just about bite size. A 130-gram muffin is about the size of a softball. A 28-gram lemon muffin contains 6 g of fat, and a 130-gram lemon muffin contains 28 g of fat. Find the difference in fat intake, in kilograms, between eating one 130-gram lemon muffin each morning for 8 weeks and eating a 28-gram lemon muffin each morning for 8 weeks. 1.232 kg

Optional Student Activity

(*Note:* These are critical thinking problems.)

1. The total weight of a box and its contents of 25 identical poker chips is 79 g. After 9 chips are removed from the box, the total weight of the box and the remaining chips is 61 g. Find the weight of the box. 29 g

2. A man and his two children must cross a river. The raft they will use can carry only 100 kg. The man weighs 100 kg and each of his children weighs 50 kg. Describe how they can cross the river.

The two children cross. One stays on the far side, while the second brings the raft back to his father. The father crosses alone, and the child on the far side returns, picks up his sibling, and they return together to the far side.

Objective B **To solve application problems**

TAKE NOTE
Although in this text we will always change units to the larger unit, it is possible to perform the calculation by changing to the smaller unit.

3 kg − 750 g
= 3000 g − 750 g
= 2250 g

Note that
2250 g = 2.250 kg.

In the application problems in this section, we perform arithmetic operations with the measurements of mass in the metric system. Remember that before measurements can be added or subtracted, they must be expressed in terms of the same unit. In this textbook, unless otherwise stated, the units should be changed to the larger unit before the arithmetic operation is performed.

To subtract 750 g from 3 kg, convert 750 g to kilograms.

$$3 \text{ kg} - 750 \text{ g} = 3 \text{ kg} - 0.750 \text{ kg}$$
$$= 2.250 \text{ kg}$$

Example 3

Find the cost of three packages of ground meat weighing 540 g, 670 g, and 890 g if the price per kilogram is $9.89. Round to the nearest cent.

Strategy
To find the cost of the meat:
• Find the total weight of the three packages.
• Convert the total weight to kilograms.
• Multiply the weight by the cost per kilogram ($9.89).

Solution
540 g + 670 g + 890 g = 2100 g

2100 g = 2.1 kg

2.1 × 9.89 = 20.769

The cost of the meat is $20.77.

You Try It 3

How many kilograms of fertilizer are required to fertilize 400 trees in an apple orchard if 300 g of fertilizer are used for each tree?

Your strategy

Your solution 120 kg

Solution on p. S24

In-Class Examples (Objective 9.2B)

1. A 10- by 10-centimeter tile weighs 350 g. Find the weight in kilograms of a box containing 72 tiles. 25.2 kg

2. A shopper purchased four packages of cheese, which weighed 850 g, 920 g, 780 g, and 760 g. Find the total weight in kilograms of the four packages of cheese. 3.31 kg

9.2 Exercises

Objective A   To convert units of mass in the metric system of measurement

For Exercises 1 to 24, convert.

1. 420 g = _0.420_ kg

2. 7421 g = _7.421_ kg

3. 127 mg = _0.127_ g

4. 43 mg = _0.043_ g

5. 4.2 kg = _4200_ g

6. 0.027 kg = _27_ g

7. 0.45 g = _450_ mg

8. 325 g = _325,000_ mg

9. 1856 g = _1.856_ kg

10. 8900 g = _8.900_ kg

11. 4057 mg = _4.057_ g

12. 1970 mg = _1.970_ g

13. 1.37 kg = _1370_ g

14. 5.1 kg = _5100_ g

15. 0.0456 g = _45.6_ mg

16. 0.2 g = _200_ mg

17. 18,000 g = _18.000_ kg

18. 0.87 kg = _870_ g

19. 3 kg 922 g = _3.922_ kg

20. 1 kg 47 g = _1.047_ kg

21. 7 g 891 mg = _7.891_ g

22. 209 g 42 mg = _209.042_ g

23. 4 kg 63 g = _4.063_ kg

24. 18 g 5 mg = _18.005_ g

Objective B   To solve application problems

25. **Consumerism**   A 1.19-kilogram container of Quaker Oats contains 30 servings. Find the number of grams in one serving of the oatmeal. Round to the nearest whole number.
40 g

Quick Quiz (Objective 9.2A)

Convert.

1. 163 g = ___ mg   163,000 mg

2. 2.74 kg = ___ g   2740 g

3. 3712 g = ___ kg   3.712 kg

4. 6 kg 38 g = ___ kg   6.038 kg

26. **Nutrition** A patient is advised to supplement her diet with 2 g of calcium per day. The calcium tablets she purchases contain 500 mg of calcium per tablet. How many tablets per day should the patient take?
    4 tablets

27. **Nutrition**
    a. One egg contains 274 mg of cholesterol. How many grams of cholesterol are in a dozen eggs?   3.288 g
    b. One glass of milk contains 33 mg of cholesterol. How many grams of cholesterol are in four glasses of milk?   0.132 g

28. **Gemology** A carat is a unit of weight equal to 200 mg. Find the weight in grams of a 10-carat precious stone.
    2 g

29. **Consumerism** The nutrition label for a corn bread mix is shown at the right.
    a. How many kilograms of mix are in the package?   0.186 kg
    b. How many grams of sodium are contained in two servings of the corn bread?   0.42 g

Nutrition Facts		
Serving Size ⅙ pkg. (31g mix)		
Servings Per Container 6		
Amount Per Serving	**Mix**	**Prepared**
Calories	110	160
Calories from Fat	10	50
		% Daily Value*
Total Fat 1g	1%	9%
Saturated Fat 0g	0%	7%
Cholesterol 0mg	0%	12%
Sodium 210mg	9%	11%
Total Carbohydrate 24g	8%	8%
Sugars 6g		
Protein 2g		

30. **Consumerism** Find the cost of three packages of ground meat weighing 470 g, 680 g, and 590 g if the price per kilogram is $8.40.
    $14.62

31. **Landscaping** Eighty grams of grass seed are used for every 100 m^2 of lawn. How many kilograms of grass seed are needed to cover 2000 m^2?
    1.6 kg

32. **Airlines** A commuter flight charges $9.95 for each kilogram or part of a kilogram over 15 kg of luggage weight. How much extra must be paid for three pieces of luggage weighing 6450 g, 5850 g, and 7500 g?
    $49.75

33. **Business** A health food store buys nuts in 10-kilogram containers and repackages the nuts for resale. The store packages the nuts in 200-gram bags, costing $.04 each, and sells them for $3.89 per bag. Find the profit on a 10-kilogram container of nuts costing $75.
    $117.50

34. **Measurements** A trailer is loaded with nine automobiles weighing 1405 kg each. Find the total weight of the automobiles.
    12,645 kg

35. **Agriculture** During 1 year the United States exported 37,141 million kg of wheat, 2680 million kg of rice, and 40,365 million kg of corn. What percent of the total of these grain exports was corn? Round to the nearest tenth of a percent.
    50.3%

APPLYING THE CONCEPTS

36. Define a metric ton. Convert the weights in Exercise 35 to metric tons.
    1 metric ton = 1000 kg; 37,141,000 metric tons; 2,680,000 metric tons; 40,365,000 metric tons

37. Discuss the advantages and disadvantages of the U.S. Customary System and the metric system of measurement.

Quick Quiz (Objective 9.2B)

1. Four hundred grams of weed killer are used for every 100 square meters of lawn. Find the amount of weed killer, in kilograms, needed to cover 1800 square meters.   7.2 kg

2. Four hundred building blocks weigh 580 kg. How much does each block weigh?   1.45 kg

9.3 Capacity

Objective A

To convert units of capacity in the metric system of measurement

The basic unit of capacity in the metric system is the liter. One **liter** is defined as the capacity of a box that is 10 cm long on each side.

The units of capacity in the metric system have the same prefixes as the units of length.

10 cm
10 cm
10 cm

Units of Capacity in the Metric System

1 kiloliter (kl) = 1000 L
1 hectoliter (hl) = 100 L
1 decaliter (dal) = 10 L
1 liter (L) = 1 L
1 deciliter (dl) = 0.1 L
1 centiliter (cl) = 0.01 L
1 milliliter (ml) = 0.001 L

1-liter bottle

1 cm
1 cm
1 cm
1 ml = 1 cm³

The milliliter is equal to 1 **cubic centimeter** (cm³).

Conversion between units of capacity in the metric system involves moving the decimal point to the right or to the left. Listing the units in order from largest to smallest will indicate how many places to move the decimal point and in which direction.

To convert 824 ml to liters, first write the units in order from largest to smallest.

kl hl dal **L** dl cl ml
        3 positions

824 ml = 0.824 L
        3 places

• Converting ml to L requires moving 3 positions to the left.

• Move the decimal point the same number of places and in the same direction.

Example 1 Convert 4 L 32 ml to liters.

Solution
32 ml = 0.032 L

4 L 32 ml = 4 L + 0.032 L
= 4.032 L

You Try It 1 Convert 2 kl 167 L to liters.

Your solution 2167 L

Solution on p. S24

Objective 9.3A

Vocabulary to Review
kilo- [9.1A]
hecto- [9.1A]
deca- [9.1A]
deci- [9.1A]
centi- [9.1A]
milli- [9.1A]

New Vocabulary
liter
cubic centimeter

Optional Student Activity
The table below shows the average fuel efficiency, in kilometers per liter of gasoline, of passenger cars in the United States in 1980, 1985, 1990, and 1996. (*Source:* U.S. Dept. of Transportation, Federal Highway Administration)

1980	6.57 km/L
1985	7.73 km/L
1990	8.93 km/L
1996	9.58 km/L

Find the difference between the number of kiloliters of gasoline needed to travel 20,000 kilometers in 1996 and the number needed in 1980. (The average American car is driven 20,000 km per year.) Round to the nearest thousandth. 0.956 kL

Instructor Note
It may be helpful to tell students that 1 L is approximately 1 qt plus $\frac{1}{4}$ c. One teaspoon is approximately 5 ml.

You might have students do the project "Name That Metric Unit" at the end of this chapter. This is a challenging activity for students unfamiliar with the metric system, but it will help them to differentiate among units of length, mass, and capacity.

In-Class Examples (Objective 9.3A)

Convert.
1. 1.34 L = ____ ml   1340 ml
2. 3600 ml = ____ L   3.6 L
3. 821 cm³ = ____ ml   821 ml
4. 6 kl 8 L = ____ kl   6.008 kl

> **Example 2** Convert 1.23 L to cubic centimeters.
>
> **Solution** 1.23 L = 1230 ml = 1230 cm³

> **You Try It 2** Convert 325 cm³ to liters.
>
> **Your solution** 0.325 L
>
> *Solution on p. S24*

Optional Student Activity

The table below shows the per capita consumption, in liters, of milk in the United States in four years that span the last century.

	Whole Milk	Low-Fat & Skim Milk
1909	101.76	27.67
1939	110.82	18.15
1969	100.81	20.69
1998	31.53	58.29

(*Source:* Economic Research Service, U.S. Dept. of Agriculture)

1. Find the difference between the number of milliliters of whole milk consumed per day in 1909 and 1998. Use a 365-day year. Round to the nearest whole number.   192 ml

2. Find the difference between the number of milliliters of low-fat and skim milk consumed per day in 1998 and 1939. Use a 365-day year. Round to the nearest whole number.   110 ml

3. Provide an explanation for the changes in milk consumption reflected in the table.
   Answers will vary. For example, the decrease in whole milk consumption and the increase in consumption of low-fat or skim milk may be a reflection of health officials' stressing the benefits of a low-fat diet. The decrease in milk consumption may be a reflection of the increase in consumption of other beverages, such as soft drinks, by the youth in this country.

Objective B   **To solve application problems**

> **TAKE NOTE**
> Although in this text we will always change units to the larger unit, it is possible to perform the calculation by changing to the smaller unit.
>
> 2.5 kl + 875 L
> = 2500 L + 875 L
> = 3375 L
>
> Note that
> 3375 L = 3.375 kl.

In the application problems in this section, we perform arithmetic operations with the measurements of capacity in the metric system. Remember that before measurements can be added or subtracted, they must be expressed in terms of the same unit. In this textbook, unless otherwise stated, the units should be changed to the larger unit before the arithmetic operation is performed.

To add 2.5 kl and 875 L, convert 875 L to kiloliters.

$$2.5 \text{ kl} + 875 \text{ L} = 2.5 \text{ kl} + 0.875 \text{ kl}$$
$$= 3.375 \text{ kl}$$

> **Example 3**
>
> A laboratory assistant is in charge of ordering acid for three chemistry classes of 30 students each. Each student requires 80 ml of acid. How many liters of acid should be ordered? (The assistant must order by the whole liter.)
>
> **Strategy**
> To find the number of liters to be ordered:
> - Find the number of milliliters of acid needed by multiplying the number of classes (3) by the number of students per class (30) by the number of milliliters of acid required by each student (80).
> - Convert milliliters to liters.
> - Round up to the nearest whole number.
>
> **Solution**
> 3(30)(80) = 7200 ml
>
> 7200 ml = 7.2 L
>
> 7.2 rounded up to the nearest whole number is 8.
>
> The assistant should order 8 L of acid.

> **You Try It 3**
>
> For $299.50, a cosmetician buys 5 L of moisturizer and repackages it in 125-milliliter jars. Each jar costs the cosmetician $.85. Each jar of moisturizer is sold for $29.95. Find the profit on the 5 L of moisturizer.
>
> **Your strategy**
>
> **Your solution**
> $864.50
>
> *Solution on p. S24*

In-Class Examples (Objective 9.3B)

1. A concession stand at a fair sold 970 medium soft drinks. Each medium soft drink cup contains 450 ml. How many liters of soft drinks were sold?
   436.5 L

2. Forty liters of maple syrup are bought and then repackaged in 250-milliliter containers. Thirty-four containers of maple syrup are sold. How many liters of the maple syrup are still in stock?   31.5 L

9.3 Exercises

Section 9.3

Objective A   **To convert units of capacity in the metric system of measurement**

Suggested Assignment
Exercises 1–33, odds
More challenging problem:
  Exercise 35

For Exercises 1 to 24, convert.

1. 4200 ml = __4.200__ L

2. 7.5 ml = __0.0075__ L

3. 3.42 L = __3420__ ml

4. 0.037 L = __37__ ml

5. 423 ml = __423__ cm^3

6. 0.32 ml = __0.32__ cm^3

7. 642 cm^3 = __642__ ml

8. 0.083 cm^3 = __0.083__ ml

9. 42 cm^3 = __0.042__ L

10. 3075 cm^3 = __3.075__ L

11. 0.435 L = __435__ cm^3

12. 2.57 L = __2570__ cm^3

13. 4.62 kl = __4620__ L

14. 0.035 kl = __35__ L

15. 1423 L = __1.423__ kl

16. 897 L = __0.897__ kl

17. 1.267 L = __1267__ cm^3

18. 4.105 L = __4105__ cm^3

19. 3 L 42 ml = __3.042__ L

20. 1 L 127 ml = __1.127__ L

21. 3 kl 4 L = __3.004__ kl

22. 6 kl 32 L = __6.032__ kl

23. 8 L 200 ml = __8.200__ L

24. 9 kl 505 L = __9.505__ kl

Objective B   **To solve application problems**

25.   **Earth Science**   The air in Earth's atmosphere is 78% nitrogen and 21% oxygen.
  a. Without calculating, determine whether the amount of oxygen in 50 L of air is more or less than 25 L.   Less
  b. Find the amount of oxygen in 50 L of air.   10.5 L

Quick Quiz (Objective 9.3A)
Convert.
1. 37 L = ____ ml   37,000 ml
2. 6.4 ml = ____ L   0.0064 L
3. 312 ml = ____ cm^3 312 cm^3
4. 18 L 376 ml = ____ L   18.376 L

26. **Consumerism**   A can of tomato juice contains 1.36 L. How many 170-milliliter servings are in one can of tomato juice?
    8 servings

27. **Measurements**   An athletic club uses 800 ml of chlorine each day for its swimming pool. How many liters of chlorine are used in a month of 30 days?
    24 L

28. **Consumerism**   The printed label from a container of milk is shown at the right. How many 230-milliliter servings are in the container? Round to the nearest whole number.
    16 servings

Dairy Hill

Skim Milk
Vitamin A & D Added
Pasteurized • Homogenized

INGREDIENTS: PASTEURIZED SKIM MILK, NONFAT MILK SOLIDS, VITAMIN A PALMITATE AND VITAMIN D3 ADDED.

0 15400 20209 1

1 GAL (3.78 L)

29. **Medicine**   A flu vaccine is being given for the coming winter season. A medical corporation buys 12 L of flu vaccine. How many patients can be immunized if each person receives 3 cm^3 of the vaccine?
    4000 patients

30. **Chemistry**   A chemistry experiment requires 12 ml of an acid solution. How many liters of acid should be ordered when 4 classes of 90 students each are going to perform the experiment? (The acid must be ordered by the whole liter.)
    5 L

31. **Consumerism**   A case of 12 one-liter bottles of apple juice costs $19.80. A case of 24 cans, each can containing 340 ml of apple juice, costs $14.50. Which case of apple juice is the better buy?
    The 12 one-liter bottles

32. **Business**   For $195, a pharmacist purchases 5 L of cough syrup and repackages it in 250-milliliter bottles. Each bottle costs the pharmacist $.55. Each bottle of cough syrup is sold for $23.89. Find the profit on the 5 L of cough syrup.
    $271.80

33. **Business**   A service station operator bought 85 kl of gasoline for $23,750. The gasoline was sold for $.379 per liter. Find the profit on the 85 kl of gasoline.
    $8465

34. **Business**   A wholesale distributor purchased 32 kl of cooking oil for $44,480. The wholesaler repackaged the cooking oil in 1.25-liter bottles. The bottles cost $.21 each. Each bottle of cooking oil was sold for $2.97. Find the distributor's profit on the 32 kl of cooking oil.
    $26,176

APPLYING THE CONCEPTS

35. After a 280-milliliter serving is taken from a 3-liter bottle of water, how much water remains in the container? Write the answer in three different ways.
    2720 ml; 2.72 L; 2 L 720 ml

36. ✎   Write an essay describing problems in trade and manufacturing that arise between the United States and Europe because they use different systems of measurement.

Answers to Writing Exercises

36. Students may discuss any number of difficulties. For example, comparing unit costs or prices is more difficult. Parts are not interchangeable within similar products. Different dimensions of machinery and tools limit trade between the United States and the European countries.

Quick Quiz (Objective 9.3B)

1. There are 24 cans of tomato soup in a case. Each can contains 305 ml of soup. How many liters of soup are in one case?   7.32 L

2. Six hundred milliliters of fluoride are added to a town's water supply each day. How many liters of fluoride are used during a 30-day period?   18 L

9.4 Energy

Objective A   **To use units of energy in the metric system of measurement**

Two commonly used units of energy in the metric system are the calorie and the watt-hour.

Heat is generally measured in units called calories or in larger units called Calories (with a capital C). A **Calorie** is 1000 calories and should be called a kilocalorie, but it is common practice in nutritional references and food labeling to simply call it a Calorie. A Calorie is the amount of heat required to raise the temperature of 1 kg of water 1 degree Celsius. One Calorie is also the energy required to lift 1 kg a distance of 427 m.

HOW TO   Swimming uses 480 Calories per hour. How many Calories are used by swimming $\frac{1}{2}$ h each day for 30 days?

Strategy   To find the number of Calories used:

- Find the number of hours spent swimming.
- Multiply the number of hours spent swimming by the Calories used per hour.

Solution $\frac{1}{2} \times 30 = 15$

$15(480) = 7200$

7200 Calories are used by swimming $\frac{1}{2}$ h each day for 30 days.

The **watt-hour** is used for measuring electrical energy. One watt-hour is the amount of energy required to lift 1 kg a distance of 370 m. A light bulb rated at 100 watts (W) will emit 100 watt-hours (Wh) of energy each hour. A **kilowatt-hour** is 1000 watt-hours.

TAKE NOTE
Recall that the prefix kilo- means 1000.

1000 watt-hours (Wh) = 1 kilowatt-hour (kWh)

HOW TO   A 150-watt bulb is on for 8 h. At 8¢ per kilowatt-hour, find the cost of the energy used.

Strategy   To find the cost:

- Find the number of watt-hours used.
- Convert to kilowatt-hours.
- Multiply the number of kilowatt-hours used by the cost per kilowatt-hour.

Solution $150 \times 8 = 1200$

$1200 \text{ Wh} = 1.2 \text{ kWh}$

$1.2 \times 0.08 = 0.096$

The cost of the energy used is $.096.

Integrating Technology
To convert watt-hours to kilowatt-hours, divide by 1000. To use a calculator to determine the number of kilowatt-hours used in the problem at the right, enter the following:

150 × 8 ÷ 1000 =

The calculator display reads 1.2.

New Vocabulary
Calorie
watt-hour
kilowatt-hour

Instructor Note
The fundamental unit of energy in the metric system is the joule, which is the energy needed each second to push an ampere of current through an ohm of resistance. However, heat is commonly measured in calories. There are about 4.19 joules in a calorie.

Concept Check
Back on July 1, 1997, the *Milwaukee Journal/Sentinel* reported that Don Gorske had, between 1973 and that date, eaten 14,837 McDonald's Big Macs. A Big Mac contains 560 Calories. Bicycling at a rate of 9 km/h burns 210 Calories/h. How many hours would Don Gorske have had to bicycle at 9 km/h to burn off the Calories contained in the 14,837 Big Macs he consumed? Round to the nearest whole number.
39,565 h

Instructor Note
Energy is sold by the joule, but it is common practice to bill electrical energy in terms of kilowatt-hours. A kilowatt-hour is 3,600,000 joules, or 3.6 megajoules.

In-Class Examples (Objective 9.4A)

1. Jogging uses 375 Calories per hour. How many Calories do you burn in 28 days of jogging if you jog for $\frac{1}{2}$ h each day?   5250 Calories

2. A black-and-white monitor is rated at 45 W. The monitor is used for 28 h in 1 week. How many kilowatt-hours of energy are used?   1.26 kWh

Optional Student Activity

The Fahrenheit temperature scale is used in the United States. Countries that use the metric system use the Celsius temperature scale. To convert from Fahrenheit to Celsius, first subtract 32 from the Fahrenheit temperature. Then multiply the difference by $\frac{5}{9}$. For example, if the Fahrenheit temperature is 104°, which is written 104°F, then

$$104 - 32 = 72$$
$$\frac{5}{9} \times 72 = 40$$

The Celsius temperature is 40°, which is written 40°C.

In each exercise below, convert the Fahrenheit temperature to Celsius. Round to the nearest whole number.

1. Normal body temperature is 98.6°F.   37°C
2. Cookies normally bake at 350°F.   177°C
3. The coldest city in the United States is International Falls, Minnesota, where the average temperature is 36.4°F. (*Source:* National Climatic Data Center)   2°C
4. The global annual temperature is 59.3°F. (*Source:* IPCC, NASA)   15°C
5. The temperature on the surface of the sun is 8700°F.   4816°C

Example 1

Walking uses 180 Calories per hour. How many Calories will you burn off by walking $5\frac{1}{4}$ h during one week?

Strategy
To find the number of Calories, multiply the number of hours spent walking by the Calories used per hour.

Solution
$$5\frac{1}{4} \times 180 = \frac{21}{4} \times 180 = 945$$

You will burn off 945 Calories.

Example 2

A clothes iron is rated at 1200 W. If the iron is used for 1.5 h, how much energy, in kilowatt-hours, is used?

Strategy
To find the energy used:
- Find the number of watt-hours used.
- Convert watt-hours to kilowatt-hours.

Solution
$1200 \times 1.5 = 1800$

$1800 \text{ Wh} = 1.8 \text{ kWh}$

1.8 kWh of energy are used.

Example 3

A TV set rated at 1800 W is on for an average of 3.5 h per day. At 7.2¢ per kilowatt-hour, find the cost of operating the set for 1 week.

Strategy
To find the cost:
- Multiply to find the total number of hours the set is used per week.
- Multiply the product by the number of watts to find the watt-hours.
- Convert watt-hours to kilowatt-hours.
- Multiply the number of kilowatt-hours by the cost per kilowatt-hour.

Solution
$3.5 \times 7 = 24.5$

$24.5 \times 1800 = 44,100$

$44,100 \text{ Wh} = 44.1 \text{ kWh}$

$44.1 \times 0.072 = 3.1752$

The cost is $3.1752.

You Try It 1

Housework requires 240 Calories per hour. How many Calories are burned off by doing $4\frac{1}{2}$ h of housework?

Your strategy

Your solution
1080 Calories

You Try It 2

Find the number of kilowatt-hours of energy used when a 150-watt light bulb burns for 200 h.

Your strategy

Your solution
30 kWh

You Try It 3

A microwave oven rated at 500 W is used an average of 20 min per day. At 8.7¢ per kilowatt-hour, find the cost of operating the oven for 30 days.

Your strategy

Your solution
43.5¢

Solutions on pp. S24–S25

9.4 Exercises

Objective A   **To use units of energy in the metric system of measurement**

1. **Health**   How many Calories can you eliminate from your diet by omitting 1 slice of bread per day for 30 days? One slice of bread contains 110 Calories.
3300 Calories

2. **Health**   How many Calories can you eliminate from your diet in 2 weeks by omitting 400 Calories per day?
5600 Calories

3. **Nutrition**   A nutrition label from a package of crisp bread is shown at the right.

   a. How many Calories are in $1\frac{1}{2}$ servings?   90 Calories

   b. How many Calories from fat are in 6 slices of the bread?   30 Calories

Nutrition Facts		
Serving Size 2 Slices (18g)		
Servings Per Container about 15		
Amount Per Serving		
Calories 60	Calories from Fat 10	
		% Daily Value*
Total Fat 1g		**2%**
Saturated Fat 0g		**0%**
Polyunsaturated Fat 0.5g		
Monounsaturated Fat 0.5g		
Cholesterol 0mg		**0%**
Sodium 60mg		**3%**
Total Carbohydrate 10g		**3%**
Dietary Fiber 3g		**10%**
Sugars 1g		
Protein 2g		
Vitamin A 0%	•	Vitamin C 0%
Calcium 0%	•	Iron 4%

* Percent Daily Values are based on a 2,000 calorie diet. Your daily values may be higher or lower depending on your calorie needs.

	Calories:	2,000	2,500
Total Fat	Less than	65g	80g
Saturated Fat	Less than	20g	25g
Cholesterol	Less than	300mg	300mg
Sodium	Less than	2,400mg	2,400mg
Total Carbohydrate		300g	375g
Dietary Fiber		25g	30g

Calories per gram:
Fat 9 • Carbohydrate 4 • Protein 4

4. **Health**   Moderately active people need 20 Calories per pound of body weight to maintain their weight. How many Calories should a 150-pound, moderately active person consume per day to maintain that weight?
3000 Calories

5. **Health**   People whose daily activity level would be described as light need 15 Calories per pound of body weight to maintain their weight. How many Calories should a 135-pound, lightly active person consume per day to maintain that weight?
2025 Calories

6. **Health**   For a healthful diet, it is recommended that 55% of the daily intake of Calories come from carbohydrates. Find the daily intake of Calories from carbohydrates that is appropriate if you want to limit your Calorie intake to 1600 Calories.
880 Calories

7. **Health**   Playing singles tennis requires 450 Calories per hour. How many Calories do you burn in 30 days playing 45 min per day?
10,125 Calories

8. **Health**   After playing golf for 3 h, Ruben had a banana split containing 550 Calories. Playing golf uses 320 Calories per hour.
   a. Without doing the calculations, did the banana split contain more or fewer Calories than Ruben burned off playing golf?   Fewer
   b. Find the number of Calories Ruben gained or lost from these two activities.   Loss of 410 Calories

9.   **Health**   Hiking requires approximately 315 Calories per hour. How many hours would you have to hike to burn off the Calories in a 375-Calorie sandwich, a 150-Calorie soda, and a 280-Calorie ice cream cone? Round to the nearest tenth.
2.6 h

10.   **Health**   Riding a bicycle requires 265 Calories per hour. How many hours would Shawna have to ride a bicycle to burn off the Calories in a 320-Calorie milkshake, a 310-Calorie cheeseburger, and a 150-Calorie apple? Round to the nearest tenth.
2.9 h

Suggested Assignment
Exercises 1–19, odds

Quick Quiz (Objective 9.4A)

1. Swimming uses 480 Calories per hour. How many Calories would you use by swimming $1\frac{1}{2}$ h each day for 30 days?   21,600 Calories

2. A 1200-watt dishwasher is operated an average of 4.5 h per week. How many kilowatt-hours of energy does it use each week?   5.4 kWh

11. **Energy**   An oven uses 500 W of energy. How many watt-hours of energy are used to cook a 5-kilogram roast for $2\frac{1}{2}$ h?
    1250 Wh

12. **Energy**   A 21-inch color TV set is rated at 90 W. The TV is used an average of $3\frac{1}{2}$ h each day for a week. How many kilowatt-hours of energy are used during the week?
    2.205 kWh

13. **Energy**   A fax machine is rated at 9 W when the machine is in standby mode and at 36 W when in operation. How many kilowatt-hours of energy are used during a week in which the fax machine is in standby mode for 39 h and in operation for 6 h?
    0.567 kWh

14. **Energy**   A 120-watt CD player is on an average of 2 h a day. Find the cost of listening to the CD player for 2 weeks at a cost of 9.4¢ per kilowatt-hour. Round to the nearest cent.
    $.32

15. **Energy**   How much does it cost to run a 2200-watt air conditioner for 8 h at 9¢ per kilowatt-hour? Round to the nearest cent.
    $1.58

16. **Energy**   A space heater is used for 3 h. The heater uses 1400 W per hour. Electricity costs 11.1¢ per kilowatt-hour. Find the cost of using the electric heater. Round to the nearest cent.
    $.47

17. **Energy**   A 60-watt Sylvania Long Life Soft White Bulb has a light output of 835 lumens and an average life of 1250 h. A 34-watt Sylvania Energy Saver Bulb has a light output of 400 lumens and an average life of 1500 h.
    a. Is the light output of the Energy Saver Bulb more or less than half that of the Long Life Soft White Bulb?   Less
    b. If electricity costs 10.8¢ per kilowatt-hour, what is the difference in cost between using the Long Life Soft White Bulb for 150 h and using the Energy Saver Bulb for 150 h? Round to the nearest cent. $.42

18. **Energy**   A house is insulated to save energy. The house used 265 kWh of electrical energy per month before insulation and saves 45 kWh of energy per month after insulation. What percent decrease does this amount represent? Round to the nearest tenth of a percent.
    17.0%

19. **Energy**   A welder uses 6.5 kWh of energy each hour. Find the cost of using the welder for 6 h a day for 30 days. The cost is 9.4¢ per kilowatt-hour.
    $109.98

APPLYING THE CONCEPTS

20. Write an essay on how to improve the energy efficiency of a home.

21. A maintenance intake of Calories allows a person to neither gain nor lose weight. Consult a book on nutrition in order to make a table of weights and the corresponding maintenance intake of Calories. Then, for each weight, add another column indicating the appropriate Calorie intake when an individual at that weight wants to lose 1 lb per week.
    Answers will vary.

Answers to Writing Exercises

20. The energy efficiency of a home can be improved in a number of ways. Students might mention increasing the amount or thickness of insulation, using weather stripping around doors and windows, using window treatments that help to retain heat or cool air in the home, and insulating the hot water heater. Also, repairing a cracked foundation or reroofing an older home can be beneficial.

9.5 Conversion Between the U.S. Customary and the Metric Systems of Measurement

Objective A   To convert U.S. Customary units to metric units

Point of Interest

The definition of 1 inch has been changed as a consequence of the wide acceptance of the metric system. One inch is now exactly 25.4 mm.

Point of Interest

Back in September of 1999, CNN reported that "NASA lost a $125 million Mars orbiter because a Lockheed Martin engineering team used English units of measurement while the agency's team used the . . . metric system for a key spacecraft operation." One team used miles, the other kilometers. This resulted in the orbiter breaking up when it hit the Martian atmosphere because the approach was too near the surface.

More than 90% of the world's population uses the metric system of measurement. Therefore, converting U.S. Customary units to metric units is essential in trade and commerce—for example, in importing foreign goods and exporting domestic goods. Approximate equivalences between the two systems follow.

Units of Length	Units of Weight	Units of Capacity
1 in. = 2.54 cm	1 oz ≈ 28.35 g	1 L ≈ 1.06 qt
1 m ≈ 3.28 ft	1 lb ≈ 454 g	1 gal ≈ 3.79 L
1 m ≈ 1.09 yd	1 kg ≈ 2.2 lb	
1 mi ≈ 1.61 km		

These equivalences can be used to form conversion rates to change one unit of measurement to another. For example, because 1 mi ≈ 1.61 km, the conversion rates $\frac{1 \text{ mi}}{1.61 \text{ km}}$ and $\frac{1.61 \text{ km}}{1 \text{ mi}}$ are both approximately equal to 1.

HOW TO   Convert 55 mi to kilometers.

$$55 \text{ mi} \approx 55 \text{ mi} \times \boxed{\frac{1.61 \text{ km}}{1 \text{ mi}}}$$

- The conversion rate must contain km in the numerator and mi in the denominator.

$$= \frac{55 \text{ mi}}{1} \times \frac{1.61 \text{ km}}{1 \text{ mi}}$$

$$= \frac{88.55 \text{ km}}{1}$$

$$55 \text{ mi} \approx 88.55 \text{ km}$$

Example 1

Convert 45 mi/h to kilometers per hour.

Solution

$$\frac{45 \text{ mi}}{h} \approx \frac{45 \text{ mi}}{h} \times \frac{1.61 \text{ km}}{1 \text{ mi}}$$

$$= \frac{72.45 \text{ km}}{1 \text{ h}}$$

$$45 \text{ mi/h} \approx 72.45 \text{ km/h}$$

- The conversion rate is $\frac{1.61 \text{ km}}{1 \text{ mi}}$ with km in the numerator and mi in the denominator.

You Try It 1

Convert 60 ft/s to meters per second. Round to the nearest hundredth.

Your solution

18.29 m/s

Solution on p. S25

In-Class Examples (Objective 9.5A)

Convert. Round to the nearest hundredth.

1. Find the weight in kilograms of a 140-pound person. (1 kg ≈ 2.2 lb)   63.64 kg

2. Find the height in meters of a person 5 ft 10 in. tall. (1 m ≈ 3.28 ft)   1.78 m

3. Express 75 mi/h in kilometers per hour. (1 mi ≈ 1.61 km)   120.75 km/h

Vocabulary to Review

conversion rate   [8.1A]

Instructor Note

Prior to teaching this lesson, you might want to review with your students the metric units of measurement (see Objectives 9.1A, 9.2A, and 9.3A) as well as the U.S. Customary units of measurement (see Chapter 8).

Instructor Note

The major focus of this section should be approximate equivalences. Ask questions such as

"Place your hands 20 cm apart."

"Estimate the weight of a desk in kilograms."

"I commute ____ km to school each day."

Optional Student Activity

The Sphinx is in Egypt, near the Pyramid of Khafre. It was constructed about 4500 years ago. It is carved out of bedrock and features the head of a man and the body of a lion.

Given below are measurements of the features of the Sphinx. Convert each measurement to the metric system. Round to the nearest tenth of a meter. (You may want to have students mark off, on the floor, the dimensions of the features of the face so that they get a better sense of its size.)

1. The mouth is 7 ft long.   2.1 m
2. The face is 13 ft 8 in. from ear to ear.   4.2 m
3. The distance from the base of the nose to the top of the face is 5 ft.   1.5 m
4. The width of each eye is 6 ft.   1.8 m
5. The Sphinx is 240 ft long.   73.2 m
6. Each paw is 50 ft long.   15.2 m
7. It is 66 ft from the base to the top of the head.   20.1 m

Example 2

The price of gasoline is $1.59/gal. Find the cost per liter. Round to the nearest tenth of a cent.

Solution

$$\frac{\$1.59}{\text{gal}} \approx \frac{\$1.59}{\text{gal}} \times \frac{1 \text{ gal}}{3.79 \text{ L}} = \frac{\$1.59}{3.79 \text{ L}} \approx \frac{\$.420}{1 \text{ L}}$$

$1.59/\text{gal} \approx \$.420/\text{L}$

You Try It 2

The price of milk is $2.69/gal. Find the cost per liter. Round to the nearest cent.

Your solution

$.71/L

Solution on p. S25

Objective B   **To convert metric units to U.S. Customary units**

Metric units are used in the United States. Cereal is sold by the gram, 35-mm film is available, and soda is sold by the liter. The same conversion rates used in Objective A are used for converting metric units to U.S. Customary units.

Example 3

Convert 200 m to feet.

Solution

$$200 \text{ m} \approx 200 \text{ m} \times \frac{3.28 \text{ ft}}{1 \text{ m}} = \frac{656 \text{ ft}}{1}$$

$200 \text{ m} \approx 656 \text{ ft}$

You Try It 3

Convert 45 cm to inches. Round to the nearest hundredth.

Your solution

17.72 in.

Example 4

Convert 90 km/h to miles per hour. Round to the nearest hundredth.

Solution

$$\frac{90 \text{ km}}{\text{h}} \approx \frac{90 \text{ km}}{\text{h}} \times \frac{1 \text{ mi}}{1.61 \text{ km}} = \frac{90 \text{ mi}}{1.61 \text{ h}} \approx \frac{55.90 \text{ mi}}{1 \text{ h}}$$

$90 \text{ km/h} \approx 55.90 \text{ mi/h}$

You Try It 4

Express 75 km/h in miles per hour. Round to the nearest hundredth.

Your solution

46.58 mi/h

Example 5

The price of gasoline is $.392/L. Find the cost per gallon. Round to the nearest cent.

Solution

$$\frac{\$.392}{1 \text{ L}} \approx \frac{\$.392}{1 \text{ L}} \times \frac{3.79 \text{ L}}{1 \text{ gal}} \approx \frac{\$1.49}{1 \text{ gal}}$$

$.392/\text{L} \approx \$1.49/\text{gal}$

You Try It 5

The price of ice cream is $1.75/L. Find the cost per gallon. Round to the nearest cent.

Your solution

$6.63/gal

Solutions on p. S25

Objective 9.5B

Discuss the Concepts

1. Name some products that are sold in metric units. (*Note:* The textbook lists 35-mm film, soda, and cereal.)

2. Why can the equivalences used to convert U.S. Customary units to metric units also be used in this objective to convert metric units to U.S. Customary units?

Concept Check

A few years ago, *The Index-Journal* of Greenwood, South Carolina, reported that the maintenance crews of the Department of Transportation had picked up 116,318.2 kg of litter during National Public Lands Month. More than 1300 highway maintenance employees picked up the trash from along 2737 km of highways across the state.

1. How many pounds of litter were picked up? Round to the nearest whole number.
   255,900 lb

2. Along how many miles of highway did the employees pick up litter?   1700 mi

3. Provide an estimate of how many pounds of litter each employee picked up.   197 lb

4. Provide an estimate of how many miles of highway each employee cleaned up.   1.3 mi

In-Class Examples (Objective 9.5B)

Convert. Round to the nearest hundredth.

1. Find the number of gallons in 6 L of antifreeze.
   (1 gal ≈ 3.79 L)   1.58 gal

2. A piece of luggage weighs 6.3 kg. Find the weight in pounds. (1 kg ≈ 2.2 lb)   13.86 lb

3. Express 88 km/h in miles per hour.
   (1 mi ≈ 1.61 km)   54.66 mi/h

9.5 Exercises

Objective A   **To convert U.S. Customary units to metric units**

For Exercises 1 to 16, convert. Round to the nearest hundredth if necessary.

1. Convert 100 yd to meters.
   91.74 m

2. Find the weight in kilograms of a 145-pound person.
   65.91 kg

3. Find the height in meters of a person 5 ft 8 in. tall.
   1.73 m

4. Find the number of liters in 2 c of soda.
   0.47 L

5. How many kilograms does a 15-pound turkey weigh?
   6.82 kg

6. Find the number of liters in 14.3 gal of gasoline.
   54.20 L

7. Find the number of milliliters in 1 c.
   235.85 ml

8. The winning long jump at a track meet was 29 ft 2 in. Convert this distance to meters.
   8.89 m

9. Express 65 mi/h in kilometers per hour.
   104.65 km/h

10. Express 30 mi/h in kilometers per hour.
    48.3 km/h

11. Fat-free hot dogs cost $3.49/lb. Find the cost per kilogram.
    $7.68/kg

12. Seedless watermelon costs $.59/lb. Find the cost per kilogram.
    $1.30/kg

13. The cost of gasoline is $1.47/gal. Find the cost per liter.
    $.39/L

14. Deck stain costs $24.99/gal. Find the cost per liter.
    $6.59/L

15. **Earth Science**   The distance around Earth is 24,887 mi. Convert this distance to kilometers.
    40,068.07 km

16. **Astronomy**   The distance from Earth to the sun is 93,000,000 mi. Convert this distance to kilometers.
    149,730,000 km

Objective B   **To convert metric units to U.S. Customary units**

For Exercises 17 to 30, convert. Round to the nearest hundredth if necessary.

17. Convert 100 m to feet.
    328 ft

18. Find the weight in pounds of an 86-kilogram person.
    189.2 lb

19. Find the number of gallons in 6 L of antifreeze.
    1.58 gal

20. Find the height in inches of a person 1.85 m tall.
    72.82 in.

Quick Quiz (Objective 9.5A)

Convert. Round to the nearest hundredth.

1. How many kilograms does a 12-pound turkey weigh? (1 kg ≈ 2.2 lb)   5.45 kg

2. Find the number of liters in 8 gal of coffee. (1 gal ≈ 3.79 L)   30.32 L

3. Express 48 mi/h in kilometers per hour. (1 mi ≈ 1.61 km)   77.28 km/h

21. Find the distance of the 1500-meter race in feet.
4920 ft

22. Find the weight in ounces of 327 g of cereal.
11.53 oz

23. How many gallons of water does a 24-liter aquarium hold?
6.33 gal

24. Find the width in inches of 35-mm film.
1.38 in.

25. Express 80 km/h in miles per hour.
49.69 mi/h

26. Express 30 m/s in feet per second.
98.4 ft/s

27. Gasoline costs $.385/L. Find the cost per gallon.
$1.46/gal

28. A 5-kilogram ham costs $10/kg. Find the cost per pound.
$4.55/lb

29. A backpack tent weighs 2.1 kg. Find its weight in pounds.
4.62 lb

30. A 2.5-kilogram bag of grass seed costs $7.89. Find the cost per pound.
$1.43/lb

31. **Health** Gary is planning a 5-day backpacking trip and decides to hike an average of 5 h each day. Hiking requires an extra 320 Calories per hour. How many pounds will Gary lose during the trip if he consumes an extra 900 Calories each day? (3500 Calories are equivalent to 1 lb.)
1 lb

32. **Health** Swimming requires 550 Calories per hour. How many pounds could be lost by swimming $1\frac{1}{2}$ h each day for 5 days if no extra calories were consumed? (3500 Calories are equivalent to 1 lb.)
1.18 lb

33. **Physics** The speed of light is 300,000 km/s. Convert this speed to miles per second.
186,335.4 mi/s

APPLYING THE CONCEPTS

34. For the following U.S. Customary units, make an estimate of the metric equivalent. Then perform the conversion and see how close you came to the actual measurement.

60 mi/h ≈   120 lb ≈   6 ft ≈

1 mi ≈   1 gal ≈   1 quarter-mile ≈

Estimates will vary.

35. Determine whether the statement is true or false.
a. A liter is more than a gallon.   False
b. A meter is less than a yard.   False
c. 30 mi/h is less than 60 km/h.   True
d. A kilogram is greater than a pound.   True
e. An ounce is less than a gram.   False

Answers to Writing Exercises
36. This is an "opinion paper." Answers will vary.

36. Should the United States keep the U.S. Customary System or convert to the metric system? Justify your position.

Quick Quiz (Objective 9.5B)

Convert. Round to the nearest hundredth.

1. How many gallons does a 92-liter tank hold?
(1 gal ≈ 3.79 L)   24.27 gal

2. Find the weight in pounds of an 86-kilogram person. (1 kg ≈ 2.2 lb)   189.2 lb

3. Express 60 m/s in feet per second. (1 m ≈ 3.28 ft)
196.8 ft/s

Focus on Problem Solving

Working Backward

Sometimes the solution to a problem can be found by *working backward*. This problem-solving technique can be used to find a winning strategy for a game called Nim.

There are many variations of this game. For our game, there are two players, Player A and Player B, who alternately place 1, 2, or 3 matchsticks in a pile. The object of the game is to place the 32nd matchstick in the pile. Is there a strategy that Player A can use to guarantee winning the game?

Working backward, if there are 29, 30, or 31 matchsticks in the pile when it is A's turn to play, A can win by placing 3 matchsticks ($29 + 3 = 32$), 2 matchsticks, ($30 + 2 = 32$), or 1 matchstick ($31 + 1 = 32$) on the pile. If there are to be 29, 30, or 31 matchsticks in the pile when it is A's turn, there must be 28 matchsticks in the pile when it is B's turn.

Working backward from 28, if there are to be 28 matches in the pile at B's turn, there must be 25, 26, or 27 at A's turn. Player A can then add 3 matchsticks, 2 matchsticks, or 1 matchstick to the pile to bring the number to 28. For there to be 25, 26, or 27 matchsticks in the pile at A's turn, there must be 24 matchsticks at B's turn.

Now working backward from 24, if there are to be 24 matches in the pile at B's turn, there must be 21, 22, or 23 at A's turn. Player A can then add 3 matchsticks, 2 matchsticks, or 1 matchstick to the pile to bring the number to 24. For there to be 21, 22, or 23 matchsticks in the pile at A's turn, there must be 20 matchsticks at B's turn.

So far, we have found that for Player A to win, there must be 28, 24, or 20 matchsticks in the pile when it is B's turn to play. Note that each time, the number is decreasing by 4. Continuing this pattern, Player A will win if there are 16, 12, 8, or 4 matchsticks in the pile when it is B's turn.

Player A can guarantee winning by making sure that the number of matchsticks in the pile is a multiple of 4. To ensure this, Player A allows Player B to go first and then adds exactly enough matchsticks to the pile to bring the total to a multiple of 4.

For example, suppose B places 3 matchsticks in the pile; then A places 1 matchstick ($3 + 1 = 4$). Now B places 2 matchsticks in the pile. The total is now 6 matchsticks. Player A then places 2 matchsticks in the pile to bring the total to 8, a multiple of 4. If play continues in this way, Player A will win.

Here are some variations of Nim. See whether you can develop a winning strategy for Player A.

1. Suppose the goal is to place the last matchstick in a pile of 30 matches.

2. Suppose the players make two piles of matchsticks, with the final number of matchsticks in each pile to be 20.

3. In this variation of Nim, there are 40 matchsticks in a pile. Each player alternately removes 1, 2, or 3 matches from the pile. The player who removes the last match wins.

Answers to Focus on Problem Solving—Working Backward

1. Player A can guarantee a win by going first and playing 2 matchsticks. Player A then adds to the pile exactly enough matchsticks to bring the total to a multiple of 4 plus 2 (2, 6, 10, 14, 18, 22, 26, 30).

2. Because there are two piles of 20, there is a total of 40 matchsticks. Player A can guarantee a win by allowing Player B to go first. Player A then adds to either pile exactly enough matchsticks to bring the total in both piles to a multiple of 4.

3. Player A can guarantee a win by allowing Player B to go first. Player A then removes from the pile exactly enough matchsticks to bring the total remaining to a multiple of 4.

Answers to Projects and Group Activities— Name That Metric Unit

1. kilometer
2. kilogram
3. centimeter
4. milliliter
5. gram
6. kiloliter
7. meter
8. centimeter
9. gram
10. kilogram
11. milliliter
12. liter
13. milligram
14. kilometer
15. millimeter
16. meter
17. kilogram
18. kiloliter
19. milliliter
20. kilogram

Projects and Group Activities

Name That Metric Unit

What unit in the metric system would be used to measure each of the following? If you are working in a group, be sure that each member agrees on the unit to be used and understands why that unit is used before going on to the next item.

1. The distance from Los Angeles to New York
2. The weight of a truck
3. A person's waist
4. The amount of coffee in a mug
5. The weight of a thumbtack
6. The amount of water in a swimming pool
7. The distance a baseball player hits a baseball
8. A person's hat size
9. The amount of protein needed daily
10. A person's weight
11. The amount of maple syrup served with pancakes
12. The amount of water in a water cooler
13. The amount of medication in an aspirin
14. The distance to the grocery store
15. The width of a hair
16. A person's height
17. The weight of a lawn mower
18. The amount of water a family uses monthly
19. The contents of a bottle of salad dressing
20. The newspapers collected at a recycling center

Metric Measurements for Computers

Other prefixes in the metric system are becoming more commonly used as a result of technological advances in the computer industry. For example, the speed of a computer used to be measured in microseconds and then in nanoseconds, but now computer speeds are measured in picoseconds.

tera-	= 1,000,000,000,000
giga-	= 1,000,000,000
mega-	= 1,000,000
micro-	= 0.000001
nano-	= 0.000000001
pico-	= 0.000000000001

1. Complete the table.

Metric System Prefix	Symbol	Magnitude	Means Multiply the Basic Unit By:
tera-	T	10^{12}	1,000,000,000,000
giga-	G	___	1,000,000,000
mega-	M	10^{6}	___
kilo-	___	___	1,000
hecto-	h	___	100
deca-	da	10^{1}	___
deci-	d	$\dfrac{1}{10}$	___
centi-	___	$\dfrac{1}{10^{2}}$	___
milli-	___	___	0.001
micro-	μ	$\dfrac{1}{10^{6}}$	___
nano-	n	$\dfrac{1}{10^{9}}$	___
pico-	p	___	0.000000000001

2. How can the Magnitude column in the table above be used to determine how many places to move the decimal point when converting to the basic unit in the metric system?

A **bit** is the smallest unit of code that computers can read; it is a binary digit, either a 0 or a 1. A bit is abbreviated b. Usually bits are grouped into **bytes** of 8 bits. Each byte stands for a letter, number, or any other symbol we might use in communicating information. For example, the letter W can be represented 01010111. A byte is abbreviated B.

The amount of memory in a computer hard drive is measured in terabytes (TB), gigabytes (GB), and megabytes (MB). Often a gigabyte is referred to as a gig, and a megabyte is referred to as a meg. Using the definitions of the prefixes given above, a kilobyte is 1000 bytes, a megabyte is 1,000,000 bytes, and a gigabyte is 1,000,000,000 bytes. However, these are not exact equivalences. Bytes are actually computed in powers of 2. Therefore, kilobytes, megabytes, gigabytes, and terabytes are powers of 2. The exact equivalences are shown below.

$$1 \text{ byte} = 2^{3} \text{ bits}$$
$$1 \text{ kilobyte} = 2^{10} \text{ bytes} = 1{,}024 \text{ bytes}$$
$$1 \text{ megabyte} = 2^{20} \text{ bytes} = 1{,}048{,}576 \text{ bytes}$$
$$1 \text{ gigabyte} = 2^{30} \text{ bytes} = 1{,}073{,}741{,}824 \text{ bytes}$$
$$1 \text{ terabyte} = 2^{40} \text{ bytes} = 1{,}099{,}511{,}627{,}776 \text{ bytes}$$

3. Find an advertisement for a computer system. What is the computer's storage capacity? Convert the capacity to bytes. Use the exact equivalences given above.

Apple iBook 900MHz
14" (M9009LL/A) Memory

SALE
Maximum Memory
640MB

Answers to Projects and Group Activities— Metric Measurements for Computers

1. Symbol column:
   k, c, m
   Magnitude column:
   10^{9}, 10^{3}, 10^{2}, $\dfrac{1}{10^{3}}$, $\dfrac{1}{10^{12}}$
   Last column:
   1,000,000
   10
   0.1
   0.01
   0.000001
   0.000000001

2. The exponent on 10 indicates the number of places to move the decimal point. For prefixes tera-, giga-, mega-, kilo-, hecto-, and deca-, move the decimal point to the right. For the other prefixes shown, move the decimal point to the left.

3. Answers will vary.

Chapter 9 Summary

Key Words	**Examples**

The *metric system* of measurement is an internationally standardized system of measurement. It is based on the decimal system. The basic unit of length in the metric system is the *meter*. [9.1A, p. 373]

1 km = 1000 m
1 kg = 1000 g
1 kl = 1000 L

The basic unit of *mass* is the *gram*. [9.2A, p. 377]

The basic unit of capacity is the *liter*. [9.3A, p. 381]

1 m = 100 cm
1 m = 1000 mm
1 g = 1000 mg
1 L = 1000 ml

Heat is commonly measured in units called *Calories*. [9.4A, p. 385]

The *watt-hour* is used in the metric system for measuring electrical energy. [9.4A, p. 385]

In the metric system, prefixes to the basic unit denote the magnitude of each unit. [9.1A, p. 373]

kilo- = 1000   *deci-* = 0.1
hecto- = 100   *centi-* = 0.01
deca- = 10   *milli-* = 0.001

Essential Rules and Procedures	**Examples**

Converting between units in the metric system involves moving the decimal point to the right or to the left. Listing the units in order from largest to smallest will indicate how many places to move the decimal point and in which direction. [9.1A, 9.2A, 9.3A, pp. 373, 377, 381]

1. When converting from a larger unit to a smaller unit, move the decimal point to the *right*.
2. When converting from a smaller unit to a larger unit, move the decimal point to the *left*.

Convert 3.7 kg to grams.

$$3.7 \text{ kg} = 3700 \text{ g}$$

Convert 2387 m to kilometers.

$$2387 \text{ m} = 2.387 \text{ km}$$

Convert 9.5 L to milliliters.

$$9.5 \text{ L} = 9500 \text{ ml}$$

Approximate equivalences between units in the U.S. Customary and the metric systems of measurement are used to form conversion rates to change one unit of measurement to another. [9.5A, p. 389]

Units of Length
1 in. = 2.54 cm
1 m ≈ 3.28 ft
1 m ≈ 1.09 yd
1 mi ≈ 1.61 km

Units of Weight
1 oz ≈ 28.35 g
1 lb ≈ 454 g
1 kg ≈ 2.2 lb

Units of Capacity
1 L ≈ 1.06 qt
1 gal ≈ 3.79 L

Convert 20 mi/h to kilometers per hour.

$$\frac{20 \text{ mi}}{\text{h}} \approx \frac{20 \text{ m\cancel{i}}}{\text{h}} \times \frac{1.61 \text{ km}}{1 \text{ m\cancel{i}}}$$

$$\approx \frac{32.2 \text{ km}}{1 \text{ h}}$$

$$\approx 32.2 \text{ km/h}$$

Convert 1000 m to yards.

$$1000 \text{ m} \approx 1000 \text{ \cancel{m}} \times \frac{1.09 \text{ yd}}{1 \text{ \cancel{m}}}$$

$$\approx 1090 \text{ yd}$$

Chapter 9 Review Exercises

1. Convert 1.25 km to meters.
1250 m   [9.1A]

2. Convert 0.450 g to milligrams.
450 mg   [9.2A]

3. Convert 0.0056 L to milliliters.
5.6 ml   [9.3A]

4. Convert the 1000-meter run to yards.
(1 m ≈ 1.09 yd)
1090 yd   [9.5B]

5. Convert 79 mm to centimeters.
7.9 cm   [9.1A]

6. Convert 5 m 34 cm to meters.
5.34 m   [9.1A]

7. Convert 990 g to kilograms.
0.990 kg   [9.2A]

8. Convert 2550 ml to liters.
2.550 L   [9.3A]

9. Convert 4870 m to kilometers.
4.870 km   [9.1A]

10. Convert 0.37 cm to millimeters.
3.7 mm   [9.1A]

11. Convert 6 g 829 mg to grams.
6.829 g   [9.2A]

12. Convert 1.2 L to cubic centimeters.
1200 cm^3 [9.3A]

13. Convert 4.050 kg to grams.
4050 g   [9.2A]

14. Convert 8.7 m to centimeters.
870 cm   [9.1A]

15. Convert 192 ml to cubic centimeters.
192 cm^3 [9.3A]

16. Convert 356 mg to grams.
0.356 g   [9.2A]

17. Convert 372 cm to meters.
3.72 m   [9.1A]

18. Convert 8.3 kl to liters.
8300 L   [9.3A]

19. Convert 2 L 89 ml to liters.
2.089 L   [9.3A]

20. Convert 5410 cm^3 to liters.
5.410 L   [9.3A]

21. Convert 3792 L to kiloliters.
3.792 kl   [9.3A]

22. Convert 468 cm^3 to milliliters.
468 ml   [9.3A]

23. **Measurements** Three pieces of wire are cut from a 50-meter roll. The three pieces measure 240 cm, 560 cm, and 480 cm. How much wire is left on the roll after the three pieces are cut?
37.2 m [9.1B]

24. **Consumerism** Find the total cost of three packages of chicken weighing 790 g, 830 g, and 655 g if the cost is $5.59 per kilogram.
$12.72 [9.2B]

25. **Consumerism** Cheese costs $3.40 per pound. Find the cost per kilogram. (1 kg ≈ 2.2 lb)
$7.48/kg [9.5A]

26. **Measurements** One hundred twenty-five guests are expected to attend a reception. Assuming that each person drinks 400 ml of coffee, how many liters of coffee should be prepared?
50 L [9.3B]

27. **Nutrition** A large egg contains approximately 90 Calories. How many Calories can you eliminate from your diet in a 30-day month by eliminating one large egg per day from your usual breakfast?
2700 Calories [9.4A]

28. **Energy** A TV uses 240 W of energy. The set is on an average of 5 h a day in a 30-day month. At a cost of 9.5¢ per kilowatt-hour, how much does it cost to run the set for 30 days?
$3.42 [9.4A]

29. **Measurements** A backpack weighs 1.90 kg. Find the weight in pounds. Round to the nearest hundredth. (1 kg ≈ 2.2 lb)
4.18 lb [9.5B]

30. **Health** Cycling burns up approximately 400 Calories per hour. How many hours of cycling are necessary to lose 1 lb? (3500 Calories are equivalent to 1 lb.)
8.75 h [9.4A]

31. **Business** Six liters of liquid soap were bought for $11.40 per liter. The soap was repackaged in 150-milliliter plastic containers. The cost of each container was $.26. Each container of soap sold for $3.29 per bottle. Find the profit on the 6 L of liquid soap.
$52.80 [9.3B]

32. **Energy** A color TV is rated at 80 W. The TV is used an average of 2 h each day for a week. How many kilowatt-hours of energy are used during the week?
1.120 kWh [9.4A]

33. **Agriculture** How many kilograms of fertilizer are necessary to fertilize 500 trees in an orchard if 250 g of fertilizer is used for each tree?
125 kg [9.2B]

Chapter 9 Test

1. Convert 2.96 km to meters.
2960 m [9.1A]

2. Convert 0.378 g to milligrams.
378 mg [9.2A]

3. Convert 0.046 L to milliliters.
46 ml [9.3A]

4. Convert 919 cm³ to milliliters.
919 ml [9.3A]

5. Convert 42.6 mm to centimeters.
4.26 cm [9.1A]

6. Convert 7 m 96 cm to meters.
7.96 m [9.1A]

7. Convert 847 g to kilograms.
0.847 kg [9.2A]

8. Convert 3920 ml to liters.
3.920 L [9.3A]

9. Convert 5885 m to kilometers.
5.885 km [9.1A]

10. Convert 1.5 cm to millimeters.
15 mm [9.1A]

11. Convert 3 g 89 mg to grams.
3.089 g [9.2A]

12. Convert 1.6 L to cubic centimeters.
1600 cm³ [9.3A]

13. Convert 3.29 kg to grams.
3290 g [9.2A]

14. Convert 4.2 m to centimeters.
420 cm [9.1A]

15. Convert 96 ml to cubic centimeters.
96 cm³ [9.3A]

16. Convert 1375 mg to grams.
1.375 g [9.2A]

17. Convert 402 cm to meters.
4.02 m [9.1A]

18. Convert 8.92 kl to liters.
8920 L [9.3A]

19. Health Sedentary people need 15 Calories per pound of body weight to maintain their weight. How many Calories should a 140-pound, sedentary person consume per day to maintain that weight?
2100 Calories [9.4A]

20. Energy A color television is rated at 100 W. The television is used an average of $4\frac{1}{2}$ h each day for a week. Find the number of kilowatt-hours of energy used during the week for operating the television.
3.15 kWh [9.4A]

21. **Carpentry** A carpenter needs 30 rafters, each 380 cm long. Find the total length of the rafters in meters.
114 m   [9.1B]

22. **Measurements** A tile measuring 20 × 20 cm weighs 250 g. Find the weight, in kilograms, of a box of 144 tiles.
36 kg   [9.2B]

23. **Medicine** The community health clinic is giving flu shots for the coming flu season. Each flu shot contains 2 cm³ of vaccine. How many liters of vaccine are needed to inoculate 2600 people?
5.2 L   [9.3B]

24. **Measurements** Convert 35 mi/h to kilometers per hour. Round to the nearest tenth. (1 mi ≈ 1.61 km)
56.4 km/h   [9.5A]

25. **Metal Works** Twenty-five rivets are used to fasten two steel plates together. The plates are 4.20 m long, and the rivets are equally spaced, with a rivet at each end. Find the distance, in centimeters, between the rivets.
17.5 cm   [9.1B]

26. **Agriculture** Two hundred grams of fertilizer are used for each tree in an orchard containing 1200 trees. At $2.75 per kilogram of fertilizer, how much does it cost to fertilize the orchard?
$660   [9.2B]

27. **Energy** An air conditioner rated at 1600 W is operated an average of 4 h per day. Electrical energy costs 8.5¢ per kilowatt-hour. How much does it cost to operate the air conditioner for 30 days?
$16.32   [9.4A]

28. **Chemistry** A laboratory assistant is in charge of ordering acid for three chemistry classes of 40 students each. Each student requires 90 ml of acid. How many liters of acid should be ordered? (The assistant must order by the whole liter.)
11 L   [9.3B]

29. ● **Sports** Three ski jumping events are held at the Olympic Games: the individual normal hill, the individual large hill, and the team large hill. The normal hill measures 90 m. The large hill measures 120 m. Convert the measure of the large hill to feet. Round to the nearest tenth. (1 m ≈ 3.28 ft)
393.6 ft   [9.5B]

30. ● **Sports** In the archery competition at the Olympic Games, the center ring, or bull's eye, of the target is approximately 4.8 in. in diameter. Convert 4.8 in. to centimeters. Round to the nearest tenth. (1 in. = 2.54 cm)
12.2 cm   [9.5A]

Cumulative Review Exercises

1. Simplify: $12 - 8 \div (6 - 4)^2 \cdot 3$
6   [1.6B]

2. Find the total of $5\frac{3}{4}$, $1\frac{5}{6}$, and $4\frac{7}{9}$.
$12\frac{13}{36}$ [2.4C]

3. Subtract: $4\frac{2}{9} - 3\frac{5}{12}$
$\frac{29}{36}$ [2.5C]

4. Divide: $5\frac{3}{8} \div 1\frac{3}{4}$
$3\frac{1}{14}$ [2.7B]

5. Simplify: $\left(\frac{2}{3}\right)^4 \cdot \left(\frac{9}{4}\right)^2$
1   [2.8B]

6. Subtract: $12.0072 - 9.937$
2.0702   [3.3A]

7. Solve the proportion $\frac{5}{8} = \frac{n}{50}$. Round to the nearest tenth.
31.3   [4.3B]

8. Write $1\frac{3}{4}$ as a percent.
175%   [5.1B]

9. 6.09 is 4.2% of what number?
145   [5.4A]

10. Convert 18 pt to gallons.
2.25 gal   [8.3A]

11. Convert 875 cm to meters.
8.75 m   [9.1A]

12. Convert 3420 m to kilometers.
3.420 km   [9.1A]

13. Convert 5.05 kg to grams.
5050 g   [9.2A]

14. Convert 3 g 672 mg to grams.
3.672 g   [9.2A]

15. Convert 6 L to milliliters.
6000 ml   [9.3A]

16. Convert 2.4 kl to liters.
2400 L   [9.3A]

17. **Finances**   The Guerrero family has a monthly income of $5244 per month. The family spends one-fourth of its monthly income on rent. How much money is left after the rent is paid?
$3933   [2.6C]

18. **Taxes**   The state income tax on a business is $620 plus 0.08 times the profit the business makes. The business made a profit of $82,340.00 last year. Find the amount of state income tax the business paid.
$7207.20   [3.4B]

19. **Real Estate**   The property tax on a $245,000 home is $4900. At the same rate, what is the property tax on a home worth $275,000?
$5500   [4.3C]

20. **Consumerism**   A car dealer offers new-car buyers a 12% rebate on some models. What rebate would a new-car buyer receive on a car that cost $23,500?
$2820   [5.2B]

21. **Investments**   Rob Akullian received a dividend of $533 on an investment of $8200. What percent of the investment is the dividend?
6.5%   [5.3B]

22. **Education**   You received grades of 78, 92, 45, 80, and 85 on five English exams. Find your mean grade.
76   [7.4A]

23. **Compensation**   Karla Perella, a ski instructor, receives a salary of $22,500. Her salary will increase by 12% next year. What will her salary be next year?
$25,200   [5.2B]

24. **Discount**   A sporting goods store has fishing rods that are regularly priced at $180 on sale for $140.40. What is the discount rate?
22%   [6.2D]

25. **Masonry**   Forty-eight bricks, each 9 in. long, are laid end to end to make the base for a wall. Find the length of the wall in feet.
36 ft   [8.1C]

26. **Packaging**   A bottle of apple juice contains 24 oz. Find the number of quarts of apple juice in a case of 16 bottles.
12 qt   [8.3C]

27. **Business**   A garage mechanic bought oil in a 40-gallon container. The mechanic bought the oil for $4.88 per gallon and sold the oil for $1.99 per quart. Find the profit on the 40-gallon container of oil.
$123.20   [8.3C]

28. **Measurements**   A school swimming pool uses 1200 ml of chlorine each school day. How many liters of chlorine are used for 20 days during the month?
24 L   [9.3B]

29. **Energy**   A 1200-watt hair dryer is used an average of 30 min a day. At a cost of 10.5¢ per kilowatt-hour, how much does it cost to operate the hair dryer for 30 days?
$1.89   [9.4A]

30. **Measurements**   Convert 60 mi/h to kilometers per hour. Round to the nearest tenth. (1.61 km ≈ 1 mi)
96.6 km/h   [9.5A]

chapter

10 Rational Numbers

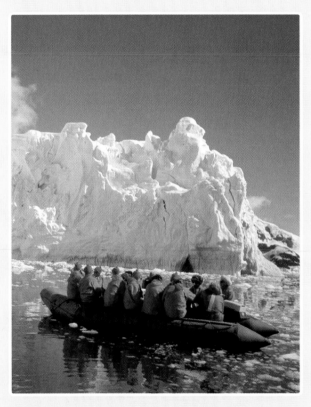

These tourists took an excursion boat in order to get a closer look at the icebergs at the end of the glaciers in Antarctica's Paradise Bay. Temperatures in places such as this fall below zero. When this happens, the temperatures are described in negative numbers. Negative numbers are used to express any value below zero, such as debt, altitude below sea level, decreases in stock prices, and golf scores under par. In **Exercises 86 to 89 on page 436,** you will be solving problems involving extreme temperature changes in parts of the United States, as well as the boiling points of certain chemical elements.

OBJECTIVES

Section 10.1

A To identify the order relation between two integers
B To evaluate expressions that contain the absolute value symbol

Section 10.2

A To add integers
B To subtract integers
C To solve application problems

Section 10.3

A To multiply integers
B To divide integers
C To solve application problems

Section 10.4

A To add or subtract rational numbers
B To multiply or divide rational numbers
C To solve application problems

Section 10.5

A To write a number in scientific notation
B To use the Order of Operations Agreement to simplify expressions

Need help? For online student resources, such as section quizzes, visit this textbook's website at **math.college.hmco.com/students.**

PREP TEST • • •

Do these exercises to prepare for Chapter 10.

1. Place the correct symbol, < or >, between the two numbers.

54   45
54 > 45   [1.1A]

2. What is the distance from 4 to 8 on the number line?
4 units   [1.3A]

For Exercises 3 to 14, add, subtract, multiply, or divide.

3. $7654 + 8193$
15,847   [1.2A]

4. $6097 - 2318$
3779   [1.3B]

5. 472×56
26,432   [1.4B]

6. $\dfrac{144}{24}$
6   [2.3B]

7. $\dfrac{2}{3} + \dfrac{3}{5}$
$1\dfrac{4}{15}$ [2.4B]

8. $\dfrac{3}{4} - \dfrac{5}{16}$
$\dfrac{7}{16}$ [2.5B]

9. $0.75 + 3.9 + 6.408$
11.058   [3.2A]

10. $5.4 - 1.619$
3.781   [3.3A]

11. $\dfrac{3}{4} \times \dfrac{8}{15}$
$\dfrac{2}{5}$ [2.6A]

12. $\dfrac{5}{12} \div \dfrac{3}{4}$
$\dfrac{5}{9}$ [2.7A]

13. 23.5×0.4
9.4   [3.4A]

14. $0.96 \div 2.4$
0.4   [3.5A]

15. Simplify: $(8 - 6)^2 + 12 \div 4 \cdot 3^2$
31   [1.6B]

GO FIGURE • • •

Super Yeast causes bread to double in volume each minute. If it takes one loaf of bread made with Super Yeast 30 minutes to fill the oven, how long does it take two loaves of bread made with Super Yeast to fill one-half the oven?   28 minutes

10.1 Introduction to Integers

Objective A   To identify the order relation between two integers

Thus far in the text, we have encountered only zero and the numbers greater than zero. The numbers greater than zero are called **positive numbers.** However, the phrases "12 degrees below zero," "$25 in debt," and "15 feet below sea level" refer to numbers less than zero. These numbers are called **negative numbers.**

The **integers** are . . . , −4, −3, −2, −1, 0, 1, 2, 3, 4, ...

Each integer can be shown on a number line. The integers to the left of zero on the number line are called **negative integers** and are represented by a negative sign (−) placed in front of the number. The integers to the right of zero are called **positive integers.** The positive integers are also called **natural numbers.** Zero is neither a positive nor a negative integer.

Point of Interest

Among the slang words for zero are *zilch*, *zip*, and *goose egg*. The word *love* for zero in scoring a tennis game comes from the French word *l'oeuf*, which means "the egg."

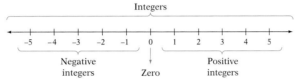

A number line can be used to visualize the order relation between two integers. A number that appears to the left of a given number is less than (<) the given number. A number that appears to the right of a given number is greater than (>) the given number.

2 is greater than negative 4.

$2 > -4$

Negative 5 is less than negative 3.

$-5 < -3$

Example 1

The temperature at the North Pole was recorded as 87° below zero. Represent this temperature as an integer.

Solution $-87°$

Example 2

Graph −2 on the number line.

Solution

Example 3

Place the correct symbol, < or >, between the numbers −5 and −7.

Solution $-5 > -7$ • −5 is to the right of −7 on the number line.

You Try It 1

The surface of the Salton Sea is 232 ft below sea level. Represent this depth as an integer.

Your solution -232 ft

You Try It 2

Graph −4 on the number line.

Your solution

You Try It 3

Place the correct symbol, < or >, between the numbers −12 and −8.

Your solution $-12 < -8$

Solutions on p. S25

Objective 10.1A

New Vocabulary
positive numbers
negative numbers
integers
negative integers
positive integers
natural numbers

Instructor Note
Within the Microsoft PowerPoint® slides available with this text is a number line that extends from −10 to 10. It can be used to create a transparency on which to graph integers.

Discuss the Concepts
Classify each number as a positive integer, a negative integer, or neither.
$-10, 17, -8, 0, \frac{2}{3}, 465$

Concept Check
A is a point on the number line halfway between −12 and 4. *B* is a point on the number line halfway between *A* and the graph of 2 on the number line. *B* is the graph of what number?
−1

Optional Student Activity
Graph each of the following pairs of numbers on the number line. Then place the correct symbol, < or >, between the two numbers.
1. −4 and 3 $-4 < 3$
2. 5 and −2 $5 > -2$
3. −1 and −3 $-1 > -3$
4. −5 and 0 $-5 < 0$
5. −7 and 1 $-7 < 1$

In-Class Examples (Objective 10.1A)

1. Represent a temperature that is 15°F below zero as an integer. $-15°F$

2. Place the correct symbol, < or >, between the two numbers. −25 −35 $-25 > -35$

3. Place the correct symbol, < or >, between the two numbers. 8 −9 $8 > -9$

Objective 10.1B

New Vocabulary
opposites
absolute value of a number

Instructor Note
The +/− key on a calculator can be used to illustrate the idea of opposite. Entering 4 +/− gives the opposite of 4. Pressing the +/− key again gives 4, the opposite of −4.

Discuss the Concepts
True or false:
1. Every negative number is to the left of every positive number on the number line.   T
2. Every positive number is to the right of every negative number on the number line.   T
3. −6 is to the left of −1 on the number line.   T
4. −8 is to the right of −12 on the number line.   T

Instructor Note
The important point for a student to understand about absolute value is magnitude. If a student runs 5 mi west or 5 mi east, the distance is the same. Only the direction is different.

Optional Student Activity
The following list provides temperatures either on the surface or in the atmospheres of bodies in our solar system. (*Continued on the next page*)

Objective B To evaluate expressions that contain the absolute value symbol

Two numbers that are the same distance from zero on the number line but on opposite sides of zero are called **opposites.**

−4 is the opposite of 4
   and
4 is the opposite of −4.

Note that a negative sign can be read as "the opposite of."

−(4) = −4   The opposite of positive 4 is negative 4.

−(−4) = 4   The opposite of negative 4 is positive 4.

The **absolute value of a number** is the distance between zero and the number on the number line. Therefore, the absolute value of a number is a positive number or zero. The symbol for absolute value is | |.

The distance from 0 to 4 is 4. Thus |4| = 4 (the absolute value of 4 is 4).

The distance from 0 to −4 is 4. Thus |−4| = 4 (the absolute value of −4 is 4).

The absolute value of a positive number is the number itself. The absolute value of a negative number is the opposite of the negative number. The absolute value of zero is zero.

Example 4 Find the absolute value of 2 and −3.

Solution |2| = 2

   |−3| = 3

You Try It 4 Find the absolute value of −7 and 21.

Your solution 7; 21

Example 5 Evaluate |−34| and |0|.

Solution |−34| = 34

   |0| = 0

You Try It 5 Evaluate |2| and |−9|.

Your solution 2; 9

Example 6 Evaluate −|−4|.

Solution   −|−4| = −4
The minus sign *in front of* the absolute value sign is not affected by the absolute value sign.

You Try It 6 Evaluate −|−12|.

Your solution −12

Solutions on p. S25

In-Class Examples (Objective 10.1B)
1. What is the opposite of −18?   18
2. Find the absolute value of −25.   25
3. Evaluate −|−9|.   −9

10.1 Exercises

Objective A   **To identify the order relation between two integers**

For Exercises 1 to 4, represent the quantity as an integer.

1. A lake 120 ft below sea level
−120 ft

2. A temperature that is 15° below zero
−15°

3. A share of stock up 2 dollars
+2 dollars

4. A loss of 324 dollars
−324 dollars

For Exercises 5 to 8, graph the numbers on the number line.

5. 3 and −3

6. −2 and 0

7. −4 and 1

8. 4 and −1

For Exercises 9 to 14, state which number on the number line is in the location given.

9. 3 units to the right of −2
1

10. 5 units to the right of −3
2

11. 4 units to the left of 3
−1

12. 2 units to the left of −1
−3

13. 6 units to the right of −3
3

14. 4 units to the right of −4
0

For Exercises 15 to 18, use the following number line.

15. a. If F is 1 and G is 2, what number is A?
−4

16. a. If G is 1 and H is 2, what number is B?
−4

b. If F is 1 and G is 2, what number is C?
−2

b. If G is 1 and H is 2, what number is D?
−2

17. a. If H is 0 and I is 1, what number is A?
−7

18. a. If G is 2 and I is 4, what number is B?
−3

b. If H is 0 and I is 1, what number is D?
−4

b. If G is 2 and I is 4, what number is E?
0

Section 10.1

Suggested Assignment
Exercises 1–113, odds
More challenging problems:
    Exercises 115–121

(*Continued from page 406*)
    Mercury's dark side:
        −346°F
    Mercury's sunlit side:
        950°F
    Mars, daytime:
        −17°F
    Saturn's moon Titan:
        −190°F
    Neptune's atmosphere:
        −360°F
    Pluto, atmospheric high:
        −163°F
    Pluto, atmospheric low:
        −397°F

a. List the temperatures given
    from coldest to warmest.
    −397, −360, −346, −190,
    −163, −17, 950

b. How many of the
    temperatures listed are below
    −300°F?
    Three of the temperatures
    listed are below −300°F
    (−397°F, −360°F, −346°F).

Quick Quiz (Objective 10.1A)

1. Represent as an integer a stock price that is down
3 dollars.   −3 dollars

2. Place the correct symbol, < or >, between the
two numbers. −5 3   −5 < 3

3. Place the correct symbol, < or >, between the
two numbers. −18 −12   −18 < −12

For Exercises 19 to 42, place the correct symbol, < or >, between the two numbers.

19. $-2 > -5$ **20.** $-6 < -1$ **21.** $-16 < 1$ **22.** $-2 < 13$

23. $3 > -7$ **24.** $5 > -6$ **25.** $-11 < -8$ **26.** $-4 > -10$

27. $35 > 28$ **28.** $42 > 19$ **29.** $-42 < 27$ **30.** $-36 < 49$

31. $21 > -34$ **32.** $53 > -46$ **33.** $-27 > -39$ **34.** $-51 < -20$

35. $-87 < 63$ **36.** $-75 < 92$ **37.** $86 > -79$ **38.** $95 > -71$

39. $-62 > -84$ **40.** $-91 < -70$ **41.** $-131 < 101$ **42.** $127 > -150$

For Exercises 43 to 51, write the given numbers in order from smallest to largest.

43. $3, -7, 0, -2$
$-7, -2, 0, 3$

44. $-4, 8, 6, -1$
$-4, -1, 6, 8$

45. $-3, 1, -5, 4$
$-5, -3, 1, 4$

46. $-6, 2, -8, 7$
$-8, -6, 2, 7$

47. $9, -4, 5, 0$
$-4, 0, 5, 9$

48. $6, -9, -12, 8$
$-12, -9, 6, 8$

49. $-10, 4, 12, -5, -7$
$-10, -7, -5, 4, 12$

50. $11, -8, -1, 7, -6$
$-8, -6, -1, 7, 11$

51. $10, -11, -2, 5, -7$
$-11, -7, -2, 5, 10$

Objective B **To evaluate expressions that contain the absolute value symbol**

For Exercises 52 to 61, find the opposite number.

52. 4
-4

53. 16
-16

54. -2
2

55. -3
3

56. 22
-22

57. 45
-45

58. -31
31

59. -59
59

60. 70
-70

61. -88
88

Quick Quiz (Objective 10.1B)
1. What is the opposite of 14? -14
2. Find the absolute value of -37. 37
3. Evaluate $-|62|$. -62

For Exercises 62 to 69, find the absolute value of the number.

62. 4
4

63. −4
4

64. −7
7

65. 9
9

66. −1
1

67. −11
11

68. 10
10

69. −12
12

For Exercises 70 to 99, evaluate.

70. $|2|$
2

71. $|-2|$
2

72. $|-6|$
6

73. $|6|$
6

74. $|8|$
8

75. $|5|$
5

76. $|-9|$
9

77. $|-1|$
1

78. $-|-1|$
−1

79. $-|-5|$
−5

80. $-|0|$
0

81. $|16|$
16

82. $|19|$
19

83. $|-12|$
12

84. $|-22|$
22

85. $-|29|$
−29

86. $-|20|$
−20

87. $-|-14|$
−14

88. $-|-18|$
−18

89. $|-15|$
15

90. $|-23|$
23

91. $-|33|$
−33

92. $-|27|$
−27

93. $|32|$
32

94. $|25|$
25

95. $-|-42|$
−42

96. $|-74|$
74

97. $|-61|$
61

98. $-|88|$
−88

99. $-|52|$
−52

For Exercises 100 to 107, place the correct symbol, $<$, $=$, or $>$, between the two numbers.

100. $|7| < |-9|$

101. $|-12| > |8|$

102. $|-5| > |-2|$

103. $|6| < |13|$

104. $|-8| > |3|$

105. $|-1| < |-17|$

106. $|-14| = |14|$

107. $|17| = |-17|$

For Exercises 108 to 113, write the given numbers in order from smallest to largest.

108. $|-8|, -3, |2|, -|-5|$
$-|-5|, -3, |2|, |-8|$

109. $-|6|, -4, |-7|, -9$
$-9, -|6|, -4, |-7|$

110. $-1, |-6|, |0|, -|3|$
$-|3|, -1, |0|, |-6|$

111. $-|-7|, -9, 5, |4|$
$-9, -|-7|, |4|, 5$

112. $-|2|, -8, 6, |1|$
$-8, -|2|, |1|, 6$

113. $-3, -|-8|, |5|, -|10|$
$-|10|, -|-8|, -3, |5|$

APPLYING THE CONCEPTS

114. **Meteorology** The graph at the right shows the lowest recorded temperatures, in degrees Fahrenheit, for selected states in the United States. Which state has the lowest recorded temperature?
New York

Lowest Recorded Temperatures

Sources: National Climatic Data Center; NESDIS; NOAA; U.S. Dept. of Commerce

115. **a.** Name two numbers that are 5 units from 3 on the number line.   −2, 8
   b. Name two numbers that are 3 units from −1 on the number line.   −4, 2

116. **a.** Find a number that is halfway between −7 and −5.   −6
   b. Find a number that is halfway between −10 and −6.   −8
   c. Find a number that is one-third of the way between −12 and −3.   −9 or −6

117. **Rocketry** Which is closer to blastoff, −12 min and counting or −17 min and counting?
−12 min and counting

118. **Investments** In the stock market, the net change in the price of a share of stock is recorded as a positive or a negative number. If the price rises, the net change is positive. If the price falls, the net change is negative. If the net change for a share of Stock A is −2 and the net change for a share of Stock B is −1, which stock showed the least net change?
Stock B

119. **Business** Some businesses show a profit as a positive number and a loss as a negative number. During the first quarter of this year, the loss experienced by a company was recorded as −12,575. During the second quarter of this year, the loss experienced by the company was −11,350. During which quarter was the loss greater?
The first quarter

120. Find the values of *a* for which $|a| = 7$.
7, −7

121. Find the values of *y* for which $|y| = 11$.
11, −11

122. **a.** Describe the *opposite of a number* in your own words.
   b. Describe the *absolute value of a number* in your own words.

Answers to Writing Exercises

122a. Students might describe the opposite of a number as the number with the sign changed, or they might rephrase the definition that states that opposite numbers are two numbers that are the same distance from zero on the number line but on opposite sides of zero.

  b. Students might rephrase the definition that states that the absolute value of a number is its distance from zero on the number line. You might check that students include in their description that the absolute value of a number cannot be negative (or that the absolute value of a number is either a positive number or zero).

10.2 Addition and Subtraction of Integers

Objective A To add integers

An integer can be graphed as a dot on a number line, as shown in the last section. An integer also can be represented anywhere along a number line by an arrow. A positive number is represented by an arrow pointing to the right. A negative number is represented by an arrow pointing to the left. The absolute value of the number is represented by the length of the arrow. The integers 5 and -4 are shown on the number lines below.

The sum of two integers can be shown on a number line. To add two integers, use arrows to represent the addends, with the first arrow starting at zero. The sum is the number directly below the tip of the arrow that represents the second addend.

$4 + 2 = 6$

$-4 + (-2) = -6$

$-4 + 2 = -2$

$4 + (-2) = 2$

The sums of the integers shown above can be categorized by the signs of the addends.

Here the addends have the same sign:

$4 + 2$ *positive* 4 plus *positive* 2
$-4 + (-2)$ *negative* 4 plus *negative* 2

Here the addends have different signs:

$-4 + 2$ *negative* 4 plus *positive* 2
$4 + (-2)$ *positive* 4 plus *negative* 2

The rule for adding two integers depends on whether the signs of the addends are the same or different.

Objective 10.2A

Vocabulary to Review
integers [10.1A]
absolute value [10.1B]
addend [1.2A]

New Rules
addition of integers

Discuss the Concepts
a. Explain the rule for adding two integers with the same sign.
b. Explain the rule for adding two integers with different signs.

Concept Check
State whether or not the operation is addition. If the operation is addition, state whether the numbers being added have the same sign or different signs.
a. $2 + (-6)$
Yes; different signs
b. $-4 + 8$ Yes; different signs
c. $7 - 13$ No
d. $(-10) + (-9)$
Yes; same sign
e. $-15 + (-3)$ Yes; same sign
f. $-5(+13)$ No

Optional Student Activity
On the number line, illustrate each of the following sums.
a. $-3 + 5$
b. $-4 + 7$
c. $2 + (-6)$
d. $5 + (-9)$
e. $-1 + (-4)$
f. $-5 + (-2)$

In-Class Examples (Objective 10.2A)
Add.
1. $-9 + 4$ -5
2. $-44 + (-71)$ -115
3. $16 + (-25) + 4$ -5

Instructor Note

There are several ways to model the addition of integers. The model on the previous page uses arrows. Another model uses money. For instance, if you are $8 in debt (−8) and you receive $5, then you are only $3 in debt (−3). This can also be related to credit card debt. If a student owes $100 (−100) and charges $25 more (−25), the student then owes $125 (−125).

Giving examples like these may help students to see that the rules are not arbitrary but are designed to model everyday experience.

Another model of addition of integers that is more manipulative in nature involves using chips: blue chips for positive and red chips for negative. One positive chip added to one negative chip gives zero. To add −8 and 5, place 8 red chips and 5 blue chips in a region. Pair as many red and blue chips as possible and remove the pairs from the region. The remaining chips give the answer—in this case, 3 red chips, or −3.

To model (−8) + (−5), place 8 red chips in the region and then 5 more red chips in the region. There are no pairs of red and blue chips, so there are 13 red chips. Therefore, the answer is −13.

Point of Interest

Although mathematical symbols are fairly standard in every country, that has not always been true. Italian mathematicians in the 15th century used a "p" to indicate plus. The "p" was from the Italian word *piu*.

Integrating Technology

To add −14 + (−47) on your calculator, enter the following:

14 +/− + 47 +/− =

Rule for Adding Two Numbers

To add numbers with the same sign, add the absolute values of the numbers. Then attach the sign of the addends.

To add numbers with different signs, find the difference between the absolute values of the numbers. Then attach the sign of the addend with the greater absolute value.

HOW TO   Add: (−4) + (−9)

$|−4| = 4, |−9| = 9$
$4 + 9 = 13$

(−4) + (−9) = −13

- Because the signs of the addends are the same, add the absolute values of the numbers.
- Then attach the sign of the addends.

HOW TO   Add: 6 + (−13)

$|6| = 6, |−13| = 13$
$13 − 6 = 7$

6 + (−13) = −7

- Because the signs of the addends are different, subtract the smaller absolute value from the larger absolute value.
- Then attach the sign of the number with the larger absolute value. Because $|−13| > |6|$, attach the negative sign.

HOW TO   Add: 162 + (−247)

162 + (−247) = −85

- Because the signs are different, find the difference between the absolute values of the numbers and attach the sign of the number with the greater absolute value.

HOW TO   Find the sum of −14 and −47.

−14 + (−47) = −61

- Because the signs are the same, add the absolute values of the numbers and attach the sign of the addends.

When adding more than two integers, start from the left and add the first two numbers. Then add the sum to the third number. Continue this process until all the numbers have been added.

HOW TO   Add: (−4) + (−6) + (−8) + 9

(−4) + (−6) + (−8) + 9 = (−10) + (−8) + 9
= (−18) + 9
= −9

- Add the first two numbers.
- Add the sum to the next number.
- Continue adding until all numbers have been added.

Example 1   What is −162 added to 98?

Solution   98 + (−162) = −64   • The signs of the addends are different.

You Try It 1   Add: −154 + (−37)

Your solution   −191

Solution on p. S25

Example 2

Add: $-2 + (-7) + 4 + (-6)$

Solution

$$
\begin{aligned}
-2 + (-7) + 4 + (-6) &= -9 + 4 + (-6) \\
&= -5 + (-6) \\
&= -11
\end{aligned}
$$

You Try It 2

Add: $-5 + (-2) + 9 + (-3)$

Your solution -1

Solution on p. S25

Objective B   **To subtract integers**

Study Tip

Be sure to do all you need to do in order to be successful at adding and subtracting integers: Read through the introductory material, work through the examples indicated by the HOW TO feature, study the paired Examples, do the You Try Its and check your solutions against those in the back of the book, and do the exercises on pages 415 to 418. See *AIM for Success* at the front of the book.

Before the rules for subtracting two integers are explained, look at the translation into words of an expression that is the difference of two integers:

$$
\begin{array}{ll}
9 - 3 & \text{positive 9 minus positive 3} \\
(-9) - 3 & \text{negative 9 minus positive 3} \\
9 - (-3) & \text{positive 9 minus negative 3} \\
(-9) - (-3) & \text{negative 9 minus negative 3}
\end{array}
$$

Note that the sign $-$ is used in two different ways. One way is as a negative sign, as in (-9), *negative* 9. The second way is to indicate the operation of subtraction, as in $9 - 3$, 9 *minus* 3.

Look at the next four subtraction expressions and decide whether the second number in each expression is a positive number or a negative number.

1. $(-10) - 8$ **2.** $(-10) - (-8)$ **3.** $10 - (-8)$ **4.** $10 - 8$

In expressions 1 and 4, the second number is a positive 8. In expressions 2 and 3, the second number is a negative 8.

> **Rule for Subtracting Two Numbers**
>
> To subtract two numbers, add the opposite of the second number to the first number.

This rule states that to subtract two integers, we rewrite the subtraction expression as the sum of the first number and the opposite of the second number.

Here are some examples:

First number	−	second number	=	first number	+	the opposite of the second number	
8	−	15	=	8	+	(-15)	$= -7$
8	−	(-15)	=	8	+	15	$= 23$
(-8)	−	15	=	(-8)	+	(-15)	$= -23$
(-8)	−	(-15)	=	(-8)	+	15	$= 7$

Integrating Technology

The $+/-$ key on your calculator is used to find the opposite of a number. The $-$ is used to perform the operation of subtraction.

HOW TO   Subtract: $(-15) - 75$

$$
\begin{aligned}
(-15) - 75 &= (-15) + (-75) \\
&= -90
\end{aligned}
$$

• To subtract, add the opposite of the second number to the first number.

In-Class Examples (Objective 10.2B)

Subtract.

1. $-4 - 17$ -21

2. $21 - (-19)$ 40

3. $-5 - 18 - (-10)$ -13

Objective 10.2B

Vocabulary to Review
minus   [1.3B]
negative   [10.1A]
opposite   [10.1B]

New Rules
subtraction of integers

Discuss the Concepts
Explain why the absolute value of -16 is greater than the absolute value of 4.

Concept Check
State whether each "$-$" sign is a minus sign or a negative sign.
a. $7 - 6$ Minus
b. $3 - (-9)$ Minus; negative
c. $-4 - 1$ Negative; minus
d. $-5 - (-2)$
Negative; minus; negative
e. $(-8) - 10$ Negative; minus

Instructor Note
Although it is more complicated than the addition model, a subtraction model using blue and red chips can be used. Blue chips represent positive numbers, and red chips represent negative numbers. To model $5 - (-3)$, place 5 blue chips in a region. Subtraction requires removing 3 red chips, but because there are no red chips in the region, add 3 pairs of a red and a blue chip (essentially adding three zeros). Now the 3 red chips can be removed. The result is 8 blue chips.

HOW TO Subtract: 27 − (−32)

$$27 - (-32) = 27 + 32$$
$$= 59$$

• To subtract, add the opposite of the second number to the first number.

When subtraction occurs several times in an expression, rewrite each subtraction as addition of the opposite and then add.

HOW TO Subtract: −13 − 5 − (−8)

$$-13 - 5 - (-8) = -13 + (-5) + 8$$
$$= -18 + 8$$
$$= -10$$

• Rewrite each subtraction as the addition of the opposite and then add.

Example 3

Find 8 less than −12.

Solution
$$-12 - 8 = -12 + (-8)$$
$$= -20$$

• Rewrite "−" as "+"; the opposite of 8 is −8.

You Try It 3

Find −8 less 14.

Your solution
−22

Example 4

Subtract: 6 − (−20)

Solution
$$6 - (-20) = 6 + 20$$
$$= 26$$

• Rewrite "−" as "+"; the opposite of −20 is 20.

You Try It 4

Subtract: 3 − (−15)

Your solution
18

Example 5

Subtract: −8 − 30 − (−12) − 7

Solution
$$-8 - 30 - (-12) - 7$$
$$= -8 + (-30) + 12 + (-7)$$
$$= -38 + 12 + (-7)$$
$$= -26 + (-7)$$
$$= -33$$

You Try It 5

Subtract: 4 − (−3) − 12 − (−7) − 20

Your solution
−18

Solutions on pp. S25–S26

Objective C To solve application problems

Example 6

Find the temperature after an increase of 9°C from −6°C.

Strategy
To find the temperature, add the increase (9°C) to the previous temperature (−6°C).

Solution
$$-6 + 9 = 3$$

The temperature is 3°C.

You Try It 6

Find the temperature after an increase of 12°C from −10°C.

Your strategy

Your solution
2°C

Solution on p. S26

Objective 10.2C

Optional Student Activity

The following list provides temperatures either on the surface or in the atmospheres of bodies in our solar system. (This is the same list provided in Objective 10.1B.)

Mercury's dark side:
−346°F

Mercury's sunlit side:
950°F

Mars, daytime:
−17°F

Saturn's moon Titan:
−190°F

Neptune's atmosphere:
−360°F

Pluto, atmospheric high:
−163°F

Pluto, atmospheric low:
−397°F

a. Find the difference between the temperatures on the sunlit side and the dark side of Mercury.   1296°F

b. Find the difference between Pluto's atmospheric high and low temperatures.   234°F

In-Class Examples (Objective 10.2C)

1. Find the temperature after a rise of 4°C from −2°C.   2°C

2. During a card game of Hearts, you had a score of 14 points before your opponent "shot the moon," subtracting a score of 26 from your total. What was your score after your opponent "shot the moon"?   −12 points

10.2 Exercises

Objective A   **To add integers**

For Exercises 1 and 2, name the negative integers in the list of numbers.

1. $-14, 28, 0, -\dfrac{5}{7}, -364, -9.5$
$-14, -364$

2. $-37, 90, -\dfrac{7}{10}, -88.8, 42, -561$
$-37, -561$

For Exercises 3 to 30, add.

3. $3 + (-5)$
-2

4. $-4 + 2$
-2

5. $8 + 12$
20

6. $16 + 23$
39

7. $-3 + (-8)$
-11

8. $-12 + (-1)$
-13

9. $-4 + (-5)$
-9

10. $-12 + (-12)$
-24

11. $6 + (-9)$
-3

12. $4 + (-9)$
-5

13. $-6 + 7$
1

14. $-12 + 6$
-6

15. $2 + (-3) + (-4)$
-5

16. $7 + (-2) + (-8)$
-3

17. $-3 + (-12) + (-15)$
-30

18. $9 + (-6) + (-16)$
-13

19. $-17 + (-3) + 29$
9

20. $13 + 62 + (-38)$
37

21. $-3 + (-8) + 12$
1

22. $-27 + (-42) + (-18)$
-87

23. $13 + (-22) + 4 + (-5)$
-10

24. $-14 + (-3) + 7 + (-6)$
-16

25. $-22 + 10 + 2 + (-18)$
-28

26. $-6 + (-8) + 13 + (-4)$
-5

27. $-16 + (-17) + (-18) + 10$
-41

28. $-25 + (-31) + 24 + 19$
-13

29. $-126 + (-247) + (-358) + 339$
-392

30. $-651 + (-239) + 524 + 487$
121

31. What is -8 more than -12?
-20

32. What is -5 more than 3?
-2

33. What is -7 added to -16?
-23

34. What is 7 added to -25?
-18

Section 10.2

Suggested Assignment
Exercises 1–97, odds
More challenging problems:
    Exercises 98–100

Quick Quiz (Objective 10.2A)
Add.
1. $8 + (-10)$ -2
2. $-15 + (-28)$ -43
3. $37 + (-12) + (-5)$ 20

35. What is −4 plus 2?
−2

36. What is −22 plus −17?
−39

37. Find the sum of −2, 8, and −12.
−6

38. Find the sum of 4, −4, and −6.
−6

39. What is the total of 2, −3, 8, and −13?
−6

40. What is the total of −6, −8, 13, and −2?
−3

Objective B **To subtract integers**

For Exercises 41 to 44, translate the expression into words. Represent each number as positive or negative.

41. −6 − 4
Negative six minus positive four

42. −6 − (−4)
Negative six minus negative four

43. 6 − (−4)
Positive six minus negative four

44. 6 − 4
Positive six minus positive four

For Exercises 45 to 48, rewrite the subtraction as the sum of the first number and the opposite of the second number.

45. 9 − (−5)
9 + 5

46. −3 − 7
−3 + (−7)

47. 1 − 8
1 + (−8)

48. −2 − (−10)
−2 + 10

For Exercises 49 to 76, subtract.

49. 16 − 8
8

50. 12 − 3
9

51. 7 − 14
−7

52. 6 − 9
−3

53. −7 − 2
−9

54. −9 − 4
−13

55. 7 − (−29)
36

56. 3 − (−4)
7

57. −6 − (−3)
−3

58. −4 − (−2)
−2

59. 6 − (−12)
18

60. −12 − 16
−28

61. −4 − 3 − 2
−9

62. 4 − 5 − 12
−13

63. 12 − (−7) − 8
11

64. −12 − (−3) − (−15)
6

65. 4 − 12 − (−8)
0

66. 13 − 7 − 15
−9

67. −6 − (−8) − (−9)
11

68. 7 − 8 − (−1)
0

69. −30 − (−65) − 29 − 4
2

70. 42 − (−82) − 65 − 7
52

71. −16 − 47 − 63 − 12
−138

72. 42 − (−30) − 65 − (−11)
18

73. 47 − (−67) − 13 − 15
86

74. −18 − 49 − (−84) − 27
−10

Quick Quiz (Objective 10.2B)
Subtract.
1. 5 − 16   −11
2. −64 − (−48)   −16
3. 13 − 25 − (−7)   −5

75. $167 - 432 - (-287) - 359$
−337

76. $-521 - (-350) - 164 - (-299)$
−36

77. Subtract −8 from −4.
4

78. Subtract −12 from 3.
15

79. What is the difference between −8 and 4?
−12

80. What is the difference between 8 and −3?
11

81. What is −4 decreased by 8?
−12

82. What is −13 decreased by 9?
−22

83. Find −2 less than 1.
3

84. Find −3 less than −5.
−2

Objective C   **To solve application problems**

85. **Temperature** Find the temperature after a rise of 7°C from −8°C.
−1°C

86. **Temperature** Find the temperature after a rise of 5°C from −19°C.
−14°C

87. **Games** During a card game of Hearts, Nick had a score of 11 points before his opponent "shot the moon," subtracting a score of 26 from Nick's total. What was Nick's score after his opponent shot the moon?
−15 points

88. **Games** In a card game of Hearts, Monique had a score of −19 before she "shot the moon," entitling her to add 26 points to her score. What was Monique's score after she shot the moon?
7 points

89. **Investments** The price of Byplex Corporation's stock fell each trading day of the first week of June. Use the figure at the right to find the change in the price of Byplex stock over the week's time.
−9 dollars

Change in Price of Byplex Corporation Stock (in dollars)

90. **Astronomy** The daytime temperature on the moon can reach 266°F, and the nighttime temperature can go as low as −292°F. Find the difference between these extremes.
558°F

91. **Earth Science** The average temperature throughout Earth's stratosphere is −70°F. The average temperature on Earth's surface is 45°F. Find the difference between these average temperatures.
115°F

Quick Quiz (Objective 10.2C)

1. Find the temperature after a rise of 6°C from −10°C.   −4°C

2. In a card game of Hearts, you had a score of −15 points before you "shot the moon," entitling you to add 26 points to your score. What was your score after you shot the moon?   11 points

Geography   The elevation, or height, of places on Earth is measured in relation to sea level, or the average level of the ocean's surface. The table below shows height above sea level as a positive number and depth below sea level as a negative number. Use the table for Exercises 92 to 94.

Mt. Everest

Continent	Highest Elevation (in meters)		Lowest Elevation (in meters)	
Africa	Mt. Kilimanjaro	5895	Lake Assal	−156
Asia	Mt. Everest	8850	Dead Sea	−411
North America	Mt. McKinley	5642	Death Valley	−28
South America	Mt. Aconcagua	6960	Valdes Peninsula	−86

92.   What is the difference in elevation between Mt. Kilimanjaro and Lake Assal?
6051 m

93.   What is the difference in elevation between Mt. Aconcagua and the Valdes Peninsula?
7046 m

94.   For which continent shown is the difference between the highest and lowest elevations greatest?
Asia

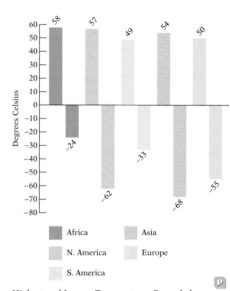

Highest and Lowest Temperatures Recorded
(in degrees Celsius)

Source: National Climatic Data Center

Meteorology   The figure at the right shows the highest and lowest temperatures ever recorded for selected regions of the world. Use this graph for Exercises 95 to 97.

95.   What is the difference between the highest and lowest temperatures recorded in Africa?
82°C

96.   What is the difference between the highest and lowest temperatures recorded in South America?
82°C

97.   What is the difference between the lowest temperature recorded in Europe and the lowest temperature recorded in Asia?
13°C

APPLYING THE CONCEPTS

98.   **Number Problems**   Consider the numbers 4, −7, −5, 13, and −9. What is the largest difference that can be obtained by subtracting one number in the list from another number in the list? Find the smallest positive difference.
22; 2

99.   **Number Problems**   Fill in the blank squares at the right with integers so that the sum of the integers along any row, column, or diagonal is zero.

−3	2	1
4	0	−4
−1	−2	3

100.   **Number Problems**   The sum of two negative integers is −8. Find the integers.
−7 and −1, −6 and −2, −5 and −3, or −4 and −4

101.   Explain the difference between the words *negative* and *minus*.

Answers to Writing Exercises

101. Students should include in their description the fact that *minus* refers to the operation of subtraction, whereas *negative* refers to the sign of a number.

10.3 Multiplication and Division of Integers

Objective A   To multiply integers

Multiplication is the repeated addition of the same number.

Several different symbols are used to indicate multiplication:

$$3 \times 2 = 6 \qquad\qquad 3 \cdot 2 = 6 \qquad\qquad (3)(2) = 6$$

When 5 is multiplied by a sequence of decreasing integers, the products decrease by 5.

$$5 \times 3 = 15$$
$$5 \times 2 = 10$$
$$5 \times 1 = 5$$
$$5 \times 0 = 0$$

The pattern developed can be continued so that 5 is multiplied by a sequence of negative numbers. The resulting products must be negative in order to maintain the pattern of decreasing by 5.

$$5 \times (-1) = -5$$
$$5 \times (-2) = -10$$
$$5 \times (-3) = -15$$
$$5 \times (-4) = -20$$

This example illustrates that the product of a positive number and a negative number is negative.

When -5 is multiplied by a sequence of decreasing integers, the products increase by 5.

$$-5 \times 3 = -15$$
$$-5 \times 2 = -10$$
$$-5 \times 1 = -5$$
$$-5 \times 0 = 0$$

The pattern developed can be continued so that -5 is multiplied by a sequence of negative numbers. The resulting products must be positive in order to maintain the pattern of increasing by 5.

$$-5 \times (-1) = 5$$
$$-5 \times (-2) = 10$$
$$-5 \times (-3) = 15$$
$$-5 \times (-4) = 20$$

This example illustrates that the product of two negative numbers is positive.

The pattern for multiplication shown above is summarized in the following rules for multiplying integers.

Integrating Technology
To multiply $(-4)(-8)$ on your calculator, enter the following:

4 `+/-` $\times$ 8 `+/-` `=`

Rule for Multiplying Two Numbers

To multiply numbers with the same sign, multiply the absolute values of the factors. The product is positive.

To multiply numbers with different signs, multiply the absolute values of the factors. The product is negative.

$$4 \cdot 8 = 32$$
$$(-4)(-8) = 32$$
$$-4 \cdot 8 = -32$$
$$(4)(-8) = -32$$

Objective 10.3A

Vocabulary to Review
product   [1.4A]

New Rules
multiplication of integers

Discuss the Concepts
Describe the rules for multiplying two integers.

Instructor Note
Another way to suggest that positive times negative equals negative is to use repeated addition. For instance, $(-5)(3)$ is $(-5) + (-5) + (-5) = -15$.
   The idea that negative times negative equals positive seems arbitrary to students. You might relate it to using a double negative in English. For example, ask them the meaning of the sentence "It is not impossible to run a 4-minute mile."

Concept Check
True or false:
a. The product of a nonzero number and its opposite is a positive number.   F
b. The product of an odd number of negative numbers is a negative number.   T
c. The product of an even number of negative numbers is a negative number.   F
d. To find the opposite of a number, multiply the number by -1.   T

In-Class Examples (Objective 10.3A)
Multiply.
1. $-5(7)$ -35
2. $-8(-12)$ 96
3. $-10(-6)(7)$ 420

420

Optional Student Activity

Illustrate each of the following products on the number line.

a. $3 \times (-3)$
b. $2 \times (-5)$
c. $3 \times (-2)$
d. $5 \times (-1)$

(*Hint:* Remember that multiplication is repeated addition.)

HOW TO   Multiply: $2(-3)(-5)(-7)$

$$2(-3)(-5)(-7) = -6(-5)(-7)$$

$$= 30(-7)$$

$$= -210$$

- To multiply more than two numbers, multiply the first two numbers.
- Then multiply the product by the third number.
- Continue until all the numbers have been multiplied.

Example 1

Multiply: $(-2)(6)$

Solution

$(-2)(6) = -12$ • The signs are different. The product is negative.

You Try It 1

Multiply: $(-3)(-5)$

Your solution

15

Example 2

Find the product of -42 and 62.

Solution

$-42 \cdot 62 = -2604$ • The signs are different. The product is negative.

You Try It 2

Find -38 multiplied by 51.

Your solution

-1938

Example 3

Multiply: $-5(-4)(6)(-3)$

Solution

$$\begin{aligned}-5(-4)(6)(-3) &= 20(6)(-3)\\ &= 120(-3)\\ &= -360\end{aligned}$$

You Try It 3

Multiply: $-7(-8)(9)(-2)$

Your solution

-1008

Solutions on p. S26

Objective 10.3B

Vocabulary to Review
quotient   [1.5A]

New Rules
division of integers

New Properties
zero and one in division

Discuss the Concepts
Describe the rules for dividing two integers.

Concept Check
What property of division is illustrated by each of the following?

a. $\dfrac{0}{-12} = 0$

Zero divided by any number other than zero is zero.

b. $\dfrac{-18}{1} = -18$

Any number divided by 1 is the number.

c. $\dfrac{-5}{0}$ is undefined.

Division by zero is undefined.

d. $\dfrac{-22}{-22} = 1$

Any number other than zero divided by itself is 1.

Objective B   To divide integers

For every division problem, there is a related multiplication problem.

Division: $\dfrac{8}{2} = 4$ Related multiplication: $4 \cdot 2 = 8$

This fact can be used to illustrate the rules for dividing signed numbers.

Rule for Dividing Two Numbers

To divide numbers with the same sign, divide the absolute values of the numbers. The quotient is positive.

To divide numbers with different signs, divide the absolute values of the numbers. The quotient is negative.

$\dfrac{8}{2} = 4$ because $4 \cdot 2 = 8$.

$\dfrac{-8}{-2} = 4$ because $4(-2) = -8$.

$\dfrac{8}{-2} = -4$ because $-4(-2) = 8$.

$\dfrac{-8}{2} = -4$ because $-4(2) = -8$.

In-Class Examples (Objective 10.3B)

Divide.

1. $15 \div (-3)$ -5

2. $-108 \div (-12)$ 9

3. $-79 \div 0$ Undefined

Note that $\frac{8}{-2}$, $\frac{-8}{2}$, and $-\frac{8}{2}$ are all equal to -4.

If a and b are two integers, then $\frac{a}{-b} = \frac{-a}{b} = -\frac{a}{b}$.

$\frac{0}{a} = 0$ because $0 \cdot a = 0$

$\frac{a}{a} = 1$ because $1 \cdot a = a$

$\frac{a}{1} = a$ because $a \cdot 1 = a$

$\frac{4}{0} = ?$ $? \cdot 0 = 4$

There is no number whose product with zero is 4.

Properties of Zero and One in Division

Zero divided by any number other than zero is zero.

Any number other than zero divided by itself is 1.

Any number divided by 1 is the number.

Division by zero is not defined.

The examples below illustrate these properties of division.

$\frac{0}{-8} = 0$ $\frac{-7}{-7} = 1$ $\frac{-9}{1} = -9$ $\frac{-3}{0}$ is undefined.

Example 4

Divide: $(-120) \div (-8)$

Solution
$(-120) \div (-8) = 15$ • The signs are the same. The quotient is positive.

You Try It 4

Divide: $(-135) \div (-9)$

Your solution
15

Example 5

Divide: $95 \div (-5)$

Solution
$95 \div (-5) = -19$ • The signs are different. The quotient is negative.

You Try It 5

Divide: $84 \div (-6)$

Your solution
-14

Example 6

Find the quotient of -81 and 3.

Solution
$-81 \div 3 = -27$

You Try It 6

What is -72 divided by 4?

Your solution
-18

Example 7

Divide: $0 \div (-24)$

Solution
$0 \div (-24) = 0$ • Zero divided by a nonzero number is zero.

You Try It 7

Divide: $-39 \div 0$

Your solution
Undefined

Solutions on p. S26

Objective 10.3C

Optional Student Activity

A **geometric sequence** is a list of numbers in which each number after the first is found by multiplying the preceding number in the list by the same number. For example, in the sequence 1, 3, 9, 27, 81, . . . , each number after the first is found by multiplying the preceding number in the list by 3. To find the multiplier in a geometric sequence, divide the second number in the sequence by the first number; for the example above, 3 ÷ 1 = 3. Geometric sequences are used in the exercises below.

1. Find the next three numbers in the geometric sequence −5, 15, −45, ...

135, −405, 1215

2. Find the next three numbers in the geometric sequence 2, −4, 8, ...   −16, 32, −64

3. Find the next three numbers in the geometric sequence −3, −12, −48, ...

−192, −768, −3072

4. Find the next three numbers in the geometric sequence −1, −5, −25, ...

−125, −625, −3125

Objective C   **To solve application problems**

Example 8

The combined scores of the top five golfers in a tournament equaled −10 (10 under par). What was the average score of the five golfers?

Strategy
To find the average score, divide the combined scores (−10) by the number of golfers (5).

Solution
−10 ÷ 5 = −2

The average score was −2.

You Try It 8

The melting point of mercury is −38°C. The melting point of argon is five times the melting point of mercury. Find the melting point of argon.

Your strategy

Your solution
−190°C

Example 9

The daily high temperatures during one week were recorded as follows: −9°F, 3°F, 0°F, −8°F, 2°F, 1°F, 4°F. Find the average daily high temperature for the week.

Strategy
To find the average daily high temperature:
- Add the seven temperature readings.
- Divide by 7.

Solution
−9 + 3 + 0 + (−8) + 2 + 1 + 4 = −7

−7 ÷ 7 = −1

The average daily high temperature was −1°F.

You Try It 9

The daily low temperatures during one week were recorded as follows: −6°F, −7°F, 1°F, 0°F, −5°F, −10°F, −1°F. Find the average daily low temperature for the week.

Your strategy

Your solution
−4°F

Solutions on p. S26

In-Class Examples (Objective 10.3C)

1. The combined scores, in relation to par, of the top seven golfers in a golf tournament equaled −98. What was the average score of the seven golfers?   −14

2. The daily high temperatures during one week were as follows: −5°, −8°, 6°, 8°, 0°, −6°, −2°. Find the average daily high temperature for the week.   −1°

10.3 Exercises

Objective A   **To multiply integers**

For Exercises 1 to 4, state whether the operation in the expression is addition, subtraction, or multiplication.

1. $5 - (-6)$
Subtraction

2. $4(-9)$
Multiplication

3. $-8(-5)$
Multiplication

4. $-3 + (-7)$
Addition

For Exercises 5 to 46, multiply.

5. 14×3
42

6. 62×9
558

7. $-4 \cdot 6$
-24

8. $-7 \cdot 3$
-21

9. $-2 \cdot (-3)$
6

10. $-5 \cdot (-1)$
5

11. $(9)(2)$
18

12. $(3)(8)$
24

13. $5(-4)$
-20

14. $4(-7)$
-28

15. $-8(2)$
-16

16. $-9(3)$
-27

17. $(-5)(-5)$
25

18. $(-3)(-6)$
18

19. $(-7)(0)$
0

20. -32×4
-128

21. -24×3
-72

22. $19 \cdot (-7)$
-133

23. $6(-17)$
-102

24. $-8(-26)$
208

25. $-4(-35)$
140

26. $-5 \cdot (23)$
-115

27. $-6 \cdot (38)$
-228

28. $9(-27)$
-243

29. $8(-40)$
-320

30. $-7(-34)$
238

31. $-4(39)$
-156

32. $4 \cdot (-8) \cdot 3$
-96

33. $5 \times 7 \times (-2)$
-70

34. $8 \cdot (-6) \cdot (-1)$
48

35. $(-9)(-9)(2)$
162

36. $-8(-7)(-4)$
-224

37. $-5(8)(-3)$
120

38. $(-6)(5)(7)$
-210

39. $-1(4)(-9)$
36

40. $6(-3)(-2)$
36

Suggested Assignment
Exercises 1–3, odds;
Exercises 5–45, every other odd;
Exercises 47–55, odds;
Exercises 57–105, every other odd;
Exercises 109–119, odds
More challenging problems:
    Exercises 121–123

Quick Quiz (Objective 10.3A)

Multiply.
1. $6(-8)$ -48
2. $-5(-4)$ 20
3. $15(-3)(5)$ -225

41. $4(-4) \cdot 6(-2)$
192

42. $-5 \cdot 9(-7) \cdot 3$
945

43. $-9(4) \cdot 3(1)$
−108

44. $8(8)(-5)(-4)$
1280

45. $(-6) \cdot 7 \cdot (-10)(-5)$
−2100

46. $-9(-6)(11)(-2)$
−1188

47. What is −5 multiplied by −4?
20

48. What is 6 multiplied by −5?
−30

49. What is −8 times 6?
−48

50. What is −8 times −7?
56

51. Find the product of −4, 7, and −5.
140

52. Find the product of −2, −4, and −7.
−56

Objective B   **To divide integers**

For Exercises 53 to 56, write the related multiplication problem.

53. $\dfrac{-36}{-12} = 3$
3(−12) = −36

54. $\dfrac{28}{-7} = -4$
−4(−7) = 28

55. $\dfrac{-55}{11} = -5$
−5(11) = −55

56. $\dfrac{-20}{-10} = 2$
2(−10) = −20

For Exercises 57 to 107, divide.

57. $12 \div (-6)$
−2

58. $18 \div (-3)$
−6

59. $(-72) \div (-9)$
8

60. $(-64) \div (-8)$
8

61. $0 \div (-6)$
0

62. $-49 \div 7$
−7

63. $45 \div (-5)$
−9

64. $-24 \div 4$
−6

65. $-36 \div 4$
−9

66. $-56 \div 7$
−8

67. $-81 \div (-9)$
9

68. $-40 \div (-5)$
8

69. $72 \div (-3)$
−24

70. $44 \div (-4)$
−11

71. $-60 \div 5$
−12

72. $-66 \div 6$
−11

73. $-93 \div (-3)$
31

74. $-98 \div (-7)$
14

75. $(-85) \div (-5)$
17

76. $(-60) \div (-4)$
15

77. $120 \div 8$
15

78. $144 \div 9$
16

79. $78 \div (-6)$
-13

80. $84 \div (-7)$
-12

81. $-72 \div 4$
-18

82. $-80 \div 5$
-16

83. $-114 \div (-6)$
19

84. $-91 \div (-7)$
13

85. $-104 \div (-8)$
13

86. $-126 \div (-9)$
14

87. $57 \div (-3)$
-19

88. $162 \div (-9)$
-18

89. $-136 \div (-8)$
17

90. $-128 \div 4$
-32

91. $-130 \div (-5)$
26

92. $(-280) \div 8$
-35

93. $(-92) \div (-4)$
23

94. $-196 \div (-7)$
28

95. $-150 \div (-6)$
25

96. $(-261) \div 9$
-29

97. $204 \div (-6)$
-34

98. $165 \div (-5)$
-33

99. $-132 \div (-12)$
11

100. $-156 \div (-13)$
12

101. $-182 \div 14$
-13

102. $-144 \div 12$
-12

103. $143 \div 11$
13

104. $168 \div 14$
12

105. $-180 \div (-15)$
12

106. $-169 \div (-13)$
13

107. $154 \div (-11)$
-14

108. Find the quotient of -132 and -11.
12

109. Find the quotient of 182 and -13.
-14

110. What is -60 divided by -15?
4

111. What is 144 divided by -24?
-6

112. Find the quotient of -135 and 15.
-9

113. Find the quotient of -88 and 22.
-4

Objective C   **To solve application problems**

114. Meteorology The daily low temperatures during one week were recorded as follows: 4°F, −5°F, 8°F, −1°F, −12°F, −14°F, −8°F. Find the average daily low temperature for the week.
−4°F

115. Meteorology The daily high temperatures during one week were recorded as follows: −6°F, −11°F, 1°F, 5°F, −3°F, −9°F, −5°F. Find the average daily high temperature for the week.
−4°F

116. **Chemistry** The graph at the right shows the boiling points of three chemical elements. The boiling point of neon is seven times the highest boiling point shown in the table.
 a. Without actually calculating the boiling point, determine whether the boiling point of neon is above 0°C or below 0°C.   Below
 b. What is the boiling point of neon?   −245°C

117. Sports The combined scores of the top ten golfers in a tournament equaled −20 (20 under par). What was the average score of the ten golfers?
−2

118. Sports The combined scores of the top four golfers in a tournament equaled −12 (12 under par). What was the average score of the four golfers?
−3

119. **Meteorology** The wind chill factor when the temperature is −20°F and the wind is blowing at 15 mph is five times the wind chill factor when the temperature is 10°F and the wind is blowing at 20 mph. If the wind chill factor at 10°F with a 20-mph wind is −9°F, what is the wind chill factor at −20°F with a 15-mph wind?
−45°F

120. Education To discourage guessing on a multiple-choice exam, an instructor graded the test by giving 5 points for a correct answer, −2 points for an answer left blank, and −5 points for an incorrect answer. How many points did a student score who answered 20 questions correctly, answered 5 questions incorrectly, and left 2 questions blank?
71

APPLYING THE CONCEPTS

121. a. Number Problems Find the greatest possible product of two negative integers whose sum is −10.   25
 b. Number Problems Find the least possible sum of two negative integers whose product is 16.   −17

122. Use repeated addition to show that the product of two integers with different signs is a negative number.
Answers will vary.

123. Determine whether the statement is true or false.
 a. The product of a nonzero number and its opposite is negative.   True
 b. The square of a negative number is a positive number.   True

124. In your own words, describe the rules for multiplying and dividing integers.

Answers to Writing Exercises

124. Students should rephrase the rules for multiplying two integers and for dividing two integers:

To multiply two integers with the same sign, multiply the absolute values of the factors; the product is positive. To multiply two integers with different signs, multiply the absolute values of the factors; the product is negative.

To divide two integers with the same sign, divide the absolute values of the numbers; the quotient is positive. To divide two integers with different signs, divide the absolute values of the numbers; the quotient is negative.

Be sure that students are describing both the signs of the numbers being multiplied or divided and the sign of the resulting product or quotient.

Quick Quiz (Objective 10.3C)

1. The combined scores, in relation to par, of the top nine golfers in a golf tournament equaled −63. What was the average score of the nine golfers?
−7

2. The daily low temperatures during one week were as follows: 4°, −6°, 8°, −2°, −9°, −11°, −5°. Find the average daily low temperature for the week.
−3°

10.4 Operations with Rational Numbers

Objective A   To add or subtract rational numbers

In this section, operations with rational numbers are discussed. A **rational number** is the quotient of two integers.

> **Rational Numbers**
>
> A rational number is a number that can be written in the form $\frac{a}{b}$, where a and b are integers and $b \neq 0$.

Each of the three numbers shown at the right is a rational number.

$$\frac{3}{4} \qquad \frac{-2}{9} \qquad \frac{13}{-5}$$

An integer can be written as the quotient of the integer and 1. Therefore, **every integer is a rational number.**

$$6 = \frac{6}{1} \qquad -8 = \frac{-8}{1}$$

A mixed number can be written as the quotient of two integers. Therefore, **every mixed number is a rational number.**

$$1\frac{4}{7} = \frac{11}{7} \qquad 3\frac{2}{5} = \frac{17}{5}$$

Recall that every fraction can be written as a decimal by dividing the numerator of the fraction by the denominator. The result is either a terminating decimal or a repeating decimal.

We can write the fraction $\frac{3}{4}$ as the terminating decimal 0.75.

TAKE NOTE

The fraction bar can be read "divided by."

$$\frac{3}{4} = 3 \div 4$$

$$
\begin{array}{r}
0.75 \\
4{\overline{\smash{\big)}\,3.00}} \\
\underline{-2\,8} \\
20 \\
\underline{-20} \\
0
\end{array}
$$

← This is a **terminating decimal.**

← The remainder is zero.

We can write the fraction $\frac{2}{3}$ as the repeating decimal $0.\overline{6}$.

TAKE NOTE

A **terminating decimal** is a decimal that has a finite number of digits after the decimal point, which means that it comes to an end and does not go on forever. A **repeating decimal** is a decimal that does not end; it has a repeating pattern of digits after the decimal point.

$$
\begin{array}{r}
0.666 = 0.\overline{6} \\
3{\overline{\smash{\big)}\,2.000}} \\
\underline{-1\,8} \\
20 \\
\underline{-18} \\
20 \\
\underline{-18} \\
2
\end{array}
$$

← This is a **repeating decimal.**

The bar over the digit 6 in $0.\overline{6}$ is used to show that this digit repeats.

← The remainder is never zero.

All terminating and repeating decimals are rational numbers.

Vocabulary to Review
integer   [10.1A]
mixed number   [2.2A]

New Vocabulary
rational number
terminating decimal
repeating decimal

Discuss the Concepts
Classify the numbers as natural numbers, integers, or rational numbers:
$-3, 7, 0, \dfrac{2}{5},$ and $-\dfrac{10}{9}$

Concept Check
Show that the number is a rational number by writing it as the quotient of two integers. (*Note:* Answers may vary. Possible answers are given.)

1. $-\dfrac{5}{6}$ $\dfrac{-5}{6}$

2. $-\dfrac{7}{25}$ $\dfrac{-7}{25}$

3. $3\dfrac{2}{9}$ $\dfrac{29}{9}$

4. -0.16 $\dfrac{-4}{25}$

5. -0.75 $\dfrac{-3}{4}$

6. 5.2 $\dfrac{26}{5}$

7. -2.4 $\dfrac{-12}{5}$

8. -6.25 $\dfrac{-25}{4}$

In-Class Examples (Objective 10.4A)

Simplify.

1. $-\dfrac{3}{4} + \left(-\dfrac{1}{6}\right)$ $-\dfrac{11}{12}$

2. $3.8 - 7.4$ -3.6

3. $6.2 - (-4.61)$ 10.81

Optional Student Activity

Have students use calculators to write $\frac{5}{99}, \frac{17}{99}, \frac{42}{99}, \frac{63}{99}$, and $\frac{81}{99}$ as repeating decimals. Have them describe the pattern. Ask them to find fractions with other denominators that have interesting decimal patterns (for example, ninths or elevenths).

Optional Student Activity

To determine whether a fraction can be written as a terminating decimal, first write the fraction in simplest form. Then look at the denominator of the fraction. If it contains prime factors of only 2s and/or 5s, then it can be expressed as a terminating decimal. If it contains prime factors other than 2s or 5s, it represents a repeating decimal. For example, if the denominator of a fraction in simplest form is 20, then it can be written as a terminating decimal because $20 = 2 \cdot 2 \cdot 5$ (prime factors of only 2 and 5). If the denominator of a fraction in simplest form is 6, it represents a repeating decimal because it contains the prime factor 3 (a number other than 2 or 5).

Assume that each of the following numbers is the denominator of a fraction written in simplest form. Does the fraction represent a terminating or a repeating decimal? (*Continued on the next page*)

To add or subtract rational numbers in fractional form, first find the least common multiple (LCM) of the denominators.

HOW TO   Add: $-\frac{7}{8} + \frac{5}{6}$

$8 = 2 \cdot 2 \cdot 2$
$6 = 2 \cdot 3$
$\text{LCM} = 2 \cdot 2 \cdot 2 \cdot 3 = 24$

- Find the LCM of the denominators.

$-\dfrac{7}{8} + \dfrac{5}{6} = -\dfrac{21}{24} + \dfrac{20}{24}$

- Rewrite each fraction using the LCM of the denominators as the common denominator.

$= \dfrac{-21 + 20}{24}$

- Add the numerators.

$= \dfrac{-1}{24} = -\dfrac{1}{24}$

TAKE NOTE
In this text, answers that are negative fractions are written with the negative sign in front of the fraction.

HOW TO   Subtract: $-\frac{7}{9} - \frac{5}{12}$

$9 = 3 \cdot 3$
$12 = 2 \cdot 2 \cdot 3$
$\text{LCM} = 2 \cdot 2 \cdot 3 \cdot 3 = 36$

- Find the LCM of the denominators.

$-\dfrac{7}{9} - \dfrac{5}{12} = -\dfrac{28}{36} - \dfrac{15}{36}$

- Rewrite each fraction using the LCM of the denominators as the common denominator.

$= \dfrac{-28}{36} + \dfrac{-15}{36}$

- Rewrite subtraction as addition of the opposite. Rewrite negative fractions with the negative sign in the numerator.

$= \dfrac{-28 + (-15)}{36}$

- Add the numerators.

$= \dfrac{-43}{36} = -\dfrac{43}{36} = -1\dfrac{7}{36}$

To add or subtract rational numbers in decimal form, use the sign rules for adding integers.

HOW TO   Add: $47.034 + (-56.91)$

$\begin{array}{r} 56.910 \\ -\ 47.034 \\ \hline 9.876 \end{array}$

- The signs are different. Find the difference between the absolute values of the numbers.

$47.034 + (-56.91)$
$= -9.876$

- Attach the sign of the number with the greater absolute value.

HOW TO Subtract: $-39.09 - 102.98$

$-39.09 - 102.98$
$= -39.09 + (-102.98)$

* Rewrite subtraction as addition of the opposite number.

$\begin{array}{r} 39.09 \\ + 102.98 \\ \hline 142.07 \end{array}$

* The signs of the addends are the same. Find the sum of the absolute values of the numbers.

$-39.09 - 102.98 = -142.07$

* Attach the sign of the addends.

Example 1

Subtract: $\dfrac{5}{16} - \dfrac{7}{40}$

Solution

$\dfrac{5}{16} - \dfrac{7}{40} = \dfrac{25}{80} - \dfrac{14}{80}$

* The LCM of 16 and 40 is 80.

$= \dfrac{25}{80} + \dfrac{-14}{80}$

* Rewrite as addition of the opposite.

$= \dfrac{25 + (-14)}{80} = \dfrac{11}{80}$

You Try It 1

Subtract: $\dfrac{5}{9} - \dfrac{11}{12}$

Your solution

$-\dfrac{13}{36}$

Example 2

Simplify: $-\dfrac{3}{4} + \dfrac{1}{6} - \dfrac{5}{8}$

Solution

$-\dfrac{3}{4} + \dfrac{1}{6} - \dfrac{5}{8} = -\dfrac{18}{24} + \dfrac{4}{24} - \dfrac{15}{24}$

* The LCM of 4, 6, and 8 is 24.

$= \dfrac{-18}{24} + \dfrac{4}{24} + \dfrac{-15}{24}$

$= \dfrac{-18 + 4 + (-15)}{24}$

$= \dfrac{-29}{24} = -\dfrac{29}{24} = -1\dfrac{5}{24}$

You Try It 2

Simplify: $-\dfrac{7}{8} - \dfrac{5}{6} + \dfrac{2}{3}$

Your solution

$-1\dfrac{1}{24}$

Example 3

Subtract: $42.987 - 98.61$

Solution

$42.987 - 98.61 = 42.987 + (-98.61)$

$\begin{array}{r} 98.610 \\ - 42.987 \\ \hline 55.623 \end{array}$

$42.987 - 98.61 = -55.623$

You Try It 3

Subtract: $16.127 - 67.91$

Your solution

-51.783

Solutions on p. S26

---

a. 4   Terminating
b. 5   Terminating
c. 7   Repeating
d. 9   Repeating
e. 10 Terminating
f. 12 Repeating
g. 15 Repeating
h. 16 Terminating
i. 18 Repeating
j. 21 Repeating
k. 24 Repeating
l. 25 Terminating
m. 28 Repeating
n. 40 Terminating

Write another number that, as the denominator of a fraction in simplest form, represents a terminating decimal.
Answers will vary. For example, 8 or 80.

Write another number that, as the denominator of a fraction in simplest form, represents a repeating decimal.
Answers will vary. For example, 3 or 30.

Example 4

Simplify: $1.02 + (-3.6) + 9.24$

Solution

$$1.02 + (-3.6) + 9.24 = -2.58 + 9.24$$
$$= 6.66$$

You Try It 4

Simplify: $2.7 + (-9.44) + 6.2$

Your solution

-0.54

Solution on p. S26

Objective B   **To multiply or divide rational numbers**

The product of two rational numbers written as fractions is the product of the numerators over the product of the denominators. Use the sign rules for multiplying integers.

> **HOW TO**   Simplify: $-\dfrac{3}{8} \times \dfrac{12}{17}$
>
> $$-\dfrac{3}{8} \times \dfrac{12}{17} = -\left(\dfrac{3 \cdot 12}{8 \cdot 17}\right) = -\dfrac{9}{34}$$
>
> • The signs are different. The product is negative.

To divide rational numbers written as fractions, invert the divisor and then multiply. Use the sign rules for dividing integers.

> **HOW TO**   Simplify: $-\dfrac{3}{10} \div \left(-\dfrac{18}{25}\right)$
>
> $$-\dfrac{3}{10} \div \left(-\dfrac{18}{25}\right) = \dfrac{3}{10} \times \dfrac{25}{18} = \dfrac{3 \cdot 25}{10 \cdot 18} = \dfrac{5}{12}$$
>
> • The signs are the same. The quotient is positive.

To multiply or divide rational numbers written in decimal form, use the sign rules for integers.

> **HOW TO**   Simplify: $(-6.89) \times (-0.00035)$
>
> $$\begin{array}{r} 6.89 \\ \times \quad 0.00035 \\ \hline 3445 \\ 2067 \\ \hline 0.0024115 \end{array}$$
>
> 2 decimal places
> 5 decimal places
>
>
> 7 decimal places
>
> $(-6.89) \times (-0.00035) = 0.0024115$
>
> • The signs are the same. Multiply the absolute values.
>
> • The product is positive.

In-Class Examples (Objective 10.4B)

Simplify.

1. $-\dfrac{5}{6}\left(\dfrac{3}{10}\right)$ $-\dfrac{1}{4}$

2. $-6.8(-2.1)$ 14.28

3. $9.44 \div (-8)$ -1.18

HOW TO Divide $1.32 \div (-0.27)$. Round to the nearest tenth.

$$
\begin{array}{r}
4.88 \approx 4.9 \\
0.27.\overline{)1.32.00} \\
-1\ 08 \\
\hline
24\ 0 \\
-21\ 6 \\
\hline
2\ 40 \\
-2\ 16 \\
\hline
24
\end{array}
$$

- Divide the absolute values. Move the decimal point two places in the divisor and then in the dividend. Place the decimal point in the quotient.

$1.32 \div (-0.27) \approx -4.9$

- The signs are different. The quotient is negative.

Example 5

Multiply: $-\dfrac{7}{12} \times \dfrac{9}{14}$

Solution   The product is negative.

$$-\frac{7}{12} \times \frac{9}{14} = -\left(\frac{7 \cdot 9}{12 \cdot 14}\right)$$

$$= -\frac{3}{8}$$

You Try It 5

Multiply: $\left(-\dfrac{2}{3}\right)\left(-\dfrac{9}{10}\right)$

Your solution $\dfrac{3}{5}$

Example 6

Divide: $-\dfrac{3}{8} \div \left(-\dfrac{5}{12}\right)$

Solution   The quotient is positive.

$$-\frac{3}{8} \div \left(-\frac{5}{12}\right) = \frac{3}{8} \times \frac{12}{5}$$

$$= \frac{3 \cdot 12}{8 \cdot 5}$$

$$= \frac{9}{10}$$

You Try It 6

Divide: $-\dfrac{5}{8} \div \dfrac{5}{40}$

Your solution -5

Example 7   Multiply: -4.29×8.2

Solution   The product is negative.

$$
\begin{array}{r}
4.29 \\
\times\ \ 8.2 \\
\hline
858 \\
3432 \\
\hline
35.178
\end{array}
$$

$-4.29 \times 8.2 = -35.178$

You Try It 7   Multiply: -5.44×3.8

Your solution -20.672

Solutions on pp. S26–S27

Example 8   Multiply: $-3.2 \times (-0.4) \times 6.9$

Solution $-3.2 \times (-0.4) \times 6.9$
    $= 1.28 \times 6.9$
    $= 8.832$

You Try It 8   Multiply: $3.44 \times (-1.7) \times 0.6$

Your solution -3.5088

Example 9   Divide: $-0.0792 \div (-0.42)$
    Round to the nearest hundredth.

Solution

$$
\begin{array}{r}
0.188 \approx 0.19 \\
0.42.\overline{)0.07.920} \\
-4\,2 \\
\hline
3\,72 \\
-3\,36 \\
\hline
360 \\
-336 \\
\hline
24
\end{array}
$$

$-0.0792 \div (-0.42) \approx 0.19$

You Try It 9   Divide: $-0.394 \div 1.7$
    Round to the nearest hundredth.

Your solution -0.23

Solutions on p. S27

Objective C   **To solve application problems**

Example 10

In Fairbanks, Alaska, the average temperature during the month of July is 61.5°F. During the month of January, the average temperature in Fairbanks is −12.7°F. What is the difference between the average temperature in Fairbanks during July and the average temperature during January?

Strategy
To find the difference, subtract the average temperature in January (−12.7°F) from the average temperature in July (61.5°F).

Solution
$61.5 - (-12.7) = 61.5 + 12.7 = 74.2$

The difference between the average temperature during July and the average temperature during January in Fairbanks is 74.2°F.

You Try It 10

On January 10, 1911, in Rapid City, South Dakota, the temperature fell from 12.78°C at 7:00 A.M. to −13.33°C at 7:15 A.M. How many degrees did the temperature fall during the 15-min period?

Your strategy

Your solution
26.11°C

Solution on p. S27

Objective 10.4C

Optional Student Activity

The table below shows the total trade balance, in billions of dollars, for the United States in the 1990s. (*Source:* U.S. Department of Commerce, Bureau of Economic Analysis)

Year	Trade Balance
1991	6.616
1992	−47.724
1993	−82.681
1994	−118.605
1995	−109.457
1996	−123.318
1997	−140.540
1998	−217.138
1999	−331.479

1. For which year was the trade balance the lowest?   1999

2. Calculate the difference between the trade balance in 1991 and that in 1999.
   $338.095 billion

3. Between which two consecutive years was the difference in the trade balance greatest?
   1998–1999

4. How many times greater was the trade balance in 1999 than in 1992? Round to the nearest whole number.
   7 times greater

5. Calculate the average trade balance per quarter in 1997.
   −$35.135 billion

6. Examine the data. Would you expect the trade balance to have increased or decreased from 1999 to 2000? Why?
   Answers will vary.

In-Class Examples (Objective 10.4C)

1. The lowest temperature ever recorded in Australia is −9.4°F. The highest temperature ever recorded is 128.0°F. (*Source:* National Climatic Data Center) Find the difference between these two extremes.   137.4°F

2. On September 15, 2003, the closing price of a share of Hewlett-Packard stock was $19.83. The change in the closing price from the previous day was −$.17. Find the closing price on the previous day. $20.00

10.4 Exercises

Objective A   **To add or subtract rational numbers**

Suggested Assignment

Exercises 1–91, odds
More challenging problems:
   Exercises 92–94

For Exercises 1 to 43, simplify.

1. $\dfrac{5}{8} - \dfrac{5}{6}$
$-\dfrac{5}{24}$

2. $\dfrac{1}{9} - \dfrac{5}{27}$
$-\dfrac{2}{27}$

3. $-\dfrac{5}{12} - \dfrac{3}{8}$
$-\dfrac{19}{24}$

4. $-\dfrac{5}{6} - \dfrac{5}{9}$
$-1\dfrac{7}{18}$

5. $-\dfrac{6}{13} + \dfrac{17}{26}$
$\dfrac{5}{26}$

6. $-\dfrac{7}{12} + \dfrac{5}{8}$
$\dfrac{1}{24}$

7. $-\dfrac{5}{8} - \left(-\dfrac{11}{12}\right)$
$\dfrac{7}{24}$

8. $-\dfrac{7}{12} - \left(-\dfrac{7}{8}\right)$
$\dfrac{7}{24}$

9. $\dfrac{5}{12} - \dfrac{11}{15}$
$-\dfrac{19}{60}$

10. $\dfrac{2}{5} - \dfrac{14}{15}$
$-\dfrac{8}{15}$

11. $-\dfrac{3}{4} - \dfrac{5}{8}$
$-1\dfrac{3}{8}$

12. $-\dfrac{2}{3} - \dfrac{5}{8}$
$-1\dfrac{7}{24}$

13. $-\dfrac{5}{2} - \left(-\dfrac{13}{4}\right)$
$\dfrac{3}{4}$

14. $-\dfrac{7}{3} - \left(-\dfrac{3}{2}\right)$
$-\dfrac{5}{6}$

15. $-\dfrac{3}{8} - \dfrac{5}{12} - \dfrac{3}{16}$
$-\dfrac{47}{48}$

16. $-\dfrac{5}{16} + \dfrac{3}{4} - \dfrac{7}{8}$
$-\dfrac{7}{16}$

17. $\dfrac{1}{2} - \dfrac{3}{8} - \left(-\dfrac{1}{4}\right)$
$\dfrac{3}{8}$

18. $\dfrac{3}{4} - \left(-\dfrac{7}{12}\right) - \dfrac{7}{8}$
$\dfrac{11}{24}$

19. $\dfrac{1}{3} - \dfrac{1}{4} - \dfrac{1}{5}$
$-\dfrac{7}{60}$

20. $\dfrac{5}{16} + \dfrac{1}{8} - \dfrac{1}{2}$
$-\dfrac{1}{16}$

21. $\dfrac{1}{2} + \left(-\dfrac{3}{8}\right) + \dfrac{5}{12}$
$\dfrac{13}{24}$

22. $-\dfrac{3}{8} + \dfrac{3}{4} - \left(-\dfrac{3}{16}\right)$
$\dfrac{9}{16}$

23. $3.4 + (-6.8)$
-3.4

24. $-4.9 + 3.27$
-1.63

25. $-8.32 + (-0.57)$
-8.89

26. $-3.5 + 7$
3.5

27. $-4.8 + (-3.2)$
-8.0

Quick Quiz (Objective 10.4A)

Simplify.

1. $-\dfrac{3}{8} - \dfrac{1}{2}$ $-\dfrac{7}{8}$

2. $5.63 - (-2.1)$ 7.73

3. $-\dfrac{5}{6} - \left(-\dfrac{2}{9}\right)$ $-\dfrac{11}{18}$

28. $6.2 + (-4.29)$
1.91

29. $-4.6 + 3.92$
−0.68

30. $7.2 + (-8.42)$
−1.22

31. $-45.71 + (-135.8)$
−181.51

32. $-35.274 + 12.47$
−22.804

33. $4.2 + (-6.8) + 5.3$
2.7

34. $6.7 + 3.2 + (-10.5)$
−0.6

35. $-4.5 + 3.2 + (-19.4)$
−20.7

36. $2.09 - 6.72 - 5.4$
−10.03

37. $-18.39 + 4.9 - 23.7$
−37.19

38. $19 - (-3.72) - 82.75$
−60.03

39. $-3.09 - 4.6 - 27.3$
−34.99

40. 🖩 $-3.89 + (-2.9) + 4.723 + 0.2$
−1.867

41. 🖩 $-4.02 + 6.809 - (-3.57) - (-0.419)$
6.778

42. 🖩 $0.0153 + (-1.0294) + (-1.0726)$
−2.0867

43. 🖩 $0.27 + (-3.5) - (-0.27) + (-5.44)$
−8.4

> **Objective B**   **To multiply or divide rational numbers**

For Exercises 44 to 79, simplify.

44. $\dfrac{1}{2} \times \left(-\dfrac{3}{4}\right)$
$-\dfrac{3}{8}$

45. $-\dfrac{2}{9} \times \left(-\dfrac{3}{14}\right)$
$\dfrac{1}{21}$

46. $\left(-\dfrac{3}{8}\right)\left(-\dfrac{4}{15}\right)$
$\dfrac{1}{10}$

47. $\left(-\dfrac{3}{4}\right)\left(-\dfrac{8}{27}\right)$
$\dfrac{2}{9}$

48. $-\dfrac{1}{2} \times \dfrac{8}{9}$
$-\dfrac{4}{9}$

49. $\dfrac{5}{12} \times \left(-\dfrac{8}{15}\right)$
$-\dfrac{2}{9}$

50. $\left(-\dfrac{5}{12}\right)\left(\dfrac{42}{65}\right)$
$-\dfrac{7}{26}$

51. $\left(\dfrac{3}{8}\right)\left(-\dfrac{15}{41}\right)$
$-\dfrac{45}{328}$

52. $\left(-\dfrac{15}{8}\right)\left(-\dfrac{16}{3}\right)$
10

53. $\left(-\dfrac{5}{7}\right)\left(-\dfrac{14}{15}\right)$
$\dfrac{2}{3}$

54. $\dfrac{5}{8} \times \left(-\dfrac{7}{12}\right) \times \dfrac{16}{25}$
$-\dfrac{7}{30}$

55. $\left(\dfrac{1}{2}\right)\left(-\dfrac{3}{4}\right)\left(-\dfrac{5}{8}\right)$
$\dfrac{15}{64}$

Quick Quiz (Objective 10.4B)
Simplify.

1. $-\dfrac{7}{12}\left(-\dfrac{4}{9}\right)$ $\dfrac{7}{27}$

2. $-9.3(12.7)$ −118.11

3. $15.33 \div (-7)$ −2.19

56. $\dfrac{1}{3} \div \left(-\dfrac{1}{2}\right)$

$-\dfrac{2}{3}$

57. $-\dfrac{3}{8} \div \dfrac{7}{8}$

$-\dfrac{3}{7}$

58. $\left(-\dfrac{3}{4}\right) \div \left(-\dfrac{7}{40}\right)$

$4\dfrac{2}{7}$

59. $\dfrac{5}{6} \div \left(-\dfrac{3}{4}\right)$

$-1\dfrac{1}{9}$

60. $-\dfrac{5}{12} \div \dfrac{15}{32}$

$-\dfrac{8}{9}$

61. $-\dfrac{5}{16} \div \left(-\dfrac{3}{8}\right)$

$\dfrac{5}{6}$

62. $\left(-\dfrac{3}{8}\right) \div \left(-\dfrac{5}{12}\right)$

$\dfrac{9}{10}$

63. $\left(-\dfrac{8}{19}\right) \div \dfrac{7}{38}$

$-2\dfrac{2}{7}$

64. $\left(-\dfrac{2}{3}\right) \div 4$

$-\dfrac{1}{6}$

65. $-6 \div \dfrac{4}{9}$

$-13\dfrac{1}{2}$

66. $-6.7 \times (-4.2)$

28.14

67. $-8.9 \times (-3.5)$

31.15

68. -1.6×4.9

−7.84

69. -14.3×7.9

−112.97

70. $(-0.78)(-0.15)$

0.117

71. $(-1.21)(-0.03)$

0.0363

72. $(-8.919) \div (-0.9)$

9.91

73. $-77.6 \div (-0.8)$

97

74. $59.01 \div (-0.7)$

−84.3

75. $(-7.04) \div (-3.2)$

2.2

76. $(-84.66) \div 1.7$

−49.8

77. $-3.312 \div (0.8)$

−4.14

78. $1.003 \div (-0.59)$

−1.7

79. $26.22 \div (-6.9)$

−3.8

 For Exercises 80 to 85, divide.

80. $(-19.08) \div 0.45$

−42.40

81. $21.792 \div (-0.96)$

−22.70

82. $(-38.665) \div (-9.5)$

4.07

83. $(-3.171) \div (-45.3)$

0.07

84. $27.738 \div (-60.3)$

−0.46

85. $(-13.97) \div (-25.4)$

0.55

436

> **Objective C**   **To solve application problems**

86.   **Meteorology**   On January 23, 1916, the temperature in Browing, Montana, was 6.67°C. On January 24, 1916, the temperature in Browing was −48.9°C. Find the difference between the temperatures in Browing on these two days.
55.57°C

87.   **Meteorology**   On January 22, 1943, in Spearfish, South Dakota, the temperature fell from 12.22°C at 9 A.M. to −20°C at 9:27 A.M. How many degrees did the temperature fall during the 27-min period?
32.22°C

88.   **Chemistry**   The boiling point of nitrogen is −195.8°C and the melting point is −209.86°C. Find the difference between the boiling point and the melting point of nitrogen.
14.06°C

89.   **Chemistry**   The boiling point of oxygen is −182.962°C. Oxygen's melting point is −218.4°C. What is the difference between the boiling point and the melting point of oxygen?
35.438°C

Investments   The chart at the right shows the closing price of a share of stock on September 15, 2003, for each of five companies. Also shown is the change in the price from the previous day. To find the closing price on the previous day, subtract the change in price from the closing price on September 15. Use this chart for Exercises 90 and 91.

Company	Closing Price	Change in Price
Del Monte Foods Co.	8.77	−0.06
General Mills, Inc.	47.10	−0.11
Hershey Foods, Inc.	72.57	+0.11
Hormel Food Corp.	22.42	−0.08
Sara Lee Corp.	19.18	−0.21

90.   **a.** Find the closing price on the previous day for General Mills. $47.21
　　　　b. Find the closing price on the previous day for Hormel Foods. $22.50

91.   **a.** Find the closing price on the previous day for Sara Lee. $19.39
　　　　b. Find the closing price on the previous day for Hershey Foods. $72.46

APPLYING THE CONCEPTS

92.   Determine whether the statement is true or false.
　　a. Every integer is a rational number.   True
　　b. Every whole number is an integer.   True
　　c. Every integer is a positive number.   False
　　d. Every rational number is an integer.   False

93.   **Number Problems**   Find a rational number between $-\frac{3}{4}$ and $-\frac{2}{3}$.
Answers will vary. For example, $-\frac{17}{24}$.

94.   **Number Problems**
　　a. Find a rational number between 0.1 and 0.2.
　　b. Find a rational number between 1 and 1.1.
　　c. Find a rational number between 0 and 0.005.
Answers will vary. For example, **a.** 0.15, **b.** 1.05, **c.** 0.001.

95.   Given any two different rational numbers, is it always possible to find a rational number between them? If so, explain how. If not, give an example of two different rational numbers for which there is no rational number between them.

Answers to Writing Exercises

95. Given any two different rational numbers, it is always possible to find a rational number between them. One method is to add the two numbers and divide by 2.

Another method is to add the numerators and add the denominators. For example, given the fractions $\frac{2}{5}$ and $\frac{3}{4}$,

$\frac{2+3}{5+4} = \frac{5}{9}$ and $\frac{2}{5} < \frac{5}{9} < \frac{3}{4}$.

Quick Quiz (Objective 10.4C)

1. The lowest temperature ever recorded in North America is −81.4°F. The highest temperature ever recorded is 134.0°F. (*Source*: National Climatic Data Center) Find the difference between these two extremes.   215.4°F

2. On September 15, 2003, the closing price of a share of Gateway stock was $6.06. The change in the closing price from the previous day was −$.09. Find the closing price on the previous day. $6.15

10.5

Scientific Notation and the Order of Operations Agreement

Objective A   **To write a number in scientific notation**

Point of Interest

The first woman mathematician for whom documented evidence exists is Hypatia (370–415). She lived in Alexandria, Egypt, and lectured at the Museum, the forerunner of our modern university. She made important contributions in mathematics, astronomy, and philosophy.

Scientific notation uses negative exponents. Therefore, we will discuss that topic before presenting scientific notation.

Look at the powers of 10 shown at the right. Note the pattern: The exponents are decreasing by 1, and each successive number on the right is one-tenth of the number above it. ($100,000 \div 10 = 10,000$; $10,000 \div 10 = 1000$; etc.)

$$10^5 = 100,000$$
$$10^4 = 10,000$$
$$10^3 = 1000$$
$$10^2 = 100$$
$$10^1 = 10$$

If we continue this pattern, the next exponent on 10 is $1 - 1 = 0$, and the number on the right side is $10 \div 10 = 1$.

$$10^0 = 1$$

The next exponent on 10 is $0 - 1 = -1$, and 10^{-1} is equal to $1 \div 10 = 0.1$.

$$10^{-1} = 0.1$$

The pattern is continued on the right. Note that a negative exponent does not indicate a negative number. Rather, each power of 10 with a negative exponent is equal to a number between 0 and 1. Also note that as the exponent on 10 decreases, so does the number it is equal to.

$$10^{-2} = 0.01$$
$$10^{-3} = 0.001$$
$$10^{-4} = 0.0001$$
$$10^{-5} = 0.00001$$
$$10^{-6} = 0.000001$$

Very large and very small numbers are encountered in the natural sciences. For example, the mass of an electron is 0.00000000000000000000000000000911 kg. Numbers such as this are difficult to read, so a more convenient system called **scientific notation** is used. In scientific notation, a number is expressed as the product of two factors, one a number between 1 and 10, and the other a power of 10.

To express a number in scientific notation, write it in the form $a \times 10^n$, where a is a number between 1 and 10 and n is an integer.

For numbers greater than 10, move the decimal point to the right of the first digit. The exponent n is positive and equal to the number of places the decimal point has been moved.

$$240,000 = 2.4 \times 10^5$$
$$93,000,000 = 9.3 \times 10^7$$

Objective 10.5A

New Vocabulary
scientific notation

Discuss the Concepts
Is the number 42.5×10^7 written in scientific notation? Why or why not?

Concept Check
Which of the following numbers represent numbers greater than 10? Which represent numbers less than 1?
a. 4.3×10^5 Greater than 10
b. 3.09×10^{-6} Less than 1
c. 8×10^{17} Greater than 10
d. 6.175×10^{-22} Less than 1

Optional Student Activity

1. Our galaxy is estimated to be 6×10^{17} mi across. How many hours would it take a spaceship to cross the galaxy traveling at 25,000 mph? Write the answer in scientific notation. 2.4×10^{13} h

2. One light year is the distance light travels in 1 year, or about 5.87×10^{12} mi. The distance to the star Alpha Centauri is 4.3 light years. That is, it would take 4.3 years traveling at the speed of light to reach this star. Astronauts travel at a speed of approximately 25,000 mph. At this speed approximately how many years would it take to reach Alpha Centauri? Round to the nearest thousand.   115,000 years

---

In-Class Examples (Objective 10.5A)
Write the number in scientific notation.
1. 87,300,000,000 8.73×10^{10}
2. 0.00000654 6.54×10^{-6}
Write the number in decimal notation.
3. 9.102×10^8 910,200,000
4. 3.7×10^{-7} 0.000000371

For numbers less than 1, move the decimal point to the right of the first nonzero digit. The exponent n is negative. The absolute value of the exponent is equal to the number of places the decimal point has been moved.

$0.0003 = 3 \times 10^{-4}$

$0.0000832 = 8.32 \times 10^{-5}$

Changing a number written in scientific notation to decimal notation also requires moving the decimal point.

When the exponent on 10 is positive, move the decimal point to the right the same number of places as the exponent.

$3.45 \times 10^9 = 3{,}450{,}000{,}000$

$2.3 \times 10^8 = 230{,}000{,}000$

When the exponent on 10 is negative, move the decimal point to the left the same number of places as the absolute value of the exponent.

$8.1 \times 10^{-3} = 0.0081$

$6.34 \times 10^{-6} = 0.00000634$

Example 1 Write 824,300,000,000 in scientific notation.

Solution The number is greater than 10. Move the decimal point 11 places to the left. The exponent on 10 is 11.

$824{,}300{,}000{,}000 = 8.243 \times 10^{11}$

You Try It 1 Write 0.000000961 in scientific notation.

Your solution 9.61×10^{-7}

Example 2 Write 6.8×10^{-10} in decimal notation.

Solution The exponent on 10 is negative. Move the decimal point 10 places to the left.

$6.8 \times 10^{-10} = 0.00000000068$

You Try It 2 Write 7.329×10^6 in decimal notation.

Your solution 7,329,000

Solutions on p. S27

Objective B **To use the Order of Operations Agreement to simplify expressions**

The Order of Operations Agreement has been used throughout this book. In simplifying expressions with rational numbers, the same Order of Operations Agreement is used. This agreement is restated here.

Objective 10.5B

Procedures to Review
Order of Operations Agreement
[1.6B]

Instructor Note
The exercises in this objective ask students to recall the Order of Operations Agreement and to practice a combination of operations with rational numbers.

Concept Check
Put the following expressions in order from the smallest value to the largest value.

a. $(-3) \div \left(-\dfrac{3}{2}\right) \div \left(-\dfrac{1}{8}\right)$ -16

b. $3 - (-4 + 2) \div 2$ 4

c. $\left(-\dfrac{1}{2}\right)^3 - \left(\dfrac{3}{4} \div \dfrac{9}{16}\right) + \dfrac{1}{6}$ $-1\dfrac{7}{24}$

d. $8 - (-7) \times (6 - 10) + 3^2$ -11

Optional Student Activity

1. What is the smallest prime number greater than $32 \div (1 - 9) + (-3)^2$? 7

2. What is the smallest integer greater than $-2^2 - (-3)^2 + 5(4) \div 10 - (-6)$? -4

3. What is the smallest prime number that divides evenly into the sum of 3^9 and 5^{11}? 2

The Order of Operations Agreement

Step 1 Do all operations inside parentheses.
Step 2 Simplify any number expressions containing exponents.
Step 3 Do multiplication and division as they occur from left to right.
Step 4 Rewrite subtraction as addition of the opposite. Then do additions as they occur from left to right.

Ⓟ

TAKE NOTE
In $(-3)^2$, we are squaring -3; multiply -3 times -3. In -3^2, we are finding the opposite of 3^2. The expression -3^2 is the same as $-(3^2)$.

Exponents may be confusing in expressions with signed numbers.

$(-3)^2 = (-3) \times (-3) = 9$
$-3^2 = -(3)^2 = -(3 \times 3) = -9$

Note that -3 is squared only when the negative sign is *inside* the parentheses.

HOW TO Simplify: $(-3)^2 - 2 \times (8 - 3) + (-5)$

$(-3)^2 - 2 \times \underbrace{(8 - 3)} + (-5)$

$\underbrace{(-3)^2} - 2 \times 5 + (-5)$ **1.** Perform operations inside parentheses.

$9 - \underbrace{2 \times 5} + (-5)$ **2.** Simplify expressions with exponents.

$\underbrace{9 - 10} + (-5)$ **3.** Do multiplications and divisions as they occur from left to right.

$\underbrace{9 + (-10)} + (-5)$ **4.** Rewrite subtraction as the addition of the opposite. Then add from left to right.

$\underbrace{(-1) + (-5)}$

-6

Integrating Technology

As shown above, the value of -3^2 is different from the value of $(-3)^2$. The keystrokes to evaluate each of these on your calculator are different. To evaluate -3^2, enter

3 x^2 $+/-$

To evaluate $(-3)^2$, enter

3 $+/-$ x^2

HOW TO Simplify: $\left(\dfrac{1}{4} - \dfrac{1}{2}\right)^2 \div \dfrac{3}{8}$

$\underbrace{\left(\dfrac{1}{4} - \dfrac{1}{2}\right)^2} \div \dfrac{3}{8}$

$\underbrace{\left(-\dfrac{1}{4}\right)^2} \div \dfrac{3}{8}$ **1.** Perform operations inside parentheses.

$\underbrace{\dfrac{1}{16} \div \dfrac{3}{8}}$ **2.** Simplify expressions with exponents.

$\underbrace{\dfrac{1}{16} \times \dfrac{8}{3}}$ **3.** Do multiplication and division as they occur from left to right.

$\dfrac{1}{6}$

Example 3 Simplify: $8 - 4 \div (-2)$

Solution
$8 - 4 \div (-2)$
$= 8 - (-2)$ • Do the division.
$= 8 + 2$ • Rewrite as
$= 10$ addition. Add.

You Try It 3 Simplify: $9 - 9 \div (-3)$

Your solution 12

Solution on p. S27

In-Class Examples (Objective 10.5B)
Simplify.
1. $(-4)^2 - 8 + 9(3)$ 35
2. $7(3 - 6) + 5$ -16
3. $(-6)^2 \times (6 - 4)^2 - (-12) \div 4$ 147

Example 4   Simplify: $12 \div (-2)^2 + 5$

Solution $12 \div (-2)^2 + 5$
$= 12 \div 4 + 5$ • Exponents
$= 3 + 5$ • Division
$= 8$ • Addition

You Try It 4   Simplify: $8 \div 4 \cdot 4 - (-2)^2$

Your solution   4

Example 5   Simplify: $12 - (-10) \div (8 - 3)$

Solution $12 - (-10) \div (8 - 3)$
$= 12 - (-10) \div 5$
$= 12 - (-2)$
$= 12 + 2$
$= 14$

You Try It 5   Simplify: $8 - (-15) \div (2 - 7)$

Your solution   5

Example 6   Simplify:
$(-3)^2 \times (5 - 7)^2 - (-9) \div 3$

Solution $(-3)^2 \times (5 - 7)^2 - (-9) \div 3$
$= (-3)^2 \times (-2)^2 - (-9) \div 3$
$= 9 \times 4 - (-9) \div 3$
$= 36 - (-9) \div 3$
$= 36 - (-3)$
$= 36 + 3$
$= 39$

You Try It 6   Simplify:
$(-2)^2 \times (3 - 7)^2 - (-16) \div (-4)$

Your solution   60

Example 7   Simplify: $3 \div \left(\dfrac{1}{2} - \dfrac{1}{4}\right) - 3$

Solution $3 \div \left(\dfrac{1}{2} - \dfrac{1}{4}\right) - 3$

$= 3 \div \dfrac{1}{4} - 3$

$= 3 \times \dfrac{4}{1} - 3$

$= 12 - 3$
$= 12 + (-3)$
$= 9$

You Try It 7   Simplify: $7 \div \left(\dfrac{1}{7} - \dfrac{3}{14}\right) - 9$

Your solution -107

Solutions on p. S27

10.5 Exercises

Objective A   To write a number in scientific notation

For Exercises 1 to 12, write the number in scientific notation.

1. 2,370,000
2.37×10^6

2. 75,000
7.5×10^4

3. 0.00045
4.5×10^{-4}

4. 0.000076
7.6×10^{-5}

5. 309,000
3.09×10^5

6. 819,000,000
8.19×10^8

7. 0.000000601
6.01×10^{-7}

8. 0.00000000096
9.6×10^{-10}

9. 57,000,000,000
5.7×10^{10}

10. 934,800,000,000
9.348×10^{11}

11. 0.000000017
1.7×10^{-8}

12. 0.0000009217
9.217×10^{-7}

For Exercises 13 to 24, write the number in decimal notation.

13. 7.1×10^5
710,000

14. 2.3×10^7
23,000,000

15. 4.3×10^{-5}
0.000043

16. 9.21×10^{-7}
0.000000921

17. 6.71×10^8
671,000,000

18. 5.75×10^9
5,750,000,000

19. 7.13×10^{-6}
0.00000713

20. 3.54×10^{-8}
0.0000000354

21. 5×10^{12}
5,000,000,000,000

22. 1.0987×10^{11}
109,870,000,000

23. 8.01×10^{-3}
0.00801

24. 4.0162×10^{-9}
0.0000000040162

25. **Physics** Light travels 16,000,000,000 mi in 1 day. Write this number in scientific notation.
1.6×10^{10} mi

26. **Earth Science** Write the mass of Earth, which is approximately 5,980,000,000,000,000,000,000,000 kg, in scientific notation.
5.98×10^{24} kg

27. **Wars** The graph at the right shows the monetary cost of four wars. Write the monetary cost of World War II in scientific notation.
$\$3.1 \times 10^{12}$

28. **Chemistry** The electric charge on an electron is 0.00000000000000000016 coulomb. Write this number in scientific notation.
1.6×10^{-19} coulomb

Monetary Cost of Wars

Source: Congressional Research Service, using numbers from the *Statistical Abstract of the United States*

Quick Quiz (Objective 10.5A)
Write the number in scientific notation.
1. 1,600,000,000,000 1.6×10^{12}
2. 0.0000000392 3.92×10^{-8}
Write the number in decimal notation.
3. 4×10^9 4,000,000,000
4. 7.5×10^{-5} 0.000075

29. ⬤ **Physics** The length of an infrared light wave is approximately 0.0000037 m. Write this number in scientific notation.
3.7×10^{-6} m

30. ⬤ **Computers** One unit used to measure the speed of a computer is the picosecond. One picosecond is 0.000000000001 of a second. Write this number in scientific notation.
1×10^{-12}

Objective B **To use the Order of Operations Agreement to simplify expressions**

For Exercises 31 to 92, simplify.

31. $8 \div 4 + 2$
4

32. $3 - 12 \div 2$
−3

33. $4 + (-7) + 3$
0

34. $-16 \div 2 + 8$
0

35. $4^2 - 4$
12

36. $6 - 2^2$
2

37. $2 \times (3 - 5) - 2$
−6

38. $2 - (8 - 10) \div 2$
3

39. $4 - (-3)^2$
−5

40. $(-2)^2 - 6$
−2

41. $4 - (-3) - 5$
2

42. $6 + (-8) - (-3)$
1

43. $4 - (-2)^2 + (-3)$
−3

44. $-3 + (-6)^2 - 1$
32

45. $3^2 - 4 \times 2$
1

46. $9 \div 3 - (-3)^2$
−6

47. $3 \times (6 - 2) \div 6$
2

48. $4 \times (2 - 7) \div 5$
−4

49. $2^2 - (-3)^2 + 2$
−3

50. $3 \times (8 - 5) + 4$
13

51. $6 - 2 \times (1 - 5)$
14

52. $4 \times 2 \times (3 - 6)$
−24

53. $(-2)^2 - (-3)^2 + 1$
−4

54. $4^2 - 3^2 - 4$
3

55. $6 - (-3) \times (-3)^2$
33

56. $4 - (-5) \times (-2)^2$
24

57. $4 \times 2 - 3 \times 7$
−13

58. $16 \div 2 - 9 \div 3$
5

59. $(-2)^2 - 5 \times 3 - 1$
−12

60. $4 - 2 \times 7 - 3^2$
−19

Quick Quiz (Objective 10.5B)
Simplify.
1. $-5(4 - 1) \div 15$ −1
2. $3 - (-5)^2 - 6 + 2(4)$ −20
3. $24 \div (5 - 7) \div (-2)^2$ −3

61. $7 \times 6 - 5 \times 6 + 3 \times 2 - 2 + 1$
17

62. $3 \times 2^2 + 5 \times (3 + 2) - 17$
20

63. $-4 \times 3 \times (-2) + 12 \times (3 - 4) + (-12)$
0

64. $3 \times 4^2 - 16 - 4 + 3 - (1 - 2)^2$
30

65. $-12 \times (6 - 8) + 1^2 \times 3^2 \times 2 - 6 \times 2$
30

66. $-3 \times (-2)^2 \times 4 \div 8 - (-12)$
6

67. $10 \times 9 - (8 + 7) \div 5 + 6 - 7 + 8$
94

68. $-27 - (-3)^2 - 2 - 7 + 6 \times 3$
−27

69. $3^2 \times (4 - 7) \div 9 + 6 - 3 - 4 \times 2$
−8

70. $16 - 4 \times 8 + 4^2 - (-18) \div (-9)$
−2

71. $(-3)^2 \times (5 - 7)^2 - (-9) \div 3$
39

72. $-2 \times 4^2 - 3 \times (2 - 8) - 3$
−17

73. $4 - 6(2 - 5)^3 \div (17 - 8)$
22

74. $5 + 7(3 - 8)^2 \div (-14 + 9)$
−30

75. $(1.2)^2 - 4.1 \times 0.3$
0.21

76. $2.4 \times (-3) - 2.5$
−9.7

77. $1.6 - (-1.6)^2$
−0.96

78. $4.1 \times 8 \div (-4.1)$
−8

79. $(4.1 - 3.9) - 0.7^2$
−0.29

80. $1.8 \times (-2.3) - 2$
−6.14

81. $(-0.4)^2 \times 1.5 - 2$
−1.76

82. $(6.2 - 1.3) \times (-3)$
−14.7

83. $4.2 - (-3.9) - 6$
2.1

84. $-\dfrac{1}{2} + \dfrac{3}{8} \div \left(-\dfrac{3}{4}\right)$

-1

85. $\left(\dfrac{3}{4}\right)^2 - \dfrac{3}{8}$

$\dfrac{3}{16}$

86. $\left(\dfrac{1}{2}\right)^2 - \left(-\dfrac{1}{2}\right)^2$

0

87. $\dfrac{5}{16} - \dfrac{3}{8} + \dfrac{1}{2}$

$\dfrac{7}{16}$

88. $\dfrac{2}{7} \div \dfrac{5}{7} - \dfrac{3}{14}$

$\dfrac{13}{70}$

89. $\dfrac{1}{2} \times \dfrac{1}{4} \times \dfrac{1}{2} - \dfrac{3}{8}$

$-\dfrac{5}{16}$

90. $\dfrac{2}{3} \times \dfrac{5}{8} \div \dfrac{2}{7}$

$1\dfrac{11}{24}$

91. $\dfrac{1}{2} - \left(\dfrac{3}{4} - \dfrac{3}{8}\right) \div \dfrac{1}{3}$

$-\dfrac{5}{8}$

92. $\dfrac{3}{8} \div \left(-\dfrac{1}{2}\right)^2 + 2$

$3\dfrac{1}{2}$

APPLYING THE CONCEPTS

93. Place the correct symbol, $<$ or $>$, between the two numbers.
   a. 3.45×10^{-14} 3.45×10^{-15} $>$
   b. 5.23×10^{18} 5.23×10^{17} $>$
   c. 3.12×10^{12} 3.12×10^{11} $>$

94.   **Astronomy**   Light travels 3×10^8 m in 1 s. How far does light travel in 1 year? (Astronomers refer to this distance as 1 light year.)
9.4608×10^{15} m

95. **a.** Evaluate $1^3 + 2^3 + 3^3 + 4^3$. 100
   b. Evaluate $(-1)^3 + (-2)^3 + (-3)^3 + (-4)^3$. -100
   c. Evaluate $1^3 + 2^3 + 3^3 + 4^3 + 5^3$. 225
   d. On the basis of your answers to parts a, b, and c, evaluate
     $(-1)^3 + (-2)^3 + (-3)^3 + (-4)^3 + (-5)^3$. -225

96. Evaluate $2^{(3^2)}$ and $(2^3)^2$. Are the answers the same? If not, which is larger?
No. $2^{(3^2)}$ is larger.

97.   Abdul, Becky, Carl, and Diana were being questioned by their teacher. One of the students had left an apple on the teacher's desk, but the teacher did not know which one. Abdul said it was either Becky or Diana. Diana said it was neither Becky nor Carl. If both those statements are false, who left the apple on the teacher's desk? Explain how you arrived at your solution.

98.   In your own words, explain how you know that a number is written in scientific notation.

99. **a.** Express the mass of the sun in kilograms using scientific notation. 1.99×10^{30} kg
   b. Express the mass of a neutron in kilograms using scientific notation. 1.67×10^{-27} kg

Answers to Writing Exercises

97. Because the first statement is false (it was either Becky or Diana), neither Becky nor Diana could have done it. Therefore, it was either Abdul or Carl. Because the second statement is false (it was neither Becky nor Carl), either Becky or Carl did it. Because the first statement ensured that it was Abdul or Carl, it must have been Carl.

98. Check that students state that, for a number to be written in scientific notation, it must be the product of a number between 1 and 10 and a power of 10, where the exponent on 10 is an integer.

Focus on Problem Solving

Drawing Diagrams How do you best remember something? Do you remember best what you hear? The word *aural* means "pertaining to the ear"; people with a strong aural memory remember best those things that they hear. The word *visual* means "pertaining to the sense of sight"; people with a strong visual memory remember best that which they see written down. Some people claim that their memory is in their writing hand—they remember something only if they write it down! The method by which you best remember something is probably also the method by which you can best learn something new.

In problem-solving situations, try to capitalize on your strengths. If you tend to understand the material better when you hear it spoken, read application problems aloud or have someone else read them to you. If writing helps you to organize ideas, rewrite application problems in your own words.

No matter what your main strength, visualizing a problem can be a valuable aid in problem solving. A drawing, sketch, diagram, or chart can be a useful tool in problem solving, just as calculators and computers are tools. A diagram can be helpful in gaining an understanding of the relationships inherent in a problem-solving situation. A sketch will help you to organize the given information and can lead to your being able to focus on the method by which the solution can be determined.

> **HOW TO** A tour bus drives 5 mi south, then 4 mi west, then 3 mi north, then 4 mi east. How far is the tour bus from the starting point?
>
> Draw a diagram of the given information.
>
> From the diagram, we can see that the solution can be determined by subtracting 3 from 5: $5 - 3 = 2$.
>
> The bus is 2 mi from the starting point.

> **HOW TO** If you roll two ordinary six-sided dice and multiply the two numbers that appear on top, how many different possible products are there?
>
> Make a chart of the possible products. In the chart below, repeated products are marked with an asterisk.

$1 \cdot 1 = 1$	$2 \cdot 1 = 2$ (*)	$3 \cdot 1 = 3$ (*)	$4 \cdot 1 = 4$ (*)	$5 \cdot 1 = 5$ (*)	$6 \cdot 1 = 6$ (*)
$1 \cdot 2 = 2$	$2 \cdot 2 = 4$ (*)	$3 \cdot 2 = 6$ (*)	$4 \cdot 2 = 8$ (*)	$5 \cdot 2 = 10$ (*)	$6 \cdot 2 = 12$ (*)
$1 \cdot 3 = 3$	$2 \cdot 3 = 6$ (*)	$3 \cdot 3 = 9$	$4 \cdot 3 = 12$ (*)	$5 \cdot 3 = 15$ (*)	$6 \cdot 3 = 18$ (*)
$1 \cdot 4 = 4$	$2 \cdot 4 = 8$	$3 \cdot 4 = 12$ (*)	$4 \cdot 4 = 16$	$5 \cdot 4 = 20$ (*)	$6 \cdot 4 = 24$ (*)
$1 \cdot 5 = 5$	$2 \cdot 5 = 10$	$3 \cdot 5 = 15$	$4 \cdot 5 = 20$	$5 \cdot 5 = 25$	$6 \cdot 5 = 30$ (*)
$1 \cdot 6 = 6$	$2 \cdot 6 = 12$	$3 \cdot 6 = 18$	$4 \cdot 6 = 24$	$5 \cdot 6 = 30$	$6 \cdot 6 = 36$

By counting the products that are not repeats, we can see that there are 18 different possible products.

Look at Sections 10.1 and 10.2. You will notice that number lines are used to help you visualize the integers, as an aid in ordering integers, to help you understand the concepts of opposite and absolute value, and to illustrate addition of integers. As you begin your work with integers, you may find that sketching a number line proves helpful in coming to understand a problem or in working through a calculation that involves integers.

Projects and Group Activities

Answers to Projects and Group Activities— Deductive Reasoning

1. 4 Ós
2. 2 Λs
3. 6 Ós
4. 24 ∫s
5. 15 Ôs
6. Chris: compact car
   Dana: sports car
   Leslie: station wagon
   Pat: sedan
7. Anna: cucumbers
   Kay: beans
   Michelle: tomatoes
   Nicole: squash
8. Ontkeans: French
   Kedrovas: Mexican
   McIvers: Chinese
   Levinsons: Italian

Deductive Reasoning

Suppose that during the last week of your math class, your instructor tells you that if you receive an A on the final exam, you will earn an A in the course. When the final exam grades are posted, you learn that you received an A on the final exam. You can then assume that you will earn an A in the course.

The process used to determine your grade in the math course is deductive reasoning. **Deductive reasoning** involves drawing a conclusion that is based on given facts. The problems below require deductive reasoning.

1. Given that $\Delta\Delta\Delta = \diamond\diamond\diamond\diamond$ and $\diamond\diamond\diamond\diamond = \acute{O}\acute{O}$, then $\Delta\Delta\Delta\Delta\Delta\Delta =$ how many Ós?

2. Given that $\ddagger\ddagger = \bullet\bullet\bullet$ and $\bullet\bullet\bullet = \Lambda$, then $\ddagger\ddagger\ddagger\ddagger =$ how many Λs?

3. Given that $\acute{O}\acute{O}\acute{O} = \Omega\Omega$ and $\square = \Omega\Omega$, then $\square\square =$ how many Ós?

4. Given that $\int\int\int\int\int = \partial\partial$ and $\partial\partial\partial\partial = ¥¥¥$, then $¥¥¥¥¥¥ =$ how many ∫s?

5. Given that $\hat{O}\hat{O}\hat{O}\hat{O}\hat{O} = \square\square\square$ and $\square\square\square\square\square\square = §§§§$, then $§§§§§§ =$ how many Ôs?

6. Chris, Dana, Leslie, and Pat are neighbors. Each drives a different type of vehicle: a compact car, a sedan, a sports car, or a station wagon. From the following statements, determine which type of vehicle each of the neighbors drives. It may be helpful to use the chart provided below.
   a. Although the vehicle owned by Chris has more mileage on it than does either the sedan or the sports car, it does not have the highest mileage of all four cars.
   b. Pat and the owner of the sports car live on one side of the street, and Leslie and the owner of the compact car live on the other side of the street.
   c. Leslie owns the vehicle with the most mileage on it.

TAKE NOTE

To use the chart to solve this problem, write an X in a box to indicate that a possibility has been eliminated. Write a √ to show that a match has been found. When a row or column has 3 X's, a √ is written in the remaining open box in that row or column of the chart.

	Compact	Sedan	Sports Car	Wagon
Chris				
Dana				
Leslie				
Pat				

7. Four neighbors, Anna, Kay, Michelle, and Nicole, each plant a different vegetable (beans, cucumbers, squash, or tomatoes) in their garden. From the following statements, determine which vegetable each neighbor plants.
   a. Nicole's garden is bigger than the one that has tomatoes but smaller than the one that has cucumbers.
   b. Anna, who planted the largest garden, didn't plant the beans.
   c. The person who planted the beans has a garden the same size as Nicole's.
   d. Kay and the person who planted the tomatoes also have flower gardens.

8. The Ontkeans, Kedrovas, McIvers, and Levinsons are neighbors. Each of the four families specializes in a different national cuisine (Chinese, French, Italian, or Mexican). From the following statements, determine which cuisine each family specializes in.
   a. The Ontkeans invited the family that specializes in Chinese cuisine and the family that specializes in Mexican cuisine for dinner last night.
   b. The McIvers live between the family that specializes in Italian cuisine and the Ontkeans. The Levinsons live between the Kedrovas and the family that specializes in Chinese cuisine.
   c. The Kedrovas and the family that specializes in Italian cuisine both subscribe to the same culinary magazine.

Chapter 10 Summary

Key Words	Examples
Positive numbers are numbers greater than zero. *Negative numbers* are numbers less than zero. The *integers* are . . . −4, −3, −2, −1, 0, 1, 2, 3, 4, ... *Positive integers* are to the right of zero on the number line. *Negative integers* are to the left of zero on the number line. [10.1A, p. 405]	9, 87, and 603 are positive numbers. They are also positive integers. −5, −41, and −729 are negative numbers. They are also negative integers.
Opposite numbers are two numbers that are the same distance from zero on the number line but on opposite sides of zero. [10.1B, p. 406]	8 is the opposite of −8. −2 is the opposite of 2.
The *absolute value of a number* is its distance from zero on a number line. The absolute value of a number is a positive number or zero. The symbol for absolute value is | |. [10.1B, p. 406]	$\|9\| = 9$ $\|-9\| = 9$ $-\|9\| = -9$
A *rational number* is a number that can be written in the form $\frac{a}{b}$, where a and b are integers and $b \neq 0$. [10.4A, p. 427]	$\frac{3}{7}$, $-\frac{5}{8}$, 9, −2, $4\frac{1}{2}$, 0.6, and $0.\overline{3}$ are rational numbers.

Essential Rules and Procedures	Examples
Order Relations [10.1A, p. 405] A number that appears to the left of a given number on the number line is less than (<) the given number. A number that appears to the right of a given number on the number line is greater than (>) the given number.	$-6 > -12$ $-8 < 4$

To add numbers with the same sign, add the absolute values of the numbers. Then attach the sign of the addends. [10.2A, p. 412]	$6 + 4 = 10$ $-6 + (-4) = -10$
To add numbers with different signs, find the difference between the absolute values of the numbers. Then attach the sign of the addend with the greater absolute value. [10.2A, p. 412]	$-6 + 4 = -2$ $6 + (-4) = 2$
To subtract two numbers, add the opposite of the second number to the first number. [10.2B, p. 413]	$6 - 4 = 6 + (-4) = 2$ $6 - (-4) = 6 + 4 = 10$ $-6 - 4 = -6 + (-4) = -10$ $-6 - (-4) = -6 + 4 = -2$
To multiply numbers with the same sign, multiply the absolute values of the factors. The product is positive. [10.3A, p. 419]	$3 \cdot 5 = 15$ $-3(-5) = 15$
To multiply numbers with different signs, multiply the absolute values of the factors. The product is negative. [10.3A, p. 419]	$-3(5) = -15$ $3(-5) = -15$
To divide two numbers with the same sign, divide the absolute values of the numbers. The quotient is positive. [10.3B, p. 420]	$15 \div 3 = 5$ $(-15) \div (-3) = 5$
To divide two numbers with different signs, divide the absolute values of the numbers. The quotient is negative. [10.3B, p. 420]	$-15 \div 3 = -5$ $15 \div (-3) = -5$
Properties of Zero and One in Division [10.3B, p. 421] Zero divided by any number other than zero is zero. Any number other than zero divided by itself is 1. Any number divided by 1 is the number. Division by zero is not defined.	$0 \div (-5) = 0$ $-5 \div (-5) = 1$ $-5 \div 1 = -5$ $-5 \div 0$ is undefined.
Scientific Notation [10.5A, pp. 437–438] To express a number in scientific notation, write it in the form $a \times 10^n$, where a is a number between 1 and 10 and n is an integer. If the number is greater than 10, the exponent on 10 will be positive. If the number is less than 1, the exponent on 10 will be negative.	$367{,}000{,}000 = 3.67 \times 10^8$ $0.0000059 = 5.9 \times 10^{-6}$
To change a number written in scientific notation to decimal notation, move the decimal point to the right if the exponent on 10 is positive and to the left if the exponent is negative. Move the decimal point the same number of places as the absolute value of the exponent on 10.	$2.418 \times 10^7 = 24{,}180{,}000$ $9.06 \times 10^{-5} = 0.0000906$

The Order of Operations Agreement [10.5B, p. 439]

Step 1 Do all operations inside parentheses.	$(-4)^2 - 3(1 - 5) = (-4)^2 - 3(-4)$
Step 2 Simplify any numerical expressions containing exponents.	$= 16 - 3(-4)$ $= 16 - (-12)$
Step 3 Do multiplication and division as they occur from left to right.	$= 16 + 12$ $= 28$
Step 4 Rewrite subtraction as addition of the opposite. Then do additions as they occur from left to right.	

Chapter 10 Review Exercises

1. Find the opposite of 22.
−22 [10.1B]

2. Subtract: $-8 - (-2) - (-10) - 3$
1 [10.2B]

3. Subtract: $\dfrac{5}{8} - \dfrac{5}{6}$
$-\dfrac{5}{24}$ [10.4A]

4. Simplify: $-0.33 + 1.98 - 1.44$
0.21 [10.4A]

5. Multiply: $\left(-\dfrac{2}{3}\right)\left(\dfrac{6}{11}\right)\left(-\dfrac{22}{25}\right)$
$\dfrac{8}{25}$ [10.4B]

6. Multiply: -0.08×16
−1.28 [10.4B]

7. Simplify: $12 - 6 \div 3$
10 [10.5B]

8. Simplify: $\left(\dfrac{2}{3}\right)^2 - \dfrac{5}{6}$
$-\dfrac{7}{18}$ [10.5B]

9. Find the opposite of −4.
4 [10.1B]

10. Place the correct symbol, $<$ or $>$, between the two numbers.
$0 > -3$ [10.1A]

11. Evaluate $-|-6|$.
−6 [10.1B]

12. Divide: $-18 \div (-3)$
6 [10.3B]

13. Add: $-\dfrac{3}{8} + \dfrac{5}{12} + \dfrac{2}{3}$
$\dfrac{17}{24}$ [10.4A]

14. Multiply: $\dfrac{1}{3} \times \left(-\dfrac{3}{4}\right)$
$-\dfrac{1}{4}$ [10.4B]

15. Divide: $-\dfrac{7}{12} \div \left(-\dfrac{14}{39}\right)$
$1\dfrac{5}{8}$ [10.4B]

16. Simplify: $16 \div 4(8 - 2)$
24 [10.5B]

17. Add: $-22 + 14 + (-18)$
−26 [10.2A]

18. Simplify: $3^2 - 9 + 2$
2 [10.5B]

19. Write 0.0000397 in scientific notation.
3.97×10^{-5} [10.5A]

20. Divide: $-1.464 \div 18.3$
-0.08 [10.4B]

21. Simplify: $-\frac{5}{12} + \frac{7}{9} - \frac{1}{3}$
$\frac{1}{36}$ [10.4A]

22. Multiply: $\frac{6}{34} \times \frac{17}{40}$
$\frac{3}{40}$ [10.4B]

23. Multiply: $1.2 \times (-0.035)$
-0.042 [10.4B]

24. Simplify: $-\frac{1}{2} + \frac{3}{8} \div \frac{9}{20}$
$\frac{1}{3}$ [10.5B]

25. Evaluate $|-5|$.
5 [10.1B]

26. Place the correct symbol, $<$ or $>$, between the two numbers.
$-2 > -40$ [10.1A]

27. Find 2 times -13.
-26 [10.3A]

28. Simplify: $-0.4 \times 5 - (-3.33)$
1.33 [10.5B]

29. Add: $\frac{5}{12} + \left(-\frac{2}{3}\right)$
$-\frac{1}{4}$ [10.4A]

30. Simplify: $-33.4 + 9.8 - (-16.2)$
-7.4 [10.4A]

31. Divide: $\left(-\frac{3}{8}\right) \div \left(-\frac{4}{5}\right)$
$\frac{15}{32}$ [10.4B]

32. Write 2.4×10^5 in decimal notation.
240,000 [10.5A]

33. **Temperatures** Find the temperature after a rise of $18°$ from $-22°$.
$-4°$ [10.2C]

34. **Education** To discourage guessing on a multiple-choice exam, an instructor graded the test by giving 3 points for a correct answer, -1 point for an answer left blank, and -2 points for an incorrect answer. How many points did a student score who answered 38 questions correctly, answered 4 questions incorrectly, and left 8 questions blank?
98 [10.3C]

35. 🔵 **Chemistry** The boiling point of mercury is $356.58°C$. The melting point of mercury is $-38.87°C$. Find the difference between the boiling point and the melting point of mercury.
$395.45°C$ [10.4C]

Chapter 10 Test

1. Subtract: $-5 - (-8)$
3   [10.2B]

2. Evaluate $-|-2|$.
-2 [10.1B]

3. Add: $-\dfrac{2}{5} + \dfrac{7}{15}$
$\dfrac{1}{15}$ [10.4A]

4. Find the product of 0.032 and -1.9.
-0.0608 [10.4B]

5. Place the correct symbol, $<$ or $>$, between the two numbers.
$-8 > -10$ [10.1A]

6. Add: $1.22 + (-3.1)$
-1.88 [10.4A]

7. Simplify: $4 \times (4 - 7) \div (-2) - 4 \times 8$
-26 [10.5B]

8. Multiply: $-5 \times (-6) \times 3$
90   [10.3A]

9. What is -1.004 decreased by 3.01?
-4.014 [10.4A]

10. Divide: $-72 \div 8$
-9 [10.3B]

11. Find the sum of -2, 3, and -8.
-7 [10.2A]

12. Add: $-\dfrac{3}{8} + \dfrac{2}{3}$
$\dfrac{7}{24}$ [10.4A]

13. Write 87,600,000,000 in scientific notation.
8.76×10^{10} [10.5A]

14. Find the product of -4 and 12.
-48 [10.3A]

15. Divide: $\dfrac{0}{-17}$
0   [10.3B]

16. Subtract: $16 - 4 - (-5) - 7$
10   [10.2B]

17. Find the quotient of $-\frac{2}{3}$ and $\frac{5}{6}$.
$-\frac{4}{5}$ [10.4B]

18. Place the correct symbol, $<$ or $>$, between the two numbers.
$0 > -4$ [10.1A]

19. Add: $16 + (-10) + (-20)$
-14 [10.2A]

20. Simplify: $(-2)^2 - (-3)^2 \div (1 - 4)^2 \times 2 - 6$
-4 [10.5B]

21. Subtract: $-\frac{2}{5} - \left(-\frac{7}{10}\right)$
$\frac{3}{10}$ [10.4A]

22. Write 9.601×10^{-8} in decimal notation.
0.00000009601 [10.5A]

23. Divide: $-15.64 \div (-4.6)$
3.4 [10.4B]

24. Find the sum of $-\frac{1}{2}, \frac{1}{3},$ and $\frac{1}{4}$.
$\frac{1}{12}$ [10.4A]

25. Multiply: $\frac{3}{8} \times \left(-\frac{5}{6}\right) \times \left(-\frac{4}{15}\right)$
$\frac{1}{12}$ [10.4B]

26. Subtract: $2.113 - (-1.1)$
3.213 [10.4A]

27. **Temperatures** Find the temperature after a rise of 11°C from −4°C.
7°C [10.2C]

28. 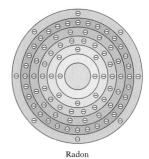 **Chemistry** The melting point of radon is −71°C. The melting point of oxygen is three times the melting point of radon. Find the melting point of oxygen.
−213°C [10.3C]

Radon

29. **Meteorology** On December 24, 1924, in Fairfield, Montana, the temperature fell from 17.22°C at noon to −29.4°C at midnight. How many degrees did the temperature fall in the 12-h period?
46.62°C [10.4C]

30. **Meteorology** The daily low temperature readings for a 3-day period were as follows: −7°F, 9°F, −8°F. Find the average low temperature for the 3-day period.
−2°F [10.3C]

Cumulative Review Exercises

1. Simplify: $16 - 4 \cdot (3 - 2)^2 \cdot 4$
0 [1.6B]

2. Find the difference between $8\frac{1}{2}$ and $3\frac{4}{7}$.
$4\frac{13}{14}$ [2.5C]

3. Divide: $3\frac{7}{8} \div 1\frac{1}{2}$
$2\frac{7}{12}$ [2.7B]

4. Simplify: $\frac{3}{8} \div \left(\frac{3}{8} - \frac{1}{4}\right) \div \frac{7}{3}$
$1\frac{2}{7}$ [2.8C]

5. Subtract: $2.907 - 1.09761$
1.80939 [3.3A]

6. Solve the proportion $\frac{7}{12} = \frac{n}{32}$.
Round to the nearest hundredth.
18.67 [4.3B]

7. 22 is 160% of what number?
13.75 [5.4A]

8. Convert: $7 \text{ qt} = \underline{\hspace{0.5cm}} \text{ gal } \underline{\hspace{0.5cm}} \text{ qt}$
1 gal 3 qt [8.3A]

9. Convert: $6692 \text{ ml} = \underline{\hspace{0.5cm}} \text{ L}$
6.692 L [9.3A]

10. Convert 4.2 ft to meters. Round to the nearest hundredth. (1 m = 3.28 ft.)
1.28 m [9.5A]

11. Find 32% of 180.
57.6 [5.2A]

12. Convert $3\frac{2}{5}$ to a percent.
340% [5.1B]

13. Add: $-8 + 5$
-3 [10.2A]

14. Add: $3\frac{1}{4} + \left(-6\frac{5}{8}\right)$
$-3\frac{3}{8}$ [10.4A]

15. Subtract: $-6\frac{1}{8} - 4\frac{5}{12}$
$-10\frac{13}{24}$ [10.4A]

16. Simplify: $-12 - (-7) - 3(-8)$
19 [10.5B]

17. What is -3.2 times -1.09?
3.488 [10.4B]

18. Multiply: $-6 \times 7 \times \left(-\frac{3}{4}\right)$
$31\frac{1}{2}$ [10.4B]

19. Find the quotient of 42 and −6.

−7 [10.3B]

20. Divide: $-2\frac{1}{7} \div \left(-3\frac{3}{5}\right)$

$\frac{25}{42}$ [10.4B]

21. Simplify: $3 \times (3 - 7) \div 6 - 2$

−4 [10.5B]

22. Simplify: $4 - (-2)^2 \div (1 - 2)^2 \times 3 + 4$

−4 [10.5B]

23. **Carpentry** A board $5\frac{2}{3}$ ft long is cut from a board 8 ft long. What is the length of the board remaining?

$2\frac{1}{3}$ ft [2.5D]

24. **Banking** Nimisha had a balance of $763.56 in her checkbook before writing checks for $135.88 and $47.81 and making a deposit of $223.44. Find her new checkbook balance.

$803.31 [6.7A]

25. **Consumerism** A suit that regularly sells for $165 is on sale for $120. Find the percent decrease in price. Round to the nearest tenth of a percent.

27.3% [6.2C]

26. **Measurement** A reception is planned for 80 guests. How many gallons of coffee should be prepared to provide 2 c of coffee for each guest?

10 gal [8.3C]

27. **Investments** A stock selling for $82.625 per share paid a dividend of $1.50 per share before the dividend was increased by 12%. Find the dividend per share after the increase.

$1.68 [6.2A]

28. **Compensation** The hourly wages for five job classifications at a company are $11.40, $9.32, $15.25, $10.73, and $13.10. Find the median hourly pay.

$11.40 [7.4A]

29. **Elections** A pre-election survey showed that 5 out of every 8 registered voters would cast ballots in a city election. At this rate, how many people would vote in a city of 960,000 registered voters?

600,000 people [4.3C]

30. **Meteorology** The daily high temperature readings for a 4-day period were recorded as follows: −19°F, −7°F, 1°F, and 9°F. Find the average high temperature for the 4-day period.

−4°F [10.3C]

This family is getting ready to head off on their vacation. They plan to drive 500 miles to their destination, and their car averages 25 miles per gallon of gasoline on the highway. How many gallons of gasoline will the car consume in driving the family there and back? We can use the formula $D = M \cdot G$, where D is the distance traveled, M is the miles per gallon, and G is the number of gallons consumed. Using this formula, we find that the car will need 40 gallons to make the trip. **Exercises 98 to 101 on pages 475 and 476** use this formula to answer other questions about gasoline consumption.

Need help? For online student resources, such as section quizzes, visit this textbook's website at **math.college.hmco.com/students.**

OBJECTIVES

Section 11.1

A To evaluate variable expressions

B To simplify variable expressions containing no parentheses

C To simplify variable expressions containing parentheses

Section 11.2

A To determine whether a given number is a solution of an equation

B To solve an equation of the form $x + a = b$

C To solve an equation of the form $ax = b$

D To solve application problems using formulas

Section 11.3

A To solve an equation of the form $ax + b = c$

B To solve application problems using formulas

Section 11.4

A To solve an equation of the form $ax + b = cx + d$

B To solve an equation containing parentheses

Section 11.5

A To translate a verbal expression into a mathematical expression given the variable

B To translate a verbal expression into a mathematical expression by assigning the variable

Section 11.6

A To translate a sentence into an equation and solve

B To solve application problems

PREP TEST • • •

Do these exercises to prepare for Chapter 11.

For Exercises 1 to 9, simplify.

1. $2 - 9$
-7 [10.2B]

2. $-5(4)$
-20 [10.3A]

3. $-16 + 16$
0 [10.2A]

4. $\dfrac{-7}{-7}$
1 [10.3B]

5. $-\dfrac{3}{8}\left(-\dfrac{8}{3}\right)$
1 [10.4B]

6. $\left(\dfrac{3}{5}\right)^{3} \cdot \left(\dfrac{5}{9}\right)^{2}$
$\dfrac{1}{15}$ [2.8B]

7. $\dfrac{2}{3} + \left(\dfrac{3}{4}\right)^{2} \cdot \dfrac{2}{9}$
$\dfrac{19}{24}$ [2.8C]

8. $-8 \div (-2)^{2} + 6$
4 [10.5B]

9. $4 + 5(2 - 7)^{2} \div (-8 + 3)$
-21 [10.5B]

GO FIGURE • • •

In the addition at the right, each letter stands for a different digit. If N = 1, I = 5, U = 7, F = 2, and T = 3, what is the value of S? 6

```
  F U N
    I N
+ T H E
-------
  S U N
```

11.1 Variable Expressions

Objective A **To evaluate variable expressions**

Often we discuss a quantity without knowing its exact value—for example, next year's inflation rate, the price of gasoline next summer, or the interest rate on a new-car loan next fall. In mathematics, a letter of the alphabet is used to stand for a quantity that is unknown or that can change, or *vary*. The letter is called a **variable.** An expression that contains one or more variables is called a **variable expression.**

A company's business manager has determined that the company will make a $10 profit on each radio it sells. The manager wants to describe the company's total profit from the sale of radios. Because the number of radios that the company will sell is unknown, the manager lets the variable n stand for that number. Then the variable expression $10 \cdot n$, or simply $10n$, describes the company's profit from selling n radios.

The company's profit from selling n radios is $10 \cdot n = \$10n$.

If the company sells 12 radios, its profit is $10 \cdot 12 = \$120$.

If the company sells 75 radios, its profit is $10 \cdot 75 = \$750$.

Replacing the variable or variables in a variable expression and then simplifying the resulting numerical expression is called **evaluating a variable expression.**

> **HOW TO**   Evaluate $3x^2 + xy - z$ when $x = -2$, $y = 3$, and $z = -4$.
>
> $3x^2 + xy - z$
>
> $3(-2)^2 + (-2)(3) - (-4)$ • Replace each variable in the expression with the number it stands for.
>
> $= 3 \cdot 4 + (-2)(3) - (-4)$ • Use the Order of Operations Agreement to simplify the resulting numerical expression.
>
> $= 12 + (-6) - (-4)$
>
> $= 12 + (-6) + 4$
>
> $= 6 + 4$
>
> $= 10$
>
> The value of the variable expression $3x^2 + xy - z$ when $x = -2$, $y = 3$, and $z = -4$ is 10.

(Continued)

2. A phone company charges a flat fee of $25 to come to your house for repair work. This charge is added to the repair bill. Let *B* represent the repair bill. Write a variable expression that represents the final bill after the flat fee has been added. $B + 25$

3. The varsity baseball coach at a high school must order jerseys for the team. The price of one jersey is $28. Let *N* represent the number of players on the baseball team. Write a variable expression that represents the total cost for jerseys for the team.
$28N$

4. A community service organization must raise a total of $500. This amount is divided by the number of members so that each member of the organization is responsible for raising the same amount of money. Let *N* represent the number of people in the organization. Write a variable expression that represents the amount of money each person is responsible for raising.
$\dfrac{500}{N}$

After each exercise in this activity, have students evaluate the variable expression written for some values of the variable that you provide.

Example 1

Evaluate $3x - 4y$ when $x = -2$ and $y = 3$.

Solution
$3x - 4y$
$3(-2) - 4(3) = -6 - 12$
$\qquad\qquad\qquad = -6 + (-12) = -18$

You Try It 1

Evaluate $6a - 5b$ when $a = -3$ and $b = 4$.

Your solution
-38

Example 2

Evaluate $-x^2 - 6 \div y$ when $x = -3$ and $y = 2$.

Solution
$-x^2 - 6 \div y$
$-(-3)^2 - 6 \div 2 = -9 - 6 \div 2$
$\qquad\qquad\qquad = -9 - 3$
$\qquad\qquad\qquad = -9 + (-3) = -12$

You Try It 2

Evaluate $-3s^2 - 12 \div t$ when $s = -2$ and $t = 4$.

Your solution
-15

Example 3

Evaluate $-\dfrac{1}{2}y^2 - \dfrac{3}{4}z$ when $y = 2$ and $z = -4$.

Solution
$-\dfrac{1}{2}y^2 - \dfrac{3}{4}z$

$-\dfrac{1}{2}(2)^2 - \dfrac{3}{4}(-4) = -\dfrac{1}{2} \cdot 4 - \dfrac{3}{4}(-4)$
$\qquad\qquad\qquad = -2 - (-3)$
$\qquad\qquad\qquad = -2 + (3) = 1$

You Try It 3

Evaluate $-\dfrac{2}{3}m + \dfrac{3}{4}n^3$ when $m = 6$ and $n = 2$.

Your solution
2

Example 4

Evaluate $-2ab + b^2 + a^2$ when $a = -\dfrac{3}{5}$ and $b = \dfrac{2}{5}$.

Solution
$-2ab + b^2 + a^2$

$-2\left(-\dfrac{3}{5}\right)\left(\dfrac{2}{5}\right) + \left(\dfrac{2}{5}\right)^2 + \left(-\dfrac{3}{5}\right)^2$

$= -2\left(-\dfrac{3}{5}\right)\left(\dfrac{2}{5}\right) + \left(\dfrac{4}{25}\right) + \left(\dfrac{9}{25}\right)$

$= \dfrac{12}{25} + \dfrac{4}{25} + \dfrac{9}{25} = \dfrac{25}{25} = 1$

You Try It 4

Evaluate $-3yz - z^2 + y^2$ when $y = -\dfrac{2}{3}$ and $z = \dfrac{1}{3}$.

Your solution
1

Solutions on p. S28

Objective B   **To simplify variable expressions containing no parentheses**

The **terms of a variable expression** are the addends of the expression. The variable expression at the right has four terms.

$$\overbrace{\underbrace{7x^2 \ + \ (-6xy) \ + \ x}_{\text{Variable terms}} \ + \ \underbrace{(-8)}_{\substack{\text{Constant} \\ \text{term}}}}^{\text{4 terms}}$$

TAKE NOTE
A variable term is a term composed of a numerical coefficient and a variable part. The numerical coefficient is the number part of the term.

Three of the terms are **variable terms:** $7x^2$, $-6xy$, and x.

One of the terms is a **constant term:** -8. A constant term has no variables.

Each variable term is composed of a **numerical coefficient** (the number part of a variable term) and a **variable part** (the variable or variables and their exponents). When the numerical coefficient is 1, the 1 is usually not written. ($1x = x$)

Like terms of a variable expression are the terms with the same variable part. (Because $y^2 = y \cdot y$, y^2 and y are not like terms.)

In variable expressions that contain constant terms, the constant terms are like terms.

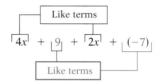

The Commutative and Associative Properties of Addition are used to simplify variable expressions. These properties can be stated in general form using variables.

TAKE NOTE
With the Commutative Property, the order in which the numbers appear changes. With the Associative Property, the order in which the numbers appear remains the same.

> **Commutative Property of Addition**
>
> If a and b are two numbers, then $a + b = b + a$.

> **Associative Property of Addition**
>
> If a, b, and c are three numbers, then $a + (b + c) = (a + b) + c$.

The phrase **simplifying a variable expression** means *combining* like terms by adding their numerical coefficients. For example, to simplify $2y + 3y$, think

$$2y + 3y = (y + y) + (y + y + y) = 5y$$

HOW TO   Simplify: $8z - 5 + 2z$

$8z - 5 + 2z = 8z + 2z - 5$ • Use the Commutative and Associative
$\qquad\qquad\quad = 10z - 5$ Properties of Addition to group like terms.
                                     Combine the like terms $8z + 2z$.

In-Class Examples (Objective 11.1B)
Simplify.
1. $11b - 2b$ $9b$
2. $8d - 7 + 9d$ $17d - 7$
3. $4a - 2b^2 - 7a + 9b^2$ $-3a + 7b^2$

Concept Check

1. Which of the following pairs of terms are like terms?

a. $3a$ and $3b$

b. $7z^2$ and $7z^3$

c. $6ab$ and $3a$

d. $-4c^2$ and $6c^2$ d

2. Classify each statement below as illustrating the Commutative Property of Addition or the Associative Property of Addition.

a. $5 + 3 = 3 + 5$
The Commutative Property

b. $(8 + 1) + 4 = 8 + (1 + 4)$
The Associative Property

c. $x + 10 = 10 + x$
The Commutative Property

d. $y + (1 + 7) = (y + 1) + 7$
The Associative Property

Instructor Note

Although students will normally say that $4x + 3$ cannot be simplified, many of these same students will try to simplify $3 + 4x$ as $7x$.

HOW TO Simplify: $12a - 4b - 8a + 2b$

$$12a - 4b - 8a + 2b = 12a + (-4)b + (-8)a + 2b$$

- Change subtraction to addition of the opposite.

$$= 12a + (-8)a + (-4)b + 2b$$
$$= 4a + (-2)b$$

- Use the Commutative and Associative Properties of Addition to group like terms. Combine like terms.

$$= 4a - 2b$$

- Recall that $a + (-b) = a - b$.

HOW TO Simplify: $6z^2 + 3 - z^2 - 7$

$$6z^2 + 3 - z^2 - 7 = 6z^2 + 3 + (-1)z^2 + (-7)$$

- Change subtraction to addition of the opposite.

$$= 6z^2 + (-1)z^2 + 3 + (-7)$$
$$= 5z^2 + (-4)$$
$$= 5z^2 - 4$$

- Use the Commutative and Associative Properties of Addition to group like terms. Combine like terms.

Example 5

Simplify: $6xy - 8x + 5x - 9xy$

Solution
$$6xy - 8x + 5x - 9xy$$
$$= 6xy + (-8)x + 5x + (-9)xy$$
$$= 6xy + (-9)xy + (-8)x + 5x$$
$$= (-3)xy + (-3)x$$
$$= -3xy - 3x$$

You Try It 5

Simplify: $5a^2 - 6b^2 + 7a^2 - 9b^2$

Your solution
$12a^2 - 15b^2$

Example 6

Simplify: $-4z^2 + 8 + 5z^2 - 3$

Solution
$$-4z^2 + 8 + 5z^2 - 3$$
$$= -4z^2 + 8 + 5z^2 + (-3)$$
$$= -4z^2 + 5z^2 + 8 + (-3)$$
$$= z^2 + 5$$

You Try It 6

Simplify: $-6x + 7 + 9x - 10$

Your solution
$3x - 3$

Example 7

Simplify: $\frac{1}{4}m^2 - \frac{1}{2}n^2 + \frac{1}{2}m^2$

Solution
$$\frac{1}{4}m^2 - \frac{1}{2}n^2 + \frac{1}{2}m^2 = \frac{1}{4}m^2 + \left(-\frac{1}{2}\right)n^2 + \frac{1}{2}m^2$$
$$= \frac{1}{4}m^2 + \frac{1}{2}m^2 + \left(-\frac{1}{2}\right)n^2$$
$$= \frac{1}{4}m^2 + \frac{2}{4}m^2 + \left(-\frac{1}{2}\right)n^2$$
$$= \frac{3}{4}m^2 + \left(-\frac{1}{2}\right)n^2$$
$$= \frac{3}{4}m^2 - \frac{1}{2}n^2$$

You Try It 7

Simplify: $\frac{3}{8}w + \frac{1}{2} - \frac{1}{4}w - \frac{2}{3}$

Your solution
$\frac{1}{8}w - \frac{1}{6}$

Solutions on p. S28

Objective C   **To simplify variable expressions containing parentheses**

The Commutative and Associative Properties of Multiplication and the Distributive Property are used to simplify variable expressions that contain parentheses. These properties can be stated in general form using variables.

Commutative Property of Multiplication

If a and b are two numbers, then $a \cdot b = b \cdot a$.

Associative Property of Multiplication

If a, b, and c are three numbers, then $a \cdot (b \cdot c) = (a \cdot b) \cdot c$.

The Associative and Commutative Properties of Multiplication are used to simplify variable expressions such as the following.

HOW TO   Simplify: $-5(4x)$

$-5(4x) = (-5 \cdot 4)x$ • Use the Associative Property of Multiplication.
$\qquad = -20x$

HOW TO   Simplify: $(6y) \cdot 5$

$(6y) \cdot 5 = 5 \cdot (6y)$ • Use the Commutative Property of Multiplication.
$\qquad = (5 \cdot 6)y = 30y$ • Use the Associative Property of Multiplication.

The **Distributive Property** is used to remove parentheses from variable expressions that contain both multiplication and addition.

Distributive Property

If a, b, and c are three numbers, then $a(b + c) = ab + ac$.

HOW TO   Simplify: $4(z + 5)$

$4(z + 5) = 4z + 4(5)$ • The Distributive Property is used to rewrite the
$\qquad = 4z + 20$ variable expression without parentheses.

HOW TO   Simplify: $-3(2x + 7)$

$-3(2x + 7) = -3(2x) + (-3)(7)$ • Use the Distributive Property.
$\qquad = -6x + (-21)$
$\qquad = -6x - 21$ • Recall that $a + (-b) = a - b$.

The Distributive Property can also be stated in terms of subtraction.

$$a(b - c) = ab - ac$$

Instructor Note

Some of the Applying the Concepts exercises at the end of this section involve using rectangles and squares to model algebraic expressions. These models can help some students better understand adding like terms and the Distributive Property.

A typical student error for the third How To is to subtract 5 from 12 and then apply the Distributive Property. Remind students that the Order of Operations Agreement requires multiplication before addition or subtraction.

Optional Student Activity

Operations such as addition and multiplication are called *binary operations*. The word *binary* means "consisting of two parts." Addition consists of two parts (the addends); multiplication consists of two parts (the factors). Other binary operations can be defined. For instance, define $\otimes$ as $a \otimes b = (a \cdot b) - (a + b)$. Then $7 \otimes 4 = (7 \cdot 4) - (7 + 4) = 28 - 11 = 17$.

a. Find $6 \otimes 8$.   34

b. Find $\dfrac{4}{3} \otimes \dfrac{3}{4}$. $-1\dfrac{1}{12}$

c. Does $a \otimes 0 = 0$?
No, for $a \neq 0$

d. Does $a \otimes 1 = 1$?   No

e. Is $\otimes$ a commutative operation?   Yes

f. Explain your answer to part e.
$(a \cdot b) - (a + b)$ is the same as $(b \cdot a) - (b + a)$ because the operations of multiplication and addition are both commutative.

HOW TO   Simplify: $8(2r - 3s)$

$$8(2r - 3s) = 8(2r) - 8(3s) \qquad \bullet \text{ Use the Distributive Property.}$$
$$= 16r - 24s$$

HOW TO   Simplify: $-5(2x - 4y)$

$$-5(2x - 4y) = (-5)(2x) - (-5)(4y) \qquad \bullet \text{ Use the Distributive Property.}$$
$$= -10x - (-20y)$$
$$= -10x + 20y \qquad \bullet \text{ Recall that } a - (-b) = a + b.$$

HOW TO   Simplify: $12 - 5(m + 2) + 2m$

$$12 - 5(m + 2) + 2m = 12 - 5m + (-5)(2) + 2m \quad \bullet \text{ Use the Distributive}$$
$$= 12 - 5m + (-10) + 2m \qquad \text{Property to simplify}$$
the expression
$-5(m + 2)$.

$$= -5m + 2m + 12 + (-10) \quad \bullet \text{ Use the Commutative and}$$
Associative Properties to
group like terms.

$$= -3m + 2 \qquad \bullet \text{ Combine like terms by}$$
adding their numerical
coefficients. Add constant
terms.

The answer $-3m + 2$ can also be written as $2 - 3m$. In this text, we will write answers with variable terms first, followed by the constant term.

Example 8

Simplify: $4(x - 3)$

Solution
$$4(x - 3) = 4x - 4(3)$$
$$= 4x - 12$$

You Try It 8

Simplify: $5(a - 2)$

Your solution
$5a - 10$

Example 9

Simplify: $5n - 3(2n - 4)$

Solution
$$5n - 3(2n - 4) = 5n - 3(2n) - (-3)(4)$$
$$= 5n - 6n - (-12)$$
$$= 5n - 6n + 12$$
$$= -n + 12$$

You Try It 9

Simplify: $8s - 2(3s - 5)$

Your solution
$2s + 10$

Example 10

Simplify: $3(c - 2) + 2(c + 6)$

Solution
$$3(c - 2) + 2(c + 6) = 3c - 3(2) + 2c + 2(6)$$
$$= 3c - 6 + 2c + 12$$
$$= 3c + 2c - 6 + 12$$
$$= 5c + 6$$

You Try It 10

Simplify: $4(x - 3) - 2(x + 1)$

Your solution
$2x - 14$

Solutions on p. S28

11.1 Exercises

Objective A   To evaluate variable expressions

For Exercises 1 to 34, evaluate the variable expression when $a = -3$, $b = 6$, and $c = -2$.

1. $5a - 3b$
-33

2. $4c - 2b$
-20

3. $2a + 3c$
-12

4. $2c + 4a$
-16

5. $-c^2$
-4

6. $-a^2$
-9

7. $b - a^2$
-3

8. $b - c^2$
2

9. $ab - c^2$
-22

10. $bc - a^2$
-21

11. $2ab - c^2$
-40

12. $3bc - a^2$
-45

13. $a - (b \div a)$
-1

14. $c - (b \div c)$
1

15. $2ac - (b \div a)$
14

16. $4ac \div (b \div a)$
-12

17. $b^2 - c^2$
32

18. $b^2 - a^2$
27

19. $b^2 \div (ac)$
6

20. $3c^2 \div (ab)$
$-\dfrac{2}{3}$

21. $c^2 - (b \div c)$
7

22. $a^2 - (b \div a)$
11

23. $a^2 + b^2 + c^2$
49

24. $a^2 - b^2 - c^2$
-31

25. $ac + bc + ab$
-24

26. $ac - bc - ab$
36

27. $a^2 + b^2 - ab$
63

28. $b^2 + c^2 - bc$
52

29. $2b - (3c + a^2)$
9

30. $\dfrac{2}{3}b + \left(\dfrac{1}{2}c - a\right)$
6

31. $\dfrac{1}{3}a + \left(\dfrac{1}{2}b - \dfrac{2}{3}a\right)$
4

32. $-\dfrac{2}{3}b - \left(\dfrac{1}{2}c + a\right)$
0

33. $\dfrac{1}{6}b + \dfrac{1}{3}(c + a)$
$-\dfrac{2}{3}$

34. $\dfrac{1}{2}c + \left(\dfrac{1}{3}b - a\right)$
4

For Exercises 35 to 38, evaluate the variable expression when $a = -\dfrac{1}{2}$, $b = \dfrac{3}{4}$, and $c = \dfrac{1}{4}$.

35. $4a + (3b - c)$
0

36. $2b + (c - 3a)$
$3\dfrac{1}{4}$

37. $2a - b^2 \div c$
$-3\dfrac{1}{4}$

38. $b \div (-c) + 2a$
-4

For Exercises 39 to 42, evaluate the variable expression when $a = 3.72$, $b = -2.31$, and $c = -1.74$.

39. $a^2 - b^2$
8.5023

40. $a^2 - b \cdot c$
9.819

41. $3ac - (c \div a)$
-18.950658

42. $2c + (b^2 - c)$
3.5961

Quick Quiz (Objective 11.1A)

Evaluate the variable expression when $a = 3$, $b = -1$, and $c = -2$.

1. $2a - (3b + 4c)$ 17

2. $3ac \div b$ 18

Section 11.1

Suggested Assignment
Exercises 1–33, every other odd
Exercises 35–129, odds
Exercises 131, 132, 134

Objective B   **To simplify variable expressions containing no parentheses**

For Exercises 43 to 46, name the terms of the variable expression. Then underline the constant term.

43. $2x^2 + 3x - 4$
$2x^2, 3x, \underline{-4}$

44. $-4y^2 + 5$
$-4y^2, \underline{5}$

45. $3a^2 - 4a + 8$
$3a^2, -4a, \underline{8}$

46. $7 - b$
$-b, \underline{7}$

For Exercises 47 to 50, name the variable terms of the expression. Then underline the coefficients of the variable terms.

47. $3x^2 - 4x + 9$
$\underline{3}x^2, \underline{-4}x$

48. $-5a^2 + a - 4$
$\underline{-5}a^2, \underline{1}a$

49. $y^2 + 6a - 1$
$\underline{1}y^2, \underline{6}a$

50. $8 - c$
$\underline{-1}c$

For Exercises 51 to 94, simplify.

51. $7z + 9z$
$16z$

52. $6x + 5x$
$11x$

53. $12m - 3m$
$9m$

54. $5y - 12y$
$-7y$

55. $5at + 7at$
$12at$

56. $12mn + 11mn$
$23mn$

57. $-4yt + 7yt$
$3yt$

58. $-12yt + 5yt$
$-7yt$

59. $-3x - 12y$
Unlike terms

60. $-12y - 7y$
$-19y$

61. $3t^2 - 5t^2$
$-2t^2$

62. $7t^2 + 8t^2$
$15t^2$

63. $6c - 5 + 7c$
$13c - 5$

64. $7x - 5 + 3x$
$10x - 5$

65. $2t + 3t - 7t$
$-2t$

66. $9x^2 - 5 - 3x^2$
$6x^2 - 5$

67. $7y^2 - 2 - 4y^2$
$3y^2 - 2$

68. $3w - 7u + 4w$
$7w - 7u$

69. $6w - 8u + 8w$
$14w - 8u$

70. $4 - 6xy - 7xy$
$-13xy + 4$

71. $10 - 11xy - 12xy$
$-23xy + 10$

72. $7t^2 - 5t^2 - 4t^2$
$-2t^2$

73. $3v^2 - 6v^2 - 8v^2$
$-11v^2$

74. $5ab - 7a - 10ab$
$-5ab - 7a$

75. $-10ab - 3a + 2ab$
$-8ab - 3a$

76. $-4x^2 - x + 2x^2$
$-2x^2 - x$

77. $-3y^2 - y + 7y^2$
$4y^2 - y$

78. $4x^2 - 8y - x^2 + y$
$3x^2 - 7y$

79. $2a - 3b^2 - 5a + b^2$
$-3a - 2b^2$

80. $8y - 4z - y + 2z$
$7y - 2z$

81. $3x^2 - 7x + 4x^2 - x$
$7x^2 - 8x$

82. $5y^2 - y + 6y^2 - 5y$
$11y^2 - 6y$

83. $6s - t - 9s + 7t$
$-3s + 6t$

84. $5w - 2v - 9w + 5v$
$-4w + 3v$

85. $4m + 8n - 7m + 2n$
$-3m + 10n$

86. $z + 9y - 4z + 3y$
$-3z + 12y$

Quick Quiz (Objective 11.1B)
Simplify.
1. $5y - 12y$ $-7y$
2. $6c - 3 + 4c$ $10c - 3$
3. $5x^2 - 6x + 7x^2 - 8x$ $12x^2 - 14x$

87. $-5ab + 7ac + 10ab - 3ac$
$5ab + 4ac$

88. $-2x^2 - 3x - 11x^2 + 14x$
$-13x^2 + 11x$

89. $\frac{4}{9}a^2 - \frac{1}{5}b^2 + \frac{2}{9}a^2 + \frac{4}{5}b^2$
$\frac{2}{3}a^2 + \frac{3}{5}b^2$

90. $\frac{6}{7}x^2 + \frac{2}{5}x - \frac{3}{7}x^2 - \frac{4}{5}x$
$\frac{3}{7}x^2 - \frac{2}{5}x$

91. $4.235x - 0.297x + 3.056x$
$6.994x$

92. $8.092y - 3.0793y + 0.063y$
$5.0757y$

93. $7.81m + 3.42n - 6.25m - 7.19n$
$1.56m - 3.77n$

94. $8.34y^2 - 4.21y - 6.07y^2 - 5.39y$
$2.27y^2 - 9.6y$

Objective C **To simplify variable expressions containing parentheses**

For Exercises 95 to 130, simplify.

95. $5(x + 4)$
$5x + 20$

96. $3(m + 6)$
$3m + 18$

97. $(y - 3)4$
$4y - 12$

98. $(z - 3)7$
$7z - 21$

99. $-2(a + 4)$
$-2a - 8$

100. $-5(b + 3)$
$-5b - 15$

101. $3(5x + 10)$
$15x + 30$

102. $2(4m - 7)$
$8m - 14$

103. $5(3c - 5)$
$15c - 25$

104. $-4(w - 3)$
$-4w + 12$

105. $-3(y - 6)$
$-3y + 18$

106. $3m + 4(m + z)$
$7m + 4z$

107. $5x + 2(x + 7)$
$7x + 14$

108. $6z - 3(z + 4)$
$3z - 12$

109. $8y - 4(y + 2)$
$4y - 8$

110. $7w - 2(w - 3)$
$5w + 6$

111. $9x - 4(x - 6)$
$5x + 24$

112. $-5m + 3(m + 4)$
$-2m + 12$

113. $-2y + 3(y - 2)$
$y - 6$

114. $5m + 3(m + 4) - 6$
$8m + 6$

115. $4n + 2(n + 1) - 5$
$6n - 3$

116. $8z - 2(z - 3) + 8$
$6z + 14$

117. $9y - 3(y - 4) + 8$
$6y + 20$

118. $6 - 4(a + 4) + 6a$
$2a - 10$

119. $3x + 2(x + 2) + 5x$
$10x + 4$

120. $7x + 4(x + 1) + 3x$
$14x + 4$

121. $-7t + 2(t - 3) - t$
$-6t - 6$

Quick Quiz (Objective 11.1C)
Simplify.
1. $3(2n - 5)$ $6n - 15$
2. $-6(w - 9)$ $-6w + 54$
3. $6w - 2(w - 5)$ $4w + 10$
4. $3y - 2(3 - y) + 5y$ $10y - 6$

122. $-3y + 2(y - 4) - y$
$-2y - 8$

123. $z - 2(1 - z) - 2z$
$z - 2$

124. $2y - 3(2 - y) + 4y$
$9y - 6$

125. $3(y - 2) - 2(y - 6)$
$y + 6$

126. $7(x + 2) + 3(x - 4)$
$10x + 2$

127. $2(t - 3) + 7(t + 3)$
$9t + 15$

128. $3(y - 4) - 2(y - 3)$
$y - 6$

129. $3t - 6(t - 4) + 8t$
$5t + 24$

130. $5x + 3(x - 7) - 9x$
$-x - 21$

APPLYING THE CONCEPTS

131. The square and the rectangle at the right can be used to illustrate algebraic expressions. The illustration below represents the expression $2x + 1$.

    $\boxed{x}\ \boxed{1}$

$\boxed{x}\ \boxed{x}\ \boxed{1}$

a. Using similar squares and rectangles, draw a figure that represents the expression $3 + 2x$. **a.** $\boxed{1}\boxed{1}\boxed{1}\ \boxed{x}\ \boxed{x}$

b. Draw a figure that represents the expression $4x + 6$. **b.** $\boxed{x}\ \boxed{x}\ \boxed{x}\ \boxed{x}\ \boxed{1}\boxed{1}\boxed{1}\boxed{1}\boxed{1}\boxed{1}$

c. Draw a figure that represents the expression $3x + 2$. **c.** $\boxed{x}\ \boxed{x}\ \boxed{x}\ \boxed{1}\boxed{1}$

d. Draw a figure that represents the expression $2x + 4$. **d.** $\boxed{x}\ \boxed{x}\ \boxed{1}\boxed{1}\boxed{1}\boxed{1}$

e. The illustration below represents the expression $3(x + 1)$. Rearrange these rectangles so that the x's are together and the 1's are together. **e.** $\boxed{x}\ \boxed{x}\ \boxed{x}\ \boxed{1}\boxed{1}\boxed{1}$

f. Write a mathematical expression for the rearranged figure. **f.** $3x + 3$

$\boxed{x}\ \boxed{1}\ \boxed{x}\ \boxed{1}\ \boxed{x}\ \boxed{1}$

132. **a.** Using squares and rectangles similar to those in Exercise 131, draw a figure that represents the expression $2 + 3x$. **a.** $\boxed{1}\boxed{1}\ \boxed{x}\ \boxed{x}\ \boxed{x}$

b. Draw a figure that represents the expression $2(2x + 3)$. **b.** $\boxed{x}\ \boxed{x}\ \boxed{1}\boxed{1}\boxed{1}\ \boxed{x}\ \boxed{x}\ \boxed{1}\boxed{1}\boxed{1}$

c. Draw a figure that represents the expression $4x + 3$. **c.** $\boxed{x}\ \boxed{x}\ \boxed{x}\ \boxed{x}\ \boxed{1}\boxed{1}\boxed{1}$

d. Draw a figure that represents the expression $4x + 6$. **d.** $\boxed{x}\ \boxed{x}\ \boxed{x}\ \boxed{x}\ \boxed{1}\boxed{1}\boxed{1}\boxed{1}\boxed{1}\boxed{1}$

e. Does the figure $2(2x + 3)$ equal the figure $4x + 6$? Explain how this equivalence is related to the Distributive Property. Yes

f. Does the figure $2 + 3x$ equal the figure $5x$? Explain how this equivalence is related to combining like terms. No

133. **a.** Simplifying variable expressions requires combining like terms. Give some examples of how this applies to everyday experience.

b. It was stated in this section that the variable terms y^2 and y are not like terms. Use measurements of area and distance to show that these terms would not be combined as measurements. Answers will vary.

134. ✎ Explain why the simplification of the expression $2 + 3(2x + 4)$ shown at the right is incorrect. What is the correct simplification?

Why is this incorrect?
$2 + 3(2x + 4) = 5(2x + 4)$
$= 10x + 20$

Answers to Writing Exercises

134. The simplification is incorrect because, by the Order of Operations Agreement, we must multiply first and then add. The correct simplification is

$2 + 3(2x + 4)$
$= 2 + 6x + 12$
$= 6x + 2 + 12$
$= 6x + 14$

11.2 Introduction to Equations

Objective A   **To determine whether a given number is a solution of an equation**

Point of Interest

Finding solutions of equations has been a principal aim of mathematics for thousands of years. However, the equals sign did not appear in any text until 1557.

An **equation** expresses the equality of two mathematical expressions. These expressions can be either numerical or variable expressions.

$$\left. \begin{array}{l} 5 + 4 = 9 \\ 3x + 13 = x - 8 \\ y^2 + 4 = 6y + 1 \\ x = -3 \end{array} \right\} \text{Equations}$$

In the equation at the right, if the variable is replaced by 4, the equation is true.

$x + 3 = 7$
$4 + 3 = 7$ A true equation

If the variable is replaced by 6, the equation is false.

$6 + 3 = 7$ A false equation

A **solution of an equation** is a number that, when substituted for the variable, results in a true equation. 4 is a solution of the equation $x + 3 = 7$. 6 is not a solution of the equation $x + 3 = 7$.

HOW TO   Is -2 a solution of the equation $-2x + 1 = 2x + 9$?

$$\begin{array}{c|c} -2x + 1 = 2x + 9 \\ \hline -2(-2) + 1 & 2(-2) + 9 \\ 4 + 1 & -4 + 9 \\ 5 = 5 \end{array}$$

• Replace the variable by the given number.

• Evaluate the numerical expressions.

• Compare the results. If the results are equal, the given number is a solution. If the results are not equal, the given number is not a solution.

Yes, -2 is a solution of the equation $-2x + 1 = 2x + 9$.

Example 1   Is $\frac{1}{2}$ a solution of $2x(x + 2) = 3x + 1$?

Solution

$$\begin{array}{c|c} 2x(x + 2) = 3x + 1 \\ \hline 2\left(\frac{1}{2}\right)\left(\frac{1}{2} + 2\right) & 3\left(\frac{1}{2}\right) + 1 \\ 2\left(\frac{1}{2}\right)\left(\frac{5}{2}\right) & \frac{3}{2} + 1 \\ \frac{5}{2} = \frac{5}{2} \end{array}$$

Yes, $\frac{1}{2}$ is a solution.

You Try It 1   Is -2 a solution of $x(x + 3) = 4x + 6$?

Your solution   Yes

Solution on p. S28

New Vocabulary
equation
solution of an equation

New Symbols
$\neq$ (is not equal to)

Discuss the Concepts
1. What is a solution of an equation?
2. How can we determine whether a number is a solution of an equation?

Concept Check
Label each of the following as either an expression or an equation.
a. $3x + 7 = 9$ Equation
b. $3x + 7$ Expression
c. $4 - 6(y + 5)$ Expression
d. $a + b = 8$ Equation
e. $a + b - 8$ Expression

Optional Student Activity
Which numbers in parentheses are solutions of the equation?
1. $x^2 - 7x = 8$
   $(-1, 2, 1, 8)$ $-1, 8$
2. $y^2 = 2y + 15$
   $(-5, -3, 5, 3)$ $-3, 5$
3. $2x(x + 3) = x + 12$
   $\left(-4, -1\frac{1}{2}, 1\frac{1}{2}, 4\right)$ $-4, 1\frac{1}{2}$
4. $2x^2 - 6x = 5x - 15$
   $\left(-3, -2\frac{1}{2}, 2\frac{1}{2}, 3\right)$ $2\frac{1}{2}, 3$

In-Class Examples (Objective 11.2A)
1. Is -2 a solution of $2x + 7 = 3$?   Yes
2. Is -3 a solution of $-9 - 2x = 9 + 4x$?   Yes
3. Is -2 a solution of $3x(x - 2) = x^2 + 15$?   No

Example 2 Is 5 a solution of
$(x - 2)^2 = x^2 - 4x + 2$?

You Try It 2 Is -3 a solution of
$x^2 - x = 3x + 7$?

Solution

$$
\begin{array}{c|c}
(x - 2)^2 = x^2 - 4x + 2 & \\
\hline
(5 - 2)^2 & 5^2 - 4(5) + 2 \\
3^2 & 25 - 4(5) + 2 \\
9 & 25 - 20 + 2 \\
 & 25 + (-20) + 2 \\
9 \neq 7 & (\neq \text{ means "is not} \\
 & \text{equal to")}
\end{array}
$$

No, 5 is not a solution.

Your solution No

Solution on p. S28

Objective 11.2B

New Vocabulary
solving an equation

Properties to Review
Addition Property of Zero   [1.2A]

New Properties
Addition Property of Equations

Discuss the Concepts
What is the solution of the equation $x = 9$? Use the answer to explain why the goal in solving the equations in this objective is to get the variable alone on one side of the equation.

Concept Check
Which of the following are equations of the form $x + a = b$?
a. $y + 7.8 = -9.2$ Yes
b. $0.3 = z + 1.4$ Yes
c. $-9 = 3d$ No
d. $-8 + n = -5.6$ Yes

Optional Student Activity
Make equations using the numbers 4, 8, and 12 to fill the boxes in

$$x + \square = \square - \square$$

a. What is the largest number solution possible?   0
b. What is the smallest number solution possible? -16

Objective B   **To solve an equation of the form $x + a = b$**

A solution of an equation is a number that, when substituted for the variable, results in a true equation. The phrase **solving an equation** means finding a solution of the equation.

The simplest equation to solve is an equation of the form *variable = constant*. The constant is the solution of the equation.

If $x = 7$, then 7 is the solution of the equation because $7 = 7$ is a true equation.

In solving an equation of the form $x + a = b$, the goal is to simplify the given equation to one of the form *variable = constant*. The Addition Properties that follow are used to simplify equations to this form.

> **Addition Property of Zero**
> The sum of a term and zero is the term.
> $a + 0 = a \qquad 0 + a = a$

> **Addition Property of Equations**
> If a, b, and c are algebraic expressions, then the equations $a = b$ and $a + c = b + c$ have the same solutions.

The Addition Property of Equations states that the same quantity can be added to each side of an equation without changing the solution of the equation.

The Addition Property of Equations is used to rewrite an equation in the form *variable = constant*. Remove a term from one side of the equation by adding the opposite of that term to each side of the equation.

In-Class Examples (Objective 11.2B)

Solve.
1. $x + 6 = 9$ 3
2. $4 + n = 10$ 6
3. $y - 2 = 14$ 16
4. $w - 2 = -6$ -4
5. $d + 3 = -8$ -11

6. $\dfrac{3}{10} + x = -\dfrac{7}{10}$ -1

TAKE NOTE

Always check the solution to an equation.

Check: $x - 7 = -2$

$$\frac{5 - 7 \mid -2}{-2 = -2 \text{ True}}$$

HOW TO   Solve: $x - 7 = -2$

$x - 7 = -2$

$x - 7 + 7 = -2 + 7$

$x + 0 = 5$

$x = 5$

The solution is 5.

- The goal is to simplify the equation to one of the form *variable = constant*.

- Add the opposite of the constant term -7 to each side of the equation. After we simplify and use the Addition Property of Zero, the equation will be in the form *variable = constant*.

Because subtraction is defined in terms of addition, the Addition Property of Equations allows the same number to be subtracted from each side of an equation.

Study Tip

When we suggest that you check a solution, substitute the solution into the original equation. For instance,

$$x + 8 = 5$$

$$\frac{-3 + 8 \mid 5}{5 = 5}$$

The solution checks.

HOW TO   Solve: $x + 8 = 5$

$x + 8 = 5$

$x + 8 - 8 = 5 - 8$

$x + 0 = -3$

$x = -3$

- The goal is to simplify the equation to one of the form *variable = constant*.

- Add the opposite of the constant term 8 to each side of the equation. This procedure is equivalent to subtracting 8 from each side of the equation.

The solution is -3. You should check this solution.

Example 3   Solve: $4 + m = -2$

Solution

$4 + m = -2$

$4 - 4 + m = -2 - 4$ • Subtract 4 from

$0 + m = -6$ each side.

$m = -6$

The solution is -6.

You Try It 3   Solve: $-2 + y = -5$

Your solution -3

Example 4   Solve: $3 = y - 2$

Solution

$3 = y - 2$

$3 + 2 = y - 2 + 2$ • Add 2 to each

$5 = y + 0$ side.

$5 = y$

The solution is 5.

You Try It 4   Solve: $7 = y + 8$

Your solution -1

Example 5   Solve: $\frac{2}{7} = \frac{5}{7} + t$

Solution

$$\frac{2}{7} = \frac{5}{7} + t$$

$$\frac{2}{7} - \frac{5}{7} = \frac{5}{7} - \frac{5}{7} + t$$ • Subtract $\frac{5}{7}$ from each side.

$$-\frac{3}{7} = 0 + t$$

$$-\frac{3}{7} = t$$

The solution is $-\frac{3}{7}$.

You Try It 5   Solve: $\frac{1}{5} = z + \frac{4}{5}$

Your solution $-\frac{3}{5}$

Solutions on p. S28

Objective 11.2C

Vocabulary to Review
reciprocal   [2.7A]

Properties to Review
Multiplication Property of One
[1.4A]

New Properties
Multiplication Property of
  Reciprocals
Multiplication Property of
  Equations

Concept Check
Which are equivalent equations?

a. $5x = -20$

b. $-2x = 8$

c. $24 = 6x$

d. $\dfrac{3}{4}x = -3$

e. $-1 = -\dfrac{1}{4}x$

f. $x = 4$

g. $x = -4$

a, b, d, and g are equivalent
equations. c, e, and f are
equivalent equations.

Optional Student Activity
Match each numbered equation
with a lettered question that can
be used to solve the equation.

1. $x + 3 = 8$ d

2. $x - 5 = 20$ b

3. $4x = 16$ a

4. $\dfrac{x}{7} = 1$ e

5. $99 = -9x$ c

a. 4 times what number is equal
to 16?

b. What number minus 5 is equal
to 20?

c. 99 is equal to −9 times what
number?

d. What number plus 3 is equal
to 8?

e. What number divided by 7 is
equal to 1?

Objective C   **To solve an equation of the form $ax = b$**

In solving an equation of the form $ax = b$, the goal is to simplify the given equation to one of the form *variable = constant*. The Multiplication Properties that follow are used to simplify equations to this form.

Multiplication Property of Reciprocals

The product of a nonzero term and its reciprocal equals 1.

Because $a = \dfrac{a}{1}$, the reciprocal of a is $\dfrac{1}{a}$. $a\left(\dfrac{1}{a}\right) = 1$ $\dfrac{1}{a}(a) = 1$

The reciprocal of $\dfrac{a}{b}$ is $\dfrac{b}{a}$. $\left(\dfrac{a}{b}\right)\left(\dfrac{b}{a}\right) = 1$ $\left(\dfrac{b}{a}\right)\left(\dfrac{a}{b}\right) = 1$

Multiplication Property of One

The product of a term and 1 is the term.
   $a \cdot 1 = a$ $1 \cdot a = a$

Multiplication Property of Equations

If a, b, and c are algebraic expressions and $c \neq 0$, then the equation $a = b$ has the same solutions as the equation $ac = bc$.

The Multiplication Property of Equations states that each side of an equation can be multiplied by the same nonzero number without changing the solutions of the equation.

Recall that the goal of solving an equation is to rewrite the equation in the form *variable = constant*. The Multiplication Property of Equations is used to rewrite an equation in this form by multiplying each side of the equation by the reciprocal of the coefficient.

HOW TO   Solve: $\dfrac{2}{3}x = 8$

$\dfrac{2}{3}x = 8$

$\left(\dfrac{3}{2}\right)\left(\dfrac{2}{3}\right)x = \left(\dfrac{3}{2}\right)8$ • Multiply each side of the equation by $\dfrac{3}{2}$, the

$1 \cdot x = 12$ reciprocal of $\dfrac{2}{3}$. After simplifying, the equation

$x = 12$ will be in the form *variable = constant*.

Check: $\dfrac{2}{3}x = 8$

$\dfrac{2}{3}(12) \;\big|\; 8$

$8 = 8$

The solution is 12.

In-Class Examples (Objective 11.2C)
Solve.

1. $6z = -30$ −5

2. $-3x = 21$ −7

3. $-8x = -32$ 4

4. $63 = 9y$ 7

5. $-\dfrac{1}{4}x = -4$ 16

6. $-\dfrac{m}{3} = 3$ −9

Because division is defined in terms of multiplication, the Multiplication Property of Equations allows each side of an equation to be divided by the same nonzero quantity.

Study Tip

When we suggest that you check a solution, substitute the solution into the original equation. For instance,

$$\begin{array}{c|c} -4x = 24 \\ \hline -4(-6) & 24 \\ 24 = 24 \end{array}$$

The solution checks.

HOW TO Solve: $-4x = 24$

$$-4x = 24$$

$$\frac{-4x}{-4} = \frac{24}{-4}$$

$$1x = -6$$

$$x = -6$$

- The goal is to rewrite the equation in the form *variable = constant*.

- Multiply each side of the equation by the reciprocal of -4. This is equivalent to dividing each side of the equation by -4. Then simplify.

The solution is -6. You should check this solution.

In using the Multiplication Property of Equations, it is usually easier to multiply each side of the equation by the reciprocal of the coefficient when the coefficient is a fraction. Divide each side of the equation by the coefficient when the coefficient is an integer or a decimal.

Example 6 Solve: $-2x = 6$

Solution

$$-2x = 6$$

$$\frac{-2x}{-2} = \frac{6}{-2}$$ • Divide each side by -2.

$$1x = -3$$

$$x = -3$$

The solution is -3.

You Try It 6 Solve: $4z = -20$

Your solution -5

Example 7 Solve: $-9 = \frac{3}{4}y$

Solution

$$-9 = \frac{3}{4}y$$

$$\left(\frac{4}{3}\right)(-9) = \left(\frac{4}{3}\right)\left(\frac{3}{4}y\right)$$ • Multiply each side by $\frac{4}{3}$.

$$-12 = 1y$$

$$-12 = y$$

The solution is -12.

You Try It 7 Solve: $8 = \frac{2}{5}n$

Your solution 20

Example 8 Solve: $6z - 8z = -5$

Solution

$$6z - 8z = -5$$

$$-2z = -5$$ • Combine like terms.

$$\frac{-2z}{-2} = \frac{-5}{-2}$$ • Divide each side by -2.

$$1z = \frac{5}{2}$$

$$z = \frac{5}{2} = 2\frac{1}{2}$$

The solution is $2\frac{1}{2}$.

You Try It 8 Solve: $\frac{2}{3}t - \frac{1}{3}t = -2$

Your solution -6

Solutions on p. S29

Objective 11.2D

New Vocabulary
formula

Optional Student Activity

Match each numbered problem with a lettered equation. Then solve the problem.

1. Since 9 A.M., the temperature has risen 16°F. It is now 4°F. What was the temperature at 9 A.M.?   c. −12°F

2. There are 16 members in the Math Club. After graduation next month, there will be 4 members in the club. How many members are graduating next month?

   a. 12 members

3. Your roommate, who borrowed $4 from you last week, repays you the $4. You put it in your pocket. You now have $16 in your pocket. How much money was in your pocket before you added the $4 to it?   d. $12

4. A couple has given their child the same allowance every week for 4 weeks. The child has saved it all and now has $16. What amount was given to the child each week?

   b. $4

a. $16 - x = 4$

b. $4x = 16$

c. $x + 16 = 4$

d. $x + 4 = 16$

Objective D To solve application problems using formulas

A **formula** is an equation that expresses a relationship among variables. Formulas are used in the examples below.

Example 9

An accountant for a greeting card store found that the weekly profit for the store was $1700 and that the total amount spent during the week was $2400. Use the formula $P = R - C$, where P is the profit, R is the revenue, and C is the amount spent, to find the revenue for the week.

Strategy
To find the revenue for the week, replace the variables P and C in the formula by the given values, and solve for R.

Solution
$$P = R - C$$
$$1700 = R - 2400$$
$$1700 + 2400 = R - 2400 + 2400$$
$$4100 = R + 0$$
$$4100 = R$$

The revenue for the week was $4100.

You Try It 9

A clothing store's sale price for a pair of slacks is $44. This is a discount of $16 off the regular price. Use the formula $S = R - D$, where S is the sale price, R is the regular price, and D is the discount, to find the regular price.

Your strategy

Your solution
$60

Example 10

A store manager uses the formula $S = R - d \cdot R$, where S is the sale price, R is the regular price, and d is the discount rate. During a clearance sale, all items are discounted 20%. Find the regular price of a jacket that is on sale for $120.

Strategy
To find the regular price of the jacket, replace the variables S and d in the formula by the given values, and solve for R.

Solution
$$S = R - d \cdot R$$
$$120 = R - 0.20R$$
$$120 = 0.80R$$
$$\frac{120}{0.80} = \frac{0.80R}{0.80}$$
$$150 = R$$

$\bullet\ R - 0.20R = 1R - 0.20R$

The regular price of the jacket is $150.

You Try It 10

Find the monthly payment when the total amount paid on a loan is $6840 and the loan is paid off in 24 months. Use the formula $A = MN$, where A is the total amount paid on a loan, M is the monthly payment, and N is the number of monthly payments.

Your strategy

Your solution
$285

Solutions on p. S29

In-Class Examples (Objective 11.2D)

1. Use the formula $P = R - C$, where P is the profit, R is the revenue, and C is the amount spent during one week, to find the revenue during a week in which the profit for the week was $1200 and the amount spent was $975. $2175

2. Use the formula $L = PN$, where L is the loan amount, P is the monthly payment, and N is the number of months, to find the monthly payment on a loan of $6400 that is to be repaid in 40 equal payments. $160

11.2 Exercises

Objective A To determine whether a given number is a solution of an equation

1. Is -3 a solution of $2x + 9 = 3$?
Yes

2. Is -2 a solution of $5x + 7 = 12$?
No

3. Is 2 a solution of $4 - 2x = 8$?
No

4. Is 4 a solution of $5 - 2x = 4x$?
No

5. Is 3 a solution of $3x - 2 = x + 4$?
Yes

6. Is 2 a solution of $4x + 8 = 4 - 2x$?
No

7. Is 3 a solution of $x^2 - 5x + 1 = 10 - 5x$?
Yes

8. Is -5 a solution of $x^2 - 3x - 1 = 9 - 6x$?
Yes

9. Is -1 a solution of $2x(x - 1) = 3 - x$?
Yes

10. Is 2 a solution of $3x(x - 3) = x - 8$?
Yes

11. Is 2 a solution of $x(x - 2) = x^2 - 4$?
Yes

12. Is -4 a solution of $x(x + 4) = x^2 + 16$?
No

13. Is $-\frac{2}{3}$ a solution of $3x + 6 = 4$?
Yes

14. Is $\frac{1}{2}$ a solution of $2x - 7 = -3$?
No

15. Is $\frac{1}{4}$ a solution of $2x - 3 = 1 - 14x$?
Yes

16. Is $-\frac{1}{3}$ a solution of $5x - 2 = 1 - 2x$?
No

17. Is $\frac{3}{4}$ a solution of $3x(x - 2) = x - 4$?
No

18. Is $\frac{2}{5}$ a solution of $5x(x + 1) = x + 3$?
No

19. Is 1.32 a solution of $x^2 - 3x = -0.8776 - x$?
No

20. Is -1.9 a solution of $x^2 - 3x = x + 3.8$?
No

21. Is 1.05 a solution of $x^2 + 3x = x(x + 3)$?
Yes

Objective B To solve an equation of the form $x + a = b$

For Exercises 22 to 57, solve.

22. $x + 3 = 9$
6

23. $x + 7 = 5$
-2

24. $y - 6 = 16$
22

25. $z - 4 = 10$
14

26. $3 + n = 4$
1

27. $6 + x = 8$
2

28. $z + 7 = 2$
-5

29. $w + 9 = 5$
-4

Section 11.2

Suggested Assignment
Exercises 1–21, every other odd
Exercises 23–105, odds
More challenging problems:
    Exercises 106–110

Quick Quiz (Objective 11.2A)
1. Is -1 a solution of $3x + 5 = 8$?   No
2. Is 3 a solution of $x^2 - x - 6 = 0$?   Yes
3. Is 6 a solution of $x^2 + 6x + 9 = (x + 3)^2$?   Yes

30. $x - 3 = -7$
-4

31. $m - 4 = -9$
-5

32. $y + 6 = 6$
0

33. $t - 3 = -3$
0

34. $v - 7 = -4$
3

35. $x - 3 = -1$
2

36. $1 + x = 0$
-1

37. $3 + y = 0$
-3

38. $x - 10 = 5$
15

39. $y - 7 = 3$
10

40. $x + 4 = -7$
-11

41. $t - 3 = -8$
-5

42. $w + 5 = -5$
-10

43. $z + 6 = -6$
-12

44. $x + 7 = -8$
-15

45. $x + 2 = -5$
-7

46. $x + \dfrac{1}{2} = -\dfrac{1}{2}$
-1

47. $x - \dfrac{5}{6} = -\dfrac{1}{6}$
$\dfrac{2}{3}$

48. $y + \dfrac{7}{11} = -\dfrac{3}{11}$
$-\dfrac{10}{11}$

49. $\dfrac{2}{5} + x = -\dfrac{3}{5}$
-1

50. $\dfrac{7}{8} + y = -\dfrac{1}{8}$
-1

51. $\dfrac{1}{3} + x = \dfrac{2}{3}$
$\dfrac{1}{3}$

52. $x + \dfrac{1}{2} = -\dfrac{1}{3}$
$-\dfrac{5}{6}$

53. $y + \dfrac{3}{8} = \dfrac{1}{4}$
$-\dfrac{1}{8}$

54. $y + \dfrac{2}{3} = -\dfrac{3}{8}$
$-1\dfrac{1}{24}$

55. $t + \dfrac{1}{4} = -\dfrac{1}{2}$
$-\dfrac{3}{4}$

56. $x + \dfrac{1}{3} = \dfrac{5}{12}$
$\dfrac{1}{12}$

57. $y + \dfrac{2}{3} = -\dfrac{5}{12}$
$-1\dfrac{1}{12}$

Objective C   To solve an equation of the form $ax = b$

For Exercises 58 to 93, solve.

58. $3y = 12$
4

59. $5x = 30$
6

60. $5z = -20$
-4

61. $3z = -27$
-9

62. $-2x = 6$
-3

63. $-4t = 20$
-5

64. $-5x = -40$
8

65. $-2y = -28$
14

66. $40 = 8x$
5

67. $24 = 3y$
8

68. $-24 = 4x$
-6

69. $-21 = 7y$
-3

70. $\dfrac{x}{3} = 5$
15

71. $\dfrac{y}{2} = 10$
20

72. $\dfrac{n}{4} = -2$
-8

73. $\dfrac{y}{7} = -3$
-21

Quick Quiz (Objective 11.2B)

Solve.

1. $c - 3 = 7$ 10

2. $m - 3 = -9$ -6

3. $w + 3 = -3$ -6

4. $\dfrac{1}{2} + x = \dfrac{5}{2}$ 2

74. $-\dfrac{x}{4} = 1$

−4

75. $-\dfrac{y}{3} = 5$

−15

76. $\dfrac{2}{3}w = 4$

6

77. $\dfrac{5}{8}x = 10$

16

78. $\dfrac{3}{4}v = -3$

−4

79. $\dfrac{2}{7}x = -12$

−42

80. $-\dfrac{1}{3}x = -2$

6

81. $-\dfrac{1}{5}y = -3$

15

82. $\dfrac{3}{8}x = -24$

−64

83. $\dfrac{5}{12}y = -16$

$-38\dfrac{2}{5}$

84. $-4 = -\dfrac{2}{3}z$

6

85. $-8 = -\dfrac{5}{6}x$

$9\dfrac{3}{5}$

86. $-12 = -\dfrac{3}{8}y$

32

87. $-9 = \dfrac{5}{6}t$

$-10\dfrac{4}{5}$

88. $\dfrac{2}{3}x = -\dfrac{2}{7}$

$-\dfrac{3}{7}$

89. $\dfrac{3}{7}y = \dfrac{5}{6}$

$1\dfrac{17}{18}$

90. $4x - 2x = 7$

$3\dfrac{1}{2}$

91. $3a - 6a = 8$

$-2\dfrac{2}{3}$

92. $\dfrac{4}{5}m - \dfrac{1}{5}m = 9$

15

93. $\dfrac{1}{3}b - \dfrac{2}{3}b = -1$

3

> **Objective D**   **To solve application problems using formulas**

Investments   In Exercises 94 to 97, use the formula $A = P + I$, where A is the value of the investment after 1 year, P is the original investment, and I is the increase in value of the investment.

94. The value of an investment in a high-tech company after 1 year was $17,700. The increase in value during the year was $2700. Find the amount of the original investment.
$15,000

95. The value of an investment in a software company after 1 year was $26,440. The increase in value during the year was $2830. Find the amount of the original investment.
$23,610

96. The original investment in a mutual fund was $8000. The value of the mutual fund after 1 year was $11,420. Find the increase in value of the investment.
$3420

97. The original investment in a money market fund was $7500. The value of the mutual fund after 1 year was $8690. Find the increase in value of the investment.
$1190

Fuel Efficiency   In Exercises 98 to 101, use the formula $D = M \cdot G$, where D is the distance, M is the miles per gallon, and G is the number of gallons. Round to the nearest tenth.

98. Julio, a sales executive, averages 28 mi/gal on a 621-mile trip. Find the number of gallons of gasoline used on the trip.
22.2 gal

Quick Quiz (Objective 11.2C)

Solve.

1. $-4y = -32$ 8

2. $-20 = 5c$ −4

3. $-\dfrac{1}{3}x = -2$ 6

4. $-\dfrac{x}{4} = 2$ −8

99. Over a 3-day weekend, you take a 592-mile trip. If you average 32 mi/gal on the trip, how many gallons of gasoline did you use?
18.5 gal

100. The manufacturer of a subcompact car estimates that the car can travel 560 mi on a 15-gallon tank of gas. Find the miles per gallon.
37.3 mi/gal

101. You estimate that your car can travel 410 mi on 12 gal of gasoline. Find the miles per gallon.
34.2 mi/gal

Markup   In Exercises 102 and 103, use the formula $S = C + M$, where S is the selling price, C is the cost, and M is the markup.

102. A computer store sells a computer for $2240. The computer has a markup of $420. Find the cost of the computer.
$1820

103. A toy store buys stuffed animals for $23.50 and sells them for $39.80. Find the markup on each stuffed animal.
$16.30

Markup   In Exercises 104 and 105, use the formula $S = C + R \cdot C$, where S is the selling price, C is cost, and R is the markup rate.

104. A store manager uses a markup rate of 24% on all appliances. Find the cost of a blender that sells for $77.50.
$62.50

105. A music store uses a markup rate of 30%. Find the cost of a compact disc that sells for $18.85.
$14.50

APPLYING THE CONCEPTS

106. Write out the steps for solving the equation $x - 3 = -5$. Identify each property of real numbers and each property of equations as you use it.
The complete solution is in the Solutions Manual.

107. Write out the steps for solving the equation $\frac{3}{4}x = 6$. Identify each property of real numbers and each property of equations as you use it.
The complete solution is in the Solutions Manual.

108. Is 2 a solution of $x = x + 4$? Try -2, 0, 3, 6, and 10. Do you think there is a solution of this equation? Why or why not?

109. Write an equation of the form $x + a = b$ that has -4 as its solution.
Answers will vary. For example, $x + 5 = 1$.

110. Write an equation of the form $a - x = b$ that has -2 as its solution.
Answers will vary. For example, $3 - x = 5$.

111. **a.** In your own words, state the Addition Property of Equations.
b. In your own words, state the Multiplication Property of Equations.

Answers to Writing Exercises

108. No, none of the numbers 2, -2, 0, 3, 6, or 10 is a solution of the equation. There is no solution of the equation because there is no number that is equal to itself plus 4.

111a. Students should rephrase the Addition Property of Equations (the same number or variable expression can be added to each side of an equation without changing the solution of the equation).

b. Students should rephrase the Multiplication Property of Equations (each side of an equation can be multiplied by the same nonzero number without changing the solution of the equation).

Quick Quiz (Objective 11.2D)

1. Use the formula $P = R - C$, where P is the profit, R is the revenue, and C is the amount spent during one week, to find the revenue for a week in which the profit was $4800 and the amount spent was $11,200. $16,000

2. Use the formula $L = PN$, where L is the loan amount, P is the monthly payment, and N is the number of months, to find the number of months in which a $3600 loan is repaid. The monthly payment on the loan is $150.   24 months

11.3 General Equations: Part I

Objective A   To solve an equation of the form $ax + b = c$

To solve an equation of the form $ax + b = c$, it is necessary to use both the Addition and the Multiplication Properties to simplify the equation to one of the form *variable = constant*.

Point of Interest

Evariste Galois, despite being killed in a duel at the age of 21, made significant contributions to solving equations. In fact, there is a branch of mathematics called Galois theory, showing what kinds of equations can be solved and what kinds cannot.

HOW TO   Solve: $\frac{x}{4} - 1 = 3$

$\frac{x}{4} - 1 = 3$ • The goal is to simplify the equation to one of the form *variable = constant*.

$\frac{x}{4} - 1 + 1 = 3 + 1$ • Add the opposite of the constant term -1 to each side of the equation. Then simplify (Addition Properties).

$\frac{x}{4} + 0 = 4$

$\frac{x}{4} = 4$

$4 \cdot \frac{x}{4} = 4 \cdot 4$ • Multiply each side of the equation by the reciprocal of the numerical coefficient of the variable term. Then simplify (Multiplication Properties).

$1x = 16$

$x = 16$

The solution is 16.   • Write the solution.

TAKE NOTE

$\frac{x}{4} = \frac{1}{4}x$

The reciprocal of $\frac{1}{4}$ is 4.

Example 1   Solve: $3x + 7 = 2$

Solution
$3x + 7 = 2$
$3x + 7 - 7 = 2 - 7$ • Subtract 7 from
$3x = -5$ each side.
$\frac{3x}{3} = \frac{-5}{3}$ • Divide each side by 3.
$x = -\frac{5}{3} = -1\frac{2}{3}$

The solution is $-1\frac{2}{3}$.

You Try It 1   Solve: $5x + 8 = 6$

Your solution $-\frac{2}{5}$

Example 2   Solve: $5 - x = 6$

Solution
$5 - x = 6$
$5 - 5 - x = 6 - 5$ • Subtract 5 from
$-x = 1$ each side.
$(-1)(-x) = (-1) \cdot 1$ • Multiply each
$x = -1$ side by -1.

The solution is -1.

You Try It 2   Solve: $7 - x = 3$

Your solution   4

Solutions on p. S29

Objective 11.3A

Properties to Review
Addition Property of Equations [11.2B]
Multiplication Property of Equations [11.2C]

Discuss the Concepts
Explain the steps you would take to solve each equation.
a. $5y + 1 = 11$
b. $2z - 9 = 11$
c. $12 = 2 + 5a$
d. $-7v + 6 = -8$
e. $8 - 5x = -12$
f. $9 = 15 - 2y$

Concept Check
Match each equation with the first step in solving that equation.
1. $3x - 7 = 5$ a
2. $4x + 7 = -5$ c
3. $7x - 5 = 2$ b
4. $-7x + 5 = -2$ d
a. Add 7 to each side.
b. Add 5 to each side.
c. Subtract 7 from each side.
d. Subtract 5 from each side.

Optional Student Activity
1. Create a three-question quiz on the material in this objective. Provide solutions for all three questions.
2. Make up an equation of the form:
    a. $x + a = b$ that has -5 as a solution
    b. $ax = b$ that has -4 as a solution
    c. $ax + b = c$ that has -2 as a solution

In-Class Examples (Objective 11.3A)
Solve.
1. $2x + 3 = 11$ 4
2. $3z - 1 = 8$ 3
3. $6w + 4 = -2$ -1
4. $7 - n = 2$ 5
5. $2y + 10 = 0$ -5
6. $-7x + 8 = -13$ 3

Objective 11.3B

New Formula

$F = 1.8C + 32$

Optional Student Activity

Two people decide to open a business to recondition toner cartridges for copy machines. They rent a building for $7000 per year and estimate that the building maintenance, taxes, and insurance will cost $6500 per year. Each person wants to make $12 per hour in the first year and will work 10 hours per day for 260 days of the year. Assume that it costs $28 to restore a cartridge and that they can sell the restored cartridge for $45.

a. How many cartridges must they restore and sell annually to break even, not including the hourly wage they wish to earn?
Approximately 794 cartridges

b. How many cartridges must they restore and sell annually just to earn the hourly wage they desire?
Approximately 3671 cartridges

c. Suppose the entrepreneurs are successful in their business and are restoring and selling 25 cartridges each day of the 260 days they are open. What would be their hourly wage for the year?
Approximately $18.65 per hour

Objective B **To solve application problems using formulas**

Anders Celsius

The Fahrenheit temperature scale was devised by Daniel Gabriel Fahrenheit (1686–1736), a German physicist and maker of scientific instruments. He invented the mercury thermometer in 1714. On the Fahrenheit scale, the temperature at which water freezes is 32°F, and the temperature at which water boils is 212°F. *Note:* The small raised circle is the symbol for degrees, and the capital F is for Fahrenheit. The Fahrenheit scale is used only in the United States.

In the metric system, temperature is measured on the Celsius scale. The Celsius temperature scale was devised by Anders Celsius (1701–1744), a Swedish astronomer. On the Celsius scale, the temperature at which water freezes is 0°C, and the temperature at which water boils is 100°C. *Note:* The small raised circle is the symbol for degrees, and the capital C is for Celsius.

On both the Celsius scale and the Fahrenheit scale, temperatures below 0° are negative numbers.

The relationship between Celsius temperatures and Fahrenheit temperatures is given by the formula

$$F = 1.8C + 32$$

where F represents degrees Fahrenheit and C represents degrees Celsius.

Integrating Technology

You can check the solution to this equation using a calculator. Evaluate the right side of the equation after substituting 37 for C. Enter

1.8 × 37 + 32 =

The display reads 98.6, the given Fahrenheit temperature. The solution checks.

HOW TO Normal body temperature is 98.6°F. Convert this temperature to degrees Celsius.

$$F = 1.8C + 32$$
$$98.6 = 1.8C + 32$$

• Substitute 98.6 for F.

$$98.6 - 32 = 1.8C + 32 - 32$$

• Subtract 32 from each side.

$$66.6 = 1.8C$$

• Combine like terms on each side.

$$\frac{66.6}{1.8} = \frac{1.8C}{1.8}$$

• Divide each side by 1.8.

$$37 = C$$

Normal body temperature is 37°C.

In-Class Examples (Objective 11.3B)

1. Convert −5°F to degrees Celsius. Use the formula $F = 1.8C + 32$, where F is the Fahrenheit temperature and C is the Celsius temperature. Round to the nearest tenth.   −20.6°C

Example 3

Find the Celsius temperature when the Fahrenheit temperature is 212°. Use the formula $F = 1.8C + 32$, where F is the Fahrenheit temperature and C is the Celsius temperature.

Strategy

To find the Celsius temperature, replace the variable F in the formula by the given value and solve for C.

Solution

$$F = 1.8C + 32$$
$$212 = 1.8C + 32$$
$$212 - 32 = 1.8C + 32 - 32$$
$$180 = 1.8C$$
$$\frac{180}{1.8C} = \frac{1.8C}{1.8}$$
$$100 = C$$

- Substitute 212 for F.
- Subtract 32 from each side.
- Combine like terms.
- Divide each side by 1.8.

The Celsius temperature is 100°.

You Try It 3

Find the Celsius temperature when the Fahrenheit temperature is −22°. Use the formula $F = 1.8C + 32$, where F is the Fahrenheit temperature and C is the Celsius temperature.

Your strategy

Your solution

−30°C

Example 4

To find the total cost of production, an economist uses the formula $T = U \cdot N + F$, where T is the total cost, U is the cost per unit, N is the number of units made, and F is the fixed cost. Find the number of units made during a week when the total cost was $8000, the cost per unit was $16, and the fixed costs were $2000.

Strategy

To find the number of units made, replace the variables T, U, and F in the formula by the given values and solve for N.

Solution

$$T = U \cdot N + F$$
$$8000 = 16 \cdot N + 2000$$
$$8000 - 2000 = 16 \cdot N + 2000 - 2000$$
$$6000 = 16 \cdot N$$
$$\frac{6000}{16} = \frac{16 \cdot N}{16}$$
$$375 = N$$

The number of units made was 375.

You Try It 4

Find the cost per unit during a week when the total cost was $4500, the number of units produced was 250, and the fixed costs were $1500. Use the formula $T = U \cdot N + F$, where T is the total cost, U is the cost per unit, N is the number of units made, and F is the fixed cost.

Your strategy

Your solution

$12

Solutions on p. S29

Section 11.3

Suggested Assignment
Exercises 1–89, every other odd
Exercises 91–101, odds
Exercise 105
More challenging problems:
    Exercises 103, 104

11.3 Exercises

> **Objective A**   To solve an equation of the form $ax + b = c$

For Exercises 1 to 90, solve.

1. $3x + 5 = 14$
  3

2. $5z + 6 = 31$
  5

3. $2n - 3 = 7$
  5

4. $4y - 4 = 20$
  6

5. $5w + 8 = 3$
  −1

6. $3x + 10 = 1$
  −3

7. $3z - 4 = -16$
  −4

8. $6x - 1 = -13$
  −2

9. $5 + 2x = 7$
  1

10. $12 + 7x = 33$
  3

11. $6 - x = 3$
  3

12. $4 - x = -2$
  6

13. $3 - 4x = 11$
  −2

14. $2 - 3x = 11$
  −3

15. $5 - 4x = 17$
  −3

16. $8 - 6x = 14$
  −1

17. $3x + 6 = 0$
  −2

18. $5x - 20 = 0$
  4

19. $-3x - 4 = -1$
  −1

20. $-7x - 22 = -1$
  −3

21. $12y - 30 = 6$
  3

22. $9b - 7 = 2$
  1

23. $3c + 7 = 4$
  −1

24. $8t + 13 = 5$
  −1

25. $-2x + 11 = -3$
  7

26. $-4x + 15 = -1$
  4

27. $14 - 5x = 4$
  2

28. $7 - 3x = 4$
  1

29. $-8x + 7 = -9$
  2

30. $-7x + 13 = -8$
  3

31. $9x + 13 = 13$
  0

32. $-2x + 7 = 7$
  0

33. $7d - 14 = 0$
  2

34. $5z + 10 = 0$
  −2

35. $4n - 4 = -4$
  0

36. $-13m - 1 = -1$
  0

37. $3x + 5 = 7$
  $\frac{2}{3}$

38. $4x + 6 = 9$
  $\frac{3}{4}$

39. $6x - 1 = 16$
  $2\frac{5}{6}$

40. $12x - 3 = 7$
  $\frac{5}{6}$

Quick Quiz (Objective 11.3A)
Solve.
1. $4x + 7 = 31$ 6
2. $5d - 8 = 27$ 7
3. $7w + 9 = -5$ −2
4. $6 - m = -1$ 7
5. $3y - 33 = 0$ 11
6. $-8x + 11 = -13$ 3

41. $2x - 3 = -8$
$-2\frac{1}{2}$

42. $5x - 3 = -12$
$-1\frac{4}{5}$

43. $-6x + 2 = -7$
$1\frac{1}{2}$

44. $-3x + 9 = -1$
$3\frac{1}{3}$

45. $-2x - 3 = -7$
2

46. $-5x - 7 = -4$
$-\frac{3}{5}$

47. $3x + 8 = 2$
-2

48. $2x - 9 = 8$
$8\frac{1}{2}$

49. $3w - 7 = 0$
$2\frac{1}{3}$

50. $7b - 2 = 0$
$\frac{2}{7}$

51. $-2d + 9 = 12$
$-1\frac{1}{2}$

52. $-7c + 3 = 1$
$\frac{2}{7}$

53. $\frac{1}{2}x - 2 = 3$
10

54. $\frac{1}{3}x + 1 = 4$
9

55. $\frac{3}{5}w - 1 = 2$
5

56. $\frac{2}{5}w + 5 = 6$
$2\frac{1}{2}$

57. $\frac{2}{9}t - 3 = 5$
36

58. $\frac{5}{9}t - 3 = 2$
9

59. $\frac{y}{3} - 6 = -8$
-6

60. $\frac{y}{2} - 2 = 3$
10

61. $\frac{x}{3} - 2 = -5$
-9

62. $\frac{x}{4} - 3 = 5$
32

63. $\frac{5}{8}v + 6 = 3$
$-4\frac{4}{5}$

64. $\frac{2}{3}v - 4 = 3$
$10\frac{1}{2}$

65. $\frac{4}{7}z + 10 = 5$
$-8\frac{3}{4}$

66. $\frac{3}{8}v - 3 = 4$
$18\frac{2}{3}$

67. $\frac{2}{9}x - 3 = 5$
36

68. $\frac{1}{2}x + 3 = -8$
-22

69. $\frac{3}{4}x - 5 = -4$
$1\frac{1}{3}$

70. $\frac{2}{3}x - 5 = -8$
$-4\frac{1}{2}$

71. $1.5x - 0.5 = 2.5$
2

72. $2.5w - 1.3 = 3.7$
2

73. $0.8t + 1.1 = 4.3$
4

74. $0.3v + 2.4 = 1.5$
-3

75. $0.4x - 2.3 = 1.3$
9

76. $1.2t + 6.5 = 2.9$
-3

77. $3.5y - 3.5 = 10.5$
4

78. $1.9x - 1.9 = -1.9$
0

79. $0.32x + 4.2 = 3.2$
-3.125

80. $5x - 3x + 2 = 8$
3

81. $6m + 2m - 3 = 5$
1

82. $4a - 7a - 8 = 4$
−4

83. $3y - 8y - 9 = 6$
−3

84. $x - 4x + 5 = 11$
−2

85. $-2y + y - 3 = 6$
−9

86. $-4y - y - 8 = 12$
−4

87. $0.032x - 0.0194 = 0.139$
4.95

88. $-3.256x + 42.38 = -16.9$
18.206388

89. $6.09x + 17.33 = 16.805$
−0.0862069

90. $1.925x + 32.87 = -16.994$
−25.903377

Objective B **To solve application problems using formulas**

Temperature Conversion In Exercises 91 and 92, use the relationship between Fahrenheit temperature and Celsius temperature, which is given by the formula $F = 1.8C + 32$, where F is the Fahrenheit temperature and C is the Celsius temperature.

91. Find the Celsius temperature when the Fahrenheit temperature is $-40°$.
−40°C

92. Find the Celsius temperature when the Fahrenheit temperature is $72°$. Round to the nearest tenth of a degree.
22.2°C

Physics In Exercises 93 and 94, use the formula $V = V_0 + 32t$, where V is the final velocity of a falling object, V_0 is the starting velocity of a falling object, and t is the time for the object to fall.

93. Find the time required for an object to increase in velocity from 8 ft/s to 472 ft/s.
14.5 s

94. Find the time required for an object to increase in velocity from 16 ft/s to 128 ft/s.
3.5 s

Manufacturing In Exercises 95 and 96, use the formula $T = U \cdot N + F$, where T is the total cost, U is the cost per unit, N is the number of units made, and F is the fixed cost.

95. Find the number of units made during a week when the total cost was $25,000, the cost per unit was $8, and the fixed costs were $5000.
2500 units

Quick Quiz (Objective 11.3B)

1. Convert 16°F to degrees Celsius. Use the formula $F = 1.8C + 32$, where F is the Fahrenheit temperature and C is the Celsius temperature. Round to the nearest tenth. −8.9°C

96. Find the cost per unit during a week when the total cost was $80,000, the total number of units produced was 500, and the fixed costs were $15,000.
$130

Taxes   In Exercises 97 and 98, use the formula $T = I \cdot R + B$, where T is the monthly tax, I is the monthly income, R is the income tax rate, and B is the base monthly tax.

97. The monthly tax that a clerk pays is $476. The clerk's monthly tax rate is 22%, and the base monthly tax is $80. Find the clerk's monthly income.
$1800

98. The monthly tax that Marcy, a teacher, pays is $770. Her monthly income is $3100, and the base monthly tax is $150. Find Marcy's income tax rate. 20%

Compensation   In Exercises 99 to 102, use the formula $M = S \cdot R + B$, where M is the monthly earnings, S is the total sales, R is the commission rate, and B is the base monthly salary.

99. A book representative earns a base monthly salary of $600 plus a 9% commission on total sales. Find the total sales during a month in which the representative earned $3480.
$32,000

100. A sales executive earns a base monthly salary of $1000 plus a 5% commission on total sales. Find the total sales during a month in which the executive earned $2800.
$36,000

101. Miguel earns a base monthly salary of $750. Find his commission rate during a month in which total sales were $42,000 and he earned $2640.
4.5%

102. Tina earns a base monthly salary of $500. Find her commission rate during a month when total sales were $42,500 and her earnings were $3560.
7.2%

APPLYING THE CONCEPTS

103. Explain in your own words the steps you would take to solve the equation $\frac{2}{3}x - 4 = 10$. State the property of real numbers or the property of equations that is used at each step.
The complete solution is in the Solutions Manual.

104. Make up an equation of the form $ax + b = c$ that has -3 as its solution.
Answers will vary. For example, $2x + 5 = -1$.

105. ✎ Does the sentence "Solve $3x + 4(x - 3)$" make sense? Why or why not?

Answers to Writing Exercises

105. No, the sentence "Solve $3x + 4(x - 3)$" does not make sense because

$$3x + 4(x - 3)$$

is an expression, and you cannot solve an expression. You can solve an equation.

Objective 11.4A

Properties to Review

Addition Property of Equations
[11.2B]
Multiplication Property of
Equations [11.2C]

Concept Check

When solving an equation of the form $ax + b = cx + d$, would it always be correct to start by adding or subtracting the variable term

a. on the right side of the equation?   Yes

b. on the left side of the equation?   Yes

c. with the smaller coefficient?   Yes

Optional Student Activity

1. Rick weighs 76 lb plus half his weight. How much does Rick weigh?   152 lb

2. At the local gym, you can either pay $75 for a year's membership and $5 per visit, or pay $150 for a year's membership plus $2 per visit. How many times during the year must you go to the gym for the two plans to be equal in price?   25 visits

3. The gauge on a water tank shows that the tank is $\frac{5}{8}$ full.

   After 18 more gallons are drained from the tank, the

   gauge shows that it is $\frac{1}{4}$ full.

   How many gallons of water were in the tank when it was

   $\frac{1}{2}$ full?   24 gallons

11.4 General Equations: Part II

Objective A   To solve an equation of the form $ax + b = cx + d$

When a variable occurs on each side of an equation, the Addition Properties are used to rewrite the equation so that variable terms are on one side of the equation and constant terms are on the other side of the equation. Then the Multiplication Properties are used to simplify the equation to one of the form *variable = constant*.

HOW TO   Solve: $4x - 6 = 8 - 3x$

$$4x - 6 = 8 - 3x$$

• The goal is to write the equation in the form *variable = constant*.

$$4x + 3x - 6 = 8 - 3x + 3x$$
$$7x - 6 = 8 + 0$$
$$7x - 6 = 8$$

• Add $3x$ to each side of the equation. Then simplify (Addition Properties). Now only one variable term occurs in the equation.

$$7x - 6 + 6 = 8 + 6$$
$$7x + 0 = 14$$
$$7x = 14$$

• Add 6 to each side of the equation. Then simplify (Addition Properties). Now only one constant term occurs in the equation.

$$\frac{7x}{7} = \frac{14}{7}$$
$$1x = 2$$
$$x = 2$$

• Divide each side of the equation by the numerical coefficient of the variable term. Then simplify (Multiplication Properties).

The solution is 2.

• Write the solution.

Study Tip

Always check the solution of an equation. For the equation at the right:

$$4x - 6 = 8 - 3x$$

$4(2) - 6$	$8 - 3(2)$
$8 - 6$	$8 - 6$
$2 = 2$	

The solution checks.

Example 1

Solve: $\frac{2}{9}x - 3 = \frac{7}{9}x + 2$

Solution

$$\frac{2}{9}x - 3 = \frac{7}{9}x + 2$$

$$\frac{2}{9}x - \frac{7}{9}x - 3 = \frac{7}{9}x - \frac{7}{9}x + 2$$

• Subtract $\frac{7}{9}x$ from each side.

$$-\frac{5}{9}x - 3 = 2$$

$$-\frac{5}{9}x - 3 + 3 = 2 + 3$$

• Add 3 to each side.

$$-\frac{5}{9}x = 5$$

$$\left(-\frac{9}{5}\right)\left(-\frac{5}{9}\right)x = \left(-\frac{9}{5}\right)5$$

• Multiply each side by $-\frac{9}{5}$.

$$x = -9$$

The solution is -9.

You Try It 1

Solve: $\frac{1}{5}x - 2 = \frac{2}{5}x + 4$

Your solution
-30

Solution on pp. S29–S30

Solve.

1. $5x + 1 = 3x + 7$ 3
2. $7x - 4 = 8x - 3$ -1
3. $2x - 3 = 15 - 4x$ 3
4. $-7 - 3x = 9 - 2x$ -16

Objective B **To solve an equation containing parentheses**

When an equation contains parentheses, one of the steps in solving the equation requires use of the Distributive Property.

$$a(b + c) = ab + ac$$

The Distributive Property is used to rewrite a variable expression without parentheses.

HOW TO Solve: $4(3 + x) - 2 = 2(x - 4)$

$4(3 + x) - 2 = 2(x - 4)$	• The goal is to write the equation in the form *variable = constant*.
$12 + 4x - 2 = 2x - 8$	• Use the Distributive Property to rewrite the equation without parentheses.
$10 + 4x = 2x - 8$	• Combine like terms.
$10 + 4x - 2x = 2x - 2x - 8$	• Use the Addition Property of Equations.
$10 + 2x = -8$	Subtract $2x$ from each side of the equation.
$10 - 10 + 2x = -8 - 10$	• Use the Addition Property of Equations.
$2x = -18$	Subtract 10 from each side of the equation.
$\dfrac{2x}{2} = \dfrac{-18}{2}$	• Use the Multiplication Property of Equations. Divide each side of the equation by the
$x = -9$	numerical coefficient of the variable term.

Check:

$$\begin{array}{c|c} 4(3 + x) - 2 = 2(x - 4) \\ \hline 4[3 + (-9)] - 2 & 2(-9 - 4) \\ 4(-6) - 2 & 2(-13) \\ -24 - 2 & -26 \\ -26 = -26 \end{array}$$

• Check the solution.

A true equation

The solution is -9.

• Write the solution.

The solution to this last equation illustrates the steps involved in solving first-degree equations.

Steps in Solving General First-Degree Equations

1. Use the Distributive Property to remove parentheses.
2. Combine like terms on each side of the equation.
3. Use the Addition Property of Equations to rewrite the equation with only one variable term.
4. Use the Addition Property of Equations to rewrite the equation with only one constant term.
5. Use the Multiplication Property of Equations to rewrite the equation so that the coefficient of the variable is 1.

Ⓟ

Properties to Review
Distributive Property [11.1C]

Discuss the Concepts
What are the steps involved in solving an equation containing parentheses?

Optional Student Activity
Solve.
1. $3(2x - 1) - (6x - 4) = -9$
   No solution
2. $2(5x - 6) - 3(x - 4)$
   $= 7x + 14$ No solution

Optional Student Activity
1. I am thinking of a number. When I subtract 3 from the number and then take 400% of the result, it is equal to the original number. What is the original number?   4
2. If $s = 4x - 3$ and $t = x + 5$, find the value of x for which $s = 2t - 1$.   6
3. The population of the town of Danville increased by 5000 people during the 1990s. In the first decade of the new millennium, the population of Danville decreased by 10%, at which time the town had 3000 more people than it had at the beginning of the 1990s. Find Danville's population at the beginning of the 1990s.
   15,000 people

In-Class Examples (Objective 11.4B)
Solve.
1. $5x + 3(x - 2) = 10$ 2
2. $5 - 4(x + 6) = 21$ -10
3. $2x + 3(x + 1) = 4(x + 7)$ 25
4. $4 - 3(x - 2) = 4(x - 1)$ 2

Example 2

Solve: $3(x + 2) - x = 11$

Solution

$$3(x + 2) - x = 11$$
$$3x + 6 - x = 11 \qquad \bullet \text{ Use the Distributive Property.}$$
$$2x + 6 = 11 \qquad \bullet \text{ Combine like terms on the left side.}$$
$$2x + 6 - 6 = 11 - 6 \qquad \bullet \text{ Use the Addition Property of Equations. Subtract 6 from each side.}$$
$$2x = 5 \qquad \bullet \text{ Combine like terms on each side.}$$
$$\frac{2x}{2} = \frac{5}{2} \qquad \bullet \text{ Use the Multiplication Property. Divide both sides by 2.}$$
$$x = 2\frac{1}{2} \qquad \bullet \text{ The solution checks.}$$

The solution is $2\frac{1}{2}$.

You Try It 2

Solve: $4(x - 1) - x = 5$

Your solution

3

Example 3

Solve: $5x - 2(x - 3) = 6(x - 2)$

Solution

$$5x - 2(x - 3) = 6(x - 2)$$
$$5x - 2x + 6 = 6x - 12 \qquad \bullet \text{ Distributive Property}$$
$$3x + 6 = 6x - 12 \qquad \bullet \text{ Combine like terms.}$$
$$3x - 6x + 6 = 6x - 6x - 12 \qquad \bullet \text{ Subtract } 6x \text{ from each side.}$$
$$-3x + 6 = -12 \qquad \bullet \text{ Combine like terms.}$$
$$-3x + 6 - 6 = -12 - 6 \qquad \bullet \text{ Subtract 6 from each side.}$$
$$-3x = -18 \qquad \bullet \text{ Combine like terms.}$$
$$\frac{-3x}{-3} = \frac{-18}{-3} \qquad \bullet \text{ Divide both sides by } -3.$$
$$x = 6 \qquad \bullet \text{ The solution checks.}$$

The solution is 6.

You Try It 3

Solve: $2x - 7(3x + 1) = 5(5 - 3x)$

Your solution

-8

Solutions on p. S30

11.4 Exercises

Objective A   **To solve an equation of the form *ax + b = cx + d***

Suggested Assignment
Exercises 1–101, every other odd
More challenging problems:
    Exercises 103–106

For Exercises 1 to 54, solve.

1. $6x + 3 = 2x + 5$
$\frac{1}{2}$

2. $7x + 1 = x + 19$
3

3. $3x + 3 = 2x + 2$
-1

4. $6x + 3 = 3x + 6$
1

5. $5x + 4 = x - 12$
-4

6. $3x - 12 = x - 8$
2

7. $7b - 2 = 3b - 6$
-1

8. $2d - 9 = d - 8$
1

9. $9n - 4 = 5n - 20$
-4

10. $8x - 7 = 5x + 8$
5

11. $2x + 1 = 16 - 3x$
3

12. $3x + 2 = -23 - 2x$
-5

13. $5x - 2 = -10 - 3x$
-1

14. $4x - 3 = 7 - x$
2

15. $2x + 7 = 4x + 3$
2

16. $7m - 6 = 10m - 15$
3

17. $c + 4 = 6c - 11$
3

18. $t - 6 = 4t - 21$
5

19. $3x - 7 = x - 7$
0

20. $2x + 6 = 7x + 6$
0

21. $3 - 4x = 5 - 3x$
-2

22. $6 - 2x = 9 - x$
-3

23. $7 + 3x = 9 + 5x$
-1

24. $12 + 5x = 9 - 3x$
$-\frac{3}{8}$

25. $5 + 2y = 7 + 5y$
$-\frac{2}{3}$

26. $9 + z = 2 + 3z$
$3\frac{1}{2}$

27. $8 - 5w = 4 - 6w$
-4

28. $9 - 4x = 11 - 5x$
2

29. $6x + 1 = 3x + 2$
$\frac{1}{3}$

30. $7x + 5 = 4x + 7$
$\frac{2}{3}$

31. $5x + 8 = x + 5$
$-\frac{3}{4}$

32. $9x + 1 = 3x - 4$
$-\frac{5}{6}$

33. $2x - 3 = 6x - 4$
$\frac{1}{4}$

Quick Quiz (Objective 11.4A)
Solve.
1. $4x + 3 = x + 9$ 2
2. $6x - 11 = x - 21$ -2
3. $4x + 5 = 23 - 2x$ 3
4. $9 - 4x = -7 - 2x$ 8

34. $4 - 3x = 4 - 5x$
0

35. $6 - 3x = 6 - 5x$
0

36. $2x + 7 = 4x - 3$
5

37. $6x - 2 = 2x - 9$
$-1\frac{3}{4}$

38. $4x - 7 = -3x + 2$
$1\frac{2}{7}$

39. $6x - 3 = -5x + 8$
1

40. $7y - 5 = 3y + 9$
$3\frac{1}{2}$

41. $-6t - 2 = -8t - 4$
-1

42. $-7w + 2 = 3w - 8$
1

43. $-3 - 4x = 7 - 2x$
-5

44. $-8 + 5x = 8 + 6x$
-16

45. $3 - 7x = -2 + 5x$
$\frac{5}{12}$

46. $3x - 2 = 7 - 5x$
$1\frac{1}{8}$

47. $5x + 8 = 4 - 2x$
$-\frac{4}{7}$

48. $4 - 3x = 6x - 8$
$1\frac{1}{3}$

49. $12z - 9 = 3z + 12$
$2\frac{1}{3}$

50. $4c + 13 = -6c + 9$
$-\frac{2}{5}$

51. $\frac{5}{7}m - 3 = \frac{2}{7}m + 6$
21

52. $\frac{4}{5}x - 1 = \frac{1}{5}x + 5$
10

53. $\frac{3}{7}x + 5 = \frac{5}{7}x - 1$
21

54. $\frac{3}{4}x + 2 = \frac{1}{4}x - 9$
-22

Objective B   **To solve an equation containing parentheses**

For Exercises 55 to 102, solve.

55. $6x + 2(x - 1) = 14$
2

56. $3x + 2(x + 4) = 13$
1

57. $-3 + 4(x + 3) = 5$
-1

58. $8b - 3(b - 5) = 30$
3

59. $6 - 2(d + 4) = 6$
-4

60. $5 - 3(n + 2) = 8$
-3

61. $5 + 7(x + 3) = 20$
$-\frac{6}{7}$

62. $6 - 3(x - 4) = 12$
2

63. $2x + 3(x - 5) = 10$
5

64. $3x - 4(x + 3) = 9$
-21

65. $3(x - 4) + 2x = 3$
3

66. $4 + 3(x - 9) = -12$
$3\frac{2}{3}$

Quick Quiz (Objective 11.4B)
Solve.
1. $2x + 5(x + 1) = -9$ -2
2. $6 - 3(x + 2) = 12$ -4
3. $4x + 2(x + 3) = 3(x - 1)$ -3
4. $3 - 2(x + 1) = 3(x - 3)$ 2

67. $2x - 3(x - 4) = 12$
0

68. $4x - 2(x - 5) = 10$
0

69. $2x + 3(x + 4) = 7$
−1

70. $3(x + 2) + 7 = 12$
$-\dfrac{1}{3}$

71. $3(x - 2) + 5 = 5$
2

72. $4(x - 5) + 7 = 7$
5

73. $3y + 7(y - 2) = 5$
$1\dfrac{9}{10}$

74. $-3z - 3(z - 3) = 3$
1

75. $4b - 2(b + 9) = 8$
13

76. $3x - 6(x - 3) = 9$
3

77. $3x + 5(x - 2) = 10$
$2\dfrac{1}{2}$

78. $3x - 5(x - 1) = -5$
5

79. $3x + 4(x + 2) = 2(x + 9)$
2

80. $5x + 3(x + 4) = 4(x + 2)$
−1

81. $2d - 3(d - 4) = 2(d + 6)$
0

82. $3t - 4(t - 1) = 3(t - 2)$
$2\dfrac{1}{2}$

83. $7 - 2(x - 3) = 3(x - 1)$
$3\dfrac{1}{5}$

84. $4 - 3(x + 2) = 2(x - 4)$
$1\dfrac{1}{5}$

85. $6x - 2(x - 3) = 11(x - 2)$
4

86. $9x - 5(x - 3) = 5(x + 4)$
−5

87. $6c - 3(c + 1) = 5(c + 2)$
$-6\dfrac{1}{2}$

88. $2w - 7(w - 2) = 3(w - 4)$
$3\dfrac{1}{4}$

89. $7 - (x + 1) = 3(x + 3)$
$-\dfrac{3}{4}$

90. $12 + 2(x - 9) = 3(x - 12)$
30

91. $2x - 3(x + 4) = 2(x - 5)$
$-\dfrac{2}{3}$

92. $3x + 2(x - 7) = 7(x - 1)$
$-3\dfrac{1}{2}$

93. $x + 5(x - 4) = 3(x - 8) - 5$
-3

94. $2x - 2(x - 1) = 3(x - 2) + 7$
$\dfrac{1}{3}$

95. $9b - 3(b - 4) = 13 + 2(b - 3)$
$-1\dfrac{1}{4}$

96. $3y - 4(y - 2) = 15 - 3(y - 2)$
$6\dfrac{1}{2}$

97. $3(x - 4) + 3x = 7 - 2(x - 1)$
$2\dfrac{5}{8}$

98. $2(x - 6) + 7x = 5 - 3(x - 2)$
$1\dfrac{11}{12}$

99. $3.67x - 5.3(x - 1.932) = 6.99$
1.9936196

100. $4.06x + 4.7(x + 3.22) = 1.774$
-1.5251142

101. $8.45(z - 10) = 3(z - 3.854)$
13.383119

102. $4(d - 1.99) - 3.92 = 3(d - 1.77)$
6.57

APPLYING THE CONCEPTS

103. If $2x - 2 = 4x + 6$, what is the value of $3x^2$?
48

104. If $3 + 2(4a - 3) = 4$ and $4 - 3(2 - 3b) = 11$, which is larger, a or b?
b

105. Explain what is wrong with the demonstration at the right, which suggests that $5 = 4$.

$5x + 7 = 4x + 7$
$5x + 7 - 7 = 4x + 7 - 7$ • Subtract 7 from each side of the equation.
$5x = 4x$
$\dfrac{5x}{x} = \dfrac{4x}{x}$ • Divide each side of the equation by x.
$5 = 4$

106. The equation $x = x + 1$ has no solution, whereas the solution of the equation $2x + 3 = 3$ is zero. Is there a difference between no solution and a solution of zero? Explain.

11.5 Translating Verbal Expressions into Mathematical Expressions

Objective A   **To translate a verbal expression into a mathematical expression given the variable**

One of the major skills required in applied mathematics is to translate a verbal expression into a mathematical expression. Doing so requires recognizing the verbal phrases that translate into mathematical operations. Following is a partial list of the verbal phrases used to indicate the different mathematical operations.

Addition	more than	5 more than x	$x + 5$
	the sum of	the sum of w and 3	$w + 3$
	the total of	the total of 6 and z	$6 + z$
	increased by	x increased by 7	$x + 7$
Subtraction	less than	5 less than y	$y - 5$
	the difference between	the difference between w and 3	$w - 3$
	decreased by	8 decreased by a	$8 - a$
Multiplication	times	3 times c	$3c$
	the product of	the product of 4 and t	$4t$
	of	two-thirds of v	$\dfrac{2}{3}v$
	twice	twice d	$2d$
Division	divided by	n divided by 3	$\dfrac{n}{3}$
	the quotient of	the quotient of z and 4	$\dfrac{z}{4}$
	the ratio of	the ratio of s to 6	$\dfrac{s}{6}$

Translating phrases that contain the words *sum, difference, product,* and *quotient* can sometimes cause a problem. In the examples at the right, note where the operation symbol is placed.

the *sum* of x and y $\qquad$ $x + y$
the *difference* between x and y $\qquad$ $x - y$
the *product* of x and y $\qquad$ $x \cdot y$
the *quotient* of x and y $\qquad$ $\dfrac{x}{y}$

Note where we place the fraction bar when translating the word *ratio*.

the *ratio* of x to y $\qquad$ $\dfrac{x}{y}$

HOW TO   Translate "the quotient of n and the sum of n and 6" into a mathematical expression.

the *quotient* of n and the *sum* of n and 6 $\qquad$ $\dfrac{n}{n + 6}$

New Vocabulary
See the list of verbal phrases that translate into mathematical operations.

Discuss the Concepts
Is the statement true or false?
1. "Five less than n" can be translated as "$5 - n$."   F
2. A variable expression contains an equals sign.   F
3. The words *quotient* and *ratio* both indicate division.   T
4. The expressions $7y - 8$ and $(7y) - 8$ are equivalent.   T
5. "Four times the difference between x and 3" can be translated as "$4x - 3$."   F

Optional Student Activity
1. Given that x is equal to -3, what is the value of x raised to the $-x$ power? -27
2. Given that y is equal to $3x$ and z is equal to $2y$, find the sum of x, y, and z in terms of x. $10x$
3. At a small college, there are three times as many commuters as boarders and four times as many boarders as faculty members. Using C to represent the number of commuters, write an expression in simplest form for the total number of commuters, boarders, and faculty at the college. $\dfrac{17}{12}C$

In-Class Examples (Objective 11.5A)
Translate into a mathematical expression.

1. y decreased by the quotient of y and 4 $y - \dfrac{y}{4}$

2. the difference between a and the product of -3 and a $a - (-3a)$

3. four times the sum of x and 8 $4(x + 8)$

Example 1   Translate "the sum of 5 and the product of 4 and n" into a mathematical expression.

Solution $5 + 4n$

You Try It 1   Translate "the difference between 8 and twice t" into a mathematical expression.

Your solution $8 - 2t$

Example 2   Translate "the product of 3 and the difference between z and 4" into a mathematical expression.

Solution $3(z - 4)$

You Try It 2   Translate "the quotient of 5 and the product of 7 and x" into a mathematical expression.

Your solution $\dfrac{5}{7x}$

Solutions on p. S30

Objective 11.5B

Objective B   **To translate a verbal expression into a mathematical expression by assigning the variable**

In most applications that involve translating phrases into mathematical expressions, the variable to be used is not given. To translate these phrases, we must assign a variable to the unknown quantity before writing the mathematical expression.

HOW TO   Translate "the difference between seven and twice a number" into a mathematical expression.

The difference between seven and twice a number
- Identify the phrases that indicate the mathematical operations.

The unknown number: n
- Assign a variable to one of the unknown quantities.

Twice the number: $2n$
- Use the assigned variable to write an expression for any other unknown quantity.

$7 - 2n$
- Use the identified operations to write the mathematical expression.

Example 3   Translate "the total of a number and the square of the number" into a mathematical expression.

Solution
The *total* of a number and the *square* of the number

The unknown number: x
The square of the number: x^2

$x + x^2$

You Try It 3   Translate "the product of a number and one-half of the number" into a mathematical expression.

Your solution

$(x)\left(\dfrac{1}{2}x\right)$

Solution on p. S30

In-Class Examples (Objective 11.5B)
Translate into a mathematical expression.

1. three times the sum of a number and five
   $3(x + 5)$

2. one less than one-fifth of a number $\dfrac{1}{5}x - 1$

3. the difference between four times a number and nine $4x - 9$

11.5 Exercises

Section 11.5

Suggested Assignment
Exercises 1–43, odds
More challenging problems:
Exercises 45, 46, 48, 49

Objective A   To translate a verbal expression into a mathematical expression given the variable

For Exercises 1 to 22, translate into a mathematical expression.

1. 9 less than y
$y - 9$

2. w divided by 7
$\dfrac{w}{7}$

3. z increased by 3
$z + 3$

4. the product of -2 and x
$-2x$

5. the sum of two-thirds of n and n
$\dfrac{2}{3}n + n$

6. the difference between the square of r and r
$r^2 - r$

7. the quotient of m and the difference between m and 3
$\dfrac{m}{m - 3}$

8. v increased by twice v
$v + 2v$

9. the product of 9 and 4 more than x
$9(x + 4)$

10. the total of a and the quotient of a and 7
$a + \dfrac{a}{7}$

11. the difference between n and the product of -5 and n
$n - (-5)n$

12. x decreased by the quotient of x and 2
$x - \dfrac{x}{2}$

13. the product of c and one-fourth of c
$c\left(\dfrac{1}{4}c\right)$

14. the quotient of 3 less than z and z
$\dfrac{z - 3}{z}$

15. the total of the square of m and twice the square of m
$m^2 + 2m^2$

16. the product of y and the sum of y and 4
$y(y + 4)$

17. 2 times the sum of t and 6
$2(t + 6)$

18. the quotient of r and the difference between 8 and r
$\dfrac{r}{8 - r}$

19. x divided by the total of 9 and x
$\dfrac{x}{9 + x}$

20. the sum of z and the product of 6 and z
$z + 6z$

21. three times the sum of b and 6
$3(b + 6)$

22. the ratio of w to the sum of w and 8
$\dfrac{w}{w + 8}$

Objective B   To translate a verbal expression into a mathematical expression by assigning the variable

For Exercises 23 to 44, translate into a mathematical expression.

23. the square of a number
x^2

24. five less than some number
$x - 5$

25. a number divided by twenty
$\dfrac{x}{20}$

26. the difference between a number and twelve
$x - 12$

27. four times some number
$4x$

28. the quotient of five and a number
$\dfrac{5}{x}$

Quick Quiz (Objective 11.5A)

Translate into a mathematical expression.

1. the quotient of c and the sum of c and 2 $\dfrac{c}{c + 2}$

2. the total of n and the square of n $n + n^2$

3. the difference between y and the product of 6 and y $y - 6y$

29. three-fourths of a number
$\dfrac{3}{4}x$

30. the sum of a number and seven
$x + 7$

31. four increased by some number
$4 + x$

32. the ratio of a number to nine
$\dfrac{x}{9}$

33. the difference between five times a number and the number
$5x - x$

34. six less than the total of three and a number
$(3 + x) - 6$

35. the product of a number and two more than the number
$x(x + 2)$

36. the quotient of six and the sum of nine and a number
$\dfrac{6}{9 + x}$

37. seven times the total of a number and eight
$7(x + 8)$

38. the difference between ten and the quotient of a number and two
$10 - \dfrac{x}{2}$

39. the square of a number plus the product of three and the number
$x^2 + 3x$

40. a number decreased by the product of five and the number
$x - 5x$

41. the sum of three more than a number and one-half of the number
$(x + 3) + \dfrac{1}{2}x$

42. eight more than twice the sum of a number and seven
$2(x + 7) + 8$

43. the quotient of three times a number and the number
$\dfrac{3x}{x}$

44. the square of a number divided by the sum of the number and twelve
$\dfrac{x^2}{x + 12}$

APPLYING THE CONCEPTS

45. **a.** Translate the expression $2x + 3$ into a phrase.
b. Translate the expression $2(x + 3)$ into a phrase.
Answers will vary. For example: **a.** 3 more than twice x; **b.** Twice the sum of x and 3

46. **a.** Translate the expression $\dfrac{2x}{7}$ into a phrase.

b. Translate the expression $\dfrac{2 + x}{7}$ into a phrase.
a. The quotient of twice x and 7.   **b.** The quotient of the sum of 2 and x and 7

47.   In your own words, explain how variables are used.

48. **Chemistry**   The chemical formula for water is H_2O. This formula means that there are two hydrogen atoms and one oxygen atom in each molecule of water. If x represents the number of atoms of oxygen in a glass of pure water, express the number of hydrogen atoms in the glass of water.
$2x$

49. **Chemistry**   The chemical formula for one molecule of glucose (sugar) is $C_6H_{12}O_6$, where C is carbon, H is hydrogen, and O is oxygen. If x represents the number of atoms of hydrogen in a sample of pure sugar, express the number of carbon atoms and the number of oxygen atoms in the sample in terms of x. $\dfrac{1}{2}x$; $\dfrac{1}{2}x$

Answers to Writing Exercises

47. Students will provide different explanations of how variables are used. Look for the idea that a variable is used to represent a number that is unknown or a number that can change, or vary.

Quick Quiz (Objective 11.5B)

Translate into a mathematical expression.

1. a number decreased by one-half of the number
$x - \dfrac{1}{2}x$

2. the sum of a number and the product of 2 and the number $x + 2x$

3. eight more than twice a number $2x + 8$

11.6 Translating Sentences into Equations and Solving

Objective A   To translate a sentence into an equation and solve

Point of Interest

Number puzzle problems similar to the one on this page have appeared in textbooks for hundreds of years. Here is one from a 1st-century Chinese textbook: "When a number is divided by 3, the remainder is 2; when it is divided by 5, the remainder is 3; when it is divided by 7, the remainder is 2. Find the number." There are actually infinitely many solutions to this problem. See whether you can find one of them.

An equation states that two mathematical expressions are equal. Therefore, to translate a sentence into an equation requires recognition of the words or phrases that mean "equals." Some of these phrases are

$$\left.\begin{array}{l} \text{equals} \\ \text{is} \\ \text{is equal to} \\ \text{amounts to} \\ \text{represents} \end{array}\right\} \text{translate to } =$$

Once the sentence is translated into an equation, the equation can be simplified to one of the form *variable = constant* and the solution can be found.

HOW TO   Translate "three more than twice a number is seventeen" into an equation and solve.

The unknown number: n

- Assign a variable to the unknown quantity.

| Three more than twice a number | is | seventeen |

- Find two verbal expressions for the same value.

$$\begin{array}{rcl} 2n + 3 & = & 17 \\ 2n + 3 - 3 & = & 17 - 3 \\ 2n & = & 14 \\ \dfrac{2n}{2} & = & \dfrac{14}{2} \\ n & = & 7 \end{array}$$

- Write a mathematical expression for each verbal expression. Write the equals sign. Solve the resulting equation.

The number is 7.

Example 1

Translate "a number decreased by six equals fifteen" into an equation and solve.

Solution
The unknown number: x

| A number decreased by six | equals | fifteen |

$$\begin{array}{rcl} x - 6 & = & 15 \\ x - 6 + 6 & = & 15 + 6 \\ x & = & 21 \end{array}$$

The number is 21.

You Try It 1

Translate "a number increased by four equals twelve" into an equation and solve.

Your solution
8

Solution on p. S30

Objective 11.6A

New Vocabulary
See the list of words and phrases that mean "equals."

Optional Student Activity

1. One bleepet plus one-fifth of a bleepet equals twenty-one. What is the value of a bleepet?
   $17\dfrac{1}{2}$

2. The difference between twelve times a number and seven is the same as the difference between four times a number and twenty-three. Find the number. -2

3. x is four more than y, and y is ten more than eight times ten. Find the mean of x, y, and $(2x + 3y)$.   214

4. The mean of x and five times x is 30. Find the value of x.
   10

5. One-half of a number plus three-fourths of the same number is 40. Find the number.
   32

6. One-half of a number equals two-thirds of the same number. Find the number.   0

7. The sum of three numbers is one hundred twenty. The second number is ten more than the first, and the third number is three times the first. Find the three numbers.
   22, 32, and 66

In-Class Examples (Objective 11.6A)

Translate into an equation and solve.

1. The sum of a number and six is fifteen. Find the number. $x + 6 = 15$; 9

2. The product of four and a number is negative twenty-four. Find the number. $4x = -24$; -6

3. The difference between nine and four times a number is three. Find the number.
   $9 - 4x = 3$; $1\dfrac{1}{2}$

4. The sum of two-fifths of a number and three is negative one. Find the number. $\dfrac{2}{5}x + 3 = -1$; -10

Example 2

The quotient of a number and six is five. Find the number.

Solution

The unknown number: z

The quotient of a number and six	is	five

$$\frac{z}{6} = 5$$

$$6 \cdot \frac{z}{6} = 6 \cdot 5$$

$$z = 30$$

The number is 30.

Example 3

Eight decreased by twice a number is four. Find the number.

Solution

The unknown number: t

Eight decreased by twice a number	is	four

$$8 - 2t = 4$$

$$8 - 8 - 2t = 4 - 8$$

$$-2t = -4$$

$$\frac{-2t}{-2} = \frac{-4}{-2}$$

$$t = 2$$

The number is 2.

Example 4

Three less than the ratio of a number to seven is one. Find the number.

Solution

The unknown number: x

Three less than the ratio of a number to seven	is	one

$$\frac{x}{7} - 3 = 1$$

$$\frac{x}{7} - 3 + 3 = 1 + 3$$

$$\frac{x}{7} = 4$$

$$7 \cdot \frac{x}{7} = 7 \cdot 4$$

$$x = 28$$

The number is 28.

You Try It 2

The product of two and a number is ten. Find the number.

Your solution

5

You Try It 3

The sum of three times a number and six equals four. Find the number.

Your solution

$-\frac{2}{3}$

You Try It 4

Three more than one-half of a number is nine. Find the number.

Your solution

12

Solutions on pp. S30–S31

Objective B   **To solve application problems**

Example 5

The cost of a television with remote control is $649. This amount is $125 more than the cost of the same television without remote control. Find the cost of the television without remote control.

Strategy

To find the cost of the television without remote control, write and solve an equation using C to represent the cost of the television without remote control.

Solution

| $649 | is | $125 more than the television without remote control |

$$649 = C + 125$$
$$649 - 125 = C + 125 - 125$$
$$524 = C$$

The cost of the television without remote control is $524.

Example 6

By purchasing a fleet of cars, a company receives a discount of $1972 on each car purchased. This amount is 8% of the regular price. Find the regular price.

Strategy

To find the regular price, write and solve an equation using P to represent the regular price of the car.

Solution

| $1972 | is | 8% of the regular price |

$$1972 = 0.08 \cdot P$$
$$\frac{1972}{0.08} = \frac{0.08P}{0.08}$$
$$24{,}650 = P$$

The regular price is $24,650.

You Try It 5

The sale price of a pair of slacks is $38.95. This amount is $11 less than the regular price. Find the regular price.

Your strategy

Your solution
$49.95

You Try It 6

At a certain speed, the engine rpm (revolutions per minute) of a car in fourth gear is 2500. This is two-thirds of the rpm of the engine in third gear. Find the rpm of the engine when it is in third gear.

Your strategy

Your solution
3750 rpm

Solutions on p. S31

Concept Check

1. A Chinese restaurant charges $11.95 for the adult buffet and $7.25 for the children's buffet. One family's bill came to $79.35. If there were three adults in the family, how many children were there?   6 children

2. A local feed store sells a 75-pound bag of feed for $10.90. If a customer buys more than one bag, each additional bag costs $10.50. A customer bought $84.40 worth of feed. How many 75-pound bags of feed did this customer purchase?   8 bags

Optional Student Activity

1. A rope is cut in half. Then one of the two pieces is cut so that the ratio of the pieces created from it is 2:1. The total length of the shortest of the three pieces and the longest of the three pieces is 6 ft. Find the length of the original piece of rope.   9 ft

2. Aaron decided to quit smoking. He started by cutting back two cigarettes a day each day for a week. He smoked 119 cigarettes during the week. How many cigarettes did he smoke the day before he started to cut back?   25 cigarettes

---

In-Class Examples (Objective 11.6B)

1. A dental assistant paid $1640 in state income tax this year. This is $86 more than last year. Find the state income tax last year. $1554

2. A plumber charged $115 to install a hot water heater. This charge included $40 for materials and $30 per hour for labor. How long did it take the plumber to install the water heater?   2.5 h

Example 7

Ron Sierra charged $1105 for plumbing repairs in an office building. This charge included $90 for parts and $35 per hour for labor. Find the number of hours he worked in the office building.

Strategy

To find the number of hours worked, write and solve an equation using N to represent the number of hours worked.

Solution

$1105	included	$90 for parts and $35 per hour for labor

$$
\begin{aligned}
1105 &= 90 + 35N \\
1105 - 90 &= 90 - 90 + 35N \\
1015 &= 35N \\
\frac{1015}{35} &= \frac{35N}{35} \\
29 &= N
\end{aligned}
$$

Ron worked 29 h.

Example 8

The state income tax for Tim Fong last month was $256. This amount is $5 more than 8% of his monthly salary. Find Tim's monthly salary.

Strategy

To find Tim's monthly salary, write and solve an equation using S to represent his monthly salary.

Solution

$256	is	$5 more than 8% of the monthly salary

$$
\begin{aligned}
256 &= 0.08 \cdot S + 5 \\
256 - 5 &= 0.08S + 5 - 5 \\
251 &= 0.08S \\
\frac{251}{0.08} &= \frac{0.08S}{0.08} \\
3137.50 &= S
\end{aligned}
$$

Tim's monthly salary is $3137.50.

You Try It 7

The total cost to make a model Z100 television is $300. The cost includes $100 for materials plus $12.50 per hour for labor. How many hours of labor are required to make a model Z100 television?

Your strategy

Your solution
16 h

You Try It 8

Natalie Adams earned $2500 last month for temporary work. This amount was the sum of a base monthly salary of $800 and an 8% commission on total sales. Find Natalie's total sales for the month.

Your strategy

Your solution
$21,250

Solutions on p. S31

11.6 Exercises

Objective A   **To translate a sentence into an equation and solve**

Suggested Assignment
Exercises 1–49, odds
More challenging problems:
   Exercises 50, 51

For Exercises 1 to 26, write an equation and solve.

1. The sum of a number and seven is twelve. Find the number.
    $x + 7 = 12$; 5

2. A number decreased by seven is five. Find the number.
    $x - 7 = 5$; 12

3. The product of three and a number is eighteen. Find the number.
    $3x = 18$; 6

4. The quotient of a number and three is one. Find the number.
    $\dfrac{x}{3} = 1$; 3

5. Five more than a number is three. Find the number.
    $x + 5 = 3$; -2

6. A number divided by four is six. Find the number.
    $\dfrac{x}{4} = 6$; 24

7. Six times a number is fourteen. Find the number.
    $6x = 14$; $2\dfrac{1}{3}$

8. Seven less than a number is three. Find the number.
    $x - 7 = 3$; 10

9. Five-sixths of a number is fifteen. Find the number.
    $\dfrac{5}{6}x = 15$; 18

10. The total of twenty and a number is five. Find the number.
    $20 + x = 5$; -15

11. The sum of three times a number and four is eight. Find the number.
    $3x + 4 = 8$; $1\dfrac{1}{3}$

12. The sum of one-third of a number and seven is twelve. Find the number.
    $\dfrac{1}{3}x + 7 = 12$; 15

13. Seven less than one-fourth of a number is nine. Find the number.
    $\dfrac{1}{4}x - 7 = 9$; 64

14. The total of a number divided by four and nine is two. Find the number.
    $\dfrac{x}{4} + 9 = 2$; -28

15. The ratio of a number to nine is fourteen. Find the number.
    $\dfrac{x}{9} = 14$; 126

16. Five increased by the product of five and a number is equal to 30. Find the number.
    $5 + 5x = 30$; 5

17. Six less than the quotient of a number and four is equal to negative two. Find the number.
    $\dfrac{x}{4} - 6 = -2$; 16

18. The product of a number plus three and two is eight. Find the number.
    $(x + 3)2 = 8$; 1

19. The difference between seven and twice a number is thirteen. Find the number.
    $7 - 2x = 13$; -3

20. Five more than the product of three and a number is eight. Find the number.
    $3x + 5 = 8$; 1

Quick Quiz (Objective 11.6A)

Translate into an equation and solve.

1. A number decreased by nine is four. Find the number.
   $x - 9 = 4$; 13

2. Three-fifths of a number is fifteen. Find the number.
   $\dfrac{3}{5}x = 15$; 25

3. Three more than the product of six and a number is twenty-one. Find the number.
   $6x + 3 = 21$; 3

4. Nine less than three-fourths of a number is six. Find the number.
   $\dfrac{3}{4}x - 9 = 6$; 20

21. Nine decreased by the quotient of a number and two is five. Find the number.

$9 - \dfrac{x}{2} = 5; 8$

22. The total of ten times a number and seven is twenty-seven. Find the number.

$10x + 7 = 27; 2$

23. The sum of three-fifths of a number and eight is two. Find the number.

$\dfrac{3}{5}x + 8 = 2; -10$

24. Five less than two-thirds of a number is three. Find the number.

$\dfrac{2}{3}x - 5 = 3; 12$

25. 📱 The difference between a number divided by 4.186 and 7.92 is 12.529. Find the number.

$\dfrac{x}{4.186} - 7.92 = 12.529; 85.599514$

26. 📱 The total of 5.68 times a number and 132.7 is the number minus 29.265. Find the number.

$5.68x + 132.7 = x - 29.265; -34.607906$

Objective B   **To solve application problems**

27. **Consumerism**   Sears has a pair of shoes on sale for $72.50. This amount is $4.25 less than the pair sells for at Target. Find the price at Target.
$76.75

28. **Compensation**   As a restaurant manager, Uechi Kim is paid a salary of $832 a week. This is $58 more a week than the salary Uechi received last year. Find the weekly salary paid to Uechi last year.
$774

29. **Depreciation**   The value of a sport utility vehicle this year is $16,000, which is four-fifths of what its value was last year. Find the value of the vehicle last year.
$20,000

30. **Real Estate**   This year the value of a lakefront summer cottage is $175,000. This amount is twice the value of the cottage 6 years ago. What was its value 6 years ago?
$87,500

31. 🔵 **Bridges**   The length of the Akashi Kaikyo Bridge is 1991 m. This is 1505 m greater than the length of the Brooklyn Bridge. Find the length of the Brooklyn Bridge.
486 m

32. 🔵 **Debt**   According to CardWeb.com, the average household credit card balance ten years ago was $3275, which is $5665 less than the average credit card balance today. Find the average credit card balance today.
$8940

Akashi Kaikyo Bridge

33. **Finances**   Each month the Manzanares family spends $1360 for their house payment and utilities, which amounts to one-fourth of the family's monthly income. Find the family's monthly income.
$5440

34. **Consumerism**   The cost of a graphing calculator is now three-fourths of what the calculator cost 5 years ago. The cost of the graphing calculator is now $72. Find the cost of the calculator 5 years ago.
$96

Quick Quiz (Objective 11.6B)

1. The value of a classic car this year is $42,000. This is five-fourths of its value two years ago. Find its value two years ago. $33,600

2. A roofing contractor charges $200 plus $75 for each square of roofing material installed. How many squares of roofing material can be installed for $1400?   16 squares

35. Business   Assume that the Dell Computer Corp. has increased its output of computers by 400 computers per month. This amount represents an 8% increase over last year's production. Find the monthly output last year.   5000 computers

36. **Sports**   The average number of home runs per major league game today is 2.21. This represents 135% of the average number of home runs per game 40 years ago. (*Source:* Elias Sports Bureau) Find the average number of home runs per game 40 years ago. Round to the nearest hundredth.
1.64 home runs

37. **Nutrition**   The nutrition label on a bag of Baked Tostitos tortilla chips lists the sodium content of one serving as 200 mg, which is 8% of the recommended daily allowance of sodium. What is the recommended daily allowance of sodium? Express the answer in grams.
2.5 g

38. Markup   The price of a pair of skis at the Solitude Ski Shop is $340. This price includes the store's cost for the skis plus a markup at the rate of 25%. Find Solitude's cost for the skis.
$272

39. Contractors   Budget Plumbing charged $400 for a water softener and installation. The charge included $310 for the water softener and $30 per hour for labor. How many hours of labor were required for the job?
3 h

40. Compensation   Sandy's monthly salary as a sales representative was $2580. This amount included her base monthly salary of $600 plus a 3% commission on total sales. Find her total sales for the month.
$66,000

41. **Conservation**   In Central America and Mexico, 1184 plants and animals are known to be at risk of extinction. This represents approximately 10.7% of all the species known to be at risk of extinction on Earth. (*Source:* World Conservation Union) Approximately how many plants and animals are known to be at risk of extinction in the world?
11,065 plants and animals

42. **Taxes**   When you use your car for business, you are able to deduct the expense on your income tax return. In 2003 the deduction was 36 cents per mile driven. How many business-related miles did a taxpayer who deducted a total of $1728 drive?
4800 mi

43. **Insecticides**   Americans spend approximately $295 million a year on remedies for cockroaches. The table at the right shows the top U.S. cities for sales of roach insecticides. What percent of the total is spent in New York? Round to the nearest tenth of a percent.
3.3%

City	Roach Insecticide Sales
Los Angeles	$16.8 million
New York	$9.8 million
Houston	$6.7 million

Source: IRI InfoScan for Combat

44. Consumerism   McPherson Cement sells cement for $75 plus $24 for each yard of cement. How many yards of cement can be purchased for $363?
12 yd

45. Conservation A water flow restrictor has reduced the flow of water to 2 gal/min. This amount is 1 gal/min less than three-fifths the original flow rate. Find the original rate.
5 gal/min

46. Temperature Conversion The Celsius temperature equals five-ninths times the difference between the Fahrenheit temperature and 32. Find the Fahrenheit temperature when the Celsius temperature is 40°.
104°F

47. Compensation Assume that a sales executive receives a base monthly salary of $600 plus an 8.25% commission on total sales per month. Find the executive's total sales during a month in which she receives total compensation of $4109.55.
$42,540

Environment The graph at the right shows projected world carbon dioxide emissions, in billions of metric tons. Use the graph for Exercises 48 and 49. Round to the nearest hundredth of a billion.

Projected World Carbon Dioxide Emissions

Source: U.S. Energy Information Administration

48. The projected world carbon dioxide emissions in 2020 are equal to 160.4% of the emissions in 1998. Find the world carbon dioxide emissions in 1998.
6.14 billion metric tons

49. The projected world carbon dioxide emissions in 2015 are equal to 152.1% of the emissions in 1990. Find the world carbon dioxide emissions in 1990.
5.83 billion metric tons

APPLYING THE CONCEPTS

50. A man's boyhood lasted $\frac{1}{6}$ of his life, he played football for the next $\frac{1}{8}$ of his life, and he married 5 years after quitting football. A daughter was born after he had been married $\frac{1}{12}$ of his life. The daughter lived $\frac{1}{2}$ as many years as her father. The man died 6 years after his daughter. How old was the man when he died? Use a number line to illustrate the time. Then write an equation and solve it.
88 years old

51. It is always important to check the answer to an application problem to be sure the answer makes sense. Consider the following problem. "A 4-quart mixture of fruit juices is made from apple juice and cranberry juice. There are 6 more quarts of apple juice than of cranberry juice. Write and solve an equation for the number of quarts of each juice used." Does the answer to this question make sense? Explain.

52. A formula is an equation that relates variables in a known way. Find two examples of formulas that are used in your college major. Explain what each of the variables represents.

Answers to Writing Exercises

51. The problem states that a 4-quart mixture of fruit juice is made from apple juice and cranberry juice. There are 6 more quarts of apple juice than of cranberry juice. If we let x = the number of quarts of cranberry juice, then $x + 6$ = the number of quarts of apple juice. The total number of quarts is 4. Therefore, we can write the equation $x + (x + 6) = 4$.

$$x + (x + 6) = 4$$
$$2x + 6 = 4$$
$$2x = -2$$
$$x = -1$$

Since x = the number of quarts of cranberry juice, there are -1 qt of cranberry juice in the mixture. We cannot add -1 qt to a mixture. The solution is not reasonable.

We see from the original problem that the answer will not be reasonable. If the total number of quarts in the mixture is 4, we cannot have more than 6 qt of apple juice in the mixture.

52. Students should provide you with information concerning their majors, as well as two formulas from the field of study and explanations of the variables used in each formula.

Focus on Problem Solving

From Concrete to Abstract

As you progress in your study of algebra, you will find that the problems become less concrete and more abstract. Problems that are concrete provide information pertaining to a specific instance. Abstract problems are theoretical; they are stated without reference to a specific instance. Let's look at an example of an abstract problem.

How many cents are in d dollars?

How can you solve this problem? Are you able to solve the same problem if the information given is concrete?

How many cents are in 5 dollars?

You know that there are 100 cents in 1 dollar. To find the number of cents in 5 dollars, multiply 5 by 100.

$100 \cdot 5 = 500$ There are 500 cents in 5 dollars.

Use the same procedure to find the number of cents in d dollars: multiply d by 100.

$100 \cdot d = 100d$ There are $100d$ cents in d dollars.

This problem might be taken a step further:

If one pen costs c cents, how many pens can be purchased with d dollars?

Consider the same problem using numbers in place of the variables.

If one pen costs 25 cents, how many pens can be purchased with 2 dollars?

To solve this problem, you need to calculate the number of cents in 2 dollars (multiply 2 by 100) and divide the result by the cost per pen (25 cents).

$\dfrac{100 \cdot 2}{25} = \dfrac{200}{25} = 8$ If one pen costs 25 cents, 8 pens can be purchased with 2 dollars.

Use the same procedure to solve the related abstract problem. Calculate the number of cents in d dollars (multiply d by 100) and divide the result by the cost per pen (c cents).

$\dfrac{100 \cdot d}{c} = \dfrac{100d}{c}$ If one pen costs c cents, $\dfrac{100d}{c}$ pens can be purchased with d dollars.

At the heart of the study of algebra is the use of variables. It is the variables in the problems above that make them abstract. But it is variables that enable us to generalize situations and state rules about mathematics.

Answers to Focus on Problem Solving— From Concrete to Abstract

1. $20d$ nickels

2. $\dfrac{100d}{c}$ minutes

3. mg miles

4. $\dfrac{60h}{x}$ miles

5. $\dfrac{25q}{5n}$ photocopies

Answers to Projects and Group Activities— Averages

1. Answers will vary.

2. Answers will vary.

3. Answers will vary.

4. 0

5. Steps 2 and 3: Answers will vary. Step 4: 0

6. The sums are the same.

7. If the activity were conducted again, the outcome would be the same.

   Students should describe the average of a set of data as a number that represents the "middle" of the data. Therefore, the data values above the average and the data values below the average will "cancel each other out."

Instructor Note

If students are working in small groups for this activity, you might assign a different pair of cities to each group.

Try the following problems.

1. How many nickels are in d dollars?

2. How long can you talk on a pay phone if you have only d dollars and the call costs c cents per minute?

3. If you travel m miles on one gallon of gasoline, how far can you travel on g gallons of gasoline?

4. If you walk a mile in x minutes, how far can you walk in h hours?

5. If one photocopy costs n nickels, how many photocopies can you make for q quarters?

Projects and Group Activities

Averages We often discuss temperature in terms of average high or average low temperature. Temperatures collected over a period of time are analyzed to determine, for example, the average high temperature for a given month in your city or state. The following activity is planned to help you better understand the concept of "average."

1. Choose two cities in the United States. We will refer to them as City X and City Y. Over an 8-day period, record the daily high temperature each day in each city.

2. Determine the average high temperature for City X for the 8-day period. (Add the eight numbers, and then divide the sum by 8.) Do not round your answer.

3. Subtract the average high temperature for City X from each of the eight daily high temperatures for City X. You should have a list of eight numbers; the list should include positive numbers, negative numbers, and possibly zero.

4. Find the sum of the list of eight differences recorded in Step 3.

5. Repeat Steps 2 through 4 for City Y.

6. Compare the two sums found in Steps 4 and 5 for City X and City Y.

7. If you were to conduct this activity again, what would you expect the outcome to be? Use the results to explain what an average high temperature is. In your own words, explain what "average" means.

Chapter 11 Summary

Key Words	Examples
A *variable* is a letter of the alphabet used to stand for a quantity that is unknown or that can change. An expression that contains one or more variables is a *variable expression*. Replacing the variable or variables in a variable expression and then simplifying the resulting numerical expression is called *evaluating the variable expression*. [11.1A, p. 457]	Evaluate $5x^3 + 2y - 6$ when $x = -1$ and $y = 4$. $5x^3 + 2y - 6$ $5(-1)^3 + 2(4) - 6 = 5(-1) + 2(4) - 6$ $\quad = -5 + 8 - 6$ $\quad = 3 - 6$ $\quad = 3 + (-6) = -3$
The *terms of a variable expression* are the addends of the expression. A *variable term* consists of a *numerical coefficient* and a *variable part*. A *constant term* has no variable part. [11.1B, pp. 458–459]	The variable expression $-3x^2 + 2x - 5$ has three terms: $-3x^2$, $2x$, and -5. $-3x^2$ and $2x$ are variable terms. -5 is a constant term. For the term $-3x^2$, the coefficient is -3 and the variable part is x^2.
Like terms of a variable expression have the same variable part. Constant terms are also like terms. [11.1B, p. 459]	$-6a^3b^2$ and $4a^3b^2$ are like terms.
An *equation* expresses the equality of two mathematical expressions. [11.2A, p. 467]	$5x + 6 = 7x - 3$ $y = 4x - 10$ $3a^2 - 6a + 4 = 0$
A *solution of an equation* is a number that, when substituted for the variable, results in a true equation. [11.2A, p. 467]	6 is a solution of $x - 4 = 2$ because $6 - 4 = 2$ is a true equation.
Solving an equation means finding a solution of the equation. The goal is to rewrite the equation in the form *variable = constant*. [11.2B, p. 468]	$x = 5$ is in the form *variable = constant*. The solution of the equation $x = 5$ is the constant 5 because $5 = 5$ is a true equation.
A *formula* is an equation that expresses a relationship among variables. [11.2D, p. 472]	The relationship between Celsius temperatures and Fahrenheit temperatures is given by the formula $F = 1.8C + 32$, where F represents degrees Fahrenheit and C represents degrees Celsius.
Some of the words and phrases that translate to *equals* are *is, is equal to, amounts to,* and *represents*. [11.6A, p. 495]	"Eight plus a number is ten" translates to $8 + x = 10$.

Essential Rules and Procedures	Examples
Commutative Property of Addition [11.1B, p. 459] $a + b = b + a$	$-9 + 5 = 5 + (-9)$

Associative Property of Addition [11.1B, p. 459]
$(a + b) + c = a + (b + c)$

$(-6 + 4) + 2 = -6 + (4 + 2)$

Commutative Property of Multiplication [11.1C, p. 461]
$a \cdot b = b \cdot a$

$-5(10) = 10(-5)$

Associative Property of Multiplication [11.1C, p. 461]
$(a \cdot b) \cdot c = a \cdot (b \cdot c)$

$(-3 \cdot 4) \cdot 6 = -3 \cdot (4 \cdot 6)$

Distributive Property [11.1C, p. 461]
$a(b + c) = ab + ac$
$a(b - c) = ab - ac$

$2(x + 7) = 2(x) + 2(7) = 2x + 14$
$5(4x - 3) = 5(4x) - 5(3) = 20x - 15$

Addition Property of Zero [11.2B, p. 468]
The sum of a term and zero is the term.

$a + 0 = a$ or $0 + a = a$

$-16 + 0 = -16$

Addition Property of Equations [11.2B, p. 468]
If a, b, and c are algebraic expressions, then the equations $a = b$ and $a + c = b + c$ have the same solutions.

The same number or variable expression can be added to each side of an equation without changing the solution of the equation.

$$x + 7 = 20$$
$$x + 7 + (-7) = 20 + (-7)$$
$$x = 13$$

Multiplication Property of Reciprocals [11.2C, p. 470]
The product of a nonzero term and its reciprocal equals 1.

$8 \cdot \dfrac{1}{8} = 1$

Multiplication Property of One [11.2C, p 470]
The product of a term and 1 is the term.

$-7(1) = -7$

Multiplication Property of Equations [11.2C, p. 470]
If a, b, and c are algebraic expressions and $c \neq 0$, then the equation $a = b$ has the same solutions as the equation $ac = bc$.

Each side of an equation can be multiplied by the same nonzero number without changing the solution of the equation.

$$\frac{3}{4}x = 24$$
$$\frac{4}{3} \cdot \frac{3}{4}x = \frac{4}{3} \cdot 24$$
$$x = 32$$

Steps in Solving General First-Degree Equations
[11.4B, p. 485]

1. Use the Distributive Property to remove parentheses.

2. Combine like terms on each side of the equation.

3. Use the Addition Property of Equations to rewrite the equation with only one variable term.

4. Use the Addition Property of Equations to rewrite the equation with only one constant term.

5. Use the Multiplication Property of Equations to rewrite the equation so that the coefficient of the variable is 1.

$$8 - 4(2x + 3) = 2(1 - x)$$
$$8 - 8x - 12 = 2 - 2x$$
$$-8x - 4 = 2 - 2x$$
$$-8x + 2x - 4 = 2 - 2x + 2x$$
$$-6x - 4 = 2$$
$$-6x - 4 + 4 = 2 + 4$$
$$-6x = 6$$
$$\frac{-6x}{-6} = \frac{6}{-6}$$
$$x = -1$$

Chapter 11 Review Exercises

1. Simplify: $-2(a - b)$
$-2a + 2b$ [11.1C]

2. Is -2 a solution of the equation $3x - 2 = -8$?
Yes [11.2A]

3. Solve: $x - 3 = -7$
-4 [11.2B]

4. Solve: $-2x + 5 = -9$
7 [11.3A]

5. Evaluate $a^2 - 3b$ when $a = 2$ and $b = -3$.
13 [11.1A]

6. Solve: $-3x = 27$
-9 [11.2C]

7. Solve: $\frac{2}{3}x + 3 = -9$
-18 [11.3A]

8. Simplify: $3x - 2(3x - 2)$
$-3x + 4$ [11.1C]

9. Solve: $6x - 9 = -3x + 36$
5 [11.4A]

10. Solve: $x + 3 = -2$
-5 [11.2B]

11. Is 5 a solution of the equation $3x - 5 = -10$?
No [11.2A]

12. Evaluate $a^2 - (b \div c)$ when $a = -2$, $b = 8$, and $c = -4$.
6 [11.1A]

13. Solve: $3(x - 2) + 2 = 11$
5 [11.4B]

14. Solve: $35 - 3x = 5$
10 [11.3A]

15. Simplify: $6bc - 7bc + 2bc - 5bc$
$-4bc$ [11.1B]

16. Solve: $7 - 3x = 2 - 5x$
$-2\frac{1}{2}$ [11.4A]

17. Solve: $-\frac{3}{8}x = -\frac{15}{32}$
$1\frac{1}{4}$ [11.2C]

18. Simplify: $\frac{1}{2}x^2 - \frac{1}{3}x^2 + \frac{1}{5}x^2 + 2x^2$
$\frac{71}{30}x^2$ [11.1B]

19. Solve: $5x - 3(1 - 2x) = 4(2x - 1)$

$-\dfrac{1}{3}$ [11.4B]

20. Solve: $\dfrac{5}{6}x - 4 = 5$

$10\dfrac{4}{5}$ [11.3A]

21. **Fuel Efficiency** A tourist drove a rental car 621 mi on 27 gal of gas. Find the number of miles per gallon of gas. Use the formula $D = M \cdot G$, where D is distance, M is miles per gallon, and G is the number of gallons.
23 mi/gal [11.2D]

22. **Temperature Conversion** Find the Celsius temperature when the Fahrenheit temperature is 100°. Use the formula $F = 1.8C + 32$, where F is the Fahrenheit temperature and C is the Celsius temperature. Round to the nearest tenth.
37.8°C [11.3B]

23. Translate "the total of n and the quotient of n and five" into a mathematical expression.
$n + \dfrac{n}{5}$ [11.5A]

24. Translate "the sum of five more than a number and one-third of the number" into a mathematical expression.
$(n + 5) + \dfrac{1}{3}n$ [11.5B]

25. The difference between nine and twice a number is five. Find the number.
2 [11.6A]

26. The product of five and a number is fifty. Find the number.
10 [11.6A]

27. **Discount** A compact disc player is now on sale for $228. This is 60% of the regular price. Find the regular price of the CD player.
$380 [11.6B]

28. **Agriculture** A farmer harvested 28,336 bushels of corn. This amount represents a 12% increase over last year's crop. How many bushels of corn did the farmer harvest last year?
25,300 bushels [11.6B]

Chapter 11 Test

1. Solve: $\frac{x}{5} - 12 = 7$
   95 [11.3A]

2. Solve: $x - 12 = 14$
   26 [11.2B]

3. Simplify: $3y - 2x - 7y - 9x$
   $-11x - 4y$ [11.1B]

4. Solve: $8 - 3x = 2x - 8$
   $3\frac{1}{5}$ [11.4A]

5. Solve: $3x - 12 = -18$
   -2 [11.3A]

6. Evaluate $c^2 - (2a + b^2)$ when $a = 3$, $b = -6$, and $c = -2$.
   -38 [11.1A]

7. Is 3 a solution of the equation $x^2 + 3x - 7 = 3x - 2$?
   No [11.2A]

8. Simplify: $9 - 8ab - 6ab$
   $-14ab + 9$ [11.1B]

9. Solve: $-5x = 14$
   $-2\frac{4}{5}$ [11.2C]

10. Simplify: $3y + 5(y - 3) + 8$
    $8y - 7$ [11.1C]

11. Solve: $3x - 4(x - 2) = 8$
    0 [11.4B]

12. Solve: $5 = 3 - 4x$
    $-\frac{1}{2}$ [11.3A]

13. Evaluate $\frac{x^2}{y} - \frac{y^2}{x}$ for $x = 3$ and $y = -2$.
    $-5\frac{5}{6}$ [11.1A]

14. Solve: $\frac{5}{8}x = -10$
    -16 [11.2C]

15. Solve: $y - 4y + 3 = 12$
    -3 [11.3A]

16. Solve: $2x + 4(x - 3) = 5x - 1$
    11 [11.4B]

17. **Finance** A loan of $6600 is to be paid in 48 equal monthly installments. Find the monthly payment. Use the formula $L = P \cdot N$, where L is the loan amount, P is the monthly payment, and N is the number of months.
$137.50 [11.2D]

18. **Manufacturing** A clock manufacturer's fixed costs per month are $5000. The unit cost for each clock is $15. Find the number of clocks made during a month in which the total cost was $65,000. Use the formula $T = U \cdot N + F$, where T is the total cost, U is the cost per unit, N is the number of units made, and F is the fixed costs.
4000 clocks [11.3B]

19. **Physics** Find the time required for a falling object to increase in velocity from 24 ft/s to 392 ft/s. Use the formula $V = V_0 + 32t$, where V is the final velocity of a falling object, V_0 is the starting velocity of a falling object, and t is the time for the object to fall.
11.5 s [11.3B]

20. Translate "the sum of x and one-third of x" into a mathematical expression.
$x + \dfrac{1}{3}x$ [11.5A]

21. Translate "five times the sum of a number and three" into a mathematical expression.
$5(x + 3)$ [11.5B]

22. Translate "three less than two times a number is seven" into an equation and solve.
$2x - 3 = 7; 5$ [11.6A]

23. The total of five and three times a number is the number minus two. Find the number.
$-3\dfrac{1}{2}$ [11.6A]

24. **Compensation** Eduardo Santos earned $3600 last month. This salary is the sum of the base monthly salary of $1200 and a 6% commission on total sales. Find his total sales for the month.
$40,000 [11.6B]

25. **Consumerism** Your mechanic charges you $278 for performing a 30,000-mile checkup on your car. This charge includes $152 for parts and $42 per hour for labor. Find the number of hours the mechanic worked on your car.
3 h [11.6B]

Cumulative Review Exercises

1. Simplify: $6^2 - (18 - 6) \div 4 + 8$
   41 [1.6B]

2. Subtract: $3\frac{1}{6} - 1\frac{7}{15}$
   $1\frac{7}{10}$ [2.5C]

3. Simplify: $\left(\frac{3}{8} - \frac{1}{4}\right) \div \frac{3}{4} + \frac{4}{9}$
   $\frac{11}{18}$ [2.8C]

4. Multiply: 9.67×0.0049
   0.047383 [3.4A]

5. Write "$84 earned in 20 hours" as a unit rate.
   $4.20/h [4.2B]

6. Solve the proportion $\frac{2}{3} = \frac{n}{40}$.
   Round to the nearest hundredth.
   26.67 [4.3B]

7. Write $5\frac{1}{3}\%$ as a fraction.
   $\frac{4}{75}$ [5.1A]

8. What percent of 30 is 42?
   140% [5.3A]

9. 8 is 125% of what number?
   6.4 [5.4A]

10. Multiply: 3 ft 9 in. $\times$ 5
    18 ft 9 in. [8.1B]

11. Convert $1\frac{3}{8}$ lb to ounces.
    22 oz [8.2A]

12. Convert 282 mg to grams.
    0.282 g [9.2A]

13. Add: $-2 + 5 + (-8) + 4$
    −1 [10.2A]

14. Find −6 less than 13.
    19 [10.2B]

15. Simplify: $(-2)^2 - (-8) \div (3 - 5)^2$
    6 [10.5B]

16. Evaluate $3ab - 2ac$ when $a = -2$, $b = 6$, and $c = -3$.
    −48 [11.1A]

17. Simplify: $3z - 2x + 5z - 8x$
    $-10x + 8z$ [11.1B]

18. Simplify: $6y - 3(y - 5) + 8$
    $3y + 23$ [11.1C]

19. Solve: $2x - 5 = -7$
    −1 [11.3A]

20. Solve: $7x - 3(x - 5) = -10$
    $-6\frac{1}{4}$ [11.4B]

21. Solve: $-\frac{2}{3}x = 5$

$-7\frac{1}{2}$ [11.2C]

22. Solve: $\frac{x}{3} - 5 = -12$

-21 [11.3A]

23. **Education**   In a mathematics class of 34 students, 6 received an A grade. Find the percent of the students in the mathematics class who received an A grade. Round to the nearest tenth of a percent.
17.6%   [5.3B]

24. **Markup**   The manager of a pottery store used a markup rate of 40%. Find the price of a piece of pottery that cost the store $28.50.
$39.90   [6.2B]

25. **Discount**   A department store has a suit regularly priced at $450 on sale for $369.
a. What is the discount? $81
b. What is the discount rate?   18%   [6.2D]

26. **Simple Interest**   A toy store borrowed $80,000 at a simple interest rate of 11% for 4 months. What is the simple interest due on the loan? Round to the nearest cent.
$2933.33   [6.3A]

27. Translate "the sum of three times a number and 4" into a mathematical expression.
$3n + 4$ [11.5B]

28. **Probability**   A tetrahedral die is one with four triangular sides numbered 1 to 4. If two tetrahedral dice are rolled, what is the probability that the sum of the upward faces is 7?
$\frac{1}{8}$ [7.5A]

29. **Compensation**   Sunah Yee, a sales executive, receives a base salary of $800 plus an 8% commission on total sales. Find the total sales during a month in which Sunah earned $3400. Use the formula $M = S \cdot R + B$, where M is the monthly earnings, S is the total sales, R is the commission rate, and B is the base monthly salary.
$32,500   [11.3B]

30. Three less than eight times a number is three more than five times the number. Find the number.
2   [11.6A]

12 Geometry

This is an aerial view of the house of William Paca, who was a Maryland Patriot and a signer of the Declaration of Independence. The house's large, formal garden has been restored to its original splendor. The best way to appreciate the shapes sculpted in the garden is to view it from above, like in this photo. Each geometric shape combines with the others to form the entire garden. **Exercise 35 on page 544** shows you how to use a geometric formula to first determine the size of an area, and then to calculate how much grass seed is needed for an area of that size.

Need help? For online student resources, such as section quizzes, visit this textbook's website at **math.college.hmco.com/students**.

OBJECTIVES

Section 12.1

A To define and describe lines and angles
B To define and describe geometric figures
C To solve problems involving angles formed by intersecting lines

Section 12.2

A To find the perimeter of plane geometric figures
B To find the perimeter of composite geometric figures
C To solve application problems

Section 12.3

A To find the area of geometric figures
B To find the area of composite geometric figures
C To solve application problems

Section 12.4

A To find the volume of geometric solids
B To find the volume of composite geometric solids
C To solve application problems

Section 12.5

A To find the square root of a number
B To find the unknown side of a right triangle using the Pythagorean Theorem
C To solve application problems

Section 12.6

A To solve similar and congruent triangles
B To solve application problems

Do these exercises to prepare for Chapter 12.

1. Solve: $x + 47 = 90$
   43   [11.2B]

2. Solve: $32 + 97 + x = 180$
   51   [11.2B]

3. Simplify: $2(18) + 2(10)$
   56   [1.6B]

4. Evaluate abc when $a = 2$, $b = 3.14$, and $c = 9$.
   56.52   [11.1A]

5. Evaluate xyz^3 when $x = \frac{4}{3}$, $y = 3.14$, and $z = 3$.
   113.04   [11.1A]

6. Solve: $\frac{5}{12} = \frac{6}{x}$
   14.4   [4.3B]

GO FIGURE • • •

Draw the figure that would come next.

Circle inside diamond inside square inside triangle

12.1 Angles, Lines, and Geometric Figures

Objective A   To define and describe lines and angles

Point of Interest

Geometry is one of the oldest branches of mathematics. Around 350 B.C., the Greek mathematician Euclid wrote the *Elements*, which contained all of the known concepts of geometry. Euclid's contribution was to unify various concepts into a single deductive system that was based on a set of axioms.

The word *geometry* comes from the Greek words for "earth" (*geo*) and "measure." The original purpose of geometry was to measure land. Today geometry is used in many sciences, such as physics, chemistry, and geology, and in applied fields such as mechanical drawing and astronomy. Geometric form is used in art and design.

Two basic geometric concepts are plane and space.

A **plane** is a flat surface, such as a table-top or a blackboard. Figures that can lie totally in a plane are called **plane figures.**

Space extends in all directions. Objects in space, such as trees, ice cubes, and doors, are called **solids.**

A **line** extends indefinitely in two directions in a plane. A line has no width.

Study Tip

A great many new vocabulary words are introduced in this chapter. There are eight new terms on this page alone: plane, plane figures, space, solids, line, line segment, parallel lines, and intersecting lines. All of these terms are in **bold type.** The bold type indicates that these are concepts you must know to learn the material. Be sure to study each new term as it is presented.

A **line segment** is part of a line and has two endpoints. The line segment *AB* is shown in the figure.

The length of a line segment is the distance between the endpoints of the line segment. The length of a line segment may be expressed as the sum of two or more shorter line segments, as shown. For this example, $AB = 5$, $BC = 3$, and $AC = AB + BC = 5 + 3 = 8$.

HOW TO   Given that $AB = 22$ and $AC = 31$, find the length of BC.

$$AC = AB + BC$$
$$31 = 22 + BC$$
$$31 - 22 = 22 - 22 + BC$$
$$9 = BC$$

• Substitute 22 for *AB* and 31 for *AC*, and solve for *BC*.

Lines in a plane can be parallel or intersecting. **Parallel lines** never meet; the distance between them is always the same. **Intersecting lines** cross at a point in the plane.

New Vocabulary

plane
plane figures
space
solids
line
line segment
parallel lines
intersecting lines
ray
angle
vertex
degree
right angle
perpendicular lines
complementary angles
straight angle
supplementary angles
acute angle
obtuse angle

New Symbols

∥ (is parallel to)
∠ (angle)
° (degrees)
∟ (right angle)
⊥ (is perpendicular to)

Instructor Note

Tell students that when no units, such as feet or meters, are given for lengths along a line segment, all the distances are assumed to be in the same unit of length.

Discuss the Concepts

1. Describe each of the following: ray, line, and line segment.
2. Does the surface of Earth lie in a plane?

(*Continued on next page*)

In-Class Examples (Objective 12.1A)

1. How many degrees are in two-thirds of a revolution?   240°
2. Find the complement of a 38° angle.   52°
3. Find the supplement of a 57° angle.   123°

(*Continued*)

3. Using any objects in the classroom, provide examples of each of the following: a right angle, an acute angle, an obtuse angle, a plane, intersecting lines, parallel lines, and perpendicular lines. (If students need some assistance, suggest that they consider the windows.)

Concept Check

1. Provide three names for the angle below.

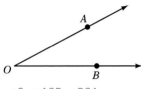

$\angle O$, $\angle AOB$, $\angle BOA$

2a. Give the number of degrees in a full circle.   360°

  b. Give the number of degrees in a straight angle.   180°

  c. Give the number of degrees in a right angle.   90°

3. How many dimensions does each of the following have?

  a. A point   0

  b. A line   1

  c. A line segment   1

  d. A ray   1

  e. An angle   2

4. What is the name given to lines in a plane that do not intersect?   Parallel lines

5. What is the name given to two lines that intersect at right angles?   Perpendicular lines

---

TAKE NOTE

For the figure below, we can refer to $\angle ABD$, $\angle ABC$, or $\angle CBD$.

Just writing $\angle B$ does not identify a specific angle.

Point of Interest

The Babylonians knew that Earth is in approximately the same position in the sky every 365 days. Historians suggest that the reason one complete revolution of a circle is 360° is that 360 is the closest number to 365 that is divisible by many numbers.

TAKE NOTE

The corner of a page of this book serves as a good model for a 90° angle.

The symbol $\parallel$ means "is parallel to." In the accompanying figure, $AB \parallel CD$ and $p \parallel q$. Note that line p contains line segment AB and that line q contains line segment CD. Parallel lines contain parallel line segments.

A **ray** starts at a point and extends indefinitely in one direction.

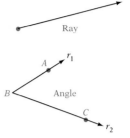
Ray

An **angle** is formed when two rays start from the same point. Rays r_1 and r_2 start from point B. The common endpoint is called the **vertex** of the angle.

Angle

If A and C are points on rays r_1 and r_2 above, respectively, then the angle is called $\angle ABC$, $\angle CBA$, or $\angle B$, where $\angle$ is the symbol for angle. Note that an angle is named by giving three points, with the vertex as the second point listed, or by giving the point at the vertex.

An angle can also be named by writing a variable between the rays close to the vertex. In the figure, $\angle x = \angle QRS = \angle SRQ$ and $\angle y = \angle SRT = \angle TRS$. Note that in this figure, more than two rays meet at the vertex. In this case, the vertex cannot be used to name the angle.

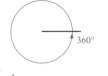

A unit in which angles are measured is the **degree.** The symbol for degree is °. One complete revolution is 360° (360 degrees).

360°

One-quarter of a revolution is 90°. A 90° angle is called a **right angle.** The symbol ∟ represents a right angle.

90°   Right angle

Perpendicular lines are intersecting lines that form right angles.

90°   90°
90°   90°

The symbol $\perp$ means "is perpendicular to." In the accompanying figure, $AB \perp CD$ and $p \perp q$. Note that line p contains line segment AB and line q contains line segment CD. Perpendicular lines contain perpendicular line segments.

Complementary angles are two angles whose sum is 90°.

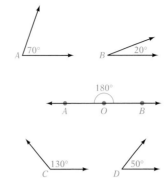

$\angle A + \angle B = 70° + 20° = 90°$
$\angle A$ and $\angle B$ are complementary angles.

One-half of a revolution is 180°. A 180° angle is called a **straight angle.** $\angle AOB$ in the figure is a straight angle.

Supplementary angles are two angles whose sum is 180°.

$\angle C + \angle D = 130° + 50° = 180°$
$\angle C$ and $\angle D$ are supplementary angles.

An **acute angle** is an angle whose measure is between 0° and 90°. $\angle D$ in the figure above is an acute angle. An **obtuse angle** is an angle whose measure is between 90° and 180°. $\angle C$ in the figure above is an obtuse angle.

In the accompanying figure, $\angle DAC = 45°$ and $\angle CAB = 55°$.

$\angle DAB = \angle DAC + \angle CAB$
$= 45° + 55° = 100°$

Example 1

Given that $MN = 15$, $NO = 18$, and $MP = 48$, find the length of OP.

Solution

$MP = MN + NO + OP$
$48 = 15 + 18 + OP$
$48 = 33 + OP$
$48 - 33 = 33 - 33 + OP$
$15 = OP$

Example 2

Find the complement of a 32° angle.

Solution

Let x represent the complement of 32°.

$x + 32° = 90°$
$x + 32° - 32° = 90° - 32°$ • The sum of
$\qquad\qquad x = 58°$ complementary
                                     angles is 90°.

58° is the complement of 32°.

You Try It 1

Given that $QR = 24$, $ST = 17$, and $QT = 62$, find the length of RS.

Your solution
21

You Try It 2

Find the supplement of a 32° angle.

Your solution
148°

Solutions on p. S31

Optional Student Activity

1. Prepare a handout on which you have drawn four angles. You might include a 40° angle, a 60° angle, a 120° angle, and a 30° angle. Orient the angles in different directions. Make enough copies of the handout so that each student in the class will have one.

   During class, provide students with protractors and instruction on how to measure angles. Give each student a copy of the handout. Have them measure each angle and then classify each angle as an acute angle or an obtuse angle.

2. On a number line, the points A, B, C, and D have coordinates −2.5, 2, 5, and 3.5, respectively. Which of these points is halfway between two others?
   D, or 3.5

3. Find the measure of the smaller angle between the hands of a clock when the time is 5 o'clock.   150°

4. On a line, Q is between P and S. R is between Q and S. S is between Q and T.
   $PT = 28$ units, $QS = 8$ units, and $PQ = QR = RS$. Find ST.
   16 units

Objective 12.1B

New Vocabulary
triangle
base of a triangle
height of a triangle
right triangle
hypotenuse
legs of a right triangle
quadrilateral
parallelogram
height of a parallelogram
rectangle
square
circle
center of a circle
diameter of a circle
radius of a circle
geometric solid
rectangular solid
cube
sphere
center of a sphere
diameter of a sphere
radius of a sphere
cylinder

New Formulas
The angles in a triangle:
$\angle A + \angle B + \angle C = 180°$
$r = \dfrac{1}{2}d$
$d = 2r$

Discuss the Concepts
1. In a right triangle, why are the two acute angles complementary?

2. Which of the following are plane figures: a square, a cube, a triangle, a sphere, a circle, a point, and a box?

Concept Check
1. The shape of Earth approximates what geometric figure?   A sphere

2. The angles of a triangle are in the ratio 2:3:7. Find the number of degrees in the largest angle.   105°

Example 3

Find the measure of $\angle x$.

Solution
$$\angle x + 47° = 90°$$
$$\angle x + 47° - 47° = 90° - 47°$$
$$\angle x = 43°$$

You Try It 3

Find the measure of $\angle a$.

Your solution
50°

Solution on p. S31

Objective B   **To define and describe geometric figures**

TAKE NOTE
Any side of a triangle can be considered the base, but generally the **base of a triangle** refers to the side that the triangle rests on. The **height of a triangle** is a line segment drawn perpendicular to the base from the opposite vertex.

A **triangle** is a closed, three-sided plane figure. Figure *ABC* is a triangle. *AB* is called the **base.** The line *CD*, perpendicular to the base, is called the **height.**

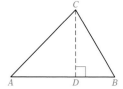

The Angles in a Triangle
The sum of the three angles in a triangle is 180°.
$$\angle A + \angle B + \angle C = 180°$$

HOW TO   In triangle *DEF*, $\angle D = 32°$ and $\angle E = 88°$. Find the measure of $\angle F$.

$\angle D + \angle E + \angle F = 180°$ • The sum of the three angles in a triangle is 180°.
$32° + 88° + \angle F = 180°$ • $\angle D = 32°$ and $\angle E = 88°$
$120° + \angle F = 180°$ • Solve for $\angle F$.
$120° - 120° + \angle F = 180° - 120°$
$\angle F = 60°$

A **right triangle** contains one right angle. The side opposite the right angle is called the **hypotenuse.** The **legs of a right triangle** are its other two sides. In a right triangle, the two acute angles are complementary.

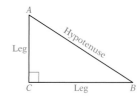

$$\angle A + \angle B = 90°$$

HOW TO   In the right triangle at the left, $\angle A = 30°$. Find the measure of $\angle B$.

$\angle A + \angle B = 90°$ • The two acute angles are complementary.
$30° + \angle B = 90°$ • $\angle A = 30°$
$30° - 30° + \angle B = 90° - 30°$ • Solve for $\angle B$.
$\angle B = 60°$

A **quadrilateral** is a closed, four-sided plane figure. Three quadrilaterals with special characteristics are described here.

A **parallelogram** has opposite sides parallel and equal. The distance *AE* between the parallel sides is called the **height.**

Parallelogram

In-Class Examples (Objective 12.1B)

1. A triangle has a 21° angle and a 64° angle. Find the measure of the third angle.   95°

2. A right triangle has a 52° angle. Find the measure of the other two angles.   90°, 38°

3. Find the radius of a circle with a diameter of 24 in.
   12 in.

4. Find the diameter of a circle with a radius of 3.5 ft.
   7 ft

A **rectangle** is a parallelogram that has four right angles.

A **square** is a rectangle that has four equal sides.

Rectangle   Square

A **circle** is a plane figure in which all points are the same distance from point *O*, which is called the **center** of the circle.

The **diameter of a circle** (*d*) is a line segment through the center of the circle with endpoints on the circle. *AB* is a diameter of the circle shown.

The **radius of a circle** (*r*) is a line segment from the center to a point on the circle. *OC* is a radius of the circle.

$$d = 2r \quad \text{or} \quad r = \frac{1}{2}d$$

Circle

HOW TO   The line segment *AB* is a diameter of the circle shown. Find the radius of the circle.

The radius is one-half the diameter. Therefore,

$$r = \frac{1}{2}d$$
$$= \frac{1}{2}(8 \text{ in.}) \qquad \bullet \quad d = 8 \text{ in.}$$
$$= 4 \text{ in.}$$

8 in.

A **geometric solid** is a figure in space, or space figure. Four common space figures are the rectangular solid, cube, sphere, and cylinder.

A **rectangular solid** is a solid in which all six faces are rectangles.

Height

Length   Width

Rectangular solid

A **cube** is a rectangular solid in which all six faces are squares.

Cube

A **sphere** is a solid in which all points on the surface are the same distance from point *O*, which is called the **center** of the sphere.

The **diameter of a sphere** is a line segment going through the center with endpoints on the sphere. *AB* is a diameter of the sphere shown.

Sphere

The **radius of a sphere** is a line segment from the center to a point on the sphere. *OC* is a radius of the sphere.

$$d = 2r \quad \text{or} \quad r = \frac{1}{2}d$$

Optional Student Activity

If time permits, you might teach your students how to prepare a circle graph. If you did the first activity described on page 517, your students know how to measure angles. You will need to provide them with instructions on drawing angles.

We suggest that you provide your students with papers on which you have drawn a few large circles for the graphs. The center of each circle should be marked with a dot. Have them prepare circle graphs for the data provided in the exercises that follow.

a. Shown below are American adults' favorite pizza toppings. (*Source:* Market Facts for Bolla wines)

Pepperoni	43%
Sausage	19%
Mushrooms	14%
Vegetables	13%
Others	7%
Onions	4%

b. According to a Pathfinder Research Group survey, more than 94% of adults have heard of the Three Stooges. The choices among those who have a favorite are as follows:

Curly	52%
Moe	31%
Larry	12%
Curly Joe	3%
Shemp	2%

HOW TO   The radius of the sphere shown at the right is 5 cm. Find the diameter of the sphere.

$d = 2r$ • The diameter equals twice the radius.
$\quad = 2(5 \text{ cm})$ • $r = 5$ cm
$\quad = 10 \text{ cm}$

The diameter is 10 cm.

The most common **cylinder** is one in which the bases are circles and are perpendicular to the side.

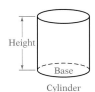

Height

Base

Cylinder

Example 4

One angle in a right triangle is equal to 50°. Find the measure of the other angles.

Solution
In a right triangle, one angle measures 90° and the two acute angles are complementary.

$$\angle A + \angle B = 90°$$
$$\angle A + 50° = 90°$$
$$\angle A + 50° - 50° = 90° - 50°$$
$$\angle A = 40°$$

The other angles measure 90° and 40°.

You Try It 4

A right triangle has one angle equal to 7°. Find the measure of the other angles.

Your solution
90° and 83°

Example 5

Two angles of a triangle measure 42° and 103°. Find the measure of the third angle.

Solution
The sum of the three angles of a triangle is 180°.

$$\angle A + \angle B + \angle C = 180°$$
$$\angle A + 42° + 103° = 180°$$
$$\angle A + 145° = 180°$$
$$\angle A + 145° - 145° = 180° - 145°$$
$$\angle A = 35°$$

The measure of the third angle is 35°.

You Try It 5

Two angles of a triangle measure 62° and 45°. Find the measure of the other angle.

Your solution
73°

Example 6

A circle has a radius of 8 cm. Find the diameter.

Solution
$d = 2r$
$\quad = 2 \cdot 8 \text{ cm} = 16 \text{ cm}$

The diameter is 16 cm.

You Try It 6

A circle has a diameter of 8 in. Find the radius.

Your solution
4 in.

Solutions on p. S32

Objective C   **To solve problems involving angles formed by intersecting lines**

Four angles are formed by the intersection of two lines. If the two lines are perpendicular, then each of the four angles is a right angle. If the two lines are not perpendicular, then two of the angles formed are acute angles and two of the angles are obtuse angles. The two acute angles are always opposite each other, and the two obtuse angles are always opposite each other.

In the figure, $\angle w$ and $\angle y$ are acute angles. $\angle x$ and $\angle z$ are obtuse angles. Two angles that are on opposite sides of the intersection of two lines are called **vertical angles.** Vertical angles have the same measure. $\angle w$ and $\angle y$ are vertical angles. $\angle x$ and $\angle z$ are vertical angles.

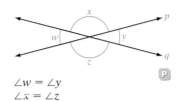

$\angle w = \angle y$
$\angle x = \angle z$

Two angles that share a common side are called **adjacent angles.** In the previous figure, $\angle x$ and $\angle y$ are adjacent angles, as are $\angle y$ and $\angle z$, $\angle z$ and $\angle w$, and $\angle w$ and $\angle x$. Adjacent angles of intersecting lines are supplementary angles.

$\angle x + \angle y = 180°$
$\angle y + \angle z = 180°$
$\angle z + \angle w = 180°$
$\angle w + \angle x = 180°$

HOW TO   In the figure at the left, $\angle c = 65°$. Find the measures of angles a, b, and d.

$\angle a = 65°$ • $\angle a = \angle c$ because $\angle c$ and $\angle a$ are vertical angles.

$\angle b + \angle c = 180°$ • $\angle c$ is supplementary to $\angle b$ because $\angle c$ and $\angle b$ are adjacent angles.

$\angle b + 65° = 180°$ • $\angle c = 65°$
$\angle b + 65° - 65° = 180° - 65°$
$\angle b = 115°$

$\angle d = 115°$ • $\angle d = \angle b$ because $\angle b$ and $\angle d$ are vertical angles.

A line intersecting two other lines at two different points is called a **transversal.**

If the lines cut by a transversal are parallel lines and the transversal is perpendicular to the parallel lines, then all eight angles formed are right angles.

Transversal

If the lines cut by a transversal are parallel lines and the transversal is not perpendicular to the parallel lines, then all four acute angles have the same measure and all four obtuse angles have the same measure. For the figure at the right,

Transversal

$\angle a = \angle c = \angle w = \angle y$ and $\angle b = \angle d = \angle x = \angle z$

Objective 12.1C

Vocabulary to Review
intersecting lines   [12.1A]
parallel lines   [12.1A]
supplementary angles   [12.1A]

New Vocabulary
vertical angles
adjacent angles
transversal
alternate interior angles
alternate exterior angles
corresponding angles

Discuss the Concepts
What is a transversal? Describe two different ways in which a transversal can intersect two other lines.

Concept Check
When a transversal intersects two parallel lines, which of the following are supplementary angles: vertical angles, adjacent angles, alternate interior angles, alternate exterior angles, or corresponding angles?
Adjacent angles

Optional Student Activity
In the figure below, $BD \parallel AC$ and $\angle BAC = \angle BCA$. Find the value of x.   65°

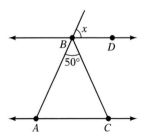

In-Class Examples (Objective 12.1C)

1. In the figure, $\angle a = 62°$. Find $\angle b$.   118°

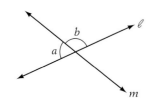

2. In the figure, $\ell_1 \parallel \ell_2$ and $\angle c = 103°$.
   a. Find $\angle a$.   103°
   b. Find $\angle b$.   77°

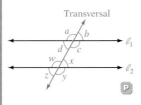

Transversal

Alternate interior angles are two nonadjacent angles that are on opposite sides of the transversal and between the parallel lines. For the figure at the left, $\angle c$ and $\angle w$ are alternate interior angles. $\angle d$ and $\angle x$ are alternate interior angles. Alternate interior angles have the same measure.

Alternate exterior angles are two nonadjacent angles that are on opposite sides of the transversal and outside the parallel lines. For the figure at the left, $\angle a$ and $\angle y$ are alternate exterior angles. $\angle b$ and $\angle z$ are alternate exterior angles. Alternate exterior angles have the same measure.

Corresponding angles are two angles that are on the same side of the transversal and are both acute angles or are both obtuse angles. For the figure at the top left, the following pairs of angles are corresponding angles: $\angle a$ and $\angle w$, $\angle d$ and $\angle z$, $\angle b$ and $\angle x$, $\angle c$ and $\angle y$. Corresponding angles have the same measure.

HOW TO In the figure at the left, $\ell_1 \parallel \ell_2$ and $\angle c = 58°$. Find the measures of $\angle f$, $\angle h$, and $\angle g$.

$\angle f = 58°$ • $\angle f = \angle c$ because $\angle f$ and $\angle c$ are alternate interior angles.

$\angle h = 58°$ • $\angle h = \angle c$ because $\angle c$ and $\angle h$ are corresponding angles.

$\angle g + \angle h = 180°$ • $\angle g$ is supplementary to $\angle h$.
$\angle g + 58° = 180°$ • $\angle h = 58°$
$\angle g = 122°$ • Subtract 58° from each side.

Example 7

In the figure, $\angle a = 75°$. Find $\angle b$.

Solution

$\angle a + \angle b = 180°$ • $\angle a$ and $\angle b$ are supplementary.

$75° + \angle b = 180°$ • $\angle a = 75°$
$\angle b = 105°$ • Subtract 75° from each side.

You Try It 7

In the figure, $\angle a = 125°$. Find $\angle b$.

Your solution

55°

Example 8

In the figure, $\ell_1 \parallel \ell_2$ and $\angle a = 70°$. Find $\angle b$.

Solution

$\angle c = \angle a = 70°$ • Corresponding angles are equal.

$\angle b + \angle c = 180°$ • $\angle b$ and $\angle c$ are supplementary.
$\angle b + 70° = 180°$ • $\angle c = 70°$
$\angle b = 110°$ • Subtract 70° from each side.

You Try It 8

In the figure, $\ell_1 \parallel \ell_2$ and $\angle a = 120°$. Find $\angle b$.

Your solution

60°

Solutions on p. S32

12.1 Exercises

Objective A   To define and describe lines and angles

Section 12.1

Suggested Assignment
Exercises 1–63, odds
More challenging problems:
  Exercises 65–67

1. The measure of an acute angle is between ___ and ___.
0°; 90°

2. The measure of an obtuse angle is between ___ and ___.
90°; 180°

3. How many degrees are in a straight angle?
180°

4. Two lines that intersect at right angles are ___ lines.
Perpendicular

5. In the figure, $EF = 20$ and $FG = 10$. Find the length of EG.   30

6. In the figure, $EF = 18$ and $FG = 6$. Find the length of EG.   24

7. In the figure, it is given that $QR = 7$ and $QS = 28$. Find the length of RS.   21

8. In the figure, it is given that $QR = 15$ and $QS = 45$. Find the length of RS.   30

9. In the figure, it is given that $AB = 12$, $CD = 9$, and $AD = 35$. Find the length of BC.   14

10. In the figure, it is given that $AB = 21$, $BC = 14$, and $AD = 54$. Find the length of CD.   19

11. Find the complement of a 31° angle.
59°

12. Find the complement of a 62° angle.
28°

13. Find the supplement of a 72° angle.
108°

14. Find the supplement of a 162° angle.
18°

15. Find the complement of a 13° angle.
77°

16. Find the complement of an 88° angle.
2°

17. Find the supplement of a 127° angle.
53°

18. Find the supplement of a 7° angle.
173°

In Exercises 19 and 20, find the measure of angle *AOB*.

19.

77°

20.

136°

Quick Quiz (Objective 12.1A)

1. How many degrees are in three-fifths of a revolution?   216°

2. Find the complement of a 54° angle.   36°

3. Find the supplement of a 22° angle.   158°

524

For Exercises 21 to 24, find the measure of angle *a*.

21.

22.

23.

24.

25. In the figure, it is given that ∠LOM = 53° and ∠LON = 139°. Find the measure of ∠MON.   86°

26. In the figure, it is given that ∠MON = 38° and ∠LON = 85°. Find the measure of ∠LOM.   47°

Objective B   **To define and describe geometric figures**

27. What is the sum of the three angles of a triangle?
180°

28. Name the side opposite the right angle in a right triangle.
Hypotenuse

29. Name a parallelogram with four right angles.
Rectangle or square

30. Name the rectangle with four equal sides.
Square

31. Name a rectangular solid in which all six faces are squares.
Cube

32. Name the solid in which all points are the same distance from the center.
Sphere

33. Name a quadrilateral in which opposite sides are parallel and equal.
Parallelogram, rectangle, or square

34. Name the plane figure in which all points are the same distance from the center.
Circle

35. Name the solid in which the bases are circular and perpendicular to the side.
Cylinder

36. Name a solid in which all the faces are rectangles.
Rectangular solid or cube

37. A triangle has a 13° angle and a 65° angle. Find the measure of the other angle.
102°

38. A triangle has a 105° angle and a 32° angle. Find the measure of the other angle.
43°

Quick Quiz (Objective 12.1B)

1. A triangle has a 110° angle and a 35° angle. Find the measure of the third angle.   35°

2. A right triangle has a 71° angle. Find the measure of the other two angles.   90°, 19°

3. Find the radius of a circle with a diameter of 15 ft.
7.5 ft

4. Find the diameter of a circle with a radius of 29 cm.   58 cm

39. A right triangle has a 45° angle. Find the measure of the other two angles.
90° and 45°

40. A right triangle has a 62° angle. Find the measure of the other two angles.
90° and 28°

41. A triangle has a 62° angle and a 104° angle. Find the measure of the other angle.
14°

42. A triangle has a 30° angle and a 45° angle. Find the measure of the other angle.
105°

43. A right triangle has a 25° angle. Find the measure of the other two angles.
90° and 65°

44. Two angles of a triangle are 42° and 105°. Find the measure of the other angle.
33°

45. Find the radius of a circle with a diameter of 16 in.
8 in.

46. Find the radius of a circle with a diameter of 9 ft.
$4\frac{1}{2}$ ft

47. Find the diameter of a circle with a radius of $2\frac{1}{3}$ ft.
$4\frac{2}{3}$ ft

48. Find the diameter of a circle with a radius of 24 cm.
48 cm

49. The radius of a sphere is 3.5 cm. Find the diameter.
7 cm

50. The radius of a sphere is $1\frac{1}{2}$ ft. Find the diameter.
3 ft

51. The diameter of a sphere is 4 ft 8 in. Find the radius.
2 ft 4 in.

52. The diameter of a sphere is 1.2 m. Find the radius.
0.6 m

Objective C   **To solve problems involving angles formed by intersecting lines**

For Exercises 53 to 56, find the measures of angles a and b.

53. 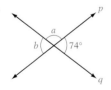 $\angle a = 106°$; $\angle b = 74°$

54. $\angle a = 131°$; $\angle b = 49°$

55. $\angle a = 112°$; $\angle b = 68°$

56. $\angle a = 131°$; $\angle b = 49°$

Quick Quiz (Objective 12.1C)

1. In the figure, $\angle a = 118°$. Find $\angle b$.   62°

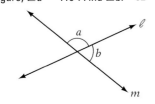

2. In the figure, $\ell_1 \parallel \ell_2$ and $\angle c = 67°$.
   a. Find $\angle a$.   67°
   b. Find $\angle b$.   113°

For Exercises 57 to 64, $\ell_1 \parallel \ell_2$. Find the measures of angles a and b.

57.

$\angle a = 38°; \angle b = 142°$

58.

$\angle a = 44°; \angle b = 44°$

59.

$\angle a = 58°; \angle b = 58°$

60.

$\angle a = 55°; \angle b = 125°$

61.

$\angle a = 152°; \angle b = 152°$

62.

$\angle a = 105°; \angle b = 75°$

63.

$\angle a = 130°; \angle b = 50°$

64.

$\angle a = 62°; \angle b = 118°$

APPLYING THE CONCEPTS

65. a. What is the smallest possible whole number of degrees in an angle of a triangle?   1°
b. What is the largest possible whole number of degrees in an angle of a right triangle?   90°

66. Determine whether the statement is always true, sometimes true, or never true.
a. Two lines that are both parallel to a third line are parallel to each other.   Always true
b. A triangle contains at least two acute angles.   Always true
c. Vertical angles are complementary angles.   Sometimes true

67. If AB and CD intersect at point O, and $\angle AOC = \angle BOC$, explain why AB is perpendicular to CD.

Answers to Writing Exercises

67. $\angle AOC$ and $\angle BOC$ are supplementary angles. Therefore, $\angle AOC + \angle BOC = 180°$. Because $\angle AOC = \angle BOC$, by substitution $\angle AOC + \angle AOC = 180°$. Therefore, $2(\angle AOC) = 180°$ and $\angle AOC = 90°$. Therefore, AB is perpendicular to CD.

12.2 Plane Geometric Figures

Objective A   **To find the perimeter of plane geometric figures**

A **polygon** is a closed figure determined by three or more line segments that lie in a plane. The **sides of a polygon** are the line segments that form the polygon. The figures below are examples of polygons.

      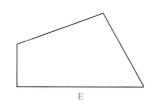

A   B   C   D   E

Point of Interest

Although a polygon is defined in terms of its *sides* (see the definition above), the word actually comes from the Latin word *polygonum*, which means having many *angles*. This is certainly the case for a polygon.

The Pentagon in Arlington, Virginia

A **regular polygon** is one in which each side has the same length and each angle has the same measure. The polygons in Figures A, C, and D above are regular polygons.

The name of a polygon is based on the number of its sides. The table below lists the names of polygons that have from 3 to 10 sides.

Number of Sides	Name of the Polygon
3	Triangle
4	Quadrilateral
5	Pentagon
6	Hexagon
7	Heptagon
8	Octagon
9	Nonagon
10	Decagon

Triangles and quadrilaterals are two of the most common types of polygons. Triangles are distinguished by the number of equal sides and also by the measures of their angles.

An **isosceles triangle** has two sides of equal length. The angles opposite the equal sides are of equal measure.
$AC = BC$
$\angle A = \angle B$

The three sides of an **equilateral triangle** are of equal length. The three angles are of equal measure.
$AB = BC = AC$
$\angle A = \angle B = \angle C$

A **scalene triangle** has no two sides of equal length. No two angles are of equal measure.

Objective 12.2A

Vocabulary to Review
plane geometric figure   [12.1A]
triangle   [12.1B]
right triangle   [12.1B]
quadrilateral   [12.1B]
parallelogram   [12.1B]
rectangle   [12.1B]
square   [12.1B]

New Vocabulary
polygon
sides of a polygon
regular polygon
isosceles triangle
equilateral triangle
scalene triangle
acute triangle
obtuse triangle
perimeter
circumference

New Symbols
π (pi)

New Formulas
Perimeter of a triangle:
  $P = a + b + c$
Perimeter of a rectangle:
  $P = 2L + 2W$
Perimeter of a square: $P = 4s$
Circumference of a circle:
  $C = \pi d$ or $C = 2\pi r$

Discuss the Concepts
1. Is every square a rectangle? Is every rectangle a square?
2. In the definition of a polygon, what does the phrase *closed figure* mean? Draw a figure that is not closed.

In-Class Examples (Objective 12.2A)
1. Find the perimeter of a square in which the sides are equal to 15 m.   60 m
2. Find the perimeter of a rectangle with a length of 5 m and a width of 1.4 m.   12.8 m
3. Find the circumference of a circle with a radius of 11 cm. Use 3.14 for π.   69.08 cm

Copyright © Houghton Mifflin Company. All rights reserved.

Instructor Note

The diagram at the right below shows the relationships among quadrilaterals. The description of each quadrilateral is within an example of that quadrilateral.

Concept Check

1. Figure A below is a rectangle. Label the length of the rectangle *L* and the width of the rectangle *W*.

Figure A

2. What is the name of a regular polygon that has three sides? Equilateral triangle

3. What is the name of a parallelogram in which all angles are the same measure?   Rectangle

An **acute triangle** has three acute angles.

An **obtuse triangle** has one obtuse angle.

A **right triangle** has a right angle.

Quadrilaterals are also distinguished by their sides and angles, as shown below. Note that a rectangle, a square, and a rhombus are different forms of a parallelogram.

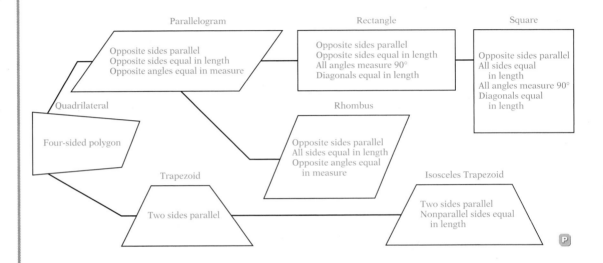

The **perimeter** of a plane geometric figure is a measure of the distance around the figure. Perimeter is used in buying fencing for a lawn or determining how much baseboard is needed for a room.

The perimeter of a triangle is the sum of the lengths of the three sides.

Perimeter of a Triangle

$P = a + b + c$

HOW TO   Find the perimeter of the triangle shown at the right.

$P = a + b + c$
$= 3 \text{ cm} + 5 \text{ cm} + 6 \text{ cm}$
$= 14 \text{ cm}$

The perimeter of the triangle is 14 cm.

The perimeter of a quadrilateral is the sum of the lengths of the four sides. The perimeter of a square is the sum of the four equal sides.

Perimeter of a Square

$P = 4s$

HOW TO   Find the perimeter of the square shown at the right.

$P = 4s$
$\quad = 4(3 \text{ ft})$ • $s = 3$ ft
$\quad = 12 \text{ ft}$

The perimeter of the square is 12 ft.

3 ft

3 ft

A rectangle is a quadrilateral with opposite sides of equal length. The length of a rectangle refers to the longer side, and the width refers to the length of the shorter side.

Perimeter of a Rectangle

$P = 2L + 2W$

L

W   W

L

HOW TO   Find the perimeter of the rectangle shown at the right.

$P = 2L + 2W$
$\quad = 2(6 \text{ m}) + 2(3 \text{ m})$ • $L = 6$ m, $W = 3$ m
$\quad = 12 \text{ m} + 6 \text{ m}$
$\quad = 18 \text{ m}$

The perimeter of the rectangle is 18 m.

3 m

6 m

The distance around a circle is called the **circumference.** The circumference of a circle is equal to the product of π (pi) and the diameter.

Circumference of a Circle

$C = \pi d$
  or
$C = 2\pi r$ • Because diameter = 2r

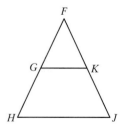

Radius

Diameter

The formula for circumference uses the number π (pi). The value of π can be approximated by a decimal or a fraction.

$$\pi \approx 3.14 \qquad \pi \approx \frac{22}{7}$$

The π key on a calculator gives a closer approximation of π than 3.14.

Point of Interest

Archimedes (c. 287–212 B.C.) was the mathematician who gave us the approximate value of π as $\frac{22}{7} = 3\frac{1}{7}$. He actually showed that π was between $3\frac{10}{71}$ and $3\frac{1}{7}$. The approximation $3\frac{10}{71}$ is closer to the exact value of π, but it is more difficult to use.

Optional Student Activity

1. The base of isosceles triangle ABC (with $AB = BC$ and $\angle B = 42°$) and the base of equilateral triangle CDE lie on line segment AE. Find the measure of $\angle BCD$.   51°

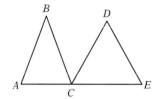

B   D

A   C   E

2. Three line segments are randomly chosen from line segments whose lengths are 1 cm, 2 cm, 3 cm, 4 cm, and 5 cm. What is the probability that a triangle can be formed from the line segments?
$\frac{3}{10}$

3. Triangle FJH is an isosceles triangle in which $FJ = FH$, $FK = KJ$, $FG = GH$. Find the perimeter of triangle FJH if $FK = 2x + 3$, $GH = 5x - 9$, and $JH = 4x$.   60 units

F

G   K

H   J

HOW TO Find the circumference of the circle shown at the right.

$C = 2\pi r$
$\approx 2 \cdot 3.14 \cdot 6 \text{ in.}$ • $r = 6$ in.
$= 37.68 \text{ in.}$

The circumference of the circle is approximately 37.68 in.

6 in.

Example 1

Find the perimeter of a rectangle with a width of $\frac{2}{3}$ ft and a length of 2 ft.

Solution

$\frac{2}{3}$ ft

2 ft

$P = 2L + 2W$
$= 2(2 \text{ ft}) + 2\left(\frac{2}{3} \text{ ft}\right)$ • $L = 2$ ft, $W = \frac{2}{3}$ ft
$= 4 \text{ ft} + \frac{4}{3} \text{ ft}$
$= 5\frac{1}{3} \text{ ft}$

The perimeter of the rectangle is $5\frac{1}{3}$ ft.

You Try It 1

Find the perimeter of a rectangle with a length of 2 m and a width of 0.85 m.

Your solution

5.7 m

Example 2

Find the perimeter of a triangle with sides 5 in., 7 in., and 8 in.

Solution

5 in.   7 in.

8 in.

$P = a + b + c$
$= 5 \text{ in.} + 7 \text{ in.} + 8 \text{ in.}$
$= 20 \text{ in.}$

The perimeter of the triangle is 20 in.

You Try It 2

Find the perimeter of a triangle with sides 12 cm, 15 cm, and 18 cm.

Your solution

45 cm

Example 3

Find the circumference of a circle with a radius of 18 cm. Use 3.14 for π.

Solution

18 cm

$C = 2\pi r$
$\approx 2 \cdot 3.14 \cdot 18 \text{ cm}$
$= 113.04 \text{ cm}$

The circumference is approximately 113.04 cm.

You Try It 3

Find the circumference of a circle with a diameter of 6 in. Use 3.14 for π.

Your solution

18.84 in.

Solutions on p. S32

Objective B — To find the perimeter of composite geometric figures

A **composite geometric figure** is a figure made from two or more geometric figures. The following composite is made from part of a rectangle and part of a circle:

$$\text{Perimeter of the composite figure} = 3 \text{ sides of a rectangle} + \tfrac{1}{2} \text{ the circumference of a circle}$$

$$\text{Perimeter of the composite figure} = 2L + W + \tfrac{1}{2}\pi d$$

The perimeter of the composite figure below is found by adding the measures of twice the length plus the width plus one-half the circumference of the circle.

$$P = 2L + W + \tfrac{1}{2}\pi d$$

$$\approx 2(12 \text{ m}) + 4 \text{ m} + \tfrac{1}{2}(3.14)(4 \text{ m})$$

• $L = 12$ m, $W = 4$ m, $d = 4$ m. *Note:* The diameter of the circle is equal to the width of the rectangle.

$$= 34.28 \text{ m}$$

The perimeter is approximately 34.28 m.

Example 4

Find the perimeter of the composite figure. Use $\tfrac{22}{7}$ for π.

Solution

$$\begin{array}{rcl}\text{Perimeter of composite figure} &=& \begin{array}{c}\text{sum of lengths of the 4 sides}\end{array} + \begin{array}{c}\tfrac{1}{2}\text{ the circumference of the circle}\end{array}\end{array}$$

$$P = 4s + \tfrac{1}{2}\pi d$$

$$\approx 4(5 \text{ cm}) + \tfrac{1}{2}\left(\tfrac{22}{7}\right)(7 \text{ cm})$$

$$= 20 \text{ cm} + 11 \text{ cm} = 31 \text{ cm}$$

The perimeter is approximately 31 cm.

You Try It 4

Find the perimeter of the composite figure. Use 3.14 for π.

Your solution
25.42 in.

Solution on p. S32

Copyright © Houghton Mifflin Company. All rights reserved.

Objective 12.2B

New Vocabulary
composite geometric figure

Optional Student Activity
(This activity is adapted from James Gray Propp's "The Slicing Game," *American Mathematical Monthly*, April 1996. Used with permission.)
  Shown below is a triangle with three "slices" through it. Note that the resulting pieces are diamonds, trapezoids, and a triangle.

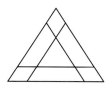

Determine how you can slice a triangle so that all the resulting pieces are triangles. Use three slices. Each slice must cut all the way through the triangle.
Solution:

(*Note:* You might also ask students to determine how to slice (a) a quadrilateral, (b) a pentagon, (c) a hexagon, and (d) a heptagon so that all the resulting pieces are triangles.)

In-Class Examples (Objective 12.2B)

Find the perimeter. Use 3.14 for π.

1. 95.98 m

2. 79.58 cm

Objective 12.2C

Optional Student Activity

1. Find the perimeter of the figure shown below. Use 3.14 for π. Round to the nearest hundredth.   32.37 cm

12 cm

40°

2. In triangle *ABC*, *AB* = 10 ft, *BC* = (*x* + 7) ft, and *AC* = (2*x* + 3) ft. The perimeter of the triangle is 32 ft. Is triangle *ABC* scalene, isosceles, or equilateral?
   Isosceles

3. The lengths of two sides of a triangle are 10 in. and 14 in. What are the possible values for *x*, the length of the third side of the triangle?
   x is greater than 4 in. and less than 24 in.

Objective C   To solve application problems

Example 5

The dimensions of a triangular sail are 18 ft, 11 ft, and 15 ft. What is the perimeter of the sail?

Strategy

11 ft   18 ft   15 ft

To find the perimeter, use the formula for the perimeter of a triangle.

Solution
$P = a + b + c$
$ = 18 \text{ ft} + 11 \text{ ft} + 15 \text{ ft}$
$ = 44 \text{ ft}$

The perimeter of the sail is 44 ft.

Example 6

If fencing costs \$4.75 per foot, how much will it cost to fence a rectangular lot that is 108 ft wide and 240 ft long?

Strategy

240 ft

108 ft

To find the cost of the fence:
• Find the perimeter of the lot.
• Multiply the perimeter by the per-foot cost of fencing.

Solution
$P = 2L + 2W$
$ = 2(240 \text{ ft}) + 2(108 \text{ ft})$
$ = 480 \text{ ft} + 216 \text{ ft}$
$ = 696 \text{ ft}$

Cost = 696 × 4.75 = 3306

The cost is \$3306.

You Try It 5

What is the perimeter of a standard piece of computer paper that measures $8\frac{1}{2}$ in. by 11 in.?

Your strategy

Your solution
39 in.

You Try It 6

Metal stripping is being installed around a workbench that is 0.74 m wide and 3 m long. At \$2.76 per meter, find the cost of the metal stripping. Round to the nearest cent.

Your strategy

Your solution
\$20.64

Solutions on p. S32

In-Class Examples (Objective 12.2C)

1. A horse trainer is planning to enclose a circular exercise area. How much fencing is needed if it has a diameter of 40 ft? Use 3.14 for π.   125.6 ft

2. The dimensions of a rectangular classroom are 30 ft by 24 ft. At \$2.20 per foot, how much does it cost to purchase an outlet strip to go around the room? Subtract 3 ft for a doorway.   \$231.00

12.2 Exercises

Objective A   **To find the perimeter of plane geometric figures**

For Exercises 1 to 8, find the perimeter or circumference of the given figures. Use 3.14 for π.

1.

12 in.   20 in.

24 in.

56 in.

2.

14 cm   13 cm

12 cm

39 cm

3.

5 ft

5 ft

20 ft

4.

2 m

2 m

8 m

5.

14 cm

32 cm

92 cm

6.

5 ft

18 ft

46 ft

7.

15 cm

47.1 cm

8.

4 in.

25.12 in.

9. Find the perimeter of a triangle with sides 2 ft 4 in., 3 ft, and 4 ft 6 in.
9 ft 10 in.

10. Find the perimeter of a rectangle with a length of 2 m and a width of 0.8 m.
5.6 m

11. Find the circumference of a circle with a radius of 8 cm. Use 3.14 for π.
50.24 cm

12. Find the circumference of a circle with a diameter of 14 in. Use $\frac{22}{7}$ for π.
44 in.

Section 12.2

Suggested Assignment
Exercises 1–37, odds
More challenging problems:
   Exercises 38–43

Quick Quiz (Objective 12.2A)

1. Find the perimeter of a square in which the sides are equal to 13.5 cm.   54 cm

2. Find the perimeter of a rectangle with a length of 3 m and a width of 0.75 m.   7.5 m

3. Find the circumference of a circle with a radius of 11 in. Use 3.14 for π.   69.08 in.

13. Find the perimeter of a square in which each side is equal to 60 m.

240 m

14. Find the perimeter of a triangle in which each side is $1\frac{2}{3}$ ft.

5 ft

15. Find the perimeter of a five-sided figure with sides of 22 cm, 47 cm, 29 cm, 42 cm, and 17 cm.

157 cm

16. Find the perimeter of a rectangular farm that is $\frac{1}{2}$ mi wide and $\frac{3}{4}$ mi long.

$2\frac{1}{2}$ mi

Objective B   **To find the perimeter of composite geometric figures**

For Exercises 17 to 24, find the perimeter. Use 3.14 for π.

17.

121 cm

18.

$20\frac{1}{6}$ ft

19.

50.56 m

20.

25.12 cm

21.

3.57 ft

22.

40.26 cm

23.

139.3 m

24.

13.107 ft

Quick Quiz (Objective 12.2B)

Find the perimeter. Use 3.14 for π.

1.

103.5 m

2.

59.98 in.

Objective C   **To solve application problems**

25. **Landscaping**   How many feet of fencing should be purchased for a rectangular garden that is 18 ft long and 12 ft wide?
    60 ft

26. **Interior Design**   Wall-to-wall carpeting is installed in a room that is 12 ft long and 10 ft wide. The edges of the carpet are nailed to the floor. Along how many feet must the carpet be nailed down?
    44 ft

27. **Quilting**   How many feet of binding are required to bind the edge of a rectangular quilt that measures 3.5 ft by 8.5 ft?
    24 ft

28. **Carpentry**   Find the length of molding needed to put around a circular table that is 3.8 ft in diameter. Use 3.14 for π.
    11.932 ft

29. **Landscaping**   The rectangular lot shown in the figure at the right is being fenced. The fencing along the road is to cost $6.20 per foot. The rest of the fencing costs $5.85 per foot. Find the total cost to fence the lot.
    $24,422.50

800 ft

1250 ft

30. **Sewing**   Bias binding is to be sewed around the edge of a rectangular tablecloth measuring 72 in. by 45 in. Each package of bias binding costs $3.50 and contains 15 ft of binding. How many packages of bias binding are needed for the tablecloth?
    2 packages

31. **Travel**   A bicycle tire has a diameter of 24 in. How many feet does the bicycle travel when the wheel makes 5 revolutions? Use 3.14 for π.
    31.4 ft

32. **Travel**   A tricycle tire has a diameter of 12 in. How many feet does the tricycle travel when the wheel makes 8 revolutions? Use 3.14 for π.
    25.12 ft

33. **Architecture**   The floor plan of a roller rink is shown in the figure at the right.
    a. Use estimation to determine whether the perimeter of the rink is more than 70 m or less than 70 m.   More than 70 m
    b. Calculate the perimeter of the roller rink. Use 3.14 for π.   81.4 m

25 m

10 m

34. **Home Improvement**   A rain gutter is being installed on a home that has the dimensions shown in the figure at the right. At a cost of $11.30 per meter, how much will it cost to install the rain gutter?
    $497.20

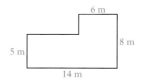
6 m

8 m

5 m

14 m

Quick Quiz (Objective 12.2C)

1. Find the length of rubber gasket needed to fit around a circular porthole that has a 20-inch diameter. Use 3.14 for π.   62.8 in.

2. A metal frame is being built to install around a rectangular sign that is 2 m wide and 3.5 m long. At $4.24 per meter, find the cost of the material for the frame. $46.64

35. **Home Improvement** Find the length of weather stripping installed around the arched door shown in the figure at the right. Use 3.14 for π.
20.71 ft

6 ft 6 in.

3 ft

36. **Astronomy** The distance from Earth to the sun is 93,000,000 mi. Approximate the distance Earth travels in making 1 revolution about the sun. Use 3.14 for π.
584,040,000 mi

37. **Earth Science** The distance from the surface to the center of Earth is 6356 km. Approximate the circumference of Earth. Use 3.14 for π.
39,915.68 km

APPLYING THE CONCEPTS

38. a. If the diameter of a circle is doubled, how many times larger is the resulting circumference? 2 times
   b. If the radius of a circle is doubled, how many times larger is the resulting circumference? 2 times

39. **Geometry** In the pattern to the right, the length of one side of a square is 1 unit. Find the perimeter of the eighth figure in the pattern.
22 units

40. **Geometry** Remove six toothpicks from the figure at the right in such a way as to leave two squares.

41. **Geometry** An equilateral triangle is placed inside an equilateral triangle as shown at the right. Now three more equilateral triangles are placed inside the unshaded equilateral triangles. The process is repeated again. Determine the perimeter of all the shaded triangles in Figure C.
$14\frac{1}{4}$ cm

|←2 cm→|
Figure A   Figure B   Figure C

42. **Metalwork** A wire whose length is given as x inches is bent into a square. Express the length of a side of the square in terms of x.
$\frac{1}{4}x$ or $\frac{x}{4}$

x

Answers to Writing Exercises

43. The ranger could measure the circumference of the trunk of the tree and then solve the equation $C = \pi d$ for d.

43. A forest ranger must determine the diameter of a redwood tree. Explain how the ranger could do this without cutting down the tree.

12.3 Area

Objective A   To find the area of geometric figures

Area is a measure of the amount of surface in a region. Area can be used to describe the size of a rug, a parking lot, a farm, or a national park. Area is measured in square units.

A square that measures 1 in. on each side has an area of 1 square inch, which is written 1 in².

A square that measures 1 cm on each side has an area of 1 square centimeter, which is written 1 cm².

Larger areas can be measured in square feet (ft²), square meters (m²), square miles (mi²), acres (43,560 ft²), or any other square unit.

The area of a geometric figure is the number of squares that are necessary to cover the figure. In the figures below, two rectangles have been drawn and covered with squares. In the figure on the left, 12 squares, each of area 1 cm², were used to cover the rectangle. The area of the rectangle is 12 cm². In the figure on the right, 6 squares, each of area 1 in², were used to cover the rectangle. The area of the rectangle is 6 in².

The area of the rectangle is 12 cm².

The area of the rectangle is 6 in².

Note from the above figures that the area of a rectangle can be found by multiplying the length of the rectangle by its width.

> **Area of a Rectangle**
>
> $A = LW$

HOW TO   Find the area of the rectangle shown at the right.

$A = LW$
$= (8 \text{ ft})(5 \text{ ft})$ • $L = 8 \text{ ft}, W = 5 \text{ ft}$
$= 40 \text{ ft}^2$

The area of the rectangle is 40 ft².

Objective 12.3A

New Vocabulary
area

New Symbols
in² (square inches)
cm² (square centimeters)
ft² (square feet)
m² (square meters)
mi² (square miles)

New Formulas
Area of a triangle: $A = \frac{1}{2} bh$
Area of a rectangle: $A = LW$
Area of a square: $A = s^2$
Area of a circle: $A = \pi r^2$

Discuss the Concepts
What is wrong with the statement?
a. The perimeter is 40 m².
b. The area is 120 ft.

Concept Check
The concepts of square units and area are difficult for students. After introducing these ideas, ask students questions such as the following:
1. Would I measure the distance from Chicago to Boston in miles or square miles?   Miles
2. Is the amount of land cultivated by a gardener measured in feet or square feet?   Square feet
3. How is the size of a state park measured?
    Acres or square miles
4. How is the length of a bedroom measured?
    Feet or meters

In-Class Examples (Objective 12.3A)

1. Find the area of a triangle with a base of 5 cm and a height of 2.6 cm.   6.5 cm²
2. Find the area of a square with a side of 8.5 ft.
    72.25 ft²
3. Find the area of a rectangle with a length of 37 in. and a width of 15 in.   555 in²
4. Find the area of a circle with a diameter of 16 in. Use 3.14 for π.   200.96 in²

538

Optional Student Activity

1. Use graph paper to draw different rectangles, each with a perimeter of 20 units. Investigate just whole-number dimensions. What dimensions will result in a rectangle with the greatest possible area?
   5×5

2. Ancient Egyptians gave the formula for the area of a circle as $\left(\frac{8}{9}d\right)^2$, where d is the diameter. Does this formula give an area that is less than or greater than the area given by the correct formula?
   Greater than

3. Choose a local pizza restaurant and a particular type of pizza. Determine the size (by diameter) and cost of a medium pizza and of a large pizza. (Use regular prices with no special discounts.) Find the ratio of the cost of a medium pizza to the size of a medium pizza and the ratio of the cost of a large pizza to the size of a large pizza. Explain what each ratio represents. Are the ratios approximately equal? Explain your findings and what they mean.

A square is a rectangle in which all sides are the same length. Therefore, both the length and the width can be represented by a side. Remember that $s \cdot s = s^2$.

Area of a Square

$A = s^2$

HOW TO Find the area of the square shown at the right.

$A = s^2$
$\quad = (14 \text{ cm})^2 \qquad \bullet \ s = 14 \text{ cm}$
$\quad = 196 \text{ cm}^2$

The area of the square is 196 cm².

The area of a circle is equal to the product of π and the square of the radius.

Area of a Circle

$A = \pi r^2$

HOW TO Find the area of the circle shown at the right.

$A = \pi r^2$
$\quad = \pi(8 \text{ in.})^2 = 64\pi \text{ in}^2$
$\quad \approx 64 \cdot 3.14 \text{ in}^2 = 200.96 \text{ in}^2$

The area is exactly $64\pi \text{ in}^2$.
The area is approximately 200.96 in².

In the figure below, AB is the base of the triangle, and CD, which is perpendicular to the base, is the height. The area of a triangle is one-half the product of the base and the height.

Area of a Triangle

$A = \frac{1}{2}bh$

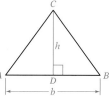

Integrating Technology

To calculate the area of the triangle shown at the right, you can enter

$20 \ \times \ 5 \ \div \ 2 \ =$

or

$.5 \ \times \ 20 \ \times \ 5 \ =$

HOW TO Find the area of the triangle shown below.

$A = \frac{1}{2}bh$

$\quad = \frac{1}{2}(20 \text{ m})(5 \text{ m}) \qquad \bullet \ b = 20 \text{ m}, h = 5 \text{ m}$

$\quad = 50 \text{ m}^2$

The area of the triangle is 50 m².

Objective 12.3B

Vocabulary to Review
composite geometric figure
[12.2B]

Optional Student Activity

Geometric probability involves using lengths or areas of geometric figures to determine the likelihood that an event will occur. Here are two exercises involving geometric probability.

1. Floating in a swimming pool measuring 8 m by 4.5 m is an inflatable raft measuring 0.8 m by 2.5 m. If an autumn leaf lands on a random point in the pool, what is the probability that it will land on the raft?

$$\frac{1}{18}$$

2. Gary Timken has a bulletin board that measures 16 in. by 24 in. in his college dorm room. On that bulletin board is a photo that measures 3.5 in. by 5.5 in. If a dart is thrown and lands at a random point on the bulletin board, what is the probability that the dart will hit the photograph? Write the answer as a decimal rounded to the nearest hundredth.   0.05

Example 1   Find the area of a circle with a diameter of 9 cm. Use 3.14 for π.

Solution $r = \frac{1}{2}d = \frac{1}{2}(9 \text{ cm}) = 4.5 \text{ cm}$

$A = \pi r^2$

$\approx 3.14(4.5 \text{ cm})^2 = 63.585 \text{ cm}^2$

The area is approximately 63.585 cm².

You Try It 1   Find the area of a triangle with a base of 24 in. and a height of 14 in.

Your solution   168 in²

Solution on p. S32

Objective B   **To find the area of composite geometric figures**

The area of the composite figure shown below is found by calculating the area of the rectangle and then subtracting the area of the triangle.

 = −

$A = LW - \frac{1}{2}bh$

$= (20 \text{ in.})(8 \text{ in.}) - \frac{1}{2}(20 \text{ in.})(3 \text{ in.}) = 160 \text{ in}^2 - 30 \text{ in}^2 = 130 \text{ in}^2$

Example 2

Find the area of the shaded portion of the figure. Use 3.14 for π.

8 m

8 m

Solution

 = −

$$\text{Area of shaded portion} = \text{area of square} - \text{area of circle}$$

$A = s^2 - \pi r^2$

$= (8 \text{ m})^2 - \pi (4 \text{ m})^2$

$\approx 64 \text{ m}^2 - 3.14(16 \text{ m}^2)$

$= 64 \text{ m}^2 - 50.24 \text{ m}^2 = 13.76 \text{ m}^2$

The area is approximately 13.76 m².

You Try It 2

Find the area of the composite figure.

6 in.

4 in.   10 in.

Your solution
48 in²

Solution on p. S33

In-Class Examples (Objective 12.3B)

Find the area. Use 3.14 for π.

1.   4 ft   390 ft²

11 ft

30 ft

2.   569.25 in²

18 in.

24 in.

Objective 12.3C

Optional Student Activity

(*Note:* For the first activity below, students will need to be able to look at an American flag.)

1. An American flag has dimensions of 65 in. by 39 in. Each short stripe has a length of 39 in. What fractional part of the area of the flag is red?
$\frac{27}{65}$

2. A rectangular box that has no top is made from pieces of cardboard. If the box is 8 in. long, 7 in. wide, and 6 in. high, what is the total area of the cardboard used to construct the box?   236 in^2

3. What is the maximum number of 3″ × 5″ index cards that will fit completely and without overlap on an $8\frac{1}{2}″ × 11″$ piece of paper?
5 index cards

Objective C   To solve application problems

Example 3

A walkway 2 m wide is built along the front and along both sides of a building, as shown in the figure. Find the area of the walkway.

Strategy

To find the area of the walkway, add the area of the front section (54 m · 2 m) and the area of the two side sections (each 35 m · 2 m).

Solution

$$\text{Area of walkway} = \underbrace{\text{area of front section}} + \underbrace{2(\text{area of one side section})}$$

$$A = (54 \text{ m})(2 \text{ m}) + 2(35 \text{ m})(2 \text{ m})$$
$$= 108 \text{ m}^2 \quad\quad + 140 \text{ m}^2$$
$$= 248 \text{ m}^2$$

The area of the walkway is 248 m^2.

You Try It 3

New carpet is installed in a room measuring 9 ft by 12 ft. Find the area of the room in square yards. (9 ft^2 = 1 yd^2)

Your strategy

Your solution
12 yd^2

Solution on p. S33

In-Class Examples (Objective 12.3C)

1. Find the area of a circular rug that is 12 ft in diameter. Use 3.14 for π.   113.04 ft^2

2. A room that measures 11 ft by 15 ft is to be carpeted. Find the number of square yards of carpet needed. (9 ft^2 = 1 yd^2) Round to the nearest tenth.   18.3 yd^2

12.3 Exercises

Objective A   **To find the area of geometric figures**

For Exercises 1 to 8, find the area of the given figures. Use 3.14 for π.

1.

6 ft

24 ft

144 ft^2

2.

8 in.

18 in.

144 in^2

3.

9 in.

9 in.

81 in^2

4.

4 in.

4 in.

16 in^2

5.

4 ft

50.24 ft^2

6.

3 cm

28.26 cm^2

7.

4 in.

10 in.

20 in^2

8.

6 m

7 m

21 m^2

9. Find the area of a right triangle with a base of 3 cm and a height of 1.42 cm.
2.13 cm^2

10. Find the area of a triangle with a base of 3 ft and a height of $\frac{2}{3}$ ft.
1 ft^2

11. Find the area of a square with a side of 4 ft.
16 ft^2

12. Find the area of a square with a side of 10 cm.
100 cm^2

Quick Quiz (Objective 12.3A)

1. Find the area of a triangle with a base of 10 ft and a height of 16 ft.   80 ft^2

2. Find the area of a square with a side of 16 cm.
256 cm^2

3. Find the area of a rectangle with a length of 64 cm and a width of 22 cm.   1408 cm^2

4. Find the area of a circle with a diameter of 26 in. Use 3.14 for π.   530.66 in^2

13. Find the area of a rectangle with a length of 43 in. and a width of 19 in.
817 in²

14. Find the area of a rectangle with a length of 82 cm and a width of 20 cm.
1640 cm²

15. Find the area of a circle with a radius of 7 in. Use $\frac{22}{7}$ for π.
154 in²

16. Find the area of a circle with a diameter of 40 cm. Use 3.14 for π.
1256 cm²

Objective B   **To find the area of composite geometric figures**

For Exercises 17 to 24, find the area. Use 3.14 for π.

17.

26 cm²

18.

50.13 in²

19.

2220 cm²

20.

1.3488 m²

21.

150.72 in²

22.

30 in²

23.

8.851323 ft²

24.

447.8208 cm²

Quick Quiz (Objective 12.3B)

Find the area. Use 3.14 for π.

1.

243 ft²

2.

343.07 m²

Objective C   **To solve application problems**

25. **Sports**   Artificial turf is being used to cover a playing field. If the field is rectangular with a length of 100 yd and a width of 75 yd, how much artificial turf must be purchased to cover the field?
7500 yd²

26. **Telescopes**   The telescope lens of the Hale telescope at Mount Palomar, California, has a diameter of 200 in. Find the area of the lens. Leave the answer in terms of π.
10,000 π in²

27. **Agriculture**   An irrigation system waters a circular field that has a 50-foot radius. Find the area watered by the irrigation system. Use 3.14 for π.
7850 ft²

28. **Interior Design**   A fabric wall hanging is to fill a space that measures 5 m by 3.5 m. Allowing for 0.1 m of the fabric to be folded back along each edge, how much fabric must be purchased for the wall hanging?
19.24 m²

29. **Home Improvement**   You plan to stain the wooden deck attached to your house. The deck measures 10 ft by 8 ft. A quart of stain will cost $9.95 and will cover 50 ft². How many quarts of stain should you buy?
2 qt

30. **Interior Design**   A carpet is to be installed in one room and a hallway, as shown in the diagram at the right. At a cost of $18.50 per square meter, how much will it cost to carpet the area?
$640.10

31. **Landscaping**   Find the area of a concrete driveway with the measurements shown in the figure.
1250 ft²

32. **Interior Design**   You want to tile your kitchen floor. The floor measures 12 ft by 9 ft. How many tiles, each a square with side $1\frac{1}{2}$ ft, should you purchase for the job?
48 tiles

33. **Interior Design**   You are wallpapering two walls of a child's room. One wall measures 9 ft by 8 ft, and the other measures 11 ft by 8 ft. The wallpaper costs $28.50 per roll, and each roll of the wallpaper will cover 40 ft². What is the cost to wallpaper the two walls?
$114

34. **Construction**   Find the area of the 2-meter boundary around the swimming pool shown in the figure.
68 m²

Quick Quiz (Objective 12.3C)

1. A circular skating rink has a 72-foot radius. Find the area of the surface of the ice rink. Use 3.14 for π.
16,277.76 ft²

2. Find the area of the floor of a rectangular greenhouse that is 24 m long and 16 m wide.
384 m²

35. **Parks** An urban renewal project involves reseeding a park that is in the shape of a square, 60 ft on each side. Each bag of grass seed costs $5.75 and will seed 1200 ft². How much money should be budgeted for buying grass seed for the park?
$17.25

36. **Architecture** The roller rink shown in the figure at the right is to be covered with hardwood floor.

80 ft

175 ft
   a. Without doing the calculations, indicate whether the area of the rink is more than 8000 ft² or less than 8000 ft². More than 8000 ft²
   b. Calculate how much hardwood floor is needed to cover the roller rink. Use 3.14 for π. 19,024 ft²

37. **Parks** Find the total area of the national park with the dimensions shown in the figure. Use 3.14 for π.
125.1492 mi²

12.7 mi

2.5 mi

4.3 mi

17.5 mi

38. **Interior Design** Find the cost of plastering the walls of a room 22 ft wide, 25 ft 6 in. long, and 8 ft high. Subtract 120 ft² for windows and doors. The cost is $2.50 per square foot.
$1600

39. a. A circle has a radius of 8 in. Find the increase in area when the radius is increased by 2 in. Use 3.14 for π. 113.04 in²
   b. A circle has a radius of 5 cm. Find the increase in area when the radius is doubled. Use 3.14 for π. 235.5 cm²

APPLYING THE CONCEPTS

40. **Geometry** What fractional part of the area of the larger of the two squares is the shaded area? Write your answer as a fraction in simplest form. This problem appeared in *Math Teacher*, vol. 86, No. 3 (September 1993). $\frac{1}{64}$

12

3

3

12

41. a. If both the length and the width of a rectangle are doubled, how many times larger is the area of the resulting rectangle? 4 times
   b. If the radius of a circle is doubled, what happens to the area? Quadrupled
   c. If the diameter of a circle is doubled, what happens to the area? Quadrupled

42. The circles at the right are identical. Is the area in the circles to the left of the line equal to, less than, or greater than the area in the circles to the right of the line? Explain your answer.

43. Determine whether the statement is always true, sometimes true, or never true.
   a. If two triangles have the same perimeter, then they have the same area. Sometimes true
   b. If two rectangles have the same area, then they have the same perimeter. Sometimes true
   c. If two squares have the same area, then the sides of the squares have the same length. Always true

44. **Geometry** All of the dots at the right are equally spaced, horizontally and vertically, 1 inch apart. What is the area of the triangle?
$2\frac{1}{2}$ in²

Answers to Writing Exercises

42. The area in the circles to the left of the line is equal to the area in the circles to the right of the line. Note that in the circle at the left in the top row, the line goes through the center of the circle; thus it is a diameter of the circle, and half the area lies on one side of the line and half lies on the other side of the line. A complete circle lies on each side of the line; the circle at the right in the top row is on one side, and the circle at the left in the bottom row lies on the other side. For the two circles at the right on the bottom row, half their combined area lies on the left side of the line, and half lies on the right side of the line.

12.4 Volume

Objective A   To find the volume of geometric solids

Volume is a measure of the amount of space inside a closed surface, or figure in space. Volume can be used to describe the amount of heating gas used for cooking, the amount of concrete delivered for the foundation of a house, or the amount of water in storage for a city's water supply.

A cube that is 1 ft on each side has a volume of 1 cubic foot, which is written 1 ft³.

A cube that measures 1 cm on each side has a volume of 1 cubic centimeter, which is written 1 cm³.

The volume of a solid is the number of cubes that are necessary to fill the solid exactly. The volume of the rectangular solid at the right is 24 cm³ because it will hold exactly 24 cubes, each 1 cm on a side. Note that the volume can be found by multiplying the length times the width times the height.

> **Volume of a Rectangular Solid**
> $V = LWH$

HOW TO   Find the volume of a rectangular solid with a length of 9 in., a width of 3 in., and a height of 4 in.

$V = LWH$
  $= (9 \text{ in.})(3 \text{ in.})(4 \text{ in.})$ • $L = 9$ in., $W = 3$ in.,
  $= 108 \text{ in}^3$ $H = 4$ in.

The volume of the rectangular solid is 108 in³.

Objective 12.4A

New Vocabulary
volume

New Symbols
in³ (cubic inches)
cm³ (cubic centimeters)
ft³ (cubic feet)
m³ (cubic meters)

New Formulas
Volume of a rectangular solid:
  $V = LWH$
Volume of a cube: $V = s^3$
Volume of a sphere: $V = \frac{4}{3}\pi r^3$
Volume of a cylinder: $V = \pi r^2 h$

Discuss the Concepts
The difficulty students have distinguishing linear measure from square measure is compounded with volume measure. Ask students to give examples of things that would be measured in, for instance, feet, square feet, and cubic feet—for example, the length of a room, the area of the floor, and the volume of air in the room. Here are two more examples.

a. The distance across a lake, the area of the surface of the lake, and the volume of water in the lake

b. The length of a driveway, the area of the driveway that needs to be plowed, and the volume of asphalt used to pave the driveway

In-Class Examples (Objective 12.4A)
Use 3.14 for π. Round to the nearest hundredth.

1. Find the volume, in cubic meters, of a rectangular solid with a length of 6 m, a width of 400 cm, and a height of 4.5 m.   108 m³

2. Find the volume of a cube with a side of 5.5 ft.
166.375 ft³

3. Find the volume of a sphere with a radius of 4 mm.
267.95 mm³

4. Find the volume of a cylinder with a radius of 15 cm and a height of 14 cm.   9891 cm³

Concept Check

1. Indicate which of the
following are rectangular
solids: a juice box, a milk
carton, a can of soup, a
compact disk, the plastic
container a compact disk is
packaged in.
A juice box and the plastic
container a compact disk is
packaged in

2. Indicate which of the
following units could not be
used to measure the volume
of a cylinder: ft³, m³, yd², cm³,
mi.   Yd², mi

The length, width, and height of a cube have the same measure. The volume of
a cube is found by multiplying the side of the cube times itself three times (side
cubed).

> **Volume of a Cube**
> $V = s^3$

HOW TO   Find the volume of the cube shown at the right.   3 ft

$$V = s^3$$
$$\quad = (3\ \text{ft})^3 \qquad \bullet\ s = 3\ \text{ft}$$
$$\quad = 27\ \text{ft}^3$$

The volume of the cube is 27 ft³.

The volume of a sphere is found by multiplying four-thirds times pi (π) times the
radius cubed.

> **Volume of a Sphere**
> $V = \dfrac{4}{3}\pi r^3$

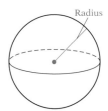

Radius

HOW TO   Find the volume of the sphere shown below. Use 3.14 for π.
Round to the nearest hundredth.

$$V = \frac{4}{3}\pi r^3$$
$$\quad \approx \frac{4}{3}(3.14)(2\ \text{in.})^3 \qquad \bullet\ r = 2\ \text{in}$$
$$\quad = \frac{4}{3}(3.14)(8\ \text{in}^3)$$
$$\quad \approx 33.49\ \text{in}^3$$

The volume is approximately 33.49 in³.

2 in.

The volume of a cylinder is found by multiplying the area of the base of the cylinder (a circle) times the height.

> **Volume of a Cylinder**
> $V = \pi r^2 h$

Height

Radius

$V = \pi r^2 h$
  $\approx 3.14(3 \text{ cm})^2(8 \text{ cm})$ • $r = 3$ cm, $h = 8$ cm
  $= 3.14(9 \text{ cm}^2)(8 \text{ cm})$
  $= 226.08 \text{ cm}^3$

8 cm

Radius = 3 cm

The volume of the cylinder is approximately 226.08 cm³.

Optional Student Activity

1. A foot is what fraction of a yard?
$\dfrac{1}{3}$

2. A square foot is what fraction of a square yard?
$\dfrac{1}{9}$

3. A cubic foot is what fraction of a cubic yard?
$\dfrac{1}{27}$

4. An inch is what fraction of a foot?
$\dfrac{1}{12}$

5. A square inch is what fraction of a square foot?
$\dfrac{1}{144}$

6. A cubic inch is what fraction of a cubic foot?
$\dfrac{1}{1728}$

7. A centimeter is what fraction of a meter?
$\dfrac{1}{100}$

8. A square centimeter is what fraction of a square meter?
$\dfrac{1}{10,000}$

9. A cubic centimeter is what fraction of a cubic meter?
$\dfrac{1}{1,000,000}$

Example 1

Find the volume of a rectangular solid with a length of 3 ft, a width of 1.5 ft, and a height of 2 ft.

Solution
$V = LWH$
  $= (3 \text{ ft})(1.5 \text{ ft})(2 \text{ ft})$
  $= 9 \text{ ft}^3$

The volume is 9 ft³.

You Try It 1

Find the volume of a rectangular solid with a length of 8 cm, a width of 3.5 cm, and a height of 4 cm.

Your solution
112 cm³

Example 2

Find the volume of a cube that has a side measuring 2.5 in.

Solution
$V = s^3$
  $= (2.5 \text{ in.})^3$
  $= 15.625 \text{ in}^3$

The volume is 15.625 in³.

You Try It 2

Find the volume of a cube with a side of 5 cm.

Your solution
125 cm³

Example 3

Find the volume of a cylinder with a radius of 12 cm and a height of 65 cm. Use 3.14 for π.

Solution
$V = \pi r^2 h$
  $\approx 3.14(12 \text{ cm})^2(65 \text{ cm})$
  $= 3.14(144 \text{ cm}^2)(65 \text{ cm})$
  $= 29,390.4 \text{ cm}^3$

The volume is approximately 29,390.4 cm³.

You Try It 3

Find the volume of a cylinder with a diameter of 14 in. and a height of 15 in. Use $\frac{22}{7}$ for π.

Your solution
2310 in³

Solutions on p. S33

Example 4

Find the volume of a sphere with a diameter of 12 in. Use 3.14 for π.

Solution

$r = \dfrac{1}{2}d = \dfrac{1}{2}(12 \text{ in.}) = 6 \text{ in.}$ • Find the radius.

$V = \dfrac{4}{3}\pi r^3$ • Use the formula for the volume of a sphere.

$\approx \dfrac{4}{3}(3.14)(6 \text{ in.})^3$

$= \dfrac{4}{3}(3.14)(216 \text{ in}^3)$

$= 904.32 \text{ in}^3$

The volume is approximately 904.32 in³.

You Try It 4

Find the volume of a sphere with a radius of 3 m. Use 3.14 for π.

Your solution

113.04 m³

Solution on p. S33

Solution on p. S33

Objective 12.4B

New Vocabulary

composite geometric solid

Optional Student Activity

The dimensions of a grocery bag are 7 in. by 12 in. by 17 in. The dimensions of a lunch bag are 4 in. by 6 in. by 12 in.

a. Find the difference between the volume of a grocery bag and a lunch bag. 1140 in³

b. If 3 in. is added to the height of a lunch bag, what is the change in the volume? 72 in³

c. How much larger is the volume of a lunch bag if 3 in. is added to the width of the base of the bag rather than to its height? 144 in³

d. If 3 in. is added to each dimension of a lunch bag, what is the change in the volume? 657 in³

e. Is this more or less than twice the original volume? More than twice the original volume

Objective B To find the volume of composite geometric solids

A **composite geometric solid** is a solid made from two or more geometric solids. The solid shown is made from a cylinder and one-half of a sphere.

Volume of the composite solid = volume of the cylinder + $\dfrac{1}{2}$ the volume of the sphere

HOW TO Find the volume of the composite solid shown above if the radius of the base of the cylinder is 3 in. and the height of the cylinder is 10 in. Use 3.14 for π.

The volume equals the volume of a cylinder plus one-half the volume of a sphere. The radius of the sphere equals the radius of the base of the cylinder.

$V = \pi r^2 h + \dfrac{1}{2}\left(\dfrac{4}{3}\pi r^3\right)$

$\approx 3.14(3 \text{ in.})^2(10 \text{ in.}) + \dfrac{1}{2}\left(\dfrac{4}{3}\right)(3.14)(3 \text{ in.})^3$

$= 3.14(9 \text{ in}^2)(10 \text{ in.}) + \dfrac{1}{2}\left(\dfrac{4}{3}\right)(3.14)(27 \text{ in}^3)$

$= 282.6 \text{ in}^3 + 56.52 \text{ in}^3$

$= 339.12 \text{ in}^3$

The volume is approximately 339.12 in³.

In-Class Examples (Objective 12.4B)

Find the volume. Use 3.14 for π.

1. 1.02 ft³

2. 1695.6 in³

Example 5

Find the volume of the solid in the figure. Use 3.14 for π.

Solution

| Volume of the solid | = | volume of rectangular solid | + | volume of cylinder |

$V = LWH + \pi r^2 h$
$\approx (8 \text{ cm})(8 \text{ cm})(2 \text{ cm}) + 3.14(1 \text{ cm})^2(2 \text{ cm})$
$= 128 \text{ cm}^3 + 6.28 \text{ cm}^3$
$= 134.28 \text{ cm}^3$

The volume is approximately 134.28 cm³.

You Try It 5

Find the volume of the solid in the figure. Use 3.14 for π.

Your solution
0.64192 m³

Example 6

Find the volume of the solid in the figure. Use 3.14 for π.

Solution

| Volume of the solid | = | volume of rectangular solid | − | volume of cylinder |

$V = LWH - \pi r^2 h$
$\approx (80 \text{ m})(40 \text{ m})(30 \text{ m}) - 3.14(14 \text{ m})^2(80 \text{ m})$
$= 96,000 \text{ m}^3 - 49,235.2 \text{ m}^3$
$= 46,764.8 \text{ m}^3$

The volume is approximately 46,764.8 m³.

You Try It 6

Find the volume of the solid in the figure. Use 3.14 for π.

Your solution
915.12 in³

Solutions on p. S33

Objective 12.4C

Objective C   **To solve application problems**

Optional Student Activity

(*Note:* You might ask students to estimate an answer to Exercise 2 below before performing the calculations. They will probably be surprised at the result.)

1. When an object is placed in water, the object displaces an amount of water that is equal to the volume of the object.

   a. A sphere with a diameter of 4 in. is placed in a rectangular tank of water. Find the volume of the water that is displaced by the sphere. Round to the nearest tenth.   33.5 in³

   b. The radius of the base of a cylinder is 2 cm, and its height is 10 cm. The cylinder is submerged in a tank of water. Find the volume of the water that is displaced by the cylinder. Round to the nearest tenth.   125.6 cm³

2. A cylindrical oil tank has a diameter of 100 ft and a height of 70 ft. How many gallons of oil can be stored in the tank? 7.48 gal of oil can be stored in a volume of 1 ft³. Round to the nearest thousand.   4,110,000 gal

---

Example 7

An aquarium is 28 in. long, 14 in. wide, and 16 in. high. Find the volume of the aquarium.

Strategy
To find the volume of the aquarium, use the formula for the volume of a rectangular solid.

Solution
$V = LWH$
$= (28 \text{ in.})(14 \text{ in.})(16 \text{ in.})$
$= 6272 \text{ in}^3$

The volume of the aquarium is 6272 in³.

Example 8

Find the volume of the bushing shown in the figure below. Use 3.14 for π.

Strategy
To find the volume of the bushing, subtract the volume of the half-cylinder from the volume of the rectangular solid.

Solution

Volume of the bushing	=	volume of rectangular solid	$-$	$\frac{1}{2}$ the volume of cylinder

$V = LWH - \frac{1}{2}\pi r^2 h$

$\approx (8 \text{ cm})(4 \text{ cm})(4 \text{ cm}) - \frac{1}{2}(3.14)(1 \text{ cm})^2(8 \text{ cm})$

$= 128 \text{ cm}^3 - 12.56 \text{ cm}^3$

$= 115.44 \text{ cm}^3$

The volume of the bushing is approximately 115.44 cm³.

You Try It 7

Find the volume of a freezer that is 7 ft long, 3 ft high, and 2.5 ft wide.

Your strategy

Your solution
52.5 ft³

You Try It 8

Find the volume of the channel iron shown in the figure below.

Your strategy

Your solution
3.4 ft³

Solutions on p. S33

In-Class Examples (Objective 12.4C)

1. A propane gas storage tank, which is in the shape of a cylinder, is 8 m high and has a 5-meter diameter. Find the volume of the gas storage tank. Use 3.14 for π.   157 m³

2. How many gallons of water will fill a fish tank that is 30 in. long, 10 in. wide, and 9 in. high? Round to the nearest tenth. (1 gal = 231 in³)   11.7 gal

12.4 Exercises

Section 12.4

Suggested Assignment
Exercises 1–35, odds
More challenging problems:
Exercises 36–39

Objective A To find the volume of geometric solids

For Exercises 1 to 8, find the volume. Round to the nearest hundredth. Use 3.14 for π.

1.

144 cm³

2.

240 ft³

3.

512 in³

4.

1728 m³

5.

2143.57 in³

6.

179.50 in³

7.

150.72 cm³

8.

157 ft³

Quick Quiz (Objective 12.4A)
Use 3.14 for π. Round to the nearest hundredth.

1. Find the volume, in cubic meters, of a rectangular solid with a length of 3 m, a width of 90 cm, and a height of 5 m.   13.5 m³

2. Find the volume of a cube with a side of 8 ft.   512 ft³

3. Find the volume of a sphere with a 5-foot diameter.   65.42 ft³

4. Find the volume of a cylinder with a radius of 30 cm and a height of 42 cm.   118,692 cm³

For Exercises 9 to 16, find the volume.

9. Find the volume, in cubic meters, of a rectangular solid with a length of 2 m, a width of 80 cm, and a height of 4 m.
6.4 m³

10. Find the volume of a cylinder with a radius of 7 cm and a height of 14 cm. Use $\frac{22}{7}$ for π.
2156 cm³

11. Find the volume of a sphere with an 11-millimeter radius. Use 3.14 for π. Round to the nearest hundredth.
5572.45 mm³

12. Find the volume of a cube with a side of 2.14 m. Round to the nearest tenth.
9.8 m³

13. Find the volume of a cylinder with a diameter of 12 ft and a height of 30 ft. Use 3.14 for π.
3391.2 ft³

14. Find the volume of a sphere with a 6-foot diameter. Use 3.14 for π.
113.04 ft³

15. Find the volume of a cube with a side of $3\frac{1}{2}$ ft.
$42\frac{7}{8}$ ft³

16. Find the volume, in cubic meters, of a rectangular solid with a length of 1.15 m, a width of 60 cm, and a height of 25 cm.
0.1725 m³

Objective B   **To find the volume of composite geometric solids**

For Exercises 17 to 22, find the volume. Use 3.14 for π.

17.
82.26 in³

18.
395.64 ft³

19.
1.6688 m³

20.
5 m³

Quick Quiz (Objective 12.4B)
Find the volume. Use 3.14 for π.

1.
19,440 in³

2.
699.12 in³

21.

69.08 in³

22.

4578.12 cm³

Objective C   **To solve application problems**

For Exercises 23 to 35, solve. Use 3.14 for π.

23. Fish Hatchery   A rectangular tank at the fish hatchery is 9 m long, 3 m wide, and 1.5 m deep. Find the volume of the water in the tank when the tank is full.
40.5 m³

24. Rocketry   A fuel tank in a booster rocket is a cylinder 10 ft in diameter and 52 ft high. Find the volume of the fuel tank.
4082 ft³

25. Ballooning   A hot air balloon is in the shape of a sphere. Find the volume of a hot air balloon that is 32 ft in diameter. Round to the nearest hundredth.
17,148.59 ft³

26. Petroleum   A storage tank for propane is in the shape of a sphere that has a diameter of 9 m. Find the volume of the tank.
381.51 m³

27. Petroleum   An oil tank, which is in the shape of a cylinder, is 4 m high and has a diameter of 6 m. The oil tank is two-thirds full. Find the number of cubic meters of oil in the tank. Round to the nearest hundredth.
75.36 m³

28. Agriculture   A silo, which is in the shape of a cylinder, is 16 ft in diameter and has a height of 30 ft. The silo is three-fourths full. Find the volume of the portion of the silo that is not being used for storage.
1507.2 ft³

29. Architecture   An architect is designing the heating system for an auditorium and needs to know the volume of the structure. Find the volume of the auditorium with the measurements shown in the figure.
809,516.25 ft³

30. Pools   A swimming pool 50 ft long and 13 ft wide contains water to a depth of 10 ft. Find the total weight of the water in the swimming pool. (1 ft³ weighs 62.4 lb.)
405,600 lb

Quick Quiz (Objective 12.4C)

1. A silo, which is in the shape of a cylinder, is 24 ft in diameter and has a height of 40 ft. Find the volume of the silo. Use 3.14 for π.   18,086.4 ft³

2. How many gallons of water will fill an aquarium that is 20 in. wide, 18 in. high, and 36 in. long? Round to the nearest tenth. (1 gal = 231 in³)
56.1 gal

31. Metal Works   Find the volume of the bushing shown at the right.
212.64 in³

32. Aquariums   How many gallons of water will fill an aquarium that is 12 in. wide, 18 in. long, and 16 in. high? Round to the nearest tenth. (1 gal = 231 in³)
15.0 gal

33. Aquariums   How many gallons of water will fill a fish tank that is 12 in. long, 8 in. wide, and 9 in. high? Round to the nearest tenth. (1 gal = 231 in³)
3.7 gal

34. Petroleum   A truck carrying an oil tank is shown in the figure at the right.
  a. Without doing the calculations, determine whether the volume of the oil tank is more than 240 ft³ or less than 240 ft³.   More than 240 ft³
  b. If the tank is half full, how many cubic feet of oil is the truck carrying? Round to the nearest hundredth.   887.57 ft³

35. Construction   The concrete floor of a building is shown in the figure at the right. At a cost of $5.85 per cubic foot, find the cost of having the floor poured. Round to the nearest cent.
$6526.41

APPLYING THE CONCEPTS

36. Half a sphere is called a hemisphere. Derive a formula for the volume of a hemisphere.
$$\frac{2}{3}\pi r^3$$

37. a. If both the length and the width of a rectangular solid are doubled, how many times larger is the resulting rectangular solid?   4 times larger
  b. If the length, width, and height of a rectangular solid are all doubled, how many times larger is the resulting rectangular solid?   8 times larger
  c. If the side of a cube is doubled, how many times larger is the resulting cube?   8 times larger

For Exercises 38 to 41, explain how you could cut through a cube so that the face of the resulting solid is the given geometric figure.

38. A square   **39.** An equilateral triangle

40. A trapezoid   **41.** A hexagon

42.   Suppose a cylinder is cut into 16 equal pieces, which are then arranged as shown at the right. The figure resembles a rectangular solid. What variable expressions could be used to represent the length, width, and height of the rectangular solid? Explain how the formula for the volume of a cylinder is derived from this approach.

12.5 The Pythagorean Theorem

Objective A To find the square root of a number

The area of a square is 36 in². What is the length of one side?

Area of the square = (side)²
36 = side · side

What number multiplied times itself equals 36?

36 = 6 · 6

The side of the square is 6 in.

The **square root** of a number is one of two identical factors of that number. The square root symbol is $\sqrt{}$.

The square root of 36 is 6.

$$\sqrt{36} = 6$$

A **perfect square** is the product of a whole number times itself.

1, 4, 9, 16, 25, and 36 are perfect squares.

1 · 1 = 1	$\sqrt{1} = 1$
2 · 2 = 4	$\sqrt{4} = 2$
3 · 3 = 9	$\sqrt{9} = 3$
4 · 4 = 16	$\sqrt{16} = 4$
5 · 5 = 25	$\sqrt{25} = 5$
6 · 6 = 36	$\sqrt{36} = 6$

The square root of a perfect square is a whole number.

If a number is not a perfect square, its square root can only be approximated. The approximate square roots of numbers can be found using a calculator. For example:

Number	Square Root
33	$\sqrt{33} \approx 5.745$
34	$\sqrt{34} \approx 5.831$
35	$\sqrt{35} \approx 5.916$

Point of Interest

The square root of a number that is not a perfect square is an irrational number. There is evidence of irrational numbers as early as 500 B.C. These numbers were not very well understood, and they were given the name *numerus surdus*. This phrase comes from the Latin word *surdus*, which means deaf or mute. Thus irrational numbers were "inaudible numbers."

Area = 36 in²

Example 1
a. Find the square roots of the perfect squares 49 and 81.
b. Find the square roots of 27 and 108. Round to the nearest thousandth.

Solution
a. $\sqrt{49} = 7$ $\sqrt{81} = 9$
b. $\sqrt{27} \approx 5.196$ $\sqrt{108} \approx 10.392$

You Try It 1
a. Find the square roots of the perfect squares 16 and 169.
b. Find the square roots of 32 and 162. Round to the nearest thousandth.

Your solution
a. $\sqrt{16} = 4$ $\sqrt{169} = 13$
b. $\sqrt{32} \approx 5.657$ $\sqrt{162} \approx 12.728$

Solution on p. S33

Objective 12.5A

New Vocabulary
square root
perfect square

New Symbols
$\sqrt{}$ (square root)

Discuss the Concepts
What is a perfect square? How can you mentally find the square root of a perfect square?

Optional Student Activity

1. Find a perfect square that is between 350 and 400.   361
2. What is the smallest integer larger than $\sqrt{3} + \sqrt{5}$?   4
3. If $x = 4$, what is the value of $\sqrt{3x - 8}$?   2
4. Find the two-digit perfect square that has exactly nine factors.   36
5. Find two whole numbers such that their difference is 10, the smaller number is a perfect square, and the larger number is two less than a perfect square.   14 and 4
6. Find the smallest integer n such that $\sqrt{1 + 2 + 3 + \cdots + n}$ simplifies to an integer.   8

In-Class Examples (Objective 12.5A)
Find the square root. Round to the nearest thousandth.
1. 43   6.557
2. 129   11.358

Objective 12.5B

Objective B   **To find the unknown side of a right triangle using the Pythagorean Theorem**

Vocabulary to Review
hypotenuse   [12.1B]
legs   [12.1B]

New Vocabulary
Pythagorean Theorem

New Formulas
Hypotenuse $= \sqrt{(\text{leg})^2 + (\text{leg})^2}$
Leg $= \sqrt{(\text{hypotenuse})^2 - (\text{leg})^2}$

Discuss the Concepts
How can you tell which is the hypotenuse and which are the legs of a right triangle?

Concept Check
1. Label the right triangle below. Include the right-angle symbol, the hypotenuse, and the legs.

2. The diagonal of a rectangle is a line drawn from one vertex to the opposite vertex. Find the length of the diagonal of a rectangle that is 5 m wide and 11 m long. Round to the nearest tenth.   12.1 m

Optional Student Activity
The numbers 3, 4, and 5 are called a **Pythagorean triple** because they are natural numbers that satisfy the equation of the Pythagorean Theorem.

In each exercise below, determine whether the numbers are a Pythagorean triple.

a. 5, 7, and 9   No

b. 8, 15, and 17   Yes

c. 11, 60, and 61   Yes

d. 28, 45, and 53   Yes
(Continued on next page)

Point of Interest

The first known proof of the Pythagorean Theorem is in a Chinese textbook that dates from 150 B.C. The book is called *Nine Chapters on the Mathematical Art*. The diagram below is from that book and was used in the proof of the theorem.

The Greek mathematician Pythagoras is generally credited with the discovery that the square of the hypotenuse of a right triangle is equal to the sum of the squares of the two legs. This is called the **Pythagorean Theorem.** However, the Babylonians used this theorem more than 1000 years before Pythagoras lived.

Square of the hypotenuse	equals	sum of the squares of the two legs
5^2	$=$	$3^2 + 4^2$
25	$=$	$9 + 16$
25	$=$	25

If the length of one side of a right triangle is unknown, one of the following formulas can be used to find it.

If the hypotenuse is unknown, use

$$\textbf{Hypotenuse} = \sqrt{(\textbf{leg})^2 + (\textbf{leg})^2}$$
$$= \sqrt{(3)^2 + (4)^2}$$
$$= \sqrt{9 + 16}$$
$$= \sqrt{25} = 5$$

If the length of a leg is unknown, use

$$\textbf{Leg} = \sqrt{(\textbf{hypotenuse})^2 - (\textbf{leg})^2}$$
$$= \sqrt{(5)^2 - (4)^2}$$
$$= \sqrt{25 - 16}$$
$$= \sqrt{9} = 3$$

Example 2

Find the hypotenuse of the triangle in the figure. Round to the nearest thousandth.

Solution
Hypotenuse $= \sqrt{(\text{leg})^2 + (\text{leg})^2}$
$= \sqrt{8^2 + 4^2}$
$= \sqrt{64 + 16}$
$= \sqrt{80} \approx 8.944$

The hypotenuse is approximately 8.944 in.

You Try It 2

Find the hypotenuse of the triangle in the figure. Round to the nearest thousandth.

Your solution
3.601 in.

Solution on p. S34

In-Class Examples (Objective 12.5B)
Find the unknown side of the triangle. Round to the nearest thousandth.

1.   12.207 cm

2.   6.708 cm

Example 3

Find the length of the leg of the triangle in the figure. Round to the nearest thousandth.

12 cm
9 cm

Solution

$\text{Leg} = \sqrt{(\text{hypotenuse})^2 - (\text{leg})^2}$
$= \sqrt{12^2 - 9^2}$
$= \sqrt{144 - 81}$
$= \sqrt{63} \approx 7.937$

The length of the leg is approximately 7.937 cm.

You Try It 3

Find the length of the leg of the triangle in the figure. Round to the nearest thousandth.

5 ft
12 ft

Your solution
10.909 ft

Solution on p. S34

Objective C   To solve application problems

Example 4

A 25-foot ladder is placed against a building at a point 21 ft from the ground, as shown in the figure. Find the distance from the base of the building to the base of the ladder. Round to the nearest thousandth.

25 ft
21 ft

Strategy

To find the distance from the base of the building to the base of the ladder, use the Pythagorean Theorem. The hypotenuse is the length of the ladder (25 ft). One leg is the distance along the building from the ground to the top of the ladder (21 ft). The distance from the base of the building to the base of the ladder is the unknown leg.

Solution

$\text{Leg} = \sqrt{(\text{hypotenuse})^2 - (\text{leg})^2}$
$= \sqrt{25^2 - 21^2}$
$= \sqrt{625 - 441}$
$= \sqrt{184} \approx 13.565$

The distance is approximately 13.565 ft.

You Try It 4

Find the distance between the centers of the holes in the metal plate in the figure. Round to the nearest thousandth.

3 cm
8 cm

Your strategy

Your solution
8.544 cm

Solution on p. S34

(Continued)

Mathematicians have investigated Pythagorean triples and have found formulas that will generate these triples. One such formula is

$a = m^2 - n^2$
$b = 2mn$
$c = m^2 + n^2$, where $m > n$

For instance, let $m = 2$ and $n = 1$. Then $a = 2^2 - 1^2 = 3$, $b = 2(2)(1) = 4$, and $c = 2^2 + 1^2 = 5$. This is the Pythagorean triple 3, 4, 5. Find the Pythagorean triple produced by each of the following.

e. $m = 3$ and $n = 1$ 6, 8, 10
f. $m = 5$ and $n = 2$ 20, 21, 29
g. $m = 4$ and $n = 2$ 12, 16, 20
h. $m = 6$ and $n = 1$ 12, 35, 37

Objective 12.5C

Concept Check

(*Note:* For these exercises, encourage students to draw a diagram.)

1. A pilot flies 90 mi east from Landown Airport to Mitchell Airfield and then 60 mi north from Mitchell Airfield to Wallace Airport. Find the distance of the return flight straight back to Landown from Wallace. Round to the nearest mile.   108 mi

2. Marta Lightfoot leaves a dock in her sailboat and sails 2.5 mi due east. She then tacks and sails 4 mi due north. The two-way radio Marta has on board has a range of 5 mi. Will she be able to call a friend on the dock from her location using the two-way radio?   Yes

Optional Student Activity

The radius of the circle shown below is 3 in. Find the length of a side of the square drawn inside the circle. Round to the nearest tenth.   4.2 in.

In-Class Examples (Objective 12.5C)

1. A ladder 10 m long is leaning against a building. How high on the building will the ladder be when the bottom of the ladder is 4 m from the base of the building? Round to the nearest thousandth.
9.165 m

2. A fence is built around the plot shown at the right. At $15.20 per meter, how much did it cost to fence the plot? $820.80

21m
9 m
9 m

Section 12.5

Suggested Assignment
Exercises 1–26, every other odd
Exercises 27–39, odds
More challenging problems:
    Exercises 42, 43, 45

12.5 Exercises

Objective A   To find the square root of a number

For Exercises 1 to 8, find the square root. Round to the nearest thousandth.

1. 7
   2.646

2. 34
   5.831

3. 42
   6.481

4. 64
   8

5. 165
   12.845

6. 144
   12

7. 189
   13.748

8. 130
   11.402

Objective B   To find the unknown side of a right triangle using the Pythagorean Theorem

For Exercises 9 to 26, find the unknown side of the triangle. Round to the nearest thousandth.

9.   5 in.
   3 in.
   4 in.

10.   13 in.
   5 in.
   12 in.

11.   8.602 cm
   5 cm
   7 cm

12.   11.402 cm
   7 cm
   9 cm

13.   11.180 ft
   15 ft
   10 ft

14.   8.718 ft
   20 ft
   18 ft

15.   4.472 cm
   4 cm   6 cm

16.   7.937 m
   9 m   12 m

17.   12.728 yd
   9 yd
   9 yd

18.   20 cm
   10 cm
   17.321 cm

19.   12 ft
   6 ft
   10.392 ft

20.   16 cm
   8 cm
   13.856 cm

21.   21.213 cm
   15 cm
   15 cm

22.   8.485 in.
   6 in.   6 in.

23.   8 m   4 m
   8.944 m

24.   8.6 cm
   4.3 cm
   7.448 cm

25.   11.3 yd
   8.1 yd
   7.879 yd

26.   13.9 ft
   8.2 ft
   11.224 ft

Quick Quiz (Objective 12.5A)
Find the square root. Round to the nearest thousandth.
1. 63   7.937
2. 111   10.536

Quick Quiz (Objective 12.5B)
Find the unknown side of the triangle. Round to the nearest thousandth.

1.
   7 cm
   12 cm
   13.892 cm

2.
   14 mi
   11 mi
   8.660 mi

Objective C   **To solve application problems**

27. **Ramps**   Find the length of the ramp used to roll barrels up to the loading dock, which is 3.5 ft high. Round to the nearest hundredth.
    9.66 ft

3.5 ft

9 ft

28. **Metal Works**   Find the distance between the centers of the holes in the metal plate in the figure at the right. Round to the nearest hundredth.
    6.32 in.

2 in.

|← 6 in. →|

29. **Travel**   If you travel 18 mi east and then 12 mi north, how far are you from your starting point? Round to the nearest tenth.
    21.6 mi

30. **Travel**   If you travel 12 mi west and 16 mi south, how far are you from your starting point?
    20 mi

31. **Geometry**   The diagonal of a rectangle is a line drawn from one vertex to the opposite vertex. Find the length of the diagonal in the rectangle shown at the right. Round to the nearest tenth.
    12.1 mi

5 mi

11 mi

32. **Geometry**   The diagonal of a rectangle is a line drawn from one vertex to the opposite vertex. (See Exercise 31.) Find the length of the diagonal in the rectangle that has a length of 8 m and a width of 3.5 m. Round to the nearest tenth.
    8.7 m

33. **Home Maintenance**   A ladder 8 m long is placed against a home in preparation for washing the windows. How high on the building does the ladder reach when the bottom of the ladder is 3 m from the home? Round to the nearest tenth.
    7.4 m

8 m

3 m

34. **Geometry**   Find the perimeter of a right triangle with legs that measure 5 cm and 9 cm. Round to the nearest tenth.
    24.3 cm

35. **Geometry**   Find the perimeter of a right triangle with legs that measure 6 in. and 10 in.
    27.7 in.

36. **Metal Works**   Find the distance between the centers of the holes in the metal plate shown in the diagram at the right. Round to the nearest tenth.
    9.8 cm

4 cm

|← 9 cm →|

37. **Parks**   An L-shaped sidewalk from the parking lot to a memorial is shown in the figure at the right. The distance directly across the grass to the memorial is 650 ft. The distance to the corner is 600 ft. Find the distance from the corner to the memorial.
    250 ft

Parking

650 ft

600 ft

Memorial

Quick Quiz (Objective 12.5C)

1. A car is driven 12 mi west and then 5 mi north. How far is the car from the starting point?   13 mi

2. Find the distance around the hiking trail.   12 km

3 km

4 km

38. Landscaping   A fence is built around the plot shown in the figure at the right. At $11.40 per meter, how much did it cost to fence the plot? (*Hint:* Use the Pythagorean Theorem to find the unknown length.)
$296.40

7 m
4 m
10 m

39. Metal Works   Four holes are drilled in the circular plate in the figure at the right. The centers of the holes are 3 in. from the center. Find the distance between the centers of adjacent holes. Round to the nearest thousandth.
4.243 in.

40. Plumbing   Find the offset distance, d, of the length of pipe shown in the diagram at the right. The total length of the pipe is 62 in.
$3\frac{3}{4}$ in.

d

$\left|\leftarrow 20\frac{3}{4}\text{ in.} \rightarrow\right|$ 9 in. $\left|\leftarrow 31\frac{1}{2}\text{ in.} \rightarrow\right|$

APPLYING THE CONCEPTS

41. Determine whether the statement is always true, sometimes true, or never true.
a. The sum of the lengths of two sides of a triangle is greater than the length of the third side of the triangle.   Always true
b. The hypotenuse is the longest side of a right triangle.   Always true

42. ✎ **Home Maintenance**   You need to clean the gutters of your home. The gutters are 24 ft above the ground. For safety, the distance a ladder reaches up a wall should be four times the distance from the bottom of the ladder to the base of the side of the house. Therefore, the ladder must be 6 ft from the base of the house. Will a 25-foot ladder be long enough to reach the gutters? Explain how you determined your answer.

24 ft
6 ft

43. ✎ Can the Pythagorean Theorem be used to find the length of side c of the triangle at the right? If so, determine c. If not, explain why the theorem cannot be used.

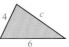
4
c
6

44. **a.** What is a Pythagorean triple?   A triple A, B, C such that $A^2 + B^2 = C^2$
b. Provide at least three examples of Pythagorean triples.
For example, 5, 12, and 13 because $5^2 + 12^2 = 13^2$. Other examples: 3, 4, and 5; 6, 8, and 10; 8, 15, and 17.

45. **Construction**   Buildings A and B are situated on opposite sides of a river. A construction company must lay a pipeline between the two buildings. The plan is to connect the buildings as shown. What is the total length of the pipe needed to connect the buildings?
8 m

A
4 m
2 m
1 m
B
6 m

12.6 Similar and Congruent Triangles

Objective A   **To solve similar and congruent triangles**

Similar objects have the same shape but not necessarily the same size. A baseball is similar to a basketball. A model airplane is similar to an actual airplane.

Similar objects have corresponding parts; for example, the propellers on the model airplane correspond to the propellers on the actual airplane. The relationship between the sizes of each of the corresponding parts can be written as a ratio, and all such ratios will be the same. If the propellers on the model plane are $\frac{1}{50}$ the size of the propellers on the actual plane, then the model wing is $\frac{1}{50}$ the size of the actual wing, the model fuselage is $\frac{1}{50}$ the size of the actual fuselage, and so on.

The two triangles ABC and DEF shown are similar. The ratios of corresponding sides are equal.

$$\frac{AB}{DE} = \frac{2}{6} = \frac{1}{3}, \frac{BC}{EF} = \frac{3}{9} = \frac{1}{3}, \text{ and } \frac{AC}{DF} = \frac{4}{12} = \frac{1}{3}$$

The ratio of corresponding sides $= \frac{1}{3}$.

Because the ratios of corresponding sides are equal, three proportions can be formed:

$$\frac{AB}{DE} = \frac{BC}{EF}, \frac{AB}{DE} = \frac{AC}{DF}, \text{ and } \frac{BC}{EF} = \frac{AC}{DF}$$

The ratio of corresponding heights equals the ratio of corresponding sides, as shown in the figure.

Ratio of corresponding sides $= \frac{1.5}{6} = \frac{1}{4}$

Ratio of heights $= \frac{2}{8} = \frac{1}{4}$

Congruent objects have the same shape *and* the same size.

The two triangles shown are congruent. They have exactly the same size.

For triangles, *congruent* means that the corresponding sides *and* angles of the triangle are equal (this contrasts with similar triangles, in which corresponding angles, but not necessarily corresponding sides, are equal).

Vocabulary to Review
ratio   [4.1A]

New Vocabulary
similar objects
similar triangles
congruent objects
congruent triangles

New Rules
side-side-side rule (SSS)
side-angle-side rule (SAS)

Discuss the Concepts
1. What is the difference between similar objects and congruent objects?
2. The text discusses two rules to determine whether two triangles are congruent, the SSS rule and the SAS rule. Are two triangles congruent if all three angles of one triangle are equal to all three angles of the second triangle?
3. What does the phrase "the included angle" mean?

Instructor Note
The concept of similar triangles and the fact that the ratios of corresponding sides are equal are not obvious to students. Nonetheless, these concepts are essential in many practical applications.
   You might introduce the topic of similar triangles by using a magnifying glass. Illustrate that an object viewed under a magnifying lens appears larger, but its shape has not changed.

In-Class Examples (Objective 12.6A)

1. Find the ratio of corresponding sides for the similar triangles.

$\frac{2}{3}$

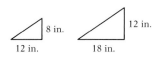

2. Triangles ABC and DEF are similar. Find side DE.

6.75 m

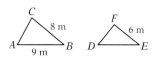

Objective 12.6B

Objective B   **To solve application problems**

Optional Student Activity

1. The lengths of the three sides of a triangle are 3 m, 4 m, and 6 m. Find the least possible perimeter of a similar triangle, one of whose sides measures 12 m.   26 m

2. In the figure below, triangle *ABC* is similar to triangle *DEF*. ∠*C* is a right angle, side *AC* measures 6 in., side *AB* measures 10 in., and side *EF* measures 3 in. Find the area of the shaded region.   20.625 in²

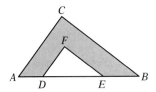

Example 5

Triangles *ABC* and *DEF* in the figure are similar. Find the area of triangle *DEF*.

Strategy

To find the area of triangle *DEF*:

• Solve a proportion to find the height of triangle *DEF*. Let h = the height.

• Use the formula $A = \frac{1}{2}bh$.

Solution

$$\frac{AB}{DE} = \frac{\text{height of triangle } ABC}{\text{height of triangle } DEF}$$

$$\frac{4 \text{ cm}}{12 \text{ cm}} = \frac{3 \text{ cm}}{h}$$

$$4h = 12 \cdot 3 \text{ cm}$$

$$4h = 36 \text{ cm}$$

$$\frac{4h}{4} = \frac{36 \text{ cm}}{4}$$

$$h = 9 \text{ cm}$$

$$A = \frac{1}{2}bh$$

$$= \frac{1}{2}(12 \text{ cm})(9 \text{ cm})$$

$$= 54 \text{ cm}^2$$

The area is 54 cm².

You Try It 5

Triangles *ABC* and *DEF* in the figure are similar right triangles. Find the perimeter of triangle *ABC*.

Your strategy

Your solution   12 in.

Solution on p. S34

In-Class Examples (Objective 12.6B)

1. Shown at right are a building and a pole casting shadows. Find the height of the building.   24 ft

2. Triangles *ABC* and *DEF* are similar. Find the area of triangle *DEF*.   3125 cm²

12.6 Exercises

Section 12.6

Suggested Assignment
Exercises 1–17, odds
More challenging problems:
 Exercises 19–21

Objective A   **To solve similar and congruent triangles**

Find the ratio of corresponding sides for the similar triangles in Exercises 1 to 4.

1. $\frac{1}{2}$

2. $\frac{1}{3}$

3. $\frac{3}{4}$

4. $\frac{1}{3}$

Determine whether the two triangles in Exercises 5 to 8 are congruent.

5.

Yes

6.

Yes

7.

Yes

8.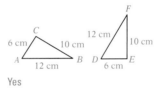

Yes

Triangles *ABC* and *DEF* in Exercises 9 to 12 are similar. Find the
indicated distance. Round to the nearest tenth.

9.   Find side *DE*.

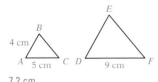

7.2 cm

10.   Find side *DE*.

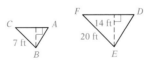

13.7 in.

11.   Find the height of triangle *DEF*.

3.3 m

12.   Find the height of triangle *ABC*.

4.9 ft

Quick Quiz (Objective 12.6A)

1. Find the ratio of corresponding sides for the
similar triangles.

$\frac{3}{4}$

2. Triangles *ABC* and *DEF* are similar. Find side *AC*.
4 cm

Objective B **To solve application problems**

Measurement The sun's rays, objects on Earth, and the shadows cast by those objects form similar triangles. For Exercises 13 and 14, find the height of the building.

13.

16 m

14.

20.8 ft

In Exercises 15 to 18, triangles *ABC* and *DEF* are similar.

15. Find the perimeter of triangle *ABC*.

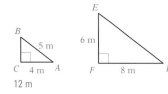

12 m

16. Find the perimeter of triangle *DEF*.

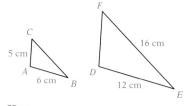

38 cm

17. Find the area of triangle *ABC*.

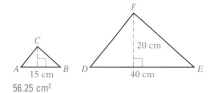

56.25 cm²

18. Find the area of triangle *DEF*.

49 m²

APPLYING THE CONCEPTS

19. Determine whether the statement is always true, sometimes true, or never true.
  a. If two angles of one triangle are equal to two angles of a second triangle, then the triangles are similar triangles.   Always true
  b. Two isosceles triangles are similar triangles.   Sometimes true
  c. Two equilateral triangles are similar triangles.   Always true

20. ✎ Are all squares similar? Are all rectangles similar? Explain. Use a drawing in your explanation.

21. Figure *ABC* is a right triangle and *DE* is parallel to *AB*. What is the perimeter of the trapezoid *ABED*?
  16.5 units

22. ✎ Explain how, by using only a yardstick, you could determine the approximate height of a tree without climbing it.

Answers to Writing Exercises

20. Yes. Given two squares, the ratios of corresponding sides are equal because the same number will be in the numerators (the length of a side of one square) and the same number will be in the denominators (the length of a side of the second square.)
  No. The lengths of the sides of a rectangle vary. Therefore, given two rectangles, the ratios of corresponding sides may vary.

22. To find the height of the tree, you could use similar triangles and a proportion. This method assumes that the sun is shining and is not directly overhead. Use the yardstick to measure the shadow of the tree. Then use the yardstick to measure the length of your own shadow. Write and solve a proportion. One possible proportion is as follows: "Your height" over "the length of your shadow" is equal to "the height of the tree" over "the length of the shadow cast by the tree."

Quick Quiz (Objective 12.6B)

1. Shown at right are a building and a pole casting shadows. Find the height of the building.   21 m

2. Triangles *ABC* and *DEF* are similar. Find the area of triangle *ABC*.   80 cm²

Focus on Problem Solving

Trial and Error   Some problems in mathematics are solved by using **trial and error.** The trial-and-error method of arriving at a solution to a problem involves performing repeated tests or experiments until a satisfactory conclusion is reached.

Many of the Applying the Concepts exercises in this text require a trial-and-error method of solution. For example, an exercise in Section 12.4 reads as follows:

Explain how you could cut through a cube so that the face of the resulting solid is **(a)** a square, **(b)** an equilateral triangle, **(c)** a trapezoid, **(d)** a hexagon.

There is no formula to apply to this problem; there is no computation to perform. This problem requires picturing a cube and the results after it is cut through at different places on its surface and at different angles. For part a, cutting perpendicular to the top and bottom of the cube and parallel to two of its sides will result in a square. The other shapes may prove more difficult.

When solving problems of this type, keep an open mind. Sometimes when using the trial-and-error method, we are hampered by our narrowness of vision; we cannot expand our thinking to include other possibilities. Then when we see someone else's solution, it appears so obvious to us! For example, for the Applying the Concepts question above, it is necessary to conceive of cutting through the cube at places other than the top surface; we need to be open to the idea of beginning the cut at one of the corner points of the cube.

One topic of the Projects and Group Activities in this chapter is symmetry. Here again, the trial-and-error method is used to determine the lines of symmetry inherent in an object. For example, in determining lines of symmetry for a square, begin by drawing a square. The horizontal line of symmetry and the vertical line of symmetry may be immediately obvious to you.

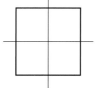

But there are two others. Do you see that a line drawn through opposite corners of the square is also a line of symmetry?

Many of the questions in this text that require an answer of "always true, sometimes true, or never true" are best solved by the trial-and-error method. For example, consider the following statement, which is presented as this type of question in Section 12.3.

If two rectangles have the same area, then they have the same perimeter.

**Answers to Focus on
Problem Solving—
Trial and Error**

No, it is not possible for two
different pairs of factors of a
number to have the same sum.

Try some numbers. Each of two rectangles, one measuring 6 units by 2 units and another measuring 4 units by 3 units, has an area of 12 square units, but the perimeter of the first is 16 units and the perimeter of the second is 14 units, so the answer "always true" has been eliminated. We still need to determine whether there is a case when it is true. After experimenting with a lot of numbers, you may come to realize that we are trying to determine whether it is possible for two different pairs of factors of a number to have the same sum. Is it?

Don't be afraid to make many experiments, and remember that *errors,* or tests that "don't work," are a part of the trial-and-*error* process.

Projects and Group Activities

**Investigating
Perimeter**

The perimeter of the square at the right is 4 units.

If two squares are joined along one of the sides, the perimeter is 6 units. Note that it does not matter which sides are joined; the perimeter is still 6 units.

If three squares are joined, the perimeter of the resulting figure is 8 units for each possible placement of the squares.

Four squares can be joined in five different ways as shown. There are two possible perimeters: 10 units for A, B, C, and D, and 8 for E.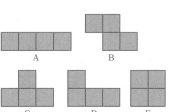

**Answers to Projects
and Group Activities—
Investigating
Perimeter**

1. 12 units
2. 10 units
3. 14 units
4. 10 units

1. If five squares are joined, what is the maximum perimeter possible?

2. If five squares are joined, what is the minimum perimeter possible?

3. If six squares are joined, what is the maximum perimeter possible?

4. If six squares are joined, what is the minimum perimeter possible?

Symmetry

Look at the letter A printed at the left. If the letter were folded along line ℓ, the two sides of the letter would match exactly. This letter has **symmetry** with respect to line ℓ. Line ℓ is called the **axis of symmetry.**

Now consider the letter H printed below at the left. Both lines ℓ_1 and ℓ_2 are axes of symmetry for this letter; the letter could be folded along either line and the two sides would match exactly.

1. Does the letter A have more than one axis of symmetry?

2. Find axes of symmetry for other capital letters of the alphabet.

3. Which lowercase letters have one axis of symmetry?

4. Do any of the lowercase letters have more than one axis of symmetry?

5. Find the number of axes of symmetry for each of the plane geometric figures presented in this chapter.

6. There are other types of symmetry. Look up the meaning of *point symmetry* and *rotational symmetry*. Which plane geometric figures provide examples of these types of symmetry?

7. Find examples of symmetry in nature, art, and architecture.

Chapter 12 Summary

Key Words

Examples

A *line* extends indefinitely in two directions. A *line segment* is part of a line and has two endpoints. The length of a line segment is the distance between the endpoints of the line segment. [12.1A, p. 515]

Line

A B Line segment

Parallel lines never meet; the distance between them is always the same. The symbol ‖ means "is parallel to." *Intersecting lines* cross at a point in the plane. *Perpendicular lines* are intersecting lines that form right angles. The symbol ⊥ means "is perpendicular to." [12.1A, pp. 515–516]

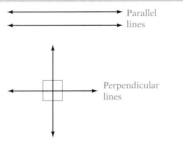

Parallel lines

Perpendicular lines

Answers to Projects and Group Activities— Symmetry

1. No

2. Besides A and H, the capital letters B, C, D, E, I, M, O, T, V, W, X, and Y have axes of symmetry. Note that the letters I, O, and X have more than one axis of symmetry. (For some letters, it depends on how the letter is written; an example is the letter U.)

3. The lowercase letters c, i, l, t, v, and w have one axis of symmetry. (For some letters, it depends on how the letter is written; an example is the letter t.)

4. The lowercase letters o and x have more than one axis of symmetry.

5. An isosceles triangle has one axis of symmetry. An equilateral triangle has three axes of symmetry. A rectangle has two axes of symmetry. A square has four axes of symmetry. A circle has an infinite number of axes of symmetry. A trapezoid has no axis of symmetry.

6. If a figure is unchanged after being rotated 180° about a point *O*, then the figure has point symmetry. A parallelogram, a rectangle, a square, and a circle have point symmetry.

   If a figure is unchanged after being rotated more than 0° and less than 360°, then the figure has rotational symmetry. An equilateral triangle, a rectangle, a square, a circle, and a parallelogram have rotational symmetry. It may be helpful to note that a regular pentagon has rotational symmetry but not point symmetry.

7. Examples of symmetry in nature include people and animals, snowflakes, starfish, and many types of flowers. Examples of symmetry in art and architecture can be found in history texts and in art and architectural magazines.

A *ray* starts at a point and extends indefinitely in one direction. An *angle* is formed when two rays start from the same point. The common point is called the *vertex* of the angle. An angle is measured in *degrees*. A 90° angle is a *right angle*. A 180° angle is a *straight angle*. *Complementary angles* are two angles whose measures have the sum 90°. *Supplementary angles* are two angles whose measures have the sum 180°. An *acute angle* is an angle whose measure is between 0° and 90°. An *obtuse angle* is an angle whose measure is between 90° and 180°. [12.1A, pp. 516–517]

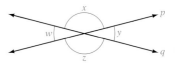

Two angles that are on opposite sides of the intersection of two lines are *vertical angles;* vertical angles have the same measure. Two angles that share a common side are *adjacent angles;* adjacent angles of intersecting lines are supplementary angles. [12.1C, p. 521]

Angles *w* and *y* are vertical angles.
Angles *x* and *y* are adjacent angles.

A line that intersects two other lines at two different points is a *transversal.* If the lines cut by a transversal are parallel lines, equal angles are formed: *alternate interior angles, alternate exterior angles,* and *corresponding angles.* [12.1C, pp. 521–522]

Parallel lines l_1 and l_2 are cut by transversal *t*. All four acute angles have the same measure. All four obtuse angles have the same measure.

A *quadrilateral* is a four-sided polygon. A *parallelogram,* a *rectangle,* and a *square* are quadrilaterals. [12.1B, pp. 518–519]

A *polygon* is a closed figure determined by three or more line segments. The line segments that form the polygon are its *sides.* A *regular polygon* is one in which each side has the same length and each angle has the same measure. Polygons are classified by the number of sides. [12.2A, p. 527]

Number of Sides	Name of the Polygon
3	Triangle
4	Quadrilateral
5	Pentagon
6	Hexagon
7	Heptagon
8	Octagon
9	Nonagon
10	Decagon

A *triangle* is a closed, three-sided plane figure. [12.1B, p. 518]

An *isosceles triangle* has two sides of equal length. The three sides of an *equilateral triangle* are of equal length. A *scalene triangle* has no two sides of equal length. An *acute triangle* has three actue angles. An *obtuse triangle* has one obtuse angle. [12.2A, pp. 527–528]

A *right triangle* contains a right angle. The side opposite the right angle is called the *hypotenuse.* The other two sides are called *legs.* [12.1B, p. 518]

Right Triangle

A *circle* is a plane figure in which all points are the same distance from the center of the circle. A *diameter* of a circle is a line segment across the cricle through the center. A *radius* of a circle is a line segment from the center of the circle to a point on the circle. [12.1B, p. 519]

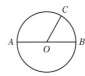

AB is a diameter of the circle.
OC is a radius.

Geometric solids are figures in space. Four common space figures are the rectangular solid, cube, sphere, and cylinder. A *rectangular solid* is a solid in which all six faces are rectangles. A *cube* is a rectangular solid in which all six faces are squares. A *sphere* is a solid in which all points on the sphere are the same distance from the center of the sphere. The most common *cylinder* is one in which the bases are circles and are perpendicular to the side. [12.1B, pp. 519–520]

Rectangular Solid

The *square root* of a number is one of two identical factors of that number. The symbol for square root is $\sqrt{\ }$. A *perfect square* is the product of a whole number times itself. The square root of a perfect square is a whole number. [12.5A, p. 555]

$1^2 = 1$, $2^2 = 4$, $3^2 = 9$, $4^2 = 16$, $5^2 = 25$, ... 1, 4, 9, 16, 25, ... are perfect squares.

$\sqrt{1} = 1$, $\sqrt{4} = 2$, $\sqrt{9} = 3$, $\sqrt{16} = 4$, ...

Similar triangles have the same shape but not necessarily the same size. The ratios of corresponding sides are equal. The ratio of corresponding heights is equal to the ratio of corresponding sides. *Congruent triangles* have the same shape and the same size. [12.6A, p. 561]

Triangles *ABC* and *DEF* are similar triangles. The ratio of corresponding sides is $\frac{1}{2}$.

Essential Rules and Procedures

Examples

Triangles [12.1B, p. 518]
Sum of three angles = 180°

Two angles of a triangle measure 32° and 48°. Find the measure of the third angle.

$\angle A + \angle B + \angle C = 180°$
$\angle A + 32° + 48° = 180°$
$\angle A + 80° = 180°$
$\angle A + 80° - 80° = 180° - 80°$
$\angle A = 100°$

The measure of the third angle is 100°.

Formulas for Perimeter (the distance around a figure)
[12.2A, pp. 528–529]
Triangle: $P = a + b + c$
Square: $P = 4s$
Rectangle: $P = 2L + 2W$
Circumference of a circle: $C = \pi d$ or $C = 2\pi r$

The length of a rectangle is 8 m. The width is 5.5 m. Find the perimeter of the rectangle.
$P = 2L + 2W$
$P = 2(8 \text{ m}) + 2(5.5 \text{ m})$
$P = 16 \text{ m} + 11 \text{ m}$
$P = 27 \text{ m}$
The perimeter is 27 m.

Formulas for Area (the amount of surface in a region)
[12.3A, pp. 537–538]
Triangle: $A = \frac{1}{2}bh$
Square: $A = s^2$
Rectangle: $A = LW$
Circle: $A = \pi r^2$

Find the area of a circle with a radius of 4 cm. Use 3.14 for π.
$A = \pi r^2$
$A \approx 3.14(4 \text{ cm})^2$
$A \approx 50.24 \text{ cm}^2$
The area is 50.24 cm².

Formulas for Volume (the amount of space inside a figure in space) [12.4A, pp. 545–546]
Rectangular solid: $V = LWH$
Cube: $V = s^3$
Sphere: $V = \frac{4}{3}\pi r^3$
Cylinder: $V = \pi r^2 h$

Find the volume of a cube that measures 3 in. on a side.
$V = s^3$
$V = 3^3$
$V = 27$
The volume is 27 in³.

Pythagorean Theorem [12.5B, p. 556]
The square of the hypotenuse of a right triangle is equal to the sum of the squares of the two legs.
If the length of one side of a triangle is unknown, one of the following formulas can be used to find it.
If the hypotenuse is unknown, use

$$\text{Hypotenuse} = \sqrt{(\text{leg})^2 + (\text{leg})^2}$$

If the length of a leg is unknown, use

$$\text{Leg} = \sqrt{(\text{hypotenuse})^2 - (\text{leg})^2}$$

Two legs of a right triangle measure 6 ft and 8 ft. Find the hypotenuse of the right triangle.

$$
\begin{aligned}
\text{Hypotenuse} &= \sqrt{(\text{leg})^2 + (\text{leg})^2} \\
&= \sqrt{6^2 + 8^2} \\
&= \sqrt{36 + 64} \\
&= \sqrt{100} \\
&= 10
\end{aligned}
$$

The length of the hypotenuse is 10 ft.

Side-Side-Side (SSS) Rule
Two triangles are congruent if three sides of one triangle equal the corresponding sides of the second triangle.

Side-Angle-Side (SAS) Rule
Two triangles are congruent if two sides and the included angle of one triangle equal the corresponding sides and included angle of the second triangle. [12.6A, p. 562]

Triangles *ABC* and *DEF* are congruent by the SAS rule.

Chapter 12 Review Exercises

1. The diameter of a sphere is 1.5 m. Find the radius of the sphere.
   0.75 m [12.1B]

2. Find the circumference of a circle with a radius of 5 cm. Use 3.14 for π.
   31.4 cm [12.2A]

3. Find the perimeter of the rectangle in the figure below.

   8 ft

   5 ft

   26 ft [12.2A]

4. Given $AB = 15$, $CD = 6$, and $AD = 24$, find the length of BC.

   3 [12.1A]

5. Find the volume of the rectangular solid shown below.

   4 ft

   5 ft   10 ft

   200 ft³ [12.4A]

6. Find the unknown side of the triangle in the figure below.

   10 cm

   24 cm

   26 cm [12.5B]

7. Find the supplement of a 105° angle.
   75° [12.1A]

8. Find the square root of 15. Round to the nearest thousandth.
   3.873 [12.5A]

9. Triangles ABC and DEF are similar. Find the height of triangle DEF.

   E

   B

   8 cm   12 cm   24 cm h

   A   C   D   F

   16 cm [12.6A]

10. Find the area of the circle shown below. Use 3.14 for π.

    9 cm

    63.585 cm² [12.3A]

11. Here $\ell_1 \parallel \ell_2$.
    a. Find the measure of angle b.
    b. Find the measure of angle a.

    t

    b

    45°   a

    ℓ_1

    ℓ_2

    a. 45° **b.** 135° [12.1C]

12. Find the area of the rectangle shown below.

    5 m

    11 m

    55 m² [12.3A]

13. Find the volume of the composite figure shown below.

    3 in.

    6 in.

    3 in.

    7 in.   8 in.

    240 in³ [12.4B]

14. Find the area of the composite figure shown below. Use 3.14 for π.

    4 in.

    8 in.

    57.12 in² [12.3B]

15. Find the volume of a sphere with a diameter of 8 ft. Use 3.14 for π. Round to the nearest tenth.
267.9 ft³   [12.4A]

16. Triangles *ABC* and *DEF* are similar. Find the area of triangle *DEF*.

64.8 m²   [12.6B]

17. Find the perimeter of the composite figure shown below. Use 3.14 for π.

10 in.

16 in.   16 in.

47.7 in.   [12.2B]

18. Here $\ell_1 \parallel \ell_2$.
   a. Find the measure of angle *b*.
   b. Find the measure of angle *a*.

a. 80°   b. 100°   [12.1C]

19. **Home Maintenance**   How high on a building will a 17-foot ladder reach when the bottom of the ladder is 8 ft from the building?
15 ft   [12.5C]

20. A right triangle has a 32° angle. Find the measures of the other two angles.
90° and 58°   [12.1B]

21. **Travel**   A bicycle tire has a diameter of 28 in. How many feet does the bicycle travel when the wheel makes 10 revolutions? Use 3.14 for π. Round to the nearest tenth of a foot.
73.3 ft   [12.2C]

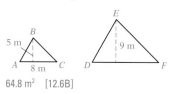

28 in.

22. **Interior Design**   New carpet is installed in a room measuring 18 ft by 14 ft. Find the area of the room in square yards. (9 ft² = 1 yd²)
28 yd²   [12.3C]

23. **Agriculture**   A silo, which is in the shape of a cylinder, is 9 ft in diameter and has a height of 18 ft. Find the volume of the silo. Use 3.14 for π.
1144.53 ft³   [12.4C]

24. Find the area of a right triangle with a base of 8 m and a height of 2.75 m.
11 m²   [12.3A]

25. **Travel**   If you travel 20 mi west and then 21 mi south, how far are you from your starting point?
29 mi   [12.5C]

Chapter 12 Test

1. Find the volume of a cylinder with a height of 6 m and a radius of 3 m. Use 3.14 for π.
169.56 m³ [12.4A]

2. Find the perimeter of a rectangle that has a length of 2 m and a width of 1.4 m.
6.8 m [12.2A]

3. Find the volume of the composite figure. Use 3.14 for π.

1406.72 cm³ [12.4B]

4. Triangles *ABC* and *FED* are congruent right triangles. Find the length of *FE*.

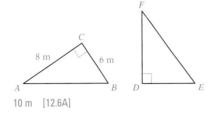

10 m [12.6A]

5. Find the complement of a 32° angle.
58° [12.1A]

6. Find the area of a circle that has a diameter of 2 m. Use $\frac{22}{7}$ for π.

$3\frac{1}{7}$ m² [12.3A]

7. In the figure below, lines ℓ_1 and ℓ_2 are parallel. Angle *x* measures 30°. Find the measure of angle *y*.

150° [12.1C]

8. Find the perimeter of the composite figure. Use 3.14 for π.

15.85 ft [12.2B]

9. Find the square root of 189. Round to the nearest thousandth.
13.748 [12.5A]

10. Find the unknown side of the triangle shown below. Round to the nearest thousandth.

9.747 ft [12.5B]

11. Find the area of the composite figure.

$1\frac{1}{2}$ ft

3 ft

$4\frac{1}{2}$ ft

10.125 ft² [12.3B]

12. In the figure below, lines ℓ_1 and ℓ_2 are parallel. Angle x measures 45°. Find the measures of angles a and b.

$\angle a = 45°; \angle b = 135°$ [12.1C]

13. Triangles *ABC* and *DEF* are similar. Find side *BC*.

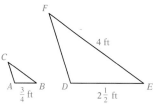

$1\frac{1}{5}$ ft   [12.6A]

14. A right triangle has a 40° angle. Find the measures of the other two angles.
90° and 50°   [12.1B]

15. **Measurement**   Use similar triangles to find the width of the canal shown in the figure at the right.
25 ft   [12.6B]

Canal

12 ft

5 ft   60 ft

16. **Consumerism**   How much more pizza is contained in a pizza with radius 10 in. than in one with radius 8 in.? Use 3.14 for π.
113.04 in²   [12.3C]

17. **Interior Design**   A carpet is to be placed as shown in the diagram at the right. At $26.80 per square yard, how much will it cost to carpet the area? Round to the nearest cent. (9 ft² = 1 yd²)
$1113.69   [12.3C]

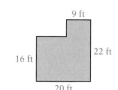

9 ft

16 ft   22 ft

20 ft

18. **Forestry**   Find the cross-sectional area of a redwood tree that is 11 ft 6 in. in diameter. Use 3.14 for π. Round to the nearest hundredth.
103.82 ft²   [12.3C]

19. **Construction**   Find the length of the rafter needed for the roof shown in the figure.
15 ft   [12.5C]

2 ft   5 ft

24 ft

20. **Measurement**   A toolbox is 1 ft 2 in. long, 9 in. wide, and 8 in. high. The sides and bottom of the toolbox are $\frac{1}{2}$ in. thick. Find the volume of the interior of the toolbox in cubic inches.
780 in³   [12.4C]

Cumulative Review Exercises

1. Find the GCF of 96 and 144.
48 [2.1B]

2. Add: $3\frac{5}{12} + 2\frac{9}{16} + 1\frac{7}{8}$
$7\frac{41}{48}$ [2.4C]

3. Find the quotient of $4\frac{1}{3}$ and $6\frac{2}{9}$.
$\frac{39}{56}$ [2.7B]

4. Simplify: $\left(\frac{2}{3}\right)^2 \div \left(\frac{1}{3} + \frac{1}{2}\right) - \frac{2}{5}$
$\frac{2}{15}$ [2.8C]

5. Simplify: $-\frac{2}{3} - \left(-\frac{5}{8}\right)$
$-\frac{1}{24}$ [10.4A]

6. Write "$348.80 earned in 40 hours" as a unit rate.
$8.72/h [4.2B]

7. Solve the proportion $\frac{3}{8} = \frac{n}{100}$.
37.5 [4.3B]

8. Write $37\frac{1}{2}\%$ as a fraction.
$\frac{3}{8}$ [5.1A]

9. Evaluate $a^2 - (b^2 - c)$ when $a = 2$, $b = -2$, and $c = -4$.
-4 [11.1A]

10. 30.94 is 36.4% of what number?
85 [5.4A]

11. Solve: $\frac{x}{3} + 3 = 1$
-6 [11.3A]

12. Solve: $2(x - 3) + 2 = 5x - 8$
$\frac{4}{3}$ [11.4B]

13. Convert 32.5 km to meters.
32,500 m [9.1A]

14. Subtract: 32 m − 42 cm
31.58 m [9.1A]

15. Solve: $\frac{2}{3}x = -10$
-15 [11.2C]

16. Solve: $2x - 4(x - 3) = 8$
2 [11.4B]

17. **Finance** You bought a car for $26,488 and made a down payment of $1000. You paid the balance in 36 equal monthly installments. Find the monthly payment.
$708 [1.5D]

18. **Taxes** The sales tax on a jacket costing $175 is $6.75. At the same rate, find the sales tax on a stereo system costing $1220.
$47.06 [4.3C]

19. Compensation A heavy-equipment operator receives an hourly wage of $16.06 an hour after receiving a 10% wage increase. Find the operator's hourly wage before the increase.

$14.60 [5.4B]

20. Discount An after-Christmas sale has a discount rate of 55%. Find the sale price of a dress that had a regular price of $120.

$54 [6.2D]

21. Investments An IRA pays 7% annual interest compounded daily. What would be the value of an investment of $25,000 after 20 years? Use the table in the Appendix.

$101,366.50 [6.3C]

22. Shipping A square tile measuring 4 in. by 4 in. weighs 6 oz. Find the weight, in pounds, of a package of 144 such tiles.

54 lb [8.2C]

23. Metal Works Twenty-five rivets are used to fasten two steel plates together. The plates are 5.4 m long, and the rivets are equally spaced with a rivet at each end. Find the distance, in centimeters, between the rivets.

22.5 cm [9.1B]

24. Integer Problems The total of four times a number and two is negative six. Find the number.

−2 [11.6A]

25. The lines ℓ_1 and ℓ_2 in the figure below are parallel.
   a. Find angle a.
   b. Find angle b.

a. 74° **b.** 106° [12.1C]

26. Find the perimeter of the composite figure. Use 3.14 for π.

29.42 cm [12.2B]

27. Find the area of the composite figure.

50 in² [12.3B]

28. Find the volume of the composite figure. Use 3.14 for π.

92.86 in³ [12.4B]

29. Find the unknown side of the triangle shown in the figure below. Round to the nearest hundredth.

10.63 ft [12.5B]

30. Triangles *ABC* and *FED* below are similar. Find the perimeter of *FED*.

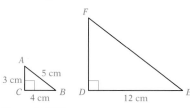

36 cm [12.6B]

Final Exam

1. Subtract: $100{,}914 - 97{,}655$
3259 [1.3B]

2. Find $34{,}821$ divided by 657.
53 [1.5C]

3. Find $90{,}001$ decreased by $29{,}796$.
60,205 [1.3B]

4. Simplify: $3^2 \cdot (5 - 3)^2 \div 3 + 4$
16 [1.6B]

5. Find the LCM of 9, 12, and 16.
144 [2.1A]

6. Add: $\dfrac{3}{8} + \dfrac{5}{6} + \dfrac{1}{5}$
$1\dfrac{49}{120}$ [2.4B]

7. Subtract: $7\dfrac{5}{12} - 3\dfrac{13}{16}$
$3\dfrac{29}{48}$ [2.5C]

8. Find the product of $3\dfrac{5}{8}$ and $1\dfrac{5}{7}$.
$6\dfrac{3}{14}$ [2.6B]

9. Divide: $1\dfrac{2}{3} \div 3\dfrac{3}{4}$
$\dfrac{4}{9}$ [2.7B]

10. Simplify: $\left(\dfrac{2}{3}\right)^3 \cdot \left(\dfrac{3}{4}\right)^2$
$\dfrac{1}{6}$ [2.8B]

11. Simplify: $\left(\dfrac{2}{3}\right)^2 \div \left(\dfrac{3}{4} + \dfrac{1}{3}\right) - \dfrac{1}{3}$
$\dfrac{1}{13}$ [2.8C]

12. Add:
$$\begin{array}{r} 4.972 \\ 28.6 \\ 1.88 \\ +\,128.725 \end{array}$$
164.177 [3.2A]

13. Multiply:
$$\begin{array}{r} 2.97 \\ \times\,0.0094 \end{array}$$
0.027918 [3.4A]

14. Divide: $0.062\overline{)0.0426}$
Round to the nearest hundredth.
0.69 [3.5A]

15. Convert 0.45 to a fraction in simplest form.
$\dfrac{9}{20}$ [3.6B]

16. Write "323.4 miles on 13.2 gallons of gas" as a unit rate.
24.5 mi/gal [4.2B]

17. Solve the proportion $\dfrac{12}{35} = \dfrac{n}{160}$.
Round to the nearest tenth.
54.9 [4.3B]

18. Write $22\dfrac{1}{2}\%$ as a fraction.
$\dfrac{9}{40}$ [5.1A]

19. Write 1.35 as a percent.
135% [5.1B]

20. Write $\frac{5}{4}$ as a percent.
125% [5.1B]

21. Find 120% of 30.
36 [5.2A]

22. 12 is what percent of 9?
$133\frac{1}{3}$% [5.3A]

23. 42 is 60% of what number?
70 [5.4A]

24. Convert $1\frac{2}{3}$ ft to inches.
20 in. [8.1A]

25. Subtract: 3 ft 2 in. − 1 ft 10 in.
1 ft 4 in. [8.1B]

26. Convert 40 oz to pounds.
2.5 lb [8.2A]

27. Find the sum of 3 lb 12 oz and 2 lb 10 oz.
6 lb 6 oz [8.2B]

28. Convert 18 pt to gallons.
2.25 gal [8.3A]

29. Divide: $3\overline{)5 \text{ gal } 1 \text{ qt}}$
1 gal 3 qt [8.3B]

30. Convert 2.48 m to centimeters.
248 cm [9.1A]

31. Convert 4 m 62 cm to meters.
4.62 m [9.1A]

32. Convert 1 kg 614 g to kilograms.
1.614 kg [9.2A]

33. Convert 2 L 67 ml to milliliters.
2067 ml [9.3A]

34. Convert 55 mi to kilometers. Round to the nearest hundredth. (1.61 km ≈ 1 mi)
88.55 km [9.5A]

35. Consumerism How much does it cost to run a 2400-watt air conditioner for 6 h at 8¢ per kilowatt-hour? Round to the nearest cent.
$1.15   [9.4A]

36. Write 0.0000000679 in scientific notation.
6.79×10^{-8} [10.5A]

37. Find the perimeter of a rectangle with a length of 1.2 m and a width of 0.75 m.
3.9 m   [12.2A]

38. Find the area of a rectangle with a length of 9 in. and a width of 5 in.
45 in²   [12.3A]

39. Find the volume of a box with a length of 20 cm, a width of 12 cm, and a height of 5 cm.
1200 cm³   [12.4A]

40. Add: $-2 + 8 + (-10)$
-4 [10.2A]

41. Subtract: $-30 - (-15)$
-15 [10.2B]

42. Multiply: $2\frac{1}{2} \times \left(-\frac{1}{5}\right)$
$-\frac{1}{2}$ [10.4B]

43. Find the quotient of $-1\frac{3}{8}$ and $5\frac{1}{2}$.
$-\frac{1}{4}$ [10.4B]

44. Simplify: $(-4)^2 \div (1 - 3)^2 - (-2)$
6   [10.5B]

45. Simplify: $2x - 3(x - 4) + 5$
$-x + 17$ [11.1C]

46. Solve: $\frac{2}{3}x = -12$
-18 [11.2C]

47. Solve: $3x - 5 = 10$
5   [11.3A]

48. Solve: $8 - 3x = x + 4$
1   [11.4A]

49. Banking You have $872.48 in your checking account. You write checks of $321.88 and $34.23 and then make a deposit of $443.56. Find your new checking account balance.
$959.93   [6.7A]

50. **Elections** On the basis of a pre-election survey, it is estimated that 5 out of 8 eligible voters will vote in an election. How many people will vote in an election with 102,000 eligible voters?
63,750 people   [4.3C]

51. **Investments** This month a company is paying its stockholders a dividend of $1.60 per share. This is 80% of what the dividend per share was 1 year ago. What was the dividend per share 1 year ago?
$2.00   [5.4B]

52. **Compensation** A sales executive received commissions of $4320, $3572, $2864, and $4420 during a 4-month period. Find the mean monthly income from commissions for the 4 months.
$3794   [7.4A]

53. **Simple Interest** A contractor borrows $120,000 for 9 months at an annual interest rate of 8%. What is the simple interest due on the loan?
$7200   [6.3A]

54. **Probability** If two dice are tossed, what is the probability that the sum of the dots on the upward faces is divisible by 3?
$\dfrac{1}{3}$ [7.5A]

55. **Wars** The top four highest death counts, by country, in World War II are shown in the circle graph. Find what percent the death count of Chinese is of the total death count in all four countries. Round to the nearest tenth of a percent.
6.7%   [7.1B]

China 1300
Japan 1100
Germany 3300
USSR 13,600

Top Four Highest Death Counts in World War II (in thousands)

Source: U.S. Department of Defense

56. **Discounts** A compact disc player that regularly sells for $314.00 is on sale for $226.08. What is the discount rate?
28%   [6.2D]

57. **Shipping** A square tile measuring 8 in. by 8 in. weighs 9 oz. Find the weight, in pounds, of a box containing 144 tiles.
81 lb   [8.2C]

58. Find the perimeter of the composite figure. Use 3.14 for π.
28.56 in.   [12.2B]

8 in.   8 in.
8 in.

59. Find the area of the composite figure. Use 3.14 for π.
16.86 cm²   [12.3B]

10 cm
2 cm

60. **Integer Problems** Five less than the quotient of a number and two is equal to three. Find the number.
16   [11.6A]

Appendix

Table of Geometric Formulas

Pythagorean Theorem

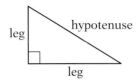

Hypotenuse $= \sqrt{(\text{leg})^2 + (\text{leg})^2}$

Leg $= \sqrt{(\text{hypotenuse})^2 - (\text{leg})^2}$

Perimeter and Area of a Triangle

$P = a + b + c$

$A = \dfrac{1}{2}bh$

Perimeter and Area of a Rectangle

$P = 2L + 2W$

$A = LW$

Perimeter and Area of a Square

$P = 4s$

$A = s^2$

Circumference and Area of a Circle

$C = 2\pi r \ \text{ or } \ C = \pi d$

$A = \pi r^2$

Area of a Trapezoid

$A = \dfrac{1}{2}h(b_1 + b_2)$

Volume of a Rectangular Solid

$V = LWH$

Volume of a Cube

$V = s^3$

Volume of a Sphere

$V = \dfrac{4}{3}\pi r^3$

Volume of a Right Circular Cylinder

$V = \pi r^2 h$

Compound Interest Table

Compounded Annually

	4%	5%	6%	7%	8%	9%	10%
1 year	1.04000	1.05000	1.06000	1.07000	1.08000	1.09000	1.10000
5 years	1.21665	1.27628	1.33823	1.40255	1.46933	1.53862	1.61051
10 years	1.48024	1.62890	1.79085	1.96715	2.15893	2.36736	2.59374
15 years	1.80094	2.07893	2.39656	2.75903	3.17217	3.64248	4.17725
20 years	2.19112	2.65330	3.20714	3.86968	4.66095	5.60441	6.72750

Compounded Semiannually

	4%	5%	6%	7%	8%	9%	10%
1 year	1.04040	1.05062	1.06090	1.07123	1.08160	1.09203	1.10250
5 years	1.21899	1.28008	1.34392	1.41060	1.48024	1.55297	1.62890
10 years	1.48595	1.63862	1.80611	1.98979	2.19112	2.41171	2.65330
15 years	1.81136	2.09757	2.42726	2.80679	3.24340	3.74531	4.32194
20 years	2.20804	2.68506	3.26204	3.95926	4.80102	5.81634	7.03999

Compounded Quarterly

	4%	5%	6%	7%	8%	9%	10%
1 year	1.04060	1.05094	1.06136	1.07186	1.08243	1.09308	1.10381
5 years	1.22019	1.28204	1.34686	1.41478	1.48595	1.56051	1.63862
10 years	1.48886	1.64362	1.81402	2.00160	2.20804	2.43519	2.68506
15 years	1.81670	2.10718	2.44322	2.83182	3.28103	3.80013	4.39979
20 years	2.21672	2.70148	3.29066	4.00639	4.87544	5.93015	7.20957

Compounded Monthly

	4%	5%	6%	7%	8%	9%	10%
1 year	1.04074	1.051162	1.061678	1.072290	1.083000	1.093807	1.104713
5 years	1.220997	1.283359	1.348850	1.417625	1.489846	1.565681	1.645309
10 years	1.490833	1.647009	1.819397	2.009661	2.219640	2.451357	2.707041
15 years	1.820302	2.113704	2.454094	2.848947	3.306921	3.838043	4.453920
20 years	2.222582	2.712640	3.310204	4.038739	4.926803	6.009152	7.328074

Compounded Daily

	4%	5%	6%	7%	8%	9%	10%
1 year	1.04080	1.05127	1.06183	1.07250	1.08328	1.09416	1.10516
5 years	1.22139	1.28400	1.34983	1.41902	1.49176	1.56823	1.64861
10 years	1.49179	1.64866	1.82203	2.01362	2.22535	2.45933	2.71791
15 years	1.82206	2.11689	2.45942	2.85736	3.31968	3.85678	4.48077
20 years	2.22544	2.71810	3.31979	4.05466	4.95217	6.04830	7.38703

To use this table:
1. Locate the section that gives the desired compounding period.
2. Locate the interest rate in the top row of that section.
3. Locate the number of years in the left-hand column of that section.
4. Locate the number where the interest-rate column and the number-of-years row meet. This is the compound interest factor.

Example An investment yields an annual interest rate of 10% compounded quarterly for 5 years.
The compounding period is "compounded quarterly."
The interest rate is 10%.
The number of years is 5.
The number where the row and column meet is 1.63862. This is the compound interest factor.

Compound Interest Table

Compounded Annually	11%	12%	13%	14%	15%	16%	17%
1 year	1.11000	1.12000	1.13000	1.14000	1.15000	1.16000	1.17000
5 years	1.68506	1.76234	1.84244	1.92542	2.01136	2.10034	2.19245
10 years	2.83942	3.10585	3.39457	3.70722	4.04556	4.41144	4.80683
15 years	4.78459	5.47357	6.25427	7.13794	8.13706	9.26552	10.53872
20 years	8.06239	9.64629	11.52309	13.74349	16.36654	19.46076	23.10560

Compounded Semiannually	11%	12%	13%	14%	15%	16%	17%
1 year	1.11303	1.12360	1.13423	1.14490	1.15563	1.16640	1.17723
5 years	1.70814	1.79085	1.87714	1.96715	2.06103	2.15893	2.26098
10 years	2.91776	3.20714	3.52365	3.86968	4.24785	4.66096	5.11205
15 years	4.98395	5.74349	6.61437	7.61226	8.75496	10.06266	11.55825
20 years	8.51331	10.28572	12.41607	14.97446	18.04424	21.72452	26.13302

Compounded Quarterly	11%	12%	13%	14%	15%	16%	17%
1 year	1.11462	1.12551	1.13648	1.14752	1.15865	1.16986	1.18115
5 years	1.72043	1.80611	1.89584	1.98979	2.08815	2.19112	2.29891
10 years	2.95987	3.26204	3.59420	3.95926	4.36038	4.80102	5.28497
15 years	5.09225	5.89160	6.81402	7.87809	9.10513	10.51963	12.14965
20 years	8.76085	10.64089	12.91828	15.67574	19.01290	23.04980	27.93091

Compounded Monthly	11%	12%	13%	14%	15%	16%	17%
1 year	1.115719	1.126825	1.138032	1.149342	1.160755	1.172271	1.183892
5 years	1.728916	1.816697	1.908857	2.005610	2.107181	2.213807	2.325733
10 years	2.989150	3.300387	3.643733	4.022471	4.440213	4.900941	5.409036
15 years	5.167988	5.995802	6.955364	8.067507	9.356334	10.849737	12.579975
20 years	8.935015	10.892554	13.276792	16.180270	19.715494	24.019222	29.257669

Compounded Daily	11%	12%	13%	14%	15%	16%	17%
1 year	1.11626	1.12747	1.13880	1.15024	1.16180	1.17347	1.18526
5 years	1.73311	1.82194	1.91532	2.01348	2.11667	2.22515	2.33918
10 years	3.00367	3.31946	3.66845	4.05411	4.48031	4.95130	5.47178
15 years	5.20569	6.04786	7.02625	8.16288	9.48335	11.01738	12.79950
20 years	9.02203	11.01883	13.45751	16.43582	20.07316	24.51534	29.94039

Monthly Payment Table

	4%	5%	6%	7%	8%	9%
1 year	0.0851499	0.0856075	0.0860664	0.0865267	0.0869884	0.0874515
2 years	0.0434249	0.0438714	0.0443206	0.0447726	0.0452273	0.0456847
3 years	0.0295240	0.0299709	0.0304219	0.0308771	0.0313364	0.0317997
4 years	0.0225791	0.0230293	0.0234850	0.0239462	0.0244129	0.0248850
5 years	0.0184165	0.0188712	0.0193328	0.0198012	0.0202764	0.0207584
15 years	0.0073969	0.0079079	0.0084386	0.0089883	0.0095565	0.0101427
20 years	0.0060598	0.0065996	0.0071643	0.0077530	0.0083644	0.0089973
25 years	0.0052784	0.0058459	0.0064430	0.0070678	0.0077182	0.0083920
30 years	0.0047742	0.0053682	0.0059955	0.0066530	0.0073376	0.0080462

	10%	11%	12%	13%
1 year	0.0879159	0.0883817	0.0888488	0.0893173
2 years	0.0461449	0.0466078	0.0470735	0.0475418
3 years	0.0322672	0.0327387	0.0332143	0.0336940
4 years	0.0253626	0.0258455	0.0263338	0.0268275
5 years	0.0212470	0.0217424	0.0222445	0.0227531
15 years	0.0107461	0.0113660	0.0120017	0.0126524
20 years	0.0096502	0.0103219	0.0110109	0.0117158
25 years	0.0090870	0.0098011	0.0105322	0.0112784
30 years	0.0087757	0.0095232	0.0102861	0.0110620

To use this table:
1. Locate the desired interest rate in the top row.
2. Locate the number of years in the left-hand column.
3. Locate the number where the interest-rate column and the number-of-years row meet. This is the monthly payment factor.

Example   A home has a 30-year mortgage at an annual interest rate of 12%.
The interest rate is 12%.
The number of years is 30.
The number where the row and column meet is 0.0102861. This is the monthly payment factor.

Table of Measurements

Prefixes in the Metric System of Measurement

kilo-	1000	deci-	0.1
hecto-	100	centi-	0.01
deca-	10	milli-	0.001

Measurement Abbreviations
U.S. Customary System

Length

in.	inches
ft	feet
yd	yards
mi	miles

Capacity

oz	fluid ounces
c	cups
qt	quarts
gal	gallons

Weight

oz	ounces
lb	pounds

Area

in^2	square inches
ft^2	square feet
yd^2	square yards
mi^2	square miles

Rate

ft/s	feet per second
mi/h	miles per hour

Time

h	hours
min	minutes
s	seconds

Metric System

Length

mm	millimeters
cm	centimeters
m	meters
km	kilometers

Capacity

ml	milliliters
cl	centiliters
L	liters
kl	kiloliters

Weight/Mass

mg	milligrams
cg	centigrams
g	grams
kg	kilograms

Area

cm^2	square centimeters
m^2	square meters
km^2	square kilometers

Rate

m/s	meters per second
km/s	kilometers per second
km/h	kilometers per hour

Time

h	hours
min	minutes
s	seconds

Table of Properties

Properties of Real Numbers

Commutative Property of Addition

If a and b are real numbers, then $a + b = b + a$.

Commutative Property of Multiplication

If a and b are real numbers, then $a \cdot b = b \cdot a$.

Associative Property of Addition

If a, b, and c are real numbers, then
$(a + b) + c = a + (b + c)$.

Associative Property of Multiplication

If a, b, and c are real numbers, then
$(a \cdot b) \cdot c = a \cdot (b \cdot c)$.

Addition Property of Zero

If a is a real number, then $a + 0 = 0 + a = a$.

Multiplication Property of Zero

If a is a real number, then $a \cdot 0 = 0 \cdot a = 0$.

Multiplication Property of One

If a is a real number, then $a \cdot 1 = 1 \cdot a = a$.

Inverse Property of Addition

If a is a real number, then $a + (-a) = (-a) + a = 0$.

Inverse Property of Multiplication

If a is a real number and $a \neq 0$, then
$a \cdot \dfrac{1}{a} = \dfrac{1}{a} \cdot a = 1$.

Distributive Property

If a, b, and c are real numbers, then
$a(b + c) = ab + ac$.

Properties of Zero and One in Division

Any number divided by 1 is the number.
Division by zero is not allowed.

Any number other than zero divided by itself is 1.
Zero divided by a number other than zero is zero.

Solutions to Chapter 1 "You Try It"

SECTION 1.1

You Try It 1

You Try It 2   **a.** $45 > 29$ **b.** $27 > 0$

You Try It 3   Thirty-six million four hundred sixty-two thousand seventy-five

You Try It 4   452,007

You Try It 5 $60,000 + 8000 + 200 + 80 + 1$

You Try It 6 $100,000 + 9000 + 200 + 7$

You Try It 7

```
            ┌──── Given place value
            │
   368,492
            │
            └──── 8 > 5
```

368,492 rounded to the nearest ten-thousand is 370,000.

You Try It 8

```
            ┌──── Given place value
            │
   3962
            │
            └──── 6 > 5
```

3962 rounded to the nearest hundred is 4000.

SECTION 1.2

You Try It 1

$$
\begin{array}{r}
\overset{1\ 1}{347} \\
+\ 12,453 \\
\hline
12,800
\end{array}
$$

- $7 + 3 = 10$
  Write the 0 in the ones column. Carry the 1 to the tens column.
  $1 + 4 + 5 = 10$
  Write the 0 in the tens column. Carry the 1 to the hundreds column.
  $1 + 3 + 4 = 8$

347 increased by 12,453 is 12,800.

You Try It 2

$$
\begin{array}{r}
\overset{2}{95} \\
88 \\
+\ 67 \\
\hline
250
\end{array}
$$

- $5 + 8 + 7 = 20$
  Write the 0 in the ones column. Carry the 2 to the tens column.

You Try It 3

$$
\begin{array}{r}
\overset{1\ 1\ 2\ 1}{392} \\
4,079 \\
89,035 \\
+\ 4,992 \\
\hline
98,498
\end{array}
$$

You Try It 4

Strategy   To find the total number of Wal-Mart stores outside of the United States, read the table to find the number of stores outside the United States in each of the four categories listed. Then add the four numbers.

Solution

$$
\begin{array}{r}
942 \\
238 \\
71 \\
+\ \ \ 37 \\
\hline
1288
\end{array}
$$

Wal-Mart has 1288 stores outside the United States.

SECTION 1.3

You Try It 1

$$
\begin{array}{r}
8925 \\
-\ 6413 \\
\hline
2512
\end{array}
\qquad
\textit{Check:}\
\begin{array}{r}
6413 \\
+\ 2512 \\
\hline
8925
\end{array}
$$

You Try It 2

$$
\begin{array}{r}
17,504 \\
-\ \ 9,302 \\
\hline
8,202
\end{array}
\qquad
\textit{Check:}\
\begin{array}{r}
9,302 \\
+\ 8,202 \\
\hline
17,504
\end{array}
$$

You Try It 3

$$
\begin{array}{r}
\overset{2\ \ 14\ \ 7\ \ 11}{\cancel{3}\ \cancel{4}\ \cancel{8}\ \cancel{1}} \\
-\ \ \ \ 8\ 6\ 5 \\
\hline
2\ 6\ 1\ 6
\end{array}
\qquad
\textit{Check:}\
\begin{array}{r}
865 \\
+\ 2616 \\
\hline
3481
\end{array}
$$

You Try It 4

$$
\begin{array}{r}
\overset{\quad\ 15}{\overset{4\ \ 5\ \ 12}{5\ 4,\cancel{3}\ \cancel{6}\ \cancel{2}}} \\
-\ 1\ 4,4\ 8\ 5 \\
\hline
4\ 0,0\ 7\ 7
\end{array}
\qquad
\textit{Check:}\
\begin{array}{r}
14,485 \\
+\ 40,077 \\
\hline
54,562
\end{array}
$$

You Try It 5

$$
\begin{array}{r}
\overset{3\ \ 10}{6\ \cancel{4},\cancel{0}\ 0\ 3} \\
-\ 5\ 4,9\ 3\ 6
\end{array}
$$

- There are 2 zeros in the minuend.
  Borrow 1 thousand from the thousands column and write 10 in the hundreds column.

(Continued)

(Continued)

$$\begin{array}{r} \overset{9}{} \\ \overset{3\ \ 10\ \ 10}{6\ \cancel{4},\cancel{0}\ \cancel{0}\ 3} \\ -\ 5\ 4,9\ 3\ 6 \\ \hline \end{array}$$

- Borrow 1 hundred from the hundreds column and write 10 in the tens column.

$$\begin{array}{r} \overset{13\ \ \ 9\ \ \ 9}{\overset{5\ \ 3\ \ 10\ \ 10\ \ 13}{\cancel{6}\ \cancel{4},\cancel{0}\ \cancel{0}\ \cancel{3}}} \\ -\ 5\ 4,9\ 3\ 6 \\ \hline 9,0\ 6\ 7 \end{array}$$

- Borrow 1 ten from the tens column and add 10 to the 3 in the ones column.

Check: $\begin{array}{r} 54{,}936 \\ +\ \ 9{,}067 \\ \hline 64{,}003 \end{array}$

You Try It 6

Strategy   To find the difference, subtract the number of personnel on active duty in the Air Force in 1945 (2,282,259) from the number of personnel on active duty in the Navy in 1945 (3,380,817).

Solution $\begin{array}{r} 3{,}380{,}817 \\ -\ 2{,}282{,}259 \\ \hline 1{,}098{,}558 \end{array}$

The difference was 1,098,558 personnel.

You Try It 7

Strategy   To find your take-home pay:
- Add to find the total of the deductions (127 + 18 + 35).
- Subtract the total of the deductions from your total salary (638).

Solution $\begin{array}{r} 127 \\ 18 \\ +\ 35 \\ \hline 180 \end{array}$ $\begin{array}{r} 638 \\ -\ 180 \\ \hline 458 \end{array}$

180 in deductions

Your take-home pay is $458.

SECTION 1.4

You Try It 1

$$\begin{array}{r} \overset{3\ 5}{648} \\ \times\ \ \ \ 7 \\ \hline 4536 \end{array}$$

- $7 \times 8 = 56$
  Write the 6 in the ones column.
  Carry the 5 to the tens column.
  $7 \times 4 = 28,\ 28 + 5 = 33$
  $7 \times 6 = 42,\ 42 + 3 = 45$

You Try It 2

$$\begin{array}{r} 756 \\ \times\ 305 \\ \hline 3780 \\ 22680 \\ \hline 230{,}580 \end{array}$$

- $5 \times 756 = 3780$
- Write a zero in the tens column for 0×756.
  $3 \times 756 = 2268$

You Try It 3

Strategy   To find the number of cars the dealer will receive in 12 months, multiply the number of months (12) by the number of cars received each month (37).

Solution $\begin{array}{r} 37 \\ \times\ \ 12 \\ \hline 74 \\ 37 \\ \hline 444 \end{array}$

The dealer will receive 444 cars in 12 months.

You Try It 4

Strategy   To find the total cost of the order:
- Find the cost of the sports jackets by multiplying the number of jackets (25) by the cost for each jacket (23).
- Add the product to the cost for the suits (4800).

Solution $\begin{array}{r} 23 \\ \times\ 25 \\ \hline 115 \\ 46 \\ \hline 575 \end{array}$ cost for jackets $\begin{array}{r} 4800 \\ +\ \ 575 \\ \hline 5375 \end{array}$

The total cost of the order is $5375.

SECTION 1.5

You Try It 1

$$9\overline{)63}^{\displaystyle 7}$$

Check: $7 \times 9 = 63$

You Try It 2

$$\begin{array}{r} 453 \\ 9\overline{)\ 4077} \\ -36 \\ \hline 47 \\ -45 \\ \hline 27 \\ -27 \\ \hline 0 \end{array}$$

Check: $453 \times 9 = 4077$

You Try It 3

$$
\begin{array}{r}
705 \\
9\overline{)\,6345} \\
-63 \\
\hline
04 \\
-\,0 \\
\hline
45 \\
-45 \\
\hline
0
\end{array}
$$

- Think $9\overline{)4}$. Place 0 in quotient.
- Subtract 0×9.
- Bring down the 5.

Check: $705 \times 9 = 6345$

You Try It 4

$$
\begin{array}{r}
870 \text{ r5} \\
6\overline{)\,5225} \\
-48 \\
\hline
42 \\
-42 \\
\hline
05 \\
-\,0 \\
\hline
5
\end{array}
$$

- Think $6\overline{)5}$. Place 0 in quotient.
- Subtract 0×6.

Check: $(870 \times 6) + 5 =$
$5220 + 5 = 5225$

You Try It 5

$$
\begin{array}{r}
3{,}058 \text{ r3} \\
7\overline{)\,21{,}409} \\
-21 \\
\hline
0\;4 \\
-\,0 \\
\hline
40 \\
-35 \\
\hline
59 \\
-56 \\
\hline
3
\end{array}
$$

- Think $7\overline{)4}$. Place 0 in quotient.
- Subtract 0×7.

Check: $(3058 \times 7) + 3 =$
$21{,}406 + 3 = 21{,}409$

You Try It 6

$$
\begin{array}{r}
109 \\
42\overline{)\,4578} \\
-42 \\
\hline
37 \\
-\,0 \\
\hline
378 \\
-378 \\
\hline
0
\end{array}
$$

- Think $42\overline{)37}$. Place 0 in quotient.
- Subtract 0×42.

Check: $109 \times 42 = 4578$

You Try It 7

$$
\begin{array}{r}
470 \text{ r29} \\
39\overline{)\,18{,}359} \\
-15\;6 \\
\hline
2\;75 \\
-2\;73 \\
\hline
29 \\
-\,0 \\
\hline
29
\end{array}
$$

- Think $3\overline{)18}$. 6×39 is too large. Try 5. 5×39 is too large. Try 4.

Check: $(470 \times 39) + 29 =$
$18{,}330 + 29 = 18{,}359$

You Try It 8

$$
\begin{array}{r}
62 \text{ r111} \\
534\overline{)\,33{,}219} \\
-32\;04 \\
\hline
1\;179 \\
-1\;068 \\
\hline
111
\end{array}
$$

Check: $(62 \times 534) + 111 =$
$33{,}108 + 111 = 33{,}219$

You Try It 9

$$
\begin{array}{r}
421 \text{ r33} \\
515\overline{)\,216{,}848} \\
-206\;0 \\
\hline
10\;84 \\
-10\;30 \\
\hline
548 \\
-515 \\
\hline
33
\end{array}
$$

Check: $(421 \times 515) + 33 =$
$216{,}815 + 33 = 216{,}848$

You Try It 10

Strategy To find the number of tires that can be stored on each shelf, divide the number of tires (270) by the number of shelves (15).

Solution

$$
\begin{array}{r}
18 \\
15\overline{)\,270} \\
-15 \\
\hline
120 \\
-120 \\
\hline
0
\end{array}
$$

Each shelf can store 18 tires.

You Try It 11

Strategy To find the number of cases produced in 8 hours:
- Find the number of cases produced in 1 hour by dividing the number of cans produced (12,600) by the number of cans to a case (24).
- Multiply the number of cases produced in 1 hour by 8.

(Continued)

(Continued)

Solution

$$\begin{array}{r} 525 \\ 24\overline{)\ 12{,}600} \\ -12\ 0 \\ \hline 60 \\ -48 \\ \hline 120 \\ -120 \\ \hline 0 \end{array}$$ cases produced in 1 hour

$$\begin{array}{r} 525 \\ \times\ \ \ 8 \\ \hline 4200 \end{array}$$

In 8 hours, 4200 cases are produced.

SECTION 1.6

You Try It 1 $2 \cdot 2 \cdot 2 \cdot 2 \cdot 3 \cdot 3 \cdot 3 = 2^4 \cdot 3^3$

You Try It 2 $10 \cdot 10 \cdot 10 \cdot 10 \cdot 10 \cdot 10 \cdot 10 = 10^7$

You Try It 3 $2^3 \cdot 5^2 = (2 \cdot 2 \cdot 2) \cdot (5 \cdot 5) = 8 \cdot 25$
$\qquad\qquad\qquad = 200$

You Try It 4 $5 \cdot (8 - 4)^2 \div 4 - 2$ • **Parentheses**
$= 5 \cdot 4^2 \div 4 - 2$ • **Exponents**
$= 5 \cdot 16 \div 4 - 2$ • **Multiplication and division**
$= 80 \div 4 - 2$
$= 20 - 2$ • **Subtraction**
$= 18$

SECTION 1.7

You Try It 1
$40 \div 1 = 40$
$40 \div 2 = 20$
$40 \div 3 \qquad$ Will not divide evenly
$40 \div 4 = 10$
$40 \div 5 = 8$
$40 \div 6 \qquad$ Will not divide evenly
$40 \div 7 \qquad$ Will not divide evenly
$40 \div 8 = 5$

1, 2, 4, 5, 8, 10, 20, and 40 are factors of 40.

You Try It 2

$$\begin{array}{c|c} \multicolumn{2}{c}{44} \\ \hline 2 & 22 \\ 2 & 11 \\ 11 & 1 \end{array}$$ • $44 \div 2 = 22$
• $22 \div 2 = 11$
• $11 \div 11 = 1$

$44 = 2 \cdot 2 \cdot 11$

You Try It 3

$$\begin{array}{c|c} \multicolumn{2}{c}{177} \\ \hline 3 & 59 \\ 59 & 1 \end{array}$$ • **Try only 2, 3, 4, 7, and 11 because $11^2 > 59$.**

$177 = 3 \cdot 59$

Solutions to Chapter 2 "You Try It"

SECTION 2.1

You Try It 1

	2	3	5
12 =	(2 · 2)	3	
27 =		(3 · 3 · 3)	
50 =	2		(5 · 5)

The LCM = $2 \cdot 2 \cdot 3 \cdot 3 \cdot 3 \cdot 5 \cdot 5$
$\qquad\quad = 2700$

You Try It 2

	2	3	5
36 =	(2 · 2)	3 · 3	
60 =	2 · 2	(3)	5
72 =	2 · 2 · 2	3 · 3	

The GCF = $2 \cdot 2 \cdot 3 = 12$.

You Try It 3

	2	3	5	11
11 =				11
24 =	2 · 2 · 2	3		
30 =	2	3	5	

Because no numbers are circled, the GCF = 1.

SECTION 2.2

You Try It 1 $4\dfrac{1}{4}$

You Try It 2 $\dfrac{17}{4}$

You Try It 3

$$\begin{array}{r} 4 \\ 5\overline{)22} \\ -20 \\ \hline 2 \end{array} \qquad \dfrac{22}{5} = 4\dfrac{2}{5}$$

You Try It 4

$$\begin{array}{r} 4 \\ 7\overline{)28} \\ -28 \\ \hline 0 \end{array} \qquad \dfrac{28}{7} = 4$$

You Try It 5 $14\dfrac{5}{8} = \dfrac{112 + 5}{8} = \dfrac{117}{8}$

SECTION 2.3

You Try It 1 $45 \div 5 = 9$ $\dfrac{3}{5} = \dfrac{3 \cdot 9}{5 \cdot 9} = \dfrac{27}{45}$

$\dfrac{27}{45}$ is equivalent to $\dfrac{3}{5}$.

You Try It 2   Write 6 as $\dfrac{6}{1}$.

$18 \div 1 = 18$ $6 = \dfrac{6 \cdot 18}{1 \cdot 18} = \dfrac{108}{18}$

$\dfrac{108}{18}$ is equivalent to 6.

You Try It 3 $\dfrac{16}{24} = \dfrac{\overset{1}{\cancel{2}} \cdot \overset{1}{\cancel{2}} \cdot \overset{1}{\cancel{2}} \cdot 2}{\underset{1}{\cancel{2}} \cdot \underset{1}{\cancel{2}} \cdot \underset{1}{\cancel{2}} \cdot 3} = \dfrac{2}{3}$

You Try It 4 $\dfrac{8}{56} = \dfrac{\overset{1}{\cancel{2}} \cdot \overset{1}{\cancel{2}} \cdot \overset{1}{\cancel{2}}}{\underset{1}{\cancel{2}} \cdot \underset{1}{\cancel{2}} \cdot \underset{1}{\cancel{2}} \cdot 7} = \dfrac{1}{7}$

You Try It 5 $\dfrac{15}{32} = \dfrac{3 \cdot 5}{2 \cdot 2 \cdot 2 \cdot 2 \cdot 2} = \dfrac{15}{32}$

You Try It 6 $\dfrac{48}{36} = \dfrac{\overset{1}{\cancel{2}} \cdot \overset{1}{\cancel{2}} \cdot 2 \cdot 2 \cdot \overset{1}{\cancel{3}}}{\underset{1}{\cancel{2}} \cdot \underset{1}{\cancel{2}} \cdot \underset{1}{\cancel{3}} \cdot 3} = \dfrac{4}{3} = 1\dfrac{1}{3}$

SECTION 2.4

You Try It 1

$\begin{array}{r} \dfrac{3}{8} \\[2mm] +\dfrac{7}{8} \\[1mm] \hline \dfrac{10}{8} = \dfrac{5}{4} = 1\dfrac{1}{4} \end{array}$

• The denominators are the same. Add the numerators. Place the sum over the common denominator.

You Try It 2

$\begin{array}{r} \dfrac{5}{12} = \dfrac{20}{48} \\[2mm] +\dfrac{9}{16} = \dfrac{27}{48} \\[1mm] \hline \dfrac{47}{48} \end{array}$

• The LCM of 12 and 16 is 48.

You Try It 3

$\begin{array}{r} \dfrac{7}{8} = \dfrac{105}{120} \\[2mm] +\dfrac{11}{15} = \dfrac{88}{120} \\[1mm] \hline \dfrac{193}{120} = 1\dfrac{73}{120} \end{array}$

• The LCM of 8 and 15 is 120.

You Try It 4

$\begin{array}{r} \dfrac{3}{4} = \dfrac{30}{40} \\[2mm] \dfrac{4}{5} = \dfrac{32}{40} \\[2mm] +\dfrac{5}{8} = \dfrac{25}{40} \\[1mm] \hline \dfrac{87}{40} = 2\dfrac{7}{40} \end{array}$

• The LCM of 4, 5, and 8 is 40.

You Try It 5 $7 + \dfrac{6}{11} = 7\dfrac{6}{11}$

You Try It 6

$\begin{array}{r} 29 \\[2mm] +17\dfrac{5}{12} \\[1mm] \hline 46\dfrac{5}{12} \end{array}$

You Try It 7

$\begin{array}{r} 7\dfrac{4}{5} = 7\dfrac{24}{30} \\[2mm] 6\dfrac{7}{10} = 6\dfrac{21}{30} \\[2mm] +13\dfrac{11}{15} = 13\dfrac{22}{30} \\[1mm] \hline 26\dfrac{67}{30} = 28\dfrac{7}{30} \end{array}$

• LCM = 30

You Try It 8

$\begin{array}{r} 9\dfrac{3}{8} = 9\dfrac{45}{120} \\[2mm] 17\dfrac{7}{12} = 17\dfrac{70}{120} \\[2mm] +10\dfrac{14}{15} = 10\dfrac{112}{120} \\[1mm] \hline 36\dfrac{227}{120} = 37\dfrac{107}{120} \end{array}$

• LCM = 120

You Try It 9

Strategy   To find the total time spent on the activities, add the three times $\left(4\dfrac{1}{2}, 3\dfrac{3}{4}, 1\dfrac{1}{3}\right)$.

Solution

$\begin{array}{r} 4\dfrac{1}{2} = 4\dfrac{6}{12} \\[2mm] 3\dfrac{3}{4} = 3\dfrac{9}{12} \\[2mm] +1\dfrac{1}{3} = 1\dfrac{4}{12} \\[1mm] \hline 8\dfrac{19}{12} = 9\dfrac{7}{12} \end{array}$

The total time spent on the three activities was $9\dfrac{7}{12}$ hours.

You Try It 10

Strategy   To find the overtime pay:
- Find the total number of overtime hours $\left(1\frac{2}{3} + 3\frac{1}{3} + 2\right)$.
- Multiply the total number of hours by the overtime hourly wage (36).

Solution

$$1\frac{2}{3}$$
$$3\frac{1}{3}$$
$$+\ 2$$
$$\overline{6\frac{3}{3}} = 7 \text{ hours}$$

$$\begin{array}{r} 36 \\ \times\ \ 7 \\ \hline 252 \end{array}$$

Jeff earned $252 in overtime pay.

SECTION 2.5

You Try It 1

$$\frac{16}{27}$$
$$-\ \frac{7}{27}$$
$$\overline{\frac{9}{27}} = \frac{1}{3}$$

- The denominators are the same. Subtract the numerators. Place the difference over the common denominator.

You Try It 2

$$\frac{13}{18} = \frac{52}{72}$$
$$-\ \frac{7}{24} = \frac{21}{72}$$
$$\overline{\frac{31}{72}}$$

- LCM = 72

You Try It 3

$$17\frac{5}{9} = 17\frac{20}{36}$$
$$-\ 11\frac{5}{12} = 11\frac{15}{36}$$
$$\overline{6\frac{5}{36}}$$

- LCM = 36

You Try It 4

$$8 = 7\frac{13}{13}$$
$$-\ 2\frac{4}{13} = 2\frac{4}{13}$$
$$\overline{5\frac{9}{13}}$$

- LCM = 13

You Try It 5

$$21\frac{7}{9} = 21\frac{28}{36} = 20\frac{64}{36}$$
$$-\ 7\frac{11}{12} = 7\frac{33}{36} = 7\frac{33}{36}$$
$$\overline{13\frac{31}{36}}$$

- LCM = 36

You Try It 6

Strategy   To find the time remaining before the plane lands, subtract the number of hours already in the air $\left(2\frac{3}{4}\right)$ from the total time of the trip $\left(5\frac{1}{2}\right)$.

Solution

$$5\frac{1}{2} = 5\frac{2}{4} = 4\frac{6}{4}$$
$$-\ 2\frac{3}{4} = 2\frac{3}{4} = 2\frac{3}{4}$$
$$\overline{2\frac{3}{4} \text{ hours}}$$

The plane will land in $2\frac{3}{4}$ hours.

You Try It 7

Strategy   To find the amount of weight to be lost during the third month:
- Find the total weight loss during the first two months $\left(7\frac{1}{2} + 5\frac{3}{4}\right)$.
- Subtract the total weight loss from the goal (24 pounds).

Solution

$$7\frac{1}{2} = 7\frac{2}{4}$$
$$+\ 5\frac{3}{4} = 5\frac{3}{4}$$
$$\overline{12\frac{5}{4}} = 13\frac{1}{4} \text{ pounds lost}$$

$$24 = 23\frac{4}{4}$$
$$-\ 13\frac{1}{4} = 13\frac{1}{4}$$
$$\overline{10\frac{3}{4} \text{ pounds}}$$

The patient must lose $10\frac{3}{4}$ pounds to achieve the goal.

SECTION 2.6

You Try It 1

$$\frac{4}{21} \times \frac{7}{44} = \frac{4 \cdot 7}{21 \cdot 44}$$

$$= \frac{\overset{1}{\cancel{2}} \cdot \overset{1}{\cancel{2}} \cdot \overset{1}{\cancel{7}}}{3 \cdot \underset{1}{\cancel{7}} \cdot \underset{1}{\cancel{2}} \cdot \underset{1}{\cancel{2}} \cdot 11} = \frac{1}{33}$$

You Try It 2

$$\frac{2}{21} \times \frac{10}{33} = \frac{2 \cdot 10}{21 \cdot 33}$$

$$= \frac{2 \cdot 2 \cdot 5}{3 \cdot 7 \cdot 3 \cdot 11} = \frac{20}{693}$$

You Try It 3 $\dfrac{16}{5} \times \dfrac{15}{24} = \dfrac{16 \cdot 15}{5 \cdot 24}$

$$= \dfrac{\overset{1}{\cancel{2}} \cdot \overset{1}{\cancel{2}} \cdot \overset{1}{\cancel{2}} \cdot 2 \cdot \overset{1}{\cancel{3}} \cdot \overset{1}{\cancel{5}}}{\underset{1}{\cancel{5}} \cdot \underset{1}{\cancel{2}} \cdot \underset{1}{\cancel{2}} \cdot \underset{1}{\cancel{2}} \cdot \underset{1}{\cancel{3}}} = \dfrac{2}{1} = 2$$

You Try It 4 $5\dfrac{2}{5} \times \dfrac{5}{9} = \dfrac{27}{5} \times \dfrac{5}{9} = \dfrac{27 \cdot 5}{5 \cdot 9}$

$$= \dfrac{\overset{1}{\cancel{3}} \cdot \overset{1}{\cancel{3}} \cdot 3 \cdot \overset{1}{\cancel{5}}}{\underset{1}{\cancel{5}} \cdot \underset{1}{\cancel{3}} \cdot \underset{1}{\cancel{3}}} = \dfrac{3}{1} = 3$$

You Try It 5 $3\dfrac{2}{5} \times 6\dfrac{1}{4} = \dfrac{17}{5} \times \dfrac{25}{4} = \dfrac{17 \cdot 25}{5 \cdot 4}$

$$= \dfrac{17 \cdot \overset{1}{\cancel{5}} \cdot 5}{\underset{1}{\cancel{5}} \cdot 2 \cdot 2} = \dfrac{85}{4} = 21\dfrac{1}{4}$$

You Try It 6 $3\dfrac{2}{7} \times 6 = \dfrac{23}{7} \times \dfrac{6}{1} = \dfrac{23 \cdot 6}{7 \cdot 1}$

$$= \dfrac{23 \cdot 2 \cdot 3}{7 \cdot 1} = \dfrac{138}{7} = 19\dfrac{5}{7}$$

You Try It 7

Strategy To find the value of the house today, multiply the old value of the house (170,000) by $2\dfrac{1}{2}$.

Solution $170{,}000 \times 2\dfrac{1}{2} = \dfrac{170{,}000}{1} \times \dfrac{5}{2}$

$$= \dfrac{170{,}000 \cdot 5}{1 \cdot 2}$$
$$= 425{,}000$$

The value of the house today is $425,000.

You Try It 8

Strategy To find the cost of the air compressor:
• Multiply to find the value of the drying chamber $\left(\dfrac{4}{5} \times 160{,}000\right)$.
• Subtract the value of the drying chamber from the total value of the two items (160,000).

Solution $\dfrac{4}{5} \times \dfrac{160{,}000}{1} = \dfrac{640{,}000}{5}$

$= 128{,}000$ • Value of the drying chamber

$160{,}000 - 128{,}000 = 32{,}000$

The cost of the air compressor was $32,000.

SECTION 2.7

You Try It 1 $\dfrac{3}{7} \div \dfrac{2}{3} = \dfrac{3}{7} \times \dfrac{3}{2} = \dfrac{3 \cdot 3}{7 \cdot 2} = \dfrac{9}{14}$

You Try It 2 $\dfrac{3}{4} \div \dfrac{9}{10} = \dfrac{3}{4} \times \dfrac{10}{9}$

$$= \dfrac{3 \cdot 10}{4 \cdot 9} = \dfrac{\overset{1}{\cancel{3}} \cdot \overset{1}{\cancel{2}} \cdot 5}{2 \cdot \underset{1}{\cancel{2}} \cdot \underset{1}{\cancel{3}} \cdot 3} = \dfrac{5}{6}$$

You Try It 3 $\dfrac{5}{7} \div 6 = \dfrac{5}{7} \div \dfrac{6}{1}$ • $6 = \dfrac{6}{1}$. The reciprocal of $\dfrac{6}{1}$ is $\dfrac{1}{6}$.

$= \dfrac{5}{7} \times \dfrac{1}{6} = \dfrac{5 \cdot 1}{7 \cdot 6}$

$$= \dfrac{5}{7 \cdot 2 \cdot 3} = \dfrac{5}{42}$$

You Try It 4 $12\dfrac{3}{5} \div 7 = \dfrac{63}{5} \div \dfrac{7}{1} = \dfrac{63}{5} \times \dfrac{1}{7}$

$$= \dfrac{63 \cdot 1}{5 \cdot 7} = \dfrac{3 \cdot 3 \cdot \overset{1}{\cancel{7}}}{5 \cdot \underset{1}{\cancel{7}}} = \dfrac{9}{5} = 1\dfrac{4}{5}$$

You Try It 5 $3\dfrac{2}{3} \div 2\dfrac{2}{5} = \dfrac{11}{3} \div \dfrac{12}{5}$

$= \dfrac{11}{3} \times \dfrac{5}{12} = \dfrac{11 \cdot 5}{3 \cdot 12}$

$$= \dfrac{11 \cdot 5}{3 \cdot 2 \cdot 2 \cdot 3} = \dfrac{55}{36} = 1\dfrac{19}{36}$$

You Try it 6 $2\dfrac{5}{6} \div 8\dfrac{1}{2} = \dfrac{17}{6} \div \dfrac{17}{2}$

$= \dfrac{17}{6} \times \dfrac{2}{17} = \dfrac{17 \cdot 2}{6 \cdot 17}$

$$= \dfrac{\overset{1}{\cancel{17}} \cdot \overset{1}{\cancel{2}}}{2 \cdot 3 \cdot \underset{1}{\cancel{17}}} = \dfrac{1}{3}$$

You Try It 7 $6\dfrac{2}{5} \div 4 = \dfrac{32}{5} \div \dfrac{4}{1}$

$= \dfrac{32}{5} \times \dfrac{1}{4} = \dfrac{32 \cdot 1}{5 \cdot 4}$

$$= \dfrac{2 \cdot 2 \cdot 2 \cdot \overset{1}{\cancel{2}} \cdot \overset{1}{\cancel{2}}}{5 \cdot \underset{1}{\cancel{2}} \cdot \underset{1}{\cancel{2}}} = \dfrac{8}{5} = 1\dfrac{3}{5}$$

You Try It 8

Strategy To find the number of products, divide the number of minutes in 1 hour (60) by the time to assemble one product $\left(7\dfrac{1}{2}\right)$.

Solution $60 \div 7\dfrac{1}{2} = \dfrac{60}{1} \div \dfrac{15}{2} = \dfrac{60}{1} \cdot \dfrac{2}{15}$

$$= \dfrac{60 \cdot 2}{1 \cdot 15} = 8$$

The factory worker can assemble eight products in 1 hour.

You Try It 9

Strategy To find the length of the remaining piece:
- Divide the total length of the board (16) by the length of each shelf $\left(3\frac{1}{3}\right)$. This will give you the number of shelves cut, with a certain fraction of a shelf left over.
- Multiply the fractional part of the result in step 1 by the length of one shelf to determine the length of the remaining piece.

Solution

$$16 \div 3\frac{1}{3} = 16 \div \frac{10}{3}$$

$$= \frac{16}{1} \times \frac{3}{10} = \frac{16 \cdot 3}{1 \cdot 10}$$

$$= \frac{\overset{1}{\cancel{2}} \cdot 2 \cdot 2 \cdot 2 \cdot 3}{\underset{1}{\cancel{2}} \cdot 5} = \frac{24}{5}$$

$$= 4\frac{4}{5}$$

There are 4 pieces $3\frac{1}{3}$ feet long. There is 1 piece that is $\frac{4}{5}$ of $3\frac{1}{3}$ feet long.

$$\frac{4}{5} \times 3\frac{1}{3} = \frac{4}{5} \times \frac{10}{3}$$

$$= \frac{4 \cdot 10}{5 \cdot 3} = \frac{2 \cdot 2 \cdot 2 \cdot \overset{1}{\cancel{5}}}{\underset{1}{\cancel{5}} \cdot 3}$$

$$= \frac{8}{3} = 2\frac{2}{3}$$

The length of the piece remaining is $2\frac{2}{3}$ feet.

SECTION 2.8

You Try It 1 $\frac{9}{14} = \frac{27}{42}$ $\frac{13}{21} = \frac{26}{42}$ $\frac{9}{14} > \frac{13}{21}$

You Try It 2

$$\left(\frac{7}{11}\right)^2 \cdot \left(\frac{2}{7}\right) = \left(\frac{7}{11} \cdot \frac{7}{11}\right) \cdot \left(\frac{2}{7}\right)$$

$$= \frac{\overset{1}{\cancel{7}} \cdot 7 \cdot 2}{11 \cdot 11 \cdot \underset{1}{\cancel{7}}} = \frac{14}{121}$$

You Try It 3

$$\left(\frac{1}{13}\right)^2 \cdot \left(\frac{1}{4} + \frac{1}{6}\right) \div \frac{5}{13} =$$

$$\left(\frac{1}{13}\right)^2 \cdot \left(\frac{5}{12}\right) \div \frac{5}{13} =$$

$$\left(\frac{1}{169}\right) \cdot \left(\frac{5}{12}\right) \div \frac{5}{13} =$$

$$\left(\frac{1 \cdot 5}{13 \cdot 13 \cdot 12}\right) \div \frac{5}{13} =$$

$$\left(\frac{1 \cdot 5}{13 \cdot 13 \cdot 12}\right) \times \frac{13}{5} =$$

$$\frac{1 \cdot \overset{1}{\cancel{5}} \cdot \overset{1}{\cancel{13}}}{\underset{1}{\cancel{13}} \cdot 13 \cdot 12 \cdot \underset{1}{\cancel{5}}} = \frac{1}{156}$$

Solutions to Chapter 3 "You Try It"

SECTION 3.1

You Try It 1 The digit 4 is in the thousandths place.

You Try It 2 $\frac{501}{1000} = 0.501$
(five hundred one thousandths)

You Try It 3 $0.67 = \frac{67}{100}$ (sixty-seven hundredths)

You Try It 4 fifty-five and six thousand eighty-three ten-thousandths

You Try It 5 806.00491   • 1 is in the hundred-thousandths place.

You Try It 6

3.675849 rounded to the nearest ten-thousandth is 3.6758.

You Try It 7

48.907 rounded to the nearest tenth is 48.9.

You Try It 8

Given place value — 31.8652

8 > 5

31.8652 rounded to the nearest whole number is 32.

You Try It 9   2.65 rounded to the nearest whole number is 3.

To the nearest inch, the average annual precipitation in Yuma is 3 inch.

SECTION 3.2

You Try It 1

```
  1 2
   4.62
  27.9
+  0.62054
  33.14054
```
• Place the decimal points on a vertical line.

You Try It 2

```
    1
   6.05
  12.
+  0.374
  18.424
```

You Try It 3

Strategy To determine the number, add the numbers of hearing-impaired Americans of ages 45 to 54, 55 to 64, and 65 to 74, and 75 and over.

Solution
```
   4.48
   4.31
   5.41
 +3.80
  18.00
```

18 million Americans ages 45 and older are hearing-impaired.

You Try It 4

Strategy To find the total income, add the four commissions (985.80, 791.46, 829.75, and 635.42) to the salary (875).

Solution 875 + 985.80 + 791.46 + 829.75 + 635.42 = 4117.43

Anita's total income was $4117.43.

SECTION 3.3

You Try It 1

```
    11 9
   6 1 10 13
   7 2.0 3 9        Check:     1 1  1
 −    8.4 7                    8.47
   6 3.5 6 9               + 63.569
                            72.039
```

You Try It 2

```
    14 9
   2 4 10 10
   3 5.0 0          Check:    1 1 1
 −   9.6 7                    9.67
   2 5.3 3                 + 25.33
                            35.00
```

You Try It 3

```
      16
   2 6 9 9 10
   3.7 0 0 0        Check:   1 111
 − 1.9 7 1 5                 1.9715
   1.7 2 8 5              + 1.7285
                            3.7000
```

You Try It 4

Strategy To find the amount of change, subtract the amount paid (6.85) from 10.00.

Solution
```
  10.00
 − 6.85
   3.15
```

Your change was $3.15.

You Try It 5

Strategy To find the new balance:
• Add to find the total of the three checks (1025.60 + 79.85 + 162.47).
• Subtract the total from the previous balance (2472.69).

Solution
```
  1025.60              2472.69
    79.85           − 1267.92
 + 162.47             1204.77
  1267.92
```

The new balance is $1204.77.

SECTION 3.4

You Try It 1

```
    870
 ×  4.6      • 1 decimal place
    522 0
   3480
   4002.0    • 1 decimal place
```

You Try It 2

```
   0.000086      • 6 decimal places
 ×    0.057      • 3 decimal places
        602
        430
   0.000004902   • 9 decimal places
```

You Try It 3

```
    4.68        • 2 decimal places
 × 6.03         • 2 decimal places
    1404
  28 080
  28.2204       • 4 decimal places
```

You Try It 4 $6.9 \times 1000 = 6900$

You Try It 5 $4.0273 \times 10^2 = 402.73$

You Try It 6

Strategy   To find the total bill:
- Find the number of gallons of water used by multiplying the number of gallons used per day (5000) by the number of days (62).
- Find the cost of water by multiplying the cost per 1000 gallons (1.39) by the number of 1000-gallon units used.
- Add the cost of the water to the meter fee (133.70).

Solution

Number of gallons = 5000(62) = 310,000

$$\text{Cost of water} = \frac{310,000}{1000} \times 1.39 = 430.90$$

Total cost = 430.90 + 133.70 = 564.60

The total bill is $564.60.

You Try It 7

Strategy   To find the cost of running the freezer for 210 hours, multiply the hourly cost (0.035) by the number of hours the freezer has run (210).

Solution

```
  0.035
×   210
  350
 7 0
 7.350
```

The cost of running the freezer for 210 hours is $7.35.

You Try It 8

Strategy   To find the total cost of the stereo:
- Multiply the monthly payment (37.18) by the number of months (18).
- Add that product to the down payment (175.00).

Solution

```
  37.18              175.00
×    18            + 669.24
 297 44              844.24
 371 8
 669.24
```

The total cost of the stereo is $844.24.

SECTION 3.5

You Try It 1

```
          2.7
0.052.)0.140.4
      -104
       36 4
      -36 4
          0
```
- Move the decimal point 3 places to the right in the divisor and the dividend. Write the decimal point in the quotient directly over the decimal point in the dividend.

You Try It 2

```
     0.4873 ≈ 0.487
76)37.0420
  -30 4
    6 64
   -6 08
     562
    -532
     300
    -228
```
- Write the decimal point in the quotient directly over the decimal point in the dividend.

You Try It 3

```
          72.73 ≈ 72.7
5.09.)370.20.00
     -356 3
       13 90
       10 18
        3 720
       -3 563
        1570
       -1527
```

You Try It 4 $309.21 \div 10,000 = 0.030921$

You Try It 5 $42.93 \div 10^4 = 0.004293$

You Try It 6

Strategy   To find how many times greater the average hourly earnings were, divide the 1990 average hourly earnings (10.02) by the 1970 average hourly earnings (3.23).

Solution $10.02 \div 3.23 \approx 3.1$

The average hourly earnings in 1990 were about 3.1 times greater than in 1970.

You Try It 7

Strategy   To find the average number of people watching TV:
- Add the number of people watching each day of the week.
- Divide the total number of people watching by 7.

Solution $91.9 + 89.8 + 90.6 + 93.9 + 78.0$
$\qquad\qquad\qquad + 77.1 + 87.7 = 609$

$$\frac{609}{7} = 87$$

An average of 87 million people watch television per day.

SECTION 3.6

You Try It 1 $\dfrac{0.56 \approx 0.6}{16\overline{)9.00}}$

You Try It 2 $4\dfrac{1}{6} = \dfrac{25}{6}$

$\dfrac{4.166 \approx 4.17}{6\overline{)25.000}}$

You Try It 3 $0.56 = \dfrac{56}{100} = \dfrac{14}{25}$

$5.35 = 5\dfrac{35}{100} = 5\dfrac{7}{20}$

You Try It 4 $0.12\dfrac{7}{8} = \dfrac{12\dfrac{7}{8}}{100} = 12\dfrac{7}{8} \div 100$

$\qquad\qquad = \dfrac{103}{8} \times \dfrac{1}{100} = \dfrac{103}{800}$

You Try It 5 $\dfrac{5}{8} = 0.625$ • Convert the fraction $\frac{5}{8}$ to a decimal.

$0.630 > 0.625$ • Compare the two decimals.

$0.63 > \dfrac{5}{8}$ • Convert 0.625 back to a fraction.

Solutions to Chapter 4 "You Try It"

SECTION 4.1

You Try It 1 $\dfrac{20 \text{ pounds}}{24 \text{ pounds}} = \dfrac{20}{24} = \dfrac{5}{6}$

20 pounds : 24 pounds = 20 : 24 = 5 : 6

20 pounds to 24 pounds = 20 to 24
$\qquad\qquad\qquad\qquad\qquad = 5 \text{ to } 6$

You Try It 2 $\dfrac{64 \text{ miles}}{8 \text{ miles}} = \dfrac{64}{8} = \dfrac{8}{1}$

64 miles : 8 miles = 64 : 8 = 8 : 1

64 miles to 8 miles = 64 to 8 = 8 to 1

You Try It 3

Strategy   To find the ratio, write the ratio of board feet of cedar (12,000) to board feet of ash (18,000) in simplest form.

Solution $\dfrac{12,000}{18,000} = \dfrac{2}{3}$

The ratio is $\dfrac{2}{3}$.

You Try It 4

Strategy   To find the ratio, write the ratio of the amount spent on radio advertising (45,000) to the amount spent on radio and television advertising (45,000 + 60,000) in simplest form.

Solution $\dfrac{\$45,000}{\$45,000 + \$60,000} = \dfrac{45,000}{105,000} = \dfrac{3}{7}$

The ratio is $\dfrac{3}{7}$.

SECTION 4.2

You Try It 1 $\dfrac{15 \text{ pounds}}{12 \text{ trees}} = \dfrac{5 \text{ pounds}}{4 \text{ trees}}$

You Try It 2 $\dfrac{260 \text{ miles}}{8 \text{ hours}}$

$\dfrac{32.5}{8\overline{)260.0}}$

32.5 miles/hour

You Try It 3

Strategy   To find Erik's profit per ounce:
• Find the total profit by subtracting the cost ($1625) from the selling price ($1720).
• Divide the total profit by the number of ounces (5).

Solution $1720 - 1625 = 95$

$95 \div 5 = 19$

The profit was $19/ounce.

SECTION 4.3

You Try It 1 $\dfrac{6}{10} \diagtimes \dfrac{9}{15}$ $\begin{array}{l} 10 \times 9 = 90 \\ 6 \times 15 = 90 \end{array}$

The cross products are equal. The proportion is true.

You Try It 2 $\dfrac{32}{6} \diagtimes \dfrac{90}{8}$ $\begin{array}{l} 6 \times 90 = 540 \\ 32 \times 8 = 256 \end{array}$

The cross products are not equal. The proportion is not true.

You Try It 3 $\dfrac{n}{14} = \dfrac{3}{7}$ • Find the cross products. Then solve for n.

$n \times 7 = 14 \times 3$
$n \times 7 = 42$
$n = 42 \div 7$
$n = 6$

Check: $\dfrac{6}{14} \diagdown\!\!\!\diagup \dfrac{3}{7}$ ➤ $14 \times 3 = 42$
➤ $6 \times 7 = 42$

You Try It 4 $\dfrac{5}{7} = \dfrac{n}{20}$ • Find the cross products. Then solve for n.

$5 \times 20 = 7 \times n$
$100 = 7 \times n$
$100 \div 7 = n$
$14.3 \approx n$

You Try It 5 $\dfrac{15}{20} = \dfrac{12}{n}$ • Find the cross products. Then solve for n.

$15 \times n = 20 \times 12$
$15 \times n = 240$
$n = 240 \div 15$
$n = 16$

Check: $\dfrac{15}{20} \diagdown\!\!\!\diagup \dfrac{12}{16}$ ➤ $20 \times 12 = 240$
➤ $15 \times 16 = 240$

You Try It 6 $\dfrac{12}{n} = \dfrac{7}{4}$

$12 \times 4 = n \times 7$
$48 = n \times 7$
$48 \div 7 = n$
$6.86 \approx n$

You Try It 7 $\dfrac{n}{12} = \dfrac{4}{1}$

$n \times 1 = 12 \times 4$
$n \times 1 = 48$
$n = 48 \div 1$
$n = 48$

Check: $\dfrac{48}{12} \diagdown\!\!\!\diagup \dfrac{4}{1}$ ➤ $12 \times 4 = 48$
➤ $48 \times 1 = 48$

You Try It 8

Strategy   To find the number of tablespoons of fertilizer needed, write and solve a proportion using n to represent the number of tablespoons of fertilizer.

Solution
$\dfrac{3 \text{ tablespoons}}{4 \text{ gallons}} = \dfrac{n \text{ tablespoons}}{10 \text{ gallons}}$ • The unit "tablespoons" is in the numerator. The unit "gallons" is in the denominator.

$3 \times 10 = 4 \times n$
$30 = 4 \times n$
$30 \div 4 = n$
$7.5 = n$

For 10 gallons of water, 7.5 tablespoons of fertilizer are required.

You Try It 9

Strategy   To find the number of jars that can be packed in 15 boxes, write and solve a proportion using n to represent the number of jars.

Solution $\dfrac{24 \text{ jars}}{6 \text{ boxes}} = \dfrac{n \text{ jars}}{15 \text{ boxes}}$

$24 \times 15 = 6 \times n$
$360 = 6 \times n$
$360 \div 6 = n$
$60 = n$

60 jars can be packed in 15 boxes.

Solutions to Chapter 5 "You Try It"

SECTION 5.1

You Try It 1   **a.** $125\% = 125 \times \dfrac{1}{100} = \dfrac{125}{100} = 1\dfrac{1}{4}$

b. $125\% = 125 \times 0.01 = 1.25$

You Try It 2 $33\dfrac{1}{3}\% = 33\dfrac{1}{3} \times \dfrac{1}{100}$

$= \dfrac{100}{3} \times \dfrac{1}{100}$

$= \dfrac{100}{300} = \dfrac{1}{3}$

You Try It 3 $0.25\% = 0.25 \times 0.01 = 0.0025$

You Try It 4 $0.048 = 0.048 \times 100\% = 4.8\%$

You Try It 5 $3.67 = 3.67 \times 100\% = 367\%$

You Try It 6 $0.62\frac{1}{2} = 0.62\frac{1}{2} \times 100\%$

$$= 62\frac{1}{2}\%$$

You Try It 7 $\frac{5}{6} = \frac{5}{6} \times 100\% = \frac{500}{6}\% = 83\frac{1}{3}\%$

You Try It 8 $1\frac{4}{9} = \frac{13}{9} = \frac{13}{9} \times 100\%$

$$= \frac{1300}{9}\% \approx 144.4\%$$

SECTION 5.2

You Try It 1
Percent $\times$ base = amount
$0.063 \times 150 = n$
$9.45 = n$

You Try It 2
Percent $\times$ base = amount

$\frac{1}{6} \times 66 = n$ • $16\frac{2}{3}\% = \frac{1}{6}$

$11 = n$

You Try It 3

Strategy
To determine the amount that came from corporations, write and solve the basic percent equation using n to represent the amount. The percent is 4%. The base is $212 billion.

Solution
Percent $\times$ base = amount
$4\% \times 212 = n$
$0.04 \times 212 = n$
$8.48 = n$

Corporations gave $8.48 billion to charities.

You Try It 4

Strategy
To find the new hourly wage:
• Find the amount of the raise. Write and solve the basic percent equation using n to represent the amount of the raise (amount). The percent is 8%. The base is $33.50.
• Add the amount of the raise to the old wage (33.50).

Solution
$8\% \times 33.50 = n$ 33.50
$0.08 \times 33.50 = n$ $\underline{+\ 2.68}$
$2.68 = n$ 36.18

The new hourly wage is $36.18.

SECTION 5.3

You Try It 1
Percent $\times$ base = amount
$n \times 32 = 16$
$n = 16 \div 32$
$n = 0.50$
$n = 50\%$

You Try It 2
Percent $\times$ base = amount
$n \times 15 = 48$
$n = 48 \div 15$
$n = 3.2$
$n = 320\%$

You Try It 3
Percent $\times$ base = amount
$n \times 45 = 30$
$n = 30 \div 45$
$n = \frac{2}{3} = 66\frac{2}{3}\%$

You Try It 4

Strategy
To find what percent of the income the income tax is, write and solve the basic percent equation using n to represent the percent. The base is $33,500 and the amount is $5025.

Solution
$n \times 33,500 = 5025$
$n = 5025 \div 33,500$
$n = 0.15 = 15\%$

The income tax is 15% of the income.

You Try It 5

Strategy
To find the percent who were women:
• Subtract to find the number of enlisted personnel who were women ($518,921 - 512,370$).
• Write and solve the basic percent equation using n to represent the percent. The base is 518,921, and the amount is the number of enlisted personnel who were women.

Solution
$518,921 - 512,370 = 6551$

$n \times 518,921 = 6551$
$n = 6551 \div 518,921$
$n \approx 0.013$

In 1950, 1.3% of the enlisted personnel in the U.S. Army were women.

SECTION 5.4

You Try It 1   Percent $\times$ base = amount
$$0.86 \times n = 215$$
$$n = 215 \div 0.86$$
$$n = 250$$

You Try It 2   Percent $\times$ base = amount
$$0.025 \times n = 15$$
$$n = 15 \div 0.025$$
$$n = 600$$

You Try It 3   Percent $\times$ base = amount
$$\frac{1}{6} \times n = 5 \qquad \bullet \ 16\frac{2}{3}\% = \frac{1}{6}$$
$$n = 5 \div \frac{1}{6}$$
$$n = 30$$

You Try It 4

Strategy   To find the original value of the car, write and solve the basic percent equation using n to represent the original value (base). The percent is 42% and the amount is $10,458.

Solution $42\% \times n = 10{,}458$
$0.42 \times n = 10{,}458$
$$n = 10{,}458 \div 0.42$$
$$n = 24{,}900$$

The original value of the car was $24,900.

You Try It 5

Strategy   To find the difference between the original price and the sale price:
- Find the original price. Write and solve the basic percent equation using n to represent the original price (base). The percent is 80% and the amount is $89.60.
- Subtract the sale price (89.60) from the original price.

Solution $80\% \times n = 89.60$
$0.80 \times n = 89.60$
$$n = 89.60 \div 0.80$$
$$n = 112.00 \quad \text{(original price)}$$

$112.00 - 89.60 = 22.40$

The difference between the original price and the sale price is $22.40.

SECTION 5.5

You Try It 1 $\dfrac{26}{100} = \dfrac{22}{n}$
$$26 \times n = 100 \times 22$$
$$26 \times n = 2200$$
$$n = 2200 \div 26$$
$$n \approx 84.62$$

You Try It 2 $\dfrac{16}{100} = \dfrac{n}{132}$
$$16 \times 132 = 100 \times n$$
$$2112 = 100 \times n$$
$$2112 \div 100 = n$$
$$21.12 = n$$

You Try It 3

Strategy   To find the number of days it snowed, write and solve a proportion using n to represent the number of days it snowed (amount). The percent is 64% and the base is 150.

Solution $\dfrac{64}{100} = \dfrac{n}{150}$
$$64 \times 150 = 100 \times n$$
$$9600 = 100 \times n$$
$$9600 \div 100 = n$$
$$96 = n$$

It snowed 96 days.

You Try It 4

Strategy   To find the percent of pens that were not defective:
- Subtract to find the number of pens that were not defective $(200 - 5)$.
- Write and solve a proportion using n to represent the percent of pens that were not defective. The base is 200, and the amount is the number of pens not defective.

Solution $200 - 5 = 195$ (number of pens not defective)
$$\frac{n}{100} = \frac{195}{200}$$
$$n \times 200 = 100 \times 195$$
$$n \times 200 = 19{,}500$$
$$n = 19{,}500 \div 200$$
$$n = 97.5$$

97.5% of the pens were not defective.

Solutions to Chapter 6 "You Try It"

SECTION 6.1

You Try It 1

Strategy   To find the unit cost, divide the total cost by the number of units.

Solution   **a.** $7.67 \div 8 = 0.95875$
$\$.959$ per battery
b. $2.29 \div 15 \approx 0.153$
$\$.153$ per ounce

You Try It 2

Strategy   To find the more economical purchase, compare the unit costs.

Solution $5.70 \div 6 = 0.95$
$3.96 \div 4 = 0.99$
$\$.95 < \$.99$

The more economical purchase is 6 cans for $5.70.

You Try It 3

Strategy   To find the total cost, multiply the unit cost (9.96) by the number of units (7).

Solution

$$\boxed{\begin{array}{c}\text{Unit}\\\text{cost}\end{array}} \times \boxed{\begin{array}{c}\text{number}\\\text{of units}\end{array}} = \boxed{\begin{array}{c}\text{total}\\\text{cost}\end{array}}$$

$$9.96 \quad \times \quad 7 \quad = \quad 69.72$$

The total cost is $69.72.

SECTION 6.2

You Try It 1

Strategy   To find the percent increase:
- Find the amount of the increase.
- Solve the basic percent equation for *percent*.

Solution

$$\boxed{\begin{array}{c}\text{New}\\\text{value}\end{array}} - \boxed{\begin{array}{c}\text{original}\\\text{value}\end{array}} = \boxed{\begin{array}{c}\text{amount of}\\\text{increase}\end{array}}$$

$$1.83 \quad - \quad 1.46 \quad = \quad 0.37$$

Percent $\times$ base = amount
$n \quad \times 1.46 = \quad 0.37$
$n = 0.37 \div 1.46$
$n \approx 0.25 = 25\%$

The percent increase was 25%.

You Try It 2

Strategy   To find the new hourly wage:
- Solve the basic percent equation for *amount*.

- Add the amount of the increase to the original wage.

Solution   Percent $\times$ base = amount
$0.14 \quad \times 12.50 = \quad n$
$1.75 = n$

$12.50 + 1.75 = 14.25$
The new hourly wage is $14.25.

You Try It 3

Strategy   To find the markup, solve the basic percent equation for *amount*.

Solution   Percent $\times$ base = amount

$$\boxed{\begin{array}{c}\text{Markup}\\\text{rate}\end{array}} \times \boxed{\text{cost}} = \boxed{\text{markup}}$$

$$0.20 \quad \times \quad 8 \quad = \quad n$$
$$1.60 = n$$

The markup is $1.60.

You Try It 4

Strategy   To find the selling price:
- Find the markup by solving the basic percent equation for *amount*.
- Add the markup to the cost.

Solution   Percent $\times$ base = amount

$$\boxed{\begin{array}{c}\text{Markup}\\\text{rate}\end{array}} \times \boxed{\text{cost}} = \boxed{\text{markup}}$$

$$0.55 \quad \times \quad 72 \quad = \quad n$$
$$39.60 = n$$

$$\boxed{\text{Cost}} + \boxed{\text{markup}} = \boxed{\text{selling price}}$$

$$72 \quad + \quad 39.60 \quad = \quad 111.60$$

The selling price is $111.60.

You Try It 5

Strategy   To find the percent decrease:
- Find the amount of the decrease.
- Solve the basic percent equation for *percent*.

Solution

$$\boxed{\begin{array}{c}\text{Original}\\\text{value}\end{array}} - \boxed{\begin{array}{c}\text{new}\\\text{value}\end{array}} = \boxed{\begin{array}{c}\text{amount of}\\\text{decrease}\end{array}}$$

$$261{,}000 \quad - 215{,}000 = \quad 46{,}000$$

Percent $\times$ base = amount
$n \quad \times 261{,}000 = 46{,}000$
$n = 46{,}000 \div 261{,}000$
$n \approx 0.176$

The percent decrease is 17.6%.

You Try It 6

Strategy To find the visibility:
- Find the amount of decrease by solving the basic percent equation for *amount*.
- Subtract the amount of decrease from the original visibility.

Solution Percent × base = amount

$$0.40 \times 5 = n$$
$$2 = n$$

$$5 - 2 = 3$$

The visibility was 3 miles.

You Try It 7

Strategy To find the discount rate:
- Find the discount.
- Solve the basic percent equation for *percent*.

Solution

Regular price	−	sale price	=	discount
12.50	−	10.99	=	1.51

Percent × base = amount

Discount rate	×	regular price	=	discount
n	×	12.50	=	1.51

$$n = 1.51 \div 12.50$$
$$n = 0.1208$$

The discount rate is 12.1%.

You Try It 8

Strategy To find the sale price:
- Find the discount by solving the basic percent equation for *amount*.
- Subtract to find the sale price.

Solution Percent × base = amount

Discount rate	×	regular price	=	discount
0.15	×	110	=	n

$$16.5 = n$$

Regular price	−	discount	=	sale price
110	−	16.5	=	93.5

The sale price is $93.50.

SECTION 6.3

You Try It 1

Strategy To find the simple interest due, multiply the principal (15,000) times the annual interest rate (8% = 0.08) times the time, in years (18 months = $\frac{18}{12}$ years = 1.5 years).

Solution

Principal	×	annual interest rate	×	time (in years)	=	interest
15,000	×	0.08	×	1.5	=	1800

The interest due is $1800.

You Try It 2

Strategy To find the maturity value:
- Use the simple interest formula to find the simple interest due.
- Find the maturity value by adding the principal and the interest.

Solution

Principal	×	annual interest rate	×	time (in years)	=	interest
3800	×	0.06	×	$\frac{90}{365}$	≈	56.22

Principal	+	interest	=	maturity value
3800	+	56.22	=	3856.22

The maturity value is $3856.22.

You Try It 3

Strategy To find the monthly payment:
- Find the maturity value by adding the principal and the interest.
- Divide the maturity value by the length of the loan in months (12).

Solution Principal + interest = maturity value

$$1900 + 152 = 2052$$

Maturity value ÷ length of the loan = payment

$$2052 \div 12 = 171$$

The monthly payment is $171.

You Try It 4

Strategy To find the finance charge, multiply the principal, or unpaid balance (1250), times the monthly interest rate (1.6%) times the number of months (1).

Solution

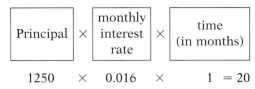

Principal × monthly interest rate × time (in months)

1250 × 0.016 × 1 = 20

The simple interest due is $20.

You Try It 5

Strategy   To find the interest earned:
- Find the new principal by multiplying the original principal (1000) by the factor found in the Compound Interest Table (3.29066).
- Subtract the original principal from the new principal.

Solution   1000 × 3.29066 = 3290.66

The new principal is $3290.66.

3290.66 − 1000 = 2290.66

The interest earned is $2290.66.

SECTION 6.4

You Try It 1

Strategy   To find the mortgage:
- Find the down payment by solving the basic percent equation for *amount*.
- Subtract the down payment from the purchase price.

Solution

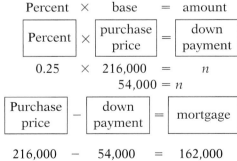

Percent × base = amount

Percent × purchase price = down payment

0.25 × 216,000 = n
54,000 = n

Purchase price − down payment = mortgage

216,000 − 54,000 = 162,000

The mortgage is $162,000.

You Try It 2

Strategy   To find the loan origination fee, solve the basic percent equation for *amount*.

Solution   Percent × base = amount
Points × mortgage = fee

0.045 × 80,000 = n
3600 = n

The loan origination fee was $3600.

You Try It 3

Strategy   To find the monthly mortgage payment:
- Subtract the down payment from the purchase price to find the mortgage.
- Multiply the mortgage by the factor found in the Monthly Payment Table.

Solution

Purchase price − down payment = mortgage

75,000 − 15,000 = 60,000
60,000 × 0.0089973 = 539.838
↑
From the table

The monthly mortgage payment is $539.84.

You Try It 4

Strategy   To find the interest:
- Multiply the mortgage by the factor found in the Monthly Payment Table to find the monthly mortgage payment.
- Subtract the principal from the monthly mortgage payment.

Solution   125,000 × 0.0083920 = 1049
↑   ↑
From the   Monthly mortgage
table   payment

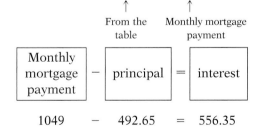

Monthly mortgage payment − principal = interest

1049 − 492.65 = 556.35

The interest on the mortgage is $556.35.

You Try It 5

Strategy   To find the monthly payment:
- Divide the annual property tax by 12 to find the monthly property tax.
- Add the monthly property tax to the monthly mortgage payment.

(Continued)

(Continued)
Solution $744 \div 12 = 62$

The monthly property tax is $62.

$415.20 + 62 = 477.20$

The total monthly payment is $477.20.

SECTION 6.5

You Try It 1

Strategy   To find the amount financed:
• Find the down payment by solving the basic percent equation for *amount*.
• Subtract the down payment from the purchase price.

Solution   Percent $\times$ base $=$ amount

$$\boxed{\text{Percent}} \times \boxed{\begin{array}{c}\text{purchase}\\\text{price}\end{array}} = \boxed{\begin{array}{c}\text{down}\\\text{payment}\end{array}}$$

$0.20 \quad \times \quad 19,200 \quad = \quad n$
$3840 = n$

The down payment is $3840.

$19,200 - 3840 = 15,360$

The amount financed is $15,360.

You Try It 2

Strategy   To find the license fee, solve the basic percent equation for *amount*.

Solution   Percent $\times$ base $=$ amount

$$\boxed{\text{Percent}} \times \boxed{\begin{array}{c}\text{purchase}\\\text{price}\end{array}} = \boxed{\text{sales tax}}$$

$0.015 \quad \times \quad 17,350 \quad = \quad n$
$260.25 = n$

The license fee is $260.25.

You Try It 3

Strategy   To find the cost, multiply the cost per mile (0.31) by the number of miles driven (23,000).

Solution $23,000 \times 0.31 = 7130$

The cost is $7130.

You Try It 4

Strategy   To find the cost per mile for car insurance, divide the cost for insurance (360) by the number of miles driven (15,000).

Solution $360 \div 15,000 = 0.024$

The cost per mile for insurance is $.024.

You Try It 5

Strategy   To find the monthly payment:
• Subtract the down payment from the purchase price to find the amount financed.
• Multiply the amount financed by the factor found in the Monthly Payment Table.

Solution $25,900 - 6475 = 19,425$

$19,425 \times 0.0244129 \approx 474.22$

The monthly payment is $474.22.

SECTION 6.6

You Try It 1

Strategy   To find the worker's earnings:
• Find the worker's overtime wage by multiplying the hourly wage by 2.
• Multiply the number of overtime hours worked by the overtime wage.

Solution $18.50 \times 2 = 37$

The hourly wage for overtime is $37.

$37 \times 8 = 296$

The construction worker earns $296.

You Try It 2

Strategy   To find the salary per month, divide the annual salary by the number of months in a year (12).

Solution $48,228 \div 12 = 4019$

The contractor's monthly salary is $4019.

You Try It 3

Strategy   To find the total earnings:
• Find the sales over $50,000.
• Multiply the commission rate by sales over $50,000.
• Add the commission to the annual salary.

Solution $175,000 - 50,000 = 125,000$

Sales over $50,000 totaled $125,000.

$125,000 \times 0.095 = 11,875$

Earnings from commissions totaled $11,875.

$27,000 + 11,875 = 38,875$

The insurance agent earned $38,875.

SECTION 6.7

You Try It 1

Strategy To find the current balance:
• Subtract the amount of the check from the old balance.
• Add the amount of each deposit.

Solution
$$
\begin{array}{rl}
302.46 & \\
-\ 20.59 & \text{check} \\
\hline
281.87 & \\
176.86 & \text{first deposit} \\
+\ 94.73 & \text{second deposit} \\
\hline
553.46 &
\end{array}
$$

The current checking account balance is $553.46.

You Try It 2

Current checkbook balance: 623.41
Check: 237 $+\ \ 78.73$
                                             702.14
Interest: $+\ \ \ 2.11$
                                             704.25
Deposit: -523.84
                                             180.41

Closing bank balance from bank statement: $180.41

Checkbook balance: $180.41

The bank statement and the checkbook balance.

Solutions to Chapter 7 "You Try It"

SECTION 7.1

You Try It 1

Strategy To find what percent of the total number of cellular phones purchased the number purchased in March represents:
• Read the pictograph to determine the number of cellular phones purchased for each month.
• Find the total cellular phone purchases for the 4-month period.
• Solve the basic percent equation for percent (n). Amount = 3000; the base is the total sales for the 4-month period.

Solution
$$
\begin{array}{r}
4,500 \\
3,500 \\
3,000 \\
1,500 \\
\hline
12,500
\end{array}
$$

$$
\begin{array}{l}
\text{Percent} \times \text{base} = \text{amount} \\
\quad n \quad \times 12,500 = 3000 \\
\qquad\qquad\qquad n = 3000 \div 12,500 \\
\qquad\qquad\qquad n = 0.24
\end{array}
$$

The number of cellular phones purchased in March represents 24% of the total number of cellular phones purchased.

You Try It 2

Strategy To find the ratio of the annual cost of fuel to the annual cost of maintenance:
• Locate the annual fuel cost and the annual maintenance cost in the circle graph.
• Write the ratio of the annual fuel cost to the annual maintenance cost in simplest form.

Solution Annual fuel cost: $700
Annual maintenance cost: $500
$$\frac{700}{500} = \frac{7}{5}$$
The ratio is $\frac{7}{5}$.

You Try It 3

Strategy   To find the amount paid for medical/dental insurance:
- Locate the percent of the distribution that is medical/dental insurance.
- Solve the basic percent equation for amount.

Solution   The percent of the distribution that is medical/dental insurance: 6%

Percent × base = amount
0.06   × 2900 = n
              174 = n

The amount paid for medical/dental insurance is $174.

SECTION 7.2

You Try It 1

Strategy   To write the ratio:
- Read the graph to find the lung capacity of an inactive female and of an athletic female.
- Write the ratio in simplest form.

Solution   Lung capacity of an inactive female: 25
Lung capacity of an athletic female: 55

$$\frac{25}{55} = \frac{5}{11}$$

The ratio is $\frac{5}{11}$.

You Try It 2

Strategy   To determine between which two years the net income of Math Associates increased the most:
- Read the line graph to determine the net income of Math Associates for each of the years shown.
- Subtract to find the difference between each two years.
- Find the greatest difference.

Solution   2000: $1 million
2001: $3 million
2002: $5 million
2003: $8 million
2004: $12 million

Between 2000 and 2001: 3 − 1 = 2
Between 2001 and 2002: 5 − 3 = 2

Between 2002 and 2003: 8 − 5 = 3
Between 2003 and 2004: 12 − 8 = 4

4 > 3 > 2

The net income of Math Associates increased the most between 2003 and 2004.

SECTION 7.3

You Try It 1

Strategy   To find the number of employees:
- Read the histogram to find the number of employees whose hourly wage is between $10 and $12 and the number whose hourly wage is between $12 and $14.
- Add the two numbers.

Solution   Number whose wage is between $10 and $12: 7
Number whose wage is between $12 and $14: 15

7 + 15 = 22

22 employees earn between $10 and $14.

You Try It 2

Strategy   To write the ratio:
- Read the graph to find the number of states with a per capita income between $28,000 and $32,000 and the number with a per capita income between $32,000 and $36,000.
- Write the ratio in simplest form.

Solution   Number of states with a per capita income between $28,000 and $32,000: 18
Number of states with a per capita income between $32,000 and $36,000: 9

$$\frac{18}{9} = \frac{2}{1}$$

The ratio is $\frac{2}{1}$.

SECTION 7.4

You Try It 1

Strategy   To find the mean amount spent by the 12 customers:

- Find the sum of the numbers.
- Divide the sum by the number of customers (12).

Solution $6.26 + 8.23 + 5.09 + 8.11 + 7.50 + 6.69 + 5.66 + 4.89 + 5.25 + 9.36 + 6.75 + 7.05 = 80.84$

$$\bar{x} = \frac{80.84}{12} \approx 6.74$$

The mean amount spent by the 12 customers was $6.74.

You Try It 2

Strategy   To find the median weight loss:
- Arrange the weight losses from smallest to largest.
- Because there is an even number of values, the median is the mean of the middle two numbers.

Solution   10, 14, 16, 16, 22, 27, 29, 31, 31, 40

$$\text{Median} = \frac{22 + 27}{2} = \frac{49}{2} = 24.5$$

The median weight loss was 24.5 pounds.

You Try It 3

Strategy   To draw the box-and-whiskers plot:
- Find the median, Q_1, and Q_3.
- Use the smallest value, Q_1, the median, Q_3, and the largest value to draw the box-and-whiskers plot.

Solution

a.

b. Answers about the spread of the data will vary. For example, in You Try It 3, the values in the interquartile range are all very close to the median. They are not

so close to the median in Example 3. The whiskers are long with respect to the box in You Try It 3, whereas they are short with respect to the box in Example 3. This shows that the data values outside the interquartile range are closer together in Example 3 than in You Try It 3.

SECTION 7.5

You Try It 1

Strategy   To find the probability:
- List the outcomes of the experiment in a systematic way. We will use a table.
- Use the table to count the number of possible outcomes of the experiment.
- Count the number of outcomes of the experiment that are favorable to the event of 2 true questions and 1 false question.
- Use the probability formula.

Solution

Question 1	Question 2	Question 3
T	T	T
T	T	F
T	F	T
T	F	F
F	T	T
F	T	F
F	F	T
F	F	F

There are 8 possible outcomes:

$S = \{\text{TTT, TTF, TFT, TFF, FTT, FTF, FFT, FFF}\}$

There are 3 outcomes favorable to the event:

$\{\text{TTF, TFT, FTT}\}$

Probability of an event
$$= \frac{\text{number of favorable outcomes}}{\text{number of possible outcomes}} = \frac{3}{8}$$

The probability of 2 true questions and 1 false question is $\frac{3}{8}$.

Solutions to Chapter 8 "You Try It"

SECTION 8.1

You Try It 1 $14 \text{ ft} = 14 \text{ ft} \times \dfrac{1 \text{ yd}}{3 \text{ ft}} = \dfrac{14 \text{ yd}}{3} = 4\dfrac{2}{3} \text{ yd}$

You Try It 2 $9240 \text{ ft} = 9240 \text{ ft} \times \dfrac{1 \text{ mi}}{5280 \text{ ft}}$

$= \dfrac{9240 \text{ mi}}{5280} = 1\dfrac{3}{4} \text{ mi}$

You Try It 3

$$\begin{array}{r} 3 \text{ ft } 6 \text{ in.} \\ 12\overline{)\;42} \\ -36 \\ \hline 6 \end{array}$$

• 12 in. = 1 ft

42 in. = 3 ft 6 in.

You Try It 4

$$\begin{array}{r} 4 \text{ yd } 2 \text{ ft} \\ 3\overline{)\;14} \\ -12 \\ \hline 2 \end{array}$$

• 3 ft = 1 yd

14 ft = 4 yd 2 ft

You Try It 5

$$\begin{array}{r} 3 \text{ ft } \;\; 5 \text{ in.} \\ + 4 \text{ ft } \;\; 9 \text{ in.} \\ \hline 7 \text{ ft } 14 \text{ in.} \end{array}$$

• 14 in. = 1 ft 2 in.

7 ft 14 in. = 8 ft 2 in.

You Try It 6

$$\begin{array}{r} {\scriptstyle 3 \text{ ft} \quad 14 \text{ in.}} \\ \cancel{4 \text{ ft } 2 \text{ in.}} \\ - 1 \text{ ft } 8 \text{ in.} \\ \hline 2 \text{ ft } 6 \text{ in.} \end{array}$$

• Borrow 1 ft (12 in.) from 4 ft and add it to 2 in.

You Try It 7

$$\begin{array}{r} 4 \text{ yd } 1 \text{ ft} \\ \times \qquad 8 \\ \hline 32 \text{ yd } 8 \text{ ft} \end{array}$$

• 8 ft = 2 yd 2 ft

32 yd 8 ft = 34 yd 2 ft

You Try It 8

$$\begin{array}{r} 3 \text{ yd } \quad 2 \text{ ft} \\ 2\overline{)\;7 \text{ yd } \quad 1 \text{ ft}} \\ -6 \text{ yd} \\ \hline 1 \text{ yd} = 3 \text{ ft} \\ 4 \text{ ft} \\ -4 \text{ ft} \\ \hline 0 \end{array}$$

You Try It 9 $6\dfrac{1}{4} \text{ ft} = 6\dfrac{3}{12} \text{ ft} = 5\dfrac{15}{12} \text{ ft}$

$$\begin{array}{r} -3\dfrac{2}{3} \text{ ft} = 3\dfrac{8}{12} \text{ ft} = 3\dfrac{8}{12} \text{ ft} \\ \hline 2\dfrac{7}{12} \text{ ft} \end{array}$$

You Try It 10

Strategy   To find the width of the storage room:
- Multiply the number of tiles (8) by the width of each tile (9 in.).
- Divide the result by the number of inches in 1 foot (12) to find the width in feet.

Solution   9 in. × 8 = 72 in.

72 ÷ 12 = 6

The width is 6 ft.

You Try It 11

Strategy   To find the length of each piece, divide the total length (9 ft 8 in.) by the number of pieces (4).

Solution

$$\begin{array}{r} 2 \text{ ft } \quad 5 \text{ in.} \\ 4\overline{)\;9 \text{ ft } \quad 8 \text{ in.}} \\ -8 \text{ ft} \\ \hline 1 \text{ ft} = 12 \text{ in.} \\ 20 \text{ in.} \\ -20 \text{ in.} \\ \hline 0 \text{ in.} \end{array}$$

Each piece is 2 ft 5 in. long.

SECTION 8.2

You Try It 1 $3 \text{ lb} = 3 \text{ lb} \times \dfrac{16 \text{ oz}}{1 \text{ lb}} = 48 \text{ oz}$

You Try It 2 $4200 \text{ lb} = 4200 \text{ lb} \times \dfrac{1 \text{ ton}}{2000 \text{ lb}}$

$= \dfrac{4200 \text{ tons}}{2000} = 2\dfrac{1}{10} \text{ tons}$

You Try It 3

$$\begin{array}{r} {\scriptstyle 6 \text{ lb} \quad 17 \text{ oz}} \\ \cancel{7 \text{ lb } 1 \text{ oz}} \\ - 3 \text{ lb } 4 \text{ oz} \\ \hline 3 \text{ lb } 13 \text{ oz} \end{array}$$

• Borrow 1 lb (16 oz) from 7 lb and add it to 1 oz.

You Try It 4

$$\begin{array}{r} 3\text{ lb }6\text{ oz} \\ \times \qquad 4 \\ \hline 12\text{ lb }24\text{ oz} = 13\text{ lb }8\text{ oz} \end{array}$$

You Try It 5

Strategy   To find the weight of 12 bars of soap:
- Multiply the number of bars (12) by the weight of each bar (7 oz).
- Convert the number of ounces to pounds.

Solution

$$\begin{array}{r} 12 \\ \times \quad 7\text{ oz} \\ \hline 84\text{ oz} \end{array} \qquad 84\ \cancel{\text{oz}} \times \frac{1\text{ lb}}{16\ \cancel{\text{oz}}} = 5\frac{1}{4}\text{ lb}$$

The 12 bars of soap weigh $5\frac{1}{4}$ lb.

SECTION 8.3

You Try It 1

$$18\text{ pt} = 18\ \cancel{\text{pt}} \times \frac{1\ \cancel{\text{qt}}}{2\ \cancel{\text{pt}}} \times \frac{1\text{ gal}}{4\ \cancel{\text{qt}}}$$
$$= \frac{18\text{ gal}}{8} = 2\frac{1}{4}\text{ gal}$$

You Try It 2

$$\begin{array}{r} 1\text{ gal}\quad 2\text{ qt} \\ 3)\ \overline{4\text{ gal}\quad 2\text{ qt}} \\ -3\text{ gal} \\ \hline 1\text{ gal} = 4\text{ qt} \\ 6\text{ qt} \\ -6\text{ qt} \\ \hline 0 \end{array}$$

You Try It 3

Strategy   To find the number of gallons of water needed:
- Find the number of quarts required by multiplying the number of quarts one student needs per day (5) by the number of students (5) by the number of days (3).

- Convert the number of quarts to gallons.

Solution $5 \times 5 \times 3 = 75$ qt

$$75\ \cancel{\text{qt}} \cdot \frac{1\text{ gal}}{4\ \cancel{\text{qt}}} = \frac{75\text{ gal}}{4} = 18\frac{3}{4}\text{ gal}$$

The students should take $18\frac{3}{4}$ gal of water.

SECTION 8.4

Your Try It 1 $18{,}000\text{ s} = 18{,}000\ \cancel{\text{s}} \times \frac{1\ \cancel{\text{min}}}{60\ \cancel{\text{s}}} \times \frac{1\text{ h}}{60\ \cancel{\text{min}}}$
$$= \frac{18{,}000\text{ h}}{3600} = 5\text{ h}$$

SECTION 8.5

You Try It 1 $4.5\text{ Btu} = 4.5\ \cancel{\text{Btu}} \times \frac{778\text{ ft}\cdot\text{lb}}{1\ \cancel{\text{Btu}}}$
$$= 3501\text{ ft}\cdot\text{lb}$$

You Try It 2

$$\text{Energy} = 800\text{ lb} \times 16\text{ ft} = 12{,}800\text{ ft}\cdot\text{lb}$$

You Try It 3 $56{,}000\text{ Btu} =$
$$56{,}000\ \cancel{\text{Btu}} \times \frac{778\text{ ft}\cdot\text{lb}}{1\ \cancel{\text{Btu}}} =$$
$$43{,}568{,}000\text{ ft}\cdot\text{lb}$$

You Try It 4 $\text{Power} = \dfrac{90\text{ ft} \times 1200\text{ lb}}{24\text{ s}}$
$$= 4500\ \frac{\text{ft}\cdot\text{lb}}{\text{s}}$$

You Try It 5 $\dfrac{3300}{550} = 6\text{ hp}$

Solutions to Chapter 9 "You Try It"

SECTION 9.1

You Try It 1 $3.07\text{ m} = 307\text{ cm}$

You Try It 2 $750\text{ m} = 0.750\text{ km}$

$$3\text{ km }750\text{ m} = 3\text{ km} + 0.750\text{ km}$$
$$= 3.750\text{ km}$$

You Try It 3

Strategy   To find the cost of the shelves:
- Multiply the length of the bookcase (175 cm) by the number of shelves (4).
- Convert centimeters to meters.
- Multiply the number of meters by the cost per meter ($15.75).

(Continued)

(Continued)

Solution   175 cm
          × ___4
            700 cm

700 cm = 7 m

            15.75
          × ____7
            110.25

The cost of the shelves is $110.25.

SECTION 9.2

You Try It 1   42.3 mg = 0.0423 g

You Try It 2   54 mg = 0.054 g

3 g 54 mg = 3 g + 0.054 g
              = 3.054 g

You Try It 3

Strategy   To find how much fertilizer is
            required:
            • Convert 300 g to kilograms.
            • Multiply the number of kilograms
              by the number of trees (400).

Solution   300 g = 0.3 kg

            400
          × _0.3 kg
            120.0 kg

To fertilize the trees, 120 kg of
fertilizer are required.

SECTION 9.3

You Try It 1   2 kl = 2000 L

2 kl 167 L = 2000 L + 167 L
              = 2167 L

You Try It 2   325 cm³ = 325 ml = 0.325 L

You Try It 3

Strategy   To find the profit:
            • Convert 5 L to milliliters.
            • Find the number of jars by
              dividing the number of milliliters
              by the number of milliliters in each
              jar (125).
            • Multiply the number of jars by the
              cost per jar ($.85).
            • Find the total cost by adding the
              cost of the jars to the cost for the
              moisturizer ($299.50).

• Find the income by multiplying the
  number of jars by the selling price
  per jar ($29.95).
• Subtract the total cost from the
  income.

Solution   5 L = 5000 ml

5000 ÷ 125 = 40   This is the
                   number of jars.

            0.85
          × __40
            34.00   This is the cost of the jars.

            299.50
          + _34.00
            333.50   This is the total cost.

            29.95
          × ___40
            1198.00   This is the income from
                       sales.

            1198.00
          − _333.50
            864.50

The profit on the 5 L of moisturizer
is $864.50.

SECTION 9.4

You Try It 1

Strategy   To find the number of Calories
            burned off, multiply the number of
            hours spent doing housework $\left(4\frac{1}{2}\right)$
            by the Calories used per hour (240).

Solution $4\frac{1}{2} \times 240 = \frac{9}{2} \times 240 = 1080$

Doing $4\frac{1}{2}$ h of housework burns off
1080 Calories.

You Try It 2

Strategy   To find the number of kilowatt-hours
            used:
            • Find the number of watt-hours
              used.
            • Convert watt-hours to kilowatt-
              hours.

Solution   150 × 200 = 30,000
            30,000 Wh = 30 kWh

30 kWh of energy are used.

You Try It 3

Strategy   To find the cost:
- Convert 20 min to hours.
- Multiply to find the total number of hours the oven is used.
- Multiply the number of hours used by the number of watts to find the watt-hours.
- Convert watt-hours to kilowatt-hours.
- Multiply the number of kilowatt-hours by the cost per kilowatt-hour.

Solution $20 \text{ min} = 20 \cancel{\text{min}} \times \dfrac{1 \text{ h}}{60 \cancel{\text{min}}}$

$= \dfrac{20}{60} \text{ h} = \dfrac{1}{3} \text{ h}$

$\dfrac{1}{3} \text{ h} \times 30 = 10 \text{ h}$

$10 \text{ h} \times 500 \text{ W} = 5000 \text{ Wh}$

$5000 \text{ Wh} = 5 \text{ kWh}$

$5 \times 8.7¢ = 43.5¢$

The cost is 43.5¢.

You Try It 2

$\dfrac{\$2.69}{\text{gal}} \approx \dfrac{\$2.69}{\cancel{\text{gal}}} \times \dfrac{1 \cancel{\text{gal}}}{3.79 \text{ L}}$

$= \dfrac{\$2.69}{3.79 \text{ L}} \approx \dfrac{\$.71}{\text{L}}$

$\$2.69/\text{gal} \approx \$.71/\text{L}$

You Try It 3 $45 \text{ cm} = \dfrac{45 \cancel{\text{cm}}}{1} \times \dfrac{1 \text{ in.}}{2.54 \cancel{\text{cm}}}$

$= \dfrac{45 \text{ in.}}{2.54} \approx 17.72 \text{ in.}$

$45 \text{ cm} \approx 17.72 \text{ in.}$

You Try It 4 $\dfrac{75 \text{ km}}{\text{h}} \approx \dfrac{75 \cancel{\text{km}}}{\text{h}} \times \dfrac{1 \text{ mi}}{1.61 \cancel{\text{km}}}$

$= \dfrac{75 \text{ mi}}{1.61 \text{ h}} \approx 46.58 \text{ mi/h}$

$75 \text{ km/h} \approx 46.58 \text{ mi/h}$

You Try It 5 $\dfrac{\$1.75}{\text{L}} \approx \dfrac{\$1.75}{\cancel{\text{L}}} \times \dfrac{3.79 \cancel{\text{L}}}{1 \text{ gal}}$

$= \dfrac{\$6.6325}{1 \text{ gal}} \approx \$6.63/\text{gal}$

$\$1.75/\text{L} \approx \$6.63/\text{gal}$

SECTION 9.5

You Try It 1

$\dfrac{60 \text{ ft}}{\text{s}} \approx \dfrac{60 \cancel{\text{ft}}}{\text{s}} \times \dfrac{1 \text{ m}}{3.28 \cancel{\text{ft}}}$

$= \dfrac{60 \text{ m}}{3.28 \text{ s}} \approx 18.29 \text{ m/s}$

$60 \text{ ft/s} \approx 18.29 \text{ m/s}$

- The conversion rate is $\dfrac{1 \text{ m}}{3.28 \text{ ft}}$ with m in the numerator and ft in the denominator.

Solutions to Chapter 10 "You Try It"

SECTION 10.1

You Try It 1 -232 ft

You Try It 2

You Try It 3 $-12 < -8$ • -12 is to the left of -8 on the number line.

You Try It 4 $|-7| = 7$; $|21| = 21$

You Try It 5 $|2| = 2$; $|-9| = 9$

You Try It 6 $-|-12| = -12$

SECTION 10.2

You Try It 1 $-154 + (-37) = -191$ • The signs of the addends are the same.

You Try It 2 $-5 + (-2) + 9 + (-3)$
$= -7 + 9 + (-3)$
$= 2 + (-3)$
$= -1$

You Try It 3 $-8 - 14$
$= -8 + (-14)$ • Rewrite "−" as "+"; the opposite of 14 is −14.
$= -22$

You Try It 4 $3 - (-15)$
$= 3 + 15$ • Rewrite "$-$" as "$+$";
$= 18$ the opposite of -15 is 15.

You Try It 5 $4 - (-3) - 12 - (-7) - 20$
$= 4 + 3 + (-12) + 7 + (-20)$
$= 7 + (-12) + 7 + (-20)$
$= -5 + 7 + (-20)$
$= 2 + (-20)$
$= -18$

You Try It 6

Strategy   To find the temperature, add the increase ($12°C$) to the previous temperature ($-10°C$).

Solution $-10 + 12 = 2$

After an increase of $12°C$, the temperature is $2°C$.

SECTION 10.3

You Try It 1 $(-3)(-5) = 15$ • The signs are the same. The product is positive.

You Try It 2   The signs are different. The product is negative.

$-38 \cdot 51 = -1938$

You Try It 3 $-7(-8)(9)(-2) = 56(9)(-2)$
$= 504(-2)$
$= -1008$

You Try It 4 $(-135) \div (-9) = 15$ • The signs are the same. The quotient is positive.

You Try It 5 $84 \div (-6) = -14$ • The signs are different. The quotient is negative.

You Try It 6 $-72 \div 4 = -18$

You Try It 7   Division by zero is undefined. $-39 \div 0$ is undefined.

You Try It 8

Strategy   To find the melting point of argon, multiply the melting point of mercury ($-38°C$) by 5.

Solution $5(-38) = -190$

The melting point of argon is $-190°C$.

You Try It 9

Strategy   To find the average daily low temperature:
• Add the seven temperature readings.
• Divide by 7.

Solution
$-6 + (-7) + 1 + 0 + (-5) + (-10) + (-1)$
$= -13 + 1 + 0 + (-5) + (-10) + (-1)$
$= -12 + 0 + (-5) + (-10) + (-1)$
$= -12 + (-5) + (-10) + (-1)$
$= -17 + (-10) + (-1)$
$= -27 + (-1)$
$= -28$

$-28 \div 7 = -4$

The average daily low temperature was $-4°F$.

SECTION 10.4

You Try It 1   The LCM of 9 and 12 is 36.

$$\frac{5}{9} - \frac{11}{12} = \frac{20}{36} - \frac{33}{36} = \frac{20}{36} + \frac{-33}{36}$$
$$= \frac{20 + (-33)}{36} = \frac{-13}{36}$$
$$= -\frac{13}{36}$$

You Try It 2   The LCM of 8, 6, and 3 is 24.

$$-\frac{7}{8} - \frac{5}{6} + \frac{2}{3}$$
$$= -\frac{21}{24} - \frac{20}{24} + \frac{16}{24}$$
$$= \frac{-21}{24} + \frac{-20}{24} + \frac{16}{24}$$
$$= \frac{-21 + (-20) + 16}{24} = \frac{-25}{24}$$
$$= -\frac{25}{24} = -1\frac{1}{24}$$

You Try It 3 $16.127 - 67.91 = 16.127 + (-67.91)$

67.910
$\underline{-16.127}$
51.783

$16.127 - 67.91 = -51.783$

You Try It 4 $2.7 + (-9.44) + 6.2$
$= -6.74 + 6.2$
$= -0.54$

You Try It 5   The product is positive.

$$\left(-\frac{2}{3}\right)\left(-\frac{9}{10}\right) = \frac{2 \cdot 9}{3 \cdot 10}$$
$$= \frac{3}{5}$$

You Try It 6   The quotient is negative.

$$-\frac{5}{8} \div \frac{5}{40} = -\left(\frac{5}{8} \div \frac{5}{40}\right)$$

$$= -\left(\frac{5}{8} \times \frac{40}{5}\right)$$

$$= -\left(\frac{5 \cdot 40}{8 \cdot 5}\right)$$

$$= -5$$

You Try It 7   The product is negative.

$$\begin{array}{r} 5.44 \\ \times\ 3.8 \\ \hline 4352 \\ 1632 \\ \hline 20.672 \end{array}$$

$$-5.44 \times 3.8 = -20.672$$

You Try It 8 $3.44 \times (-1.7) \times 0.6$
$= (-5.848) \times 0.6$
$= -3.5088$

You Try It 9

$$\begin{array}{r} 0.231 \\ 1.7\overline{)0.3.940} \\ -3\ 4 \\ \hline 54 \\ -51 \\ \hline 30 \\ -17 \\ \hline 13 \end{array}$$

$$-0.394 \div 1.7 \approx -0.23$$

You Try It 10

Strategy   To find how many degrees the temperature fell, subtract the lower temperature ($-13.33°C$) from the higher temperature ($12.78°C$).

Solution $12.78 - (-13.33) = 12.78 + 13.33$
$= 26.11$

The temperature fell $26.11°C$ in the 15-min period.

SECTION 10.5

You Try It 1   The number is less than 1. Move the decimal point 7 places to the right. The exponent on 10 is -7.

$$0.000000961 = 9.61 \times 10^{-7}$$

You Try It 2   The exponent on 10 is positive. Move the decimal point 6 places to the right.

$$7.329 \times 10^6 = 7,329,000$$

You Try It 3 $9 - 9 \div (-3)$
$= 9 - (-3)$ • Do the division.
$= 9 + 3$ • Rewrite as addition. Add.
$= 12$

You Try It 4 $8 \div 4 \cdot 4 - (-2)^2$
$= 8 \div 4 \cdot 4 - 4$ • Exponents
$= 2 \cdot 4 - 4$ • Division
$= 8 - 4$ • Multiplication
$= 8 + (-4)$ • Subtraction
$= 4$

You Try It 5 $8 - (-15) \div (2 - 7)$
$= 8 - (-15) \div (-5)$
$= 8 - 3$
$= 8 + (-3)$
$= 5$

You Try It 6 $(-2)^2 \times (3 - 7)^2 - (-16) \div (-4)$
$= (-2)^2 \times (-4)^2 - (-16) \div (-4)$
$= 4 \times 16 - (-16) \div (-4)$
$= 64 - (-16) \div (-4)$
$= 64 - 4$
$= 64 + (-4)$
$= 60$

You Try It 7 $7 \div \left(\frac{1}{7} - \frac{3}{14}\right) - 9$

$$= 7 \div \left(-\frac{1}{14}\right) - 9$$

$$= 7(-14) - 9$$
$$= -98 - 9$$
$$= -98 + (-9)$$
$$= -107$$

Solutions to Chapter 11 "You Try It"

SECTION 11.1

You Try It 1 $6a - 5b$

$$6(-3) - 5(4) = -18 - 20$$
$$= -18 + (-20)$$
$$= -38$$

You Try It 2 $-3s^2 - 12 \div t$

$$-3(-2)^2 - 12 \div 4 = -3(4) - 12 \div 4$$
$$= -12 - 12 \div 4$$
$$= -12 - 3$$
$$= -12 + (-3)$$
$$= -15$$

You Try It 3 $-\dfrac{2}{3}m + \dfrac{3}{4}n^3$

$$-\dfrac{2}{3}(6) + \dfrac{3}{4}(2)^3 = -\dfrac{2}{3}(6) + \dfrac{3}{4}(8)$$
$$= -4 + 6$$
$$= 2$$

You Try It 4 $-3yz - z^2 + y^2$

$$-3\left(-\dfrac{2}{3}\right)\left(\dfrac{1}{3}\right) - \left(\dfrac{1}{3}\right)^2 + \left(-\dfrac{2}{3}\right)^2$$
$$= -3\left(-\dfrac{2}{3}\right)\left(\dfrac{1}{3}\right) - \dfrac{1}{9} + \dfrac{4}{9}$$
$$= \dfrac{2}{3} - \dfrac{1}{9} + \dfrac{4}{9} = \dfrac{6}{9} - \dfrac{1}{9} + \dfrac{4}{9} = \dfrac{9}{9}$$
$$= 1$$

You Try It 5 $5a^2 - 6b^2 + 7a^2 - 9b^2$

$$= 5a^2 + (-6)b^2 + 7a^2 + (-9)b^2$$
$$= 5a^2 + 7a^2 + (-6)b^2 + (-9)b^2$$
$$= 12a^2 + (-15)b^2$$
$$= 12a^2 - 15b^2$$

You Try It 6 $-6x + 7 + 9x - 10$

$$= (-6)x + 7 + 9x + (-10)$$
$$= (-6)x + 9x + 7 + (-10)$$
$$= 3x + (-3)$$
$$= 3x - 3$$

You Try It 7 $\dfrac{3}{8}w + \dfrac{1}{2} - \dfrac{1}{4}w - \dfrac{2}{3}$

$$= \dfrac{3}{8}w + \dfrac{1}{2} + \left(-\dfrac{1}{4}\right)w + \left(-\dfrac{2}{3}\right)$$
$$= \dfrac{3}{8}w + \left(-\dfrac{1}{4}\right)w + \dfrac{1}{2} + \left(-\dfrac{2}{3}\right)$$
$$= \dfrac{3}{8}w + \left(-\dfrac{2}{8}\right)w + \dfrac{3}{6} + \left(-\dfrac{4}{6}\right)$$
$$= \dfrac{1}{8}w + \left(-\dfrac{1}{6}\right)$$
$$= \dfrac{1}{8}w - \dfrac{1}{6}$$

You Try It 8 $5(a - 2) = 5a - 5(2)$
$$= 5a - 10$$

You Try It 9 $8s - 2(3s - 5)$
$$= 8s - 2(3s) - (-2)(5)$$
$$= 8s - 6s + 10$$
$$= 2s + 10$$

You Try It 10 $4(x - 3) - 2(x + 1)$
$$= 4x - 4(3) - 2x - 2(1)$$
$$= 4x - 12 - 2x - 2$$
$$= 4x - 2x - 12 - 2$$
$$= 2x - 14$$

SECTION 11.2

You Try It 1

$x(x + 3) = 4x + 6$	
$(-2)(-2 + 3)$	$4(-2) + 6$
$(-2)(1)$	$(-8) + 6$
$-2 = -2$	

Yes, -2 is a solution.

You Try It 2

$x^2 - x = 3x + 7$	
$(-3)^2 - (-3)$	$3(-3) + 7$
$9 + 3$	$-9 + 7$
$12 \neq -2$	

No, -3 is not a solution.

You Try It 3

$$-2 + y = -5$$
$$-2 + 2 + y = -5 + 2 \qquad \bullet \text{ Add 2 to}$$
$$0 + y = -3 \qquad\qquad \text{each side.}$$
$$y = -3$$

The solution is -3.

You Try It 4

$$7 = y + 8$$
$$7 - 8 = y + 8 - 8 \qquad \bullet \text{ Subtract 8}$$
$$-1 = y + 0 \qquad\qquad \text{from each}$$
$$-1 = y \qquad\qquad\quad \text{side.}$$

The solution is -1.

You Try It 5

$$\dfrac{1}{5} = z + \dfrac{4}{5}$$
$$\dfrac{1}{5} - \dfrac{4}{5} = z + \dfrac{4}{5} - \dfrac{4}{5} \qquad \bullet \text{ Subtract } \dfrac{4}{5} \text{ from}$$
$$-\dfrac{3}{5} = z + 0 \qquad\qquad\qquad \text{each side.}$$
$$-\dfrac{3}{5} = z$$

The solution is $-\dfrac{3}{5}$.

You Try It 6
$$4z = -20$$
$$\frac{4z}{4} = \frac{-20}{4} \quad \bullet \text{ Divide each side by 4.}$$
$$1z = -5$$
$$z = -5$$

The solution is -5.

You Try It 7
$$8 = \frac{2}{5}n$$
$$\left(\frac{5}{2}\right)(8) = \left(\frac{5}{2}\right)\frac{2}{5}n \quad \bullet \text{ Multiply each side by } \frac{5}{2}.$$
$$20 = 1n$$
$$20 = n$$

The solution is 20.

You Try It 8
$$\frac{2}{3}t - \frac{1}{3}t = -2$$
$$\frac{1}{3}t = -2 \quad \bullet \text{ Combine like terms.}$$
$$\left(\frac{3}{1}\right)\frac{1}{3}t = \left(\frac{3}{1}\right)(-2) \quad \bullet \text{ Multiply each side by 3.}$$
$$1t = -6$$
$$t = -6$$

The solution is -6.

You Try It 9

Strategy To find the regular price, replace the variables S and D in the formula by the given values and solve for R.

Solution
$$S = R - D$$
$$44 = R - 16$$
$$44 + 16 = R - 16 + 16$$
$$60 = R$$

The regular price is $60.

You Try It 10

Strategy To find the monthly payment, replace the variables A and N in the formula by the given values and solve for M.

Solution
$$A = MN$$
$$6840 = M \cdot 24$$
$$6840 = 24M$$
$$\frac{6840}{24} = \frac{24M}{24}$$
$$285 = M$$

The monthly payment is $285.

SECTION 11.3

You Try It 1
$$5x + 8 = 6$$
$$5x + 8 - 8 = 6 - 8 \quad \bullet \text{ Subtract 8 from each side.}$$
$$5x = -2$$

$$\frac{5x}{5} = \frac{-2}{5} \quad \bullet \text{ Divide each side by 5.}$$
$$x = -\frac{2}{5}$$

The solution is $-\frac{2}{5}$.

You Try It 2
$$7 - x = 3$$
$$7 - 7 - x = 3 - 7 \quad \bullet \text{ Subtract 7 from each side.}$$
$$-x = -4$$
$$(-1)(-x) = (-1)(-4) \quad \bullet \text{ Multiply each side by } -1.$$
$$x = 4$$

The solution is 4.

You Try It 3

Strategy To find the Celsius temperature, replace the variable F in the formula by the given value and solve for C.

Solution
$$F = 1.8C + 32$$
$$-22 = 1.8C + 32 \quad \bullet \text{ Substitute } -22 \text{ for } F.$$
$$-22 - 32 = 1.8C + 32 - 32 \quad \bullet \text{ Subtract 32 from each side.}$$
$$-54 = 1.8C \quad \bullet \text{ Combine like terms.}$$
$$\frac{-54}{1.8} = \frac{1.8C}{1.8} \quad \bullet \text{ Divide each side by 1.8.}$$
$$-30 = C$$

The temperature is $-30°C$.

You Try It 4

Strategy To find the cost per unit, replace the variables T, N, and F in the formula by the given values and solve for U.

Solution
$$T = U \cdot N + F$$
$$4500 = U \cdot 250 + 1500$$
$$4500 = 250U + 1500$$
$$4500 - 1500 = 250U + 1500 - 1500$$
$$3000 = 250U$$
$$\frac{3000}{250} = \frac{250U}{250}$$
$$12 = U$$

The cost per unit is $12.

SECTION 11.4

You Try It 1
$$\frac{1}{5}x - 2 = \frac{2}{5}x + 4$$
$$\frac{1}{5}x - \frac{2}{5}x - 2 = \frac{2}{5}x - \frac{2}{5}x + 4 \quad \bullet \text{ Subtract } \frac{2}{5}x \text{ from each side.}$$
$$-\frac{1}{5}x - 2 = 4$$

$$-\frac{1}{5}x - 2 + 2 = 4 + 2$$ • Add 2 to each side.

$$-\frac{1}{5}x = 6$$

$$(-5)\left(-\frac{1}{5}x\right) = (-5)6$$ • Multiply each side by −5.

$$x = -30$$

The solution is −30.

You Try It 2

$$4(x - 1) - x = 5$$
$$4x - 4 - x = 5$$
$$3x - 4 = 5$$
$$3x - 4 + 4 = 5 + 4$$
$$3x = 9$$
$$\frac{3x}{3} = \frac{9}{3}$$
$$x = 3$$

The solution is 3.

You Try It 3

$$2x - 7(3x + 1) = 5(5 - 3x)$$
$$2x - 21x - 7 = 25 - 15x$$ • Distributive Property
$$-19x - 7 = 25 - 15x$$ • Combine like terms.
$$-19x + 15x - 7 = 25 - 15x + 15x$$ • Add 15x to both sides.
$$-4x - 7 = 25$$ • Combine like terms.
$$-4x - 7 + 7 = 25 + 7$$ • Add 7 to both sides.
$$-4x = 32$$ • Combine like terms.
$$\frac{-4x}{-4} = \frac{32}{-4}$$ • Divide each side by −4.
$$x = -8$$

The solution is −8.

SECTION 11.5

You Try It 1 $8 - 2t$

You Try It 2 $\dfrac{5}{7x}$

You Try It 3   The product of a number and one-half of the number
The unknown number: x

One-half the number: $\dfrac{1}{2}x$

$$(x)\left(\frac{1}{2}x\right)$$

SECTION 11.6

You Try It 1   The unknown number: x

$$x + 4 = 12$$
$$x + 4 - 4 = 12 - 4$$
$$x = 8$$

The number is 8.

You Try It 2   The unknown number: x

The product of two and a number is ten

$$2x = 10$$
$$\frac{2x}{2} = \frac{10}{2}$$
$$x = 5$$

The number is 5.

You Try It 3   The unknown number: x

The sum of three times a number and six equals four

$$3x + 6 = 4$$
$$3x + 6 - 6 = 4 - 6$$
$$3x = -2$$
$$\frac{3x}{3} = \frac{-2}{3}$$
$$x = -\frac{2}{3}$$

The number is $-\dfrac{2}{3}$.

You Try It 4   The unknown number: x

Three more than one-half of a number is nine

$$\frac{1}{2}x + 3 = 9$$
$$\frac{1}{2}x + 3 - 3 = 9 - 3$$
$$\frac{1}{2}x = 6$$
$$2 \cdot \frac{1}{2}x = 2 \cdot 6$$
$$x = 12$$

The number is 12.

You Try It 5

Strategy   To find the regular price, write and solve an equation using R to represent the regular price of the slacks.

Solution

$38.95	is	$11 less than the regular price

$$38.95 = R - 11$$
$$38.95 + 11 = R - 11 + 11$$
$$49.95 = R$$

The regular price of the slacks is $49.95.

You Try It 6

Strategy   To find the rpm of the engine when it is in third gear, write and solve an equation using R to represent the rpm of the engine in third gear.

Solution

2500	is	two-thirds of the rpm of the engine in third gear

$$2500 = \frac{2}{3}R$$

$$\frac{3}{2}(2500) = \left(\frac{3}{2}\right)\frac{2}{3}R$$

$$3750 = R$$

The rpm of the engine when in third gear is 3750 rpm.

You Try It 7

Strategy   To find the number of hours of labor, write and solve an equation using H to represent the number of hours of labor required.

Solution

$300	includes	$100 for materials plus $12.50 per hour of labor

$$300 = 100 + 12.50H$$
$$300 - 100 = 100 - 100 + 12.50H$$
$$200 = 12.50H$$
$$\frac{200}{12.50} = \frac{12.50H}{12.50}$$
$$16 = H$$

16 h of labor are required.

You Try It 8

Strategy   To find the total sales, write and solve an equation using S to represent the total sales.

Solution

$2500	is	the sum of $800 and an 8% commission on total sales

$$2500 = 800 + 0.08S$$
$$2500 - 800 = 800 - 800 + 0.08S$$
$$1700 = 0.08S$$
$$\frac{1700}{0.08} = \frac{0.08S}{0.08}$$
$$21{,}250 = S$$

Natalie's total sales for the month were $21,250.

Solutions to Chapter 12 "You Try It"

SECTION 12.1

You Try It 1

$$QT = QR + RS + ST$$
$$62 = 24 + RS + 17$$
$$62 = 41 + RS$$
$$62 - 41 = 41 - 41 + RS$$
$$21 = RS$$

You Try It 2   Let x represent the supplement of a 32° angle. The sum of supplementary angles is 180°.

$$x + 32° = 180°$$
$$x + 32° - 32° = 180° - 32°$$
$$x = 148°$$

148° is the supplement of 32°.

You Try It 3

$$\angle a + 68° = 118°$$
$$\angle a + 68° - 68° = 118° - 68°$$
$$\angle a = 50°$$

You Try It 4 In a right triangle, one angle measures 90° and the two acute angles are complementary.

$$\angle A + \angle B = 90°$$
$$\angle A + 7° = 90°$$
$$\angle A + 7° - 7° = 90° - 7°$$
$$\angle A = 83°$$

The other angles measure 90° and 83°.

You Try It 5 The sum of the three angles of a triangle is 180°.

$$\angle A + \angle B + \angle C = 180°$$
$$\angle A + 62° + 45° = 180°$$
$$\angle A + 107° = 180°$$
$$\angle A + 107° - 107° = 180° - 107°$$
$$\angle A = 73°$$

The measure of the other angle is 73°.

You Try It 6 $r = \frac{1}{2}d$

$r = \frac{1}{2}(8 \text{ in.}) = 4 \text{ in.}$

The radius is 4 in.

You Try It 7 Angles a and b are supplementary angles.

$$\angle a + \angle b = 180°$$
$$125° + \angle b = 180°$$
$$125° - 125° + \angle b = 180° - 125°$$
$$\angle b = 55°$$

You Try It 8 $\angle c$ and $\angle a$ are corresponding angles. Corresponding angles are equal.

$\angle c = \angle a = 120°$

$\angle b$ and $\angle c$ are supplementary angles.

$$\angle b + \angle c = 180°$$
$$\angle b + 120° = 180°$$
$$\angle b + 120° - 120° = 180° - 120°$$
$$\angle b = 60°$$

SECTION 12.2

You Try It 1 $P = 2L + 2W$
$= 2(2 \text{ m}) + 2(0.85 \text{ m})$
$= 4 \text{ m} + 1.7 \text{ m}$
$= 5.7 \text{ m}$

The perimeter of the rectangle is 5.7 m.

You Try It 2 $P = a + b + c$
$= 12 \text{ cm} + 15 \text{ cm} + 18 \text{ cm}$
$= 45 \text{ cm}$

The perimeter of the triangle is 45 cm.

You Try It 3 $C = \pi d$
$\approx 3.14 \cdot 6 \text{ in.}$
$= 18.84 \text{ in.}$

The circumference is approximately 18.84 in.

You Try It 4

$$\begin{matrix} \text{Perimeter} \\ \text{of} \\ \text{composite} \\ \text{figure} \end{matrix} = \begin{matrix} \text{two} \\ \text{lengths} \\ \text{of a} \\ \text{rectangle} \end{matrix} + \begin{matrix} \text{the} \\ \text{circumference} \\ \text{of a} \\ \text{circle} \end{matrix}$$

$P = 2L + \pi d$
$\approx 2(8 \text{ in.}) + 3.14(3 \text{ in.})$
$= 16 \text{ in.} + 9.42 \text{ in.}$
$= 25.42 \text{ in.}$

The perimeter is approximately 25.42 in.

You Try It 5

Strategy To find the perimeter, use the formula for the perimeter of a rectangle.

Solution $P = 2L + 2W$
$= 2(11 \text{ in.}) + 2\left(8\frac{1}{2} \text{ in.}\right)$
$= 22 \text{ in.} + 17 \text{ in.}$
$= 39 \text{ in.}$

The perimeter of the computer paper is 39 in.

You Try It 6

Strategy To find the cost:
• Find the perimeter of the workbench.
• Multiply the perimeter by the perimeter cost of the stripping.

Solution $P = 2L + 2W$
$= 2(3 \text{ m}) + 2(0.74 \text{ m})$
$= 6 \text{ m} + 1.48 \text{ m}$
$= 7.48 \text{ m}$

$\$2.76 \times 7.48 = \$20.6448$

The cost is \$20.64.

SECTION 12.3

You Try It 1 $A = \frac{1}{2}bh = \frac{1}{2}(24 \text{ in.})(14 \text{ in.}) = 168 \text{ in}^2$

The area is 168 in².

You Try It 2

A = area of rectangle − area of triangle

$A = LW - \dfrac{1}{2}bh$

$= (10 \text{ in.} \times 6 \text{ in.}) - \left(\dfrac{1}{2} \times 6 \text{ in.} \times 4 \text{ in.}\right)$

$= 60 \text{ in}^2 - 12 \text{ in}^2$

$= 48 \text{ in}^2$

The area is 48 in^2.

You Try It 3

Strategy   To find the area of the room:
- Find the area in square feet.
- Convert to square yards.

Solution $A = LW$

$= 12 \text{ ft} \cdot 9 \text{ ft}$

$= 108 \text{ ft}^2$

$108 \text{ ft}^2 \times \dfrac{1 \text{ yd}^2}{9 \text{ ft}^2} = \dfrac{108}{9} \text{ yd}^2$

$= 12 \text{ yd}^2$

The area of the room is 12 yd^2.

SECTION 12.4

You Try It 1 $V = LWH$

$= (8 \text{ cm})(3.5 \text{ cm})(4 \text{ cm})$

$= 112 \text{ cm}^3$

The volume is 112 cm^3.

You Try It 2 $V = s^3$

$= (5 \text{ cm})^3$

$= 125 \text{ cm}^3$

The volume is 125 cm^3.

You Try It 3 $r = \dfrac{1}{2}d = \dfrac{1}{2}(14 \text{ in.}) = 7 \text{ in.}$

$V = \pi r^2 h$

$\approx \dfrac{22}{7}(7 \text{ in.})^2(15 \text{ in.})$

$= 2310 \text{ in}^3$

The volume is approximately 2310 in^3.

You Try It 4 $V = \dfrac{4}{3}\pi r^3$

$\approx \dfrac{4}{3}(3.14)(3 \text{ m})^3$

$= 113.04 \text{ m}^3$

The volume is approximately 113.04 m^3.

You Try It 5

V = volume of rectangular solid
     + volume of cylinder

$V = LWH + \pi r^2 h$

$\approx (1.5 \text{ m})(0.4 \text{ m})(0.4 \text{ m}) + 3.14(0.8 \text{ m})^2(0.2 \text{ m})$

$= 0.24 \text{ m}^3 + 0.40192 \text{ m}^3$

$= 0.64192 \text{ m}^3$

The volume is approximately 0.64192 m^3.

You Try It 6

V = volume of rectangular solid
     + $\dfrac{1}{2}$ the volume of cylinder

$V = LWH + \dfrac{1}{2}\pi r^2 h$

$\approx (24 \text{ in.})(6 \text{ in.})(4 \text{ in.}) + \dfrac{1}{2}(3.14)(3 \text{ in.})^2(24 \text{ in.})$

$= 576 \text{ in}^3 + 339.12 \text{ in}^3$

$= 915.12 \text{ in}^3$

The volume is approximately 915.12 in^3.

You Try It 7

Strategy   To find the volume of the freezer, use the formula for the volume of a rectangular solid.

Solution $V = LWH$

$= (7 \text{ ft})(2.5 \text{ ft})(3 \text{ ft})$

$= 52.5 \text{ ft}^3$

The volume of the freezer is 52.5 ft^3.

You Try It 8

Strategy
To find the volume of the channel iron, subtract the volume of the cut-out rectangular solid from the large rectangular solid.

Solution

$V = LWH - LWH$

$= (10 \text{ ft})(0.5 \text{ ft})(0.8 \text{ ft}) - (10 \text{ ft})(0.3 \text{ ft})(0.2 \text{ ft})$

$= 4 \text{ ft}^3 - 0.6 \text{ ft}^3$

$= 3.4 \text{ ft}^3$

The volume of the channel iron is 3.4 ft^3.

SECTION 12.5

You Try It 1   **a.** $\sqrt{16} = 4$

$\sqrt{169} = 13$

b. $\sqrt{32} \approx 5.657$

$\sqrt{162} \approx 12.728$

You Try It 2
$$\begin{aligned} \text{Hypotenuse} &= \sqrt{(\text{leg})^2 + (\text{leg})^2} \\ &= \sqrt{8^2 + 11^2} \\ &= \sqrt{64 + 121} \\ &= \sqrt{185} \\ &\approx 13.601 \end{aligned}$$

The hypotenuse is approximately 13.601 in.

You Try It 3
$$\begin{aligned} \text{Leg} &= \sqrt{(\text{hypotenuse})^2 - (\text{leg})^2} \\ &= \sqrt{12^2 - 5^2} \\ &= \sqrt{144 - 25} \\ &= \sqrt{119} \\ &\approx 10.909 \end{aligned}$$

The length of the leg is approximately 10.909 ft.

You Try It 4

Strategy To find the distance between the holes, use the Pythagorean Theorem. The hypotenuse is the distance between the holes. The length of each leg is given (3 cm and 8 cm).

Solution
$$\begin{aligned} \text{Hypotenuse} &= \sqrt{(\text{leg})^2 + (\text{leg})^2} \\ &= \sqrt{3^2 + 8^2} \\ &= \sqrt{9 + 64} \\ &= \sqrt{73} \\ &\approx 8.544 \end{aligned}$$

The distance is approximately 8.544 cm.

SECTION 12.6

You Try It 1 $\dfrac{4 \text{ cm}}{7 \text{ cm}} = \dfrac{4}{7}$

You Try It 2
$$\frac{AB}{DE} = \frac{AC}{DF}$$
$$\frac{7 \text{ cm}}{14 \text{ cm}} = \frac{3 \text{ cm}}{x}$$
$$7x = 14 \cdot 3 \text{ cm}$$
$$7x = 42 \text{ cm}$$
$$\frac{7x}{7} = \frac{42 \text{ cm}}{7}$$
$$x = 6 \text{ cm}$$

Side *DF* is 6 cm.

You Try It 3 $AC = DF$, $\angle ACB = \angle DFE$, but $BC \neq EF$ because $\angle CAB \neq \angle FDE$. Therefore, the triangles are not congruent.

You Try It 4
$$\frac{AC}{DF} = \frac{\text{height } CH}{\text{height } FG}$$
$$\frac{10 \text{ m}}{15 \text{ m}} = \frac{7 \text{ m}}{h}$$
$$10h = 15 \cdot 7 \text{ m}$$
$$10h = 105 \text{ m}$$
$$\frac{10h}{10} = \frac{105 \text{ m}}{10}$$
$$h = 10.5 \text{ m}$$

The height *FG* is 10.5 m.

You Try It 5

Strategy To find the perimeter of triangle *ABC*:
- Solve a proportion to find the lengths of sides *BC* and *AC*.
- Use the formula $P = a + b + c$.

Solution
$$\frac{BC}{EF} = \frac{AB}{DE}$$
$$\frac{BC}{10 \text{ in.}} = \frac{4 \text{ in.}}{8 \text{ in.}}$$
$$8(BC) = 10 \text{ in.}(4)$$
$$8(BC) = 40 \text{ in.}$$
$$\frac{8(BC)}{8} = \frac{40 \text{ in.}}{8}$$
$$BC = 5 \text{ in.}$$

$$\frac{AC}{DF} = \frac{AB}{DE}$$
$$\frac{AC}{6 \text{ in.}} = \frac{4 \text{ in.}}{8 \text{ in.}}$$
$$8(AC) = 6 \text{ in.}(4)$$
$$8(AC) = 24 \text{ in.}$$
$$\frac{8(AC)}{8} = \frac{24 \text{ in.}}{8}$$
$$AC = 3 \text{ in.}$$

$$\begin{aligned} \text{Perimeter} &= 4 \text{ in.} + 5 \text{ in.} + 3 \text{ in.} \\ &= 12 \text{ in.} \end{aligned}$$

The perimeter of triangle *ABC* is 12 in.

Answers to Chapter 1 Selected Exercises

PREP TEST

1. 8   **2.** 1 2 3 4 5 6 7 8 9 10   **3.** a and D; b and E; c and A; d and B; e and F; f and C

SECTION 1.1

1. **3.** **5.** $37 < 49$ **7.** $101 > 87$
9. $245 > 158$ **11.** $0 < 45$ **13.** $815 < 928$ **15.** Millions   **17.** Hundred-thousands   **19.** Three thousand seven
hundred ninety   **21.** Fifty-eight thousand four hundred seventy-three   **23.** Four hundred ninety-eight thousand five
hundred twelve   **25.** Six million eight hundred forty-two thousand seven hundred fifteen   **27.** 357   **29.** 63,780
31. 7,024,709   **33.** $6000 + 200 + 90 + 5$ **35.** $400,000 + 50,000 + 3000 + 900 + 20 + 1$
37. $300,000 + 1000 + 800 + 9$ **39.** $3,000,000 + 600 + 40 + 2$ **41.** 850   **43.** 4000   **45.** 53,000   **47.** 630,000
49. 250,000   **51.** 72,000,000   **53.** 999; 10,000

SECTION 1.2

1. 28   **3.** 125   **5.** 102   **7.** 154   **9.** 1489   **11.** 828   **13.** 1584   **15.** 1219   **17.** 102,317   **19.** 79,326
21. 1804   **23.** 1579   **25.** 19,740   **27.** 7420   **29.** 120,570   **31.** 207,453   **33.** 24,218   **35.** 11,974
37. 9323   **39.** 77,139   **41.** 14,383   **43.** 9473   **45.** 33,247   **47.** 5058   **49.** 1992   **51.** 68,263
53. Cal.: 17,754   **55.** Cal.: 2872   **57.** Cal.: 101,712   **59.** Cal.: 158,763   **61.** Cal.: 261,595   **63.** Cal.: 946,718
Est.: 17,700   Est.: 2900   Est.: 101,000   Est.: 158,000   Est.: 260,000   Est.: 940,000
65. Cal.: 32,691,621   **67.** Cal.: 34,420,922   **69.** There were 118,295 multiple births during the year.
Est.: 33,000,000   Est.: 34,000,000
71. The estimated income from the four *Star Wars* movies was $1,500,000,000.   **73. a.** The income from the two movies
with the lowest box-office returns is $599,300,000.   **b.** Yes, this income exceeds the income from the 1977 *Star Wars*
production.   **75. a.** During the three days, 1285 miles will be driven.   **b.** At the end of the trip, the odometer will read
69,977 miles.   **77.** Americans ages 16 to 34 have invested $1796 in these three investments.   **79.** The sum of the
average amounts invested in home equity and retirement for all Americans is greater than the sum of all categories for
Americans aged 16 and 34.   **81.** 11 different sums   **83.** No. $0 + 0 = 0$ **85.** 10 numbers

SECTION 1.3

1. 4   **3.** 4   **5.** 10   **7.** 4   **9.** 9   **11.** 22   **13.** 60   **15.** 66   **17.** 31   **19.** 901   **21.** 791   **23.** 1125
25. 3131   **27.** 47   **29.** 925   **31.** 4561   **33.** 3205   **35.** 1222   **37.** 3021   **39.** 3022   **41.** 3040
43. 212   **45.** 60,245   **47.** 65   **49.** 17   **51.** 23   **53.** 456   **55.** 57   **57.** 375   **59.** 3139   **61.** 3621
63. 738   **65.** 3545   **67.** 749   **69.** 5343   **71.** 66,463   **73.** 16,590   **75.** 52,404   **77.** 38,777   **79.** 4638
81. 3612   **83.** 2913   **85.** 2583   **87.** 5268   **89.** 71,767   **91.** 11,239   **93.** 8482   **95.** 625   **97.** 76,725
99. 23   **101.** 4648   **103.** Cal.: 29,837   **105.** Cal.: 36,668   **107.** Cal.: 101,998
Est.: 30,000   Est.: 40,000   Est.: 100,000
109. There was an increase of 75,621 complaints from 2001 to 2002.   **111.** The amount that remains to be paid is
$10,275.   **113.** The Giant erupts 25 feet higher than Old Faithful.   **115.** The expected increase over 10 years is
106,000.   **117.** The amount remaining is $436.   **119. a.** True   **b.** False   **c.** False

SECTION 1.4

1. 6×2 or $6 \cdot 2$ **3.** 4×7 or $4 \cdot 7$ **5.** 12   **7.** 35   **9.** 25   **11.** 0   **13.** 72   **15.** 198   **17.** 335
19. 2492   **21.** 5463   **23.** 4200   **25.** 6327   **27.** 1896   **29.** 5056   **31.** 1685   **33.** 46,963   **35.** 59,976
37. 19,120   **39.** 19,790   **41.** 108   **43.** 3664   **45.** 20,036   **47.** 71,656   **49.** 432   **51.** 1944   **53.** 41,832
55. 43,620   **57.** 335,195   **59.** 594,625   **61.** 321,696   **63.** 342,171   **65.** 279,220   **67.** 191,800
69. 463,712   **71.** 180,621   **73.** 478,800   **75.** 158,422   **77.** 4,696,714   **79.** 5,542,452   **81.** 198,423
83. 18,834   **85.** 260,178   **87.** 315,109,895   **89.** Cal.: 440,076   **91.** Cal.: 6,491,166   **93.** Cal.: 18,728,744
Est.: 450,000   Est.: 6,300,000   Est.: 18,000,000

95. Cal.: 57,691,192   **97.** The plane used 5190 gallons of fuel on a 6-hour flight.   **99.** The area is 256 square miles.
Est.: 54,000,000

101. Company A offers the lower total price.   **103.** The estimated cost of the electricians' labor is $5100.   **105.** The total cost is $2138.   **107.** There are 12 accidental deaths each hour; 288 deaths each day; and 105,120 deaths each year.

SECTION 1.5

1. 2   **3.** 6   **5.** 7   **7.** 16   **9.** 210   **11.** 44   **13.** 703   **15.** 910   **17.** 21,560   **19.** 3580   **21.** 482
23. 1075   **25.** 52   **27.** 34   **29.** 2 r1   **31.** 5 r2   **33.** 13 r1   **35.** 10 r3   **37.** 90 r2   **39.** 230 r1
41. 204 r3   **43.** 1347 r3   **45.** 1720 r2   **47.** 409 r2   **49.** 6214 r2   **51.** 8708 r2   **53.** 1080 r2   **55.** 4200
57. 19,600   **59.** 1 r38   **61.** 1 r26   **63.** 21 r21   **65.** 30 r22   **67.** 5 r40   **69.** 9 r17   **71.** 200 r21
73. 303 r1   **75.** 67 r13   **77.** 176 r13   **79.** 1086 r7   **81.** 403   **83.** 12 r456   **85.** 4 r160   **87.** 160 r27
89. 1669 r14   **91.** 7950   **93.** Cal.: 2225   **95.** Cal.: 11,016   **97.** Cal.: 26,656   **99.** Cal.: 504   **101.** Cal.: 541
Est.: 2000   Est.: 10,000   Est.: 30,000   Est.: 500   Est.: 500

103. Cal.: 20,621   **105.** The average monthly expense for housing is $976.   **107.** The average monthly claim for theft
Est.: 20,000

is $25,000.   **109.** The average number of hours worked by employees in Britain is 35 hours.   **111.** 380 pennies are in
circulation for each person.   **113.** 198,400,000 cases of eggs are produced during the year.   **115.** The average U.S.
household will spend $130 on gasoline each month.   **117.** The average starting salary of an accounting major is $13,092
greater than that of a psychology major.   **119.** The difference is $10,848.   **121.** The total pay is $491.   **123.** 212

SECTION 1.6

1. 2^3 **3.** $6^3 \cdot 7^4$ **5.** $2^3 \cdot 3^3$ **7.** $5 \cdot 7^5$ **9.** $3^3 \cdot 6^4$ **11.** $3^3 \cdot 5 \cdot 9^3$ **13.** 8   **15.** 400   **17.** 900   **19.** 972
21. 120   **23.** 360   **25.** 0   **27.** 90,000   **29.** 540   **31.** 4050   **33.** 11,025   **35.** 25,920   **37.** 4,320,000
39. 5   **41.** 10   **43.** 47   **45.** 8   **47.** 5   **49.** 8   **51.** 6   **53.** 53   **55.** 44   **57.** 19   **59.** 67   **61.** 168
63. 27   **65.** 14   **67.** 10   **69.** 9   **71.** 12   **73.** 32   **75.** 39   **77.** 1024

SECTION 1.7

1. 1, 2, 4   **3.** 1, 2, 5, 10   **5.** 1, 7   **7.** 1, 3, 9   **9.** 1, 13   **11.** 1, 2, 3, 6, 9, 18   **13.** 1, 2, 4, 7, 8, 14, 28, 56
15. 1, 3, 5, 9, 15, 45   **17.** 1, 29   **19.** 1, 2, 11, 22   **21.** 1, 2, 4, 13, 26, 52   **23.** 1, 2, 41, 82   **25.** 1, 3, 19, 57
27. 1, 2, 3, 4, 6, 8, 12, 16, 24, 48   **29.** 1, 5, 19, 95   **31.** 1, 2, 3, 6, 9, 18, 27, 54   **33.** 1, 2, 3, 6, 11, 22, 33, 66
35. 1, 2, 4, 5, 8, 10, 16, 20, 40, 80   **37.** 1, 2, 3, 4, 6, 8, 12, 16, 24, 32, 48, 96   **39.** 1, 2, 3, 5, 6, 9, 10, 15, 18, 30, 45, 90
41. $2 \cdot 3$ **43.** Prime   **45.** $2 \cdot 2 \cdot 2 \cdot 3$ **47.** $3 \cdot 3 \cdot 3$ **49.** $2 \cdot 2 \cdot 3 \cdot 3$ **51.** Prime   **53.** $2 \cdot 3 \cdot 3 \cdot 5$
55. $5 \cdot 23$ **57.** $2 \cdot 3 \cdot 3$ **59.** $2 \cdot 2 \cdot 7$ **61.** Prime   **63.** $2 \cdot 31$ **65.** $2 \cdot 11$ **67.** Prime   **69.** $2 \cdot 3 \cdot 11$
71. $2 \cdot 37$ **73.** Prime   **75.** $5 \cdot 11$ **77.** $2 \cdot 2 \cdot 2 \cdot 3 \cdot 5$ **79.** $2 \cdot 2 \cdot 2 \cdot 2 \cdot 2 \cdot 5$ **81.** $2 \cdot 2 \cdot 2 \cdot 3 \cdot 3 \cdot 3$
83. $5 \cdot 5 \cdot 5 \cdot 5$ **85.** 3, 5; 5, 7; 11, 13; 41, 43; 71, 73; and 17, 19 are all the twin primes less than 100.

CHAPTER 1 REVIEW EXERCISES*

1. 600 [1.6A]   **2.** 10,000 + 300 + 20 + 7 [1.1C]   **3.** 1, 2, 3, 6, 9, 18 [1.7A]   **4.** 12,493 [1.2A]
5. 1749 [1.3B]   **6.** 2135 [1.5A]   **7.** 101 > 87 [1.1A]   **8.** $5^2 \cdot 7^5$ [1.6A]   **9.** 619,833 [1.4B]
10. 5409 [1.3B]   **11.** 1081 [1.2A]   **12.** 2 [1.6B]   **13.** 45,700 [1.1D]   **14.** Two hundred seventy-six thousand
fifty-seven [1.1B]   **15.** 1306 r59 [1.5C]   **16.** 2,011,044 [1.1B]   **17.** 488 r2 [1.5B]   **18.** 17 [1.6B]
19. 32 [1.6B]   **20.** $2 \cdot 2 \cdot 2 \cdot 3 \cdot 3$ [1.7B]   **21.** 2133 [1.3A]   **22.** 22,761 [1.4B]   **23.** The total pay for last
week's work is $768. [1.4C]   **24.** He drove 27 miles per gallon of gasoline. [1.5D]   **25.** Each monthly car payment is
$310. [1.5D]   **26.** The total income from commissions is $2567. [1.2B]   **27.** The total amount deposited is $301.

Note: The numbers in brackets following the answers in the Chapter Review are a reference to the objective that
corresponds to that problem. For example, the reference [1.2A] stands for Section 1.2, Objective A. This notation will be
used for all Prep Tests, Chapter Reviews, Chapter Tests, and Cumulative Reviews throughout the text.

The new checking account balance is $817.   [1.2B]   **28.** The total of the car payments is $2952.   [1.4C]
29. More males were involved in college sports in 2001 than in 1972.   [1.1A]   **30.** The difference between the numbers of male and female athletes in 1972 was 140,407 students.   [1.3C]   **31.** The number of female athletes increased by 120,939 students from 1972 to 2001.   [1.3C]   **32.** 159,421 more students were involved in ahtletics in 2001 than in 1972.   [1.3C]

CHAPTER 1 TEST

1. 432   [1.6A]   **2.** Two hundred seven thousand sixty-eight   [1.1B]   **3.** 9333   [1.3B]   **4.** 1, 2, 4, 5, 10, 20   [1.7A]
5. 6,854,144   [1.4B]   **6.** 9   [1.6B]   **7.** $900,000 + 6000 + 300 + 70 + 8$ [1.1C]   **8.** 75,000   [1.1D]
9. 1121 r27   [1.5C]   **10.** $3^3 \cdot 7^2$ [1.6A]   **11.** 54,915   [1.2A]   **12.** $2 \cdot 2 \cdot 3 \cdot 7$ [1.7B]   **13.** 4   [1.6B]
14. 726,104   [1.4A]   **15.** 1,204,006   [1.1B]   **16.** 8710 r2   [1.5B]   **17.** $21 > 19$ [1.1A]   **18.** 703   [1.5A]
19. 96,798   [1.2A]   **20.** 19,922   [1.3B]   **21.** The difference in projected total enrollment between 2012 and 2009 is 154,000 students.   [1.3C]   **22.** The average enrollment in each of the grades 9 through 12 in 2012 is 3,858,500 students.   [1.5D]   **23.** 3000 boxes were needed to pack the lemons.   [1.5D]   **24.** The investor receives $2844 over the 12-month period.   [1.4C]   **25. a.** 855 miles were driven during the 3 days.   [1.2B]   **b.** The odometer reading at the end of the 3 days is 48,481 miles.   [1.2B]

Answers to Chapter 2 Selected Exercises

PREP TEST

1. 20   [1.4A]   **2.** 120   [1.4A]   **3.** 9   [1.4A]   **4.** 10   [1.2A]   **5.** 7   [1.3A]   **6.** 2 r3   [1.5C]   **7.** 1, 2, 3, 4, 6, 12   [1.5B]   **8.** 59   [1.6B]   **9.** 7   [1.3A]   **10.** $44 < 48$ [1.1A]

SECTION 2.1

1. 40   **3.** 24   **5.** 30   **7.** 12   **9.** 24   **11.** 60   **13.** 56   **15.** 9   **17.** 32   **19.** 36   **21.** 660
23. 9384   **25.** 24   **27.** 30   **29.** 24   **31.** 576   **33.** 1680   **35.** 1   **37.** 3   **39.** 5   **41.** 25   **43.** 1
45. 4   **47.** 4   **49.** 6   **51.** 4   **53.** 1   **55.** 7   **57.** 5   **59.** 8   **61.** 1   **63.** 25   **65.** 7   **67.** 8
69. Two composite numbers are relatively prime if they do not have any common factor except the factor 1. Examples: 4 and 5, 8 and 9, and 16 and 21.

SECTION 2.2

1. Improper fraction   **3.** Proper fraction   **5.** $\frac{3}{4}$ **7.** $\frac{7}{8}$ **9.** $1\frac{1}{2}$ **11.** $2\frac{5}{8}$ **13.** $3\frac{3}{5}$ **15.** $\frac{5}{4}$ **17.** $\frac{8}{3}$
19. $\frac{28}{8}$ **21.**   **23.**   **25.**   **27.** $2\frac{3}{4}$ **29.** 5   **31.** $1\frac{1}{8}$ **33.** $2\frac{3}{10}$ **35.** 3
37. $1\frac{1}{7}$ **39.** $2\frac{1}{3}$ **41.** 16   **43.** $2\frac{1}{8}$ **45.** $2\frac{2}{5}$ **47.** 1   **49.** 9   **51.** $\frac{7}{3}$ **53.** $\frac{13}{2}$ **55.** $\frac{41}{6}$ **57.** $\frac{37}{4}$
59. $\frac{21}{2}$ **61.** $\frac{73}{9}$ **63.** $\frac{58}{11}$ **65.** $\frac{21}{8}$ **67.** $\frac{13}{8}$ **69.** $\frac{100}{9}$ **71.** $\frac{27}{8}$ **73.** $\frac{85}{13}$

SECTION 2.3

1. $\frac{5}{10}$ **3.** $\frac{9}{48}$ **5.** $\frac{12}{32}$ **7.** $\frac{9}{51}$ **9.** $\frac{12}{16}$ **11.** $\frac{27}{9}$ **13.** $\frac{20}{60}$ **15.** $\frac{44}{60}$ **17.** $\frac{12}{18}$ **19.** $\frac{35}{49}$ **21.** $\frac{10}{18}$
23. $\frac{21}{3}$ **25.** $\frac{35}{45}$ **27.** $\frac{60}{64}$ **29.** $\frac{21}{98}$ **31.** $\frac{30}{48}$ **33.** $\frac{15}{42}$ **35.** $\frac{102}{144}$ **37.** $\frac{153}{408}$ **39.** $\frac{340}{800}$ **41.** $\frac{1}{3}$ **43.** $\frac{1}{2}$
45. $\frac{1}{6}$ **47.** $1\frac{1}{9}$ **49.** 0   **51.** $\frac{9}{22}$ **53.** 3   **55.** $\frac{4}{21}$ **57.** $\frac{12}{35}$ **59.** $\frac{7}{11}$ **61.** $1\frac{1}{3}$ **63.** $\frac{3}{5}$ **65.** $\frac{1}{11}$

67. 4   **69.** $\frac{1}{3}$ **71.** $\frac{3}{5}$ **73.** $2\frac{1}{4}$ **75.** $\frac{1}{5}$ **77.** $\frac{3}{1}, \frac{6}{2}, \frac{9}{3}, \frac{12}{4}, \frac{15}{5}$ are fractions that are equal to 3.   **79. a.** $\frac{4}{25}$

b. $\frac{4}{25}$

SECTION 2.4

1. $\frac{3}{7}$ **3.** 1   **5.** $1\frac{4}{11}$ **7.** $3\frac{2}{5}$ **9.** $2\frac{4}{5}$ **11.** $2\frac{1}{4}$ **13.** $1\frac{3}{8}$ **15.** $1\frac{7}{15}$ **17.** $\frac{15}{16}$ **19.** $1\frac{4}{11}$ **21.** 1

23. $1\frac{7}{8}$ **25.** $1\frac{1}{6}$ **27.** $\frac{13}{14}$ **29.** $\frac{53}{60}$ **31.** $1\frac{1}{56}$ **33.** $\frac{23}{60}$ **35.** $\frac{56}{57}$ **37.** $1\frac{17}{18}$ **39.** $1\frac{11}{48}$ **41.** $1\frac{9}{20}$

43. $1\frac{109}{180}$ **45.** $1\frac{91}{144}$ **47.** $2\frac{5}{72}$ **49.** $\frac{39}{40}$ **51.** $1\frac{19}{24}$ **53.** $1\frac{65}{72}$ **55.** $3\frac{2}{3}$ **57.** $10\frac{1}{12}$ **59.** $9\frac{2}{7}$ **61.** $6\frac{7}{40}$

63. $9\frac{47}{48}$ **65.** $8\frac{3}{13}$ **67.** $16\frac{29}{120}$ **69.** $24\frac{29}{40}$ **71.** $33\frac{7}{24}$ **73.** $10\frac{5}{36}$ **75.** $10\frac{5}{12}$ **77.** $14\frac{73}{90}$ **79.** $10\frac{13}{48}$

81. $42\frac{13}{54}$ **83.** $8\frac{1}{36}$ **85.** $14\frac{1}{12}$ **87.** $8\frac{61}{72}$ **89.** The length of the shaft is $1\frac{5}{16}$ inches.   **91.** The total thickness

is $1\frac{5}{16}$ inches.   **93. a.** A total of 20 hours was worked.   **b.** Your total salary for the week is $220.   **95.** The total

length of wood needed is $55\frac{3}{4}$ feet.   **97.** Those who changed homes outside the county were $\frac{1}{3}$ of the people who

changed homes.

SECTION 2.5

1. $\frac{2}{17}$ **3.** $\frac{1}{3}$ **5.** $\frac{1}{10}$ **7.** $\frac{5}{13}$ **9.** $\frac{1}{3}$ **11.** $\frac{4}{7}$ **13.** $\frac{1}{4}$ **15.** $\frac{9}{23}$ **17.** $\frac{1}{4}$ **19.** $\frac{1}{2}$ **21.** $\frac{19}{56}$ **23.** $\frac{1}{2}$

25. $\frac{11}{60}$ **27.** $\frac{15}{56}$ **29.** $\frac{37}{51}$ **31.** $\frac{17}{70}$ **33.** $\frac{49}{120}$ **35.** $\frac{8}{45}$ **37.** $\frac{11}{21}$ **39.** $\frac{23}{60}$ **41.** $\frac{1}{18}$ **43.** $5\frac{1}{5}$ **45.** $10\frac{9}{17}$

47. $4\frac{7}{8}$ **49.** $\frac{16}{21}$ **51.** $5\frac{1}{2}$ **53.** $5\frac{4}{7}$ **55.** $7\frac{5}{24}$ **57.** $48\frac{31}{70}$ **59.** $85\frac{2}{9}$ **61.** $9\frac{5}{13}$ **63.** $15\frac{11}{20}$ **65.** $4\frac{23}{24}$

67. $2\frac{11}{15}$ **69.** The missing dimension is $9\frac{1}{2}$ inches.   **71.** The difference between Meyfarth's distance and Coachman's

distance was $9\frac{3}{8}$ inches. The difference between Kostadinova's distance and Meyfarth's distance was $5\frac{1}{4}$ inches.

73. a. The distance from the starting point to the second checkpoint is $7\frac{17}{24}$ miles.   **b.** The distance from the second

checkpoint to the finish line is $4\frac{7}{24}$ miles.   **75. a.** Yes   **b.** The wrestler needs to lose $3\frac{1}{4}$ pounds to reach the desired

weight.   **77.** $2\frac{5}{6}$ **79.**

$\frac{3}{8}$	$\frac{3}{4}$	$\frac{3}{4}$
1	$\frac{5}{8}$	$\frac{1}{4}$
$\frac{1}{2}$	$\frac{1}{2}$	$\frac{7}{8}$

SECTION 2.6

1. $\frac{7}{12}$ **3.** $\frac{7}{48}$ **5.** $\frac{1}{48}$ **7.** $\frac{11}{14}$ **9.** $\frac{1}{7}$ **11.** $\frac{1}{8}$ **13.** 6   **15.** $\frac{5}{12}$ **17.** 6   **19.** $\frac{2}{3}$ **21.** $\frac{3}{16}$ **23.** $\frac{3}{80}$

25. 10   **27.** $\frac{1}{15}$ **29.** $\frac{2}{3}$ **31.** $\frac{7}{26}$ **33.** 4   **35.** $\frac{100}{357}$ **37.** $\frac{5}{24}$ **39.** $\frac{1}{12}$ **41.** $\frac{4}{15}$ **43.** $1\frac{1}{2}$ **45.** 4

47. $\frac{4}{9}$ **49.** $\frac{1}{2}$ **51.** $16\frac{1}{2}$ **53.** 10   **55.** $6\frac{3}{7}$ **57.** $18\frac{1}{3}$ **59.** $1\frac{5}{7}$ **61.** $3\frac{1}{2}$ **63.** $25\frac{5}{8}$ **65.** $1\frac{11}{16}$ **67.** 16

69. 3   **71.** $8\frac{37}{40}$ **73.** $6\frac{19}{28}$ **75.** $7\frac{2}{5}$ **77.** 16   **79.** $32\frac{4}{5}$ **81.** $28\frac{1}{2}$ **83.** 9   **85.** $\frac{5}{8}$ **87.** $3\frac{1}{40}$

89. The salmon costs $11.   **91. a.** No. $\frac{1}{3}$ of 9 feet is approximately 3 ft.   **b.** The length of the board cut off is $3\frac{1}{12}$ feet.

93. The area of the square is $27\frac{9}{16}$ square feet.   **95. a.** The amount budgeted for housing and utilities is $1680.

b. The amount remaining for other than housing and utilities is \$2520. **97.** The total cost is \$363. **99.** The weight of the $12\frac{7}{12}$-foot steel rod is $54\frac{19}{36}$ pounds. **101.** $\frac{3}{16}$ of the total portfolio is invested in corporate bonds. **105.** *A*

SECTION 2.7

1. $\frac{5}{6}$ **3.** 1 **5.** 0 **7.** $\frac{1}{2}$ **9.** $2\frac{73}{256}$ **11.** $1\frac{1}{9}$ **13.** $\frac{1}{6}$ **15.** $\frac{7}{10}$ **17.** 2 **19.** 2 **21.** $\frac{1}{6}$ **23.** 6

25. $\frac{1}{15}$ **27.** 2 **29.** $2\frac{1}{2}$ **31.** 3 **33.** $1\frac{1}{6}$ **35.** $3\frac{1}{3}$ **37.** 6 **39.** $\frac{1}{2}$ **41.** $\frac{1}{30}$ **43.** $1\frac{4}{5}$ **45.** 13

47. $\frac{25}{288}$ **49.** 3 **51.** $\frac{1}{5}$ **53.** $\frac{11}{28}$ **55.** $\frac{5}{86}$ **57.** 120 **59.** $\frac{11}{40}$ **61.** $\frac{33}{40}$ **63.** $4\frac{4}{9}$ **65.** $\frac{13}{32}$ **67.** $10\frac{2}{3}$

69. $9\frac{39}{40}$ **71.** $\frac{12}{53}$ **73.** $4\frac{62}{191}$ **75.** 68 **77.** $8\frac{2}{7}$ **79.** $3\frac{13}{49}$ **81.** 4 **83.** $1\frac{3}{5}$ **85.** $\frac{9}{34}$ **87.** There are

12 servings in 16 ounces of cereal. **89.** Each acre costs \$24,000. **91.** The nut will make 12 turns in moving $1\frac{7}{8}$ inches.

93. a. The total weight of the fat and bone is $1\frac{5}{12}$ pounds. **b.** The chef can cut 28 servings from the roast.

95. The actual length of wall a is $12\frac{1}{2}$ feet. The actual length of wall b is 18 feet. The actual length of wall c is $15\frac{3}{4}$ feet.

97. $\frac{17}{50}$ of the money borrowed is spent on home improvement, cars and tuition. **99.** The capacity of the music center is 1800 people. **101.** $\frac{1}{6}$ of the puzzle is left to complete. **103.** Your total earnings for last week's work are \$111.

105. The average teenage boy consumes 3500 calories each week in soda. **107.** 740 miles **109. a.** $\frac{2}{3}$ **b.** $2\frac{5}{8}$

111. The quotient

SECTION 2.8

1. $\frac{11}{40} < \frac{19}{40}$ **3.** $\frac{2}{3} < \frac{5}{7}$ **5.** $\frac{5}{8} > \frac{7}{12}$ **7.** $\frac{7}{9} < \frac{11}{12}$ **9.** $\frac{13}{14} > \frac{19}{21}$ **11.** $\frac{7}{24} < \frac{11}{30}$ **13.** $\frac{9}{64}$ **15.** $\frac{8}{729}$ **17.** $\frac{1}{24}$

19. $\frac{8}{245}$ **21.** $\frac{1}{121}$ **23.** $\frac{81}{625}$ **25.** $\frac{7}{36}$ **27.** $\frac{27}{49}$ **29.** $\frac{5}{6}$ **31.** $1\frac{5}{12}$ **33.** $\frac{7}{48}$ **35.** $\frac{29}{36}$ **37.** $\frac{55}{72}$ **39.** $\frac{35}{54}$

41. 2 **43.** $\frac{9}{19}$ **45.** $\frac{7}{32}$ **47.** $\frac{64}{75}$

CHAPTER 2 REVIEW EXERCISES

1. $\frac{2}{3}$ [2.3B] **2.** $\frac{5}{16}$ [2.8B] **3.** $\frac{13}{4}$ [2.2A] **4.** $1\frac{13}{18}$ [2.4B] **5.** $\frac{11}{18} < \frac{17}{24}$ [2.8A] **6.** $14\frac{19}{42}$ [2.5C]

7. $\frac{5}{36}$ [2.8C] **8.** $9\frac{1}{24}$ [2.6B] **9.** 2 [2.7B] **10.** $\frac{25}{48}$ [2.5B] **11.** $3\frac{1}{3}$ [2.7B] **12.** 4 [2.1B]

13. $\frac{24}{36}$ [2.3A] **14.** $\frac{3}{4}$ [2.7A] **15.** $\frac{32}{44}$ [2.3A] **16.** $16\frac{1}{2}$ [2.6B] **17.** 36 [2.1A] **18.** $\frac{4}{11}$ [2.3B]

19. $1\frac{1}{8}$ [2.4A] **20.** $10\frac{1}{8}$ [2.5C] **21.** $18\frac{13}{54}$ [2.4C] **22.** 5 [2.1B] **23.** $3\frac{2}{5}$ [2.2B] **24.** $\frac{1}{15}$ [2.8C]

25. $5\frac{7}{8}$ [2.4C] **26.** 54 [2.1A] **27.** $\frac{1}{3}$ [2.5A] **28.** $\frac{19}{7}$ [2.2B] **29.** 2 [2.7A] **30.** $\frac{1}{15}$ [2.6A]

31. $\frac{1}{8}$ [2.6A] **32.** $1\frac{7}{8}$ [2.2A] **33.** The total rainfall for the 3 months was $21\frac{7}{24}$ inches. [2.4D] **34.** The cost per acre was \$36,000. [2.7C] **35.** The second checkpoint is $4\frac{3}{4}$ miles from the finish line. [2.5D] **36.** The car can travel 243 miles. [2.6C]

CHAPTER 2 TEST

1. $\frac{4}{9}$ [2.6A] **2.** 8 [2.1B] **3.** $1\frac{3}{7}$ [2.7A] **4.** $\frac{7}{24}$ [2.8C] **5.** $\frac{49}{5}$ [2.2B] **6.** 8 [2.6B] **7.** $\frac{5}{8}$ [2.3B]

8. $\frac{3}{8} < \frac{5}{12}$ [2.8A] **9.** $\frac{5}{6}$ [2.8C] **10.** 120 [2.1A] **11.** $\frac{1}{4}$ [2.5A] **12.** $3\frac{3}{5}$ [2.2B] **13.** $2\frac{2}{19}$ [2.7B]

14. $\frac{45}{72}$ [2.3A] **15.** $1\frac{61}{90}$ [2.4B] **16.** $13\frac{81}{88}$ [2.5C] **17.** $\frac{7}{48}$ [2.5B] **18.** $\frac{1}{6}$ [2.8B] **19.** $1\frac{11}{12}$ [2.4A]

20. $22\frac{4}{15}$ [2.4C] **21.** $\frac{11}{4}$ [2.2A] **22.** The electrician earns $840. [2.6C] **23.** 11 lots were available for sale.
[2.7C] **24.** The developer plans to build 30 houses on the property. [2.7C] **25.** The total rainfall for the 3-month

period was $21\frac{11}{24}$ inches. [2.4D]

CUMULATIVE REVIEW EXERCISES

1. 290,000 [1.1D] **2.** 291,278 [1.3B] **3.** 73,154 [1.4B] **4.** 540 r12 [1.5C] **5.** 1 [1.6B]

6. $2 \cdot 2 \cdot 11$ [1.7B] **7.** 210 [2.1A] **8.** 20 [2.1B] **9.** $\frac{23}{3}$ [2.2B] **10.** $6\frac{1}{4}$ [2.2B] **11.** $\frac{15}{48}$ [2.3A]

12. $\frac{2}{5}$ [2.3B] **13.** $1\frac{7}{48}$ [2.4B] **14.** $14\frac{11}{48}$ [2.4C] **15.** $\frac{13}{24}$ [2.5B] **16.** $1\frac{7}{9}$ [2.5C] **17.** $\frac{7}{20}$ [2.6A]

18. $7\frac{1}{2}$ [2.6B] **19.** $1\frac{1}{20}$ [2.7A] **20.** $2\frac{5}{8}$ [2.7B] **21.** $\frac{1}{9}$ [2.8B] **22.** $5\frac{5}{24}$ [2.8C]

23. The amount in the checkbook account at the end of the week was $862. [1.3C] **24.** The total income from the
tickets was $1410. [1.4C] **25.** The total weight is $12\frac{1}{24}$ pounds. [2.4D] **26.** The length of the remaining piece is

$4\frac{17}{24}$ feet. [2.5D] **27.** The car travels 225 miles on $8\frac{1}{3}$ gallons of gas. [2.6C] **28.** 25 parcels can be sold from the
remaining land. [2.7C]

Answers to Chapter 3 Selected Exercises

PREP TEST

1. $\frac{3}{10}$ [2.2A] **2.** 36,900 [1.1D] **3.** Four thousand seven hundred ninety-one [1.1B] **4.** 6842 [1.1B]

5. 9394 [1.2A] **6.** 1638 [1.3B] **7.** 76,804 [1.4B] **8.** 278 r18 [1.5C]

SECTION 3.1

1. Thousandths **3.** Ten-thousandths **5.** Hundredths **7.** 0.3 **9.** 0.21 **11.** 0.461 **13.** 0.093 **15.** $\frac{1}{10}$

17. $\frac{47}{100}$ **19.** $\frac{289}{1000}$ **21.** $\frac{9}{100}$ **23.** Thirty-seven hundredths **25.** Nine and four tenths **27.** Fifty-three
ten-thousandths **29.** Forty-five thousandths **31.** Twenty-six and four hundredths **33.** 3.0806 **35.** 407.03
37. 246.024 **39.** 73.02684 **41.** 5.4 **43.** 30.0 **45.** 413.60 **47.** 6.062 **49.** 97 **51.** 5440 **53.** 0.0236
55. 0.18 ounce **57.** 26.2 miles **59.** For example, **a.** 0.15 **b.** 1.05 **c.** 0.001

SECTION 3.2

1. 150.1065 **3.** 95.8446 **5.** 69.644 **7.** 92.883 **9.** 113.205 **11.** 0.69 **13.** 16.305 **15.** 110.7666
17. 104.4959 **19.** Cal.: 234.192 **21.** Cal.: 781.943 **23.** The total length of the shaft is 5.65 inches.
     Est.: 234   Est.: 782
25. The amount in the checking account is $3664.20. **27.** The combined populations of Asia and Africa in 2050 are
expected to be 7.1 billion people. **29.** The number of self-employed people who earn more than $5000 is 6.5 million.
31. No, $10 is not enough. **33.** No, a 4-foot rope cannot be wrapped around the box.

SECTION 3.3

1. 5.627 **3.** 113.6427 **5.** 6.7098 **7.** 215.697 **9.** 53.8776 **11.** 72.7091 **13.** 0.3142 **15.** 1.023
17. 261.166 **19.** 655.32 **21.** 342.9268 **23.** 8.628 **25.** 4.685 **27.** 10.383
29. Cal.: 2.74506 **31.** Cal.: 7.14925 **33.** The missing dimension is 7.55 inches. **35.** **a.** The total amount of
     Est.: 3   Est.: 7

the checks is $607.36.   **b.** Your new balance is $422.38.   **37.** The increase is 0.39 million births.   **39.** The growth in online shopping from 2000 to 2003 is 22.6 million households.   **41. a.** 0.1   **b.** 0.01   **c.** 0.001

SECTION 3.4

1. 0.36   **3.** 0.30   **5.** 0.25   **7.** 0.45   **9.** 6.93   **11.** 1.84   **13.** 4.32   **15.** 0.74   **17.** 39.5   **19.** 2.72
21. 0.603   **23.** 0.096   **25.** 13.50   **27.** 79.80   **29.** 4.316   **31.** 1.794   **33.** 0.06   **35.** 0.072   **37.** 0.1323
39. 0.03568   **41.** 0.0784   **43.** 0.076   **45.** 34.48   **47.** 580.5   **49.** 20.148   **51.** 0.04255   **53.** 0.17686
55. 0.19803   **57.** 14.8657   **59.** 0.0006608   **61.** 53.9961   **63.** 0.536335   **65.** 0.429   **67.** 2.116   **69.** 0.476
71. 1.022   **73.** 37.96   **75.** 2.318   **77.** 3.2   **79.** 6.5   **81.** 6285.6   **83.** 3200   **85.** 35,700   **87.** 6.3
89. 3.9   **91.** 49,000   **93.** 6.7   **95.** 0.012075   **97.** 0.0117796   **99.** 0.31004   **101.** 0.082845   **103.** 5.175
105. Cal.: 91.2   **107.** Cal.: 1.0472   **109.** Cal.: 3.897   **111.** Cal.: 11.2406   **113.** Cal.: 0.371096
Est.: 90   Est.: 0.8   Est.: 4.5   Est.: 12   Est.: 0.32
115. Cal.: 31.8528   **117.** Cal.: 1941.069459   **119.** Cal.: 0.00004351152   **121.** The motor costs $1.51 to operate.
Est.: 30   Est.: 2000   Est.: 0.00005
123. a. The estimated amount received is $25.00.   **b.** The amount received was $23.40.   **125.** The area is 23.625
square feet.   **127. a.** The amount of overtime pay is $650.25.   **b.** The nurse's total income is $1806.25.
129. The cost is $341.25.   **131. a.** The total cost for grade 1 is $21.12.   **b.** The total cost for grade 2 is $29.84.
c. The total cost for grade 3 is $201.66.   **d.** The total cost is $252.62.   **133.** Cars 2 and 5 would fail the test.
137. $1\frac{3}{10} \times 2\frac{31}{100} = \frac{13}{10} \times \frac{231}{100} = \frac{3003}{1000} = 3\frac{3}{1000} = 3.003$

SECTION 3.5

1. 0.82   **3.** 4.8   **5.** 89   **7.** 60   **9.** 84.3   **11.** 32.3   **13.** 5.06   **15.** 1.3   **17.** 0.11   **19.** 3.8   **21.** 6.3
23. 0.6   **25.** 2.5   **27.** 1.1   **29.** 130.6   **31.** 0.81   **33.** 0.09   **35.** 40.70   **37.** 0.46   **39.** 0.019
41. 0.087   **43.** 0.360   **45.** 0.103   **47.** 0.009   **49.** 1   **51.** 3   **53.** 1   **55.** 57   **57.** 0.407   **59.** 4.267
61. 0.01037   **63.** 0.008295   **65.** 0.82537   **67.** 0.032   **69.** 0.23627   **71.** 0.000053   **73.** 0.0018932
75. 18.42   **77.** 16.07   **79.** 0.0135   **81.** 0.023678   **83.** 0.112   **85.** Cal.: 11.1632   **87.** Cal.: 884.0909
Est.: 10   Est.: 1000
89. Cal.: 1.8269   **91.** Cal.: 58.8095   **93.** Cal.: 72.3053   **95.** Cal.: 0.0023   **97.** 6.23 yards are gained per carry.
Est.: 1.5   Est.: 50   Est.: 100   Est.: 0.0025
99. The cost per can is $.28.   **101.** The cost per mile is $.04.   **103.** The monthly payment is $58.65.
105. 2.9 million more women are expected to be attending institutions of higher learning in 2010.   **107.** The Army's
budget was 4.2 times greater than the Navy's advertising budget.   **109.** The population of this segment is expected to be
2.1 times greater in 2030 than in 2000.   **111.** The cigarette consumption in 2000 was 2.5 times greater than in 1960.
113. A total of 9.7 million acres was burned.   **119.** $\times$ **121.** $\times$ **123.** $\div$ **125.** 2.53

SECTION 3.6

1. 0.625   **3.** 0.667   **5.** 0.167   **7.** 0.417   **9.** 1.750   **11.** 1.500   **13.** 4.000   **15.** 0.003   **17.** 7.080
19. 37.500   **21.** 0.160   **23.** 8.400   **25.** $\frac{4}{5}$ **27.** $\frac{8}{25}$ **29.** $\frac{1}{8}$ **31.** $1\frac{1}{4}$ **33.** $16\frac{9}{10}$ **35.** $8\frac{2}{5}$ **37.** $8\frac{437}{1000}$
39. $2\frac{1}{4}$ **41.** $\frac{23}{150}$ **43.** $\frac{703}{800}$ **45.** $7\frac{19}{50}$ **47.** $\frac{57}{100}$ **49.** $\frac{2}{3}$ **51.** 0.6 > 0.45   **53.** 3.89 < 3.98
55. 0.025 < 0.105   **57.** $\frac{4}{5}$ < 0.802   **59.** 0.85 < $\frac{7}{8}$ **61.** $\frac{7}{12}$ > 0.58   **63.** $\frac{11}{12}$ < 0.92   **65.** 0.623 > 0.6023
67. 0.87 > 0.087   **69.** 0.033 < 0.3   **71.** The population ages 0 to 19 is more than $\frac{1}{4}$ of the total population.
73. Yes, the digits 538461 repeat.

CHAPTER 3 REVIEW EXERCISES

1. 54.5 [3.5A]   **2.** 833.958 [3.2A]   **3.** 0.055 < 0.1 [3.6C]   **4.** Twenty-two and ninety-two ten-thousandths [3.1A]
5. 0.05678 [3.1B]   **6.** 2.33 [3.6A]   **7.** $\frac{3}{8}$ [3.6B]   **8.** 36.714 [3.2A]   **9.** 34.025 [3.1A]   **10.** $\frac{5}{8}$ > 0.62 [3.6C]

11. 0.778 [3.6A]   **12.** $\frac{2}{3}$ [3.6B]   **13.** 22.8635 [3.3A]   **14.** 7.94 [3.1B]   **15.** 8.932 [3.4A]

16. Three hundred forty-two and thirty-seven hundredths [3.1A]   **17.** 3.06753 [3.1A]   **18.** 25.7446 [3.4A]

19. 6.594 [3.5A]   **20.** 4.8785 [3.3A]   **21.** The new balance in your account is $661.51. [3.3B]   **22.** There are 53.466 million children in grades K–12. [3.2B]   **23.** There are 40.49 million more children in public school than in private school. [3.3B]   **24.** During a 5-day school week, 9.5 million gallons of milk are served. [3.4B]   **25.** The number who drove is 6.4 times greater than the number who flew. [3.5B]

CHAPTER 3 TEST

1. $0.66 < 0.666$ [3.6C]   **2.** 4.087 [3.3A]   **3.** Forty-five and three hundred two ten-thousandths [3.1A]

4. 0.692 [3.6A]   **5.** $\frac{33}{40}$ [3.6B]   **6.** 0.0740 [3.1B]   **7.** 1.538 [3.5A]   **8.** 27.76626 [3.3A]   **9.** 7.095 [3.1B]

10. 232 [3.5A]   **11.** 458.581 [3.2A]   **12.** The missing dimension is 1.37 inches. [3.3B]   **13.** 0.00548 [3.4A]

14. 255.957 [3.2A]   **15.** 209.07086 [3.1A]   **16.** Each payment is $395.40. [3.5B]   **17.** Your total income is $3087.14. [3.2B]   **18.** The cost of the call is $4.63. [3.4B]   **19.** The yearly average computer use by a 10th-grade student is 348.4 hours. [3.4B]   **20.** On average a 2nd-grade student uses the computer 36.4 hours more per year than a 5th-grade student. [3.4B]

CUMULATIVE REVIEW EXERCISES

1. 235 r17 [1.5C]   **2.** 128 [1.6A]   **3.** 3 [1.6B]   **4.** 72 [2.1A]   **5.** $4\frac{2}{5}$ [2.2B]   **6.** $\frac{37}{8}$ [2.2B]

7. $\frac{25}{60}$ [2.3A]   **8.** $1\frac{17}{48}$ [2.4B]   **9.** $8\frac{35}{36}$ [2.4C]   **10.** $5\frac{23}{36}$ [2.5C]   **11.** $\frac{1}{12}$ [2.6A]   **12.** $9\frac{1}{8}$ [2.6B]

13. $1\frac{2}{9}$ [2.7A]   **14.** $\frac{19}{20}$ [2.7B]   **15.** $\frac{3}{16}$ [2.8B]   **16.** $2\frac{5}{18}$ [2.8C]   **17.** Sixty-five and three hundred nine ten-thousandths [3.1A]   **18.** 504.6991 [3.2A]   **19.** 21.0764 [3.3A]   **20.** 55.26066 [3.4A]   **21.** 2.154 [3.5A]

22. 0.733 [3.6A]   **23.** $\frac{1}{6}$ [3.6B]   **24.** $\frac{8}{9} < 0.98$ [3.6C]   **25.** Sweden mandates 14 days more vacation than Germany. [1.3C]   **26.** The patient must lose $7\frac{3}{4}$ pounds the third month to achieve the goal. [2.5D]

27. Your checking account balance is $617.38. [3.3B]   **28.** The resulting thickness is 1.395 inches. [3.3B]   **29.** You paid $6008.80 in income tax last year. [3.4B]   **30.** The amount of each payment is $23.87. [3.5B]

Answers to Chapter 4 Selected Exercises

PREP TEST

1. $\frac{4}{5}$ [2.3B]   **2.** $\frac{1}{2}$ [2.3B]   **3.** 24.8 [3.6A]   **4.** 4×33 [1.4A]   **5.** 4 [1.5A]

SECTION 4.1

1. $\frac{1}{5}$ 1:5   1 to 5   **3.** $\frac{2}{1}$ 2:1   2 to 1   **5.** $\frac{3}{8}$ 3:8   3 to 8   **7.** $\frac{37}{24}$ 37:24   37 to 24   **9.** $\frac{1}{1}$ 1:1   1 to 1

11. $\frac{7}{10}$ 7:10   7 to 10   **13.** $\frac{1}{2}$ 1:2   1 to 2   **15.** $\frac{2}{1}$ 2:1   2 to 1   **17.** $\frac{3}{4}$ 3:4   3 to 4   **19.** $\frac{5}{3}$ 5:3   5 to 3

21. $\frac{2}{3}$ 2:3   2 to 3   **23.** $\frac{2}{1}$ 2:1   2 to 1   **25.** The ratio is $\frac{1}{3}$.   **27.** The ratio is $\frac{3}{8}$.   **29.** The ratio is $\frac{2}{77}$.

31. The ratio is $\frac{1}{12}$.   **33. a.** The amount of increase is $20,000.   **b.** The ratio is $\frac{2}{9}$.   **35.** The ratio is $\frac{11}{31}$.

37. No, the ratio $= \frac{830}{1389} \approx 0.5976$ is greater than $\frac{2}{5}$ (0.4).

SECTION 4.2

1. $\dfrac{3\ \text{pounds}}{4\ \text{people}}$ **3.** $\dfrac{\$20}{3\ \text{boards}}$ **5.** $\dfrac{20\ \text{miles}}{1\ \text{gallon}}$ **7.** $\dfrac{5\ \text{children}}{2\ \text{families}}$ **9.** $\dfrac{8\ \text{gallons}}{1\ \text{hour}}$ **11.** 2.5 feet/second **13.** $975/week
15. 110 trees/acre **17.** $18.84/hour **19.** 52.4 miles/hour **21.** 28 miles/gallon **23.** $1.65/pound **25.** The gas mileage was 28.4 miles/gallon. **27.** The rocket uses 213,600 gallons/minute. **29. a.** 175 pounds of beef was packaged. **b.** The beef cost $2.09/pound. **31. a.** The camera goes through film at the rate of 336 feet/minute. **b.** The camera uses the film at a rate of 89 seconds/roll. **33. a.** The price of the computer hardware would be 109,236 euros. **b.** The cost of the car would be 3,978,000 yen.

SECTION 4.3

1. True **3.** Not true **5.** Not true **7.** True **9.** True **11.** Not true **13.** True **15.** True **17.** True
19. Not true **21.** True **23.** Not true **25.** 3 **27.** 6 **29.** 9 **31.** 5.67 **33.** 4 **35.** 4.38 **37.** 88
39. 3.33 **41.** 26.25 **43.** 96 **45.** 9.78 **47.** 3.43 **49.** 1.34 **51.** 50.4 **53.** A 0.5-ounce serving contains 50 calories. **55.** Ron used 70 pounds of fertilizer. **57.** There were 375 wooden bats produced.
59. The distance is 16 miles. **61.** 1.25 ounces are required. **63.** 160,000 people would vote. **65.** The monthly payment is $176.75. **67.** You will own 400 shares. **69.** A bowling ball would weigh 2.67 pounds on the moon.
71. The dividend would be $1071.

CHAPTER 4 REVIEW EXERCISES

1. True [4.3A] **2.** $\dfrac{2}{5}$ 2:5 2 to 5 [4.1A] **3.** 62.5 miles/hour [4.2B] **4.** True [4.3A] **5.** 68 [4.3B]
6. $7.50/hour [4.2B] **7.** $1.75/pound [4.2B] **8.** $\dfrac{2}{7}$ 2:7 2 to 7 [4.1A] **9.** 36 [4.3B] **10.** 19.44 [4.3B]
11. $\dfrac{2}{5}$ 2:5 2 to 5 [4.1A] **12.** Not true [4.3A] **13.** $\dfrac{\$15}{4\ \text{hours}}$ [4.2A] **14.** 27.2 miles/gallon [4.2B]
15. $\dfrac{1}{1}$ 1:1 1 to 1 [4.1A] **16.** True [4.3A] **17.** 65.45 [4.3B] **18.** $\dfrac{100\ \text{miles}}{3\ \text{hours}}$ [4.2A] **19.** The ratio
is $\dfrac{2}{5}$. [4.1B] **20.** The property tax is $6400. [4.3C] **21.** The ratio is $\dfrac{2}{1}$. [4.1B] **22.** The cost per phone
is $37.50. [4.2C] **23.** 1344 blocks would be needed. [4.3C] **24.** The ratio is $\dfrac{5}{2}$. [4.1B] **25.** The turkey costs
$.93/pound. [4.2C] **26.** The average was 56.8 miles/hour. [4.2C] **27.** The cost is $493.50. [4.3C] **28.** The cost
is $44.75/share. [4.2C] **29.** 22.5 pounds of fertilizer will be used. [4.3C] **30.** The ratio is $\dfrac{1}{2}$. [4.1B]

CHAPTER 4 TEST

1. $3836.40/month [4.2B] **2.** $\dfrac{1}{6}$ 1:6 1 to 6 [4.1A] **3.** $\dfrac{9\ \text{supports}}{4\ \text{feet}}$ [4.2A] **4.** Not true [4.3A]
5. $\dfrac{3}{2}$ 3:2 3 to 2 [4.1A] **6.** 144 [4.3B] **7.** 30.5 miles/gallon [4.2B] **8.** $\dfrac{1}{3}$ 1:3 1 to 3 [4.1A]
9. True [4.3A] **10.** 40.5 [4.3B] **11.** $\dfrac{\$27}{4\ \text{boards}}$ [4.2A] **12.** $\dfrac{3}{5}$ 3:5 3 to 5 [4.1A] **13.** The dividend is
$625. [4.3C] **14.** The ratio is $\dfrac{43}{56}$. [4.1B] **15.** The plane's speed is 538 miles/hour. [4.2C] **16.** The college
student's body contains 132 pounds of water. [4.3C] **17.** The cost of the lumber is $1.73/foot. [4.2C] **18.** The amount
of medication required is 0.875 ounce. [4.3C] **19.** The ratio is $\dfrac{4}{5}$. [4.1B] **20.** 36 defective hard drives are expected
to be found in the production of 1200 hard drives. [4.3C]

CUMULATIVE REVIEW EXERCISES

1. 9158 [1.3B] **2.** $2^4 \cdot 3^3$ [1.6A] **3.** 3 [1.6B] **4.** $2 \cdot 2 \cdot 2 \cdot 2 \cdot 2 \cdot 5$ [1.7B] **5.** 36 [2.1A] **6.** 14 [2.1B]
7. $\dfrac{5}{8}$ [2.3B] **8.** $8\dfrac{3}{10}$ [2.4C] **9.** $5\dfrac{11}{18}$ [2.5C] **10.** $2\dfrac{5}{6}$ [2.6B] **11.** $4\dfrac{2}{3}$ [2.7B] **12.** $\dfrac{23}{30}$ [2.8C]

13. Four and seven hundred nine ten-thousandths [3.1A]   **14.** 2.10 [3.1B]   **15.** 1.990 [3.5A]   **16.** $\frac{1}{15}$ [3.6B]

17. $\frac{1}{8}$ [4.1A]   **18.** $\frac{29¢}{2 \text{ pencils}}$ [4.2A]   **19.** 33.4 miles/gallon [4.2B]   **20.** 4.25 [4.3B]   **21.** The car's speed is 57.2 miles/hour. [4.2C]   **22.** 36 [4.3B]   **23.** Your new balance is $744. [1.3C]   **24.** The monthly payment is $570. [1.5D]   **25.** 105 pages remain to be read. [2.6C]   **26.** The cost per acre was $36,000. [2.7C]   **27.** The change was $17.62. [3.3B]   **28.** Your monthly salary is $3468.25. [3.5B]   **29.** 25 inches will erode in 50 months. [4.3C]   **30.** 1.6 ounces are required. [4.3C]

Answers to Chapter 5 Selected Exercises

PREP TEST

1. $\frac{19}{100}$ [2.6B]   **2.** 0.23 [3.4A]   **3.** 47 [3.4A]   **4.** 2850 [3.4A]   **5.** 4000 [3.5A]   **6.** 32 [2.7B]

7. 62.5 [3.6A]   **8.** $66\frac{2}{3}$ [2.2B]   **9.** 1.75 [3.5A]

SECTION 5.1

1. $\frac{1}{4}$, 0.25   **3.** $1\frac{3}{10}$, 1.30   **5.** 1, 1.00   **7.** $\frac{73}{100}$, 0.73   **9.** $3\frac{83}{100}$, 3.83   **11.** $\frac{7}{10}$, 0.70   **13.** $\frac{22}{25}$, 0.88

15. $\frac{8}{25}$, 0.32   **17.** $\frac{2}{3}$ **19.** $\frac{5}{6}$ **21.** $\frac{1}{9}$ **23.** $\frac{5}{11}$ **25.** $\frac{3}{70}$ **27.** $\frac{1}{15}$ **29.** 0.065   **31.** 0.123   **33.** 0.0055

35. 0.0825   **37.** 0.0505   **39.** 0.02   **41.** 0.804   **43.** 0.049   **45.** 73%   **47.** 1%   **49.** 294%   **51.** 0.6%

53. 310.6%   **55.** 70%   **57.** 37%   **59.** 40%   **61.** 12.5%   **63.** 150%   **65.** 166.7%   **67.** 87.5%   **69.** 48%

71. $33\frac{1}{3}\%$ **73.** $166\frac{2}{3}\%$ **75.** $87\frac{1}{2}\%$ **77.** 6% of those surveyed named something other than corn on the cob, cole slaw, corn bread, or fries.   **79.** This represents $\frac{1}{2}$ off the regular price.   **81. a.** False   **b.** For example, $200\% \times 4 = 2 \times 4 = 8$

SECTION 5.2

1. 8   **3.** 10.8   **5.** 0.075   **7.** 80   **9.** 51.895   **11.** 7.5   **13.** 13   **15.** 3.75   **17.** 20   **19.** 210

21. 5% of 95   **23.** 79% of 16   **25.** 72% of 40   **27.** 25,805.0324   **29.** About 13.2 million people aged 18 to 24 do not have health insurance.   **31.** 12,151 more faculty members described their political views as liberal than described their views as far left.   **33. a.** The sales tax is $1770.   **b.** The total cost of the car is $31,270.   **35.** 6232 respondents did not answer yes to the question.   **37.** Employees spend 48.2 hours with family and friends.   **39.** There are approximately 3.4 hours difference between the actual and preferred amounts of time the employees spent on self.

SECTION 5.3

1. 32%   **3.** $16\frac{2}{3}\%$ **5.** 200%   **7.** 37.5%   **9.** 18%   **11.** 0.25%   **13.** 20%   **15.** 400%   **17.** 2.5%

19. 37.5%   **21.** 0.25%   **23.** 2.4%   **25.** 9.6%   **27.** 70% of couples disagree about financial matters.   **29.** Approximately 25.4% of the vegetables were wasted.   **31.** The typical American household spends 6% of the total amount spent on energy utilities on lighting.   **33.** 27% of the food produced in the United States is wasted.   **35.** Approximately 27.4% of the total expenses is spent for food.   **37.** 91.8% of the total is spent on all categories except training.

SECTION 5.4

1. 75   **3.** 50   **5.** 100   **7.** 85   **9.** 1200   **11.** 19.2   **13.** 7.5   **15.** 32   **17.** 200   **19.** 80   **21.** 9

23. 504 **25.** 108 **27.** There were 15.8 million travelers who allowed their children to miss school to go along on a trip. **29.** 3364 people responded to the survey. **31. a.** 3000 boards were tested. **b.** 2976 boards tested were not defective. **33.** 200,935,308 people in the United States were age 20 and older in 2000. **35.** The recommended daily allowance of copper for an adult is 2 milligrams.

SECTION 5.5

1. 65 **3.** 25% **5.** 75 **7.** 12.5% **9.** 400 **11.** 19.5 **13.** 14.8% **15.** 62.62 **17.** 5 **19.** 45 **21.** 15 **23.** The total amount collected was $24,500. **25.** The world's total land area is 58,750,000 square miles. **27.** 22,577 hotels in the United States are located along highways. **29.** 14,000,000 ounces of gold were mined in the United States that year. **31.** 46.8% of the deaths were due to traffic accidents. **33.** The 108th House of Representatives had the larger percent of Republicans.

CHAPTER 5 REVIEW EXERCISES

1. 60 [5.2A] **2.** 20% [5.3A] **3.** 175% [5.1B] **4.** 75 [5.4A] **5.** $\frac{3}{25}$ [5.1A] **6.** 19.36 [5.2A]

7. 150% [5.3A] **8.** 504 [5.4A] **9.** 0.42 [5.1A] **10.** 5.4 [5.2A] **11.** 157.5 [5.4A] **12.** 0.076 [5.1A]

13. 77.5 [5.2A] **14.** $\frac{1}{6}$ [5.1A] **15.** 160% [5.5A] **16.** 75 [5.5A] **17.** 38% [5.1B] **18.** 10.9 [5.4A]

19. 7.3% [5.3A] **20.** 613.3% [5.3A] **21.** The student answered 85% of the questions correctly. [5.5B] **22.** The company spent $4500 for newspaper advertising. [5.2B] **23.** 31.7% of the cost is for electricity. [5.3B] **24.** The total cost of the video camera is $1041.25. [5.2B] **25.** Approximately 78.6% of the women wore sunscreen often. [5.3B] **26.** The world's population in 2000 was approximately 6,100,000,000 people. [5.4B] **27.** The cost of the computer 4 years ago was $3000. [5.5B] **28.** The dollar value of all the online transactions made that year was $52 billion. [5.4B]

CHAPTER 5 TEST

1. 0.973 [5.1A] **2.** $\frac{5}{6}$ [5.1A] **3.** 30% [5.1B] **4.** 163% [5.1B] **5.** 150% [5.1B] **6.** $66\frac{2}{3}$% [5.1B]

7. 50.05 [5.2A] **8.** 61.36 [5.2A] **9.** 76% of 13 [5.2A] **10.** 212% of 12 [5.2A] **11.** The amount spent for for advertising is $4500. [5.2B] **12.** 1170 pounds of vegetables were not spoiled. [5.2B] **13.** 14.7% of the daily recommended amount of potassium is provided. [5.3B] **14.** 9.1% of the daily recommended number of calories is provided. [5.3B] **15.** 16% of the permanent employees are hired. [5.3B] **16.** The student answered approximately 91.3% of the questions correctly. [5.3B] **17.** 80 [5.4A] **18.** 28.3 [5.4A] **19.** 32,000 PDAs were tested. [5.4B] **20.** The increase was 60% of the original price. [5.3B] **21.** 143.0 [5.5A] **22.** 1000% [5.5A] **23.** The dollar increase is $1.74. [5.5B] **24.** The population now is 220% of what it was 10 years ago. [5.5B] **25.** The value of the car is $12,500. [5.5B]

CUMULATIVE REVIEW EXERCISES

1. 4 [1.6B] **2.** 240 [2.1A] **3.** $10\frac{11}{24}$ [2.4C] **4.** $12\frac{41}{48}$ [2.5C] **5.** $12\frac{4}{7}$ [2.6B] **6.** $\frac{7}{24}$ [2.7B]

7. $\frac{1}{3}$ [2.8B] **8.** $\frac{13}{36}$ [2.8C] **9.** 3.08 [3.1B] **10.** 1.1196 [3.3A] **11.** 34.2813 [3.5A] **12.** 3.625 [3.6A]

13. $1\frac{3}{4}$ [3.6B] **14.** $\frac{3}{8} < 0.87$ [3.6C] **15.** 53.3 [4.3B] **16.** $9.60/hour [4.2B] **17.** $\frac{11}{60}$ [5.1A]

18. $83\frac{1}{3}$% [5.1B] **19.** 19.56 [5.2A/5.5A] **20.** $133\frac{1}{3}$% [5.3A/5.5A] **21.** 9.92 [5.4A/5.5A]

22. 342.9% [5.3A/5.5A] **23.** Sergio's take-home pay is $592. [2.6C] **24.** Each monthly payment is $204.25. [3.5B] **25.** 420 gallons were used during the month. [3.5B] **26.** The real estate tax is $5000. [4.3C] **27.** The sales tax is 6% of the purchase price. [5.3B/5.5B] **28.** 45% of the people did not favor the candidate. [5.3B/5.5B] **29.** The approximate average number of hours spent watching TV in a week is 61.3 hours. [5.2B/5.5B] **30.** 18% of the children tested had levels of lead that exceeded federal standards. [5.3B/5.5B]

Answers to Chapter 6 Selected Exercises

PREP TEST

1. 0.75 [3.5A] **2.** 52.05 [3.4A] **3.** 504.51 [3.3A] **4.** 9750 [3.4A] **5.** 45 [3.4A] **6.** 1417.24 [3.2A]
7. 3.33 [3.5A] **8.** 0.605 [3.5A] **9.** 0.379 < 0.397 [3.6C]

SECTION 6.1

1. The unit cost is $.055 per ounce. **3.** The unit cost is $.374 per ounce. **5.** The unit cost is $.080 per tablet.
7. The unit price is $6.975 per clamp. **9.** The unit cost is $.199 per ounce. **11.** The unit cost is $.119 per screw.
13. The Sutter Home pasta sauce is the more economical purchase. **15.** 20 ounces is the more economical purchase.
17. 400 tablets is the more economical purchase. **19.** Land O' Lakes cheddar cheese is the more economical purchase.
21. Maxwell House coffee is the more economical purchase. **23.** Friskies Chef's Blend is the more economical purchase.
25. The total cost is $13.77. **27.** The total cost is $1.84. **29.** The total cost is $6.37. **31.** The total cost is $2.71.
33. The total cost is $5.96.

SECTION 6.2

1. The percent increase is 19.6%. **3.** The percent increase is 117.6%. **5.** The percent increase is 457.1%.
7. The population 8 years later was 157,861 people. **9.** The average age of American mothers giving birth in 2000 was
24.9 years. **11.** The markup is $35.70. **13.** The markup rate is 30%. **15. a.** The markup is $77.76. **b.** The
selling price is $239.76. **17. a.** The markup is $1.10. **b.** The selling price is $3.10. **19.** The selling price is $74.
21. The amount represents a decrease of 40%. **23.** There is a decrease of 540 employees. **25. a.** The amount of
decrease is 13 minutes. **b.** The amount represents a decrease of 25%. **27. a.** The amount of decrease is $.60.
b. The new dividend is $1.00. **29.** The amount represents a decrease of 6.6%. **31.** The discount rate is $33\frac{1}{3}$%.
33. The discount is $68. **35.** The discount rate is 30%. **37. a.** The discount is $.25 per pound. **b.** The sale price
is $1.00 per pound. **39. a.** The amount of the discount is $4. **b.** The discount rate is 20%. **41.** Yes

SECTION 6.3

1. a. $10,000 **b.** $850 **c.** 4.25% **d.** 2 years **3.** The simple interest owed is $960. **5.** The simple interest
due is $3375. **7.** The simple interest due is $1320. **9.** The simple interest due is $92.47. **11.** The simple interest
due is $84.76. **13.** The maturity value is $5120. **15.** The maturity value is $164,250. **17.** The total amount due
on the loan is $12,875. **19.** The maturity value is $14,543.70. **21.** The monthly payment is $6187.50.
23. a. The interest charged is $1080. **b.** The monthly payment is $545. **25.** The monthly payment is $1332.50.
27. The finance charge is $1.48. **29.** The finance charge is $185.53. **31.** The difference in finance charges is $10.07.
33. The value of the investment after 1 year is $780.60. **35.** The value of the investment after 15 years is $7281.78.
37. a. The value of the investment in 5 years will be $111,446.25. **b.** The amount of interest earned will be $36,446.25.
39. The amount of interest earned is $5799.48. **41.** The interest owed is $16.

SECTION 6.4

1. The mortgage is $82,450. **3.** The down payment is $7500. **5.** The down payment is $212,500. **7.** The loan
origination fee is $3750. **9. a.** The down payment is $7500. **b.** The mortgage is $142,500. **11.** The mortgage is
$189,000. **13.** The monthly mortgage payment is $1157.73. **15.** No, the couple cannot afford to buy the house.
17. The monthly property tax is $112.35. **19. a.** The monthly mortgage payment is $1678.40. **b.** The interest
payment is $736.68. **21.** The total monthly payment for the mortgage and property tax is $982.70. **23.** The monthly
mortgage payment is $1645.53. **25.** By using the 20-year loan, the couple will save $63,408.

SECTION 6.5

1. No, Amanda does not have enough money for the down payment. **3.** The sales tax is $1192.50. **5.** The license fee is $450. **7. a.** The sales tax is $1120. **b.** The total cost of the sales tax and the license fee is $1395. **9. a.** The down payment is $4050. **b.** The amount financed is $12,150. **11.** The amount financed is $28,000. **13.** The monthly truck payment is $552.70. **15.** The cost is $5120. **17.** The cost is about $.11 per mile. **19.** The amount of interest is $74.75. **21. a.** The amount financed is $153,200. **b.** The monthly payment is $2961.78. **23.** The monthly payment is $649.63. **25.** The 7% loan has the lesser loan cost.

SECTION 6.6

1. Lewis earns $380. **3.** The real estate agent's commission is $3930. **5.** The stockbroker's commission is $84. **7.** The teacher's monthly salary is $3244. **9.** The overtime wage is $51.60/hour. **11.** The golf pro's commission was $112.50. **13.** The typist earns $618.75. **15.** Maxine's hourly wage is $85. **17. a.** Mark's hourly wage on Saturday is $23.85. **b.** Mark earns $190.80. **19. a.** The nurse's increase in hourly pay is $2.15. **b.** The nurse's hourly pay is $23.65. **21.** Nicole's earnings were $475. **23.** The starting salary for a chemical engineer increased $922 over that of the previous year. **25.** The starting salary for a political science major decreased $4115 from that of the previous year.

SECTION 6.7

1. Your current checking account balance is $486.32. **3.** The real estate firm's current checking account balance is $1222.47. **5.** The nutritionist's current balance is $825.27. **7.** The current checking account balance is $3000.82. **9.** Yes, there is enough money in the carpenter's account to purchase the refrigerator. **11.** Yes, there is enough money in the account to make the two purchases. **13.** The bank statement and checkbook balance. **15.** The bank statement and checkbook balance. **17.** added **19.** subtract

CHAPTER 6 REVIEW EXERCISES

1. The unit cost is $.195 per ounce or 19.5¢ per ounce. [6.1A] **2.** The cost is $.164 or 16.4¢ per mile. [6.5B] **3.** The percent increase is 30.4%. [6.2A] **4.** The markup is $72. [6.2B] **5.** The simple interest due is $3000. [6.3A] **6.** The value of the investment after 10 years is $45,550.75. [6.3C] **7.** The percent increase is 15%. [6.2A] **8.** The total monthly payment for the mortgage and property tax is $578.53. [6.4B] **9.** The monthly payment is $518.02. [6.5B] **10.** The value of the investment will be $53,593. [6.3C] **11.** The down payment is $18,750. [6.4A] **12.** The total cost of the sales tax and license fee is $1471.25. [6.5A] **13.** The selling price is $2079. [6.2B] **14.** The interest paid is $97.33. [6.5B] **15.** The commission was $3240. [6.6A] **16.** The sale price is $141. [6.2D] **17.** The current checkbook balance is $943.68. [6.7A] **18.** The maturity value is $31,200. [6.3A] **19.** The origination fee is $1875. [6.4A] **20.** The more economical purchase is 33 ounces for $6.99. [6.1B] **21.** The monthly mortgage payment is $934.08. [6.4B] **22.** The total income was $655.20. [6.6A] **23.** The donut shop's checkbook balance is $8866.58. [6.7A] **24.** The monthly payment is $14,093.75. [6.3A] **25.** The finance charge is $7.20. [6.3B]

CHAPTER 6 TEST

1. The cost per foot is $6.92. [6.1A] **2.** The more economical purchase is 3 pounds for $7.49. [6.1B] **3.** The total cost is $14.53. [6.1C] **4.** The percent increase in the cost of the exercise bicycle is 20%. [6.2A] **5.** The selling price of the compact disc player is $301. [6.2B] **6.** The percent decrease is 7.7%. [6.2C] **7.** The percent decrease is 20%. [6.2C] **8.** The sale price of the corner hutch is $209.30. [6.2D] **9.** The discount rate is 40%. [6.2D] **10.** The simple interest due is $2000. [6.3A] **11.** The maturity value is $26,725. [6.3A] **12.** The finance charge is $4.50. [6.3B] **13.** The amount of interest earned in 10 years will be $24,420.60. [6.3C] **14.** The loan origination fee is $3350. [6.4A] **15.** The monthly mortgage payment is $1713.44. [6.4B] **16.** The amount financed is $19,000. [6.5A] **17.** The monthly truck payment is $482.68. [6.5B] **18.** Shaney earns $1071. [6.6A] **19.** The current checkbook balance is $6612.25. [6.7A] **20.** The bank statement and the checkbook balance. [6.7B]

CUMULATIVE REVIEW EXERCISES

1. 13 [1.6B] **2.** $8\frac{13}{24}$ [2.4C] **3.** $2\frac{37}{48}$ [2.5C] **4.** 9 [2.6B] **5.** 2 [2.7B] **6.** 5 [2.8C] **7.** 52.2 [3.5A]

8. 1.417 [3.6A] **9.** $51.25/hour [4.2B] **10.** 10.94 [4.3B] **11.** 62.5% [5.1B] **12.** 27.3 [5.2A]

13. 0.182 [5.1A] **14.** 42% [5.3A] **15.** 250 [5.4A] **16.** 154.76 [5.4A/5.5A] **17.** The total rainfall is $13\frac{11}{12}$

inches. [2.4D] **18.** The amount paid in taxes is $970. [2.6C] **19.** The ratio is $\frac{3}{5}$. [4.1B] **20.** 33.4 miles are

driven per gallon. [4.2C] **21.** The cost is $.93 per pound. [4.2C] **22.** The dividend is $280. [4.3C] **23.** The

sale price is $720. [6.2D] **24.** The selling price of the disc player is $119. [6.2B] **25.** The percent increase in Sook

Kim's salary is 8%. [6.2A] **26.** The simple interest due is $2700. [6.3A] **27.** The monthly car payment is $791.81.

[6.5B] **28.** The family's new checking account balance is $2243.77. [6.7A] **29.** The cost per mile is $.20. [6.5B]

30. The monthly mortgage payment is $1232.26. [6.4B]

Answers to Chapter 7 Selected Exercises

PREP TEST

1. 48.0% was bill-related mail. [5.3B] **2. a.** The greatest cost increase is between 2009 and 2010. **b.** Between those

years, there was an increase of $5318. [1.3C] **3. a.** The ratio is $\frac{5}{3}$. **b.** The ratio is $1:1$. [4.1B]

4. a. 3.9, 3.9, 4.2, 4.5, 5.2, 5.5, 7.1 [3.6C] **b.** The average is 4.9 million viewers per night. [3.5B]

5. a. 4500 women are in the Marine Corps. [5.2B] **b.** $\frac{1}{20}$ of the women in the military are in the Marine Corps. [5.1A]

SECTION 7.1

1. The gross revenue is $750 million. **3.** The percent is 40%. **5.** 50 more people agreed that humanity should

explore space than agreed that space exploration impacts daily life. **7.** 150 children said they hid their vegetables under

a napkin. **11.** The ratio is $\frac{1}{3}$. **13.** The percent is 9.4%. **15.** The number of people surveyed is 150 people.

17. The percent is 28%. **19.** Americans spent $279,000,000 on portable game machines. **21.** Yes.

23. The number of homeless aged 25 to 34 is more than twice the number of homeless under the age of 25.

25. Out of every 100,000 homeless people, there are 8000 people over age 54. **27.** North America is 2,550,000 square

miles larger than South America. **29.** Australia is 5.2% of the total land area. **31.** $2027.50 is spent on health care.

33. No, the amount spent on housing is not more than twice the amount spent on transporation.

SECTION 7.2

1. 39 million passenger cars were produced worldwide. **3.** The percent is 28%. **5.** The Mini Cooper gets

approximately 10 more miles per gallon while traveling on the highway. **7.** The maximum salary of police officers in the

suburbs is $16,000 higher than the maximum salary of police officers in the city. **9.** The greatest difference in salaries is

in Philadelphia. **11.** The average snowfall during January is 20 inches. **13.** The snowfall during March and April

was 25 inches. **15.** The difference is 800 Calories. **17.** The ratio is $\frac{7}{6}$.

19.

Year	Difference Between the Amount Spent for Foreign and for Domestic Aid (in billions of dollars)
1991	$1.3
1992	$1.1
1993	$1.0
1994	$0.9
1995	$0.8
1996	$0.8
1997	$1.1
1998	$0.9
1999	$1.0

SECTION 7.3

1. a. 44 students have a tuition that is between $3000 and $6000.　**3.** 18 students paid more than $12,000.　**5.** There are 410 cars between 6 and 12 years old.　**7.** 230 cars are more than 12 years old.　**9.** 54 adults spend between 1 and 2 hours at the mall.　**11.** The percent is 22%.　**13.** There were 24 entrants in the discus finals.　**15.** 37.5% of the entrants had distances between 160 feet and 170 feet.　**17.** The percent is 10.8%.　**21.** The percent is 32.4%.　**23.** 900,000 students scored above 800.

SECTION 7.4

1. a. Median　**b.** Mean　**c.** Mode　**d.** Median　**e.** Mode　**f.** Mean　**3.** The mean number of seats filled is 381.5625 seats. The median number of seats filled is 394.5 seats. Since each number occurs only once, there is no mode.　**5.** The mean cost is $45.615. The median cost is $45.855.　**7.** The mean monthly rate is $403.625. The median monthly rate is $404.50.　**9.** The mean life expectancy is 70.8 years. The median life expectancy is 72 years.　**13. a.** 25%　**b.** 75%　**c.** 75%　**d.** 25%　**15.** Lowest is $46,596. Highest is $82,879. $Q_1 = $56,067$. $Q_3 = $66,507$. Median = $61,036. Range = $36,283. Interquartile range = $10,440.　**17. a.** There were 40 adults who had cholesterol levels above 217.　**b.** There were 60 adults who had cholesterol levels below 254.　**c.** There are 20 cholesterol levels in each quartile.　**d.** 25% of the adults had cholesterol levels not more than 198.　**19. a.** Range = 4.39 million metric tons. $Q_1 = 0.56$ million metric tons. $Q_3 = 2.10$ million metric tons. Interquartile range = 1.54 million metric tons

b.

```
        ┌──────┬──┐
  ●─────┤      │  ├──────────────●
        └──────┴──┘
0.41  |    2.10            4.80
   0.56  0.815
```

c. 4.80　**21. a.** No, the difference in the means is not greater than 1 inch.

b. The difference in medians is 0.3 inch.　**c.**

23. Answers will vary. For example, 55, 55, 55, 55, 55, or 50, 55, 55, 55, 60

SECTION 7.5

1. {(HHHH), (HHHT), (HHTT), (HHTH), (HTTT), (HTHH), (HTTH), (HTHT), (TTTT), (TTTH), (TTHH), (THHH), (TTHT), (THHT), (THTT), (THTH)}　**3.** {(1, 1), (1, 2), (1, 3), (1, 4), (2, 1), (2, 2), (2, 3), (2, 4), (3, 1), (3, 2), (3, 3), (3, 4), (4, 1), (4, 2), (4, 3), (4, 4)}　**5. a.** {1, 2, 3, 4, 5, 6, 7, 8}　**b.** {1, 2, 3}　**7. a.** The probability that the sum is 5 is $\frac{1}{9}$.

b. The probability that the sum is 15 is 0.　**c.** The probablity that the sum is less than 15 is 1.　**d.** The probability that the sum is 2 is $\frac{1}{36}$.　**9. a.** The probability that the number is divisible by 4 is $\frac{1}{4}$.　**b.** The probability that the number is a multiple of 3 is $\frac{1}{3}$.   **11.** The probability of throwing a sum of 5 is greater.   **13. a.** The probability is $\frac{4}{11}$ that the letter *I* is drawn.   **b.** The probability of choosing an *S* is greater.   **15. a.** The probability is $\frac{1}{3}$ that the marble chosen is green.   **b.** The probability of choosing a red marble is greater.   **17.** The probability is $\frac{8}{47}$ that the paper has a B grade.   **19.** The probability is 0.81 that an employee has a group health insurance plan.

CHAPTER 7 REVIEW EXERCISES

1. The agencies spent $349 million on maintaining websites.   [7.1B]   **2.** The ratio is $\frac{9}{8}$.   [7.1B]   **3.** 8.9% of the total amount of money was spent by NASA.   [7.1B]   **4.** Texas had the larger population.   [7.2B]   **5.** The population of California is 12.5 million people more than the population of Texas.   [7.2B]   **6.** The Texas population increased the least from 1925 to 1950.   [7.2B]   **7.** There were 54 games in which the Knicks scored fewer than 100 points.   [7.3B]   **8.** The ratio is $\frac{31}{8}$.   [7.3B]   **9.** The percent is 11.3%.   [7.3B]   **10.** From the pictograph, O'Hare airport has 10 million more passengers than Los Angeles airport.   [7.1A]   **11.** The ratio is 2 : 3.   [7.1A]   **12.** The difference was 50 days.   [7.2A]   **13.** The percent is 50%.   [7.2A]   **14. a.** The Southeast had the lowest number of days of full operation.   **b.** This region had 30 days of full operation.   [7.2A]   **15.** The probability of one tail and three heads is $\frac{1}{4}$.   [7.5A]   **16.** There were 15 people who slept 8 or more hours.   [7.3A]   **17.** The percent is 28.3%.   [7.3A]

18. a. The mean heart rate is 91.6 heartbeats per minute. The median heart rate is 93.5 heartbeats per minute. The mode is 96 heartbeats per minute.   [7.4A]   **b.** The range is 36 heartbeats per minute. The interquartile range is 15 heartbeats per minute.   [7.4B]

CHAPTER 7 TEST

1. 19 students spent between $15 and $25 each week.   [7.3B]   **2.** The ratio is $\frac{2}{3}$.   [7.3B]   **3.** The percent is 45%.   [7.3B]   **4.** There were 36 people that were surveyed for the Gallup poll.   [7.1A]   **5.** The ratio is $\frac{5}{2}$.   [7.1A]   **6.** The percent is 58.3%.   [7.1A]   **7.** During 1995 and 1996, the number of fatalities was the same.   [7.2A]   **8.** There were 32 fatal accidents from 1991 to 1999.   [7.2A]   **9.** There were 4 more fatalities from 1995 to 1998.   [7.2A]   **10.** There were 355 more films rated R.   [7.1B]   **11.** There were 16 times more films rated PG-13.   [7.1B]   **12.** The percent of films rated G was 5.6%.   [7.1B]   **13.** There are 24 states that have a median income between $40,000 and $60,000.   [7.3A]   **14.** The percent is 72%.   [7.3A]   **15.** The percent is 18%.   [7.3A]   **16.** The probability is $\frac{3}{10}$ that the ball chosen is red.   [7.5A]   **17.** The student enrollment increased the least during the 1990s.   [7.2B]   **18.** The increase in the enrollment was 11 million students.   [7.2B]   **19. a.** The mean time is 2.53 days.   [7.4A]   **b.** The median time is 2.55 days.   [7.4A]   **c.**

[7.4B]

CUMULATIVE REVIEW EXERCISES

1. 540 [1.6A]   **2.** 14 [1.6B]   **3.** 120 [2.1A]   **4.** $\frac{5}{12}$ [2.3B]   **5.** $12\frac{3}{40}$ [2.4C]   **6.** $4\frac{17}{24}$ [2.5C]

7. 2 [2.6B]   **8.** $\frac{64}{85}$ [2.7B]   **9.** $8\frac{1}{4}$ [2.8C]   **10.** 209.305 [3.1A]   **11.** 2.82348 [3.4A]   **12.** 16.67 [3.6A]

13. 26.4 miles/gallon [4.2A]   **14.** 3.2 [4.3B]   **15.** 80% [5.1B]   **16.** 80 [5.4A]   **17.** 16.34 [5.2A]

18. 40% [5.3A]   **19.** The salesperson's income for the week was $650. [6.6A]   **20.** The cost is $407.50. [4.3C]

21. The interest due is $3750. [6.3A]   **22.** The markup rate is 55%. [6.2B]   **23.** The amount budgeted for food is $570. [7.1B]   **24.** The difference in the number answered correctly is 12 problems. [7.2B]   **25.** The mean high temperature is 69.6°F. [7.4A]   **26.** The probability is $\frac{5}{36}$ that the sum of the dots on the two dice is 8. [7.5A]

Answers to Chapter 8 Selected Exercises

PREP TEST

1. 702 [1.2A] **2.** 58 [1.3B] **3.** 4 [2.6B] **4.** 10 [2.6B] **5.** 25 [2.6B] **6.** 46 [2.6B] **7.** 238 [1.5A]
8. 1.5 [3.5A]

SECTION 8.1

1. 72 in. **3.** $2\frac{1}{2}$ ft **5.** 39 ft **7.** $5\frac{1}{3}$ yd **9.** 84 in. **11.** $3\frac{1}{3}$ yd **13.** 10,560 ft **15.** $\frac{5}{8}$ ft **17.** 1 mi 1120 ft

19. 9 ft 11 in. **21.** 2 ft 9 in. **23.** 14 ft 6 in. **25.** 2 ft 8 in. **27.** $11\frac{1}{6}$ ft **29.** 4 mi 2520 ft **31.** 14 tiles can

be placed along one row. **33.** The missing dimension is $1\frac{5}{6}$ ft. **35.** The length of material needed is $7\frac{1}{2}$ in.

37. The length of each piece is $1\frac{2}{3}$ ft. **39.** The length of framing needed is 6 ft 6 in. **41.** The total length of the wall

is $33\frac{3}{4}$ ft. **43.** Yes.

SECTION 8.2

1. 4 lb **3.** 128 oz **5.** $1\frac{3}{5}$ tons **7.** 12,000 lb **9.** $4\frac{1}{8}$ lb **11.** 24 oz **13.** 2600 lb **15.** $\frac{1}{4}$ ton

17. $11\frac{1}{4}$ lb **19.** 4 tons 1000 lb **21.** 2 lb 8 oz **23.** 5 tons 400 lb **25.** 1 ton 1700 lb **27.** 33 lb

29. 14 lb **31.** 9 oz **33.** 1 lb 7 oz **35.** The load of bricks weighs 2000 lb. **37.** The package of tiles weighs 63 lb.
39. The weight of the case of soft drinks is 9 lb. **41.** Each container holds 1 lb 5 oz of shampoo. **43.** The ham roast
costs $27. **45.** The cost of mailing the manuscript is $8.75. **47.** Answers will vary.

SECTION 8.3

1. $7\frac{1}{2}$ c **3.** 24 fl oz **5.** 4 pt **7.** 7 c **9.** $5\frac{1}{2}$ gal **11.** 9 qt **13.** $3\frac{3}{4}$ qt **15.** $1\frac{1}{4}$ pt **17.** $4\frac{1}{4}$ qt

19. 3 gal 2 qt **21.** 2 qt 1 pt **23.** 7 qt **25.** 6 fl oz **27.** $17\frac{1}{2}$ pt **29.** $\frac{7}{8}$ gal **31.** 5 gal 1 qt

33. 1 gal 2 qt **35.** $2\frac{3}{4}$ gal **37.** $7\frac{1}{2}$ gal of coffee should be prepared. **39.** There are $4\frac{1}{4}$ qt of final solution.

41. The farmer used $8\frac{3}{4}$ gal of oil. **43.** The more economical purchase is $1.59 for 1 qt. **45.** The profit made was

$83.50. **47.** 1 grain = 0.002286 oz; 1 dram = 0.0625 oz; 1 furlong = $\frac{1}{8}$ mi; 1 rod = 16.5 ft

SECTION 8.4

1. 14 weeks **3.** 150 h **5.** $9\frac{1}{4}$ h **7.** 1110 s **9.** $3\frac{1}{2}$ h **11.** 23,400 s **13.** $3\frac{1}{2}$ days **15.** 3600 min

17. 4 weeks **19.** 504 h **21.** 2 days **23.** 259,200 s **25.** Yes **27.** Yes **29.** 7 years **31.** 2190 days

33. 30,660 h

SECTION 8.5

1. 19,450 ft · lb   **3.** 19,450,000 ft · lb   **5.** 1500 ft · lb   **7.** 29,700 ft · lb   **9.** 30,000 ft · lb   **11.** 25,500 ft · lb

13. 35,010,000 ft · lb   **15.** 9,336,000 ft · lb   **17.** 2 hp   **19.** 8 hp   **21.** $4950 \frac{\text{ft} \cdot \text{lb}}{\text{s}}$ **23.** $3850 \frac{\text{ft} \cdot \text{lb}}{\text{s}}$

25. $500 \frac{\text{ft} \cdot \text{lb}}{\text{s}}$ **27.** $4800 \frac{\text{ft} \cdot \text{lb}}{\text{s}}$ **29.** $1440 \frac{\text{ft} \cdot \text{lb}}{\text{s}}$ **31.** 9 hp   **33.** 12 hp

CHAPTER 8 REVIEW EXERCISES

1. 48 in.   [8.1A]   **2.** 2 ft 6 in.   [8.1B]   **3.** 1600 ft · lb   [8.5A]   **4.** 40 fl oz   [8.3A]   **5.** $4\frac{2}{3}$ yd   [8.1A]

6. $1\frac{1}{5}$ tons   [8.2A]   **7.** 2 lb 7 oz   [8.2B]   **8.** 54 oz   [8.2A]   **9.** 9 ft 3 in.   [8.1B]   **10.** 1 ton 1000 lb   [8.2B]

11. 7 c 2 fl oz   [8.3B]   **12.** 1 yd 2 ft   [8.1B]   **13.** 3 qt   [8.3A]   **14.** $6\frac{1}{4}$ h   [8.4A]   **15.** $1375 \frac{\text{ft} \cdot \text{lb}}{\text{s}}$ [8.5B]

16. 44 lb   [8.2B]   **17.** 38,900 ft · lb   [8.5A]   **18.** 7 hp   [8.5B]   **19.** The length of the remaining piece is 3 ft 6 in.

[8.1C]   **20.** The cost of mailing the book is $8.40.   [8.2C]   **21.** There are $13\frac{1}{2}$ qt in a case.   [8.3C]

22. 16 gal of milk were sold that day.   [8.3C]   **23.** 27,230,000 ft · lb   [8.5A]   **24.** $480 \frac{\text{ft} \cdot \text{lb}}{\text{s}}$ [8.5B]

CHAPTER 8 TEST

1. 30 in.   [8.1A]   **2.** 2 ft 5 in.   [8.1B]   **3.** $1\frac{1}{3}$ ft   [8.1C]   **4.** The wall is 48 ft long.   [8.1C]   **5.** 46 oz   [8.2A]

6. 2 lb 8 oz   [8.2B]   **7.** 17 lb 1 oz   [8.2B]   **8.** 1 lb 11oz   [8.2B]   **9.** The total weight of the workbooks is 750 lb.

[8.2C]   **10.** The amount the class received for recycling was $28.13.   [8.2C]   **11.** $3\frac{1}{4}$ gal   [8.3A]   **12.** 28 pt   [8.3A]

13. $12\frac{1}{4}$ gal   [8.3B]   **14.** 8 gal 1 qt   [8.3B]   **15.** $4\frac{1}{2}$ weeks   [8.4A]   **16.** 4680 min   [8.4A]   **17.** There are 60 c in

a case.   [8.3C]   **18.** Nick's profit is $144.   [8.3C]   **19.** 3750 ft · lb   [8.5A]   **20.** 31,120,000 ft · lb   [8.5A]

21. $160 \frac{\text{ft} \cdot \text{lb}}{\text{s}}$ [8.5B]   **22.** 4 hp   [8.5B]

CUMULATIVE REVIEW EXERCISES

1. 180   [2.1A]   **2.** $5\frac{3}{8}$ [2.2B]   **3.** $3\frac{7}{24}$ [2.5C]   **4.** 2   [2.7B]   **5.** $4\frac{3}{8}$ [2.8C]   **6.** 2.10   [3.1B]

7. 0.038808   [3.4A]   **8.** 8.8   [4.3B]   **9.** 1.25   [5.2A]   **10.** 42.86   [5.4A]   **11.** The unit cost is $5.15/lb.   [6.1A]

12. $8\frac{11}{15}$ in.   [8.1B]   **13.** 1 lb 8 oz   [8.2B]   **14.** 31 lb 8 oz   [8.2B]   **15.** $2\frac{1}{2}$ qt   [8.3B]   **16.** 1 lb 12 oz   [8.2B]

17. The dividend would be $280.   [4.3C]   **18.** Anna's balance is $642.79.   [6.7A]   **19.** The executive's monthly

income is $4100.   [6.6A]   **20.** The amount of carrots that could be sold is 2425 lb.   [5.2B]   **21.** The percent is 18%.

[7.3A]   **22.** The selling price of a compact disc player is $308.   [6.2B]   **23.** The interest paid on the loan is $8000.

[6.3A]   **24.** Each student received $633.   [8.2C]   **25.** The cost of mailing the books is $20.16.   [8.2C]   **26.** The

better buy is 36 oz for $2.98.   [6.1B]   **27.** The probability is $\frac{1}{9}$ that the sum of the dots on the two dice is 9.   [7.5A]

28. 3200 ft · lb   [8.5A]   **29.** $400 \frac{\text{ft} \cdot \text{lb}}{\text{s}}$ [8.5B]

Answers to Chapter 9 Selected Exercises

PREP TEST

1. 37,320   [3.4A]   **2.** 659,000   [3.4A]   **3.** 0.04107   [3.5A]   **4.** 28.496   [3.5A]   **5.** 5.125   [3.3A]
6. 5.96   [3.2A]   **7.** 0.13   [3.4A]   **8.** 56.35   [2.6A, 3.4A]   **9.** 0.5   [2.6B, 3.5A]   **10.** 675   [2.6B]

SECTION 9.1

1. 420 mm   **3.** 8.1 cm   **5.** 6.804 km   **7.** 2109 m   **9.** 4.32 m   **11.** 88 cm   **13.** 7.038 km   **15.** 3500 m
17. 2.60 m   **19.** 168.5 cm   **21.** 148 mm   **23.** 62.07 m   **25.** 31.9 cm   **27.** 8.075 km   **29.** The missing
dimension is 7.8 cm.   **31.** The distance between the rivets is 17.9 cm.   **33.** The amount of fencing left on the roll is
15.6 m.   **35.** It takes 500 s for light to travel from the sun to Earth.   **37.** Light travels 25,920,000,000 km in 1 day.

SECTION 9.2

1. 0.420 kg   **3.** 0.127 g   **5.** 4200 g   **7.** 450 mg   **9.** 1.856 kg   **11.** 4.057 g   **13.** 1370 g   **15.** 45.6 mg
17. 18.000 kg   **19.** 3.922 kg   **21.** 7.891 g   **23.** 4.063 kg   **25.** There are 40 g in 1 serving.   **27. a.** There are
3.288 g of cholesterol in 12 eggs.   **b.** There are 0.132 g of cholesterol in four glasses of milk.   **29. a.** There is 0.186 kg
of mix in the package.   **b.** There is 0.42 g of sodium in two servings.   **31.** The amount of seed needed is 1.6 kg.
33. The profit from repackaging the nuts is $117.50.   **35.** Corn was 50.3% of the total exports.

SECTION 9.3

1. 4.2 L   **3.** 3420 ml   **5.** 423 cm^3 **7.** 642 ml   **9.** 0.042 L   **11.** 435 cm^3 **13.** 4620 L   **15.** 1.423 kl
17. 1267 cm^3 **19.** 3.042 L   **21.** 3.004 kl   **23.** 8.200 L   **25. a.** There is less than 25 L of oxygen in 50 L of air.
b. There are 10.5 L of oxygen in 50 L of air.   **27.** 24 L of chlorine were used in a month.   **29.** 4000 patients can be
immunized.   **31.** The 12 one-liter bottles are the better buy.   **33.** The profit on the gasoline was $8465.   **35.** 2.72 L;
2720 ml; 2L 720 ml

SECTION 9.4

1. 3300 Calories can be omitted from your diet.   **3. a.** There are 90 Calories in $1\frac{1}{2}$ servings.   **b.** There are 30 fat
Calories in 6 slices of bread.   **5.** 2025 Calories would be needed.   **7.** You burn 10,125 Calories.   **9.** You would have
to hike for 2.6 h.   **11.** 1250 Wh are used.   **13.** The fax machine used 0.567 kWh.   **15.** The cost of running an air
conditioner is $1.58.   **17. a.** 400 lumens is less than half the output of Soft White bulb.   **b.** The energy saver bulb
costs $.42 less to operate.   **19.** The cost of using the welder is $109.98.   **21.** Answers will vary.

SECTION 9.5

1. 91.74 m   **3.** 1.73 m   **5.** 6.82 kg   **7.** 235.85 ml   **9.** 104.65 km/h   **11.** $7.68/kg   **13.** $.39/L
15. 40,068.07 km   **17.** 328 ft   **19.** 1.58 gal   **21.** 4920 ft   **23.** 6.33 gal   **25.** 49.69 mi/h   **27.** $1.46/gal
29. 4.62 lb   **31.** Gary will lose 1 lb.   **33.** 186,335.40 mi/s   **35. a.** False   **b.** False   **c.** True   **d.** True
e. False

CHAPTER 9 REVIEW EXERCISES

1. 1250 m   [9.1A]   **2.** 450 mg   [9.2A]   **3.** 5.6 ml   [9.3A]   **4.** 1090 yd   [9.5B]   **5.** 7.9 cm   [9.1A]
6. 5.34 m   [9.1A]   **7.** 0.990 kg   [9.2A]   **8.** 2.550 L   [9.3A]   **9.** 4.870 km   [9.1A]   **10.** 3.7 mm   [9.1A]
11. 6.829 g   [9.2A]   **12.** 1200 cm^3 [9.3A]   **13.** 4050 g   [9.2A]   **14.** 870 cm   [9.1A]   **15.** 192 cm^3 [9.3A]
16. 0.356 g   [9.2A]   **17.** 3.72 m   [9.1A]   **18.** 8300 L   [9.3A]   **19.** 2.089 L   [9.3A]   **20.** 5.410 L   [9.3A]
21. 3.792 kl   [9.3A]   **22.** 468 ml   [9.3A]   **23.** There are 37.2 m of wire left on the roll.   [9.1B]   **24.** The total cost
of the chicken is $12.72.   [9.2B]   **25.** $7.48/kg   [9.5A]   **26.** The amount of coffee that should be prepared is 50 L.
[9.3B]   **27.** You can eliminate 2700 Calories.   [9.4A]   **28.** The cost of running the TV set is $3.42.   [9.4A]

29. 4.18 lb   [9.5B]   **30.** 8.75 h of cycling are needed.   [9.4A]   **31.** The profit was $52.80.   [9.3B]
32. The color TV used 1.120 kWh of electricity.   [9.4A]   **33.** The amount of fertilizer used was 125 kg.   [9.2B]

CHAPTER 9 TEST

1. 2960 m   [9.1A]   **2.** 378 mg   [9.2A]   **3.** 46 ml   [9.3A]   **4.** 919 ml   [9.3A]   **5.** 4.26 cm   [9.1A]
6. 7.96 m   [9.1A]   **7.** 0.847 kg   [9.2A]   **8.** 3.920 L   [9.3A]   **9.** 5.885 km   [9.1A]   **10.** 15 mm   [9.1A]
11. 3.089 g   [9.2A]   **12.** 1600 cm³   [9.3A]   **13.** 3290 g   [9.2A]   **14.** 420 cm   [9.1A]   **15.** 96 cm³   [9.3A]
16. 1.375 g   [9.2A]   **17.** 4.02 m   [9.1A]   **18.** 8920 L   [9.3A]   **19.** A 140-pound sedentary person should consume
2100 Calories per day to maintain that weight.   [9.4A]   **20.** 3.15 kWh of energy is used during the week for operating
the television.   [9.4A]   **21.** The total length of the rafters is 114 m.   [9.1B]   **22.** The weight of the box is 36 kg.   [9.2B]
23. The amount of vaccine needed is 5.2 L.   [9.3B]   **24.** 56.4 km/h   [9.5A]   **25.** The distance between the rivets is
17.5 cm.   [9.1B]   **26.** The cost to fertilize the trees is $660.   [9.2B]   **27.** The total cost is $16.32.   [9.4A]   **28.** The
assistant should order 11 L of acid.   [9.3B]   **29.** The measure of the large hill is 393.6 ft.   [9.5B]
30. 4.8 in. is approximately 12.2 cm.   [9.5A]

CUMULATIVE REVIEW EXERCISES

1. 6   [1.6B]   **2.** $12\frac{13}{36}$ [2.4C]   **3.** $\frac{29}{36}$ [2.5C]   **4.** $3\frac{1}{14}$ [2.7B]   **5.** 1   [2.8B]   **6.** 2.0702   [3.3A]

7. 31.3   [4.3B]   **8.** 175%   [5.1B]   **9.** 145   [5.4A]   **10.** 2.25 gal   [8.3A]   **11.** 8.75 m   [9.1A]
12. 3.420 km   [9.1A]   **13.** 5050 g   [9.2A]   **14.** 3.672 g   [9.2A]   **15.** 6000 ml   [9.3A]   **16.** 2400 L   [9.3A]
17. $3933 is left after the rent is paid.   [2.6C]   **18.** The business paid $7207.20 in income tax.   [3.4B]   **19.** The
property tax is $5500.   [4.3C]   **20.** The car buyer will receive a rebate of $2820.   [5.2B]   **21.** The percent is 6.5%.
[5.3B]   **22.** Your mean grade is 76.   [7.4A]   **23.** Karla's salary next year will be $25,200.   [5.2B]   **24.** The discount
rate is 22%.   [6.2D]   **25.** The length of the wall is 36 ft.   [8.1C]   **26.** There are 12 qt of apple juice in the case.
[8.3C]   **27.** The profit was $123.20.   [8.3C]   **28.** 24 L of chlorine was used.   [9.3B]   **29.** The total cost of
operating the hair dryer is $1.89.   [9.4A]   **30.** 96.6 km/h   [9.5A]

Answers to Chapter 10 Selected Exercises

PREP TEST

1. 54 > 45   [1.1A]   **2.** 4 [1.3A]   **3.** 15,847   [1.2A]   **4.** 3779   [1.3B]   **5.** 26,432   [1.4B]   **6.** 6   [2.3B]
7. $1\frac{4}{15}$ [2.4B]   **8.** $\frac{7}{16}$ [2.5B]   **9.** 11.058   [3.2A]   **10.** 3.781   [3.3A]   **11.** $\frac{2}{5}$ [2.6A]   **12.** $\frac{5}{9}$ [2.7A]
13. 9.4   [3.4A]   **14.** 0.4   [3.5A]   **15.** 31   [1.6B]

SECTION 10.1

1. −120 ft   **3.** +2 dollars   **5.**   **7.**
9. 1   **11.** −1   **13.** 3   **15. a.** A is −4.   **b.** C is −2.   **17. a.** A is −7.   **b.** D is −4.   **19.** −2 > −5
21. −16 < 1   **23.** 3 > −7   **25.** −11 < −8   **27.** 35 > 28   **29.** −42 < 27   **31.** 21 > −34   **33.** −27 > −39
35. −87 < 63   **37.** 86 > −79   **39.** −62 > −84   **41.** −131 < 101   **43.** −7, −2, 0, 3   **45.** −5, −3, 1, 4
47. −4, 0, 5, 9   **49.** −10, −7, −5, 4, 12   **51.** −11, −7, −2, 5, 10   **53.** −16   **55.** 3   **57.** −45   **59.** 59
61. 88   **63.** 4   **65.** 9   **67.** 11   **69.** 12   **71.** 2   **73.** 6   **75.** 5   **77.** 1   **79.** −5   **81.** 16   **83.** 12
85. −29   **87.** −14   **89.** 15   **91.** −33   **93.** 32   **95.** −42   **97.** 61   **99.** −52   **101.** |−12| > |8|
103. |6| < |13| **105.** |−1| < |−17| **107.** |17| = |−17| **109.** −9, −|6|, −4, |−7| **111.** −9, −|−7|, |4|, 5
113. −|10|, −|−8|, −3, |5| **115. a.** 8 and −2 are 5 units from 3.   **b.** 2 and −4 are 3 units from −1.   **117.** −12 min
and counting is closer to blastoff.   **119.** The loss was greater during the first quarter.   **121.** 11, −11

SECTION 10.2

1. $-14, -364$ **3.** -2 **5.** 20 **7.** -11 **9.** -9 **11.** -3 **13.** 1 **15.** -5 **17.** -30 **19.** 9 **21.** 1
23. -10 **25.** -28 **27.** -41 **29.** -392 **31.** -20 **33.** -23 **35.** -2 **37.** -6 **39.** -6
41. Negative six minus positive four **43.** Positive six minus negative four **45.** $9 + 5$ **47.** $1 + (-8)$ **49.** 8
51. -7 **53.** -9 **55.** 36 **57.** -3 **59.** 18 **61.** -9 **63.** 11 **65.** 0 **67.** 11 **69.** 2 **71.** -138
73. 86 **75.** -337 **77.** 4 **79.** -12 **81.** -12 **83.** 3 **85.** The temperature is $-1°C$. **87.** Nick's score
was -15 points after his opponent shot the moon. **89.** The change in the price of the stock is -9 dollars. **91.** The
temperature is 115°F. **93.** The difference in elevation is 7046 m. **95.** The difference is 82°C. **97.** The difference in
temperature is 13°C. **99.**

-3	2	1
4	0	-4
-1	-2	3

SECTION 10.3

1. Subtraction **3.** Multiplication **5.** 42 **7.** -24 **9.** 6 **11.** 18 **13.** -20 **15.** -16 **17.** 25 **19.** 0
21. -72 **23.** -102 **25.** 140 **27.** -228 **29.** -320 **31.** -156 **33.** -70 **35.** 162 **37.** 120
39. 36 **41.** 192 **43.** -108 **45.** -2100 **47.** 20 **49.** -48 **51.** 140 **53.** $3(-12) = -36$
55. $-5(11) = -55$ **57.** -2 **59.** 8 **61.** 0 **63.** -9 **65.** -9 **67.** 9 **69.** -24 **71.** -12 **73.** 31
75. 17 **77.** 15 **79.** -13 **81.** -18 **83.** 19 **85.** 13 **87.** -19 **89.** 17 **91.** 26 **93.** 23 **95.** 25
97. -34 **99.** 11 **101.** -13 **103.** 13 **105.** 12 **107.** -14 **109.** -14 **111.** -6 **113.** -4
115. The average high temperature was $-4°F$. **117.** The average score was -2. **119.** The wind chill factor is $-45°F$.
121. a. 25 **b.** -17 **123. a.** True **b.** True

SECTION 10.4

1. $-\dfrac{5}{24}$ **3.** $-\dfrac{19}{24}$ **5.** $\dfrac{5}{26}$ **7.** $\dfrac{7}{24}$ **9.** $-\dfrac{19}{60}$ **11.** $-1\dfrac{3}{8}$ **13.** $\dfrac{3}{4}$ **15.** $-\dfrac{47}{48}$ **17.** $\dfrac{3}{8}$ **19.** $-\dfrac{7}{60}$ **21.** $\dfrac{13}{24}$
23. -3.4 **25.** -8.89 **27.** -8.0 **29.** -0.68 **31.** -181.51 **33.** 2.7 **35.** -20.7 **37.** -37.19
39. -34.99 **41.** 6.778 **43.** -8.4 **45.** $\dfrac{1}{21}$ **47.** $\dfrac{2}{9}$ **49.** $-\dfrac{2}{9}$ **51.** $-\dfrac{45}{328}$ **53.** $\dfrac{2}{3}$ **55.** $\dfrac{15}{64}$ **57.** $-\dfrac{3}{7}$
59. $-1\dfrac{1}{9}$ **61.** $\dfrac{5}{6}$ **63.** $-2\dfrac{2}{7}$ **65.** $-13\dfrac{1}{2}$ **67.** 31.15 **69.** -112.97 **71.** 0.0363 **73.** 97 **75.** 2.2
77. -4.14 **79.** -3.8 **81.** -22.70 **83.** 0.07 **85.** 0.55 **87.** The temperature fell 32.22°C in 27 min.
89. The difference between the boiling point and the melting point of oxygen is 35.438°C. **91. a.** The closing price was
$19.39. **b.** The closing price was $72.46. **93.** Answers will vary. $-\dfrac{17}{24}$ is one example.

SECTION 10.5

1. 2.37×10^6 **3.** 4.5×10^{-4} **5.** 3.09×10^5 **7.** 6.01×10^{-7} **9.** 5.7×10^{10} **11.** 1.7×10^{-8} **13.** 710,000
15. 0.000043 **17.** 671,000,000 **19.** 0.00000713 **21.** 5,000,000,000,000 **23.** 0.00801 **25.** 1.6×10^{10} mi
27. $\$3.1 \times 10^{12}$ **29.** 3.7×10^{-6} m **31.** 4 **33.** 0 **35.** 12 **37.** -6 **39.** -5 **41.** 2 **43.** -3 **45.** 1
47. 2 **49.** -3 **51.** 14 **53.** -4 **55.** 33 **57.** -13 **59.** -12 **61.** 17 **63.** 0 **65.** 30 **67.** 94
69. -8 **71.** 39 **73.** 22 **75.** 0.21 **77.** -0.96 **79.** -0.29 **81.** -1.76 **83.** 2.1 **85.** $\dfrac{3}{16}$ **87.** $\dfrac{7}{16}$
89. $-\dfrac{5}{16}$ **91.** $-\dfrac{5}{8}$ **93. a.** $3.45 \times 10^{-14} > 3.45 \times 10^{-15}$ **b.** $5.23 \times 10^{18} > 5.23 \times 10^{17}$
c. $3.12 \times 10^{12} > 3.12 \times 10^{11}$ **95. a.** 100 **b.** -100 **c.** 225 **d.** -225 **99. a.** 1.99×10^{30} kg
b. 1.67×10^{-27} kg

CHAPTER 10 REVIEW EXERCISES

1. -22 [10.1B] **2.** 1 [10.2B] **3.** $-\dfrac{5}{24}$ [10.4A] **4.** 0.21 [10.4A] **5.** $\dfrac{8}{25}$ [10.4B] **6.** -1.28 [10.4B]

7. 10 [10.5B] **8.** $-\dfrac{7}{18}$ [10.5B] **9.** 4 [10.1B] **10.** $0 > -3$ [10.1A] **11.** -6 [10.1B] **12.** 6 [10.3B]

13. $\dfrac{17}{24}$ [10.4A] **14.** $-\dfrac{1}{4}$ [10.4B] **15.** $1\dfrac{5}{8}$ [10.4B] **16.** 24 [10.5B] **17.** -26 [10.2A] **18.** 2 [10.5B]

19. 3.97×10^{-5} [10.5A] **20.** -0.08 [10.4B] **21.** $\dfrac{1}{36}$ [10.4A] **22.** $\dfrac{3}{40}$ [10.4B] **23.** -0.042 [10.4B]

24. $\dfrac{1}{3}$ [10.5B] **25.** 5 [10.1B] **26.** $-2 > -40$ [10.1A] **27.** -26 [10.3A] **28.** 1.33 [10.5B]

29. $-\dfrac{1}{4}$ [10.4A] **30.** -7.4 [10.4A] **31.** $\dfrac{15}{32}$ [10.4B] **32.** 240,000 [10.5A] **33.** $-4°$ [10.2C]

34. The student's score was 98. [10.3C] **35.** The difference between the boiling and melting points is 395.45°C. [10.4C]

CHAPTER 10 TEST

1. 3 [10.2B] **2.** -2 [10.1B] **3.** $\dfrac{1}{15}$ [10.4A] **4.** -0.0608 [10.4B] **5.** $-8 > -10$ [10.1A]

6. -1.88 [10.4A] **7.** -26 [10.5B] **8.** 90 [10.3A] **9.** -4.014 [10.4A] **10.** -9 [10.3B] **11.** -7 [10.2A]

12. $\dfrac{7}{24}$ [10.4A] **13.** 8.76×10^{10} [10.5A] **14.** -48 [10.3A] **15.** 0 [10.3B] **16.** 10 [10.2B]

17. $-\dfrac{4}{5}$ [10.4B] **18.** $0 > -4$ [10.1A] **19.** -14 [10.2A] **20.** -4 [10.5B] **21.** $\dfrac{3}{10}$ [10.4A]

22. 0.00000009601 [10.5A] **23.** 3.4 [10.4B] **24.** $\dfrac{1}{12}$ [10.4A] **25.** $\dfrac{1}{12}$ [10.4B] **26.** 3.213 [10.4A]

27. The temperature is 7°C. [10.2C] **28.** The melting point of oxygen is -213°C. [10.3C] **29.** The temperature
fell 46.62°C. [10.4C] **30.** The average low temperature was -2°F. [10.3C]

CUMULATIVE REVIEW EXERCISES

1. 0 [1.6B] **2.** $4\dfrac{13}{14}$ [2.5C] **3.** $2\dfrac{7}{12}$ [2.7B] **4.** $1\dfrac{2}{7}$ [2.8C] **5.** 1.80939 [3.3A] **6.** 18.67 [4.3B]

7. 13.75 [5.4A] **8.** 1 gal 3 qt [8.3A] **9.** 6.692 L [9.3A] **10.** 1.28 m [9.5A] **11.** 57.6 [5.2A]

12. 340% [5.1B] **13.** -3 [10.2A] **14.** $-3\dfrac{3}{8}$ [10.4A] **15.** $-10\dfrac{13}{24}$ [10.4A] **16.** 19 [10.5B]

17. 3.488 [10.4B] **18.** $31\dfrac{1}{2}$ [10.4B] **19.** -7 [10.3B] **20.** $\dfrac{25}{42}$ [10.4B] **21.** -4 [10.5B] **22.** -4 [10.5B]

23. The length remaining is $2\dfrac{1}{3}$ ft. [2.5D] **24.** Nimisha's new balance is \$803.31. [6.7A] **25.** The percent decrease
is 27.3%. [6.2C] **26.** The amount of coffee that should be prepared is 10 gal. [8.3C] **27.** The dividend per share
after the increase was \$1.68. [6.2A] **28.** The median hourly pay is \$11.40. [7.4A] **29.** 600,000 people would
vote. [4.3C] **30.** The average high temperature is $-4°$. [10.3C]

Answers to Chapter 11 Selected Exercises

PREP TEST

1. -7 [10.2B] **2.** -20 [10.3A] **3.** 0 [10.2A] **4.** 1 [10.3B] **5.** 1 [10.4B] **6.** $\dfrac{1}{15}$ [2.8B]

7. $\dfrac{19}{24}$ [2.8C] **8.** 4 [10.5B] **9.** -21 [10.5B]

SECTION 11.1

1. -33 **3.** -12 **5.** -4 **7.** -3 **9.** -22 **11.** -40 **13.** -1 **15.** 14 **17.** 32 **19.** 6 **21.** 7

23. 49 **25.** -24 **27.** 63 **29.** 9 **31.** 4 **33.** $-\frac{2}{3}$ **35.** 0 **37.** $-3\frac{1}{4}$ **39.** 8.5023

41. -18.950658 **43.** $2x^2, 3x, \underline{-4}$ **45.** $3a^2, -4a, \underline{8}$ **47.** $\underline{3x^2}, \underline{-4x}$ **49.** $\underline{1}y^2, 6a$ **51.** $16z$ **53.** $9m$ **55.** $12at$

57. $3yt$ **59.** Unlike terms   **61.** $-2t^2$ **63.** $13c - 5$ **65.** $-2t$ **67.** $3y^2 - 2$ **69.** $14w - 8u$

71. $-23xy + 10$ **73.** $-11v^2$ **75.** $-8ab - 3a$ **77.** $4y^2 - y$ **79.** $-3a - 2b^2$ **81.** $7x^2 - 8x$ **83.** $-3s + 6t$

85. $-3m + 10n$ **87.** $5ab + 4ac$ **89.** $\frac{2}{3}a^2 + \frac{3}{5}b^2$ **91.** $6.994x$ **93.** $1.56m - 3.77n$ **95.** $5x + 20$ **97.** $4y - 12$

99. $-2a - 8$ **101.** $15x + 30$ **103.** $15c - 25$ **105.** $-3y + 18$ **107.** $7x + 14$ **109.** $4y - 8$ **111.** $5x + 24$

113. $y - 6$ **115.** $6n - 3$ **117.** $6y + 20$ **119.** $10x + 4$ **121.** $-6t - 6$ **123.** $z - 2$ **125.** $y + 6$

127. $9t + 15$ **129.** $5t + 24$ **131. a.** | 1 | 1 | 1 | x | x | **b.** | x | x | x | x | 1 | 1 | 1 | 1 | 1 | 1 |

c. | x | x | x | 1 | 1 | **d.** | x | x | 1 | 1 | 1 | 1 | **e.** | x | x | x | 1 | 1 | 1 | **f.** $3x + 3$

133. Answers will vary.

SECTION 11.2

1. Yes   **3.** No   **5.** Yes   **7.** Yes   **9.** Yes   **11.** Yes   **13.** Yes   **15.** Yes   **17.** No   **19.** No   **21.** Yes

23. -2 **25.** 14 **27.** 2 **29.** -4 **31.** -5 **33.** 0 **35.** 2 **37.** -3 **39.** 10 **41.** -5 **43.** -12

45. -7 **47.** $\frac{2}{3}$ **49.** -1 **51.** $\frac{1}{3}$ **53.** $-\frac{1}{8}$ **55.** $-\frac{3}{4}$ **57.** $-1\frac{1}{12}$ **59.** 6 **61.** -9 **63.** -5 **65.** 14

67. 8 **69.** -3 **71.** 20 **73.** -21 **75.** -15 **77.** 16 **79.** -42 **81.** 15 **83.** $-38\frac{2}{5}$ **85.** $9\frac{3}{5}$

87. $-10\frac{4}{5}$ **89.** $1\frac{17}{18}$ **91.** $-2\frac{2}{3}$ **93.** 3 **95.** The original investment was \$23,610.   **97.** The value of the fund increased by \$1190.   **99.** 18.5 gal of gasoline was used.   **101.** The car gets 34.2 mi/gal.   **103.** The markup on each stuffed animal is \$16.30.   **105.** The compact disc costs \$14.50.   **109.** Answers will vary. For example, $x + 5 = 1$.

SECTION 11.3

1. 3 **3.** 5 **5.** -1 **7.** -4 **9.** 1 **11.** 3 **13.** -2 **15.** -3 **17.** -2 **19.** -1 **21.** 3 **23.** -1

25. 7 **27.** 2 **29.** 2 **31.** 0 **33.** 2 **35.** 0 **37.** $\frac{2}{3}$ **39.** $2\frac{5}{6}$ **41.** $-2\frac{1}{2}$ **43.** $1\frac{1}{2}$ **45.** 2 **47.** -2

49. $2\frac{1}{3}$ **51.** $-1\frac{1}{2}$ **53.** 10 **55.** 5 **57.** 36 **59.** -6 **61.** -9 **63.** $-4\frac{4}{5}$ **65.** $-8\frac{3}{4}$ **67.** 36 **69.** $1\frac{1}{3}$

71. 2 **73.** 4 **75.** 9 **77.** 4 **79.** -3.125 **81.** 1 **83.** -3 **85.** -9 **87.** 4.95 **89.** -0.0862069

91. The temperature is $-40°\text{C}$.   **93.** The time is 14.5 s.   **95.** 2500 units were made.   **97.** The clerk's monthly income is \$1800.   **99.** The total sales were \$32,000.   **101.** Miguel's commission rate was 4.5%.

SECTION 11.4

1. $\frac{1}{2}$ **3.** -1 **5.** -4 **7.** -1 **9.** -4 **11.** 3 **13.** -1 **15.** 2 **17.** 3 **19.** 0 **21.** -2 **23.** -1

25. $-\frac{2}{3}$ **27.** -4 **29.** $\frac{1}{3}$ **31.** $-\frac{3}{4}$ **33.** $\frac{1}{4}$ **35.** 0 **37.** $-1\frac{3}{4}$ **39.** 1 **41.** -1 **43.** -5 **45.** $\frac{5}{12}$

47. $-\frac{4}{7}$ **49.** $2\frac{1}{3}$ **51.** 21 **53.** 21 **55.** 2 **57.** -1 **59.** -4 **61.** $-\frac{6}{7}$ **63.** 5 **65.** 3 **67.** 0

69. -1 **71.** 2 **73.** $1\frac{9}{10}$ **75.** 13 **77.** $2\frac{1}{2}$ **79.** 2 **81.** 0 **83.** $3\frac{1}{5}$ **85.** 4 **87.** $-6\frac{1}{2}$ **89.** $-\frac{3}{4}$

91. $-\frac{2}{3}$ **93.** -3 **95.** $-1\frac{1}{4}$ **97.** $2\frac{5}{8}$ **99.** 1.9936196 **101.** 13.383119 **103.** 48

SECTION 11.5

1. $y - 9$ **3.** $z + 3$ **5.** $\frac{2}{3}n + n$ **7.** $\frac{m}{m - 3}$ **9.** $9(x + 4)$ **11.** $n - (-5)n$ **13.** $c\left(\frac{1}{4}c\right)$ **15.** $m^2 + 2m^2$

17. $2(t + 6)$ **19.** $\frac{x}{9 + x}$ **21.** $3(b + 6)$ **23.** x^2 **25.** $\frac{x}{20}$ **27.** $4x$ **29.** $\frac{3}{4}x$ **31.** $4 + x$ **33.** $5x - x$

35. $x(x + 2)$ **37.** $7(x + 8)$ **39.** $x^2 + 3x$ **41.** $(x + 3) + \frac{1}{2}x$ **43.** $\frac{3x}{x}$ **45. a.** Answers will vary. For example,
the sum of twice x and 3. **b.** Answers will vary. For example, twice the sum of x and 3. **49.** carbon: $\frac{1}{2}x$; oxygen: $\frac{1}{2}x$

SECTION 11.6

1. $x + 7 = 12; 5$ **3.** $3x = 18; 6$ **5.** $x + 5 = 3; -2$ **7.** $6x = 14; 2\frac{1}{3}$ **9.** $\frac{5}{6}x = 15; 18$ **11.** $3x + 4 = 8; 1\frac{1}{3}$

13. $\frac{1}{4}x - 7 = 9; 64$ **15.** $\frac{x}{9} = 14; 126$ **17.** $\frac{x}{4} - 6 = -2; 16$ **19.** $7 - 2x = 13; -3$ **21.** $9 - \frac{x}{2} = 5; 8$

23. $\frac{3}{5}x + 8 = 2; -10$ **25.** $\frac{x}{4.186} - 7.92 = 12.529; 85.599514$ **27.** The price at Target is $76.75. **29.** The value of
the SUV last year was $20,000. **31.** The length of the Brooklyn Bridge is 486 m. **33.** The family's monthly income is
$5440. **35.** The monthly output a year ago was 5000 computers. **37.** The recommended daily allowance of sodium
is 2.5 g. **39.** It took 3 h to install the water softener. **41.** About 11,065 plants and animals are known to be at risk of
extinction in the world. **43.** The percent is 3.3%. **45.** The original water flow rate was 5 gal/min. **47.** The total
sales for the month were $42,540. **49.** The world carbon dioxide emissions in 1990 were 5.83 billion metric tons.

CHAPTER 11 REVIEW EXERCISES

1. $-2a + 2b$ [11.1C] **2.** Yes [11.2A] **3.** -4 [11.2B] **4.** 7 [11.3A] **5.** 13 [11.1A] **6.** -9 [11.2C]
7. -18 [11.3A] **8.** $-3x + 4$ [11.1C] **9.** 5 [11.4A] **10.** -5 [11.2B] **11.** No [11.2A] **12.** 6 [11.1A]

13. 5 [11.4B] **14.** 10 [11.3A] **15.** $-4bc$ [11.1B] **16.** $-2\frac{1}{2}$ [11.4A] **17.** $1\frac{1}{4}$ [11.2C] **18.** $\frac{71}{30}x^2$ [11.1B]

19. $-\frac{1}{3}$ [11.4B] **20.** $10\frac{4}{5}$ [11.3A] **21.** The car averaged 23 mi/gal. [11.2D] **22.** The temperature

is 37.8°C. [11.3B] **23.** $n + \frac{n}{5}$ [11.5A] **24.** $(n + 5) + \frac{1}{3}n$ [11.5B] **25.** The number is 2. [11.6A] **26.** The

number is 10. [11.6A] **27.** The regular price of the CD player is $380. [11.6B] **28.** Last year's crop was 25,300
bushels. [11.6B]

CHAPTER 11 TEST

1. 95 [11.3A] **2.** 26 [11.2B] **3.** $-11x - 4y$ [11.1B] **4.** $3\frac{1}{5}$ [11.4A] **5.** -2 [11.3A] **6.** -38 [11.1A]

7. No [11.2A] **8.** $-14ab + 9$ [11.1B] **9.** $-2\frac{4}{5}$ [11.2C] **10.** $8y - 7$ [11.1C] **11.** 0 [11.4B]

12. $-\frac{1}{2}$ [11.3A] **13.** $-5\frac{5}{6}$ [11.1A] **14.** -16 [11.2C] **15.** -3 [11.3A] **16.** 11 [11.4B] **17.** The monthly

payment is $137.50. [11.2D] **18.** 4000 clocks were made during the month. [11.3B] **19.** The time required is

11.5 s. [11.3B] **20.** $x + \frac{1}{3}x$ [11.5A] **21.** $5(x + 3)$ [11.5B] **22.** $2x - 3 = 7; 5$ [11.6A]

23. The number is $-3\frac{1}{2}$. [11.6A] **24.** Santos's total sales for the month were $40,000. [11.6B] **25.** The mechanic
worked for 3 h. [11.6B]

CUMULATIVE REVIEW EXERCISES

1. 41 [1.6B]   **2.** $1\frac{7}{10}$ [2.5C]   **3.** $\frac{11}{18}$ [2.8C]   **4.** 0.047383 [3.4A]   **5.** $4.20/h [4.2B]   **6.** 26.67 [4.3B]

7. $\frac{4}{75}$ [5.1A]   **8.** 140% [5.3A]   **9.** 6.4 [5.4A]   **10.** 18 ft 9 in. [8.1B]   **11.** 22 oz [8.2A]

12. 0.282 g [9.2A]   **13.** -1 [10.2A]   **14.** 19 [10.2B]   **15.** 6 [10.5B]   **16.** -48 [11.1A]

17. $-10x + 8z$ [11.1B]   **18.** $3y + 23$ [11.1C]   **19.** -1 [11.3A]   **20.** $-6\frac{1}{4}$ [11.4B]   **21.** $-7\frac{1}{2}$ [11.2C]

22. -21 [11.3A]   **23.** The percent is 17.6%. [5.3B]   **24.** The price of the piece of pottery is $39.90. [6.2B]
25. a. The discount is $81.   **b.** The discount rate is 18%. [6.2D]   **26.** The simple interest due on the loan is

$2933.33. [6.3A]   **27.** $3n + 4$ [11.5B]   **28.** The probability is $\frac{1}{8}$ that the sum of the upward faces on the two dice

is 7. [7.5A]   **29.** The total sales were $32,500. [11.3B]   **30.** The number is 2. [11.6A]

Answers to Chapter 12 Selected Exercises

PREP TEST

1. 43 [11.2B]   **2.** 51 [11.2B]   **3.** 56 [1.6B]   **4.** 56.52 [11.1A]   **5.** 113.04 [11.1A]   **6.** 14.4 [4.3B]

SECTION 12.1

1. 0°; 90°   **3.** 180°   **5.** 30   **7.** 21   **9.** 14   **11.** 59°   **13.** 108°   **15.** 77°   **17.** 53°   **19.** 77°   **21.** 118°
23. 133°   **25.** 86°   **27.** 180°   **29.** Rectangle or square   **31.** Cube   **33.** Parallelogram, rectangle, or square
35. Cylinder   **37.** 102°   **39.** 90° and 45°   **41.** 14°   **43.** 90° and 65°   **45.** 8 in.   **47.** $4\frac{2}{3}$ ft   **49.** 7 cm
51. 2 ft 4 in.   **53.** $\angle a = 106°$, $\angle b = 74°$ **55.** $\angle a = 112°$, $\angle b = 68°$ **57.** $\angle a = 38°$, $\angle b = 142°$
59. $\angle a = 58°$, $\angle b = 58°$ **61.** $\angle a = 152°$, $\angle b = 152°$ **63.** $\angle a = 130°$, $\angle b = 50°$ **65. a.** 1°   **b.** 90°

SECTION 12.2

1. 56 in.   **3.** 20 ft   **5.** 92 cm   **7.** 47.1 cm   **9.** 9 ft 10 in.   **11.** 50.24 cm   **13.** 240 m   **15.** 157 cm
17. 121 cm   **19.** 50.56 m   **21.** 3.57 ft   **23.** 139.3 m   **25.** The amount of fencing needed is 60 ft.   **27.** The
amount of binding needed is 24 ft.   **29.** The total cost of the fence is $24,422.50.   **31.** The bicycle travels 31.4 ft.
33. a. The perimeter of the rink is larger than the perimeter of the rectangle. The perimeter of the rink is more than 70 m.
b. The perimeter of the rink is 81.4 m.   **35.** Approximately 20.71 ft of weather stripping are installed.   **37.** The
circumference of Earth is approximately 39,915.68 km.   **39.** 22 units   **41.** $14\frac{1}{4}$ cm

SECTION 12.3

1. 144 ft²   **3.** 81 in²   **5.** 50.24 ft²   **7.** 20 in²   **9.** 2.13 cm²   **11.** 16 ft²   **13.** 817 in²   **15.** 154 in²
17. 26 cm²   **19.** 2220 cm²   **21.** 150.72 in²   **23.** 8.851323 ft²   **25.** The area of the playing field is 7500 yd².
27. The area watered by the irrigation system is approximately 7850 ft².   **29.** You should buy 2 qt.   **31.** The area of
the driveway is 1250 ft².   **33.** The cost to wallpaper the two walls is $114.   **35.** $17.25 should be budgeted for grass
seed.   **37.** The total area of the park is 125.1492 mi².   **39. a.** The amount of the increase is about 113.04 in².
b. The amount of the increase is about 235.5 cm².   **41. a.** If the length and width are doubled, the area is increased
4 times.   **b.** If the radius is doubled, the area is quadrupled.   **c.** If the diameter is doubled, the area is quadrupled.
43. a. Sometimes true   **b.** Sometimes true   **c.** Always true

SECTION 12.4

1. 144 cm³ **3.** 512 in³ **5.** 2143.57 in³ **7.** 150.72 cm³ **9.** 6.4 m³ **11.** 5572.45 mm³ **13.** 3391.2 ft³

15. $42\frac{7}{8}$ ft³ **17.** 82.26 in³ **19.** 1.6688 m³ **21.** 69.08 in³ **23.** The volume of the water in the tank is 40.5 m³.

25. The volume is approximately 17,148.59 ft³. **27.** The amount of oil is approximately 75.36 m³. **29.** The volume of the auditorium is approximately 809,516.25 ft³. **31.** The volume of the bushing is approximately 212.64 in³. **33.** The tank will hold 3.7 gal. **35.** The cost is approximately $6526.41. **37. a.** 4 times larger **b.** 8 times larger **c.** 8 times larger

SECTION 12.5

1. 2.646 **3.** 6.481 **5.** 12.845 **7.** 13.748 **9.** 5 in. **11.** 8.602 cm **13.** 11.180 ft **15.** 4.472 cm **17.** 12.728 yd **19.** 10.392 ft **21.** 21.213 cm **23.** 8.944 m **25.** 7.879 yd **27.** The ramp is 9.66 ft long. **29.** The distance is 21.6 mi. **31.** The length of the diagonal is 12.1 mi. **33.** The ladder is 7.4 m high on the building. **35.** The perimeter is 27.7 in. **37.** The distance is 250 ft. **39.** The distance is 4.243 in. **41. a.** Always true **b.** Always true **45.** 8 m

SECTION 12.6

1. $\frac{1}{2}$ **3.** $\frac{3}{4}$ **5.** congruent **7.** congruent **9.** 7.2 cm **11.** 3.3 m **13.** The height of the building is 16 m. **15.** The perimeter is 12 m. **17.** The area is 56.25 cm². **19. a.** Always true **b.** Sometimes true **c.** Always true **21.** 16.5 units

CHAPTER 12 REVIEW EXERCISES

1. 0.75 m [12.1B] **2.** 31.4 cm [12.2A] **3.** 26 ft [12.2A] **4.** 3 [12.1A] **5.** 200 ft³ [12.4A] **6.** 26 cm [12.5B] **7.** 75° [12.1A] **8.** 3.873 [12.5A] **9.** 16 cm [12.6A] **10.** 63.585 cm² [12.3A] **11. a.** 45° **b.** 135° [12.1C] **12.** 55 m² [12.3A] **13.** 240 in³ [12.4B] **14.** 57.12 in² [12.3B] **15.** 267.9 ft³ [12.4A] **16.** 64.8 m² [12.6B] **17.** 47.7 in. [12.2B] **18. a.** 80° **b.** 100° [12.1C] **19.** The ladder will reach 15 ft up the building. [12.5C] **20.** The other angles of the triangle are 90° and 58°. [12.1B] **21.** The bicycle travels approximately 73.3 ft in 10 revolutions. [12.2C] **22.** The area of the room is 28 yd². [12.3C] **23.** The volume of the silo is approximately 1144.53 ft³. [12.4C] **24.** The area is 11 m². [12.3A] **25.** The distance from the starting point is 29 mi. [12.5C]

CHAPTER 12 TEST

1. 169.56 m³ [12.4A] **2.** 6.8 m [12.2A] **3.** 1406.72 cm³ [12.4B] **4.** 10 m [12.6A] **5.** 58° [12.1A] **6.** $3\frac{1}{7}$ m² [12.3A] **7.** 150° [12.1C] **8.** 15.85 ft [12.2B] **9.** 13.748 [12.5A] **10.** 9.747 ft [12.5B] **11.** 10.125 ft² [12.3B] **12.** $\angle a = 45°$; $\angle b = 135°$ [12.1C] **13.** $1\frac{1}{5}$ ft [12.6A] **14.** 90° and 50° [12.1B] **15.** The width of the canal is 25 ft. [12.6B] **16.** The amount of extra pizza is 113.04 in². [12.3C] **17.** It will cost $1113.69 to carpet the area. [12.3C] **18.** The area is 103.82 ft². [12.3C] **19.** The length of the rafter is 15 ft. [12.5C] **20.** The volume of the interior of the toolbox is 780 in³. [12.4C]

CUMULATIVE REVIEW EXERCISES

1. 48 [2.1B] **2.** $7\frac{41}{48}$ [2.4C] **3.** $\frac{39}{56}$ [2.7B] **4.** $\frac{2}{15}$ [2.8C] **5.** $-\frac{1}{24}$ [10.4A] **6.** $8.72/h [4.2B] **7.** 37.5 [4.3B] **8.** $\frac{3}{8}$ [5.1A] **9.** -4 [11.1A] **10.** 85 [5.4A] **11.** -6 [11.3A] **12.** $\frac{4}{3}$ [11.4B] **13.** 32,500 m [9.1A] **14.** 31.58 m [9.1A] **15.** -15 [11.2C] **16.** 2 [11.4B] **17.** The monthly payment

is $708. [1.5D]   **18.** The sales tax on the stereo is $47.06. [4.3C]   **19.** The original wage was $14.60. [5.4B]
20. The sale price of the dress is $54. [6.2D]   **21.** The value of the investment after 20 years would be $101,366.50.
[6.3C]   **22.** The weight of the package is 54 lb. [8.2C]   **23.** The distance between the rivets is 22.5 cm. [9.1B]
24. The number is -2. [11.6A]   **25. a.** 74°   **b.** 106° [12.1C]   **26.** 29.42 cm [12.2B]   **27.** 50 in^2 [12.3B]
28. 92.86 in^3 [12.4B]   **29.** 10.63 ft [12.5B]   **30.** 36 cm [12.6B]

FINAL EXAM

1. 3259 [1.3B]   **2.** 53 [1.5C]   **3.** 60,205 [1.3B]   **4.** 16 [1.6B]   **5.** 144 [2.1A]   **6.** $1\frac{49}{120}$ [2.4B]

7. $3\frac{29}{48}$ [2.5C]   **8.** $6\frac{3}{14}$ [2.6B]   **9.** $\frac{4}{9}$ [2.7B]   **10.** $\frac{1}{6}$ [2.8B]   **11.** $\frac{1}{13}$ [2.8C]   **12.** 164.177 [3.2A]

13. 0.027918 [3.4A]   **14.** 0.69 [3.5A]   **15.** $\frac{9}{20}$ [3.6B]   **16.** 24.5 mi/gal [4.2B]   **17.** 54.9 [4.3B]

18. $\frac{9}{40}$ [5.1A]   **19.** 135% [5.1B]   **20.** 125% [5.1B]   **21.** 36 [5.2A]   **22.** $133\frac{1}{3}$% [5.3A]   **23.** 70 [5.4A]
24. 20 in. [8.1A]   **25.** 1 ft 4 in. [8.1B]   **26.** 2.5 lb [8.2A]   **27.** 6 lb 6 oz [8.2B]   **28.** 2.25 gal [8.3A]
29. 1 gal 3 qt [8.3B]   **30.** 248 cm [9.1A]   **31.** 4.62 m [9.1A]   **32.** 1.614 kg [9.2A]   **33.** 2067 ml [9.3A]
34. 88.55 km [9.5A]   **35.** The cost is $1.15. [9.4A]   **36.** 6.79×10^{-8} [10.5A]   **37.** 3.9 m [12.2A]

38. 45 in^2 [12.3A]   **39.** 1200 cm^3 [12.4A]   **40.** -4 [10.2A]   **41.** -15 [10.2B]   **42.** $-\frac{1}{2}$ [10.4B]

43. $-\frac{1}{4}$ [10.4B]   **44.** 6 [10.5B]   **45.** $-x + 17$ [11.1C]   **46.** -18 [11.2C]   **47.** 5 [11.3A]   **48.** 1 [11.4A]
49. Your new balance is $959.93. [6.7A]   **50.** 63,750 people will vote. [4.3C]   **51.** One year ago the dividend
per share was $2.00. [5.4B]   **52.** The average monthly income is $3794. [7.4A]   **53.** The simple interest due is
$7200. [6.3A]   **54.** The probability is $\frac{1}{3}$. [7.5A]   **55.** China has 6.7% of the death count of the four
countries. [7.1B]   **56.** The discount rate for the compact disc player is 28%. [6.2D]   **57.** The weight of the box is
81 lb. [8.2C]   **58.** The perimeter is approximately 28.56 in. [12.2B]   **59.** The area of the composite figure is
approximately 16.86 cm^2. [12.3B]   **60.** The number is 16. [11.6A]

Glossary

absolute value of a number The distance between zero and the number on the number line. [10.1]

acute angle An angle whose measure is between 0° and 90°. [12.1]

acute triangle A triangle that has three acute angles. [12.2]

addend In addition, one of the numbers added. [1.2]

addition The process of finding the total of two numbers. [1.2]

Addition Property of Zero Zero added to a number does not change the number. [1.2]

adjacent angles Two angles that share a common side. [12.1]

alternate exterior angles Two nonadjacent angles that are on opposite sides of the transversal and outside the parallel lines. [12.1]

alternate interior angles Two nonadjacent angles that are on opposite sides of the transversal and between the parallel lines. [12.1]

angle An angle is formed when two rays start at the same point; it is measured in degrees. [12.1]

approximation An estimated value obtained by rounding an exact value. [1.1]

area A measure of the amount of surface in a region. [12.3]

Associative Property of Addition Numbers to be added can be grouped (with parentheses, for example) in any order; the sum will be the same. [1.2]

Associative Property of Multiplication Numbers to be multiplied can be grouped (with parentheses, for example) in any order; the product will be the same. [1.4]

average The sum of all the numbers divided by the number of those numbers. [1.5]

average value The sum of all values divided by the number of those values; also known as the mean value. [7.4]

balancing a checkbook Determining whether the checking account balance is accurate. [6.7]

bank statement A document showing all the transactions in a bank account during the month. [6.7]

bar graph A graph that represents data by the height of the bars. [7.2]

base of a triangle The side that the triangle rests on. [12.1]

basic percent equation Percent times base equals amount. [5.2]

borrowing In subtraction, taking a unit from the next larger place value in the minuend and adding it to the number in the given place value in order to make that number larger than the number to be subtracted from it. [1.3]

box-and-whiskers plot A graph that shows the smallest value in a set of numbers, the first quartile, the median, the third quartile, and the greatest value. [7.4]

British thermal unit A unit of energy. 1 British thermal unit = 778 foot-pounds. [8.5]

broken-line graph A graph that represents data by the position of the lines and shows trends and comparisons. [7.2]

Calorie A unit of energy in the metric system. [9.4]

capacity A measure of liquid substances. [8.3]

carrying In addition, transferring a number to another column. [1.2]

center of a circle The point from which all points on the circle are equidistant. [12.1]

center of a sphere The point from which all points on the surface of the sphere are equidistant. [12.1]

centi- The metric system prefix that means one-hundredth. [9.1]

check A printed form that, when filled out and signed, instructs a bank to pay a specified sum of money to the person named on it. [6.7]

checking account A bank account that enables you to withdraw money or make payments to other people, using checks. [6.7]

circle A plane figure in which all points are the same distance from point O, which is called the center of the circle. [12.1]

circle graph A graph that represents data by the size of the sectors. [7.1]

circumference The distance around a circle. [12.2]

class frequency The number of occurrences of data in a class interval on a histogram; represented by the height of each bar. [7.3]

class interval Range of numbers represented by the width of a bar on a histogram. [7.3]

class midpoint The center of a class interval in a frequency polygon. [7.3]

commission That part of the pay earned by a salesperson that is calculated as a percent of the salesperson's sales. [6.6]

common factor A number that is a factor of two or more numbers is a common factor of those numbers. [2.1]

common multiple A number that is a multiple of two or more numbers is a common multiple of those numbers. [2.1]

Commutative Property of Addition Two numbers can be added in either order; the sum will be the same. [1.2]

Commutative Property of Multiplication Two numbers can be multiplied in either order; the product will be the same. [1.4]

complementary angles Two angles whose sum is 90°. [12.1]

composite geometric figure A figure made from two or more geometric figures. [12.2]

composite geometric solid A solid made from two or more geometric solids. [12.4]

composite number A number that has whole-number factors besides 1 and itself. For instance, 18 is a composite number. [1.7]

compound interest Interest computed not only on the original principal but also on interest already earned. [6.3]

congruent objects Objects that have the same shape and the same size. [12.6]

congruent triangles Triangles that have the same shape and the same size. [12.6]

constant term A term that has no variables. [11.1]

conversion rate A relationship used to change one unit of measurement to another. [8.1]

corresponding angles Two angles that are on the same side of the transversal and are both acute angles or are both obtuse angles. [12.1]

cost The price that a business pays for a product. [6.2]

cross product In a proportion, the product of the numerator on the left side of the proportion times the denominator on the right, and the product of the denominator on the left side of the proportion times the numerator on the right. [4.3]

cube A rectangular solid in which all six faces are squares. [12.1]

cubic centimeter A unit of capacity equal to 1 milliliter. [9.3]

cup A U.S. Customary measure of capacity. 2 cups = 1 pint. [8.3]

cylinder A geometric solid in which the bases are circles and are perpendicular to the height. [12.1]

data Numerical information. [7.1]

day A unit of time. 24 hours = 1 day. [8.4]

decimal A number written in decimal notation. [3.1]

decimal notation Notation in which a number consists of a whole-number part, a decimal point, and a decimal part. [3.1]

decimal part In decimal notation, that part of the number that appears to the right of the decimal point. [3.1]

decimal point In decimal notation, the point that separates the whole-number part from the decimal part. [3.1]

degree Unit used to measure angles; one complete revolution is 360°. [12.1]

denominator The part of a fraction that appears below the fraction bar. [2.2]

deposit slip A form for depositing money in a checking account. [6.7]

diameter of a circle A line segment with endpoints on the circle and going through the center. [12.1]

diameter of a sphere A line segment with endpoints on the sphere and going through the center. [12.1]

difference In subtraction, the result of subtracting two numbers. [1.3]

discount The difference between the regular price and the sale price. [6.2]

discount rate The percent of a product's regular price that is represented by the discount. [6.2]

dividend In division, the number into which the divisor is divided to yield the quotient. [1.5]

division The process of finding the quotient of two numbers. [1.5]

divisor In division, the number that is divided into the dividend to yield the quotient. [1.5]

double-bar graph A graph used to display data for purposes of comparison. [7.2]

down payment The percent of a home's purchase price that the bank, when issuing a mortgage, requires the borrower to provide. [6.4]

empirical probability The ratio of the number of observations of an event to the total number of observations. [7.5]

energy The ability to do work. [8.5]

equation A statement of the equality of two mathematical expressions. [11.2]

equilateral triangle A triangle that has three sides of equal length; the three angles are also of equal measure. [12.2]

equivalent fractions Equal fractions with different denominators. [2.3]

evaluating a variable expression Replacing the variable or variables with numbers and then simplifying the resulting numerical expression. [11.1]

event One or more outcomes of an experiment. [7.5]

expanded form The number 46,208 can be written in expanded form as 40,000 + 6000 + 200 + 0 + 8. [1.1]

experiment Any activity that has an observable outcome. [7.5]

exponent In exponential notation, the raised number that indicates how many times the number to which it is attached is taken as a factor. [1.6]

exponential notation The expression of a number to some power, indicated by an exponent. [1.6]

factors In multiplication, the numbers that are multiplied. [1.4]

factors of a number The whole-number factors of a number divide that number evenly. [1.7]

favorable outcomes The outcomes of an experiment that satisfy the requirements of a particular event. [7.5]

finance charges Interest charges on purchases made with a credit card. [6.3]

first quartile In a set of numbers, the number below which one-quarter of the data lie. [7.4]

fixed-rate mortgage A mortgage in which the monthly payment remains the same for the life of the loan. [6.4]

fluid ounce A U.S. Customary measure of capacity. 8 fluid ounces = 1 cup. [8.3]

foot A U.S. Customary unit of length. 3 feet = 1 yard. [8.1]

foot-pound A U.S. Customary unit of energy. One foot-pound is the amount of energy required to lift 1 pound a distance of 1 foot. [8.5]

foot-pounds per second A U.S. Customary unit of power. [8.5]

formula An equation that expresses a relationship among variables. [11.2]

fraction The notation used to represent the number of equal parts of a whole. [2.2]

fraction bar The bar that separates the numerator of a fraction from the denominator. [2.2]

frequency polygon A graph that displays information similarly to a histogram. A dot is placed above the center of each class interval at a height corresponding to that class's frequency. [7.3]

gallon A U.S. Customary measure of capacity. 1 gallon = 4 quarts. [8.3]

geometric solid A figure in space. [12.1]

gram The basic unit of mass in the metric system. [9.2]

graph A display that provides a pictorial representation of data. [7.1]

graph of a whole number A heavy dot placed directly above that number on the number line. [1.1]

greater than A number that appears to the right of a given number on the number line is greater than the given number. [1.1]

greatest common factor (GCF) The largest common factor of two or more numbers. [2.1]

height of a parallelogram The distance between parallel sides. [12.1]

height of a triangle A line segment perpendicular to the base from the opposite vertex. [12.1]

histogram A bar graph in which the width of each bar corresponds to a range of numbers called a class interval. [7.3]

horsepower The U.S. Customary unit of power. 1 horsepower = 550 foot-pounds per second. [8.5]

hourly wage Pay calculated on the basis of a certain amount for each hour worked. [6.6]

hypotenuse The side opposite the right angle in a right triangle. [12.1]

improper fraction A fraction greater than or equal to 1. [2.2]

inch A U.S. Customary unit of length. 12 inches = 1 foot. [8.1]

integers The numbers . . . , −3, −2, −1, 0, 1, 2, 3, ... [10.1]

interest Money paid for the privilege of using someone else's money. [6.3]

interest rate The percent used to determine the amount of interest. [6.3]

interquartile range The difference between the third quartile and the first quartile. [7.4]

intersecting lines Lines that cross at a point in the plane. [12.1]

inverting a fraction Interchanging the numerator and denominator. [2.7]

isosceles triangle A triangle that has two sides of equal length; the angles opposite the equal sides are of equal measure. [12.2]

kilo- The metric system prefix that means one thousand. [9.1]

kilowatt-hour A unit of electrical energy in the metric system equal to 1000-watt hours. [9.4]

least common denominator (LCD) The least common multiple of denominators. [2.4]

least common multiple (LCM) The smallest common multiple of two or more numbers. [2.1]

legs of a right triangle The two shortest sides of a right triangle. [12.1]

length A measure of distance. [8.1]

less than A number that appears to the left of a given number on the number line is less than the given number. [1.1]

license fees Fees charged for authorization to operate a vehicle. [6.5]

like terms Terms of a variable expression that have the same variable part. [11.1]

line A line extends indefinitely in two directions in a plane; it has no width. [12.1]

line segment Part of a line; it has two endpoints. [12.1]

liter The basic unit of capacity in the metric system. [9.3]

loan origination fee The fee a bank charges for processing mortgage papers. [6.4]

markup The difference between selling price and cost. [6.2]

markup rate The percent of a product's cost that is represented by the markup. [6.2]

mass The amount of material in an object. On the surface of Earth, mass is the same as weight. [9.2]

maturity value of a loan The principal of a loan plus the interest owed on it. [6.3]

mean The sum of all values divided by the number of those values; also known as the average value. [7.4]

measurement A measurement has both a number and a unit. Examples include 7 feet, 4 ounces, and 0.5 gallon. [8.1]

median The value that separates a list of values in such a way that there is the same number of values below the median as above it. [7.4]

meter The basic unit of length in the metric system. [9.1]

metric system A system of measurement based on the decimal system. [9.1]

mile A U.S. Customary unit of length. 5280 feet = 1 mile. [8.1]

milli- The metric system prefix that means one-thousandth. [9.1]

minuend In subtraction, the number from which another number (the subtrahend) is subtracted. [1.3]

minute A unit of time. 60 minutes = 1 hour. [8.4]

mixed number A number greater than 1 that has a whole-number part and a fractional part. [2.2]

mode In a set of numbers, the value that occurs most frequently. [7.4]

monthly mortgage payment One of 12 payments due each year to the lender of money to buy real estate. [6.4]

mortgage The amount borrowed to buy real estate. [6.4]

multiples of a number The products of that number and the numbers 1, 2, 3, ... [2.1]

multiplication The process of finding the product of two numbers. [1.4]

Multiplication Property of One The product of a number and one is the number. [1.4]

Multiplication Property of Zero The product of a number and zero is zero. [1.4]

negative integers The numbers . . . , −5, −4, −3, −2, −1. [10.1]

natural numbers The numbers 1, 2, 3, 4, 5, . . . ; also called the positive integers. [10.1]

negative numbers Numbers less than zero. [10.1]

number line A line on which a number can be graphed. [1.1]

numerator The part of a fraction that appears above the fraction bar. [2.2]

numerical coefficient The number part of a variable term. When the numerical coefficient is 1 or −1, the 1 is usually not written. [11.1]

obtuse angle An angle whose measure is between 90° and 180°. [12.1]

obtuse triangle A triangle that has one obtuse angle. [12.2]

opposite numbers Two numbers that are the same distance from zero on the number line, but on opposite sides. [10.1]

Order of Operations Agreement A set of rules that tells us in what order to perform the operations that occur in a numerical expression. [1.6]

ounce A U.S. Customary unit of weight. 16 ounces = 1 pound. [8.2]

parallel lines Lines that never meet; the distance between them is always the same. [12.1]

parallelogram A quadrilateral that has opposite sides equal and parallel. [12.1]

percent Parts per hundred. [5.1]

percent decrease A decrease of a quantity, expressed as a percent of its original value. [6.2]

percent increase An increase of a quantity, expressed as a percent of its original value. [6.2]

perfect square The product of a whole number and itself. [12.5]

perimeter The distance around a plane figure. [12.2]

period In a number written in standard form, each group of digits separated from other digits by a comma or commas. [1.1]

perpendicular lines Intersecting lines that form right angles. [12.1]

pictograph A graph that uses symbols to represent information. [7.1]

pint A U.S. Customary measure of capacity. 2 pints = 1 quart. [8.3]

place value The position of each digit in a number written in standard form determines that digit's place value. [1.1]

place-value chart A chart that indicates the place value of every digit in a number. [1.1]

plane A flat surface. [12.1]

plane figures Figures that lie totally in a plane. [12.1]

points A term banks use to mean percent of a mortgage; used to express the loan origination fee. [6.4]

polygon A closed figure determined by three or more line segments that lie in a plane. [12.2]

positive integers The numbers 1, 2, 3, 4, 5, . . . ; also called the natural numbers. [10.1]

positive numbers Numbers greater than zero. [10.1]

pound A U.S. Customary unit of weight. 1 pound = 16 ounces. [8.2]

power The rate at which work is done or energy is released. [8.5]

prime factorization The expression of a number as the product of its prime factors. [1.7]

prime number A number whose only whole-number factors are 1 and itself. For instance, 13 is a prime number. [1.7]

principal The amount of money originally deposited or borrowed. [6.3]

probability A number from 0 to 1 that tells us how likely it is that a certain outcome of an experiment will happen. [7.5]

product In multiplication, the result of multiplying two numbers. [1.4]

proper fraction A fraction less than 1. [2.2]

property tax A tax based on the value of real estate. [6.4]

proportion An expression of the equality of two ratios or rates. [4.3]

Pythagorean Theorem The square of the hypotenuse of a right triangle is equal to the sum of the squares of the two legs. [12.5]

quadrilateral A four-sided closed figure. [12.1]

quart A U.S. Customary measure of capacity. 4 quarts = 1 gallon. [8.3]

quotient In division, the result of dividing the divisor into the dividend. [1.5]

radius of a circle A line segment going from the center to a point on the circle. [12.1]

radius of a sphere A line segment going from the center to a point on the sphere. [12.1]

range In a set of numbers, the difference between the largest and smallest values. [7.4]

rate A comparison of two quantities that have different units. [4.2]

ratio A comparison of two quantities that have the same units. [4.1]

rational number A number that can be written as the ratio of two integers, where the denominator is not zero. [10.4]

ray A ray starts at a point and extends indefinitely in one direction. [12.1]

reciprocal of a fraction The fraction with the numerator and denominator interchanged. [2.7]

rectangle A parallelogram that has four right angles. [12.1]

rectangular solid A solid in which all six faces are rectangles. [12.1]

regular polygon A polygon in which each side has the same length and each angle has the same measure. [12.2]

remainder In division, the quantity left over when it is not possible to separate objects or numbers into a whole number of equal groups. [1.5]

repeating decimal A decimal in which a block of one or more digits repeats forever. [10.4]

right angle A 90° angle. [12.1]

right triangle A triangle that contains one right angle. [12.1]

rounding Giving an approximate value of an exact number. [1.1]

salary Pay based on a weekly, biweekly, monthly, or annual time schedule. [6.6]

sale price The reduced price. [6.2]

sales tax A tax levied by a state or municipality on purchases. [6.5]

sample space All the possible outcomes of an experiment. [7.5]

scalene triangle A triangle that has no sides of equal length; no two of its angles are of equal measure. [12.2]

scientific notation Notation in which a number is expressed as a product of two factors, one a number between 1 and 10 and the other a power of 10. [10.5]

second A unit of time. 60 seconds = 1 minute. [8.4]

sector of a circle One of the "pieces of the pie" in a circle graph. [7.1]

selling price The price for which a business sells a product to a customer. [6.2]

service charge A sum of money charged by a bank for handling a transaction. [6.7]

sides of a polygon The line segments that form the polygon. [12.2]

similar objects Objects that have the same shape but not necessarily the same size. [12.6]

similar triangles Triangles that have the same shape but not necessarily the same size. [12.6]

simple interest Interest computed on the original principal. [6.3]

simplest form of a fraction A fraction is in simplest form when there are no common factors in the numerator and denominator. [2.3]

simplest form of a rate A rate is in simplest form when the numbers that make up the rate have no common factor. [4.2]

simplest form of a ratio A ratio is in simplest form when the two numbers do not have a common factor. [4.1]

simplifying a variable expression Combining like terms by adding their numerical coefficients. [11.1]

solids Objects in space. [12.1]

solution of an equation A number that, when substituted for the variable, results in a true equation. [11.2]

solving an equation Finding a solution of the equation. [11.2]

sphere A solid in which all points are the same distance from point *O*, which is called the center of the sphere. [12.1]

square A rectangle that has four equal sides. [12.1]

square root A square root of a number is one of two identical factors of that number. [12.5]

standard form A whole number is in standard form when it is written using the digits 0, 1, 2, . . . , 9. An example is 46,208. [1.1]

statistics The branch of mathematics concerned with data, or numerical information. [7.1]

straight angle A 180° angle. [12.1]

subtraction The process of finding the difference between two numbers. [1.3]

subtrahend In subtraction, the number that is subtracted from another number (the minuend). [1.3]

sum In addition, the total of the numbers added. [1.2]

supplementary angles Two angles whose sum is 180°. [12.1]

terminating decimal A decimal that has a finite number of digits after the decimal point, which means that it comes to an end and does not go on forever. [10.4]

terms of a variable expression The addends of the expression. [11.1]

theoretical probability A fraction with the number of favorable outcomes of an experiment in the numerator and the total number of possible outcomes of the experiment in the denominator. [7.5]

third quartile In a set of numbers, the number above which one-quarter of the data lie. [7.4]

ton A U.S. Customary unit of weight. 1 ton = 2000 pounds. [8.2]

total cost The unit cost multiplied by the number of units purchased. [6.1]

transversal A line intersecting two other lines at two different points. [12.1]

triangle A three-sided closed figure. [12.1]

true proportion A proportion in which the fractions are equal. [4.3]

unit cost The cost of one item. [6.1]

unit rate A rate in which the number in the denominator is 1. [4.2]

variable A letter used to stand for a quantity that is unknown or that can change. [11.1]

variable expression An expression that contains one or more variables. [11.1]

variable part In a variable term, the variable or variables and their exponents. [11.1]

variable term A term composed of a numerical coefficient and a variable part. [11.1]

vertex The common endpoint of two rays that form an angle. [12.1]

vertical angles Two angles that are on opposite sides of the intersection of two lines. [12.1]

volume A measure of the amount of space inside a closed surface. [12.4]

watt-hour A unit of electrical energy in the metric system. [9.4]

week A unit of time. 7 days = 1 week. [8.4]

weight A measure of how strongly Earth is pulling on an object. [8.2]

whole numbers The whole numbers are 0, 1, 2, 3, ... [1.1]

whole-number part In decimal notation, that part of the number that appears to the left of the decimal point. [3.1]

yard A U.S. Customary unit of length. 36 inches = 1 yard. [8.1]

Index

Index of Applications

(Continued from inside front cover)